MANUAL OF

Commercial
Methods
in Clinical
Microbiology

M A N U A L   O F

Commercial Methods in Clinical Microbiology

EDITOR

Allan L. Truant

Clinical Microbiology, Immunology, and Virology Laboratories
and
Departments of Pathology and Laboratory Medicine,
Microbiology and Immunology, and Internal Medicine
Temple University Hospital and School of Medicine
Philadelphia, Pennsylvania

ASM
PRESS

WASHINGTON, D.C.

Address editorial correspondence to ASM Press, 1752 N Street NW,
Washington, DC 20036-2904, USA

Send orders to ASM Press, P.O. Box 605, Herndon, VA 20172, USA
Phone: (800) 546-2416 or (703) 661-1593
Fax: (703) 661-1501
E-mail: books@asmusa.org
Online: www.asmpress.org

Copyright © 2002 ASM Press
American Society for Microbiology
1752 N Street NW
Washington, DC 20036-2904

Library of Congress Cataloging-in-Publication Data

Manual of commercial methods in clinical microbiology / edited by Allan L. Truant.
    p. ; cm.
Includes bibliographical references and index.
ISBN 1-55581-189-2
1. Medical microbiology—Equipment and supplies—Handbooks, manuals, etc. I. Truant, Allan L.
[DNLM: 1. Equipment and Supplies—Catalogs. 2. Microbiological Techniques. 3. Catalogs,
Commercial. QW 26 M294 2002]
QR46.M433 2002
616′.01—dc21

00-065062

10 9 8 7 6 5 4 3 2 1

To my father, the late Joseph Paul Truant, Ph.D. (1923–1988);

my mother, Flora Rina (Lenardon) Truant;

my children, Patti, Kathleen, and Steven;

and my sisters, Linda Ann Thompson and Anita Louise Ryan, and their families

Contents

Coordinating Authors

STEPHEN D. ALLEN   CHAPTER 4

Department of Pathology and Laboratory Medicine, Indiana University School of Medicine, Clarian Health Partners, and Methodist-IU-Riley Hospitals, Indianapolis, IN 46228

ALAN T. EVANGELISTA   CHAPTERS 3 AND 17

Ortho-McNeil Pharmaceutical, Inc., 1000 Route 202, Room 3046, Raritan, NJ 08869, and Department of Pathology and Laboratory Medicine, MCP Hahnemann University, Philadelphia, PA 19129

LYNNE S. GARCIA   CHAPTER 11

LSG & Associates, 512 12th St., Santa Monica, CA 90402-2908

SHARON L. HANSEN   CHAPTER 1

11 Ole Grist Run, Milton, DE 19968
Retired from the Microbiology Branch, Division of Clinical Laboratory Devices, Center for Devices and Radiological Health, Food and Drug Administration

RICHARD L. HODINKA   CHAPTERS 6 AND 13

Departments of Pediatrics and Pathology and Clinical Virology Laboratory, Children's Hospital of Philadelphia and University of Pennsylvania School of Medicine, Philadelphia, PA 19104

THOMAS J. INZANA   CHAPTER 14

Center for Molecular Medicine and Infectious Diseases, Virginia-Maryland Regional College of Veterinary Medicine, Virginia Polytechnic Institute and State University, Blacksburg, VA 24061

ROBERT C. JERRIS   CHAPTER 7

Microbiology, DeKalb Medical Center, Decatur, GA 30033, and Department of Pathology, Emory University School of Medicine, Atlanta, GA 30322

DONALD JUNGKIND   CHAPTER 12

Departments of Pathology and Microbiology and Clinical Microbiology Laboratories, Thomas Jefferson University, 207 Pavilion Building, 11th and Walnut Sts., Philadelphia, PA 19107-4998

CHARLES T. LADOULIS   CHAPTER 15

Spartan Consulting Group, Inc., 5 Grand Tour, Locust, NJ 07760-2343, and TNB Laboratories, Inc., 300 Prince Philip Drive, Spencer Hall, Memorial University of Newfoundland, St. John's, Newfoundland, Canada

MARILYN A. MENEGUS   CHAPTER 5

Clinical Microbiology Laboratories and Departments of Microbiology and Immunology, Pathology and Laboratory Medicine, and Pediatrics, University of Rochester School of Medicine, Rochester, NY 14642

JAMSHID MOGHADDAS   APPENDIX

Clinical Microbiology and Immunology Laboratory, Temple University Hospital, Philadelphia, PA 19140

DAVID H. PERSING   CHAPTER 18

Corixa Corporation, Suite 200, 1124 Columbia St., Seattle, WA 98104

GLENN D. ROBERTS   CHAPTERS 9 AND 10

Clinical Mycology and Mycobacteriology Laboratories, Division of Clinical Microbiology, Mayo Clinic and Mayo Foundation, 200 First St. SW, Rochester, MN 55905

BYUNGSE SUH   CHAPTER 16

Section of Infectious Diseases, Temple University Hospital, Philadelphia, PA 19140

KEN B. WAITES   CHAPTER 8

Department of Pathology, University of Alabama at Birmingham, 619 S. 19th St., WP P230, Birmingham, AL 35233-7331

MELVIN P. WEINSTEIN   CHAPTER 2

University of Medicine and Dentistry of New Jersey-Robert
Wood Johnson Medical School and Microbiology Laboratory,
Robert Wood Johnson University Hospital, New Brunswick,
NJ 08901-0019

Contributors

RAY D. ALLER
Medical Affairs and Informatics, MDS Laboratory Services
(US), 5217 Maryland Way, Suite 303, Brentwood, TN 37027

CÉCILE M. BÉBÉAR
Laboratoire de Bactériologie, Université Victor Segalen
Bordeaux 2, 33076 Bordeaux, France

CAROLYN M. BLACK
Scientific Resources Program, National Center for Infectious
Diseases, Centers for Disease Control and Prevention,
Atlanta, GA 30333

PAUL P. BOURBEAU
Division of Laboratory Medicine, Geisinger Medical Center,
Danville, PA 17822

CHRISTOPHER L. EMERY
Department of Pathology and Laboratory Medicine, MCP
Hahnemann University School of Medicine and Medical
College of Pennsylvania-Hahnemann Hospitals, Philadelphia,
PA 19102

LESLIE HALL
Clinical Mycology and Mycobacteriology Laboratories,
Division of Clinical Microbiology, Mayo Clinic and Mayo
Foundation, 200 First St. SW, Rochester, MN 55905

HARALD H. KESSLER
Institute of Hygiene and Molecular Diagnostics Laboratory,
Karl-Franzens-Universität Graz, Universitätsplatz 4, A-8010
Graz, Austria

JAY R. KOSTMAN
Presbyterian Medical Center, University of Pennsylvania,
Philadelphia, PA 19104

DAVID S. LINDSAY
Center for Molecular Medicine and Infectious Diseases,
Virginia-Maryland Regional College of Veterinary Medicine,
Virginia Polytechnic Institute and State University,
Blacksburg, VA 24061

RICHARD MANIGLIA
Presbyterian Medical Center, University of Pennsylvania,
Philadelphia, PA 19104

XIANG-JIN MENG
Center for Molecular Medicine and Infectious Diseases,
Virginia-Maryland Regional College of Veterinary Medicine,
Virginia Polytechnic Institute and State University,
Blacksburg, VA 24061

KAREN W. POST
Rollins Animal Disease Diagnostic Laboratory, 2101 Blue
Ridge Rd., Raleigh, NC 27607

L. BARTH RELLER
Departments of Pathology and Medicine and Clinical
Microbiology Laboratory, Duke University Medical Center,
Durham, NC 27710

JEAN A. SIDERS
Department of Pathology and Laboratory Medicine, Clarian
Health Partners, Methodist-IU-Riley Hospitals, Indianapolis,
IN 46228

DEBORAH F. TALKINGTON
Division of Bacterial and Mycotic Diseases, National Center
for Infectious Diseases, Centers for Disease Control and
Prevention, Atlanta, GA 30333

ALLAN L. TRUANT
Clinical Microbiology, Immunology, and Virology Laboratories
and Departments of Pathology and Laboratory Medicine,
Microbiology and Immunology, and Internal Medicine,
Temple University Hospital and School of Medicine, 2 Park
Ave. Pavilion, Broad and Ontario Sts., Philadelphia, PA
19140

DONNA M. WOLK
Molecular Diagnostics Laboratories, Southern Arizona VA
Health Care System, Laboratory Services (6-113), 3601 S. 6th
Ave., Tucson, AZ 85723

Foreword

Over 3 years ago, I received a phone call from Allan Truant asking me whether I would be willing to write a foreword to a new clinical microbiology book he was editing. I was immediately skeptical; I had not seen a truly "new" book in clinical microbiology during all of the 19 years I have been practicing in this discipline. I grew increasingly skeptical as I listened to Dr. Truant explain what he planned: a comprehensive, scholarly multiauthor look at commercial test systems in clinical microbiology. I did not think that this was possible, but I was wrong.

In this work, Dr. Truant and his coauthors have indeed created something "brand new." The information provided in this important work simply is not available anywhere else, at least not in as organized and comprehensive a fashion. More importantly, this text will have immense value for practicing clinical microbiologists, especially those faced with making decisions as to what systems should be used in specific laboratory settings.

The text is divided into 18 chapters; 12 address specific functional areas of clinical microbiology (e.g., blood cultures, bacterial identification and susceptibility testing, virology, mycobacteriology, etc.). In addition, there are separate chapters on licensure and regulation of commercial products, veterinary microbiology, laboratory information systems, immunoassay systems, emerging infectious diseases, and the future of technology in clinical microbiology. The text concludes with an appendix containing a comprehensive listing of commercial sources for the systems described.

Most chapters consist of an introductory section, a comprehensive review of salient test systems, including a comparison of performance characteristics, and summary remarks. Dr. Truant relies heavily on practicing clinical microbiologists as authors of this text, and as a result the information presented herein is of real practical utility to laboratorians. The text is very well-written, comprehensive, well-organized, and appropriately referenced. One of the major pitfalls of most multiauthored texts is lack of consistency in style, scope, and approach between different chapters and/or sections. These problems have been scrupulously avoided in this manual. As a result of careful editing, there is conspicuous balance and clear continuity across all of the chapters of this text.

One of the principal advantages of writing forewords for books such as this is that one gets to see the information before anyone else does. In my position as the director of a clinical microbiology laboratory in a large academic medical center, I find myself consulting the chapters nearly every day. I have not been disappointed.

Increasingly, clinical microbiology laboratories have become reliant on new technologies in their day-to-day provision of diagnostic services. It is estimated that more technology has been introduced into the clinical microbiology laboratory during the past decade than during the entire century that preceded it. This trend will only continue, if not accelerate. Keeping up with new technologies becomes nearly impossible. In this respect, the *Manual of Commercial Methods in Clinical Microbiology* will undoubtedly come to occupy a readily accessible and prominent spot on the bookshelves of practicing clinical microbiologists everywhere. This is one of those rare books that is destined for a tattered and worn existence soon after it is published. This reality further ensures that the first edition of the manual must be followed by an updated second edition, and then a third edition, and a fourth. . . .

One final (and perhaps political) comment. Beyond its practical application, reading this text from cover to cover leads to a clear perspective on what a truly enormous impact new technologies have had on the provision of clinical microbiology services over a very short time. Obviously, the same has been true in all areas of health care. What a difference a decade makes. Nonetheless, I suspect that 100 years from now, when future workers in our field peer back through their retrospectoscopes and ask the question, "What was the single most defining element in medicine during the beginning of the new millennium, at least in the United States?", the answer will not be technology advances. It probably will not be managed care either, or, for that matter, novel approaches to diagnosing and treating diseases like diabetes, heart disease, AIDS, and cancer. It will not even be genomics. I strongly suspect the answer will be acceptance of evidence-based medicine as a valid paradigm.

In view of the myriad of technologies that now confront us, we must begin to ask serious questions about their real clinical value, their cost-effectiveness, and how they can benefit us in ways that old approaches could not. The studies need to be systematic, objective, thoughtful, and comprehensive. And then we must carefully analyze and evaluate the data. It is only then that we can make valid decisions as to the most prudent applications of these new technologies. In this respect, the *Manual of Commercial Methods in Clinical Microbiology* serves another very useful purpose. It serves as a valuable starting point for the design and implementation of the many investigations that must certainly follow. Dr. Truant and his coauthors have done all of us who practice clinical microbiology a lasting service.

GARY V. DOERN

Preface

The *Manual of Commercial Methods in Clinical Microbiology* is a natural progression of the texts currently available in the discipline of clinical microbiology. The 3 decades since the first edition of the *Manual of Clinical Microbiology* (MCM) was published by the American Society for Microbiology have brought tremendous strides in the availability, quality, and clinical usefulness of commercial products in clinical microbiology, particularly bacteriology. In fact, in the first edition of MCM (1970), other than the description of culture media, reagents and stains, one would be hard pressed to find information on commercially available products of any kind. Commercially available microscopes and illuminators (since many light sources were separate from the microscopes at that time) were briefly mentioned, and a commercially available latex particle agglutination test for *Coccidioides immitis* was reviewed. This serves to emphasize the paucity of commercially available kits which the clinical microbiologist could use for the identification and antimicrobial susceptibility of significant pathogens at that time. In addition, in recent years, the advent of molecular methods in clinical microbiology has revolutionized the evaluation of patients with presumed infectious diseases. We are moving from an era of phenotypic identification of organisms and experiencing a paradigm shift to a true genotypic identification. This has refined and sometimes even changed the taxonomic placement of organisms which were previously categorized primarily by phenotypic characteristics.

We are just beginning to realize the value of our discipline in evaluating patients with infectious diseases and contributing to the use of therapeutic modalities in their treatment and cure. If one measures the discipline of clinical microbiology from the initiation of ASM's clinical microbiology section (now a division), it is barely middle-aged. As clinical microbiologists, we are only at the beginning of our contributions to infectious diseases and clinical medicine.

Bacteriologists and microbiologists practiced this discipline long before it was known as clinical microbiology. Indeed, Pasteur, Koch, and others used the science of microbiology to help their fellow humans to improve their collective health and quality of life. Some may consider them the first clinical microbiologists. And, although many discussions ensued between those arguing that the discipline should be called clinical microbiology and those arguing that it should be called diagnostic microbiology, the discipline is generally known today as clinical microbiology, since those who practice this specialty perform services which are both diagnostic and consultative. We consult with clinicians (indeed, some people are both microbiologists and clinicians), interpret laboratory results, and work very closely with specialists in infectious diseases and other medical disciplines, infection control practitioners, and many other health care providers.

This reference is not intended to provide an exhaustive review of clinical microbiology procedures. The *Manual of Clinical Microbiology*, the *Manual of Clinical Laboratory Immunology*, the *Clinical Microbiology Procedures Handbook*, *Essential Procedures for Clinical Microbiology*, and other fine texts provide this resource for those of us practicing clinical microbiology. However, to date, there has been no general resource for all subdisciplines of clinical microbiology to use when evaluating commercial methods, tests or products. Sources such as ASM books, the *Cumitech* series, periodicals (such as the *Journal of Clinical Microbiology*), reviews and scientific journal articles, and package inserts, which occasionally contain in-house data from the manufacturer, have been the main resource for those needing specific information on commercial methods and products.

In this book, we attempt to review the commercial tests (both manual and automated) in the discipline of clinical microbiology. Descriptions of their sensitivities and specificities and predictive values from peer-reviewed sources, when available, are included. The authors have attempted to include the most recent peer-reviewed data from respected journals. We also try to predict new tests or methods which may be used in the next several years. Additional information which will be of value is included in chapters devoted to molecular microbiology, information management, emerging infectious deseases, and veterinary clinical microbiology.

This manual, then, can be used as a starting point when one needs to evaluate which test, kit, procedure, or instrument may be considered or chosen for an organism detection, organism identification, or antibiotic susceptibility test. Current pricing information, newly published peer-reviewed product evaluations, new products, and other information will supplement the information provided in this manual.

The authors have attempted not to duplicate recent information, methods, and procedures or product reviews which are readily available in other references from ASM or other sources. Among the topics which are not reviewed in this text are taxonomy; routine culture information; and general issues such as normal floras, microscopy, laboratory management, infection control, and a comprehensive description of all of the major groups of organisms. Additionally, if substantive product reviews have been included in recent reviews or texts, they are not duplicated in this manual, and the reader is referred to the authoritative reference. This text also does not review the plethora of antibody tests which are covered authoritatively and extensively in the *Manual of Clinical Laboratory Immunology*. However, there is some unavoidable and necessary overlap of selected information. For example, the topic of collection, storage, and transportation is addressed by some authors if deemed significantly different from previous descriptions or unique to a specific commercial test. The exclusion of antibody tests will not pertain to the chapters on human immunodeficiency virus and mycoplasmas, since many of the significant methods are antibody based. In addition, for hepatitis and rickettsia methods, the reader is referred to the *Manual of Clinical Microbiology* and the *Manual of Clinical Laboratory Immunology* for current authoritative reviews.

The contributors to this book consist of a wonderful group of practicing clinical microbiologists, physicians, and scientists with a wealth of experience. I thank all of the authors for their diligence and completeness. I apologize for any unintentional omission of tests or data, and I encourage the reader to contact me or the ASM Press editorial office with any suggestions or corrections for future editions. The authors have done their best to describe the most commonly used products in clinical microbiology. As all clinical microbiologists readily understand, comparisons between products are quite difficult and scientifically challenging. Among these difficulties are changing reagents, pricing, software, hardware, and databases; different groups of products which are evaluated and described in the literature at different times, in different laboratories, or with different reference procedures; new products; and products which have been removed from the marketplace. The contributors have attempted to sort through this maze and have tried to include only the most recent and pertinent references. Of course, each laboratory and director must thoroughly evaluate each method for its usefulness in a specific setting before implementation, keeping in mind the complexity of the method and technical expertise required, review of the test by the Food and Drug Administration, the cost of the method (including reagents, personnel and equipment), and its clinical usefulness.

I hope the information presented in this manual proves useful to clinical microbiologists, pathologists, infectious disease specialists, and other health care providers. And I especially hope that it serves as an incentive for improvement of this reference in future editions and an inspiration for future accomplishments in clinical microbiology and infectious diseases.

ALLAN L. TRUANT

Acknowledgments

I thank Jeff Holtmeier, director of ASM Press, for his guidance and wisdom. Without his foresight and diligence, this project would not have been possible. Many thanks are also given to Ken April of ASM Press for his persistence and skills in bringing this project to completion. Thanks also are extended to Paul Edelstein, Irv Nachamkin, Nancy Strockbine, Washington Winn, Marianna Wilson, and Helen Buckley for review of selected portions of this manual. I also thank the College of American Pathologists for providing information to the authors of this manual and *CAP Today* for providing information and data for chapter 15.

I also thank a number of family members, to whom this book is dedicated. I thank my father, the late Joseph Paul Truant, Ph.D. (1923–1988), the founder and coeditor of the first three editions of the *Manual of Clinical Microbiology*, and the first chairperson of the Clinical Microbiology Section of the American Society for Microbiology. The Clinical Microbiology Section, which later became a division, is now the largest division of ASM. He also served as secretary, vice-chairman, and chairman of the Medical Division of ASM and as an editorial board member for journals published by ASM and the Canadian Society for Microbiology. His selfless dedication and time devoted to ASM and to helping establish, with the collaboration of many others, clinical microbiology as an accepted discipline have paved the way for those of us who practice this specialty and strive to minimize the morbidity and mortality associated with infectious diseases. I also am grateful to my mother, Flora Rina (Lenardon) Truant, who practiced infection control in the home before it was a well-understood principle, who taught her children persistence and determination, and who is the epitome of America's work ethic and moral strength; my children, Patti, Kathleen, and Steven, who give me an emotional compass and who have loved and supported me throughout this project and well beyond; and my sisters, Linda Ann Thompson and Anita Louise Ryan, and their families for their love and support.

ALLAN L. TRUANT

Important Notice

The authors and editor have taken great care to confirm the accuracy of the information presented in this manual. However, the authors, editor, and publisher make no warranty, expressed or implied, that the information in this book is accurate or appropriate for any particular facility or environment or any individual employee's personal situation, and they are not responsible for any consequences of application of any of the information in this book by any reader. The inclusion of specific products, instruments, reagents, or kits by the contributors of this manual does not represent any endorsement of any such product by the American Society for Microbiology or ASM Press or its contributors or editors, nor does the inclusion or inadvertent exclusion of any product, instrument, reagent, or kit reflect a preference for any product over other similar competitive products. The comments included in this manual are those of the authors and do not necessarily reflect the views of their employers or institutions.

Some of the tests and methods discussed in this manual have Food and Drug Administration (FDA) clearance for selected uses. It is the responsibility of the laboratorian or health care provider to ascertain the FDA status of each product which is considered for use in his or her hospital, setting, or practice.

Role of the U.S. Food and Drug Administration in Regulation of Commercial Clinical Microbiology Products

SHARON L. HANSEN

1

INTRODUCTION AND HISTORICAL REVIEW OF LEGISLATION LEADING TO FEDERAL REGULATION OF DRUGS, BIOLOGICS, AND MEDICAL DEVICES

Because the U.S. Food and Drug Administration (FDA) approval process has a major impact on how and when new in vitro diagnostic devices (IVDs) become available for use by the health care professional or the layperson, it is important to examine how the FDA and the Health Care Finance Administration (HCFA) regulatory processes operate.

Congress has a long legislative history of reaction to a series of catastrophes (emergencies) which led to food and drug laws enacted to promote and protect the public health.

- In 1848, the Drug Importation Act passed by Congress required U.S. Customs Service inspection to stop the entry of adulterated drugs from foreign countries. This was the first attempt to regulate the quality of drugs and was a direct result of adulterated quinine, which was being supplied to U.S. troops in Mexico.
- The passage of the Biologics Control Act in 1902 occurred after 13 children in St. Louis, Mo., died from tetanus after being injected with diphtheria antitoxin. The equine serum containing the diphtheria antitoxin was found to be contaminated with *Clostridium tetani*. The intent of this law was to ensure the purity and safety of serums, vaccines, and similar products used to prevent or treat diseases in humans.
- The original Food and Drugs Act was passed in 1906 and signed by President Theodore Roosevelt. It required food and drugs marketed via interstate commerce to meet their professed minimal standards of strength, purity, and quality. If they did not meet their labeling claims, the product would be considered adulterated. This law gave "policing" authority to the federal bureau charged with the policing but did not grant enforcement authority. Compliance by the industry was therefore voluntary.
- The 1912 Sherley Amendment to the Food and Drugs Act coined the term "misbranded," which refers to fraudulent or false claims of therapeutic effect on labeled medicines. Thus, the concept of a food or drug having to be safe and effective before it could be sold via interstate commerce was born.

- The concept of safety was solidified in 1938 when 107 people were killed after ingesting a cough syrup containing sulfanilamide. The poisonous solvent used in the elixir to dissolve the sulfanilamide was diethylene glycol. Through President Roosevelt's efforts and commitment to protection of the public from unsafe food and drugs, the Federal Food, Drug and Cosmetic (FFDC) Act became law. The FFDC Act authorized the creation of what later became the FDA and gave it direct regulatory authority over the food, drug, device, and cosmetic manufacturers selling products via interstate commerce. The law required that before a new drug could be placed in interstate commerce it must be shown to be safe. It also authorized factory inspections and added the remedy of court injunctions to the previous penalties of seizures and prosecutions.
- In 1944, the Public Health Service Act was passed, covering a broad spectrum of health concerns, including regulation of biological products and control of communicable diseases. Provisions in this act eventually led to the umbrella organization that became the Department of Health and Human Services, under which the FDA, Centers for Disease Control and Prevention, and HCFA now exist.
- In 1955, after 264 cases of polio occurred in individuals who had received a poliovirus vaccine containing incompletely inactivated virus, the first regulations for the manufacture of biologic products were enacted.
- Although thalidomide, a new sleeping pill, had not been approved for use in the United States, the drug was found to be the cause of birth defects in thousands of babies born in Western Europe. News reports of the role of Frances Kelsey, an FDA medical officer, in keeping the drug off the U.S. market aroused public support for stronger drug regulation. The public outcry led to an amendment to the FFDC Act in 1962. This amendment requires that a drug manufacturer prove to the FDA the effectiveness of a product before it can be introduced to the marketplace.
- The major source of FDA authority over medical devices is the FFDC Act with its Medical Device Amendments (MDA) of 1976. The enactment of these amendments was a direct result of a retrospective 10-year search of the medical literature by the then Department of Health,

Education, and Welfare. The search revealed that over 10,000 verified injuries were directly related to medical devices, 751 of which were fatal. A plethora of commercially available "quack" devices, injuries due to implantable devices, failures of devices to perform as intended, and user injuries showed the need for regulatory control. The amendments clarified drug and device status, required that the FDA be notified before a medical device could be placed in interstate commerce, provided two pathways for the industry to gain entry into the marketplace, established a classification system to assign the level of control necessary to ensure the safety and effectiveness of each device, and, for the first time, authorized mechanisms of action the agency could take against products in violation of the regulations. Another significant effect of the amendments was to establish criteria for and regulation of "good manufacturing practices." The law also required registration of all manufacturers of medical devices sold via interstate commerce in the United States.

- The Safe Medical Devices Act (SMDA) became law in 1990. A major revision of the 1976 amendments, it added new provisions to better ensure that devices entering the marketplace are safe and effective, provided means for the FDA to learn quickly about serious device problems, and provided means to remove defective devices from the market.

- The Prescription Drug User Fee Act of 1992 requires drug and biologics manufacturers to pay fees for product applications and supplements and other services. The act also requires the FDA to use these funds to hire more reviewers to assess the safety and effectiveness of new drug applications at the Center for Drug Evaluation and Research (CDER).

- The Food and Drug Administration Modernization Act of 1997 (FDAMA) reauthorizes the Prescription Drug User Fee Act of 1992 and mandates the most wide-ranging reforms in agency practices since 1938. The provisions include measures to accelerate FDA review of devices, regulate advertising of unapproved drugs and devices, and regulate health claims for foods.

FDA REGULATION OF CLINICAL MICROBIOLOGY IVDs

For the purpose of this chapter, examples and cited regulations will for the most part be confined to the FDA process which permits commercial interstate sale and distribution of clinical microbiology IVDs by the Center for Devices and Radiological Health (CDRH).

MDA of 1976

Definition (Title 21, *Code of Federal Regulations* [21 *CFR*] section 809.3)

IVDs are those reagents, instruments, and systems intended for the diagnosis of disease or other conditions, including a determination of the state of health, in order to cure, mitigate, treat, or prevent disease or its sequelae. Such products are intended for use in the collection, preparation, and examination of specimens taken from the human body. These products are devices as defined in section 201(h) of the FFDC Act and may also be biological products subject to section 351 of the Public Health Service Act.

Classification

The 1976 MDA required the FDA to classify all medical devices in the marketplace intended for human use in one of three regulatory categories. These regulatory categories (classes) are assigned according to the extent of control necessary to ensure the safety and effectiveness of each device.

- Class I, general controls, includes devices which appear to pose relatively little risk to human health and for which a series of general controls are believed sufficient to ensure safety and effectiveness. The controls include regulations that (i) prohibit the sale of adulterated or misbranded devices; (ii) require domestic device manufacturers and initial distributors to register their establishments with the FDA and provide the FDA with a list of all devices being sold before the passage of the law (preamendment devices); (iii) grant the FDA authority to ban certain devices; (iv) provide for notification to the FDA of risks and of repair, replacement, or refund (recall); (v) restrict the sale, distribution, or use of certain devices; and (vi) govern good manufacturing practices, records and reports, and inspections. These requirements also apply to class II and class III devices.

- Class II, performance standards, includes devices for which general controls alone are believed insufficient to ensure safety and effectiveness and for which existing information is sufficient to establish a performance standard that provides this assurance. Class II devices must comply not only with general controls but also with any applicable mandatory standards.

- Class III, premarket approval, includes devices for which insufficient information exists to ensure that general controls and performance standards provide reasonable assurance of safety and effectiveness. Generally, class III devices are those represented as life sustaining or life supporting, those implanted in the body, and those representing unreasonable risk of illness or injury. New devices are assigned to class III until they are found to be substantially equivalent to a preamendment device (class I or II) or are reclassified as class I or II devices through petition. New class III devices must have either approved premarket approval applications (PMAs) or completed product development protocols before sale in the United States.

Under the amendments, a provision titled Premarket Notification [section 510(k) of the FFDC Act] requires that a manufacturer notify the FDA 90 days before it begins marketing a device that was not sold before 28 May 1976. The purpose of this requirement was to ensure that manufacturers do not begin marketing new devices until such devices either receive premarket approval or are reclassified as class I or II.

Any new device is automatically placed in class III unless it is found to be substantially equivalent to either a preamendment device or a postamendment device that has already been reclassified in class I or II (a "predicate device"). If the new device is found to be substantially equivalent to a preamendment device or a reclassified device through the 510(k) notification process, then the new device automatically assumes the classification of the predicate device and enjoys the same level of freedom for commercialization, producing a kind of marketing equity and avoiding de facto monopolies.

After the passage of the amendments, a series of classification panels convened to establish classification categories for all medical devices already in the marketplace. These panels represented product categories like those described for IVDs and were divided into subpanels based on the laboratory discipline they were concerned with (Clinical Chemistry and Clinical Toxicology Devices, Hematology and Pathology Devices, and Immunology and Microbiology Devices). Subject matter experts for each panel were selected from academia, public health services, industry, professional clinical groups, and consumers to determine the level of regulatory control (class I, class II, or class III) required to provide reasonable assurance of the safety and effectiveness of each device before the panel.

Microbiology device regulations and their classifications were divided into four subparts, which can be found in 21 CFR, Microbiology Regulation, sections 866.1 through 866.3930. The reader is encouraged to review these regulations to gain an understanding of where the science of clinical microbiology was in 1976 and where we are today, particularly when we contrast the intended uses and natures of the devices then and now. The following are examples of class I and class II microbiology devices, quoted from 21 CFR.

§866.2660 Microorganism Differentiation and Identification Device.
Identification. A microorganism differentiation and identification device is a device intended for medical purposes that consists of one or more components, such as differential culture media, biochemical reagents, and paper discs or strips impregnated with test reagents, that are usually contained in individual compartments and used to differentiate and identify selected microorganisms. The device aids in the diagnosis of disease.
Classification. Class I (general controls).
§866.3165 *Cryptococcus neoformans* Serological Reagents
Identification. *C. neoformans* serologic reagents are devices that consist of antigens and antisera used in serologic tests to identify antibodies to *C. neoformans* in serum. Additionally, some of these reagents consist of antisera conjugated with a fluorescent dye (immunofluorescent reagents) and are used to identify *C. neoformans* directly from clinical specimens or from cultured isolates derived from clinical specimens. The identification aids in the diagnosis of cryptococcosis and provides epidemiological information on this type of disease. Cryptococcosis infections are found most often as chronic meningitis (inflammation of brain membranes) and, if not treated, are usually fatal.
Classification. Class II (performance standards).

Detection and Identification

Detection and identification of microorganisms directly from clinical material were in their infancy in 1976 with the exception of antisera conjugated with a fluorescent dye and directed at fastidious organisms, such as *Francisella tularensis*, *Toxoplasma gondii*, rickettsiae, or rabies virus. Culture isolation of bacteria and fungi with subsequent identification of organisms through growth-dependent phenotypic or serologic characterization of isolated colonies, identification of specific microbial antibodies in sera, and agar disk diffusion susceptibility testing were the most commonly used commercial processes. All media, with the exception of susceptibility test media, media for isolation of pathogenic *Neisseria* spp., and media for collection and transport of clinical specimens, are class I and have been exempted from 510(k) premarket notification (effective date of exemption, 1989).

After the medical device regulations were promulgated, administrative responsibility was given to the new FDA CDRH, created over the Office of Device Evaluation, which includes the Division of Clinical Laboratory Devices (DCLD). The Microbiology Branch has the responsibility for reviewing premarket submissions for all in vitro microbiology and infectious disease diagnostic devices, with the exception of those for the human immunodeficiency virus and detection of infectious disease etiological agents in blood and blood products, which are under the regulatory authority of the FDA Center for Biologics Evaluation and Research.

Microbiology Device Review Post-MDA

In the 1980s, 510(k) review for microbiology devices was essentially a notification of intent to market a device in 90 days. Few if any well-controlled clinical or laboratory studies were required to establish the performance characteristics of the product before it was cleared for the marketplace. Little or no information or data to substantiate labeling claims were generated in the clinical setting where the device was intended to be used or from patient specimens in the patient population for which the device would have an indicated use (the target patient population). Diagnostic products often reached the marketplace before the disease being diagnosed was well understood, e.g., assays for detecting antibodies to *Borrelia burgdorferi*, the causative organism of Lyme disease. The addition of antimicrobials to commercial susceptibility test panels was cleared or approved without assessing their abilities to detect resistance in the organisms the drug had been approved for use against. Product inserts for estimations of accuracy, sensitivity, specificity, 95% confidence intervals, and predictive values in low- or high-prevalence target patient populations, when present, were often not defined following appropriate clinical or statistical criteria.

SMDA OF 1990

The SMDA of 1990, a major revision of the 1976 amendments, added new provisions to better ensure that devices entering the marketplace were safe and effective, and it provided means for the FDA to learn quickly about serious device problems and to remove defective devices from the marketplace. The SMDA gave the FDA authority to ask for clinical data deemed necessary to support product claims before clearing 510(k) submissions. Before passage of the SMDA, manufacturers of IVDs that were class I or II sent 510(k) notification of their intent to enter the marketplace in 90 days. Product review had essentially been an administrative review and seldom required scientific data and evidence to provide reasonable assurance that the new device was as safe and effective as the predicate device. The SMDA prohibited manufacturers from marketing products until they had received an order from the FDA to do so. This act authorizes the FDA to order device product recalls and requires nursing homes, hospitals, and other facilities that use medical devices to report to the FDA incidents that suggest that a medical device (including IVDs) probably caused or contributed to the death, serious injury, or serious illness of a patient.

At the request of the Commissioner of Food and Drugs, a committee, chaired by Robert Temple and composed of experienced clinical and statistical reviewers from the CDER, was asked to perform reviews of selected pending and approved applications for medical devices. In the final report, the committee's findings for deficiencies in the clinical data submitted to the CDRH in support of 510(k) and

PMA submissions were as follows: (i) the fundamental problem leading to inadequate data in most applications was lack of attention to basic clinical study design; (ii) there were deficiencies in the reporting and statistical analysis of data generated during the studies. The CDRH and the chairs of the CDRH Advisory Panels reviewed these findings. The final recommendations and conclusions for improving the quality of the data submitted in device applications were as follows: there was a need for (i) significant outreach programs to educate the device industry on study design, control, and analysis; (ii) integration of biostatisticians into the review process; (iii) development of guidance to the industry on design and analysis in general and for specific device classes; (iv) specific training and guidance on trial design and analysis for CDRH reviewers; and (v) consistent and reliable advice throughout the testing and review process.

The SMDA, the final report of the Temple Committee in 1993, and new leadership in the CDRH and the Office of Device Evaluation led to a number of initiatives directed at raising the scientific and statistical basis for 510(k) and PMA device review. Additional scientists, medical officers, and biostatisticians were added to the review staff in the DCLD and the Microbiology Branch.

In an effort to provide the microbiology IVD industry with direction about the kinds and amount of data and information necessary to obtain clearance or approval of devices for use in detecting, monitoring, or assisting in therapeutic decisions for a patient with an infectious disease and to establish a level playing field for the industry, the Microbiology Branch developed guidance for review criteria for a number of products. In addition, the industry was encouraged to interact with the branch early in the product development phase and before they initiated their outside clinical studies. The branch developed a generic outline for writing a study protocol to establish performance characteristics for devices intended to identify, diagnose, monitor, and treat etiologic agents of an infectious disease. Sound scientific and statistical principles were stressed as the foundation for "truth in labeling" of the final product insert. Product limitations were to be clearly stated in the labeling so that the laboratory user and clinician could avoid the risk resulting from false or misleading claims.

Microbiology Device Review Post-SMDA

510(k) Premarket Notification

Premarket notification under the 510(k) section of the MDA creates the largest volume of review activity for the Microbiology Branch in the DCLD (about 200 submissions per year). One of the revisions in the 1990 law concerns substantial equivalence determinations by the FDA based on the information or finding that the device presented in a 510(k) notification is considered as safe and effective as, and substantially equivalent to, a legally marketed device. Typically, in a 510(k) notification, the manufacturer must describe the device methodology or technology, intended use, indications for use, and performance characteristics as shown in labeling (the product insert) and in promotional material and advertisements; compare and contrast the new device with a similar legally marketed device with supporting data; or, in the case of modified devices, show that the modification could affect the safety and effectiveness of the device and provide documentation and data to address those effects.

The FDA assesses the information and scientific data in the 510(k) notification to determine if the device is substantially equivalent to and as safe and effective as a legally marketed device, in which case the device is cleared and the manufacturer is free to market it in the United States. If the determination is made that the device has a new intended use; that the technology raises different issues of safety and effectiveness, such as over-the-counter use rather than use by a professional; or that the analyte being detected was not known before 1976 (no predicate device), then the device is found not substantially equivalent and a PMA is required. In assessing a 510(k) premarket notification, the FDA considers the nature of the analyte and its intended use, how it compares with existing device technologies or methodologies, the adequacy of the manufacturer's evaluation, the establishment of the performance characteristics of the device, and benefits and risks associated with the device's clinical utility.

The 510(k) review in DCLD is entirely a paper review; the CDRH Office of Device Evaluation does not submit the actual products to direct laboratory evaluation, and the agency therefore has no hands-on experience with the vast majority of products under review. The DCLD is continually challenged by the need to establish appropriate criteria to determine safety and effectiveness and for the substantial-equivalence decision, since performance standards for class II devices have never been established as required by statute. The National Committee for Clinical Laboratory Standards (NCCLS) consensus process has resulted in a number of NCCLS standards, reference methods, and definitions, many of which are used by the FDA and industry as guidance for certain aspects of product evaluation. As an example, the Microbiology Branch has embraced the NCCLS reference antimicrobial susceptibility testing (AST) standards for assessment of commercial AST products since 1991. A component of the performance evaluation of commercial AST IVDs requires comparison to the appropriate NCCLS method, with the results of that comparison being summarized in the FDA-cleared labeling of the device product insert in addition to the limitations of the commercial method with certain organism-drug combinations.

The 510(k) review is usually done by an individual scientist and occasionally by a team of FDA scientists and a statistician. Completion of the review is targeted for 90 days but may be delayed due to the need for additional information and data from the manufacturer. As will be discussed below, the center has adopted policies to reduce the review time for certain 510(k) submissions.

PMAs

A PMA is required when a device is determined not to be substantially equivalent to a preamendment device (i.e., class III). In addition to an in-depth scientific review [as is the case with moderate- and high-risk class I and II 510(k) devices] and before a PMA can be approved, the firm must undergo a comprehensive good manufacturing practice inspection following the Quality System Regulation [not required for 510(k) submissions], and the PMA is reviewed by the FDA microbiology advisory panel of outside experts, who provide recommendations to the FDA for approval or disapproval of the PMA. Examples of microbiology devices requiring a PMA are devices directed at detection and typing of human papillomavirus and hepatitis A, B, and C viruses; parvovirus B19 diagnostic, detection, and monitoring devices; and devices for direct detection of *Mycobacterium tuberculosis* from clinical material by nucleic acid amplification techniques. The internal review team for PMAs is composed of agency scientists, medical officers, statisticians, and compliance and administrative personnel. Device PMA review is frequently lengthy, but the review time is seldom equal to the time required to either approve a new drug (CDER) or license and approve a new biologic (Center for Biologics Evaluation and Research).

IVD PRODUCT LABELING

IVDs are unique among medical devices, as they have their own collective labeling regulation, found in 21 CFR section 809.10. Federal IVD regulations apply to all legally marketed commercial IVDs [21 CFR, section 809.10(a) and (b)], whether class I exempt or nonexempt, class II, or class III; devices for investigative use [21 CFR, section 809.10(c)(2)(ii)] and research use [21 CFR, section 809.10(c)(2)(I)]; and "home brew" devices which fall under the Analyte Specific Reagent (ASR) Regulation [21 CFR, section 809.30] and are labeled under 21 CFR, section 809.10(e).

Products for In Vitro Diagnostic Use

The proposed package insert is part of the 510(k) notification or PMA. The final step of the Microbiology Branch review after all analytical and clinical data have been critically reviewed is to clear or approve the proposed product labeling. The branch pays particular attention to the intended use, specimen collection, transport and storage recommendations, warnings and limitations, expected values, validation of cutoff, results and their interpretation, quality control recommendations, and specific performance characteristics. The reviewer verifies that all claims made in the product insert have been substantiated in the 510(k) review. The nature of the studies to establish product performance characteristics must be clearly stated. Truth in labeling is critical to alert users to the kind of results and performance they can reasonably expect to attain under the conditions specified in the product insert for the claimed intended use of the product. All microbiology products which have undergone a scientific evaluation of data to substantiate product performance claims (all nonexempt devices) as stated in the product insert should be expected to maintain that performance throughout the life of the product. Failure to maintain the expected performance, if known and reported to the FDA, could result in compliance or regulatory action. The statement "for in vitro diagnostic use" can legally be found in a product insert when the device is exempt from 510(k) notification or when the device is FDA cleared through the 510(k) notification or FDA approved through the PMA.

As mentioned previously, devices that are exempt from 510(k) review must also have a product insert that contains the required labeling information. The FDA does not review labeling in the product inserts for exempt devices before the products reach the marketplace.

Products for Research Use Only
(Test Kits and Instrument Assays)

Products which are in the laboratory research phase of development and are not represented as effective IVDs must bear the statement "for research use only; not for use in diagnostic procedures" on the product labeling. If the research use involves the participation of human subjects, institution review board approval and informed consent apply. While research tests may be performed using either clinical or nonclinical materials, research use products have no intended clinical use, and the testing performed is not designed to provide data addressing or demonstrating safety and effectiveness.

Devices labeled "for research use" are mislabeled if the device is being used for investigational purposes, i.e., in a clinical study, even if it involves only one subject, where the diagnostic or prognostic measurement will be reported to the patient's physician or in medical records or will be used to assess the patient's condition, regardless of whether confirmatory tests or procedures are used. Research use is limited to the initial research phase of product development that is necessary to identify test kit methods, components, and analytes to be measured or to laboratory research that is entirely unrelated to product development [21 CFR, section 809.10(c)(2)(1)].

Labeling of research use only devices may not contain any reference to expected values or specific performance characteristics or a specific intended use with a target patient population. Tests performed with in vitro products intended for research use are used in a preclinical or nonclinical setting. By definition, research devices have no intended use, and the testing performed is not designed to provide data addressing or demonstrating safety and effectiveness.

Products for Investigational Use Only
(Test Kits and Instrument Assays)

For a product being shipped or delivered for product testing before full commercial marketing (for example, for use on specimens derived from humans to compare the usefulness of the product with that of other products or procedures which are in current use or recognized as useful), all labeling must bear the statement, prominently placed, "for investigational use only; the performance characteristics of this product have not been established." An investigational device is a device that is the object of an investigation. Investigation means clinical investigation or research involving one or more subjects to determine the safety or effectiveness of a device for its intended clinical use, to develop performance characteristics for the product, and to establish the expected ranges of values when the device is used. Unless they are exempt from premarket notification requirements, it is illegal to commercially market IVDs that have not been approved or cleared by the FDA. The FFDC Act and the FDA regulations prohibit promotion and misrepresentation of an investigational device.

Labeling for investigational devices in addition to the statement cited in the previous paragraph must describe all relevant contraindications, hazards, adverse effects, interfering substances or devices, warnings, and precautions. The investigational device labeling must not bear any statement that is false or misleading or include any information on performance characteristics or reference ranges, and the device must not be represented as safe or effective for the purposes for which it is being investigated.

ASRs

ASRs are regulated under 21 CFR, section 864.4020, and are defined as antibodies, both polyclonal and monoclonal; specific receptor proteins; ligands; nucleic acid sequences; and similar reagents which, through specific binding or chemical reaction with substrates in a specimen, are intended for use in a diagnostic application for identification and quantification of an individual chemical substance or ligand in biological specimens.

ASRs may be thought of as the "active ingredients" of tests that are used to identify one specific disease or condition. ASRs are purchased by manufacturers, who use them as components of tests that have been approved or cleared by the FDA. Clinical laboratories also use ASRs to develop in-house tests used exclusively by each laboratory. These in-house-developed tests (sometimes referred to as home brew tests) include those that measure a wide variety of antibodies used in the diagnosis of infectious disease, cancer, and genetic and various other conditions.

The final rule for ASRs was published in the *Federal Register* on 21 November 1997 and became effective on 23 November 1998. The final rule classified or reclassified the majority of ASRs as class I medical devices. The final rule also exempted these class I devices from premarket notification requirements of section 510(k) of the FFDC Act. A few ASRs that are used in blood-banking tests were designated class II devices, e.g., certain cytomegalovirus serological and *Treponema pallidum* nontreponemal test reagents. Provisions were established for classification of ASRs as class III when (i) the analyte is intended as a component in a test intended for use in the diagnosis of a contagious condition that is highly likely to result in a fatal outcome and prompt, accurate diagnosis offers the opportunity to mitigate the public health impact of the condition (e.g., human immunodeficiency virus and AIDS or tuberculosis) or (ii) the analyte is intended for use in donor screening for conditions for which the FDA has recommended or required testing in order to safeguard the blood supply or establish the safe use of blood or blood products (e.g., tests for hepatitis or for identifying blood groups).

The regulation further imposes restrictions on the sale, use, labeling, and distribution of ASRs. Sale of ASRs to IVD manufacturers as a component of an FDA-approved or -cleared test or to organizations that use the reagents to make tests for purposes other than providing diagnostic information to patients and practitioners (e.g., forensic, academic research, and other nonclinical laboratories) are not regulated or restricted (21 CFR, section 809.30).

Labeling of ASRs

ASRs must be labeled in accordance with 21 CFR, section 809.10(e). For class I exempt ASRs, the statement "analyte-specific reagent; analytical and performance characteristics are not established" must be incorporated in the labeling. Labeling of class II and III ASRs must contain the statement "analyte-specific reagent; except as a component of the approved/cleared test (name of approved/cleared test), analytical and performance characteristics of this ASR are not established."

High-complexity (Clinical Laboratory Improvement Amendments of 1988 [CLIA 88]) clinical laboratories that develop an in-house test using a class II or III ASR must inform the ordering person of the test result by appending to the test report the statement "this test was developed and its performance characteristics determined by (laboratory name); it has not been cleared or approved by the U.S. Food and Drug Administration." This statement is not applicable or required when test results are generated using a test that was cleared or approved in conjunction with the FDA review of the class II or III ASR.

THE FDAMA

Congress amended the FFDC Act in an effort to streamline the process of bringing safe and effective drugs, medical devices, and other therapies to the U.S. marketplace.

With respect to medical devices, the FDA is directed to focus its resources on the regulation of those devices that pose the greatest risk to the public and those that offer the most significant benefits. The FDA must base its decisions on clearly defined criteria and provide for appropriate interaction with the regulated industry. The new legislation assumes that enhanced collaboration between the FDA and regulated industry will accelerate the introduction of safe and effective devices to the U.S. marketplace. It remains to be seen if the legislative intent is realized.

The majority of statutory changes that the FDAMA requires for medical devices will have an impact on FDA and industry interactions, agreements, review timeframes, and data requirements to support submissions regulated under the Investigational Device Exemption, Premarket Approval, Premarket Notification [510(k)], and Humanitarian Device Exemption sections of the act. The FDAMA will also have an impact on the Classification, Labeling, and Advertising; Postmarket Surveillance; Medical Device Reporting and Recall Reports; and Compliance and Enforcement sections of the law. The new law establishes a system for recognizing national and international standards in product reviews and permits a declaration of conformity to a standard to satisfy a premarket submission requirement when applicable.

Much of the progress achieved by the FDA's medical device program during the 1990s was focused on enhancing the quality and timeliness of the review of new devices before marketing. This was because backlogs of pending submissions had reached levels that were unacceptable to both the industry and the agency. Through a number of management changes over the past 3 years, this situation has improved dramatically. The 510(k) review backlog is essentially zero. In order to maintain an efficient and effective program that maintains maximum public health benefit in light of diminishing current and anticipated budgets, the CDRH initiated a risk-based approach to restructure the workload. That approach has been codified with the passage of the FDAMA.

One mandate of the FDAMA that likely will have an unintended direct impact on clinical laboratories is the exemption of 159 class I and 26 class II IVDs, subject to certain limitations, from premarket notification requirements. By definition, class I devices are considered low-risk, well-understood products, and in the case of IVDs, a misdiagnosis as a result of using the device would not be associated with high morbidity or mortality. The class I and II microbiology IVDs exempted from premarket notification requirements generally include all microorganism differentiation and identification devices intended for use on culture-isolated organisms, culture media (except for automated blood-culturing systems, susceptibility testing, and isolation of pathogenic *Neisseria*), serologic reagents for determination of immune status or those used for epidemiological purposes, unassayed quality control material, colony counters, incubators, and gas-generating devices. A product that is exempt from 510(k) notification requirements, such as a commercial system for identification of microorganisms, can now proceed to the marketplace without any review by the FDA.

A further consideration for the laboratories that use commercial products that are exempt from FDA review and have a waived CLIA complexity categorization is the fact that these products essentially have no scientific regulatory oversight. Laboratories with certificates of waiver for these tests must follow the manufacturers' instructions for performing the tests without the need to validate or verify the stated performance characteristics of the products. It would be prudent for the test users to critically evaluate how the performance characteristics of products exempt from FDA review and categorized as CLIA waived were established.

Because the FDA cannot predict all possible intended uses or changes in fundamental scientific technologies that may significantly affect safety and effectiveness, limitations to the exemption of these class I and II devices are in the best interest of public health because they ensure that devices incorporating such changes will be reviewed for safety and effectiveness by the agency before they go to market. In order to efficiently allocate review resources, the agency has developed a risk-based approach to use of the limitations on exemptions to ensure that high-risk devices remain subject to premarket review. These limitations to the exemption of class I and II microbiology devices are directed at IVDs whose results could be used as sole or major determinants for the diagnosis or monitoring of a disease or condition and could have an impact on acute or chronic treatment decisions by the clinician. The limitations to the exemption from 510(k) notification for microbiology devices can be found in 21 CFR, section 866.9. The following is a brief synopsis of the circumstances under which a 510(k) notification must be submitted for a class I or II exempt device.

A. The device is intended for a use different from the intended use of a legally marketed device of the same generic type, e.g., the device is intended for a different medical purpose, or the device is intended for lay use where the former intended use was by health care professionals only.
B. The modified device operates using a different fundamental scientific technology than a legally marketed device of the same generic type, e.g., a surgical instrument cuts tissue with a laser beam rather than with a sharpened metal blade, or an in vitro diagnostic device detects or identifies infectious agents by using DNA probe or nucleic acid hybridization technology rather than culture or immunoassay technology.
C. The device is an IVD that is intended for one of the following purposes:
  1. Use in the diagnosis, monitoring, or screening of neoplastic diseases, with the exception of immunohistochemical devices
  2. Use in screening or diagnosis of familial and acquired genetic disorders, including inborn errors of metabolism
  3. Measuring an analyte that serves as a surrogate marker for screening, diagnosis, or monitoring of life-threatening diseases, such as AIDS, chronic or active hepatitis, tuberculosis, or myocardial infarction or to monitor therapy
  4. Assessing the risk of cardiovascular diseases
  5. Diabetes management
  6. Identifying or inferring the identity of a microorganism directly from clinical material
  7. Detection of antibodies to microorganisms other than immunoglobulin G (IgG) and IgG assays when the results are not qualitative or are used to deter-

mine immunity or when the assay is intended for use in matrices other than serum or plasma
  8. Noninvasive testing as defined in 21 CFR, section 812.3(k)
  9. Near patient testing (point of care)

Microbiology Device Review Post-FDAMA

Microbiology devices requiring a 510(k) notification and marketing clearance with intended uses as defined above will continue to undergo scientific, statistical, and labeling review by the agency. However, it may be difficult for the clinical laboratory to discern which devices are exempt from FDA review and which are not. In addition, devices cleared before the adoption of the SMDA provisions probably did not receive the same level of scientific, statistical, and labeling review as new devices with the same intended use cleared after the SMDA.

Although the IVD labeling regulation clearly requires that accuracy, precision, specificity, and sensitivity be included under the performance characteristics in the package insert when appropriate, that information may not always have a valid scientific and statistical basis. Users of laboratory IVDs are encouraged to critically assess stated performance characteristics as found in product package inserts. As a part of the substantial-equivalence decision, the DCLD has a long history of accepting new-device–to–predicate-device comparison testing using preselected specimens to establish performance characteristics of new devices. In fact, comparison testing of a new device and a predicate has been the most common approach used by the industry to support the intended use of the new device and to obtain 510(k) clearance. Frequently, estimates of sensitivity and specificity have been inappropriately included in the device product insert. Device-to-device comparison provides only estimates of the agreement of the results from one device with those from the other and does not estimate the "truth" of the result. The performance, therefore, would be better described in terms of agreement or concordance without claims of implied clinical or analytic sensitivity and specificity.

The most desirable attribute of a laboratory test is its ability to consistently produce accurate and precise results over an extended period. Device-to-device comparisons cannot address these issues when the clinical accuracy of the result is unknown. For the past several years, the microbiology device industry has been encouraged to provide new 510(k) device performance characteristics based on data that is traceable to the best available approximation of the truth, commonly termed a "gold standard," a reference method or a clinical diagnosis (disease or condition present or absent). Once a product has progressed through the research and development phase, the investigational-phase study should have the objective of characterizing the assay performance in the setting and population for which the test is intended to be used. The study should be carefully designed to provide information and data to establish the clinical and analytical accuracy of the assay. Estimates of sensitivity and specificity with 95% confidence intervals based on the prevalence of the disease or condition in the population studied would lead to meaningful positive and negative predictive values for the assay. The limitations of the assay should be clearly stated. When such information is not in the product insert, CLIA-certified laboratories are required to establish verification and validation procedures for the performance characteristics of the test. When product labeling performance claims are reflective of the prevalence of the disease in the user laboratory clinical setting, verification and validation procedures for product performance before adaptation of the new assay can be significantly reduced (1).

In June 1999, the agency proposed DCLD IVD draft labeling guidance to the industry and FDA reviewers and staff in an effort to ensure truth in labeling. The document acknowledges two categories of endpoints for assessing the performance of an assay before clearance for the marketplace. These categories are as follows. (i) Operational truth is a comparison of the results of the new device to the patient disease state as defined by well-established case definitions or well-defined diagnostic algorithms (reference methods or consensus standards). The definitions being used as truth in establishing device accuracy, sensitivity, specificity, and positive and negative predictive values should be clearly stated and explained in tables and text in the product labeling. Studies to establish performance characteristics should be performed in the target patient population and should include patients with and without the disease, patients with and without comorbidities, and patients for whom the test would most likely be ordered. (ii) Laboratory equivalence is the characterization of the new test in terms of comparison to a predicate. The conditions of performance should be defined and should include specimen acceptability, test site setting, controls applied, and the experience of the individual performing the test. A test that has been characterized in relation to a predicate and not to true diagnostic states should be labeled without estimates of sensitivity or specificity claims. Relative performance should be described in terms of agreement, copositivity and conegativity, concordance, or other such terms.

The draft proposal was sent to the IVD industry, industry trade groups, laboratory and other health care professionals, and professional societies and was placed on the CDRH website for public comments. If truth in labeling criteria as described above are adopted and required by the DCLD, the benefit to the user community and health care providers will be considerable. It is my belief and experience that laboratory and other health care providers are skeptical of IVD labeling claims. The draft proposal, if required for IVD device package inserts, would allow users to better understand and evaluate the studies done to support product claims and ensure appropriate use of estimates of sensitivity and specificity.

The CDRH Division of Biostatistics has provided guidance to the IVD industry and the DCLD in the form of written documents, public workshops, and consultation. Biostatisticians consult with division scientists on the appropriate statistical principles that should be applied to establish meaningful scientific and valid statistical performance claims in product labeling. These principles are incorporated into numerous guidance documents to provide microbiology review criteria for specific technologies, such as nucleic acid amplification methods, or specific infectious disease assays, such as viral IgM antibody assays, assays for *Helicobacter pylori*, chlamydiae, *M. tuberculosis*, and others. All are freely accessible on the CDRH website.

Briefly, the following are examples for evaluating the performance of a new qualitative microbiology laboratory test.

Operational Truth Study (Assumes Scientific and Statistically Sound Multicenter Study Design)

The key information needed to accurately describe the performance characteristics of a diagnostic laboratory test begins with the purpose of the test. The purpose of a diagnostic test is to perform a specific function in a specific population that is believed to have a specific disease or condition. In order to design a study, each of these components must be clearly defined. The patient population for which the diagnostic test is intended is defined as the target population (e.g., all newborns, sexually active individuals, or individuals suspected of having a disease or condition based on specific clinical signs and symptoms). The target population represents two distinct groups, those with the disease or condition and those without, based on the best approximation of truth using a well-defined reference method or clinical criteria.

General Study Design Considerations

Major study design considerations are as follows.

1. A prospective study is more desirable than a retrospective study in order to minimize any potential bias.

2. Specimens should be from patients representative of the target population and belonging to one of two distinct groups, those with the disease or condition and those without.

3. The sample size should be based on a single specimen per patient. If multiple specimens are obtained from the same patient, the analysis should account for the correlation of results for that patient.

4. The new test and the determination of truth should be applied independently of one another. Masking should be used whenever possible.

5. The test should be evaluated at multiple testing sites, which is not the same as multiple patient enrollment sites. The possibility of a site-by-site performance interaction should be evaluated.

6. Case categorization as positive or negative must be possible without error.

7. Case definitions of positive and negative must be applied uniformly to every patient or specimen.

8. In order to determine appropriate sample sizes for estimating sensitivity and specificity, the performance goals and requirements for the test must be known.

Assessing the Performance of a New Test

In order to assess the performance of a new test, it is necessary to determine the following.

- Performance in both the diseased and nondiseased target patient population
- Proportion of positive test results in the diseased group of patients (sensitivity)
- Proportion of negative test results in the nondiseased group of patients (specificity)
- Proportion of indeterminate (equivocal) test results in the target patient population based on the new assay result when the new assay has an indeterminate interpretive cutoff point or range; receiver operating curves should be presented to justify the establishment of assay cutoff points

Two-by-Two Truth Table Is Representative of Results

The estimation of sensitivity and specificity for a new test involves comparing results from the new test to the true disease status in the target patient specimen. For this purpose, a two-by-two truth table is constructed as shown in

Table 1. (The entries in the truth table are used in the equations below.) Indeterminate results must never be used for sensitivity and specificity estimates.

Estimated Performance Characteristics of Test

Clinical sensitivity. Clinical sensitivity is given by the following equation: clinical sensitivity = $a/(a + c) \times 100$ = percent test positivity. It is the measure of the probability that a test result will be positive if the disease or condition being investigated is present.

Clinical specificity. Clinical specificity is given by the following equation: clinical specificity = $d/(b + d) \times 100$ = percent test negativity. It is the measure of the probability that a test result will be negative if the disease or condition being investigated is not present.

Ninety-five percent confidence intervals using the exact method. The larger the number of specimens tested, the greater the precision of the mean, or percentage, and correspondingly, the narrower the 95% confidence interval for the mean or percentage.

Prevalence. The prevalence of the condition or disease in the target population is the pretest probability of a particular clinical state in a specified population; it is the frequency of a disease in the population of interest at a given time. The prevalence is calculated as follows: (number of patients in the target population with disease/total number of patients in the target population) $\times 100$.

Positive predictive value. The positive predictive value is the probability that a new test-positive patient has the disease. It equals $a/(a + b)$ based on disease prevalence in the study.

Negative predictive value. The negative predictive value is the probability that a new test-negative patient will not have the disease. It equals $d/(c + d)$ based on disease prevalence in the study.

The positive and negative predictive values of a test indicate the ability of a test to predict the presence or absence of disease and are influenced by the sensitivity and specificity of the test and by the prevalence of the disease in the population studied.

Test accuracy. The accuracy of a test is given by the following equation: accuracy of the test = $(a + d)/(a + b + c + d) \times 100$. It is the overall ability of the test to discriminate between two or more clinical states.

TABLE 1 Truth table for sensitivity and specificity

New test result	Evaluation of result[a]	
	Diseased	Nondiseased
Positive	a	b
Negative	c	d
Indeterminate	No	No

[a] Evaluation of the results of the new test when the true disease status of patients within the target population studied is as shown. a, true positive; b, false positive; c, false negative; d, true negative.

Sensitivity and specificity are useful performance characteristics only if the case definitions for diseased and nondiseased are useful. In practice, operationally defining two clinically meaningful groups (diseased and nondiseased) is not always straightforward. While the choice of an appropriate algorithm defining the two groups is a scientific issue, not a statistical issue, certain principles should be followed so that the performance characteristics can be interpreted statistically. The following are some suggestions.

1. The algorithm divides the target population into only two groups.
2. The categorization as case positive or case negative can be made without error (i.e., the algorithm always gives the same answer when applied to the same subject or specimen).
3. The algorithm does not use the outcome of the new test being evaluated (this can be one of the problems with "discrepant resolution") (2–5).
4. Every patient or specimen is evaluated using the same case definition. (This may not be the case in a retrospective study or when banked specimens are used.)

The case definitions used in the study should also be provided in the test kit labeling, since estimates of sensitivity and specificity will most likely differ, depending on the case definitions used. However, if the same case definitions are used for verification and validation of a commercial test cleared through the 510(k) process or approved through the PMA process before its introduction into the clinical laboratory, the user should achieve the same estimates for sensitivity and specificity. It would also be prudent to apply these principles before introducing a home brew test into the clinical setting.

Laboratory Equivalence Study (Predicate–to–New-Device Comparison)

Historically, device-to-device comparison has been the most common approach used by the industry and accepted by the agency to gain market clearance for class I and II IVDs. In the absence of clinical or analytical truth of the patient disease status or known presence of the analyte in the specimens to be tested, the objective of such a study would be to determine the ability of the new test to produce results in agreement with results using the predicate device. Although a two-by-two truth table can be used to demonstrate positive, negative, and indeterminate result agreement, such a comparison would not meet statistical requirements for estimates of sensitivity, specificity, or accuracy as explained above. The labeling of many legally marketed products contains such erroneous estimates. Occasionally the terms relative sensitivity and relative specificity have also been used, with or without the following explanation. "Please be advised that 'relative' refers to the comparison of this assay's result to that of a similar device. No judgement can be made as to the comparison assay's accuracy to predict presence or absence of disease."

When a new device is compared to a predicate, the conditions of the testing should be clearly defined in the package labeling. The nature and source of the specimens, the criteria for their acceptability, how and where testing was performed for the new as well as the predicate device, and the results of the comparison testing should also be clearly stated in terms of agreement. Table 2 is an example of how comparative-testing data should be presented in a package insert.

TABLE 2 Comparative-testing data presentation in a package insert[a]

New device result	Predicate device result	
	Positive	Negative
Positive	a	b
Negative	c	d

[a] Percent agreement is given by the following equations: percent positive agreement = $[a/(a + c)] \times 100$; percent negative agreement = $[d/(b + d)] \times 100$; percent overall agreement = $[(a + d)/(a + b + c + d)] \times 100$.

Users of commercial devices are encouraged to carefully critique the package insert to assess whether appropriate scientific and statistical principles were followed to establish the described performance characteristics of the device. There undoubtedly are wide variations in the representation of product performance characteristics for similar and diverse products. These inconsistencies are likely reflective of changing DCLD internal policies and statutory and regulatory changes with no control in place to ensure consistency in the application of IVD labeling requirements.

The "least burdensome" provision of the FDAMA, once implemented by the CDRH, likely will have the most impact on future premarket submissions. Least burdensome is currently defined as a successful means of addressing a premarket issue that involves the smallest investment of time, effort, and money on the part of the submitter. It is my opinion that the goal for addressing IVD review issues will be to reduce the amount and kinds of data needed to make a substantial-equivalence decision. The IVD industry has lobbied long and hard to reduce the burden of reasonable evidence (data and information necessary) to demonstrate that a device is safe and effective. A further confounding belief among many is that IVDs are low-risk, safe products and that like products with the same intended use and technological characteristics present no new issues of safety and effectiveness. Should these views be adopted by the CDRH, the burden of establishing accurate device effectiveness (performance characteristics) will probably fall on the user clinical laboratories.

CLIA 88

The 1988 CLIA to the 1967 Clinical Laboratory Improvement Act were the result of highly publicized reports of inaccurate test results of Pap smears for detection of cervical carcinoma. These reports raised questions as to how testing laboratories functioned and what quality control existed. CLIA 88 extended federal regulation to cover all laboratories that examine human specimens for the diagnosis, prevention, or treatment of any disease or impairment or for the assessment of the health of human beings. Categorization of laboratory tests based on the complexity of performing the testing and quality control requirements for tests in the moderate- or high-complexity category were set forth by regulation in 1992. In addition, all new tests introduced into the laboratory after 1992 which use a method developed in house, a modification of a manufacturer's test procedure, or a commercially available test that has not been cleared or approved by the FDA must have their performance characteristics and adequacy of quality control procedures verified before results from patient material are reported. The HCFA was authorized to administer the pro-

gram and was charged with establishing a laboratory inspection program for regulatory compliance. The Centers for Disease Control and Prevention were initially designated to perform complexity categorization. That function was transferred to the FDA effective 31 January 2000.

Laboratorians should be aware that most CLIA-waived FDA class I and II exempt devices reach the marketplace without undergoing FDA scientific assessment of the performance characteristics of the product. Additionally, these same waived products are not subject to CLIA quality control, verification, and validation requirements. CLIA-designated moderate- and high-complexity tests, if not exempt from FDA 510(k) notification, will continue to undergo FDA review before they are legally commercially available. However, a user likely will not know whether a product is exempt from FDA notification. The FDA has no regulatory authority to require a manufacturer to put such logical information in the product labeling (insert). Such products continue to be subject to CLIA quality control, verification, and validation requirements when appropriate.

The FDA retains postmarket enforcement and compliance authority over all commercial IVDs regardless of review status (exempt or nonexempt) or product class (I, II, or III). Failure of a device to meet the stated performance characteristics could by definition misbrand the product. Failure of a manufacturer to adhere to good manufacturing practices and the manufacturing quality system regulations can lead, and has led, to monetary penalties and removal of the product from the marketplace. Documented postmarket product performance failures should be reported to the manufacturer and to the FDA through the MedWatch program. Such product failure reports have in the recent past led to the issuance of public health advisories, product safety alerts, and notices from the CDRH. The FDA has no regulatory authority over clinical microbiology laboratories; the role of the FDA is to promote and protect public health in concert with the public it serves and the industry it regulates.

An FDAMA provision requires that the FDA establish better communication and information pathways for the public and the regulated industry. In keeping with that edict, I have included a listing of FDA, Centers for Disease Control and Prevention, and HCFA websites, which should prove useful for individual, laboratory, or institutional understanding of the federal government's role in the regulation of commercially available products intended for the diagnosis, cure, mitigation, treatment, or prevention of human infectious disease or its sequelae.

In this emerging age of diagnostic molecular techniques, current and future challenges to the FDA scientific review process and postmarketing surveillance activities and to the clinical community are to clearly define what is meant by and used as the gold standard for the approximation of truth. Molecular techniques, such as nucleic acid amplification (NAA), may replace culture as a more sensitive, specific, and rapid method for the identification of specific microorganisms directly from clinical material. Such methods are particularly appealing when the pathogenic microorganism is difficult to isolate in vitro, there is no culture method for isolation, or in vitro culture isolation requires days or weeks. Accurate assessment of truth to establish estimates of clinical sensitivity and specificity using NAA technology may require both laboratory methods and clinical criteria to establish the best approximation of truth. The DCLD took such an approach to assess the performance of commercially available NAA methods for direct detection of *Chlamydia trachomatis* and *M. tuberculosis*. The criteria used for each of these products are clearly summarized in the package insert. Evaluation criteria to establish the performance characteristics for a new NAA-based assay likely will vary depending on the organism and the nature of the infectious disease. It therefore becomes prudent to attempt to define the performance of an NAA test during all stages of the disease.

APPENDIX

Useful Websites

CDRH home page	http://www.fda.gov/cdrh/
CDRH Referral List	http://www.fda.gov/cdrh/referral.html
CDRH Guidance Documents	http://www.fda.gov/cdrh/topindx.html
CLIA program and payment	http://hcfa.gov/medicaid/clia/cliahome.htm
CFR	http://www.access.gpo.gov/nara/cfr
21 CFR (FDA)	http://www.fda.gov.cdrh.devadvice/365.html
42 CFR, section 493 (CLIA), revised 1999	http://www.phppo.cdc.gov/dls/clia/docs/42cfr49399.htm
Consumer information (CDRH)	http://www.fda.gov/cdrh/consumer/index.html
DSMA home page	http://www.fda.gov/cdrh/dsma/dsmamain.html
Device advice	http://www.fda.gov/cdrh/devadvice/
Staff directory	http://www.fda.gov/cdrh/dsma/dsmastaf.html#contents
FFDC Act	http://www.fda.gov/opacom/laws/fdcact/fdctoc.html
FDA and HCFA (CLIA)	http://www.fda.gov/cdrh/hcfa
FDA home page	http://www.fda.gov/
FDAMA	http://www.fda.gov/cdrh/modact/modguid.html
Freedom of information	http://www.fda.gov/opacom/backgrounders/foiahand.html
Global harmonization	http://www.ghtf.org
Guidance documents (Office of Device Evaluation and DCLD)	http://www.fda.gov/scripts/cdrh/cfdocs/cfggp/results.cfm
Medical device reporting home page	http://www.fda.gov/cdrh/mdr.html
Medical device exemptions	http://www.fda.gov/cdrh/devadvice/315.html
MedWatch	http://www.fda.gov/medwatch/revise.htm
Premarket approval information	http://www.accessdata.fda.gov/scripts/cdrh/cfdocs/cfpma/search.cfm
Safety alerts, public health advisories, and notices from CDRH	http://www.cdrh/dsma/dsmamain.html
510(k) Information and releasable database	http://www.accessdata.fda.gov/scripts/cdrh/cfdocs/cfpmn/search.cfm
Product approvals and warning letters	http://www.fda.gov/foi/electrr.html
Product classification database	http://www.accessdata.fda.gov/scripts/cdrh/cfdocs/cfpcd/search.cfm
Recalls	http://www.fda.gov/cdrh/safety/recall.html
Registration searchable database	http://www.accessdata.fda.gov/scripts/cdrh/cfdocs/cfrl/registra/search.cfm
Listing searchable database	http://www.accessdata.fda.gov/scripts/cdrh/cfdocs/cfrl/Listing/search.cfm
Reengineering efforts in CDRH	http://www.fda.gov/cdrh/reengine.html
CDRH standards program	http://www.fda.gov/cdrh/stdsprog.html
Third-party review program	http://www.accessdata.fda.gov/scripts/cdrh/year2000/y2k_search.cfm

REFERENCES

1. **Elder, B. L., S. A. Hansen, J. A. Kellogg, F. J. Marsik, and R. J. Zabransky.** 1997. *Cumitech 31, Verification and Validation of Procedures in the Clinical Microbiology Laboratory.* Coordinating ed., B. W. McCurdy. American Society for Microbiology, Washington, D.C.
2. **Hagdu, A.** 1996. The discrepancy in discrepant analysis. *Lancet* **348:**592–593.
3. **Hagdu, A.** 1997. Bias in the evaluation of DNA-amplification tests for detecting *Chlamydia trachomatis. Stat. Med.* **16:**1391–1399.
4. **Hayden, C. L., and M. L. Feldstein.** 2000. Dealing with discrepancy analysis. Part 1. The problem of bias. *IVD Technol.* **6**(1):37–42.
5. **Hayden, C. L., and M. L. Feldstein.** 2000. Dealing with discrepancy analysis. Part 2. Alternative analytical strategies. *IVD Technol.* **6**(2):51–58.

Commercial Blood Culture Systems and Methods

MELVIN P. WEINSTEIN AND L. BARTH RELLER

2

During the last third of the 20th century in the United States, the techniques for detecting microorganisms in the bloodstream evolved from manual methods using media prepared in individual clinical laboratories to more widely available commercial systems and media prepared and sold by manufacturers to hospitals and clinical laboratories. Currently, both manual and automated systems are available, with the market being dominated increasingly by instrument-based automated systems and methods. This chapter will emphasize those manual and automated systems and media cleared for diagnostic use in the United States at the start of the 21st century as well as comment on possible developments in this rapidly evolving area of clinical microbiology. Additionally, comparative performance data for automated systems will be provided.

MANUAL BLOOD CULTURES

The fundamental concept of blood culturing has been based on inoculating a defined volume of blood, preferably obtained by a sterile venipuncture, into a broth culture medium that will support the growth of most pathogenic bacteria and fungi. Manual methods have employed this concept in its most basic form. Although there formerly were multiple commercial sources for bottled blood culture media that could be processed manually, there now are relatively few, due in part to consolidations and mergers among the various manufacturers and in part to the declining demand for manual systems. Indeed, some familiar commercial products and manufacturers (e.g., Difco Laboratories) and systems (e.g., Vacutainer Blood Culture System [Becton Dickinson]) either no longer exist or no longer are marketed because of declining demand. However, broth-based manual blood culture systems and a manual system based on the lysis-centrifugation method still are marketed in the United States.

Septi-Chek

The Septi-Chek Blood Culture System (BD Biosciences, Sparks, Md.) (Fig. 1) consists of a conventional blood culture bottle containing broth medium to which is attached an agar-coated plastic paddle, creating a biphasic system similar to that of the classic biphasic Castaneda bottle. Blood is inoculated into the culture bottle through a threaded plastic cap with a rubber diaphragm, which can be unscrewed and replaced with the agar-coated paddle after the culture specimen has been received by the laboratory. After the paddle is attached, the bottle is inverted so that the blood-broth mixture floods the paddle, thereby inoculating the agar with microorganisms that may be present. Aerobic and anaerobic bottles, the latter without the agar paddle device which is permeable to atmospheric oxygen, are then incubated with or without agitation and inspected for evidence of microbial growth once or twice daily. The agar-coated paddle can be removed for better inspection by unscrewing it from the top of its plastic container. Following each examination of the agar paddle, the bottle is inverted, in effect repeating the subculture of the blood-broth mixture to the agar.

The Septi-Chek system originally was developed as a labor-saving alternative to conventional blood cultures, which had to be subcultured manually after the first overnight incubation in the laboratory and, in some laboratories, at the end of the incubation period. Several medium formulations are available, including brain heart infusion, tryptic soy, tryptic soy with 10% sucrose, Columbia, thioglycolate, and Schaedler broths. The paddles contain both nonselective and differential selective agars: chocolate agar, MacConkey agar, and malt agar. If growth occurs on the paddles first or on the paddles and in broth at the same time, colonies are available immediately for identification and susceptibility testing. Moreover, preliminary microbiologic information may be available (e.g., lactose-positive or lactose-negative colonies on the MacConkey agar). In a number of controlled clinical evaluations, the Septi-Chek system performed well (2, 21, 36–38).

Oxoid Signal

The Oxoid Signal Blood Culture System (Oxoid Inc., Ogdensburg, N.Y.) is marketed as a one-bottle system and, like the Septi-Chek system, was developed as a labor-saving alternative to conventional manual blood culture systems. Similar to the Septi-Chek system, the Oxoid bottle is inoculated with blood through a conventional rubber diaphragm. Once in the laboratory, the diaphragm is removed and the plastic signal device is attached to the top of the bottle where it is anchored by a plastic outer sleeve (Fig. 2). The device consists of a long needle that goes below the level of the blood-broth mixture and extends up into the clear plastic cylinder of the signal device. When

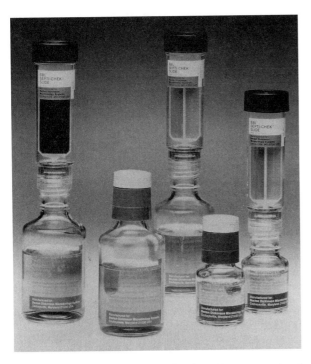

FIGURE 1 Septi-Chek aerobic bottle with agar-coated paddle attached. Courtesy of Becton-Dickinson Microbiology Systems.

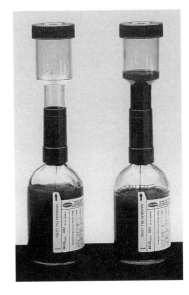

FIGURE 2 Oxoid Signal system showing positive (right) and negative (left) bottles. Courtesy of Oxoid Ltd.

microbial growth occurs in the bottle, gases are released into the bottle headspace, increasing the atmospheric pressure within the bottle and forcing some of the blood-broth mixture up through the needle and into the signal device. Thus, in addition to the conventional macroscopic examinations of blood culture bottles (looking for hemolysis, gas formation, colonies growing in the liquid medium), the Oxoid Signal system offers the potential advantage of easy visualization of a positive culture. The fluid in the signal cylinder can be seen without difficulty by the microbiologist when the blood cultures are inspected each day, and positive cultures can then be worked up in the traditional manner.

Only one medium formulation of the Oxoid Signal system has been marketed. In published controlled clinical evaluations done in the United States comparing both manual (19) and automated (31, 32, 34) blood culture systems, the one-bottle Signal system performed less well than the comparators.

Isolator

The Isolator Blood Culture System (Wampole Laboratories, Cranbury, N.J.) is based on the principle of lysis-centrifugation and is the only commercially available blood culture system that does not utilize broth culture medium. Blood is inoculated to Isolator tubes (Fig. 3) that contain a lysing solution consisting of saponin, an anticoagulant (EDTA), and a fluorocarbon that acts as a cushion during the centrifugation step of blood processing. Once the lysis and centrifugation steps have taken place, the rubber stopper is removed from the Isolator tube and the supernatant is removed by aspiration with a disposable pipette. The pellet is then resuspended and can be inoculated directly to culture media that will support the various potential pathogens for which detection is desired. The system can be used for detection of routine bacterial pathogens but has been

reported to have reduced detection of anaerobes, *Haemophilus* species, and pneumococci if specimens are not processed within 8 h of receipt in the laboratory (10, 11, 14, 30). The Isolator has been shown to be an excellent system for detecting yeasts and dimorphic fungi, mycobacteria, and *Bartonella* spp. (1). The system is labor-intensive, particularly in the initial processing of specimens and in comparison to the newer automated continuous-monitoring systems now available.

AUTOMATED BLOOD CULTURE SYSTEMS

Automated blood culture systems were introduced in the United States in the 1970s. The first clinically and commercially successful system was the BACTEC 460 radiometric system (Becton Dickinson). This system was suc-

FIGURE 3 Isolator blood culture system. Courtesy of Wampole.

ceeded by the BACTEC 660, 730, and 860 nonradiometric systems in the mid-1980s. The BACTEC 460, 660, 730, and 860 systems are not available for use in the United States now. However, the detection principle utilized in these early systems remains important because it forms the underlying basis for the continuous-monitoring systems that are used widely now. In the BACTEC radiometric system, the broth culture medium contained carbohydrate substrates that incorporated radiolabeled ^{14}C. When microbial growth occurred, the carbohydrate substrates were utilized, and the ^{14}C was released as CO_2 in the headspace of the bottle. As bottles passed through the instrument's detection system, needles sampled the headspace of each bottle. If the amount of ^{14}C detected exceeded an arbitrary threshhold, the bottle was flagged as positive, alerting the technologist to remove it from the system and to perform Gram stain and subculture. The nonradiometric system replaced the radiometric system in the mid-1980s, in part due to concerns about disposal of radioactive waste. In this system, the increase of CO_2 in the bottle headspace was detected by infrared spectrophotometry.

In the early 1990s, the first of the continuous-monitoring blood culture systems, BacT/Alert (Organon Teknika Corp., Durham, N.C.) was introduced; it was followed shortly by the BACTEC 9000 system (BD Biosciences) and the ESP Blood Culture System (Trek Diagnostic Systems, Westlake, Ohio). The availability of these systems propelled many laboratories toward the use of automated blood culture technology and rendered the previously marketed automated BACTEC systems virtually obsolete.

The continuous-monitoring systems have a number of characteristics in common. The systems are modular, allowing a single computer to coordinate the function of one to as many as 50 incubator units. Blood culture bottles are placed in individual cells of an incubator unit, and testing is performed without further manipulation of the bottle (until a bottle is flagged as positive and removed for Gram stain and subcultures), thereby reducing technologist hands-on time. In each system, blood culture bottles are individually monitored at intervals of 10 to 15 min for evidence of microbial growth. The individual bottle readings are transmitted to the system's computer for storage and analysis, and growth curves are calculated according to sophisticated instrument algorithms. Because of the frequent testing that takes place round-the-clock, it is possible to detect microbial growth approximately 1 to 1.5 days earlier than was the case for the BACTEC radiometric and nonradiometric systems (17, 20, 44). The more frequent measurements and greater number of data points also seem to be associated with fewer instrument false-positive signals than was the case with the earlier automated systems (44). Table 1 summarizes some of the relevant information on the commercially available continuous-monitoring systems.

BacT/Alert

In 1991, the BacT/Alert Blood Culture System was introduced as the first commercial continuous-monitoring blood culture system (Fig. 4). In 1999, the BacT/Alert 3D system was introduced. The 3D system is more compact and incorporates a touch screen for the instrument's computer, enabling the technologist to more easily load and unload bottles as well as to perform quality control functions on the instrument. Each BacT/Alert incubation module holds up to 240 bottles.

In the BacT/Alert system, a carbon dioxide sensor is incorporated at the base of each culture bottle, separated from the blood-broth mixture by a CO_2 semipermeable membrane that monitors the amount of CO_2 in the bottle. The sensor changes color as the concentration of CO_2 increases in the bottle. At the base of the cell in which the bottle resides in the incubating unit are light-emitting and light-sensing diodes. When microbial growth occurs the sensor changes color, and the amount of light reflected from the sensor increases. The increased reflectance is measured by the instrument as increased voltage, and the data are transmitted to the central computer. The computer's detection algorithms have the ability to flag a culture as positive in several ways: (i) by recognition that an arbitrary threshold has been exceeded, (ii) by recognition of a linear increase in CO_2, or (iii) by recognition of a change in the rate of CO_2 production.

Several medium formulations are available for use in the BacT/Alert system. These include standard aerobic and anaerobic media that contain 40 ml of tryptic soy broth and accept up to 10 ml of blood per culture vial; aerobic and anaerobic FAN media, which contain 40 ml of brain heart infusion broth, activated charcoal, and Fuller's earth (Ecosorb) designed to inactivate or bind antimicrobial agents in the blood and accept up to 10 ml of blood per culture vial; and the PF Pediatric bottle, which contains 20 ml of tryptic soy broth supplemented with brain heart infusion solids and activated charcoal, which accepts up to 4 ml of blood and which has been marketed for use with pediatric patients and those elderly patients from whom it is difficult to obtain larger volumes of blood. The MB blood bottle for detection of mycobacteria in blood has recently been introduced.

Many clinical evaluations of the BacT/Alert blood culture system and medium formulations have been published, and some but not all will be mentioned herein. The first controlled clinical field evaluation of the BacT/Alert system compared it to the BACTEC NR system, with 5 ml of blood inoculated into the standard aerobic and anaerobic bottles of each system, at three university medical centers (44). The study found the systems to be equivalent for overall recovery of microorganisms causing sepsis, with earlier detection of positives and fewer instrument false-positive signals in BacT/Alert. Further analysis of the data from this study revealed that BacT/Alert bottles could be processed using a 5-day rather than a 7-day incubation period without a significant decrease in the detection of microorganisms causing sepsis (42). Similar findings were reported independently by others (8) in a study that also showed that terminal subcultures of bottles shown to be negative by the instrument were not necessary. Another study showed that the BacT/Alert aerobic standard bottle inoculated with 10 ml of blood detected more microorganisms than the same bottle inoculated with 5 ml of blood (35). Rohner and colleagues (25) compared the BacT/Alert system (5 ml inoculated into standard aerobic and anaerobic bottles) with the Oxoid Signal (10 ml inoculated into one bottle) and demonstrated a higher yield and faster detection of microorganisms by the BacT/Alert system. Hellinger et al. (9) compared the BacT/Alert standard aerobic bottle and the Isolator and found no significant difference in overall detection of microorganisms. However, the Isolator recovered significantly more staphylococci and yeasts, whereas the BacT/Alert system recovered more aerobic and facultatively anaerobic gram-negative rods.

Several studies have evaluated the BacT/Alert FAN media. Weinstein et al. (33) compared the aerobic FAN bottles and standard bottles and found superior detection of

TABLE 1 Commercially available continuous-monitoring blood culture systems[c]

System (manufacturer)	Method(s) for detecting growth	Bottle capacity per module	Maximum no. of modules (no. of bottles[b]) per system	Test cycle (min)	Agitation type (speed [speed/min])	Dimensions (cm)
BacT/Alert 240 (Organon TeknikaCorp.)	CO_2 detection, colorimetric	240	6 (1,440)	10	Rocking (34)	175 × 87 × 66
BacT/Alert 120	CO_2 detection, colorimetric	120	6 (720)	10	Rocking (34)	87 × 87 × 55
BacT/Alert 3D	CO_2 detection, colorimetric	240	12 (2,880)	10	Rocking (34)	90 × 49 × 61
BACTEC 9240 (BD Biosciences)	CO_2 detection, O_2 detection,[a] fluorescence	240	5 (1,200) (core) 20 (4,800) (Vision) 50 (12,000) (EpiCenter)	10	Rocking (30)	93 × 128 × 55
BACTEC 9120	CO_2 detection, O_2 detection,[a] fluorescence	120	5 (600) (core) 20 (2,400) (Vision) 50 (6,000) (EpiCenter)	10	Rocking (30)	61 × 129 × 56
BACTEC 9050	CO_2 detection, O_2 detection,[a] fluorescence	50	1 (50)	10	Continuous rotation	61 × 72 × 65
ESP 128 (Trek Diagnostic Systems)	Manometric, pressure change	128	5 (640)	12 (aerobic) or 24 (anaerobic)	Rotary (aerobic only) (160)	90 × 86 × 65
ESP 256	Manometric, pressure change	256	5 (1,280)	12 (aerobic) or 24 (anaerobic)	Rotary (aerobic only) (160)	199 × 86 × 65
ESP 384	Manometric, pressure change	384	5 (1,792)	12 (aerobic) or 24 (anaerobic)	Rotary (aerobic only) (160)	199 × 86 × 65
Vital (bioMérieux)	CO_2 detection, fluorescence	400	3 (1,200)	15	Sinusoidal (150)	108 × 78 × 114

[a] O_2 detection is for Myco/F Lytic medium only.
[b] Maximum number of bottles accommodated depends on data management system selected (core, Vision, or EpiCenter).
[c] Adapted from references 24 and 45.

FIGURE 4 BacT/Alert system showing BacT/Alert 240 module (left), MB module (mycobacterial detection module) (right center), and 3D module (far right). Courtesy of Organon Teknika.

staphylococci, fungi, and all microorganisms combined in the FAN system. Wilson and colleagues (43) compared the anaerobic FAN and standard bottles and demonstrated superior recovery of staphylococci, *Escherichia coli*, and all microorganisms combined but inferior recovery of nonfermentative gram-negative rods and yeasts in FAN bottles. A subsequent report assessed the clinical importance of these two studies and concluded that use of FAN bottles detected significantly more episodes of septicemia, especially recurrences and episodes in patients receiving theoretically effective antimicrobial therapy (16). Potential disadvantages associated with use of FAN bottles in the studies included the greater cost of these bottles as well as enhanced detection of (and need to work up) contaminating coagulase-negative staphylococci.

The FAN bottle has been compared with culture media from both the ESP and BACTEC continuous-monitoring systems. In a study with pediatric patients, the FAN and ESP 80A bottles were compared (40). Overall detection rates were equivalent, but the FAN bottle detected more *Staphylococcus aureus* organisms whereas the ESP 80A bottle detected more streptococci and enterococci. The FAN bottle detected more microorganisms from patients already receiving antimicrobial agents. Doern et al. (5) compared the aerobic FAN bottle to the aerobic 80A bottle in the ESP system in an adult population and found higher recovery rates of clinically important microorganisms in FAN bottles, in particular *S. aureus*, members of the family *Enterobacteriaceae*, and gram-negative rods other than *Pseudomonas aeruginosa* from patients receiving antimicrobial therapy. The 80A bottle detected more beta-hemolytic streptococci. In this study, more contaminants were detected in the FAN bottle, and the ESP system had more instrument false-positive readings. Jorgensen and colleagues (12) conducted a multicenter comparison of the aerobic FAN and BACTEC Plus resin media and found an equivalent overall yield of clinically important microorganisms and septic episodes. *Histoplasma capsulatum* was detected more frequently in the BACTEC resin bottle. Streptococci were detected approximately 7 h earlier in the resin bottle, whereas *Candida albicans* was detected approximately 8 h earlier in the FAN bottle.

Bactec 9000 Series

In 1992, Becton Dickinson marketed the BACTEC 9000 series continuous-monitoring blood culture system (Fig. 5). Two instrument formats were marketed, the smaller-capacity BACTEC 9120 (120 bottles per incubating unit) and the larger-capacity BACTEC 9240 (240 bottles per unit). In the late 1990s, the smaller BACTEC 9050 bench top instrument (50-bottle capacity) was marketed for laboratories that processed relatively few blood cultures but wanted to use an automated rather than a manual blood culture system (Fig. 6). Like the BacT/Alert system, there is a carbon dioxide sensor at the base of each culture bottle. Rather than using colorimetric detection of increased CO_2 as in the BacT/Alert, the BACTEC system utilizes a fluorescence-sensing mechanism to detect microbial growth. When the amount of CO_2 increases, increased fluorescence is detected by the instrument and transmitted to its computer. The principle detection criteria are a linear increase in fluorescence and an increase in the rate of fluorescence.

The system has multiple medium formulations that can be used. There are Standard Aerobic/F and Anaerobic/F media, which contain 40 ml of tryptic soy broth; Plus Aerobic/F and Plus Anaerobic/F media, which contain 25 ml of tryptic soy broth plus antibiotic-binding resins on glass beads; the Lytic/10 Anaerobic/F medium, which contains 40 ml of tryptic soy broth plus a lysing agent; the Peds Plus/F medium, which contains 40 ml of tryptic soy broth and resins; and the Myco/F Lytic medium, designed for improved detection of fungi and mycobacteria. All of the culture vials for these formulations accept up to 10 ml of blood with the exception of the Peds Plus/F vials, which are designed to accept 1 to 3 ml of blood from infants and young children.

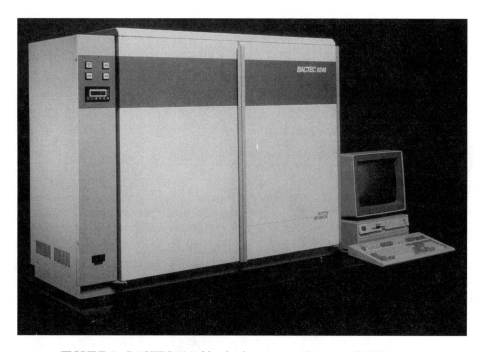

FIGURE 5   BACTEC 9240 blood culture system. Courtesy of BD Biosciences.

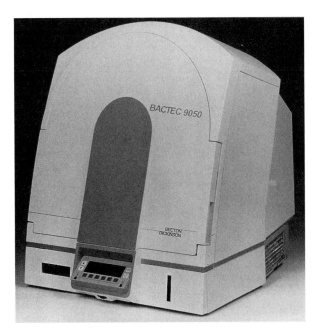

FIGURE 6 BACTEC 9050 blood culture system. Courtesy of BD Biosciences.

Numerous studies have evaluated the BACTEC 9000 series instruments and medium formulations; selected reports will be highlighted herein. An initial evaluation compared the Standard Aerobic/F and Anaerobic/F bottles processed in the BACTEC 9240 instrument to the standard aerobic and anaerobic bottles in the BACTEC nonradiometric 660 (660 NR) system (20). More staphylococci and all microorganisms combined were detected in the Aerobic/F bottles, and more coagulase-negative staphylococci, members of the family *Enterobacteriaceae*, and all microorganisms combined were detected in the Anaerobic/F bottles than in the respective comparator bottles used in the 660 NR system. Moreover, positive cultures were detected earlier in the 9240 system than in the 660 NR system.

Most published studies of the BACTEC 9000 system have assessed the Plus/F resin media. Schwabe and colleagues (27) compared BACTEC resin media formulated for and processed in the 9240 instrument versus that formulated for and processed in the 660 NR instrument. There were no differences in yield of microorganisms, but microorganisms were detected earlier by the 9240 instrument. A more recent multicenter study (18) evaluated the aerobic and anaerobic Plus/F media processed in the 9240 instrument versus the 9050 instrument. Phase 1 evaluated the aerobic Plus/F medium in the two instruments and found no differences in the recovery of individual microorganisms or groups thereof, with the exception of *Streptococcus pneumoniae*, which was isolated more frequently in the bottles processed in the 9050 instrument. The times to detection of positive cultures were similar in both systems. Phase 2 evaluated the anaerobic Plus/F medium and found no significant differences in the recovery of microorganisms in the two BACTEC instruments, and overall times to detection of positive cultures were similar except for those for anaerobes, which were detected more quickly in the 9240 instrument (mean, 35 h) than in the 9050 instrument (mean, 61 h). In both phases of the study, instrument false-positive signals occurred more frequently in the 9240 instrument than in the 9050 instrument.

Several studies have compared the Plus/F medium processed in the BACTEC 9240 and 9120 instruments to other commercially available media and systems. Smith et al. (28) compared the aerobic resin media to the BacT/Alert standard media and found superior detection of staphylococci and all microorganisms combined in the BACTEC resin bottles. The times to detection of positive cultures were similar in the two systems. In a comparison of the Plus/F medium with the Septi-Chek Release medium (brain heart infusion broth supplemented with a lytic agent, saponin), Rohner and colleagues (26) reported enhanced detection of *Haemophilus* spp. in Plus/F medium but greater detection of members of the family *Enterobacteriaceae*, gram-negative anaerobes, and all microorganisms combined in Septi-Chek medium. Positive cultures were detected earlier by the BACTEC system. Two studies have compared the Plus/F medium and the Isolator system. In one, Pohlman and colleagues (23) reported improved detection of staphylococci, members of the family *Enterobacteriaceae*, *P. aeruginosa*, and all microorganisms combined in Plus/F medium, especially if patients were receiving antimicrobial agents at the time the blood cultures were obtained. In the second study, Cockerill et al. (3) reported better detection of staphylococci and all microorganisms combined in the resin bottle and more contaminating bacteria in the Isolator. In both studies, the BACTEC system detected positive cultures earlier than the Isolator and required less technologist time for specimen processing and examination.

Two studies have directly compared the BACTEC Plus/F aerobic blood culture bottle and the BacT/Alert FAN aerobic bottle. In a single-center study, Pohlman et al. (22) reported equivalent detection of clinically important microorganisms by the two systems. However, when *Pseudomonas* and the *Enterobacteriaceae* were analyzed together, the FAN bottle detected significantly more strains. In this study, more false-positive signals were detected in the Plus/F bottle than in the FAN bottle. There were no differences in times to detection or in the detection of positive cultures in patients receiving antimicrobial agents at the time of blood culture. In a multicenter study, Jorgensen and colleagues (12) reported equivalent detection of microorganisms except for enhanced detection of *H. capsulatum* in the Plus/F medium. More contaminants were detected in the FAN medium. Times to detection of positive cultures were similar overall, but streptococci were detected on average 7 h earlier by the BACTEC system, whereas yeasts were detected on average 8 h earlier by the BacT/Alert system.

A new medium, MYCO/F Lytic, has been developed and marketed recently for use in the BACTEC 9000 instruments. This medium is formulated to enhance the detection of mycobacteria and fungi while retaining the ability to detect bacteria. Unlike other BACTEC media formulated for the 9000 series instruments, the MYCO/F Lytic medium's fluorescent sensor detects decreasing oxygen concentrations rather than increasing carbon dioxide concentrations in the presence of microbial growth. A single published study (29) assessed the MYCO/F Lytic medium in AIDS patients with suspected mycobacteremia and fungemia. Since the number of patients with positive cultures was small, it was difficult to draw conclusions from this study. Compared with the Isolator for detection of fungemia, the MYCO/F Lytic bottle detected fewer *H. capsulatum* organisms but more *Cryptococcus neoformans* organisms. Compared with the ESP II bottle for the detection of mycobacteremia, there was superior detection of *Mycobacterium avium* complex in MYCO/F Lytic medium. Further published studies of this medium are needed.

ESP Blood Culture System

The third continuous-monitoring blood culture system is the ESP system, introduced by Difco Laboratories and now marketed by Trek Diagnostic Systems (Fig. 7). This system differs from both the BacT/Alert and BACTEC 9000 instruments in several ways. After bottles are fitted with an adapter and loaded into the instrument, growth is detected by monitoring pressure changes within the headspace of each bottle as gases (oxygen, hydrogen, nitrogen, and carbon dioxide) are either consumed or produced by metabolizing microorganisms. Aerobic bottles are agitated orbitally at 160 rpm (versus the gentle rocking motion of bottles in the other systems); anaerobic bottles are not agitated. Lastly, the basal medium in the ESP aerobic bottle is soy-casein peptone broth, and the medium in the ESP anaerobic bottle is proteose-peptone broth. Trek Diagnostic Systems also markets a medium that will support the growth and detection of mycobacteria in the blood. Similar to the other continuous-monitoring systems, bottles are monitored frequently, every 12 min for aerobic bottles and every 24 min for anaerobic bottles, and pressure changes are plotted against time to yield growth curves. Positive cultures are signaled based on the instrument's proprietary algorithms.

In an initial multicenter comparative clinical study, the ESP system was compared with the BACTEC 660 NR system (17). ESP aerobic bottles recovered more pneumococci and all microorganisms combined than did the BACTEC aerobic bottles. ESP anaerobic bottles recovered more aerobic gram-positive bacteria, *Candida* species, and all microorganisms combined than did the BACTEC anaerobic bottles. Unfortunately, this study suffered from a number of design flaws that made the results difficult to interpret. For example, five different BACTEC media were used in this comparison, and the volume of blood inoculated into the culture bottles was not controlled.

Zwadyk et al. (47) compared the ESP system to the BacT/Alert system and found that ESP aerobic bottles recovered significantly more *S. aureus* organisms and all microorganisms combined and that ESP anaerobic bottles recovered more anaerobes. However, significantly more bacteremias caused by streptococci were detected by the BacT/Alert system. The speed of detecting microorganisms in the two systems was variable; some microorganisms were detected earlier by the ESP system, whereas others were detected more quickly by the BacT/Alert system.

Several studies have compared the ESP system to the Isolator system. Kirkley and colleagues (15) found that the Isolator detected more *S. aureus*, enterococci, *E. coli*, *Alcaligenes xylosoxidans*, *Stenotrophomonas maltophilia*, *C. albicans*, and *Candida glabrata* than the ESP 80A (aerobic) blood culture bottle. Kellogg et al. (13) compared the ESP aerobic and anaerobic bottles to the Isolator plus a manual Thiol broth bottle. In this system-versus-system comparison, the Isolator-Thiol broth combination recovered significantly more staphylococci, pneumococci, members of the family *Enterobacteriaceae*, and all microorganisms combined. Cockerill and colleagues (4) compared the ESP 80A bottle to the Isolator and found that the former recovered more coagulase-negative staphylococci causing bacteremia, whereas the latter recovered more *S. aureus*, *Candida* spp., and all microorganisms combined. In this study, the ESP 80A bottle also was compared to the manual Septi-Chek bottle and found to be equivalent in the recovery of microorganisms.

In a comparison of the ESP 80A and BacT/Alert FAN aerobic bottles, Doern and colleagues (5) reported higher rates of recovery of *S. aureus*, members of the family *Enterobacteriaceae*, and miscellaneous gram-negative bacilli in the FAN bottles and enhanced recovery of beta-hemolytic streptococci in the ESP 80A bottles. More contaminants grew in the FAN bottles, and fewer false-positive signals were noted in the BacT/Alert instrument.

Welby et al. (39) compared the ESP system to the Septi-Chek system for the detection of bloodstream infections in children and found the systems to be equivalent in recovering pathogens. More contaminants were recovered in the ESP system, and the detection of positive cultures occurred earlier in the ESP system. A second study reported by the same group (40) compared the ESP 80A bottle to the BacT/Alert FAN aerobic bottle for use with pediatric patients. The FAN bottle detected more *S. aureus* organisms, whereas the ESP 80A bottle detected more streptococci and enterococci. The FAN bottle also detected more positives in patients receiving antibiotic therapy.

In a multicenter retrospective evaluation, Doern et al. (6) reported that a 4-day incubation period could be used in the ESP system without substantial loss of detection of positives. Subsequently, Han and Truant (7) suggested that the incubation period in the ESP system might even be shortened to 3 days.

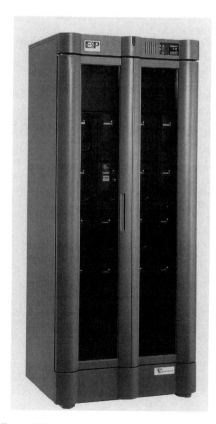

FIGURE 7 ESP 384 blood culture system. Courtesy of Trek Diagnostic Systems, Inc.

Vital Blood Culture System

The Vital Blood Culture System (Fig. 8) was developed in France and is marketed in Europe but not cleared by the Food and Drug Administration for use in the United States. It differs from other continuous-monitoring systems in several aspects. In this system, a fluorescent liquid compound is incorporated into the culture broth. When microbial growth occurs, the pH and redox potential in the culture bottle decrease and fluorescence is quenched. This change is detected by the instrument's sensors and transmitted to a computer which in turn signals positive results according to a proprietary algorithm. In the Vital instrument, bottles are placed in one of four drawers, each of which is capable of holding 100 bottles, giving the instrument a capacity of 400 bottles. For smaller laboratories, the manufacturer has marketed a bench top mini-Vital that has a capacity of 100 bottles. Vital aerobic and anaerobic culture vials contain supplemented soybean-casein digest broth and can accept up to 10 ml of blood per bottle.

The Vital system has been compared to the BACTEC NR 660 system in two large multicenter controlled clinical trials, one with adult patients and the other with children. Wilson and colleagues (41) collected 9,446 20-ml samples from adult inpatients and inoculated 5-ml aliquots into Vital aerobic and anaerobic bottles and BACTEC aerobic and anaerobic standard media without resins. Although the Vital system detected significantly more *S. aureus* organisms, BACTEC detected significantly more microorganisms overall. Whereas both systems detected growth, Vital did so earlier, but the time difference was small. The authors inferred from the data that Vital would not detect positive results as quickly as the other continuous-monitoring blood culture systems. In a study of children with suspected bacteremia or fungemia, Zaidi and colleagues (46) compared the yield and speed of detection of microorganisms in the Vital aerobic bottle with those of the BACTEC Peds Plus

bottle as formulated for the BACTEC NR 660 system. Bottles were weighed before and after they were filled with blood, and sample volumes were judged to be adequate in 6,276 paired sets of bottles. Staphylococci, yeasts, and all microorganisms combined were detected significantly more often in the Peds Plus bottles. The BACTEC advantage was present regardless of whether patients were receiving antimicrobial therapy.

FUTURE DIRECTIONS

Although current blood culture systems have reduced the time to detecting positive cultures, clinicians and patients still would benefit from accurate assay methods that provide results in minutes to hours as opposed to days. Thus, there continues to be a need for further evolution and development of new commercial blood culture systems.

As molecular methods become used more commonly in clinical microbiology laboratories, it is reasonable to speculate that blood culture methodology will evolve in this direction as well. The sensitivity of modern blood culture systems is a single viable microorganism in the volume of blood inoculated for culture. Detection systems, however, although greatly improved by continuous monitoring, still require on the order of 10^4 to 10^5 microorganisms per ml for signaling positivity. A fruitful initial approach may be to couple molecular amplification with an available culture system to detect positive cultures earlier when lower numbers of replicating organisms are present. Substantial hurdles will still remain, however, including the problem of substances in blood that inhibit molecular amplification assays, the need for determining antimicrobial susceptibility results when assays are positive, difficulties in interpreting the clinical significance of positive results if the analytical sensitivity of the new tests are greatly enhanced compared to current techniques, and added costs. In addition to possible earlier detection of common pathogens, much research needs to be done to delineate the role of other pathogens known to be present in blood, such as fungi, viruses, chlamydiae, and mycoplasmas that heretofore have not been detectable with current technology.

FIGURE 8 Vital blood culture system. Courtesy of bioMérieux, Inc.

REFERENCES

1. **Brenner, S. A., J. A. Rooney, P. Manzewitsch, and R. L. Regnery.** 1997. Isolation of *Bartonella* (*Rochalimaea*) *henselae*: effects of methods of blood collection and handling. *J. Clin. Microbiol.* **35:**544–547.
2. **Bryan, L. E.** 1981. Comparison of a slide blood culture system with a supplemented peptone broth culture method. *J. Clin. Microbiol.* **14:**389–392.
3. **Cockerill, F. R., III, G. S. Reed, J. G. Hughes, C. A. Torgerson, E. A. Vetter, W. S. Harmsen, J. C. Dale, G. D. Roberts, D. M. Ilstrup, and N. K. Henry.** 1997. Clinical comparison of BACTEC 9240 Plus Aerobic/F resin bottles and the Isolator aerobic culture system for detection of bloodstream infections. *J. Clin. Microbiol.* **35:**1469–1472.
4. **Cockerill, F. R., III, C. A. Torgerson, G. S. Reed, E. A. Vetter, A. L. Weaver, J. C. Dale, G. D. Roberts, N. K. Henry, D. M. Ilstrup, and J. E. Rosenblatt.** 1996. Clinical comparison of Difco ESP, Wampole Isolator, and Becton Dickinson Septi-Chek aerobic blood culture systems. *J. Clin. Microbiol.* **34:**20–24.

5. **Doern, G. V., A. Barton, and S. Rao.** 1998. Controlled comparative evaluation of BacT/Alert FAN and ESP 80A aerobic media as means for detecting bacteremia and fungemia. *J. Clin. Microbiol.* **36**:2686–2689.

6. **Doern, G. V., A. B. Brueggemann, W. M. Dunne, S. G. Jenkins, D. C. Halstead, and J. C. McLaughlin.** 1997. Four-day incubation period for blood culture bottles processed with the Difco ESP blood culture system. *J. Clin. Microbiol.* **35**:1290–1292.

7. **Han, X. Y., and A. L. Truant.** 1999. The detection of positive blood cultures by the AccuMed ESP-384 system: the clinical significance of three-day testing. *Diagn. Microbiol. Infect. Dis.* **33**:1–6.

8. **Hardy, D. J., B. B. Hulbert, and P. C. Migneault.** 1992. Time to detection of positive BacT/Alert blood cultures and lack of need for routine subculture of 5- to 7-day negative cultures. *J. Clin. Microbiol.* **30**:2743–2745.

9. **Hellinger, W. C., J. J. Zawley, S. Alvarez, S. F. Hogan, W. S. Harensen, D. M. Ilstrup, and F. R. Cockerill.** 1995. Clinical comparison of the Isolator and BacT/Alert aerobic blood culture systems. *J. Clin. Microbiol.* **33**:1787–1790.

10. **Henry, N. K., C. M. Grewell, P. E. Van Grevenhof, D. M. Ilstrup, and J. A. Washington II.** 1984. Comparison of lysis-centrifugation with a biphasic blood culture medium for the recovery of aerobic and facultatively anaerobic bacteria. *J. Clin. Microbiol.* **20**:413–416.

11. **Henry, N. K., C. A. McLimans, A. J. Wright, R. L. Thompson, W. R. Wilson, and J. A. Washington II.** 1983. Microbiological and clinical evaluation of the Isolator lysis-centrifugation blood culture tube. *J. Clin. Microbiol.* **17**:864–869.

12. **Jorgensen, J. H., S. Mirrett, L. C. McDonald, P. R. Murray, M. P. Weinstein, J. Fune, C. W. Trippy, M. Masterson, and L. B. Reller.** 1997. Controlled clinical laboratory comparison of BACTEC Plus Aerobic/F resin medium with BacT/Alert Aerobic FAN medium for detection of bacteremia and fungemia. *J. Clin. Microbiol.* **35**:53–58.

13. **Kellogg, J. A., D. A. Bankert, J. P. Manzella, K. S. Parsey, S. L. Scott, and S. H. Cavanaugh.** 1994. Clinical comparison of Isolator and thiol broth with ESP aerobic and anaerobic bottles for recovery of pathogens from blood. *J. Clin. Microbiol.* **32**:2050–2055.

14. **Kiehn, T. E., B. Wong, F. F. Edwards, and D. Armstrong.** 1983. Comparative recovery of bacteria and yeasts from lysis-centrifugation and a conventional blood culture system. *J. Clin. Microbiol.* **18**:300–304.

15. **Kirkley, B. A., D. A. Easley, and J. A. Washington.** 1994. Controlled clinical evaluation of Isolator and ESP aerobic blood culture systems for detection of bloodstream infections. *J. Clin. Microbiol.* **32**:1547–1549.

16. **McDonald, L. C., J. Fune, L. B. Gaido, M. P. Weinstein, L. G. Reimer, T. M. Flynn, M. L. Wilson, S. Mirrett, and L. B. Reller.** 1996. Clinical importance of increased sensitivity of BacT/Alert FAN aerobic and anaerobic blood culture bottles. *J. Clin. Microbiol.* **34**:2180–2184.

17. **Morello, J. A., C. Leitsh, S. Nitz, J. W. Dyke, M. Andruszewski, G. Maier, W. Landau, and M. A. Beard.** 1994. Detection of bacteremia by Difco ESP blood culture system. *J. Clin. Microbiol.* **32**:811–818.

18. **Murray, P. R., G. E. Hollick, R. C. Jerris, and M. L. Wilson.** 1998. Multicenter comparison of BACTEC 9050 and BACTEC 9240 blood culture systems. *J. Clin. Microbiol.* **36**:1601–1603.

19. **Murray, P. R., A. C. Niles, R. L. Heeren, M. M. Curren, L. E. James, and J. E. Hoppe-Bauer.** 1988. Comparative evaluation of the Oxoid Signal and Roche Septi-Chek blood culture systems. *J. Clin. Microbiol.* **26**:2526–2530.

20. **Nolte, F. S., J. M. Williams, R. C. Jerris, and J. A. Morello.** 1993. Multicenter clinical evaluation of a continuous monitoring blood culture system using fluorescent-sensor technology (BACTEC 9240). *J. Clin. Microbiol.* **31**:552–557.

21. **Pfaller, M. A., T. K. Sibley, L. M. Westfall, J. E. Hoppe-Bauer, M. A. Keating, and P. R. Murray.** 1982. Clinical laboratory comparison of a slide blood culture system with a conventional broth system. *J. Clin. Microbiol.* **16**:525–530.

22. **Pohlman, J. K., B. A. Kirkley, K. A. Easley, B. A. Basille, and J. A. Washington.** 1995. Controlled clinical evaluation of BACTEC Plus Aerobic/F and BacT/Alert Aerobic FAN bottles for detection of bloodstream infections. *J. Clin. Microbiol.* **33**:2856–2858.

23. **Pohlman, J. K., B. A. Kirkley, K. A. Easley, and J. A. Washington.** 1995. Controlled clinical comparison of Isolator and BACTEC 9240 Aerobic/F resin bottle for detection of bloodstream infections. *J. Clin. Microbiol.* **33**:2525–2529.

24. **Reimer, L. G., M. L. Wilson, and M. P. Weinstein.** 1997. Update on the detection of bacteremia and fungemia. *Clin. Microbiol. Rev.* **10**:444–465.

25. **Rohner, P., B. Pepey, and R. Auckenthaler.** 1995. Comparison of BacT/Alert with Signal blood culture system. *J. Clin. Microbiol.* **33**:313–317.

26. **Rohner, P., B. Pepey, and R. Auckenthaler.** 1996. Comparative evaluation of BACTEC Aerobic Plus/F and Septi-Chek Release blood culture media. *J. Clin. Microbiol.* **34**:126–129.

27. **Schwabe, L. D., R. B. Thomson, Jr., K. K. Flint, and F. P. Koontz.** 1995. Evaluation of BACTEC 9240 blood culture system using high-volume resin media. *J. Clin. Microbiol.* **33**:2451–2453.

28. **Smith, J. A., E. A. Bryce, J. H. Ngui-Yen, and F. J. Roberts.** 1995. Comparison of BACTEC 9240 and BacT/Alert blood culture systems in an adult hospital. *J. Clin. Microbiol.* **33**:1905–1908.

29. **Waite, R. T., and G. L. Woods.** 1998. Evaluation of BACTEC MYCO/F Lytic medium for recovery of mycobacteria and fungi from blood. *J. Clin. Microbiol.* **36**:1176–1179.

30. **Washington, J. A., II, and D. M. Ilstrup.** 1986. Blood cultures: issues and controversies. *Rev. Infect. Dis.* **8**:792–802.

31. **Weinstein, M. P., S. Mirrett, L. G. Reimer, and L. B. Reller.** 1989. Effect of agitation and terminal subcultures on yield and speed of detection of the Oxoid Signal blood culture system versus the BACTEC radiometric system. *J. Clin. Microbiol.* **27**:427–430.

32. **Weinstein, M. P., S. Mirrett, L. G. Reimer, and L. B. Reller.** 1990. Effect of altered headspace atmosphere on yield and speed of detection of the Oxoid Signal blood culture system versus the BACTEC radiometric system. *J. Clin. Microbiol.* **28**:795–797.

33. **Weinstein, M. P., S. Mirrett, L. G. Reimer, M. L. Wilson, S. Smith-Elekes, C. R. Chuard, K. L. Joho, and L. B. Reller.** 1995. Controlled evaluation of BacT/Alert standard aerobic and FAN aerobic blood culture bottles for detection of bacteremia and fungemia. *J. Clin. Microbiol.* **33**:978–981.

34. **Weinstein, M. P., S. Mirrett, and L. B. Reller.** 1988. Comparative evaluation of the Oxoid Signal and

BACTEC radiometric blood culture systems for the detection of bacteremia and fungemia. *J. Clin. Microbiol.* **26:**962–964.

35. **Weinstein, M. P., S. Mirrett, M. L. Wilson, L. G. Reimer, and L. B. Reller.** 1994. Controlled evaluation of 5 versus 10 milliliters of blood cultured in aerobic BacT/Alert blood culture bottles. *J. Clin. Microbiol.* **32:**2103–2106.

36. **Weinstein, M. P., L. B. Reller, S. Mirrett, W.-L. L. Wang, and D. V. Alcid.** 1985. Controlled evaluation of trypticase soy broth in agar slide and conventional blood culture systems. *J. Clin. Microbiol.* **21:**626–629.

37. **Weinstein, M. P., L. B. Reller, S. Mirrett, W.-L. L. Wang, and D. V. Alcid.** 1985. Clinical comparison of an agar slide blood culture system with trypticase soy broth and a conventional blood culture bottle with supplemented peptone broth. *J. Clin. Microbiol.* **21:**815–818.

38. **Weinstein, M. P., L. B. Reller, S. Mirrett, C. W. Stratton, L. G. Reimer, and W.-L. L. Wang.** 1986. Controlled evaluation of the agar slide and radiometric blood culture systems for the detection of bacteremia and fungemia. *J. Clin. Microbiol.* **23:**221–225.

39. **Welby, P. L., D. S. Keller, and G. A. Storch.** 1995. Comparison of automated Difco ESP blood culture system with biphasic BBL Septi-Chek system for detection of bloodstream infections in pediatric patients. *J. Clin. Microbiol.* **33:**1084–1088.

40. **Welby-Sellenriek, P. L., D. S. Keller, R. J. Ferrett, and G. A. Storch.** 1997. Comparison of the BacT/Alert FAN aerobic and the Difco ESP 80A aerobic bottles for pediatric blood cultures. *J. Clin. Microbiol.* **35:**1166–1171.

41. **Wilson, M. L., S. Mirrett, L. C. McDonald, M. P. Weinstein, J. Fune, and L. B. Reller.** 1999. Controlled clinical comparison of bioMerieux VITAL and BACTEC NR-660 blood culture systems for detection of bacteremia and fungemia in adults. *J. Clin. Microbiol.* **37:**1709–1713.

42. **Wilson, M. L., S. Mirrett, L. B. Reller, M. P. Weinstein, and L. G. Reimer.** 1993. Recovery of clinically important microorganisms from the BacT/Alert blood culture system does not require 7 day testing. *Diagn. Microbiol. Infect. Dis.* **16:**31–34.

43. **Wilson, M. L., M. P. Weinstein, S. Mirrett, L. G. Reimer, S. Smith-Elekes, C. R. Chirard, and L. B. Reller.** 1995. Controlled evaluation of BacT/Alert standard anaerobic and FAN anaerobic blood culture bottles for the detection of bacteremia and fungemia. *J. Clin. Microbiol.* **33:**2265–2270.

44. **Wilson, M. L., M. P. Weinstein, L. G. Reimer, S. Mirrett, and L. B. Reller.** 1992. Controlled comparison of the BacT/Alert and BACTEC 660/730 nonradiometric blood culture systems. *J. Clin. Microbiol.* **30:**323–329.

45. **Wilson, M. L., M. P. Weinstein, and L. B. Reller.** 1994. Automated blood culture systems. *Clin. Lab. Med.* **14:**149–169.

46. **Zaidi, A. K. M., S. Mirrett, J. C. McDonald, E. E. Rubin, L. C. McDonald, M. P. Weinstein, M. Gupta, and L. B. Reller.** 1997. Controlled comparison of bioMerieux VITAL and BACTEC NR-660 systems for detection of bacteremia and fungemia in pediatric patients. *J. Clin. Microbiol.* **35:**2007–2012.

47. **Zwadyk, P., Jr., C. L. Pierson, and C. Young.** 1994. Comparison of Difco ESP and Organon Teknika BacT/Alert continuous-monitoring blood culture systems. *J. Clin. Microbiol.* **32:**1273–1279.

Rapid Systems and Instruments for the Identification of Bacteria

ALAN T. EVANGELISTA, ALLAN L. TRUANT, AND PAUL P. BOURBEAU

3

During the past 10 years, the clinical microbiology laboratory has experienced a relative explosion in the number and variety of tests available for the initial detection and identification of bacterial pathogens. Indeed, even the usage of terms such as "conventional tests" and "gold standard" must constantly be reevaluated in the presence of the ever-increasing sensitivities and specificities of new testing modalities. Conventional tests from a generation ago consisted exclusively of tubed media (both solid and broth), plated media, tubed biochemical tests, and selected bacterial stains. The earliest kit methodologies of the 1970s consisted primarily of miniaturization of the standard biochemical tests with the occasional addition of rare proprietary tests. Preliminary detection of bacterial pathogens and isolation in pure culture on solid media was then followed by culture confirmation using one of the early-generation kits for bacterial identification.

One of the distinct advances in the second- and third-generation identification kits of the 1980s and 1990s was the incorporation of a large number of isolates in the respective kit databases and the addition of increasing numbers of unusual and rare isolates. This significantly increased the accuracy of bacterial identifications. The reader is referred to the *Color Atlas and Textbook of Diagnostic Microbiology*, 5th ed., by Koneman et al. (58) for an overview of packaged systems, computer-assisted methods, and descriptions of numeric coding systems, frequencies, and percent probabilities used for the identification of bacteria. The initial step of kit miniaturization was rapidly followed by semiautomated approaches for the identification and susceptibility testing of bacterial pathogens of human interest. The incorporation of both rapid kit methodologies and semiautomated approaches assisted the clinical microbiologist in decreasing the use of many of the manual techniques which were previously required.

The clinical microbiology laboratory has not yet reached the level of full automation, and it does not appear close to the implementation of robotics, which is available today in a few select clinical chemistry laboratories for analyte evaluations. The detection of most bacterial pathogens is still most easily performed with solid media, followed by the use of kit methods and semiautomated methods for identification and antibiotic profile determinations. Occasionally, extensive conventional testing must be performed to identify unusual or fastidious organisms. Public health laboratories, such as city and state laboratories, those at the Centers for Disease Control and Prevention (CDC), and other specialty reference laboratories, continue to play a pivotal role in selected microbiological circumstances. These laboratories often are involved with infections of high public interest or concern (e.g., sexually transmitted infections and mycobacterial disease), epidemiological investigations, and the study of very rare or unusual organisms or those with important emerging antimicrobial resistance trends.

In more recent years, the development and, in some cases, very common use of direct antigen testing has revolutionized the algorithms for specimen testing which are used today in physicians' office laboratories, community hospitals, and tertiary-care medical centers. Early antigen tests sometimes lacked the sensitivity which was offered by conventional testing. Therefore, many early antigen tests were useful when the results were positive, but negative results needed to be further evaluated by conventional testing. More recently developed direct antigen tests have improved significantly, and some even approach the sensitivity and specificity of tests previously considered conventional tests. However, many still need culture backup to assure a high level of sensitivity.

A discussion by Miller and O'Hara of manual and automated systems for microbial identification was recently included in the *Manual of Clinical Microbiology*, 7th ed., (68). The reader is referred to that chapter for a review of criteria for selecting a system, methods for evaluating an instrument or system, and other general considerations. Additional information can be found in other authoritative references which are mentioned in this chapter (24, 45, 46, 59). Here, we attempt to review those tests and methods, which have not been discussed in detail elsewhere.

It is particularly important that a pure culture of an unknown isolate be used as the inoculum in a commercial identification system and that a purity plate be used for each test to ensure that the identification rendered by the test system reflects the profile of one organism. As reviewed by Farmer in the *Manual of Clinical Microbiology*, 7th ed. (23), when an unusual organism identification is given by a commercial identification system, the clinical microbiologist must consider several possibilities. First, the identification may be correct and just unusual. Second, the identification may be incorrect for several reasons: the inoculum may be mixed or handling or coding errors were made during the testing process. If an unusual organism identification is rendered by the commercial system, it is scientifically prudent to confirm the purity of the isolate and to repeat the test in the same commercial system and possibly in a second commercial or manual system. Comparison of known antibiograms for the organism may also prove helpful.

When considering the purchase and implementation of a product, it is extremely important to evaluate the technical performance of the kit or method in addition to other factors, such as reagent and kit cost, personnel time, technical requirements, and clinical usefulness. Peer-reviewed evaluations and comparisons are very helpful as an important adjunct to the manufacturer's in-house performance data. In addition, review and clearance by the Food and Drug Administration (FDA), when required and indicated, gives further evidence for and support of stated claims and clinical applications. It is the user's ultimate responsibility to be knowledgeable about the products used in his or her laboratory and to use the products only within their stated shelf lives and only for the approved clinical applications. Additionally, as emphasized by LaClaire and Facklam (60), it is important for both the users of commercial kits and the manufacturers themselves to continually monitor and update the databases of their products to ensure that newly described organisms and redefined species are included.

We have attempted to include a comprehensive, updated list of tests, manufacturers, and references. We also recognize that in the clinical microbiology marketplace, the introduction and discontinuation of commercial kits are ongoing processes. Therefore, the omission of any tests or kits in this chapter, while likely to occur, is unintentional.

IDENTIFICATION OF GRAM-POSITIVE ORGANISMS

Staphylococcus spp.

Staphylococci are catalase-positive, gram-positive cocci. The genus Staphylococcus contains 32 species and 15 subspecies. While staphylococci are often normal skin flora, many staphylococcal species are significant human pathogens. Staphylococcus aureus is associated with a wide spectrum of disease, from skin and soft-tissue infections to endocarditis, pneumonia, and osteomyelitis, and causes significant morbidity and mortality. Coagulase-negative Staphylococcus spp. are associated with infections from foreign bodies, such as catheters and prosthetic devices.

Commercial Identification Methods for S. aureus

A variety of tests are utilized to differentiate S. aureus from the other species of staphylococci. Conventional tests for the identification of S. aureus have included the slide coagulase test, which detects bound coagulase, or clumping factor, and the tube coagulase test, which detects free coagulase. The slide coagulase test produces a rapid result, while the tube coagulase test requires 4 to 24 h for completion; the overall accuracy of the tube coagulase test is better than the accuracy of the slide coagulase test. In an attempt to produce a test with both high sensitivity and high specificity, manufacturers have developed a variety of commercial agglutination test kits designed to permit rapid and accurate differentiation of S. aureus from other species of staphylococci.

Personne and colleagues (80) have described agglutination assays for the identification of S. aureus as representing three generations of products (Table 1). The first-generation product, Staphyloslide (marketed by bioMerieux [Hazelwood, Mo.] as Staphyslide outside the United States), utilizes sheep red blood cells sensitized with fibrinogen to detect clumping factor. The second-generation products (Fig. 1) utilize latex particles coated with fibrinogen and immunoglobulin G to detect clumping factor and protein A, respectively. Based upon the observation that some strains of methicillin-resistant S. aureus mask clumping factor and protein A with capsular antigens, third-generation products have been developed with individual antibodies (monoclonal or polyclonal) which are specific for S. aureus capsular antigens.

The performances of several agglutination assays for identification of S. aureus are summarized in Table 2. Each of the assays included in Table 2 has had at least three recently published evaluations. The performances of individual kits vary significantly depending upon the strains included in the evaluation. In general, the third-generation

TABLE 1 Characteristics of agglutination assays for identification of S. aureus[a]

Commercial product	Manufacturer	Test type	Factors utilized[c]		
			Clumping factor	Protein A	Capsular antigens
BBL Staphyloslide	BD Diagnostic Systems	Hemagglutination	+	−	−
BACTi Staph[b]	Remel	Latex agglutination	+	+	−
Pastorex Staph	Sanofi Diagnostics Pasteur	Latex agglutination	+	+	−
Prolex Staph Latex	Pro-Lab Diagnostics	Latex agglutination	+	+	−
Staphaurex	Murex/Remel	Latex agglutination	+	+	−
Staphytect	Oxoid	Latex agglutination	+	+	−
Pastorex Staph-Plus	Sanofi Diagnostics Pasteur	Latex agglutination	+	+	+
Slidex Staph	bioMerieux	Hemagglutination and latex agglutination	+	+	+
Slidex Staph-Plus	bioMerieux	Latex agglutination	+	+	+
Staphaurex Plus	Murex/Remel	Latex agglutination	+	+	+
Staphytect Plus	Oxoid	Latex agglutination	+	+	+
Dryspot Staphytect Plus	Oxoid	Latex agglutination	+	+	+

[a]Modified from reference 32.
[b]Figure 1.
[c]+, utilized; −, not utilized.

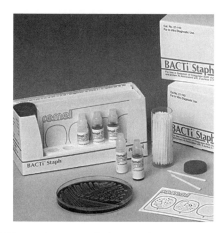

FIGURE 1 BACTi Staph latex agglutination test kit (courtesy of Remel).

assays have higher sensitivity but lower specificity than first- or second-generation assays. The decision as to which kit, if any, should be used in a specific laboratory should be based upon an evaluation which includes the strains routinely tested in that laboratory.

Commercial Identification Methods for *Staphylococcus* spp. and Other Gram-Positive Organisms

A variety of commercial products are available for the identification of *Staphylococcus* spp. In addition to staphylococci, these products often have the capability of identifying a variety of gram-positive cocci, including micrococci, streptococci, and enterococci (Table 3). Only a limited number of recent publications have evaluated staphylococcal identification kits. Meaningful interpretation of product accuracy is often complicated by software and codebook updates, including nomenclature changes. These updates often make evaluations that were published more than 5 to 10 years ago very difficult to interpret. Thus, considerable care should be taken when reviewing a published evaluation. In general, most human strains of staphylococci are identified by commercial systems with an accuracy of 70 to 90% (55).

API RAPIDEC Staph

The API RAPIDEC Staph kit (bioMerieux) consists of six cupules in a strip format. The kit identifies *S. aureus*, *Staphylococcus epidermidis*, and *Staphylococcus saprophyticus* after a 2-h incubation period. A UV lamp is required to read the fluorescent coagulase substrate in the second cupule, which is compared to the negative-control substrate in the first cupule. The remaining cupules contain chromogenic substrates, which are read using a non-UV (white) light source. An evaluation of the RAPIDEC kit reported a sensitivity of 100% for the identification of *S. aureus* but only 70% for *S. epidermidis* and 81% for *S. saprophyticus* (50).

API STAPH

The API STAPH kit (bioMerieux) consists of 20 cupules in a strip format. The kit identifies staphylococci and micrococci and requires overnight incubation. After the strip is read manually, a seven-digit code is generated and is then translated into the organism identification from a numerical database. An evaluation of the API STAPH kit yielded a sensitivity of 94% for the identification of *S. epidermidis*, 85% for *Staphylococcus haemolyticus*, and 75% for *Staphylococcus hominis* (64).

ID 32 Staph

The ID 32 Staph kit (bioMerieux) consists of 32 cupules in a strip format. It requires 24 h of incubation and can be read manually to generate a numerical profile, which can then be matched with an extensive database. The strip may also be read automatically by the bioMerieux ATB system, which is available in Europe. In the United States, the ID 32 Staph kit is available for industrial use but not for use by clinical laboratories. The test is available for clinical use in many countries outside the United States. In an evaluation of the ID 32 Staph kit, Ieven and colleagues tested 440 consecutive staphylococcal isolates, of which 95.2% were correctly identified (44). Of the isolates of *S. epidermidis*, *S. haemolyticus*, and *S. hominis*, which together made up 396 (90%) of the 440 isolates tested, 99.3, 98.1, and 75%, respectively, were correctly identified. The 44 remaining isolates, representing 10 species, were correctly identified 88.6% of the time. The authors reported that the ID 32 Staph kit was user friendly. An earlier evaluation of the ID

TABLE 2 Performance of agglutination assays for identification of *S. aureus*

Commercial product	Sensitivity (%)			Specificity (%)	References
	MRSA[a]	MSSA[b]	All strains		
Pastorex Plus	95.1–100	100	98.6–100	75.5–96.1	25, 64, 80
Slidex			96.5–99.6	73.2–96.4	80, 90, 110
Slidex Staph Plus	98.1	100	98.7–100	81.8–99.3	32, 80, 90
Staphaurex	49.2–100	98.1–100	84.5–99.6	91.7–100	25, 32, 64, 82, 88, 90, 98, 106, 110
Staphaurex Plus	100	98.9–100	98.7–100	72.7–97.9	32, 80, 98, 110; Kloos et al., *Abstr. 98th Gen. Meet. Am. Soc. Microbiol.*
Staphyloslide	72.1–96.8	98.7–100	87.7–99.1	72.5–100	25, 32, 82, 88, 90, 98, 110; Kloos et al., *Abstr. 98th Gen. Meet. Am. Soc. Microbiol.*

[a]MRSA, methicillin-resistant *S. aureus*.
[b]MSSA, methicillin-susceptible *S. aureus*.

TABLE 3 Systems for identification of *Staphylococcus* spp. and other gram-positive cocci[a]

Product	Manufacturer	Speed[d] (h)	Genera detected[e]										
			Staphylococcus	*Micrococcus*	*Streptococcus*	*Enterococcus*	*Aerococcus*	*Alloiococcus*	*Gemella*	*Lactococcus*	*Leuconostoc*	*Pediococcus*	*Stomatococcus*
API RAPIDEC Staph[b]	bioMerieux	R (2)	X										
API Staph	bioMerieux	O	X	X									
ID 32 Staph[c]	bioMerieux	O	X	X			X						X
BBL Crystal Gram-Pos ID system	BD Diagnostic Systems	O	X	X	X	X	X	X	X	X	X	X	X
BBL Crystal Rapid Gram-Pos ID system	BD Diagnostic Systems	R (4)	X	X	X	X	X		X	X	X	X	X
Phoenix Gram-Pos panel	BD Diagnostic Systems			X	X	X	X		X	X		X	X
MicroScan Pos ID panels	Dade Behring	O	X	X	X	X	X				X	X	
MicroScan Rapid Gram Pos ID panels	Dade Behring	R (2)	X	X	X	X	X		X				
Vitek GPI (Vitek 1)	bioMerieux	R (2–15)	X	X	X	X	X		X				
Vitek ID-GPC (Vitek 2)	bioMerieux	R (2–15)	X	X	X	X	X		X				
MIDI Sherlock	MIDI	O	X	X	X	X	X		X	X	X	X	X
Biolog GP MicroPlate *Streptococcus* ID panels	Biolog	O	X	X	X	X							
API 20 Strep	bioMerieux	R and O			X	X	X	X	X	X			
Rapid ID 32 Strep[c]	bioMerieux	R (4)			X	X	X	X	X	X			
IDS RapID STR	Remel	R (4)			X	X							

[a]Some products also include some genera of gram-positive rods in their databases.
[b]API RAPIDEC Staph is designed to identify *S. aureus*, *S. epidermidis*, and *S. saprophyticus*. Presumptive identification of several of the species is also made.
[c]Currently available for industrial use but not clinical use in the United States.
[d]R, rapid; O, overnight.
[e]X, detected.

32 Staph kit by Brun and colleagues against a variety of staphylococci yielded a sensitivity of 95.5% (12).

BBL Crystal Gram-Positive ID System

The BBL Crystal Gram-Positive ID system (BD Diagnostic Systems, Sparks, Md.) consists of 30 cupules in a three-row format. Test units require 18 to 24 h of incubation and must be read with a BBL Crystal Panel viewer. The system is capable of identifying a variety of gram-positive cocci but has not been recently evaluated for staphylococci.

BBL Crystal Rapid Gram-Positive ID System

The BBL Crystal Rapid Gram-Positive ID system (BD Diagnostic Systems) is identical in format to the Crystal system described above but requires only a 4-h incubation period. This system has not been recently evaluated for staphylococci.

MicroScan Pos ID Panels

The MicroScan Pos ID (conventional) panels (Dade Behring, West Sacramento, Calif.) are available for identification only or in combination with MIC or breakpoint susceptibility test formats. The MicroScan panels for antimicrobial susceptibility testing are discussed in chapter 17. The ID component includes 26 tests in a dried microtiter format. The plates require a minimum of 16 to 20 h of incubation, with some requiring an additional 24 h. The microtiter plates can be read manually or with MicroScan

instrumentation and contain substrates for the identification of a variety of gram-positive organisms. Weinstein and colleagues (107) tested 285 recent staphylococcal blood culture isolates using MicroScan Dried Pos ID panels with version 20.30 software. The results were assessed for isolates based upon a high probability (HP) (≥85%) or low probability (LP) (<85%) of identification. *S. epidermidis*, *S. hominis*, and *S. haemolyticus* isolates were correctly identified (HP versus LP) 85.6 versus 12.8, 42.9 versus 46.9, and 50 versus 50% of the time, respectively. The remaining 23 isolates belonged to eight species and were correctly identified (HP versus LP) 69.6 versus 17.3% of the time. In the same study, a revised conventional panel (CPID-2) was also evaluated. *S. epidermidis*, *S. hominis*, and *S. haemolyticus* isolates were correctly identified (HP versus LP) 95.1 versus 4.3, 85.7 versus 12.2, and 88.5 versus 3.8% of the time, respectively. The remaining 23 isolates were correctly identified (HP versus LP) 78.3 versus 8.7% of the time. The authors concluded that the prototype CPID-2 panel was a substantial improvement over the conventional MicroScan panels. Kloos and colleagues (W. E. Kloos, C. G. George, J. M. Miller, S. K. McCallister, M. F. Sierra, L. VanPelt, S. Connell, and B. L. Zimmer, *Abstr. 98th Gen. Meet. Am. Soc. Microbiol.*, abstr. C-176, p. 160, 1998) also evaluated the revised conventional MicroScan panel (CPID-2), testing 150 staphylococcal isolates. Overall (HP versus LP), 91.9 versus 5.6% of the isolates were correctly identified. Among specific species, 94% of the *S. aureus* isolates, 100%

of the *S. epidermidis*, *S. saprophyticus*, *S. haemolyticus*, *S. hominis*, and *Staphylococcus schleiferi* species, and 80% of the *Staphylococcus lugdunensis* isolates were correctly identified.

MicroScan Rapid Pos ID Panels

The MicroScan Rapid Pos ID panels (Dade Behring) include 34 tests in a microtiter format that use fluorogenic substrates, which must be read with the MicroScan Walk-Away instrument. The incubation time is 2 h. Weinstein and colleagues (107) tested 285 recent staphylococcal blood culture isolates using MicroScan Rapid Gram Pos ID panels with version 20.30 software. The results were assessed for isolates based upon an HP (≥85%) or LP (<85%) of identification. *S. epidermidis*, *S. hominis*, and *S. haemolyticus* isolates were correctly identified (HP versus LP) 95.7 versus 2.7, 32.7 versus 28.6, and 50.0 versus 19.2% of the time, respectively. These three species accounted for 262 (92%) of the isolates tested. The remaining 23 isolates belonged to eight species and were correctly identified (HP versus LP) 82.6 versus 8.7% of the time. McCallister and colleagues (S. K. McCallister, C. M. O'Hara, and J. M. Miller, *Abstr. 99th Gen. Meet. Am. Soc. Microbiol.*, abstr. C-440, p. 195, 1999) tested 302 staphylococcal strains representing 24 species using MicroScan Rapid Gram Pos ID panels with version 22 software. The evaluation yielded correct identification of 72% of the strains after the routine 2-h incubation. Following the performance of additional required off-line tests, another 65 strains (19.3%) were correctly identified for a total of 91.9% of the strains correct at 24 h. All isolates of *S. aureus*, *S. lugdunensis*, and *Staphylococcus xylosus* were correctly identified at 24 h. Only four human subspecies, *Staphylococcus schleiferi* subsp. *schleiferi*, *Staphylococcus cohnii* subsp. *urealyticum*, *Staphylococcus auricularis*, and *Staphylococcus cohnii* subsp. *cohnii*, were correct at less than 90% at 24 h.

Vitek GPI Card

The Vitek GPI card (bioMerieux) is a plastic card containing 30 microwells that is designed for use in the automated Vitek system (bioMerieux). Additional cards are also available for use on the Vitek system for the identification of gram-negative organisms, which are discussed below. A number of Vitek cards are also available for antimicrobial susceptibility testing; they are discussed in chapter 17. The GPI card contains substrates for the identification of a variety of gram-positive organisms. The identifications require an incubation period of 3 to 15 h, with coagulase-negative staphylococci usually requiring 10 to 13 h. An evaluation of 500 clinical gram-positive isolates by using GPI cards was reported by Bannerman and colleagues (5). The GPI card correctly identified 92% of *S. epidermidis*, 95% of *S. haemolyticus*, 100% of *S. saprophyticus*, and 88% of *Staphylococcus capitis* subsp. *capitis* isolates. The overall agreement between the GPI card identifications and identifications by conventional methods was 89%.

Vitek ID-GPC Card

The Vitek ID-GPC card (bioMerieux) is a 64-microwell card that is designed for use in the automated Vitek 2 instrument (bioMerieux). The card uses fluorogenic substrates and will identify most clinically significant gram-positive cocci within 2 h. Evaluations for the identification of staphylococci have not yet been published.

MIDI Sherlock System

The Sherlock System (MIDI, Inc., Newark, Del.) performs identification of a variety of gram-positive organisms by the technique of cellular fatty acid analysis using computerized high-resolution gas-liquid chromatography. The cellular fatty acid profile of the individual isolate is compared to a computerized database of profiles of known organisms. In an evaluation of 470 gram-positive isolates, the MIDI system correctly identified all strains of *S. epidermidis*, *S. cohnii*, *Staphylococcus intermedius*, *S. lugdunensis*, *S. schleiferi*, *Staphylococcus simulans*, *Staphylococcus sciuri*, and *S. xylosis* (95). Misidentified strains were noted among *S. hominis* and *S. saprophyticus* isolates. The overall agreement between the MIDI system identifications and identifications by conventional methods was 87.8%.

Biolog GP MicroPlate

The GP MicroPlate panel (Biolog, Inc., Hayward, Calif.) is a 96-well microtiter panel for the identification of gram-positive organisms based on the oxidation of substrates. The tests are read visually after a 4- to 24-h incubation period, depending on the organism. An evaluation of 113 clinical isolates found the overall accuracy to be 69% at a CDC laboratory site and 73% at a New Mexico site (69). The study conclusions were that improvements to the system were required to increase the accuracy of the identification of coagulase-negative staphylococci. See "*Corynebacterium* spp." below for an additional description of the Biolog system.

Streptococcus spp.

Streptococci are gram-positive, catalase-negative, facultatively anaerobic bacteria which grow as cocci in pairs or chains. Significant taxonomic changes have occurred in recent years to the genus *Streptococcus*; however, the primary division of the genus is still based upon hemolytic reactions on sheep blood agar. The reader is referred to a recent review of streptococcal taxonomy by Ruoff and colleagues in the *Manual of Clinical Microbiology*, 7th ed. (89).

Commercial Identification Methods for Beta-Hemolytic Streptococci

Streptococcus spp. producing a beta-hemolytic reaction on sheep blood agar may be further classified serologically based on their cell wall carbohydrate antigens. Rebecca Lancefield first developed an identification schema for these organisms, which is referred to as the Lancefield grouping system for beta-hemolytic streptococci (61). Streptococci causing human infections are serologically classified as Lancefield groups A, B, C, D, F, and G. The grouping method originally used by Lancefield, the capillary precipitin test, has been replaced by commercial methods that extract and solubilize the cell wall antigens by either acid extraction (nitrous acid) or enzymatic extraction. The solubilized cell wall antigens are detected and visualized by their interaction with group-specific antisera fixed on *S. aureus* cells (coagglutination) or on polystyrene latex beads (latex agglutination). Selected commercial methods for the serogrouping of beta-hemolytic streptococci are listed in Table 4.

Direct Detection of Pharyngeal Group A Streptococci

During the past 10 years, a number of rapid tests have been developed for the direct detection of group A streptococcal antigen from throat swabs. Compared to culture, the direct-detection methods have ranged in sensitivity from 75 to 85% (3, 35, 114), and as a result, the manufacturers of the rapid tests have recommended that negative test results be

TABLE 4 Commercial serogrouping assays for identification of beta-hemolytic streptococci

Commercial product	Manufacturer	Test type	Extraction method
Directigen	BD Diagnostic Systems	Latex agglutination	Enzyme
Meritec Strep	Meridian Diagnostics	Coagglutination	Enzyme
Oxoid Strep Grouping Kit	Oxoid	Latex agglutination	Enzyme
Oxoid Strep Plus	Oxoid	Latex agglutination	Acid
Patho-Dx	Diagnostic Products Corp.	Latex agglutination	Acid
Phadebact *Streptococcus*	Karo Bio Diagnostics	Coagglutination	Enzyme
Slidex Strep	bioMerieux	Latex agglutination	Acid
Strep Grouping Kit	The Binding Site	Latex agglutination	Enzyme
Streptex	Murex/Remel	Latex agglutination	Enzyme or acid

backed up with culture. Within the past 5 years, an increase in self-contained direct-detection tests using an immunochromatographic test format have been introduced to the clinical market. Many of these newer tests have been granted waived status under the Clinical Laboratory Improvement Act of 1988 (CLIA 88), and they are used in physicians' offices, clinics, and emergency departments (Fig. 2 and 3). With a sensitivity similar to that of the standard direct-detection methods (14, 33; S. Hoffman and K. Witt, *Abstr. 99th Gen. Meet. Am. Soc. Microbiol.*, abstr. C-22, p. 109, 1999), these waived tests have an important utility in point-of-care testing, with results available in less than 10 min. Early detection of group A streptococcal pharyngitis allows the appropriate and timely use of antimicrobial agents and an earlier return to work for parents and guardians of pediatric patients. It is not uncommon for a single manufacturer of a rapid waived test to enter into marketing agreements with a few different distributors, resulting in the test being sold under different names (Table 5). The sensitivities of these waived tests are highly dependent on the sampling technique and organism inoculum and improve with organism quantities of greater than 10^5 CFU per swab (A. T. Evangelista, P. Marciano, K. VanDerhei, and M. Dhand, *Abstr. Integr. Point-of-Care Test. Continuity Care: Eff. Outcome*, abstr. 23, p. 18, 1997; L. Morgan, S. B. Overman, and J. A. Ribes, *Abstr. 99th Gen. Meet. Am. Soc. Microbiol.*, abstr. C-23, p. 109, 1999).

In addition to the waived tests, rapid tests of greater complexity, which are not waived under CLIA 88, may be used by clinical laboratories for the rapid detection of group A streptococci (Fig. 4). The more commonly used tests which are not waived are listed in Table 6. Most of the antigen detection tests are completed in 10 to 20 min, while the DNA probe test requires about 75 min for completion. The DNA probe test has a reported sensitivity of 86 to 92% in comparison to culture (36, 84).

Direct Detection of GBS

In 1997, the FDA issued a safety alert warning of the risks of using devices for the direct detection of group B streptococcal (GBS) antigens (15, 73). The FDA warned of both false-negative and false-positive test results when these devices were employed. Specifically, the FDA cited the danger of relying on the devices for the detection of GBS colonization in pregnant women, recommending culture confirmation of all negative direct-test results. In addition, testing with these devices for GBS antigen from the urine of infants is not recommended. Since the publication of the FDA recommendations, the use of these

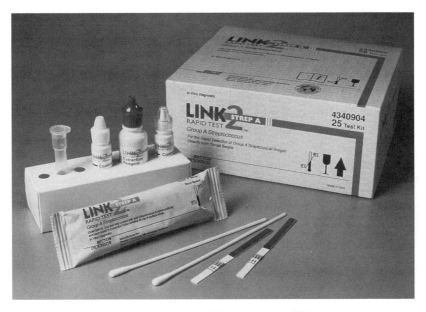

FIGURE 2 BD LINK 2 Strep A Rapid Test (courtesy of BD Diagnostic Systems).

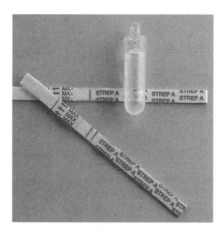

FIGURE 3 RIM A.R.C. Strep A test strips with negative test (top) and positive test (bottom) results (courtesy of Remel).

devices has substantially declined. Therefore, published evaluations of direct GBS tests are not included in this chapter.

Direct Detection of *Streptococcus pneumoniae* Urinary Antigen

The rapid diagnosis of invasive disease caused by *S. pneumoniae* is of great importance clinically, since the pneumococcus is the major cause of community-acquired bacterial pneumonia. The Binax NOW *S. pneumoniae* test is a rapid immunochromatographic test that detects pneumococcal antigen in a patient's urine specimen. The result is read in 15 min, with a reported sensitivity of 94% compared to culture (E. Molokova, D. Gentile, M. Fent, N. Moore, and V. Koulchin, *Abstr. 99th Gen. Meet. Am. Soc. Microbiol.*, abstr. C-21, p. 109, 1999).

Commercial Identification Methods for *Streptococcus* spp.

The commercial products available for the identification of streptococci have various numbers of species in their databases. Descriptions of the products, databases, and intended uses can be found in "*Staphylococcus* spp." above, and the product names are listed in Table 3. There have been few

recent published evaluations of streptococcal identification products.

BBL Crystal Gram-Positive ID System

von Baum and colleagues (104) reported that the BBL Crystal Gram-Positive ID system correctly identified 51 of 63 (81%) streptococcal isolates to the species level, with the remaining 12 isolates correct to the genus level.

Rapid ID 32 Strep

Freney and colleagues (26) evaluated the Rapid ID 32 Strep with 293 streptococcal isolates, representing over 30 species. Overall performance was excellent, with 71% of the isolates identified correctly with no supplemental tests. An additional 23% were correctly identified with supplemental tests, while 1% were misidentified and 4% were not identified.

Vitek GPI Card

Hinnebusch and colleagues (37) evaluated the performance of the Vitek GPI card in the identification of 203 viridans group streptococci. Overall, 61% were correctly identified without supplemental testing. An additional 21% were correctly identified with supplemental tests, while 10% were misidentified and 8% were not identified.

MicroScan Rapid Gram Pos ID Panels

Hinnebusch and colleagues (37) evaluated the performance of the MicroScan Rapid Gram Pos ID panels for the identification of 203 viridans group streptococci. Overall, 66% were correctly identified without supplemental testing. An additional 11% were correctly identified with supplemental testing, while 17% were misidentified and 5% were not identified.

Enterococcus spp.

Enterococci are catalase-negative, gram-positive cocci which are facultatively anaerobic. A pseudocatalase is occasionally produced, which must be differentiated from a true positive catalase test reaction. There are 16 species in the genus *Enterococcus*. Other catalase-negative, gram-positive cocci are recovered in clinical specimens, including *Lactococcus*, *Leuconostoc*, *Pediococcus*, and *Vagococcus*. For information on differentiating enterococci from other similar gram-positive cocci, see the discussion by Facklam

TABLE 5 Rapid, direct group A streptococcal detection tests granted waived status

Commercial product	Manufacturer
Abbott Signify Strep A test	Wyntek Diagnostics, Inc.
Applied Biotech SureStep Strep A (II)	Applied Biotech, Inc.
Beckman Coulter ICON Fx Strep A test	Binax, Inc.
BioStar Acceava Strep A test	Wyntek Diagnostics, Inc.
BD LINK2 Strep A Rapid test[a]	Applied Biotech, Inc.
Genzyme Contrast Strep A	Genzyme Diagnostics
Jant Pharmacal AccuStrip Strep A (II)	Applied Biotech, Inc.
Mainline Confirms Strep A Dots test	Applied Biotech, Inc.
Meridian Diagnostics ImmunoCard STAT! Strep A	Applied Biotech, Inc.
QuickVue In-Line One-Step Strep A	Quidel
RIM A.R.C. Strep A test[b]	Remel
Wyntek Diagnostics OSOM Strep A test	Wyntek Diagnostics, Inc.

[a]Figure 2.
[b]Figure 3.

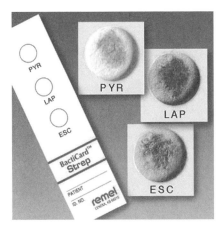

FIGURE 4 BactiCard Strep with positive test results (courtesy of Remel).

and colleagues in the *Manual of Clinical Microbiology*, 7th ed. (22).

Commercial Identification Methods for *Enterococcus* spp.

Identification of enterococci to the species level has acquired increased significance with the emergence of vancomycin-resistant enterococci. A specific species identification can be critical to determining the clinical, epidemiological, and microbiological significance of potential vancomycin-resistant isolates. With the exception of *Enterococcus faecalis* isolates, commercially available products have been variable in the identification of *Enterococcus* spp. (22). Descriptions of the products and databases can be found in Table 3 in "*Staphylococcus* spp." above.

BBL Crystal Gram-Positive ID System

von Baum and colleagues (104) reported that 100% of *Enterococcus* spp. tested were correctly identified with the BBL Crystal Gram-Positive ID system. However, only 39 enterococcal isolates were included in the evaluation, including 23 *E. faecalis*, 14 *Enterococcus faecium*, and 2 *Enterococcus avium* isolates.

TABLE 6 Rapid, direct group A streptococcal detection tests which do not have waived status

Commercial product	Manufacturer
TestPack	Abbott Laboratories
TestPack Plus/OBC	Abbott Laboratories
TestPack Plus/OBC II	Abbott Laboratories
Directigen 1-2-3	BD Diagnostic Systems
Q Test Strep	BD Diagnostic Systems
Strep A OIA	BioStar, Inc.
Strep A OIA Max	BioStar, Inc.
QuickVue Flex Strep A	Quidel
CARDS Q.S. Strep A	Quidel
BactiCard Strep[a]	Remel
Clearview Strep A	Unipath-Wampole Laboratories
Group A Strep Direct test	Gen-Probe, Inc.

[a]Figure 4.

MicroScan Pos ID Panel

A revised MicroScan Dried Pos ID panel (CPID2) was evaluated by Iwen and colleagues (47). They tested 202 isolates representing eight species of enterococci. Excluding *Enterococcus gallinarum* isolates, only 1 of 143 isolates was misidentified; 16 of 30 *Enterococcus casseliflavus* isolates were identified with LP, requiring motility and colony pigmentation tests for confirmation. Of the 59 *E. gallinarum* isolates tested, 33 were incorrectly identified as *E. faecium*. The authors suggest that confirmation by off-line tests of all isolates with an identification of *E. faecium* is required. Styers and colleagues (D. A. Styers, B. M. Brill, M. M. Henry, and P. E. Oefinger, *Abstr. 99th Gen. Meet. Am. Soc. Microbiol.*, abstr. C-441, p. 196, 1999) also evaluated the CPID2 panel in a limited study of 40 isolates. The overall accuracy was 80%. However, the number of species of non-*E. faecium* and non-*E. faecalis* isolates, including *E. gallinarum*, was not specified.

Rapid ID 32 Strep

The Rapid ID 32 Strep kit was evaluated by Freney and colleagues (26). Of the 71 enterococcal isolates tested, 67 were initially correctly identified, 3 required extra tests, and 1 was not identified. All 12 *E. faecium* and 27 *E. faecalis* isolates were correctly identified. Six of eight *E. gallinarum* isolates were correctly identified without additional testing, while two isolates required extra tests for correct identification. This test is currently not available for use in clinical laboratories in the United States.

Vitek GPI Card

Styers and colleagues (Styers et al., Abstr. 99th Gen. Meet. Am. Soc. Microbiol., 1999) evaluated the Vitek GPI card in a limited study of 40 isolates. The overall accuracy was 92.5% when supplemental tests were included for non-*E. faecalis* isolates. The exact number of non-*E. faecium* and non-*E. faecalis* species was not specified in the study.

Corynebacterium spp.

Corynebacteria are aerobic, gram-positive, non-spore-forming members of the normal skin flora of humans and many animals. Occasional corynebacteria are normal inhabitants of the environment. This group of organisms is of particular and growing significance as opportunistic pathogens, especially in immunocompromised and hospitalized patients (29). Currently, 46 species are described, with 31 having documented medical relevance (30). Evaluating the clinical relevance of *Corynebacterium* spp. from human specimens can present the clinical microbiologist and clinician with significant challenges due to the omnipresence of the organism on the skin and mucous membranes. Particularly helpful characteristics in the determination of clinical significance may be (i) isolation of the organism from acceptable specimens, especially if from multiple specimens; (ii) isolation of the organism from normally sterile body sites, particularly multiple specimens from blood; (iii) isolation of the organism in pure culture from the urine in numbers greater than 10^4/ml and from properly collected material; and (iv) elevated leukocyte reactivity and consistent clinical presentation.

Commercial Identification Methods for *Corynebacterium* spp.

The available commercial systems for the identification of *Corynebacterium* spp. are listed in Table 7. To date, evaluations have been published only for the API Coryne system

TABLE 7 Commercial identification systems for *Corynebacterium* spp.

Commercial product	Manufacturer
API Coryne system	bioMerieux
IDS RapID CB Plus system	Remel
BBL Crystal Gram-Pos ID system	BD Diagnostic Systems
BBL Crystal Rapid Gram-Pos ID system	BD Diagnostic Systems
MicroScan Pos ID panels	Dade Behring
MicroScan Rapid Gram Pos ID panels	Dade Behring
Vitek GPI card (Vitek 1)	bioMerieux
Vitek ID-GPC card (Vitek 2)	bioMerieux
MIDI Sherlock	MIDI
MCN Gram-Pos Plate	Merlin Diagnostics
Biolog GP MicroPlate	Biolog

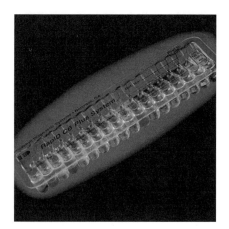

FIGURE 6 IDS RapID CB Plus system (courtesy of Remel).

(bioMerieux), the IDS RapID CB Plus system (Remel, Lenexa, Kans. [formerly Innovative Diagnostic Systems, Norcross, Ga.]), and the Biolog GP MicroPlate system (27, 28, 42, 62).

API Coryne System

The API Coryne system with database version 2.0 was recently evaluated by Funke et al. (28) (Fig. 5). The 20 biochemical reactions included in database version 2.0 were identical to those in the previous database version 1.0. Database version 2.0, however, included more taxa and additional differential tests. In the multicenter study by Funke et al., 407 strains of coryneform bacteria were tested (24 h at 37°C); 390 strains belonged to the 49 taxa included in database version 2.0, and 17 strains belonged to taxa not included in the new database. The updated version 2.0 API Coryne system in the study correctly identified 90.5% of the strains belonging to the taxa which were included in the database. Additional tests were needed for correct identification of 55.1% of all strains tested. Of all strains in the study, 5.6% were not identified and 3.8% were misidentified.

IDS RapID CB Plus System

The IDS RapID CB Plus system (Remel) consists of 4 carbohydrate and 14 preformed enzyme tests which are evaluated after 4 h in 37°C ambient air (Fig. 6). It has been evaluated recently by Hudspeth et al. (42) and Funke et al. (27). Hudspeth et al. tested 98 clinical isolates of *Corynebacterium spp.*, other coryneforms, *Listeria monocyto-*

genes, and 17 ATCC strains; 95% (40 of 42) of *Corynebacterium* spp. were accurately identified to the species level by the RapID CB Plus system, and 75% (27 of 36) of coryneform bacteria were correctly identified to the species level. Funke et al. (27) tested 345 strains of coryneform bacteria and 33 strains of *Listeria* spp. (representing 49 taxa). In this study, 80.9% were identified to the species level and 12.2% were identified to the genus level, with 3.7% of the strains misidentified and 3.2% of the strains not identified. Funke et al. found that the RapID CB Plus system correctly identified 88.5% of the most frequently encountered clinical isolates of coryneform bacteria. Further information on the API Coryne system and the RapID CB Plus system can be found in a review by Janda in the *Clinical Microbiology Newsletter* (48).

Biolog GP MicroPlate

The Biolog format for the identification of corynebacteria is a proprietary carbon source utilization test methodology in a MicroPlate test format (Fig. 7); the biochemical results may be automatically read and recorded. The MicroStation (Fig. 8), developed by Biolog, can identify over 1,400 species of both aerobic and anaerobic bacteria and yeasts. In addition to clinical and veterinary applications, the MicroStation may have applications for the pharmaceutical, cosmetics, environmental, food and beverage, agriculture, plant pathology, marine, and other industries. Biolog

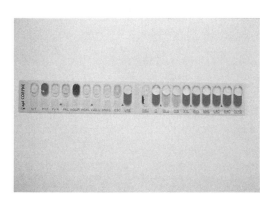

FIGURE 5 API Coryne system (courtesy of bioMerieux).

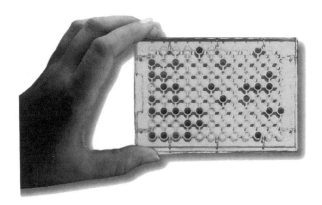

FIGURE 7 MicroPlate test format (courtesy of Biolog).

FIGURE 8 MicroStation (courtesy of Biolog).

offers the fully automated MicroStation and two manual MicroLog systems for laboratories with low testing volumes or limited financial resources, the MicroLog 1 and the MicroLog 2. The automated MicroStation system includes (i) instrumentation and software to read the MicroPlates, (ii) a turbidimeter for preparation of inocula, (iii) a colony magnifier lamp, and (iv) an eight-channel repeating electronic pipetter. In the manual systems, the automated reading capability is replaced with visual reading and manual reaction entry. The MicroLog 2 system is identical to the MicroStation system except that it does not include the MicroStation reader and pipetter. The MicroLog 1 system is identical to the MicroLog 2 system except that the system software does not have the user database-building features or the advanced identification analysis features. In all of the systems, consumables and databases are purchased separately. Lindenmann et al. evaluated the Biolog system for identification of asporogenous aerobic gram-positive rods by testing 174 strains belonging to 42 different taxa (62) and found overall that 50 and 60.3% of all strains were correctly identified to the genus or species level after 4 and 24 h, respectively. Of all strains, 38.5 and 32.2% were incorrectly identified to these levels, and for 11.4 and 7.5%, the system did not provide any identification. The database used for this study was version 3.50; it has recently been updated to version 4.0. Lindenmann et al. noted in their evaluation that the Biolog system has the advantage that some strains are identified within 4 h.

IDENTIFICATION OF GRAM-NEGATIVE BACTERIA

Prepackaged commercially available kit systems for the identification of clinically significant gram-negative bacteria have been available for a number of years and offer the clinical microbiologist several distinct advantages over previously available methodologies. Most kits have a relatively long shelf life and require little storage space. The availability of commercial kits or automated methods obviates the need to maintain large amounts of fresh media or reagents for conventional testing. Most of the kits are relatively easy to use and to read, and they have large databases, which has significantly improved their accuracy and reproducibility. Some of the methods described in this chapter have auto-

mated capabilities, which further improve their usefulness in a busy laboratory, and some have computer-assisted functions which act as expert systems in both identification and antimicrobial susceptibility testing. Many of these characteristics are particularly useful in small laboratories or in laboratories where there is minimal technical expertise. In many respects, the advances in commercial kits and automated methods have to a large extent "leveled the playing field" in microbiological identification, especially with regard to gram-negative bacteria. The confidence of clinical microbiologists in the vast majority of commercial kit systems has been fully justified, since many of the kits have demonstrated 95% or greater agreement with conventional methods.

Some of the disadvantages of commercial kits and automated methods are still present to some degree today. Fastidious or slowly growing organisms may exhibit weak or delayed reactions, which may lead to incorrect or LP identifications. Also, organisms either not in the database or present only in low numbers may be misidentified or unidentified. As we move closer to a genotypic identification of microorganisms and as commercial kits and automated methods incorporate more genetic markers (rather than phenotypic characteristics), we may see an ever-increasing refinement of commercially available methods for both identification and antimicrobial susceptibility testing.

Automated methods offer unique advantages over commercially available kits and conventional methods. The ability to interface with computers and download information, demographics, and both preliminary and final results saves the laboratory significant time and effort and minimizes clerical and transcriptional errors. One must always verify, however, that transmitted data are both complete and accurate. Automated methods also give the clinical microbiologist access to many computerized functions, such as summary results, epidemiological reports, antibiograms, budgetary and personnel information, expert analytic capability, such as linkage of the microbiology results with the pharmacy, and other functions either unavailable or minimally available in conventional or kit systems.

Unique disadvantages of automated systems are both few and relatively infrequent. Occasional power outages, surges, or fluctuations can severely disrupt the operations of automated systems. The tests in some automated systems can be read and evaluated manually when the instrument fails. In other automated systems, the tests cannot be read manually and the entire test must be repeated. The maintenance and repair record of each automated system and its reliability must be evaluated prior to purchase or lease, and it must be continually monitored. The initial costs associated with automated systems and the costs of maintenance and required reagents and supplies may be similar to or exceed those of other commercially available kit systems. In the evaluation of systems, however, one must include the total costs associated with reagents and technical time when comparing manual and automated systems. Although the initial cost or lease fee for an automated system may be higher than the cost of a comparable manual test, the associated personnel costs may bring the tests closer to fiscal equity. Each clinical facility must carefully compare and evaluate its own choice of test methods, since local costs for instrumentation, reagents, maintenance contracts, and personnel may vary significantly.

Selected information on manual, semiautomated, and automated identification systems for members of the family *Enterobacteriaceae* and other gram-negative bacteria is included in two authoritative texts edited by Koneman et al. (*Color Atlas and Textbook of Diagnostic Microbiology*, 5th ed., chapters 4 and 5) (58) and Isenberg (*Clinical Microbiology Procedures Handbook*, vol. I, section 1) (45). Listed in Table 8 are the available commercial identification systems for gram-negative bacteria, the manufacturer of each system, the organism groups identified, and the appropriate text reference where the detailed description of that system can be found. Information relating to the biochemical tests

TABLE 8 Commercial systems for identification of members of the family *Enterobacteriaceae* and gram-negative nonmembers

Commercial product	Manufacturer	Speed[b] (h)	Organisms in database[c]	Reference
API 20E[a]	bioMerieux	O	*Enterobacteriaceae* and GNNE	45, 58
API Rapid 20E[a]	bioMerieux	R (4)	*Enterobacteriaceae* and GNNE	45, 58
API NFT (20NE)[a]	bioMerieux	O	GNNE	45, 58
BBL Crystal Enteric/ Nonfermenter ID	BD Diagnostic Systems	R (4)	*Enterobacteriaceae* and GNNE	58
BBL Crystal Rapid Stool/Enteric ID	BD Diagnostic Systems	R (3)	*Enterobacteriaceae* including stool pathogens	
Enterotube II	BD Diagnostic Systems	O	*Enterobacteriaceae*	45, 58
Oxi/Ferm II	BD Diagnostic Systems	O	GNNE	45, 58
Phoenix Gram-Neg	BD Diagnostic Systems	R (2–4)	*Enterobacteriaceae* and GNNE	
Fox Extra GNI	Medical Specialties	O	*Enterobacteriaceae* and GNNE	
GN MicroPlate	Biolog	R (4) and O	*Enterobacteriaceae* and GNNE	58
MIDI Sherlock	MIDI	O	*Enterobacteriaceae* and GNNE	
MicroScan NEG ID Type 2	Dade Behring	O	*Enterobacteriaceae* and GNNE	58
MicroScan Rapid NEG ID Types 2 and 3	Dade Behring	R (2–2.5)	*Enterobacteriaceae* and GNNE	58
BACTiCard *E. coli*	Remel	R (4)	*E. coli*	
Micro ID	Remel	R (4)	*Enterobacteriaceae*	45, 58
N/F System	Remel	O	GNNE	45, 58
RapID NF Plus	Remel	R (4)	GNNE	58
RapID ONE	Remel	R (4)	*Enterobacteriaceae* and other oxidase-negative gram-negative bacteria	58
RapID SS/u	Remel	R (2)	Common UTI bacteria	
r/b System	Remel	O	*Enterobacteriaceae*	45
Uni-N/F Tek	Remel	O	GNNE	
Sensititre AP80	Trek	R (5) and O	*Enterobacteriaceae* and GNNE	58
Vitek EPS (Enteric Pathogen Screen)	bioMerieux	R (4–8)	*Salmonella*, *Shigella*, *Yersinia*	58
Vitek GNI Plus[a]	bioMerieux	R (2) and O	*Enterobacteriaceae* and GNNE	58
Vitek UID and UID-3	bioMerieux	R (1) and O	Common UTI bacteria	58
Vitek 2 ID-GNB	bioMerieux	R (2) and O	*Enterobacteriaceae* and GNNE	

[a]Figure 9.
[b]R, rapid; O, overnight.
[c]GNNE, gram-negative non-*Enterobacteriaceae*; UTI, urinary tract infection.

available in the manual kits and their uses in differentiating gram-negative bacilli can be found in appendix 2-1 of *Essential Procedures in Clinical Microbiology* (46).

Manual Systems for the Identification of Gram-Negative Bacteria

Most of the published evaluations of manual systems for the identification of gram-negative bacteria were performed and published over 5 years ago, and many are 10 to 15 or more years old. Therefore, a review of the many scientific comparison studies evaluating these manual tests will not be included in this chapter. Most of the manual tests have been shown to perform at a level of ≥90% accuracy, and many have performed at or above 95% accuracy (58), and therefore are considered comparable to conventional identification protocols. Recently published evaluations for the automated identification systems are discussed below.

Automated Systems for the Identification of Gram-Negative Bacteria

MicroScan Rapid Neg ID Panels

York and colleagues performed an evaluation of a MicroScan Rapid Negative Combo Type 2 panel (Dade Behring) that provides a 2-h identification, using a MicroScan WalkAway system with version 17.02 software (113). A total of 400 gram-negative fermentative bacteria were tested, and a sensitivity of 96% was reported. Pfaller and colleagues conducted a study of nonenteric bacilli with a MicroScan WalkAway-96 system and reported a sensitivity of 92.3% for isolates, with an identification probability of greater than 85% (81). Rhoads and coworkers, using a MicroScan WalkAway-96 system with version 20.20 software, reported a sensitivity of 97.5% for *Acinetobacter baumannii* and 82.1% for *Pseudomonas aeruginosa* (86).

Bascomb and colleagues performed a multicenter evaluation of the MicroScan Rapid Neg ID3 panel, which was designed as an improvement over the Rapid Neg ID2 panel with the replacement of 10 biochemical tests and a time to identification of 2.5 h (7). A total of 405 common gram-negative rods belonging to 54 species were tested, with a correct identification of 96.8%. A second series of 247 isolates representing new species added to the database yielded a correct identification of 89.5%. O'Hara and Miller recently evaluated the MicroScan Rapid Neg ID3 panel in a MicroScan WalkAway system using version 22.01 software (75). A total of 511 isolates of the family *Enterobacteriaceae*, along with some common gram-negative glucose nonfermenters, were tested, and the results were compared to those of conventional tube biochemical tests. At the end of the initial 2.5 h of incubation, 79.3% of the identifications were correct, increasing to 88.8% correct at 24 h after additional off-line biochemical tests were performed.

Vitek GNI

Robinson and coworkers examined 381 enteric and 131 nonenteric gram-negative rods with the Vitek GNI card (bioMerieux) and reported correct identification of 95.5% of the isolates without supplemental testing (87). Rhoads and colleagues, using a GNI card with Vitek 1 version R08.2 software, reported correct identification of 100% for 80 isolates of *A. baumannii* and 84.6% for 39 isolates of *P. aeruginosa* (28).

Vitek GNI Plus

An evaluation of the Vitek GNI Plus card (Fig. 9) was performed with 619 enteric and nonenteric gram-negative rods by O'Hara and colleagues using Vitek 1 version 5.01 software (76). After additonal tests for nonenteric bacteria suggested by the Vitek software, the accuracy was reported as 87.6%. The average time for enteric identifications was 4.1 h, and that for nonenteric identifications was 6.8 h.

Vitek 2 ID-GNB card

An evaluation of the new Vitek 2 system (bioMerieux) was performed with 845 isolates of the family *Enterobacteriaceae* and nonenteric gram-negative rods by Funke and colleagues (31). The ID-GNB card contains 64 wells, and the new fluorescence-based technology of the Vitek 2 system yields final results within 3 h. Initially, 84.7% of the isolates were correctly identified, increasing to 88.5% with the performance of simple manual off-line tests on 32 isolates.

Sensititre Gram-Negative AP80 Panel

The Gram-Negative AP80 panel of the Sensititre system (Trek Diagnostics) uses a 96-microwell format with fluorogenic substrates and contains three identical sections of 32 tests, thus allowing the identification of three organisms per panel. Staneck and colleagues evaluated the Sensititre AP80 panels by testing 879 isolates of the family *Enterobacteriaceae* and 144 nonenteric organisms (93). Results were available to the species level after 5 h of incubation for 90% of the isolates, and the remaining results were finalized at 18 h. Correct identification was reported for 92.5% of the enteric organisms and 84.7% of the nonenteric organisms.

Phoenix Gram-Negative Panel

The Gram-Negative panel of the Phoenix system (BD Biosciences) consists of 45 biochemical substrates comprising carbohydrates, fluorogens, carbon sources, and chromogens and provides identification results in less than 4 h. The Gram-Negative panel was evaluated by Salomon and coworkers by testing 75 isolates of the family *Enterobacteriaceae* and 50 gram-negative glucose nonfermenters (J. E. Saloman, T. Wiles, C. Yu, and T. Dunk, *Abstr. 99th Gen. Meet. Am. Soc. Microbiol.*, abstr. C-448, p. 197, 1999). In this early evaluation of the Phoenix system, the majority of the substrates showed good reactivity with enteric organisms in 2 to 3 h and with nonenteric organisms in 2 to 4 h. Identification of both groups of organisms was made without additional off-line tests. The evaluation did not report the percent accuracy of the identification results compared to those of conventional methods. The Phoenix system is expected to be available for use in clinical laboratories in the United States in 2001 or 2002.

SELECTED MEMBERS OF THE FAMILY *ENTEROBACTERIACEAE*

Commercial Identification Methods for *Escherichia coli*

The common gram-negative rod *E. coli* is usually motile, ferments D-glucose, is asporogenous, is oxidase negative, and grows well on MacConkey agar (11). There are five species in the genus *Escherichia*: *Escherichia blattae*, *E. coli*, *Escherichia fergusonnii*, *Escherichia hermannii*, and *Escherichia vulneris*. *E. coli* is a common inhabitant of the intestines of humans and animals. *E. coli* (and occasionally other species) may cause intestinal disease, bacteremia, and

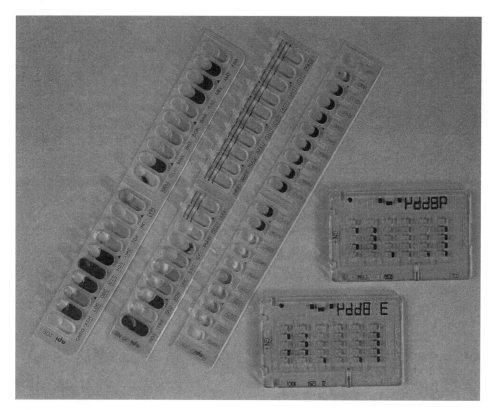

FIGURE 9   API 20E, API Rapid 20E, API 20 NE, and Vitek GNI Plus (courtesy of bioMerieux).

meningitis, and it is a common etiologic agent of urinary tract infections.

In addition to biochemical "spot tests," there is a wide spectrum of commercial kits, manual and automated, that are capable of identifying *E. coli* (Table 8). Moreover, there are several commercial kits available for the sole identification of *E. coli* isolates. Among the tests available are the RAPID DETECT *E. coli* test system (Organon Teknika), which employs three enzymatic reactions and takes approximately 30 min, and the 4-methylumbelliferyl-β-D-glucuronide (MUG) test (Remel), which tests for the enzyme β-D-glucuronidase and acts on the substrate MUG in the MUG disk. If the result is positive, 4-methyl-umbelliferone, which fluoresces blue under long-wave UV light, is released. This test can be performed using either a broth dilution method (1 h) or the direct disk method (1/2 h). Some species of *Salmonella*, *Shigella*, and *Yersinia* may also contain β-glucuronidase; however, since these organisms are not lactose fermenters, they can be easily differentiated from *E. coli*. BactiCard *E. coli* is also available from Remel for the presumptive identification of *E. coli* using enzyme technology (Fig. 10).

Commercial Identification Methods for Diarrhea-Producing *E. coli*

There are at least four categories of diarrhea-producing *E. coli*: (i) Shiga-like toxin-producing (or verocytotoxin-producing) *E. coli* (STEC), which are also called enterohemorrhagic *E. coli* (EHEC), (ii) enteropathogenic *E. coli*, (iii) enteroinvasive *E. coli*, and (iv) enterotoxigenic *E. coli*. Other possible categories which have been proposed are enteroaggregative *E. coli* and enteroadherent *E. coli*. There are dozens of serotypes of *E. coli* which have been isolated from human diarrheal sources and which belong to one of

these categories. *E. coli* O157:H7 is one of the most important serotypes which cause human disease in the United States (37) and belongs to the (STEC) group. Many laboratories screen stools for O157:H7 *E. coli* or limit the screening to only those patients with bloody stools.

Immunofluorescence-Antibody Method

An immunofluorescent-antibody test incorporating a cell wall-directed fluorescein isothiocyanate-labeled O157 antibody (Kirkegaard and Perry Laboratories, Gaithersburg, Md.) stains organisms directly from stool specimens. In a study by Park et al. (77), a total of 336 abnormal fecal sam-

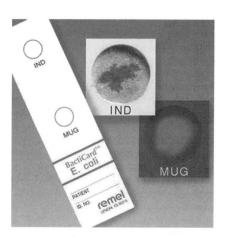

FIGURE 10   BactiCard *E. coli* with positive test results (courtesy of Remel).

ples were examined by immunofluorescence; 12 of 12 bleach-treated and concentrated samples of *E. coli* O157:H7 that were positive by culture were also positive by immunofluorescence. Two untreated specimens failed to reveal the organism by immunofluorescence testing.

Premier EHEC and Premier *E. coli* O157

An enzyme immunoassay (Premier EHEC; Meridian Diagnostics, Inc. Cincinnati, Ohio) (Fig. 11) which detects Shiga toxin (VT1 and VT2) in stool filtrates or broth cultures and an immunoassay (Premier *E. coli* O157; Meridian Diagnostics) which detects the presence of O157 are available. Mackenzie et al. evaluated the performances of these two assays (66) and classified the results as field results (those obtained at the site laboratories) and resolved results (those obtained after repeat testing in the central laboratory). They found that the field sensitivity of the Premier *E. coli* O157 test was 86% and the sensitivity of the Premier EHEC test was 89%. The specificity of the Premier *E. coli* O157 test was 98%. The performance by field versus resolved results for the Premier *E. coli* O157 test from this study is contained in Table 9.

Kehl and coworkers (54), in an evaluation of the Premier EHEC assay, found a sensitivity of 100% and a specificity of 99.7% for *E. coli* O157:H7. In this study, additional non-O157:H7 STEC isolates were detected. Table 10 contains the results of this study compared to those of culture.

In a different study of 270 specimens by Park et al. (78), 11 were positive using the Premier EHEC test (6 were positive for O157:H7; 5 were positive for non-O157:H7 isolates). Although found to perform with good sensitivity and specificity, the Premier EHEC has been reported to be negative when testing culture supernatants from Stx2e-producing *E. coli* O101 (2), and false-positive reactions have been reported with *P. aeruginosa* (8).

ImmunoCard STAT! *E. coli* O157 Plus

Meridian also offers a 10-min, one-step test for the detection of antigens from STEC O157 (ImmunoCard STAT! *E. coli* O157 Plus) (Fig. 12). It can be used to test stool specimens directly or as a confirmatory test for cultures grown in MacConkey broth or sorbitol MacConkey agar plates. In a multicenter study by MacKenzie et al. (65) 14 of 14 prospectively studied specimens positive by culture for *E. coli* O157:H7 were positive with the ImmunoCard STAT! O157:H7 test. No false positives were observed with 263 culture-negative specimens. In a retrospective study with 417 culture-positive specimens, MacKenzie and coworkers found a sensitivity of 81% and a specificity of 95% with the rapid 10-min test.

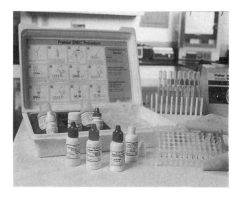

FIGURE 11 Premier EHEC (courtesy of Meridian Diagnostics).

E. coli O157 Antigen Detection Kit

An immunoassay (*E. coli* O157 Antigen Detection Kit; LMD Laboratories, Carlsbad, Calif.) to detect O157 strains in stool specimens is available. In a study by Dylla et al. of 185 stool specimens (19) to evaluate the immunoassay, nine of nine culture-positive *E. coli* O157:H7 isolates were also positive by the LMD test (sensitivity, 100%). The specificity on initial testing was found to be 98.9%. In another study by Park et al. (78) using the immunoassay, 31 of 34 samples were detected, yielding a sensitivity of 91.2% and a specificity of 99.5%.

ProSpecT Shiga Toxin *E. coli* (STEC)

An enzyme immunoassay [ProSpecT Shiga Toxin *E. coli* (STEC) Microplate Assay] which detects Stx1 and Stx2 in stool specimens or in broth-enriched cultures is available from Alexon-Trend, Inc. (Ramsey, Minn.).

RIM Immuno Shiga Toxin *E. coli* (STEC)

A direct qualitative enzyme immunoassay for the detection of *E. coli* Shiga toxins 1 and 2 in fecal specimens and broth-enriched fecal cultures [RIM (Rapid Identification Method) Immuno Shiga Toxin *E. coli* (STEC) Microplate Assay] is available from Remel (see "Commercial Identification Methods for *Campylobacter* spp." below for figures).

Reveal *E. coli*

A colloidal-gold-labeled antibody (specific to *E. coli* O157:H7) test is available (Reveal *E. coli* test system; Neogen Corp., Lexington, Ky.) which produces a line in a test window due to the concentration of the gold label. The Reveal *E. coli* test is used for the evaluation of food products and is not for human or veterinary diagnostic use.

TABLE 9 Performance of Premier *E. coli* O157 test in the field and in resolved results after repeat testing

Category	Result[a]			
	Sensitivity	Specificity	PPV	NPV
Field	57 / 66 (86)	798 / 811 (98)	57 / 70 (81)	798 / 807 (99)
Resolved	60 / 68 (88)	806 / 809 (99)	60 / 63 (95)	806 / 814 (99)

[a]Number of specimens with correct result/total number of specimens (%). Sensitivity = [number of true positives/(number of true positives + number of false negatives)] × 100; specificity = [number of true negatives/(number of true negatives + number of false positives)] × 100; positive predictive value (PPV) = [number of true positives/(number of true positives + number of false positives)] × 100; negative predictive value (NPV) = [number of true negatives/(number of true negatives + number of false negatives)] × 100. (Taken with permission from reference 66.)

TABLE 10 Sensitivities, specificities, and predictive values for the detection of *E. coli* O157:H7[a]

Method	Sensitivity (%)[b]	Specificity (%)[c]	Predictive value (%)	
			Positive[d]	Negative[e]
sMac culture[f]	60	100	100	99.6
Premier EHEC assay	100	99.7	81.3	100

[a]Taken with permission from reference 54.

[b]Sensitivity was calculated as follows: [number of true positives/(number of true positives + number of false negatives)] × 100.

[c]Specificity was calculated as follows: [number of true negatives/(number of true negatives + number of false positives)] × 100.

[d]Calculated as follows: [number of true positives/(number of true positives + number of false positives)] × 100.

[e]Calculated as follows: [number of true negatives/(number of true negatives + number of false negatives)] × 100.

[f]An additional two specimens containing *E. coli* O157:H7 by culture were not tested by Premier EHEC due to insufficient sample size and are not considered in this analysis. sMac, sorbitol MacConkey agar.

Binax EH *E. coli* O157

The Binax EH *E. coli* O157 and O157:H7 test is a rapid immunochromatographic test for use in food products which is not for human or veterinary use.

E. coli O157 Latex Agglutination Assays

Three latex agglutination tests have been evaluated by Sowers et al. (91) for detecting *E. coli* O157 antigen. The *E. coli* O157 Latex Test (Oxoid Diagnostic Reagents, Basingstoke Hampshire, England) was evaluated along with a test from Pro-Lab, Inc. (Richmond Hill, Ontario, Canada). A third assay from Remel Microbiology Products for detecting *E. coli* O157 antigen and an assay for detecting H7 antigen (RIM *E. coli* O157-H7 latex test [Fig. 13]) was also included in the study. Using 159 strains of *E. coli* and related organisms, the Oxoid, Pro-Lab, and Remel O157 latex reagents all had sensitivities and specificities of 100% compared to CDC reference antiserum. The Remel H7 latex reagent performed with a sensitivity of 97% and a specificity of 100% compared to the standard tube agglutination method using CDC H7 antiserum. An additional latex agglutination test (*E. coli*-stat) for *E. coli* O157:H7 is available from Novamed, Ltd. (Taepiot, Israel); it can be used to identify the organism from colonies obtained from MacConkey-sorbitol agar or other selective agar. In this test, no pretreatment of the samples is required, and results are obtained in approximately 2 min.

E. coli O157 Antiserum

The BBL *E. coli* O Antiserum O157 and H Antiserum H7 tests are available from BD Diagnostic Systems for the identification of *E. coli* O157:H7.

Verotox F Assay

A microplate latex agglutination assay (Verotox-F assay; Denka Seiken Co., Ltd., Tokyo, Japan) is available for the detection of VT1, VT2, and VT2c. Karmali et al. evaluated the Verotox-F assay and compared it to the Vero cell assay (52). They studied 68 verotoxin-positive *E. coli* strains (33 O157 strains, 32 non-O157 strains, and 3 reference strains) and 104 verotoxin-negative strains (100 isolates and 4 reference strains). In their study, the Verotox-F assay was determined to be 100% sensitive and 100% specific for the detection of verotoxins in culture filtrates. The sensitivity of the Verotox-F assay was determined to be 14 pg (0.7 ng/ml) for VT1, 12 pg (0.6 ng/ml) for VT2, and

FIGURE 12 ImmunoCard STAT! *E. coli* O157 Plus (courtesy of Meridian Diagnostics).

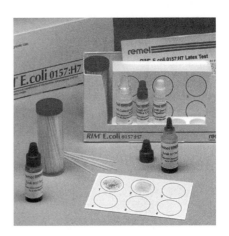

FIGURE 13 RIM *E. coli* O157:H7 latex test kit (courtesy of Remel).

350 pg (17.5 ng/ml) for VT2c. The authors report that this level of sensitivity is comparable to that of bioassay.

E. coli ST EIA

Enterotoxins produced by enterotoxigenic *E. coli* strains may be detected by the *E. coli* ST EIA kit (Denka Seiken Co., Ltd.). Stavric et al. (94) evaluated the *E. coli* ST EIA, a competitive enzyme immunoassay kit, and compared it to an infant mouse assay reference test. Of 46 strains tested, 31 were positive with both assays and 15 were negative. The investigators reported that the sensitivity of the ST EIA commercial kit was up to 64-fold lower than that of the infant mouse assay.

VET-RPLA

Yam and coworkers evaluated and optimized a reverse passive latex agglutination assay, VET-RPLA (Oxoid, Ogdensburg, N.Y.), which detects both heat-labile enterotoxin and cholera toxin. The investigators found that the sensitivity of the assay could be improved from 90.6% (for heat-labile toxin) and 75% (for cholera toxin) to 100% with the selection of appropriate culture media (112). The results of this study were confirmed with bioassays and DNA hybridization assays.

Other Assays

No other commercial assays are available for enteropathogenic, enteroinvasive, enteroaggregative, or enteroadhesive *E. coli*, although tissue culture, immunologic, or nucleic acid-based assays may be considered for further evaluations.

SELECTED BACTERIA THAT ARE NOT MEMBERS OF THE FAMILY *ENTEROBACTERIACEAE*

Commercial Identification Methods for *Campylobacter* spp.

Campylobacter spp. are curved, gram-negative, asporogenous rods which are usually microaerophilic. The genus currently consists of 18 species (71). *Campylobacter jejuni* and *Campylobacter coli* are considered the most common members of the genus *Campylobacter* associated with diarrheal disease, and they may be among the most common enteric pathogens associated with diarrhea.

INDX-Campy (jcl)

One of the commercial tests available for the identification of isolated colonies of *Campylobacter* spp. is the INDX-Campy (jcl) (Integrated Diagnostics, Baltimore, Md.). The INDX-Campy (jcl) is a latex agglutination test for the confirmatory identification to the genus level of *C. jejuni*, *C. coli*, and *Campylobacter laridis* (now called *Campylobacter lari*) from culture. In 1990, Nachamkin and Barbagallo reported 100% sensitivity in detecting *C. jejuni* and *C. coli* but a low sensitivity (zero of six isolates tested) with *C. laridis* isolates when evaluating the Campy (jcl) test (72). In their study, two of two *Campylobacter upsaliensis* isolates reacted with the test. In their hands, the test had 100% specificity for 101 non-*Campylobacter* isolates.

Campyslide

A latex slide agglutination test for the confirmatory identification of *C. jejuni*, *C. coli*, *C. laridis*, and *Campylobacter fetus* subsp. *fetus* from cultures is the Campyslide test (BD Diagnostic Systems) (Fig. 14). The test takes approximately 3 to 10 min from extraction to results. In 1988, Hodinka and Gilligan (38) evaluated the Campyslide test and found 100% of the stock cultures (27 *C. jejuni* and 3 *C. coli*) and fresh clinical isolates (45 *C. jejuni* and 5 *C. coli*) tested were correctly identified. In their study, only one rough strain of *P. aeruginosa* gave a positive result in the Campyslide test (of 173 non-*Campylobacter* isolates).

API Campy

The API Campy system (API-bioMerieux SA, Marcy l'Etoile, France) is marketed for industrial use but is not yet available for human clinical use and is not yet marketed in the United States. It is a strip consisting of 20 microtubes containing dehydrated substrates. It is made up of two parts, (i) enzymatic and conventional tests and (ii) assimilation or inhibition tests. The strip is read after 24 h of incubation at 35 to 37°C. In 1995, Huysmans et al. evaluated the API Campy system and found 92% agreement with the results of conventional methods (43). There was 100% correlation

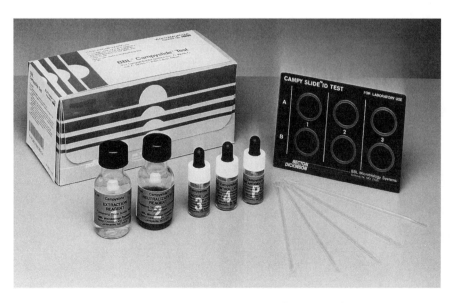

FIGURE 14 Campyslide (courtesy of BD Diagnostic Systems).

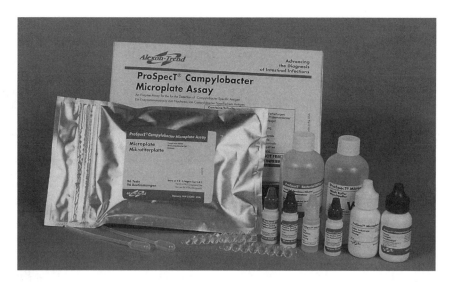

FIGURE 15 ProSpecT *Campylobacter* Microplate Assay (courtesy of Alexon-Trend).

for the identification of *C. jejuni* isolates (78 of 78), while 14 of 19 (74%) *C. coli* and 2 of 3 (67%) *C. laridis* isolates were correctly identified.

ProSpecT *Campylobacter* Microplate Assay

A test, the ProSpecT *Campylobacter* Microplate assay (Fig. 15), is available from Alexon-Trend for the rapid detection of *Campylobacter* antigens directly from stool. The Alexon-Trend assay can be completed in 2 h, and all incubations are done at room temperature. The results can be read either visually or spectrophotometrically. The assay is available for use in various batch sizes by using a 96-test format with breakaway wells.

RIM Immuno *Campylobacter* Microplate Assay

A qualitative immunoassay is available from Remel for the detection of *Campylobacter*-specific antigen from fecal specimens or from broth-enriched fecal cultures (Fig. 16 and 17).

AccuProbe *Campylobacter*

DNA probe methodology (AccuProbe; Gen-Probe Inc., San Diego, Calif.) is available for culture confirmation of

C. jejuni, *Campylobacter jejuni* subsp. *doylei*, *C. coli*, and *C. laridis*. Several studies have documented 100% sensitivity with this probe (33, 34). However, in one of the studies (85), the probe hybridized with 2 of 17 isolates of *Campylobacter hyointestinalis*. A recommendation has been made that this method be used in confirming *Campylobacter* spp. if other tests are not conclusive (71).

Commercial Identification Methods for *Helicobacter* spp.

The genus *Helicobacter* consists of gram-negative, oxidase-positive, non-spore-forming spiral or curved bacilli and contains 19 species (103). Present evidence has associated *Helicobacter pylori* with peptic ulcer disease and cancers of the human gastrointestinal system. Two primary methods available for the evaluation of *Helicobacter* infections are (i) detection of extracellular urease, since *H. pylori* produces large amounts of this enzyme, and (ii) detection of *H. pylori* antigens by enzyme immunoassay. Other diagnostic methods include (i) biopsy followed by hematoxylin and eosin

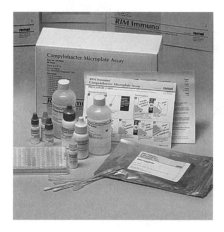

FIGURE 16 RIM Immuno *Campylobacter* Microplate Assay (courtesy of Remel).

FIGURE 17 RIM Immuno *Campylobacter* Microplate Assay tray with positive wells (courtesy of Remel).

staining, (ii) imprint cytologic analysis with a rapid Giemsa or Gram stain, (iii) DNA amplification occasionally available in research settings, (iv) serologic analyses, and (v) culture, which is becoming increasingly popular to determine antimicrobial susceptibilities for isolates from patients with recurrent *H. pylori* disease. A more detailed discussion of some of these methods can be found in chapter 51 of the *Manual of Clinical Microbiology*, 7th ed. (103).

Premier Platinum HpSA test

An available antigen test is the Premier Platinum HpSA test (Meridian Diagnostics) (Fig. 18). The Meridian test is performed with a stool specimen and polyclonal antibodies and requires approximately 1 h to perform. The plate is read at a dual wavelength (450 and 630 nm) on a plate reader. The HpSA test does not require diet modification. In a 1997 abstract (K. Kozak, C. Larka, A. Nickol, and A. Yi, *Abstr. 97th Gen. Meet. Am. Soc. Microbiol.*, abstr. 271, 1997), Kozak et al. evaluated the HpSA test and reported a sensitivity of 97.4% and a specificity of 100% compared to the urea breath test. In another 1997 study, Trevisani et al. reported a sensitivity of 93.9% and a specificity of 96.3% with the HpSA test as defined by concordance of histology and the rapid urease test (100). In a recent prospective multicenter study by Vaira et al. (101), stools from 501 patients were tested by the HpSA test and evaluated using the carbon-13–urea breath test, histology (hematoxylin and eosin), Giemsa stain of the antrum and corpus, culture, and the rapid urease test. In this multicenter study, the sensitivity, specificity, positive predictive value, and negative predictive value were 94.1, 91.8, 93.4, and 92.6%, respectively. Nonculture methods, such as histology, serology, urease

testing, and the urea breath test, are described elsewhere (10, 57, 102). DNA amplification and nucleic acid detection of *H. pylori* is only available in research settings, since the reagents are not yet available commercially in kit form.

IDENTIFICATION OF SELECTED FASTIDIOUS BACTERIA

Clinical microbiology laboratories have used commercially available identification systems for clinically significant fastidious bacteria, such as *Haemophilus* spp. and *Neisseria* spp., for the past 25 years. The identification systems have evolved over time and have been based on either growth-dependent or enzyme-mediated substrate utilization or on the detection of specific nucleic acid sequences. Selected commercial systems are listed in Table 11 by the test name, manufacturer, time to result, and organism groups identified. The fastidious bacteria that are discussed in this section include *Haemophilus* spp., *Neisseria* spp., *Gardnerella vaginalis*, *Bordetella* spp., and *Legionella* spp.

Haemophilus spp.

Haemophilus spp., a group of pleomorphic, oxidase-positive, gram-negative rods, can be responsible for a variety of human infections, although a dramatic reduction in the incidence of *Haemophilus influenzae* type b infections has occurred recently, largely due to the introduction of an effective vaccine (9, 13). The clinically important species include *H. influenzae*, *Haemophilus parainfluenzae*, *Haemophilus ducreyi*, *Haemophilus aphrophilus*, *Haemophilus*

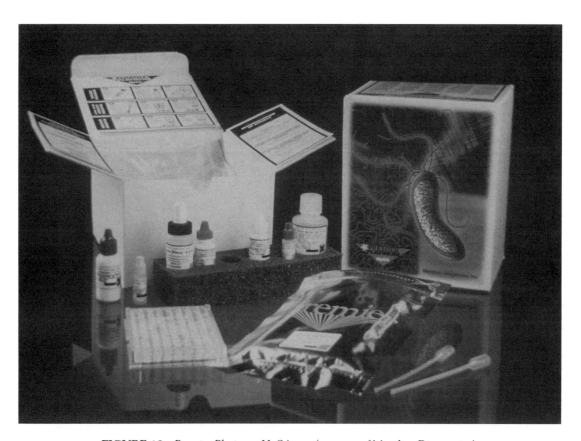

FIGURE 18 Premier Platinum HpSA test (courtesy of Meridian Diagnostics).

TABLE 11 Commercial systems for the identification of *Haemophilus* spp., *Neisseria* spp., and other fastidious bacteria

Commercial product	Manufacturer	Speed[e]	Organisms in database
API-NH	bioMerieux	O	*Neisseria* spp. and other fastidious gram-negative bacteria
API Quad-Ferm+[a]	bioMerieux	R (2 h)	*Neisseria* spp. and M. *catarrhalis*
Vitek NHI card	bioMerieux	R (4 h)	*Neisseria* spp., *Haemophilus* spp., and other fastidious gram-negative bacteria
BactiCard *Neisseria*[b]	Remel	R (4 h)	*Neisseria* spp. and M. *catarrhalis*
IDS RapID NH panel	Remel	R (4 h)	*Neisseria* spp., *Haemophilus* spp., and other fastidious gram-negative bacteria
Catarrhalis test disk[c]	Remel	R (5 min)	M. *catarrhalis*
Haemophilus identification test kit	Remel	O	*Haemophilus* spp.
BBL Crystal *Neisseria/Haemophilus* panel	BD Diagnostic Systems	R (4)	*Neisseria* spp., *Haemophilus* spp., and other fastidious gram-negative bacteria
Gonobio test	IAF Production	R	*N. gonorrhoeae*
Gonochek II	E-Y Laboratories	R (30 min)	*Neisseria* spp. and M. *catarrhalis*
GonoGen I and II	New Horizons	R	*N. gonorrhoeae*
Gonostat II	Sierra Diagnostics	O	*N. gonorrhoeae*
Gonozyme	Abbott Laboratories	R	*N. gonorrhoeae*
LCx *N. gonorrhoeae*	Abbott Laboratories	R (5 h)	*N. gonorrhoeae*
Meritec GC	Meridian Diagnostics	R	*N. gonorrhoeae*
MicroScan HNID	Dade Behring	R (4 h)	*Neisseria* spp., *Haemophilus* spp., and other fastidious gram-negative bacteria
Neisseria-Kwik[d]	Micro-Bio-Logics	R (4 h)	*Neisseria* spp.
Neisstrip	Lab M Ltd.	R	*Neisseria* spp.
Pace II DNA Probe	Gen-Probe	R	*N. gonorrhoeae*
AccuProbe *Neisseria gonorrhoeae* test	Gen-Probe	R	*N. gonorrhoeae*
Phadebact GC Omni Test	Karo-Bio	R	*N. gonorrhoeae*
Amplicore PCR *N. gonorrhoeae*	Roche Diagnostic Systems	R (5 h)	*N. gonorrhoeae*

[a]Figure 20.
[b]Figure 22.
[c]Figure 23.
[d]Figure 21.
[e]R, rapid; O, overnight.

paraphrophilus, *Haemophilus haemolyticus*, *Haemophilus parahaemolyticus*, and *Haemophilus segnis*.

Commercially available kits or methods to identify *Haemophilus* spp. are listed in Table 11. Several investigators have also reported the use of the Micro ID (Remel) (Fig. 19) and API 20E (bioMerieux) kits, commonly used for enteric bacteria, in biotyping *Haemophilus* spp. (20, 39). Color photographs of many of these kits may be found in Koneman et al. (59).

IDS RapID NH

Doern and Chapin evaluated the IDS RapID NH system (Remel) using 187 clinical isolates of *H. influenzae* (16). They found that the RapID NH system correctly identified 168 of 187 (89.8%) isolates and that there was no difference observed between strains possessing and lacking capsular type b antigen.

Vitek NHI Card

Janda et al. evaluated the Vitek NHI (*Neisseria-Haemophilus* Identification) card (bioMerieux) using a large number (480) of clinical isolates and stock strains of *Neisseria* spp., *Haemophilus* spp., and other fastidious bacteria (51). They found that with no further testing, the NHI card identified 83.2% of 244 *Neisseria* spp. and *Moraxella (Branhamella) catarrhalis*, 54.9% of 164 *Haemophilus* spp., and 84.7% of 72 fastidious gram-negative bacteria. In their study, the correct identifications increased to 97.1% (*Neisseria* and *Moraxella*), 92.7% (*Haemophilus*), and 94.4% (fastidious gram-negative bacteria) when they included isolates which produced good-confidence marginal-separation identifications. They also observed that for common clinical isolates, such as *Neisseria gonorrhoeae*, *Neisseria meningitidis*, *H. influenzae*, and *H. parainfluenzae*, the NHI card correctly identified 99.1, 98.5, 93.9, and 95.6%, respectively. In their large

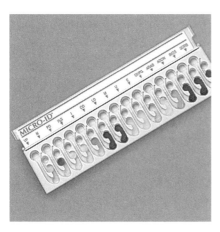

FIGURE 19 Micro ID system with inoculated wells, used primarily for *Enterobacteriaceae* (courtesy of Remel).

study, *H. aphrophilus*, *Eikenella corrodens*, *G. vaginalis*, and *Kingella denitrificans* were also consistently identified.

MicroScan HNID panel

In another large study, Janda et al. evaluated the MicroScan HNID panel (Dade Behring) using 423 clinical isolates and stock strains of *Haemophilus* spp. and *Neisseria* spp., *M. catarrhalis*, and *G. vaginalis* (49). In their laboratory, the HNID panel correctly identified 95.3% of *N. gonorrhoeae*, 96% of *G. vaginalis*, 100% of *Neisseria lactamica* and *M. catarrhalis*, 64.7% of *N. meningitidis*, 98.8% of *H. influenzae*, 97.1% of *H. parainfluenzae*, and 80% of *H. aphrophilus* and *H. paraphrophilus* isolates.

API-NH

The API-NH test (bioMerieux) is used for the identification of most *Neisseria* spp., *Haemophilus* spp., and *M. catarrhalis*. In a study by Barbe et al. (6), 318 strains belonging to these species were tested. A total of 305 strains were included in the database of the API-NH test, and 225 (73.8%) were accurately identified without additional tests. Another 79 (25.9%) were correctly identified after extra tests were performed, and 1 isolate (0.3%) was misidentified.

H. *influenzae* Antigen Detection

Latex agglutination assays for *H. influenzae* antigen detection are not reviewed in this chapter, since most laboratories have either discontinued or severely restricted testing of body fluids for *H. influenzae* type b and other common etiologic agents of bacterial meningitis. This is in part because of the significant reduction in *H. influenzae* type b infection due to vaccine availability and highly effective prophylactic and therapeutic antibiotic regimens. In addition, the sensitivity and highly contributory nature of the Gram stain has to a high degree obviated the need for routine bacterial antigen testing. Of course, selected patients previously treated with antibiotics may benefit from rapid antigen detection methods under rare circumstances. Further information concerning antigen detection assays for *H. influenzae* can be found in the *Manual of Clinical Microbiology*, 7th ed. (13).

Neisseria spp. and M. catarrhalis

Neisseria spp. and *M. catarrhalis* are gram-negative, oxidase-positive, kidney bean-shaped organisms which may either be common inhabitants of the oro- and nasopharyngeal sites (e.g., *Neisseria*) or have an as-yet-undetermined status as possible normal flora (56). Among *Neisseria* spp., *N. gonorrhoeae* and *N. meningitidis* are the most common human pathogens, with *N. gonorrhoeae* always considered pathogenic when present. *M. catarrhalis* may cause infections ranging from local infections, e.g., otitis media and sinusitis, to serious systemic infections, e.g., meningitis.

Commercially available tests used for the identification of culture isolates of *Neisseria* spp. include carbohydrate utilization and degradation tests, chromogenic enzyme substrate tests, and immunologic methods (21). In addition, multitest identification systems are available which can be used for the identification of *Neisseria* spp. as well as for that of other related organisms. In the last decade, DNA probe technologies have become available for the direct detection of *N. gonorrhoeae* from clinical specimens and for culture confirmation.

Carbohydrate Utilization Tests

As listed in Table 11, commercially available carbohydrate utilization tests include the API Quad-Ferm+ (bioMerieux) (Fig. 20), the *Neisseria*-Kwik test kit (Micro-Bio-Logics) (Fig. 21), and the Gonobio test (I.A.F. Production, Inc., Laval, Quebec, Canada). The RIM-*Neisseria* test (Remel) was discontinued in July 2000.

Chromogenic Enzyme Tests

Chromogenic enzyme substrate tests include Gonocheck II (Dupont or E-Y Laboratories), BactiCard *Neisseria* (Remel) (Fig. 22), and the Neisstrip (Lab M Ltd., Bury, United Kingdom).

Immunologic Tests

Immunologic methods include a fluorescent monoclonal antibody test (FA Test *Neisseria gonorrhoeae* Culture Confirmation Test; Syva). Immunologic methods also include coagglutination tests, such as the Phadebact GC OMNI test (Karo-Bio, Huddinge, Sweden), GonoGen I and II tests (New Horizons Diagnostics), and Meritec GC (Meridian). The Gonozyme assay (Abbott Laboratories) is an enzyme immunoassay methodology.

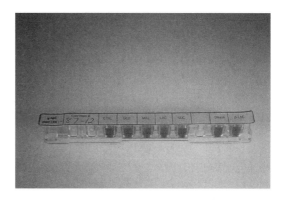

FIGURE 20 API Quad-Ferm+ (courtesy of bioMerieux).

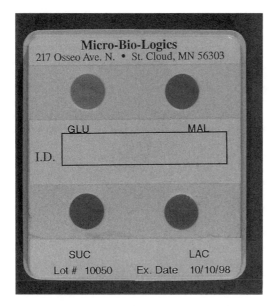

FIGURE 21 *Neisseria*-Kwik tray (courtesy of Micro-Bio-Logics).

Multitest Systems

Multitest systems, which can be used for the identification of *Neisseria* spp., *Haemophilus* spp., and other similar fastidious organisms, include the RapID NH test (Remel), the Vitek NHI card (bioMerieux), the MicroScan HNID panel (Dade Behring), and the API-NH system (bioMerieux). Many of the tests have been described in detail, and selected color photographs can be found in reference 59. Therefore, an extensive review of their performance characteristics will not be included in this chapter.

N. gonorrhoeae Nucleic Acid Tests

Other available methods include a nucleic acid probe culture confirmation test for *N. gonorrhoeae* (AccuProbe; Gen-Probe), a nonculture direct-detection nucleic acid probe test (Pace II; Gen-Probe), and a nucleic acid amplification test (ligase chain reaction) (LCx; Abbott Laboratories). Other tests which have been developed include a PCR assay (AmpliCore; Roche Diagnostic Systems) and a tran-

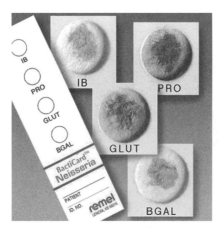

FIGURE 22 BactiCard *Neisseria* with positive test results (courtesy of Remel).

FIGURE 23 Catarrhalis test disk with positive test results (courtesy of Remel).

scription-mediated amplification test (TMA; Gen-Probe). Additional information on this topic can be found in the chapters on *Chlamydia* and molecular microbiology in this manual and in reference 59.

N. meningitidis Antigen Detection

Commercially available tests for the rapid detection of *N. meningitidis* bacterial antigens in body fluids include coagglutination tests (Pharmacia) and latex agglutination tests (Wellcogen/Remel and BD Biosciences). Bacterial antigen tests will not be discussed in detail in this chapter; see "*Haemophilus* spp." above for further comments. At the present time, there are no commercial tests available for the detection of *N. gonorrhoeae* antigens in body fluids.

G. vaginalis

G. vaginalis is the only named species in the genus *Gardnerella*; however, eight *G. vaginalis* biotypes have been proposed (30). Gram stain may show thin gram-variable rods or coccobacilli. The cell wall of *G. vaginalis* is similar to those of other gram-positive bacteria, but the peptidoglycan layer is very thin, contributing to the observation of gram-variable staining with this organism. *G. vaginalis* does not have a defined phylogenetic relationship to other coryneformlike bacteria. Phenotypically, *G. vaginalis* does not produce catalase and is nonmotile. *G. vaginalis* is considered a member of the normal vaginal flora of adult women and can be isolated from the normal anorectal areas of both children and adults (male and female). The organism is commonly associated with a clinical entity known as bacterial vaginosis, which can progress to more serious complications, such as premature rupture of membranes, preterm delivery, chorioamnionitis, and postpartum fevers. It may also be associated with neonatal infections. The recommended method for diagnosis of bacterial vaginosis is the direct examination by Gram stain or wet preparation of vaginal secretions for "clue cells" and not the culture of *G. vaginalis* (92). Clue cells are vaginal squamous epithelial cells that are covered to the margin with large numbers of small gram-negative and gram-variable bacteria, observed in specimens with markedly reduced numbers of lactobacilli (74, 92).

In those selected situations where *G. vaginalis* identification is required or to rule out other pathogens, the organism can be identified with rapid kits, such as the API Coryne system (bioMerieux), the Vitek NHI card (bioMerieux), the

MicroScan Rapid HNID panel (Dade Behring), and the BBL Crystal *Neisseria/Haemophilus* panel (BD Diagnostic Systems) (Table 11). In the multicenter evaluation by Funke et al., the API Coryne system with database 2.0 was used to test six strains of *G. vaginalis*, of which four were correctly identified, one was not identified, and one was misidentified (28).

Bordetella spp.

Bordetella spp. are catalase-positive, strictly aerobic, small gram-negative coccobacilli, with seven species currently identified: *Bordetella pertussis*, *Bordetella parapertussis*, *Bordetella bronchiseptica*, *Bordetella avium*, *Bordetella hinzii*, *Bordetella holmesii*, and *Bordetella trematum* (41). This group of organisms appears to be responsible primarily for respiratory infections; however, members have been isolated from blood cultures, wounds, and ear infections. *B. pertussis* may be solely a human pathogen; however, other *Bordetella* spp. have been observed in multiple host species (with the possible exception of *B. avium*, which has not been documented yet in human hosts).

At the present time, the isolation of *Bordetella* spp. from respiratory specimens followed by growth in culture is considered the "gold standard" due to its high level of specificity (70). *B. pertussis* is a rather fastidious organism, and it may be the most fastidious of all the *Bordetella* spp.; therefore, culture suffers from low sensitivity. *B. pertussis* is the primary etiologic agent of whooping cough (or pertussis), but other organisms, such as *B. parapertussis*, *B. bronchiseptica*, viruses, and *Chlamydia* spp., may also be responsible agents causing a similar syndrome.

DFA Assays

Other than culture, the most commonly used antigen-based tests for the diagnosis of pertussis are direct fluorescent antibody (DFA) and PCR assays. The use of DFA testing for the diagnosis of pertussis became common in the United States in the early 1960s (40, 109). Although DFA testing offers a rapid diagnostic approach, its sensitivity and specificity have been limiting factors (4, 34). Most authorities agree that evaluation of specimens by DFA testing should be performed with concomitant culturing.

The available DFA tests include a polyclonal (species specific for *B. pertussis* and *B. parapertussis*), fluorescein-labeled reagent from Difco/BD Diagnostic Systems and a dual-fluorochrome DFA test (Cytovax Biotechnologies, Edmonton, Alberta, Canada) combining monoclonal antibody 1H2 (*B. pertussis*) and monoclonal antibody A7D12 (*B. parapertussis*). Both monoclonal antibodies are isotype

immunoglobulin G 2a. This dual-fluorochrome reagent contains a fluorescein-conjugated murine monoclonal antibody, which is specific for the lipooligosaccharide of *B. pertussis*, and a rhodamine-conjugated murine monoclonal antibody, which is specific for the lipopolysaccharide of *B. parapertussis*. With this test, *B. pertussis* will appear brilliant green and *B. parapertussis* will appear brilliant orange-red. This reagent is known to cross-react with some strains of *B. bronchiseptica* which may occasionally infect humans. At this time, there are no published reports documenting the use of the Cytovax dual-fluorochrome reagent. However, McNicol et al. (67) have evaluated a monoclonal antibody (BL-5) which may be similar to the 1H2 antibody produced by Cytovax. BL-5 recognizes the lipooligosaccharide of *B. pertussis* and does not cross-react with *B. parapertussis*. BL-5 does, however, show weak reactivity with *B. bronchiseptica*. McNicol et al. have evaluated this monoclonal reagent versus culture and found the sensitivity, specificity, and positive and negative predictive values to be 65.1, 99.6, 87.2, and 98.5%, respectively (67). In their study, the sensitivity of culture compared with that of PCR was 45.5%. McNicol et al. also demonstrated that an "expanded gold standard" of positivity defined by culture or PCR resulted in sensitivity, specificity, and positive and negative predictive values for DFA of 32.3, 97.1, 83.3, and 76.1%, respectively.

PCR

A recent comparison of in-house PCR, culture, and DFA testing (using fluorescein isothiocyanate-conjugated *B. pertussis* or *B. parapertussis*) from Difco/BD Diagnostic Systems for detection of *B. pertussis* has been reported by Loeffelholz et al. (63). The sensitivity, specificity, and positive and negative predictive values are contained in Table 12. In this study, a culture sensitivity of only 15.2% was reported. Wadowsky et al. (105), however, reported a higher culture sensitivity of 73.4% in a study comparing PCR and culture (Table 13). Loeffelholz et al. have suggested that these differences in culture sensitivity may be attributed to the different transport conditions defined for each study and the different patient populations tested.

A previous study (1993) by Strebel et al. using Difco reagents reported the sensitivity, specificity, and positive and negative predictive values for the commercial polyclonal reagents to be 61, 95, 53, and 96%, respectively (97). Unless improvements are made in culture and DFA approaches, or other methods are developed, in the future PCR or other molecularly based assays may be the methods of choice to detect

TABLE 12 Performance characteristics of culture, DFA, and PCR for the detection of *B. pertussis* in nasopharyngeal-swab specimens[a]

Gold standard	No. (%) of patients positive	Test	Sensitivity (%)	Specificity (%)	PPV[c] (%)	NPV[d] (%)
Culture	7 (2.2)	DFA	71.4	92.3	17.2	99.3
		PCR	100	85.9	13.7	100
Expanded gold standard[b]	46 (14.4)	Culture	15.2	100	100	87.5
		DFA	52.2	98.2	82.8	92.4
		PCR	93.5	97.1	84.3	98.9

[a]Taken with permission from reference 63.
[b]Either (i) culture positive, (ii) PCR and DFA positive, or (iii) PCR or DFA positive with clinical features indicating pertussis.
[c]PPV, positive predictive value.
[d]NPV, negative predictive value.

TABLE 13 Comparison of the sensitivities and specificities of the multiplex PCR-based assay and culture for *B. pertussis* following the resolution of discrepant results[a]

Test and result	No. of specimens resolved as:		Sensitivity (%)	Specificity (%)
	Positive	Negative		
PCR-based assay				
Positive	93	1	98.9	99.7
Negative	1	394		
Culture				
Positive	69	0	73.4	100
Negative	25	395		

[a]Includes 489 of the 496 original pairs. (Taken with permission from reference 105.)

Bordetella spp. due to the speed of detection and their greater sensitivity. In the meantime, if DFA approaches are used, it is recommended that simultaneous cultures be performed.

Legionella spp.

The genus *Legionella* consists of slender, weakly gram-negative, asporogenous rods, with approximately 40 named species (111). They are weakly oxidase and catalase positive and are common inhabitants of the environment, particularly surface and potable water. Pneumonia is considered the most common serious presentation of legionella infection,

but subclinical, nonpneumonic, and extrapulmonary disease can also be observed. In addition to culture, common diagnostic tests include direct immunofluorescence, enzyme immunoassay, latex agglutination, and serology (96, 111).

Direct Immunofluorescence Tests

Direct immunofluorescence tests suffer from relatively low sensitivity (25 to 70%) and false-positive results, due in part to cross-reactions with other bacteria (111). Selected immunofluorescence tests and other diagnostic methods used for *Legionella* evaluation are shown in Table 14. Winn

TABLE 14 Diagnostic methods for *Legionella* evaluation[a]

Test (manufacturer)	Specimen type	Sensitivity (%)	Specificity (%)	Comments
Culture	Lower respiratory tract, wound, blood, tissue	70	100	Widely available, detects all species
Direct immunofluorescence (Zeus, Remel, Sanofi/ Bio-Rad, M-tech, MarDx, SciMedx)	Lower respiratory tract, wound, blood, tissue, and cultures	25–70	>95	Limitations of low sensitivity in low-prevalence areas
Legionella differentiation disk (Remel)	Culture	NA[b]	NA	For presumptive identification of *Legionella* spp. by satellite phenomena
Radioimmunoassay	Urine	70–90	>99	*Legionella pneumophila* serogroup 1 only; not commercially available
Enzyme immunoassay (Binax, Bartels/Intracel)	Urine	70–90[c] 64–89[d]	>99	*L. pneumophila* serogroup 1 only
Enzyme immunoassay (Biotest)	Urine	67–87[d]		For research use only in the United States; company plans to submit for FDA review; available in Europe
Immunochromatography (Binax)	Urine	56–97	>99	*L. pneumophila*-serogroup 1
Latex agglutination	Urine	55–90	85–99	*L. pneumophila* serogroup 1 only; not commercially available
Serology[e]	Serum	70–80	>95	Limited reagent availability; poor discrimination with single titers
Nucleic acid amplification[f]	NA	NA	NA	Not commercially available; "home-brew" assays

[a]Adapted with permission from reference 111.
[b]NA, not available.
[c]Reference 111.
[d]Reference 18.
[e]Indirect immunofluorescence.
[f]Reference 108.

has recommended (111) that this test should not be performed in most clinical situations in a low-prevalence environment due to its low sensitivity and false-positive performance characteristics. Antibody titers also may not clearly discriminate between true and false cases of *Legionella* infection. Plouffe et al. (83) have found that acute-phase antibody titers of ≥256 did not discriminate between true (10%) and false (6%) cases.

Enzyme Immunoassays

Two enzyme assays (Biotest *Legionella* Urin Antigen enzyme immunoassay [Biotest AG, Dreieich, Germany] and *Legionella* urinary antigen enzyme immunoassay [Binax, Portland, Maine]) have recently been evaluated by Dominguez et al. (18), using both concentrated and nonconcentrated urine samples. The sensitivity of the Biotest enzyme immunoassay was found to be 67% in nonconcentrated urine and 87% in concentrated urine. The sensitivity of the Binax enzyme immunoassay was reported to be 64 and 89% in nonconcentrated and concentrated urines, respectively. The specificity was reported to be 100% for both assays when either nonconcentrated or concentrated samples were tested.

In a 1997 study evaluating the Binax test, Kazandjian et al. studied 159 patients with suspected or proven legionellosis and 209 controls (53). They found that the sensitivity of the test was significantly greater than that of the direct fluorescence test (83 versus 42%) and similar to those of culture (85%) and serology (91%). The specificity in this study was 99.5% for the Binax test.

Immunochromatographic Test

In a 1999 evaluation, Dominguez et al. (17) compared a new immunochromatographic assay (Binax Now *Legionella* Urinary Antigen Test) with enzyme immunoassay using nonconcentrated and concentrated urine samples. The authors report the overall agreement between the two tests to be 98.1% and the specificities of both to be 100%. They found the sensitivity of the immunochromatographic test to be 55.5% (nonconcentrated urine) and 97.2% (concentrated urine) in comparison to the concentrated urine enzyme immunoassay.

REFERENCES

1. **Acheson, D., and J. Jaeger.** 1999. Shiga toxin-producing *Escherichia coli. Clin. Microbiol. Newsl.* **21:**183–188.
2. **Allerberger, F., D. Rossboth, M. P. Dierich, S. Aleksic, H. Schmidt, and H. Karch.** 1996. Prevalence and clinical manifestations of shiga toxin-producing *Escherichia coli* infections in Austrian children. *Eur. J. Clin. Microbiol. Infect. Dis.* **15:**545–550.
3. **Anhalt, J. P., B. J. Heiter, D. W. Naumovitz, and P. P. Bourbeau.** 1992. Comparison of three methods for detection of group A streptococci in throat swabs. *J. Clin. Microbiol.* **30:**2135–2138.
4. **Annick, C. A., L. W. L. Chui, M. G. Paranchych, M. S. Peppler, R. G. Marusyk, and W. L. Albritton.** 1993. Major outbreak of pertussis in Northern Alberta, Canada: analysis of discrepant direct fluorescent-antibody and culture results by using polymerase chain reaction methodology. *J. Clin. Microbiol.* **31:**1715–1725.
5. **Bannerman, T. L., K. T. Kleeman, and W. E. Kloos.** 1993. Evaluation of the Vitek Systems gram-positive identification card for species identification of coagu-

lase-negative staphylococci. *J. Clin. Microbiol.* **31:** 1322–1325.
6. **Barbe, G., M. Babolat, J. M. Boeufgras, D. Monget, and J. Freney.** 1994. Evaluation of API NH, a new 2-hour system for identification of *Neisseria* and *Haemophilus* species and *Moraxella catarrhalis* in a routine clinical laboratory. *J. Clin. Microbiol.* **32:**187–189.
7. **Bascomb, S., S. L. Abbott, J. D. Bobolis, D. A. Bruckner, S. J. Connell, S. K. Cullen, M. Daughterty, D. Glenn, J. M. Janda, S. L. Lentsch, D. Lindquist, P. B. Mayhew, D. M. Nothaft, J. R. Skinner, G. B. Williams, J. Wong, and B. L. Zimmer.** 1997. Multicenter evaluation of the MicroScan Rapid gram-negative identification type 3 panel. *J. Clin. Microbiol.* **35:**2531–2536.
8. **Beutin, L., S. Zimmermann, and K. Gleier.** 1996. *Pseudomonas aeruginosa* can cause false-positive identification of verotoxin (Shiga-like toxin) production by a commercial enzyme immune assay system for the detection of Shiga-like toxins (SLTs) infection. *Infection* **24:**267–268.
9. **Bisgard, K. M., A. Kao, J. Leake, P. M. Strebel, B. A. Perkins, and M. Wharton.** 1998. *Haemophilus influenzae* invasive disease in the United States, 1994–1995: near disappearance of a vaccine-preventable childhood disease. *Emerg. Infect. Dis.* **4:**229–237.
10. **Bonagura, A. F., and M. A. Dabezies.** 1996. *Helicobacter pylori* infection. The importance of eradication in patients with gastric diseases. *Postgrad. Med.* **100:**115–129.
11. **Bopp, C. A., F. W. Brenner, J. G. Wells, and N. A. Stockbine.** 1999. *Escherichia, Shigella,* and *Salmonella,* p. 459–474. *In* P. R. Murray, E. J. Baron, M. A. Pfaller, F. C. Tenover, and R. H. Yolken (ed.), *Manual of Clinical Microbiology,* 7th ed. American Society for Microbiology, Washington, D.C.
12. **Brun, Y., M. Bes, J. M. Boeufgras, D. Monget, J. Fleurette, R. Auckenthaler, L. A. Devriese, M. Kocur, R. R. Marples, and Y. Piemont.** 1990. International collaborative evaluation of the ATB 32 Staph gallery for identification of the *Staphylococcus* species. *Zentbl. Bakteriol.* **273:**319–326.
13. **Campos, J. M.** 1999. *Haemophilus,* p. 604–613. *In* P. R. Murray, E. J. Baron, M. A. Pfaller, F. C. Tenover, and R. H. Yolken (ed.), *Manual of Clinical Microbiology,* 7th ed. American Society for Microbiology, Washington, D.C.
14. **Carroll, K., and L. Reimer.** 1996. Microbiology and laboratory diagnosis of upper respiratory tract infections. *Clin. Infect. Dis.* **23:**442–448.
15. **Department of Health and Human Services.** 24 March 1997. FDA alert: risks of devices for direct detection of Group B streptococcal antigen. Department of Health and Human Services, Washington, D.C.
16. **Doern, G., and K. C. Chapin.** 1984. Laboratory identification of *Haemophilus influenzae.* Effects of basal media on the results of the satellitism test and evaluation of the RapID NH system. *J. Clin. Microbiol.* **20:**599–601.
17. **Dominguez, J., N. Gali, L. Matas, P. Pedroso, A. Hernandez, E. Padilla, and V. Ausina,** 1999. Evaluation of a rapid immunochromatographic assay for the detection of *Legionella* antigen in urine samples. *Eur. J. Clin. Microbiol. Infect. Dis.* **18:**896–898.
18. **Dominguez, J. A., N. Gali, P. Pedroso, A. Fargas, E. Padilla, J. M. Manterola, and L. Matas.** 1998. Comparison of the Binax *Legionella* urinary antigen enzyme immunoassay (EIA) with the Biotest *Legionella* Urine Antigen EIA for detection of *Legionella* antigen in both

concentrated and nonconcentrated urine samples. *J. Clin. Microbiol.* **36:**2718–2722.

19. **Dylla, B. L., E. A. Vetter, J. G. Hughes, and F. R. Cockerill III.** 1995. Evaluation of an immunoassay for direct detection of *Escherichia coli* O157 in stool specimens. *J. Clin. Microbiol.* **33:**222–224.

20. **Edberg, S. E., E. Melton, and J. M. Singer.** 1980. Rapid biochemical characterization of *Haemophilus* species by using the Micro-ID. *J. Clin. Microbiol.* **11:**22–26.

21. **Evangelista, A. T., and H. R. Beilstein.** 1993. Cumitech 4A, laboratory diagnosis of gonorrhea. Coordinating ed., C. Abramson. American Society for Microbiology, Washington, D.C.

22. **Facklam, R. R., D. F. Sahm, and L. M. Teixeira.** 1999. *Enterococcus*, p. 297–305. *In* P. R. Murray, E. J. Baron, M. A. Pfaller, F. C. Tenover, and R. H. Yolken (ed.), *Manual of Clinical Microbiology,* 7th ed. American Society for Microbiology, Washington, D.C.

23. **Farmer, J. J., III.** 1999. *Enterobacteriaceae*: introduction and identification, p. 442–458. *In* P. R. Murray, E. J. Baron, M. A. Pfaller, F. C. Tenover, and R. H. Yolken (ed.), *Manual of Clinical Microbiology,* 7th ed. American Society for Microbiology, Washington, D.C.

24. **Forbes, B. A., D. F. Sahm, and A. S. Weissfeld.** 1998. *Bailey and Scott's Diagnostic Microbiology,* 10th ed. Mosby, St. Louis, Mo.

25. **Fournier, J. M., A. Bouvet, D. Mathieu, F. Noto, A. Boutonnier, R. Gerbel, P. Brunengo, C. Saulnier, N. Sagot, B. Slizewicz, and J. C. Mazie.** 1993. New latex reagent using monoclonal antibodies to capsular polysaccharide for reliable identification of both oxacillin-susceptible and oxacillin-resistant *Staphylococcus aureus. J. Clin. Microbiol.* **31:**1342–1344.

26. **Freney, J., S. Bland, J. Etienne, M. Desmonceaux, J. M. Boeufgras, and J. Fleurette.** 1992. Description and evaluation of the semiautomated 4-hour Rapid ID 32 Strep method for identification of streptococci and members of related genera. *J. Clin. Microbiol.* **30:**2657–2661.

27. **Funke, G., K. Peters, and M. Aravena-Roman.** 1998. Evaluation of the RapID CB Plus System for identification of coryneform bacteria and *Listeria* spp. *J. Clin. Microbiol.* **36:**2439–2442.

28. **Funke, G., F. N. R. Renaud, J. Freney, and P. Riegel.** 1997. Multicenter evaluation of the updated and extended API (RAPID) Coryne Database 2.0. *J. Clin. Microbiol.* **35:**3122–3126.

29. **Funke, G., A. von Graevenitz, J. E. Clarridge, and K. A. Bernard.** 1997. Clinical microbiology of coryneform bacteria. *Clin. Microbiol. Rev.* **10:** 125–159.

30. **Funke, G., and K. A. Bernard.** 1999. Coryneform gram-positive rods, p. 319–345. *In* P. R. Murray, E. J. Baron, M. A. Pfaller, F. C. Tenover, and R. H. Yolken (ed.), *Manual of Clinical Microbiology,* 7th ed. American Society for Microbiology, Washington, D.C.

31. **Funke, G., D. Monnet, C. deBernardis, A. von Graevenitz, and J. Freney.** 1998. Evaluation of the Vitek 2 System for rapid indentification of medically relevant gram-negative rods. *J. Clin. Microbiol.* **36:**1948–1952.

32. **Gupta, H., N. McKinnon, L. Louie, M. Louie, and A. E. Simor.** 1998. Comparison of six rapid agglutination tests for the identification of *Staphylococcus aureus,* including methicillin-resistant strains. *Diagn. Microbiol. Infect. Dis.* **31:**333–336.

33. **Gutman, S.** 1996. Rapid streptococcal tests. *Pediatrics* **97:**783–784.

34. **Halperin, S. A., R. Bortolussi, and A. J. Wort.** 1989. Evaluation of culture, immunofluorescence, and serology for the diagnosis of pertussis. *J. Clin. Microbiol.* **27:**752–757.

35. **Harbeck, R. J., J. Teague, G. R. Crossen, D. M. Maul, and P. L. Childers.** 1993. Novel, rapid optical immunoassay technique for detection of group A streptococci from pharyngeal specimens: comparison with standard culture methods. *J. Clin. Microbiol.* **31:** 839–844.

36. **Heiter, B. J., and P. Bourbeau.** 1994. Comparison of the GenProbe Group A Streptococcus direct test with culture and a rapid streptococcal antigen assay for diagnosis of streptococcal pharyngitis. *J. Clin. Microbiol.* **31:**2070–2073.

37. **Hinnebusch, C. J., D. M. Nikolai, and D. A. Bruckner.** 1991. Comparison of API Rapid Strep, Baxter MicroScan Rapid Pos ID Panel, BBL Minitek Differential Identification System, IDS Rapid STR System and Vitek GPI to conventional biochemical tests for identification of viridans streptococci. *Am. J. Clin. Pathol.* **96:**459–463.

38. **Hodinka, R., and P. H. Gilligan.** 1988. Evaluation of the Campyslide Agglutination test for confirmatory identification of selected *Campylobacter* species. *J. Clin. Microbiol.* **26:**47–49.

39. **Holmes, R. L., L. M. DeFranco, and M. Otto.** 1982. Novel method of biotyping *Haemophilus influenzae* that uses API 20E. *J. Clin. Microbiol.* **15:**1150–1152.

40. **Holwerda, J., and G. Eldering.** 1963. Culture and fluorescent antibody methods in diagnosis of whooping cough. *J. Bacteriol.* **86:**449–451.

41. **Hope, J. E.** 1999. *Bordetella*, p. 614–624. *In* P. R. Murray, E. J. Baron, M. A. Pfaller, F. C. Tenover, and R. H. Yolken (ed.), *Manual of Clinical Microbiology,* 7th ed. American Society for Microbiology, Washington, D.C.

42. **Hudspeth, M. K., S. H. Gerardo, D. M. Citron, and E. J. C. Goldstein.** 1998. Evaluation of the RapID CB Plus System for Identification of *Corynebacterium* species and other gram-positive rods. *J. Clin. Microbiol.* **36:**543–547.

43. **Huysmans, M. B., J. D. Turnidge, and J. H. Williams.** 1995. Evaluation of API Campy in comparison with conventional methods for identification of thermophilic campylobacters. *J. Clin. Microbiol.* **33:** 3345–3346.

44. **Ieven, M., J. Verhoeven, S. R. Pattyn, and H. Goossens.** 1995. Rapid and economical method for species identification of clinically significant coagulase-negative staphylococci. *J. Clin. Microbiol.* **33:**1060–1063.

45. **Isenberg, H. D.** 1994. *Clinical Microbiology Procedures Handbook.* American Society for Microbiology, Washington, D.C.

46. **Isenberg, H. D.** 1998. *Essential Procedures in Clinical Microbiology.* American Society for Microbiology, Washington, D.C.

47. **Iwen, P. C., M. A. Rupp, P. C. Schrekenberger, and S. H. Hinricks.** 1999. Evaluation of the revised Microscan Dried Overnight Gram-Positive Identification panel to identify *Enterococcus* species. *J. Clin. Microbiol.* **38:**3756–3758.

48. **Janda, W. M.** 1999. The corynebacteria revisited: new species, identification kits and antimicrobial susceptibility testing. *Clin. Microbiol. Newsl.* **21:**175–182.

49. **Janda, W. M., J. J. Bradna, and P. Ruther.** 1989. Identification of *Neisseria* spp., *Haemophilus* spp., and other fastidious gram-negative bacteria with the Micro-Scan *Haemophilus-Neisseria* identification panel. *J. Clin. Microbiol.* **27:**869–873.

50. **Janda, W. M., K. Ristow, and D. Novak.** 1994. Evaluation of RAPIDEC Staph for identification of *Staphylococcus aureus*, *Staphylococcus epidermidis*, and *Staphylococcus saprophyticus*. *J. Clin. Microbiol.* **32:**2056–2059.

51. **Janda, W. M., P. J. Malloy, and P. C. Schreckenberger.** 1987. Clinical evaluation of the Vitek *Neisseria-Haemophilus* Identification Card. *J. Clin. Microbiol.* **25:**37–41.

52. **Karmali, M., M. Petric, and M. Bielaszewska.** 1999. Evaluation of a microplate latex agglutination method (Verotox-F Assay) for detecting and characterizing verotoxins (Shiga toxins) in *Escherichia coli*. *J. Clin. Microbiol.* **37:**396–399.

53. **Kazandjian, D., R. Chiew, and G. L. Gilbert.** 1997. Rapid diagnosis of *Legionella pneumophila* serogroup 1 infection with the Binax enzyme immunoassay urinary antigen test. *J. Clin. Microbiol.* **35:**954–956.

54. **Kehl, K. S., P. Havens, C. E. Behnke, and D. W. K. Acheson.** 1997. Evaluation of the Premier EHEC assay for detection of Shiga toxin-producing *Escherichia coli*. *J. Clin. Microbiol.* **35:**2051–2054.

55. **Kloos, W. E., and T. L. Bannerman.** 1994. Update on clinical significance of coagulase-negative staphylococci. *Clin. Microbiol. Rev.* **7:**117–140.

56. **Knapp, J. S., and E. H. Koumans.** 1999. *Neisseria* and *Branhamella*, p. 586–603. *In* P. R. Murray, E. J. Baron, M. A. Pfaller, F. C. Tenover, and R. H. Yolken (ed.), *Manual of Clinical Microbiology*, 7th ed. American Society for Microbiology, Washington, D.C.

57. **Koepke, J. A.** 1999. Tips From the clinical experts: diagnostic tests for *Helicobacter pylori*. *Med. Lab. Obs.* **31:**8–10.

58. **Koneman, E. W., S. D. Allen, W. M. Janda, P. C. Schreckenberger, and W. C. Winn.** 1997. *Color Atlas and Textbook of Diagnostic Microbiology*, 5th ed., p. 230–242 and 287–309. Lippincott, Philadelphia, Pa.

59. **Koneman, E. W., S. D. Allen, W. M. Janda, P. C. Schreckenberger, and W. C. Winn.** 1997. *Color Atlas and Textbook of Diagnostic Microbiology*, 5th ed., p. 510–515. Lippincott. Philadelphia, Pa.

60. **LaClaire, L. L., and R. R. Facklam.** 2000. Comparison of three commercial rapid identification systems for the unusual gram-positive cocci *Dolodigranulum pigrum*, *Ignavigranum ruoffiae*, and *Facklamia* species. *J. Clin. Microbiol.* **38:**2037–2042.

61. **Lancefield, R. C.** 1933. A serological differentiation of human and other groups of beta-hemolytic streptococci. *J. Exp. Med.* **57:**571–595.

62. **Lindenmann, K., A. von Graevenitz, and G. Funke.** 1995. Evaluation of the Biolog system for the identification of asporogenous, aerobic gram-positive rods. *Med. Microbiol. Lett.* **4:**287–296.

63. **Loeffelholz, M., C. J. Thompson, K. S. Long, and M. Gilchrist.** 1999. Comparison of PCR, culture, and direct fluorescent-antibody testing for detection of *Bordetella pertussis*. *J. Clin. Microbiol.* **37:**2872–2876.

64. **Luijendijk, A., A. van Belkum, H. Verburgh, and J. Kluytmans.** 1996. Comparison of five tests for identification of *Staphylococcus aureus* from clinical specimens. *J. Clin. Microbiol.* **34:**2267–2269.

65. **MacKenzie, A., E. Orrbine, L. Hyde, M. Benoit, F. Chan, C. Park, J. Alverson, A. Lembke, D. Hoban, and W. Kennedy.** 2000. Performance of the Immuno-Card STAT! E. coli O157-H7 Test for detection of *Escherichia coli* O157:H7 in stools. *J. Clin. Microbiol.* **38:**1866–1868.

66. **Mackenzie, A., P. Lebel, E. Orrbine, P. C. Rowe, L. Hyde, F. Chan, W. Johnson, P. N. McLaine, and the Synsorb PK Study Investigators.** 1998. Sensitivities and specificities of Premier *E. coli* O157 and Premier EHEC enzyme immunoassays for diagnosis of infection with verotoxin (Shiga-like toxin)-producing *Escherichia coli*. *J. Clin. Microbiol.* **36:**1608–1611.

67. **McNicol, P., S. M. Giercke, M. Gray, D. Martin, B. Brodeur, M. S. Peppler, T. Williams, and G. Hammond.** 1995. Evaluation and validation of a monoclonal immunofluorescent reagent for direct detection of *Bordetella pertussis*. *J. Clin. Microbiol.* **33:**2868–2871.

68. **Miller, J. M., and C. M. O'Hara.** 1999. Manual and automated systems for microbial identification, p. 193–201. *In* P. R. Murray, E. J. Baron, M. A. Pfaller, F. C. Tenover, and R. H. Yolken (ed.), *Manual of Clinical Microbiology*, 7th ed. American Society for Microbiology, Washington, D.C.

69. **Miller J. M., J. W. Biddle, V. K. Quenzer, and J. C. McLaughlin.** 1993. Evaluation of the Biolog for identification of members of the family *Micrococcaceae*. *J. Clin. Microbiol.* **31:**3170–3173.

70. **Müller, F., J. Hope, and C. Wirsing von König.** 1997. Laboratory diagnosis of pertussis: state of the art in 1997. *J. Clin. Microbiol.* **35:**2435–2443.

71. **Nachamkin, I.** 1999. *Campylobacter* and *Arcobacter*, p. 716–726. *In* P. R. Murray, E. J. Baron, M. A. Pfaller, F. C. Tenover, and R. H. Yolken (ed.), *Manual of Clinical Microbiology*, 7th ed. American Society for Microbiology, Washington, D.C.

72. **Nachamkin, I., and S. Barbagallo.** 1990. Culture confirmation of *Campylobacter* spp. by latex agglutination. *J. Clin. Microbiol.* **28:**817–818.

73. **Nguyen, T. M., D. W. Gauthier, T. D. Myles, B. S. Nuwayhid, M. A. Viana, and P. C. Schreckenberger.** 1998. Detection of group B streptococcus: comparison of an optical immunoassay with direct plating and broth-enhanced culture methods. *J. Matern. Fetal Med.* **7(4):**172–176.

74. **Nugent, R. P., M. A. Krohn, and S. L. Hillier.** 1991. Reliability of diagnosing bacterial vaginosis is improved by a standardized method of Gram stain interpretation. *J. Clin. Microbiol.* **29:**297–301.

75. **O'Hara, C. M., and J. M. Miller.** 2000. Evaluation of the MicroScan Rapid Neg ID3 Panel for identification of *Enterobacteriaceae* and some common gram-negative nonfermenters. *J. Clin. Microbiol.* **38:**3577–3580.

76. **O'Hara, C. M., G. L. Westbrook, and J. M. Miller.** 1997. Evaluation of Vitek GNI+ and Becton Dickinson Microbiology Systems Crystal E/NF identification systems for identification of members of the family *Enterobacteriaceae* and other gram-negative, glucose-fermenting and non-glucose-fermenting bacilli. *J. Clin. Microbiol.* **35:**3269–3273.

77. **Park, C. H., D. L. Hixon, W. L. Morrison, and C. B. Cook.** 1994. Rapid diagnosis of enterohemorrhagic *Escherichia coli* O157:H7 directly from fecal specimens using immunofluorescence stain. *Am. J. Clin. Pathol.* **101:**91–94.

78. **Park, C. H., K. M. Gates, N. M. Vandel, and D. L. Hixon.** 1996. Isolation of Shiga-like toxin producing

Escherichia coli (O157 and non-O157) in a community hospital. *Diagn. Microbiol. Infect. Dis.* **26**:69–72.

79. **Park, C. H., N. M. Vandel, and D. L. Hixon.** 1996. Rapid immunoassay for detection of *Escherichia coli* O157 directly from stool specimens. *J. Clin. Microbiol.* **34**:988–990.

80. **Personne, P., M. Bes, G. Lina, F. Vandenesch, Y. Brun, and J. Atienne.** 1997. Comparative performance of six agglutination kits assessed by using typical and atypical strains of *Staphylococcus aureus. J. Clin. Microbiol.* **35**:1138–1140.

81. **Pfaller, M. A., D. Sahm, C. O'Hara, C. Ciaglia, M. Yu, N. Yamane, G. Scharnweber, and D. Roden.** 1991. Comparison of the AutoSCAN-W/A rapid bacterial identification system and the Vitek AutoMicrobic system for identification of gram-negative bacilli. *J. Clin. Microbiol.* **29**:1422–1428.

82. **Piper, J., T. Hadfield, F. McCleskey, M. Evans, S. Friedstrom, P. Lauderdale, and R. Winn.** 1986. Efficacies of rapid agglutination tests for identification of methicillin-resistant staphylococcal strains as *Staphylococcus aureus. J. Clin. Microbiol.* **26**:1907–1909.

83. **Plouffe, J. F., T. M. File, R. F. Breiman, B. A. Hackman, S. J. Salstrom, B. J. Marston, and B. S. Fields.** 1995. Reevaluation of the definition of Legionnaire's Disease: use of the urinary antigen assay. *Clin. Infect. Dis.* **20**:1286–1291.

84. **Pokorski, S. J., E. A. Vetter, P. C. Wollan, and F. R. Cockerill III.** 1994. Comparison of Gen-Probe Group A Streptococcus Direct Test with culture for diagnosing streptococcal pharyngitis. *J. Clin. Microbiol.* **32**:1440–1443.

85. **Popovic-Uroic, T., C. M. Patton, I. K. Wachsmuth, and R. Roeder.** 1991. Evaluation of an oligonucleotide probe for identification of *Campylobacter* species. *Lab. Med.* **22**:533–539.

86. **Rhoads, S., L. Marinelli, C. A. Imperatrice, and I. Nachamkin.** 1995. Comparison of MicroScan Walk-Away system and Vitek system for identification of gram-negative bacteria. *J. Clin. Microbiol.* **33**:3044–3046.

87. **Robinson, A., Y. S. McCarter, and J. Tetreault.** 1995. Comparison of Crystal Enteric/Nonfermenter system, API 20E system, and Vitek AutoMicrobic system for identification of gram-negative bacilli. *J. Clin. Microbiol.* **33**:364–370.

88. **Ruane, P., M. A. Morgan, D. M. Citron, and M. E. Mulligan.** 1986. Failure of rapid agglutination methods to detect oxacillin-resistant *Staphylococcus aureus. J. Clin. Microbiol.* **24**:490–492.

89. **Ruoff, K. L., R. A. Whiley, and D. Beighton.** 1999. *Streptococcus*, p. 283–296. *In* P. R. Murray, E. J. Baron, M. A. Pfaller, F. C. Tenover and R. H. Yolken (ed.), *Manual of Clinical Microbiology*, 7th ed. American Society for Microbiology, Washington, D.C.

90. **Smole, S. C., E. Aronson, A. Durbin, S. M. Brecher, and R. D. Arbeit.** 1998. Sensitivity and specificity of an improved latex agglutination test for identification of methicillin-sensitive and -resistant *Staphylococcus aureus* isolates. *J. Clin. Microbiol.* **36**:1109–1112.

91. **Sowers, E. G., J. G. Wells, and N. A. Strockbine.** 1996. Evaluation of commercial latex reagents for identification of O157 and H7 antigens of *Escherichia coli. J. Clin. Microbiol.* **34**:1286–1289.

92. **Spiegel, C. A.** 1999. Bacterial vaginosis: changes in laboratory practice. *Clin. Microbiol. Newsl.* **21**:33–37.

93. **Staneck, J. L., L. S. Weckbach, R. C. Tilton, R. J. Zabransky, L. Bayola-Mueller, C. M. O'Hara, and J. M. Miller.** 1993. Collaborative evaluation of the Radiometer Sensititre AP80 for identification of gram-negative bacilli. *J. Clin. Microbiol.* **31**:1179–1184.

94. **Stavric, S., B. Buchanan, and J. Speirs.** 1992. Comparison of a competitive enzyme immunoassay kit and the infant mouse assay for detecting *Escherichia coli* heat stable enterotoxin. *Lett. Appl. Microbiol.* **14**:47–50.

95. **Stoakes L., M. A. John, R. Lannigan, B. C. Schieven, M. Ramos, D. Harley, and Z. Hussain.** 1993. Gas-liquid chromatography of cellular fatty acids for identification of staphylococci. *J. Clin. Microbiol.* **32**:1908–1910.

96. **Stout, J. E.** 2000. Laboratory diagnosis of Legionnaires' Disease: the expanding role of the Legionella urinary antigen test. *Clin. Microbiol. Newsl.* **22**:62–64.

97. **Strebel, P. M., S. L. Cochi, K. M. Farizo, B. J. Payne, S. D. Hanauer, and A. L. Baughman.** 1993. Pertussis in Missouri: evaluation of nasopharyngeal culture, direct fluorescent antibody testing, and clinical case definitions in the diagnosis of pertussis. *Clin. Infect. Dis.* **16**:276–285.

98. **Summers, W. C., E. S. Brookings, and K. B. Waites.** 1998. Identification of oxacillin-susceptible and oxacillin-resistant *Staphylococcus aureus* using commercial latex agglutination tests. *Diagn. Microbiol. Infect. Dis.* **30**:131–134.

99. **Tenover, F. C., L. Carlson, S. Barbagallo, and I. Nachamkin.** 1990. DNA probe culture confirmation assay for identification of thermophilic *Campylobacter* species. *J. Clin. Microbiol.* **28**:1284–1287.

100. **Trevisani, L., S. Sartori, F. Galvani, M. R. Rossi, M. Ruina, and M. Caselli.** 1997. Two unusual techniques for diagnosing *Helicobacter pylori* infection, p. 58–60. *In* II International Meeting, Developing Knowledge on *Helicobacter pylori*. Gastroenterology International Congress Proceedings. Gastroenterology International Congress, Ferrara, Italy.

101. **Vaira, D., P. Malfertheiner, F. M Graud, A. T. R. Axon, M. Deltenre, A. M. Hirschl, G. Gasbarrini, C. O'Morain, J. M. P. Garcia, M. Quina, G. N. J. Tytgat, and the HpSA European Study Group.** 1999. Diagnosis of Helicobacter pylori infection with a new non-invasive antigen-based assay. *Lancet* **354**:30–33.

102. **Vakil, N.** 1998. Diagnostic testing for *H. pylori. Adv. Admin. Lab.* **7**:39–41.

103. **Versalovic, J., and J. G. Fox.** 1999. *Helicobacter*, p. 727–738. *In* P. R. Murray, E. J. Baron, M. A. Pfaller, F. C. Tenover, and R. H. Yolken (ed.), *Manual of Clinical Microbiology*, 7th ed. American Society for Microbiology, Washington, D.C.

104. **von Baum, H., F. R. Klemme, H. K. Geiss, and H. G. Sonntag.** 1998. Comparative evaluation of a commercial system for identification of gram-positive cocci. *Eur. J. Clin. Microbiol. Infect. Dis.* **17**:849–852.

105. **Wadowsky, R. M., R. H. Michaels, T. Libert, L. Kingsley, and G. D. Ehrlich.** 1996. Multiplex PCR-based assay for detection of *Bordetella pertussis* in nasopharyngeal swab specimens. *J. Clin. Microbiol.* **34**:2645–2649.

106. **Weers-Pothoff, G., C. E. M. Moolhuijzen, and G. P. A. Bongaerts.** 1987. Comparison of seven coagulase tests for identification of *Staphylococcus aureus. Eur. J. Clin. Microbiol.* **6**:589–591.

107. **Weinstein, M. P., S. Mirrett, L. Van Pelt, M. Mc-Kinnon, B. L. Zimmer, W. Kloos, and L. B. Reller.** 1998. Clinical importance of identifying coagulase-negative staphylococci isolated from blood cultures: evaluation of MicroScan rapid and dried overnight gram-positive panels versus a conventional reference method. *J. Clin. Microbiol.* **36:**2089–2092.

108. **Weir, S. C., S. H. Fischer, F. Stock, and V. J. Gill.** 1998. Detection of *Legionella* by PCR in respiratory specimens using a commercially available kit. *Am. J. Clin. Pathol.* **110:**295–300.

109. **Whitaker, J. A., P. Donaldson, and J. D. Nelson.** 1960. Diagnosis of pertussis by the fluorescent antibody method. *N. Engl. J. Med.* **26:**850–851.

110. **Wilkerson, M., S. McAllister, J. M. Miller, B. J. Heiter, and P. P. Bourbeau.** 1997. Comparison of five agglutination tests for identification of *Staphylococcus aureus*. *J. Clin. Microbiol.* **35:**148–151.

111. **Winn, W.** 1999. *Legionella*, p. 572–585. *In* P. R. Murray, E. J. Baron, M. A. Pfaller, F. C. Tenover, and R. H. Yolken (ed.), *Manual of Clinical Microbiology*, 7th ed. American Society for Microbiology, Washington, D.C.

112. **Yam, W. C., M. L. Lung, and M. H. Ng.** 1992. Evaluation and optimization of a latex agglutination assay for detection of cholera toxin and *Escherichia coli* heat-labile toxin. *J. Clin. Microbiol.* **30:**2518–2520.

113. **York, M. K., G. F. Brooks, and E. H. Fiss.** 1992. Evaluation of the autoSCAN-W/A rapid system for identification and susceptibility testing of gram-negative fermentative bacilli. *J. Clin. Microbiol.* **30:**2903–2910.

114. **Yu, P. K., J. J. Germer, C. A. Torgerson, and J. P. Anhalt.** 1988. Evaluation of TestPack Strep A for the detection of group A streptococci in throat swabs. *Mayo Clin. Proc.* **63:**33–36.

Anaerobic Bacteriology

STEPHEN D. ALLEN, CHRISTOPHER L. EMERY, AND JEAN A. SIDERS

4

INTRODUCTION AND CLINICAL CONSIDERATIONS

Anaerobic bacteria, from a practical standpoint, are defined here as those bacteria that do not multiply on the surface of nutritionally adequate solid media incubated in air or in a CO_2 incubator (10% CO_2). They do not use molecular oxygen as a terminal electron acceptor. While many ferment a wide variety of carbohydrates, many others do not (i.e., they are asaccharolytic) but obtain their energy through fermentation of amino acids or various other organic compounds. Vegetative cells of anaerobic bacteria, but not the spores of clostridia, are killed by exposure to molecular oxygen in ambient air. However, as can be shown by aerotolerance testing, anaerobes vary considerably in their sensitivity or ability to tolerate oxygen (77). Some, referred to as moderate obligate anaerobes (e.g., *Bacteroides fragilis* and *Clostridium perfringens)*, can remain viable for many hours when their colonies on agar plates are left out in room air. Others, referred to as strict obligate anaerobes (e.g., *Clostridium novyi* type B and *Clostridium haemolyticum)*, cannot survive exposure to oxygen longer than a few minutes. Although most of the anaerobes associated with diseases in humans tolerate inspection and subculture of colonies on the open laboratory bench, the rapidity and amount of growth upon subculture can be improved through use of an anaerobic holding box or jar (7, 77). When working with strict obligate anaerobes, such manipulations are most likely to be successful only when done in the absence of oxygen (i.e., within an anaerobic chamber).

The taxonomic classification of the anaerobes continues to change, in part due to their broad range of biological diversity. Included among them are all morphologic forms of bacteria—both spore formers and non-spore formers; many are gram-positive and many are gram-negative. In the early 1970s when the eighth edition of *Bergey's Manual of Determinative Bacteriology* was published, about 20 different genera of anaerobic bacteria were recognized (23). Differentiation of these genera was weighted heavily upon morphologic features, short-chain acid metabolic products determined using gas-liquid chromatography (GLC), and relatively crude nucleic acid data (e.g., percent guanine plus cytosine content). Further differentiation to species level was based on conventional culture and biochemical data. Although the taxonomic classification of anaerobes is beyond the scope of this text, there have been extensive changes in recent years, largely based on extensive chemotaxonomic work and more sophisticated molecular genetic studies (in contrast to the nucleic studies done in the past) (120). Summaries of recent taxonomic changes regarding anaerobic bacteria are available elsewhere (4, 69, 70, 114). Keeping up with the many changes in the nomenclature and classification of anaerobes, with increased numbers of new genera and species in recent years, has been a challenge for manufacturers of commercial identification systems, as well as clinical microbiologists and clinicians alike (14). The databases of most of the commercial identification systems are out of date with respect to the current nomenclature for anaerobic bacteria.

Anaerobic bacteria predominate within the indigenous microflora of humans (e.g., upper respiratory tract, oral cavity, gastrointestinal tract, genitourinary tract, and skin), and they take part in essentially all types of infections that involve bacteria. Although a few diseases involving anaerobes can be of exogenous origin (e.g., food-borne botulism, tetanus, and pseudomembranous colitis [caused by *Clostridium difficile])*, the vast majority of anaerobic infections are endogenous, arising from the body's own microbial flora. These types of infections are often polymicrobial and commonly involve anaerobes mixed with aerobes, facultative anaerobes, or other anaerobes. However, some anaerobic infections involve only a single species. Anaerobic infections are often difficult to treat and life-threatening, associated with high mortality rates (3, 42).

Although clinicians and microbiologists generally know more about anaerobic infections today than they did 30 years ago, anaerobic infections probably are still among the most commonly underdiagnosed bacterial infections (42). Factors that influence the selection of specimens to submit to the laboratory for anaerobic culture include the probability of whether or not anaerobes are likely to be a significant component of the infection and the perceived clinical importance of documenting the presence of anaerobes in the infectious process. Other important factors include the ability to collect an uncontaminated specimen, the cost of empirical treatment alone without anaerobic culture data, and the perceived cost benefit of having anaerobic culture data to aid in management of the patient. Several types of infections involve anaerobes relatively frequently. These include abscesses of all anatomic regions of the body (e.g., central nervous system, respiratory tract, intra-abdominal, pelvic, and other sites). In addition, other conditions in which anaerobes are commonly involved include aspiration pneumonia, appendicitis, bacteremia, cholecystitis, chronic otitis media, chronic sinusitis, crepitant and noncrepitant cellulitis, dental-oral infections, endocarditis, endometritis, myonecrosis, necrotizing fasciitis, neutropenic enterocolitis (caused by C. septicum), peritonitis, thoracic empyema, septic arthritis, subdural empyema, and many other types of infections (18, 42, 43). Examples of infections in which anaerobes are seldom important include acute cholecystitis, acute otitis media, acute osteomyelitis, acute sinusitis, bacterial meningitis, bronchitis, cystitis, pharyngitis, pyelonephritis, superficial skin (surface) infections, and spontaneous (primary) peritonitis.

Even though there is a vast and diverse assortment of obligately anaerobic bacteria in the normal human microbiota, only a few species or groups of anaerobes are commonly isolated from properly selected and properly collected specimens from patients with disease (42, 77). Awareness of this fact has aided microbiologists and manufacturers of commercial products in influencing the selection of tests to be used in identification of anaerobes or to be included in commercial packaged kits for identification. Some of the more frequently isolated and/or clinically important anaerobes from properly selected and collected specimens are the B. fragilis group, Prevotella and Porphyromonas species, Fusobacterium nucleatum and Fusobacterium necrophorum, the anaerobic cocci [e.g., Peptostreptococcus anaerobius and Finegoldia (Peptostreptococcus) magna], Actinomyces israelii, and certain Clostridium species (e.g., C. perfringens, C. ramosum, C. clostridioforme, C. septicum, and C. difficile) (43, 77, 108, 125). The incidence of anaerobic bacteria in infectious diseases has been reviewed in detail elsewhere (3, 18, 42, 43).

A number of clinical clues can suggest that anaerobes are probably involved in an infection (18, 42). For example, a putrid odor from an infected lesion or purulent discharge is strong evidence of infection with anaerobes. No other types of bacterial, mycobacterial, fungal, parasitic, or viral infections emit such foul odors. Presumably various products of anaerobic metabolism, including short-chain fatty acids (e.g., butyrate, isocaproate, and succinate), amines, hydrogen sulfide, and probably other sulfides can contribute to the unpleasant odors. However, in many if not the majority of cases, these types of odors go unnoticed. Thus, a lack of odor does not rule out infection with anaerobes. Lesions that arise in close proximity to mucosal surfaces where anaerobes reside within the flora, or from inocula derived from these sites (e.g., dental/oral infections, necrotizing pulmonary infections following aspiration of oral contents, and infections secondary to penetrating wounds to the abdomen or pelvis), are likely to involve anaerobes. Gas in tissue; underlying disease with tissue necrosis, impaired blood supply, or both (e.g., due to malignancy); gangrenous necrosis; abscess formation; previous antibiotic treatment; septic thrombophlebitis; infections following animal or human bites; and "sulfur granules" from suspected actinomycosis also are important clues. As has been illustrated in detail elsewhere, the direct microscopic examination of Gram-stained smears frequently provides presumptive evidence consistent with anaerobic infections (77, 91). Multiple different morphologic forms of bacteria within aspirated pus from a deep tissue site could represent polymicrobial infection involving a combination of anaerobes and facultative anaerobes or anaerobes mixed with anaerobes. Anaerobic infections are commonly polymicrobial, as mentioned previously. In smears of abdominal exudate, gram-negative,

pale, irregularly staining, pleomorphic rods of varying sizes with bipolar staining would be consistent with Bacteroides. Pale, gram-negative, filamentous, slim rods with pointed ends, particularly in material from a lower respiratory tract infection, could suggest infection with F. nucleatum (91). From a patient with suspected gas gangrene, relatively large, broad, gram-positive rods with blunted ends in a smear showing a necrotic background with few or rare white blood cells could suggest the presence of C. perfringens. In addition, the inspection of colonies on primary isolation plates following anaerobic incubation aids in the preliminary identification of a number of different anaerobes (e.g., Actinomyces species, the B. fragilis group, C. perfringens, F. nucleatum, and others) (77).

Steps in the Diagnosis of Anaerobic Bacterial Infections

The diagnosis of anaerobic bacterial infections involves the following steps:

1. Selection, collection, and transportation of specimens
2. Rejection of unacceptable specimens; rapid processing of acceptable specimens
3. Direct microscopic examination of clinical materials
4. Selection and use of selective and nonselective primary isolation media
5. Use of anaerobic systems for incubation
6. Proper incubation conditions
7. Use of an anaerobic holding jar or anaerobic holding box procedure
8. Inspection and subculture of colonies
9. Identification of anaerobe isolates
10. Antimicrobial susceptibility testing of significant isolates
11. Issuing preliminary reports at appropriate times (e.g., steps 3, 8, 9, and 10) prior to the final report

Physicians' decisions regarding specimens to select and collect for culture can be critically important for both the patient and the laboratory. As emphasized by Finegold, some types of diseases involving anaerobes (e.g., tetanus) should be diagnosed on clinical grounds alone; thus, in such cases, anaerobic cultures are not relevant (42). In other instances, it could be misleading to culture for anaerobes even though the likelihood of their presence is high (e.g., surface of a decubitus ulcer or eschar of a burn). In recent years, Bilophila wadsworthia has been isolated from appendices and identified (15, 17). While this is an interesting discovery, anaerobic cultures of appendices or peritoneal surfaces at the time of appendectomy do not aid physicians in managing individual patients with acute appendicitis (42). On the other hand, if a patient subsequently develops peritonitis, perhaps with intraabdominal abscess formation and clinical findings suggesting sepsis, anaerobic cultures of abdominal pus and blood cultures could become important (e.g., a week or two postoperatively). More information on collection, transport, and storage of specimens is provided in the section below. Upon receipt of the specimen in the laboratory, the direct microscopic examination of clinical materials can provide important information rapidly. The processing of specimens, selection and inoculation of media for primary isolation, use of anaerobic systems for incubation, use of anaerobic holding jars and boxes, procedures for inspection and subculture of colonies, identification, and anaerobic susceptibility testing, while important topics as well, are beyond the scope of this text. These topics are covered extensively elsewhere (31, 59, 61, 77, 110, 132).

COLLECTION, TRANSPORT, AND STORAGE OF SPECIMENS

Some of the most important considerations involved in the laboratory diagnosis of anaerobic infections include (i) selecting, collecting, and transporting specimens for microbiologic examination and (ii) processing and examining the specimens in the laboratory as rapidly as possible after they are received. General information on specimen collection, transport, and storage is provided by Miller and Holmes (196). More detailed information on the selection, collection, and transport of specimens related to anaerobic infections is available elsewhere (77, 89, 132).

Careful selection of materials to be examined for anaerobic bacteria is extremely important. The specimen should be collected from the active site of infection and should not be contaminated with extraneous flora (7, 31). Because anaerobes are abundant components of the indigenous flora of mucous membrane surfaces and present on skin, several types of clinical materials should not be cultured for anaerobic bacteria. These include the following examples: throat or nasopharyngeal swabs; gingival swabs; expectorated sputum; bronchoscopic specimens not collected by a protective, double-lumen catheter (5, 144); gastric contents; small bowel contents (except in blind-loop and similar syndromes); feces (except in the workup of diseases due to C. difficile and certain other clostridia); rectal swabs; colocutaneous fistulae; colostomy stomata; surface material from decubitus ulcers; swab samples of other surfaces; sinus tract specimens; eschars; materials adjacent to skin or mucous membranes other than the above which have not been properly decontaminated; voided urine; and vaginal or cervical swabs (7).

Guidelines for collection of specimens for anaerobic culture are summarized in Table 1. Examples of acceptable specimens for isolation of anaerobes include an adequate sample volume of pus aspirated from an abscess or a deep wound (closed), tissue (biopsy, surgically removed, or autopsy specimen), normally sterile body fluids (e.g., cerebrospinal, pleural, pericardial, paracentesis, and synovial), blood, bone marrow, fine-needle aspirates of lung or other body site specimens, and sulfur granules. Aspiration of liquid samples by syringe and needle or biopsy of infected tissue are the methods of choice for collecting specimens to be processed for anaerobes. Swabs are much less desirable because they retain only a relatively small volume of sample, they tend to dry out, and they probably subject anaerobes to undue oxygen exposure. Prior to aspiration or biopsy of clinical material, the skin or mucous membrane surface should be properly decontaminated to prevent contamination with extraneous microorganisms at the time of specimen collection (77).

After aspiration of pus or other liquid, the sample should be carefully injected into an oxygen-free, anaerobic transport vial or anaerobic transport tube. The container should contain an oxidation-reduction potential indicator (77, 132). Because of the potential danger for needle stick injury, the syringe and needle should not be transported to the lab-

TABLE 1 Examples of specimens and collection procedures in anaerobic bacteriology[a]

Site	Specimens and methods of collection
Cardiovascular system	Blood cultures
Central nervous system	Abscess material, tissue obtained surgically, cerebrospinal fluid
Dental area, ear, nose, throat, and sinuses	Carefully aspirated or biopsied material from abscesses after surface decontamination with povidone-iodine, needle aspirates and surgical specimens from sinuses in chronic sinusitis patients
Pulmonary area	Transtracheal aspiration, fine-needle aspirate of lung, thoracotomy specimen, thoracentesis (pleural fluid), bronchoscopic specimen obtained with protective, double-lumen catheter[b]
Abdominal area	Paracentesis fluid, needle-and-syringe aspiration of deep abscesses under ultrasound or at surgery, surgical specimen if not contaminated with intestinal flora, bile
Female genital tract	Culdocentesis after surface decontamination of the vagina with povidone-iodine; laparoscopy specimens, surgical specimens, endometrial cavity specimen with double-lumen catheter[c] and microbiologic brush after cervical os is decontaminated
Urinary tract	Suprapubic aspirate of urine
Bone and joint	Aspirate of joint (in suppurative arthritis), deep aspirate of drainage material after surgery (e.g., in osteomyelitis)
Soft tissue	Open wounds (deep aspirate of margin or biopsy specimen of the depths of wound only after careful surface decontamination with povidone-iodine); sinus tracts (aspiration by syringe and small plastic catheter after careful decontamination of skin orifice); deep abscess, anaerobic cellulitis, infected vascular gangrene, clostridial myonecrosis (needle aspirate after surface decontamination); surgical specimens, including curettings and biopsy material; decubiti and other surface ulcers (thoroughly cleanse area with povidone-iodine by surgical scrub technique, and aspirate pus from deep pockets or obtain biopsy specimen from deep tissue at margin)

[a]Adapted from references 7 and 31.
[b]Reviewed elsewhere by Allen and Siders (5).
[c]Technique uses a telescoping double-catheter assembly similar to that available for bronchoscopy. This procedure is inadequate to diagnose postpartum endometritis (132).

oratory, and appropriate safety guidelines for prevention of needle stick injuries should be followed (96). A second reason not to transport aspirates in plastic syringes is that oxygen diffuses through the plastic. Tissue can be placed in a sterile container (capped loosely) and transported to the laboratory in an anaerobic bag that removes oxygen from the atmosphere inside the container (89, 132). Although swabs are the least desirable (of the methods described above for specimen collection), at times it is impossible to obtain an aspirate or tissue sample, and a swab may be used. Swabs are best transported in anaerobic transport containers that are commercially available for this purpose (Table 2) (77).

Optimally, the viability and relative proportions of anaerobic bacteria or other microorganisms should not change while the specimen is being transported to the laboratory. In order to minimize loss of viability of anaerobes or overgrowth of aerobic microorganisms within the specimen container, rapid transportation is highly recommended. Under certain circumstances, some anaerobes lose viability during refrigeration (e.g., *B. fragilis* and *C. perfringens* at 4°C) (52, 132). However, at room temperature or warmer temperatures, some aerobic or facultatively anaerobic bacteria may overgrow (e.g., *Escherichia coli* or *Pseudomonas* species), making it difficult or impossible to recover any anaerobes that could originally have been present when the sample was collected (47, 52). For these reasons, specimens are best maintained at 15 to 25°C during transportation to the laboratory.

INCUBATION, INSPECTION, AND SUBCULTURE OF COLONIES

Once a properly collected and transported specimen has been received, appropriate selective and nonselective media are inoculated and incubated in an anaerobic system. Of course, the direct microscopic examination of clinical materials may suggest the presence of anaerobes long before culture results become available. Satisfactory commercially available primary isolation media and anaerobic incubation systems are discussed elsewhere along with detailed recommendations for their use (77, 132). Several commercially available sources of systems for anaerobic incubation are provided in Tables 3 and 4. Relatively current illustrations, discussions, and recommendations on the use of a number of these products are available elsewhere (37, 77, 132). As mentioned previously in the introduction, anaerobes vary considerably in their oxygen tolerance. Therefore, we again emphasize the value of using the holding jar or holding box procedure at the time plate cultures are inspected and subcultured (31, 77). Because many anaerobes have distinctive colony characteristics, it is wise to inspect and subculture colonies carefully under a stereoscopic dissecting micro-

scope (magnification, ×7 to ×15). An important second reason is to improve the microbiologist's ability to avoid contamination and ensure purity of isolates that are to be identified.

IDENTIFICATION OF ANAEROBES

Characteristics of colonies obtained through dissecting microscope examination, along with the results of Gram's stain observations, and aerotolerance testing are especially valuable for initial presumptive identification of isolates. Photographs of anaerobe colonies on commercially available CDC Anaerobe Blood Agar and selective media, along with their microscopic features of anaerobe isolates, are shown in the *Color Atlas and Textbook of Diagnostic Microbiology* of Koneman and colleagues (77). Appearances of colonies on other media (e.g., brucella blood agar or bacteroides bile esculin agar may be different) (132). The characteristics in the list that follows are especially useful for presumptive identification of common anaerobes involved in illnesses of humans:

1. Relation to oxygen (e.g., obligate anaerobe, aerotolerant anaerobe, facultative anaerobe)
2. Colony characteristics (e.g., size, form, elevation, margin, surface, consistency, pigment)
3. Pitting of medium
4. Hemolysis
5. Microscopic features (e.g., Gram stain reaction, morphology, and presence or absence of spores)
6. Motility
7. Growth in liquid medium (e.g., rapidity, appearance, and gas production)
8. Growth in presence of 20% bile
9. Catalase production on a medium containing hemin
10. Reaction on egg yolk agar (e.g., lecithinase, lipase, and proteolysis)
11. Production of indole, H_2S, and urease
12. Hydrolysis of esculin and starch
13. Fermentation of key carbohydrates (e.g., glucose, mannitol, lactose, sucrose, maltose, salicin, glycerol, xylose, arabinose, mannose, rhamnose, trehalose)
14. Inhibition by sodium polyanethol sulfonate (SPS) and certain antibiotics (e.g., penicillin, kanamycin, rifampin, vancomycin, etc.)
15. Short-chain acid metabolic products determined using GLC (Table 4 provides lists of manufacturers and vendors of gas chromatographs)

The above characteristics formed the basis of a practical approach for identifying anaerobic bacteria by Dowell, Lombard, and a number of colleagues formerly at the Centers for Disease Control and Prevention (CDC), and also by

TABLE 2 Commercially available anaerobe transport devices

Manufacturer (product)	Manufacturer no.	Package contents	List price ($)
Starplex (Anaerobic Transport System)	S120	Pack of 10	35.12
Starplex (Starswab II)	SP130X	Pack of 50	23.22
BBL (Anaerobic Culturette)	4362218	Pack of 100	157.00
BD[a] (Port-A-Cul Tube with swab, sterile)	4321607	Pack of 10	42.40
BD (Port-A-Cul Tube)	4321606	Pack of 10	28.30
BBL (Vacutainer Anaerobic Specimen collector)	4336500	Pack of 25	105.00

[a]BD, Becton Dickinson.

TABLE 3 Commercially available anaerobic incubation systems: manufacturers and catalog list prices

Manufacturer (product and description)	Manufacturer no.	Package contents	List price ($)
BD[a] (Type A Bio-Bag: holds one 100-by-15-mm petri dish)	4361214	Pack of 100	364.00
BD (Type A Bio-Bag: holds three 100-by-15-mm petri dishes)	4361216	Pack of 50	291.00
BD (GasPak Pouch Environmental System)	4360651	Pack of 25	91.80
BD (GasPak 100)	4360626	Each	369.00
BD (GasPak 150)	4360628	Each	550.00
BD (GasPak Plus generator [envelopes for jar])	4371040	Pack of 10	23.50
Difco (Anaerobic Jar, Complete)	219511	Each	361.20
Difco (Anaerobic System [envelopes for jar])	219521	Pack of 10	20.20
EM Science (Anaerobic Jar)	53-13681	Each	260.71
EM Science (Anaerocult A [gas generator for jar])	53-13677	Pack of 10	22.23
EM Science (Anaerocult A-Mini [individual bag etc.])	53-01661	Pack of 25	106.31
Oxoid (Anaerobic Jar Complete)	65-11	Each	550.50
Oxoid (AnaeroGen anaerobic atmosphere generator)	65-35	Pack of 10	14.50

[a]BD, Becton Dickinson.

us at the Indiana University Medical Center. Of course, reference laboratories and research laboratories may use additional tests if they are developing new databases or performing chemotaxonomic research (16, 32, 34, 77).

COMMERCIALLY AVAILABLE MANUAL TESTING PROCEDURES

Many products are available for the characterization and identification of anaerobic bacteria. These products include prereduced anaerobically sterilized (PRAS) media for conventional biochemical testing, the Presumpto plate system, various alternative tests for presumptive identification, and commerical packaged micromethod kits that require 24 to 72 h of incubation. In addition, several packaged kits for rapid identification within 4 h or less of incubation are available. An automated GLC system for cellular fatty acid analysis is also available.

Conventional Identification Procedures Using PRAS Biochemicals

The "gold standard" for phenotypic characterization of anaerobic isolates (to aid in definitive identification of anaerobes) has long been based on the use of conventional biochemical tests using PRAS media according to the directions in the *Virginia Polytechnic Institute Anaerobe Laboratory Manual (hereafter referred to as the VPI Manual)* (59, 61). Accordingly, the results of (i) Gram's stains, (ii) morphologic features, and (iii) short-chain acid metabolic product analyses determined using GLC are key characteristics for identification of isolates. These results permit differentiation of most clinical isolates to the genus or group level and differentiation of a few to the species level. The results of conventional biochemical tests, obtained using PRAS media, aid in further differentiation of anaerobe isolates to the species level. In practice, the biochemical tests to be selected for identification of an anaerobe isolate are determined after observing its colony characteristics, its Gram reaction and microscopic features, and findings based on GLC (Table 5 below). Commercial suppliers of conventional PRAS liquid media, prepared according to the *VPI Manual* methods (59); conventional liquid media; and agar differential media, prepared according to the recommendations of Dowell and others (formerly of the CDC) (32, 36), are listed in Table 6.

For biochemical testing with liquid media, the CDC has traditionally used a thioglycolate formulation with bromthymol blue indicator (i.e., CHO media, [Table 6]) (32). To our knowledge, biochemical test media prepared commercially according to instructions in the CDC manual are currently available only from Smith River Biologicals [9388 Charity Highway, Ferrum, VA 24088; phone, (540) 930-2369]. Through the years, we have worked with both the Virginia Polytechnic Institute (VPI) and the CDC conventional procedures for identification (2, 7). In general, the results of identifying anaerobe isolates obtained using the VPI procedures and tables compared with results obtained using the CDC procedures and tables have shown excellent agreement (Siders and Allen, unpublished information). In the taxonomic literature, however, conventional phenotypic characteristics of various newly described and reclassified anaerobes have been based on VPI procedures far more frequently than based on CDC procedures. Detailed directions regarding the use of conventional identification procedures are given elsewhere (32, 59, 61, 122).

Tubes of PRAS biochemical test media can be inoculated with the aid of a special apparatus described in the *VPI Manual* (60). Alternatively, depending upon how they are supplied by the manufacturer, they can be inoculated through a Vacutainer-type closure using a syringe and needle (Fig. 1). Following incubation for an appropriate time, a pH meter is used to determine the pH of fermentation test media. For some rapidly growing anaerobes (e.g., some of the *B. fragilis* group, and many *Clostridium* species), biochemicals can be read after overnight incubation. Some of the so-called rapid identification systems available commercially, which require heavy inocula (e.g., turbidity of resuspended colonies equivalent to a McFarland no. 3 to no. 5 turbidity standard) of a pure culture isolate prior to inoculation, are no more rapid than the conventional systems (130).

Presumpto Quadrant Plates

During the mid-1970s at the CDC, a number of individuals (including one of us, S.D.A.), began developing what later was to become the Presumpto plate procedure for identifying commonly encountered anaerobic bacteria based on reactions obtained using three quadrant plates. Referred to as Presumpto 1, 2, and 3, the quadrant plates contained four differential agar media per plate (Fig. 2). The media are considered conventional, and prepared using Lombard-

TABLE 4 Manufacturers and vendors[a] of anaerobe chambers and GCs

Manufacturer or vendor contact information
Anaerobe chambers
Coy, Anaerobe Chamber
14500 Coy Dr.
Grass Lake, MI 49137
(734) 475-2200
Forma, Anaerobe Chamber
Forma Scientific
P.O. Box 649
Marietta, OH 45750
(800) 848-3080, ext. 2052
Bactron Anaerobe Chamber
Sheldon Manufacturing, Inc.
300 North 26th Ave.
Cornelius, OR 97113
Toucan, Anaerobic Workstations
Toucan Technologies, Inc.
1158 Altadena Dr.
Cincinnati, OH 45230-3817
(800) 506-2266
Gas chromatograph vendors
Agilent Technologies (formerly part of Hewlett Packard)
P.O. Box 10395
Palo Alto, CA 94306
(800) 227-9770
http://www.agilent.com
Gow-Mac Instrument Co.
277 Brodhead Rd.
Bethlehem, PA 18017
(610) 954-9000
http://www.gow-mac.com
Shimadzu Scientific Instruments, Inc.
7102 Riverwood Dr.
Columbia, MD 21046
(800) 477-1227
http://www.ssi.shimadzu.com
SRI Instruments
20720 Earl St.
Torrance, CA 90503-2162
(310) 214-5092
http://www.srigc.com
Varian Instruments
2700 Mitchell Dr.
Walnut Creek, CA 94598
(800) 926-3000
http://www.varianinc.com

[a] Because of the numerous sizes and configurations of these kinds of equipment, it is suggested that the interested reader contact the manufacturer or vendor for sizes and models and other descriptive information and prices.

Dowell agar as the basal medium (35). Following inoculation, the Presumpto plates are incubated anaerobically for 24 to 48 h. Thus, compared with the newer enzymatic systems that are incubated only 4 h (or less), the Presumpto method is not a rapid procedure. Characteristics of anaerobic bacteria that can be obtained using the Presumpto quadrant plates are listed in Table 7. Detailed instructions pertaining to the Presumpto quadrant plates, along with identification tables based on an extensive database developed at CDC, are available elsewhere (77, 143).

In addition to the 21 test parameters available from using the Presumpto quadrant plates, other characteristics of an anaerobe isolate can be obtained by including a tube of enriched thioglycolate medium (BBL-0135C) and CDC Anaerobe Blood agar (Table 7). The latter is used for determining relationships to oxygen, colony characteristics, and reactions obtained with various disk tests (77). The disk tests include inhibition by penicillin (2-U disk), rifampin (15-μg disk), kanamycin (1,000-μg disk) and SPS disk test. Inhibition by SPS is a key characteristic of *P. anaerobius* that can be used to differentiate this species from many other species of anaerobic cocci that are not inhibited by SPS. A disk test format can also be used to test for reduction of nitrate (77).

When commonly encountered anaerobes were tested, a high degree of agreement between the Presumpto plate procedure and the CDC reference method was found (143). Whaley and colleagues commented further that the "Presumpto plate [procedure] is as accurate as commercially available enzyme systems for the identification of many anaerobic species but is less expensive to perform" (143). The Presumpto quadrant plates are currently available from Smith River Biologicals and from Remel (Lenexa, Kans.) (Table 6). The Smith River Biologicals list price is $1.20 per quadrant plate. The Remel list price is $9.32 per quadrant plate. We agree with the statement of Whaley and colleagues that the time required for an experienced microbiologist to inoculate and prepare three quadrant plates for incubation is less than 2 min (143).

Alternative Procedures for Characterization of Anaerobes

As practical alternatives to the conventional VPI and CDC reference laboratory methods described above, and in addition to the Presumpto quadrant plate procedure, a variety of other methods and schema for characterizing and identifying anaerobes have been described (Table 8) (16, 26, 37). Compared to the more traditional approaches for identifying anaerobes, the motivation for developing alternative procedures, at least in part, has often been to decrease materials costs, save time, simplify the work involved, or find a more accurate and reproducible way to differentiate between anaerobe isolates. Among the approaches taken have been to test for certain key characteristics of isolates on differential agar media (e.g., bacteroides bile esculin agar; Anaerobe Systems, San Jose, Calif.) or to use relatively small volumes of media in small containers that can be easily and rapidly manipulated at the laboratory bench. Other convenient test formats have been to use disks to test for growth stimulation or growth inhibition and to use disks or tablets for determining certain enzyme-substrate reactions. Supplies and other materials for performing several practical and simple tests are described in the current edition of the *Wadsworth Anaerobic Bacteriology Manual* (132) and are commercially available (Table 9).

TABLE 5 Selection of conventional PRAS biochemical tests to differentiate anaerobe isolates[a]

Gram stain[b]	Spore-forming rods	GPR[c]	GPR	GPR	GNR	GNR	GNR	GPC	GPC	GPC GNC or GPR	GPC or GNC	GPC or pitting GNR
GLC[d]	Varied	aibBivlCab	ALS	Varied	Varied	aivS	aB	AL	a	aP[e]	aIC[f], aC[g]	aB[h], a[i]
Genus	Clostridium	C. difficile and C. perfringens	Actinomyces	Varied	Varied	B. fragilis group	Fusobacterium	Streptococcus		Veillonella and Propionibacterium	Peptostreptococcus	Peptostreptococcus or B. ureolyticus
Biochemicals:												
PY base[j]	+	+	+	+	+	+	+	+				+
PY amygdalin	+		+	+	+			+	+			
PY arabinose	+		+	+	+	+		+				
PY cellobiose	+		+	+	+	+	+	+	+			
PY erythritol	+		+	+				+				
PY glucose	+	+	+	+	+	+	+	+	+			+
PY lactose	+	+	+	+	+	+	+	+	+			
PY maltose	+	+	+	+	+	+	+	+	+			
PY mannitol	+	+	+	+				+				
PY mannose	+	+	+	+	+		+	+				
PY melezitose			+			+						
PY melibiose			+		+	+						
PY raffinose			+	+								
PY rhamnose			+	+	+	+						
PY ribose			+	+	+	+						
PY salicin	+	+	+	+	+	+						
PY starch			+									
PY sucrose	+	+	+	+	+	+	+	+	+			
PY trehalose	+	+	+	+	+	+	+	+				
PY xylose	+		+	+	+	+	+	+				
PY xylan												
PY glycerol	+		+	+								
PY galactose	+		+									
PY lactate							+					
PY threonine							+					

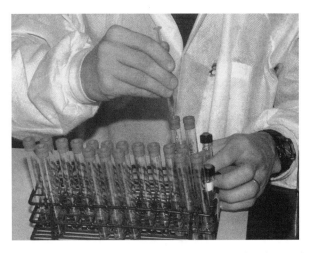

FIGURE 1 Inoculation of conventional PRAS biochemical test media.

Many anaerobic bacteria produce constitutive enzymes which rapidly hydrolyze various chromogenic substrates. WEE-TABS (Key Scientific Products, Round Rock, TX 78681) are stand-alone chromogenic tests for several enzyme-substrate reactions which can be used as alternative and supplemental tests for differentiation of certain anaerobes (Table 10; Fig. 3). Hudspeth and colleagues found WEE-TABS to be a rapid and more accurate alternative to the API ZYM and RapID ANA II systems (described later) for identification of *Porphyromonas* species isolated from infected dog and cat bite wounds in humans (64). WEE-TABS are available in single-test and dual-test formats with nitrophenol- and naphthylamide-bound substrates. According to the manufacturer's instructions, inocula are prepared by harvesting a sufficient number of colonies into 1 to 2 ml of distilled water to make a suspension equal to a no. 5 McFarland standard. At least 5 or 6 drops of this suspension is placed into each tube containing a WEE-TAB tablet, and the tube is mixed vigorously or vortexed to disintegrate the

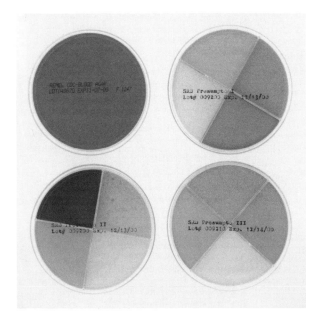

FIGURE 2 Presumpto quadrant plates 1, 2, and 3. One plate of CDC Anaerobe Blood Agar is also shown. See text and Table 8 for details.

Esculin
Indole-nitrate
Gelatin
PYG agar slant

PYG bile
Urease
Milk

a Table based on an algorithm developed at Indiana University by Siders and Allen, as modified from Holdeman et al., by Siders (59, 122).

b GPR, gram-positive rods; GNR, gram-negative rods; GPC, gram-positive cocci; GNC, gram-negative cocci.

c GPR having the characteristic colonial morphology of *C. difficile* or *C. perfringens*.

d Acid products of metabolism: A, acetic acid; P, propionic acid; IB, isobutyric acid; B, butyric acid; IV, isovaleric acid; V, valeric acid; IC, isocaproic acid; C, caproic acid.

e Identifies *Veilonella* species and/or *Propionibacterium acnes* (any two positive reactions). A GPR series of biochemical tests is set up after ruling out *P. acnes*.

f Identifies *P. anaerobius*.

g Identifies *Megasphera elsdeni*.

h Identifies *Peptostreptococcus asaccharolyticus*, *Peptostreptococcus prevotii*, and *Peptostreptococcus tetradius*.

i Identifies *B. ureolyticus* versus *Campylobacter gracilis* that have a typical pitting colonial morphology.

j PY, peptone-yeast.

57

TABLE 6 Commercially available products for anaerobe identification and list prices

Test	Manufacturer contact information	Product	Package	Catalog list price ($)
Conventional	Anaerobe Systems 15906 Concord Cir. Morgan Hill, CA 95037	PRAS media	Each tube	1.30
	Randolph Biomedical 21 McElroy St. West Warwick, RI 02893 (401) 826-1407	PRAS media	Each tube	1.50
	Remel 12076 Santa Fe Dr. Lenexa, KS 66215	PRAS media	Each tube	5.00
	Smith River Biologicals 9388 Charity Highway Ferrum, VA 24088	CDC-based CHO media	Each tube	0.90
Agar based	Remel	CDC-based QUAD plate	Each plate	9.32
	Smith River Biologicals 9388 Charity Highway Ferrum, Va 24088	CDC-based QUAD plate	Each tube	1.20

TABLE 7 Media and characteristics of anaerobe isolates that can be determined using the Presumpto quadrant plate procedure for anaerobe identification[d]

Medium	Characteristics
Blood agar	Relation to oxygen,[a] colony characteristics, hemolysis, pigment, fluorescence with UV light (Wood's lamp), pitting of agar, cellular morphology, Gram reaction, spores, motility (wet mount); inhibition by penicillin, rifampin, and kanamycin (disk tests)
Enriched thioglycolate	Appearance of growth, rapidity of growth, gas production, odor, cellular morphology
Presumpto 1 plate	
LD[c] agar	Growth on LD medium, production of indole, indole derivative, catalase[b]
LD esculin agar	Esculin hydrolysis, hydrogen sulfide, catalase[b]
LD egg yolk agar	Lipase, lecithinase, proteolysis on egg yolk agar
LD bile agar	Growth in presence of 20% bile (2% oxgall), insoluble precipitate under and immediately surrounding growth
Presumpto 2 plate	
LD glucose agar	Glucose fermentation; stimulation of growth by fermentable carbohydrate
LD starch agar	Starch hydrolysis
LD milk agar	Proteolysis of milk
LD DNA agar	Deoxyribonuclease activity
Presumpto 3 plate	
LD gelatin	Gelatin hydrolysis
LD mannitol agar	Mannitol fermentation
LD lactose agar	Lactose fermentation
LD rhamnose agar	Rhamnose fermentation

[a]By comparing growth on anaerobe plate with that on blood agar (or chocolate agar) incubated in 5 to 10% CO_2 incubator (or candle jar) or in room air.

[b]The catalase test can be performed by adding 3% hydrogen peroxide to the growth on Lombard-Dowell (LD) agar, but the reactions of catalase-positive cultures are more vigorous on LD esculin agar.

[c]LD, Lombard-Dowell.

[d]Modified from reference 77.

TABLE 8 Practical and simple manual methods for characterization and presumptive identification of anaerobes[a]

Test or procedure	Rationale, comments, commercial sources
Observations of colony characteristics on anaerobe blood agar and anaerobe selective media; Gram stain reactions and microscopic morphology.	Morphologic observations are necessary for identification of all anaerobes. Commercially prepared formulations of anaerobe blood agar are available from several manufacturers, including Anaerobe Systems, BD Biosciences Microbiology Products, PML Microbiologicals, Remel, and Springs River Biologicals.
Aerotolerance tests on anaerobe blood agar.	Necessary to define relationships to O_2. See above for sources of prepared media.
Egg yolk agar reactions	Lipase positive species include *Clostridium sporogenes, Clostridium botulinum, Clostridium novyi* type A, *Prevotella intermedia*, and *Fusobacterium necrophorum*. Lecithinase positive species include *C. perfringens, Clostridium sordellii, Clostridium bifermentans, Clostridium limosum, Clostridium subterminale*, and *Clostridium baratii*. Commercially prepared egg yolk agar is available from a number of manufacturers, including Anaerobe Systems, BD Biosciences Microbiology Products, PML Microbiologicals, Remel, and Springs River Biologicals.
Catalase test on hemin-supplemented medium (e.g., 3% H_2O_2 on LD[b] esculin agar, or slide catalase using 15% H_2O_2 with added Tween 80)	A positive catalase test is characteristic of some species of *Bacteroides, Bilophila wadsworthia*, some *Peptostreptococcus prevotii* strains, *Staphylococcus saccharolyticus, Veillonella parvula, Propionibacterium* spp. (*P. propionicus* is negative and an exception), and *Actinomyces viscosus*.
Spot indole test using *p*-dimethylaminocinamaldehyde	Useful for all groups of anaerobes. Reagent available from several sources, including Anaerobe Systems, BD Biosciences Microbiology Products, and Remel.
Enhanced growth in presence of 20% bile in medium or around disks containing bile	Key characteristic of the *B. fragilis* group, *Bilophila* spp., *Fusobacterium varium, Fusobacterium ulcerans, Leptotrichia buccalis*, and *Mitsuokella multiacida*. Available commercially in quadrant plates from Remel and Smith River Biologicals.
Antibiotic differentiation disks (e.g., kanamycin, 1 mg; rifampin, 15 μg; penicillin, 2 U; colistin, 10 μg; vancomycin, 5 μg.) (used to detect growth in presence of drug or growth inhibition).	Used by Sutter and colleagues to aid in distinguishing between gram-positive and gram-negative rods; also aids in presumptive differentiation between selected *Bacteroides, Prevotella, Porphyromonas*, and *Fusobacterium* spp.
SPS disk test	SPS inhibits *P. anaerobius*; other anaerobic cocci are resistant to SPS. Alternatively, *P. anaerobius* is the only anaerobic coccus that produces a major isocaproic acid peak on GLC analysis. SPS disks are available from Anaerobe Systems and BD Biosciences Microbiology Products.
Disk test for nitrate reduction	Most if not all *Bacteroides ureolyticus, Veillonella* spp. and *Eubacterium lentum* strains are positive; test helpful for other cocci, non-spore-forming rods, and clostridia as well. Available from Anaerobe Systems and Becton Dickinson.
Urease test (various procedures can be used)	A positive urease reaction is a key characteristic of *B. ureolyticus, Bilophila wadsworthia, C. sordellii, Peptostreptococcus tetradius, S. saccharolyticus* and certain *Actinomyces* spp. Available from Anaerobe Systems and BD Biosciences Microbiology Products.
Rapid glutamic acid decarboxylase tube test	Positive reactions given with most strains of the *B. fragilis* group, many strains of *Fusobacterium* spp., *P. micros* (53% of strains), most *C. perfringens* and *C. sordellii* strains, some *C. difficile* strains, and all *E. limosum* strains tested. Available from Remel.
Esculin hydrolysis (on LD esculin agar or using a spot test)	Useful for all groups of anaerobes. Available in quadrant plates from Remel and Smith River Biologicals.
Fluorescence under long-wavelength (366-mm) UV light	Useful for the pigmented *Prevotella-Porphyromonas* group; *P. gingivalis* does not fluoresce, whereas *Porphyromonas asaccharolytica* is red or yellow; *Prevotella intermedia* is bright red, whereas *P. melaninogenica* is red-orange. Also useful for *Veillonella* spp. (red). Observations made on anaerobe blood agar.
Lack of constitutive β-glucosidase (esculinase)	*Fusobacterium* species lack the enzyme, but it is produced by *Bacteroides* spp. and *Prevotella* spp.

[a]Modified from reference 7.
[b]LD, Lombard-Dowell.

tablet and mix the suspension. The tubes are incubated aerobically at 37°C for 2 h. For the glycosidase tests, the absence of change at 2 h is a negative result. The development of a yellow color at any time during the 2 h is considered a positive glycosidase test. After reading the glycosidase results (dual-test format), or after incubating a single naphthylamide test, 2 drops of *p*-dimethylaminocinnamaldehyde reagent are added. The tube is then incubated for 15 min for color development. Positive tests will be red, while negative tests will be yellow or a light peach color. A

TABLE 9 Source of supplies for disk tests used in presumptive identification of anaerobes

Disk test	Manufacturer no.	Package contents	List price ($)
SPS	AS-702[a]	50 tests	20.20
Nitrate	AS-703	50 tests	27.40
Colistin	AS-705	50 tests	10.85
Kanamycin	AS-706	50 tests	10.85
Vancomycin	AS-707	50 tests	10.85

[a]AS, Anaerobe Systems, Morgan Hill, Calif.

triple-test format is also available with nitrophenol- methylumbelliferyl (MU)-, and naphthylamide-bound substrates in each tube. Hydrolysis of MU-bound substrates produces a bright, blue-green fluorescence that is detected using a long-wavelength UV light (366 nm). The manufacturer's list price for WEE-TABS is $13.00 for 28 tests. When only a small battery of enzyme tests is needed for characterizing an anaerobe isolate (e.g., six or fewer tests), WEE-TABS probably are more cost-effective than the prepackaged commercial kits for detecting various preformed enzymes (in terms of time and materials costs). Diagnostic tablets for bacterial enzyme testing that are sold in Europe, but not in the United States, are manufactured by Rosco Diagnostica (Taastrup, Denmark).

The API 20A and Minitek Commercial Packaged Microsystems

For nearly 3 decades, the API 20A (bioMérieux-Vitek) (Fig. 4) and the Minitek (Becton Dickinson Microbiology Systems, Cockeysville, Md.) (Fig. 5) micromethod kits have been marketed for the identification of anaerobes isolated from clinical materials. These miniaturized systems, while smaller than the conventional biochemical identification systems discussed previously, require turbid inocula and depend upon growth of anaerobes during incubation in an anaerobic system. The API 20A kit offers a fixed set of biochemical substrates. The Minitek kit offers a larger variety of substrates that could be selected by the user; however, most of those that might be relevant to anaerobe identification are carbohydrates. Laboratory personnel working with either of these micromethod systems should use only the manufacturer's identification tables, numerical profiles, or code books for interpreting test results obtained with these kits.

The construction of the API 20A kit along with the composition of substrates and reagents continues to be the same today as it was previously, when the manufacturer was Analytab Products, Inc. (personal communication, David Pincus, bioMérieux-Vitek). Briefly, the API 20A consists of a plastic strip with 20 microtubules of dehydrated substrates to test for the following: indole; catalase; and hydrolysis of urea, gelatin, and esculin, plus the fermentation of glucose and 15 other carbohydrates. The microcupules of the strip are inoculated with a turbid suspension (i.e., ≥ no. 3 McFarland standard) of fresh colonies prepared in Lombard-Dowell broth and incubated anaerobically for 24 to 48 h. Databases exist for reading the strips at 24 or 48 h. The indicator for the fermentation tests is bromcresol purple. An acid reaction is yellow (pH 5.2) or yellow-green; a negative reaction is purple. Various anaerobes (e.g., a number of *Clostridium* species) reduce the indicator to colorless, shades of green-brown, or other colors, sometimes making reactions difficult to interpret. According to the manufacturer's instructions, "other tests such as colonial and microscopic morphology, Gram stain, etc., should be performed and the results used to confirm or complete the identification" (bioMérieux-Vitek API 20A manual). Additional recommendations for use of the API 20A are supplied by bioMérieux along with identification tables and a numerical Analytical Profile Index.

The Minitek Anaerobe II Set provides paper disks impregnated with various biochemical substrates that the

TABLE 10 Tests available in single-tablet format for assessing preformed enzymatic activity of anaerobes[a]

Test	Substrate (abbreviation)	Enzyme or comment
Glycosidase test (WEE-TABS)[b]	*p*-Nitrophenol-α-D-glucopyranoside (α-GLU)	α-Glucosidase
	p-Nitrophenol-β-D-glucopyranoside (β-GLU)	β-Glucosidase
	o-Nitrophenol-α-D-glucopyranoside (α-GAL)	α-Galactosidase[d]
	o-Nitrophenol-β-D-galactopyranoside (β-GAL or ONPG)	β-Galactosidase[d]
	p-Nitrophenol-β-*N*-acetyl-D-glucosaminide (β-NAG)	β-*N*-Acetylglucosaminidase[d]
	p-Nitrophenol-α-L-fucopyranoside (α-FUC)	α-Fucosidase[d]
	p-Nitrophenol phosphate (PNP[e])	Alkaline phosphatase
Naphythylamide-based aminopeptidase test tablets (ADD-A-TEST TABLETS)[c]	Na-benzoyl-DL-arginine β-naphthylamine (BANA or TRY)	Trypsin-like activity[d]
	N-glutaryl-gly-gly-phe-β-naphthylamide (CHY)	Chymotrypsin-like activity[d]
	L-Phenylalanine α-naphthylamide (PAL)	
	L-Arginine α-naphthylamide (ARG)	
	L-Serine α-naphthylamide (SER)	
	L-Pyroglutamic acid α-naphthylamide (PYR)	
	L-Proline α-naphthylamide (PRO)	
	Leucyl-glycine α-naphthylamide (GLY)	

[a]In the United States, WEE-TABS and ADD-A-TEST are available commercially from Key Scientific Products. List price is $13.00 for 28 tests. In Europe, tablets containing many of the substrates listed above are available from Rosco Diagnostica.

[b]Enzymes are glycosidases (listed in far right column). When bound to nitrophenol, hydrolysis of a colorless aryl-substituted glycoside substrate or phosphoester releases the nitrophenol base. The absence of color change at 2 h is a negative result; yellow color at any time during the 2 h of aerobic incubation is a positive glycosidase test.

[c]Naphthylamide-based tests. Enzyme hydrolysis of the arylamide in the tablet releases free α-naphthylamine, which is detected and shown by the color change after adding PEP reagent (i.e., *p*-dimethyl-amino-cinnamaldehyde in HCl). After reading the glycosidase test, 2 drops of PEP is added, followed by 15 min of incubation; positive tests are red, while negative tests are yellow.

[d]Test aids in differentiation of *Porphyromonas* spp.

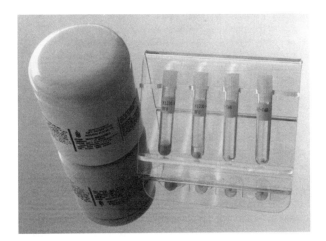

FIGURE 3 WEE-TABS stand-alone chromogenic tests. Each tube contains one tablet and a small volume of bacterial suspension.

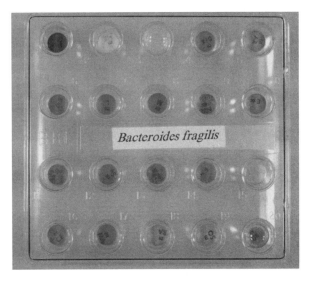

FIGURE 5 Minitek Anaerobe II tray containing paper disks impregnated with biochemical substrates. After inoculation with a bacterial suspension and 48 h of anaerobic incubation, positive reactions are indicated by color changes (see text).

user dispenses into wells of disposable plastic plates. Thus, biochemical characteristics of anaerobes that can be determined using this system include indole production, esculin hydrolysis, nitrate reduction, and the fermentation of glucose plus several other carbohydrates. As is true for the API 20A system, the bacterial inoculum for the Minitek system is prepared from fresh colonies on anaerobe blood agar to achieve a dense suspension (i.e., ≥ no. 5 McFarland standard) in Lombard-Dowell broth. The Minitek plates must be incubated anaerobically for 48 h. The phenol red indicator for detection of fermentation reactions turns yellow at pH 6.8 and lower (positive reactions), while orange or orange-red is interpreted as negative. Interpretation of color reactions can be difficult with this system also, owing to reduction of the indicator. In both the Minitek and the API 20A systems, exposure of the esculin well to long-wave UV light is recommended for determination of esculin hydrolysis. The manufacturer's instructions should be read for further details. A numerical identification code book is available for the Minitek system.

An excellent review of the older literature describing the development and evaluations of these packaged micromethod kits was published by Stargel and colleagues (130). A number of other reports have discussed the performance, advantages, and limitations of these systems (19, 54–56, 63, 65, 115, 129, 133). In terms of performance, two of the largest and most recent reports (published in the 1980s) indicate that the API 20A system correctly identified only 50 to 56% of approximately 800 anaerobe isolates to the species level (115, 133). The Minitek Anaerobe II system identified only 48% of 330 anaerobes to the species level. Although these systems can accurately identify some commonly encountered species such as *B. fragilis* and *C.*

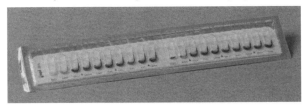

FIGURE 4 API 20A test strip. Miniaturized cupules are inoculated with a bacterial suspension and after 24 to 48 h of incubation, biochemical reactions are visualized.

perfringens, many of the anaerobes that are less reactive or more fastidious (e.g., many species of *Prevotella, Porphyromonas, Fusobacterium,* other clostridia, anaerobic non-spore-forming rods, and most anaerobic cocci) require additional tests and GLC to achieve correct identification. The need for supplemental tests decreases the usefulness of these systems and increases the time for identifications by another day or two. Thus, although both the API 20A and the Minitek systems can be set up relatively rapidly, they do not necessarily enable the user to identify anaerobes any faster than can be done using the conventional tests described previously. The lack of sufficient tests in these kits prompted Dowell, Lombard, and others of the CDC to develop the Presumpto quadrant plate system described previously (33). An additional issue is the extent to which the databases for these systems have not kept up with the current nomenclature. In this regard, Summanen and Jousimes-Somer noted in 1988 that the API 20A database was far behind (133), and, to our knowledge, this is still the case today. The Minitek database for anaerobes has not been addressed, at least in the refereed literature, in recent years. Current list prices of these products are listed in Table 11.

Commercial Packaged Microsystems for Characterization and Identification of Anaerobes after 4 h of Aerobic Incubation

Commercially available microsystems in kit form that have been used for characterization and identification of anaerobes following aerobic incubation for 4 h include the following:

1. API ZYM (bioMérieux, Inc.)
2. An-IDENT (bioMérieux, Inc.)
3. RapID-ANA II (Remel)
4. Vitek ANI (bioMérieux, Inc.)
5. MicroScan Rapid Anaerobe Panel (Dade Behring)
6. Rapid ID 32A (bioMérieux, Inc.)
7. BBL Crystal Anaerobe Identification System (Becton Dickinson Microbiology Systems)

These growth-independent systems are designed to test for various miniaturized conventional reactions and

TABLE 11 Miniaturized and rapid commercial packaged kit systems for identification of anaerobes

Packaged kit system	Manufacturer no.	Package	List price ($)
Miniaturized			
Minitek Anaerobe II set	225146	50 tests	447.75
API 20A	20300	25 tests	168.00
Rapid (≤4 h)			
Crystal ANR ID	4345010	20 determinations	150.00
RapID ANA II	83-11002	20 panels	103.50
MicroScan Panels; Rapid anaerobe ID	B1017-2	20 panels	165.50
API An-IDENT	60900	25 tests	160.00
API ZYM	25200	25 tests	151.00
Vitek ANI	V1309	20 tests	117.00

preformed enzymes (e.g., aminopeptidases and glycosidases). All of them require that inocula be prepared using dense cell suspensions from the surfaces of purity plate cultures. The short incubation time of 4 h or less for each of the systems is in sharp contrast to the longer incubation times required for the growth-dependent identification systems discussed previously. Except for the API ZYM system, which was designed for research purposes, all of the other packaged systems listed above provide numerical codes or computerized databases that aid in identification of isolates characterized using these systems. (The manufacturer does not provide identification tables or a database for the API ZYM system.)

API ZYM

The API ZYM system is a semiquantitative micromethod that uses chromogenic substrates to detect 19 different enzymatic reactions (Table 12; Fig. 6). The manufacturer recommends that the inoculum suspension have a turbidity between that of a McFarland no. 5 and a no. 6 standard, which can be difficult to achieve for fastidious isolates. Enzymatic cleavage of the substrates produces color changes that can be compared against a color chart and graded in intensity. The enzyme reactions available in this system have been reported to aid in the differentiation of certain *Bacteroides, Fusobacterium, Prevotella, Porphyromonas, Veillonella, Peptostreptococcus, Streptococcus, Actinomyces, Bifidobacterium, Eubacterium, Propionibacterium,* and *Clostridium* species (21, 36, 44, 56, 58, 75, 81, 82, 90, 123, 135, 136). In our experience, it was possible to differentiate between 41 of 46 (89%) strains of anaerobic gram-negative bacilli, 19 of 32 (59%) strains of anaerobic cocci, all 23 strains of anaerobic gram-positive non-spore-forming bacilli, and all of 50 strains of the *Clostridium* species we tested using only the API ZYM system plus a few supplemental tests (e.g., Gram reaction, microscopic morphology, colony characteristics, and relation to oxygen) (90). The manufacturer's list price for the API ZYM system in the United States is $151 per box of 25 strips, or $6.04 per strip.

Rapid Commercial Micromethod Systems with Numerical Codes or Computerized Databases for Identification Provided by the Manufacturer

The direct costs (manufacturers' list prices) of several packaged micromethod kits used in identification of anaerobes are summarized in Table 11. The time involved in prepara-

TABLE 12 Test substrates available in the API ZYM system for detecting preformed enzymatic activities

Microcupule no.	Substrate	Enzyme
1		Control
2	2-Naphthyl phosphate	Alkaline phosphatase
3	2-Naphthyl butyrate	Esterase (C-4)
4	2-Naphthyl caprylate	Esterase lipase (C-8)
5	2-Naphthyl myristate	Lipase (C-14)
6	L-Leucyl-2-naphthylamide	Leucine arylamidase
7	L-Valyl-2-naphthylamide	Valine arylamidase
8	L-Cystyl-2-naphthylamide	Cystine arylamidase
9	N-Benzoyl-DL-arginine-2-naphthylamide	Trypsin
10	N-Glutaryl-phenylalanine-2-naphthylamide	Chymotrypsin
11	2-Naphthyl phosphate	Acid phosphatase
12	Naphthol-AS-Bl-phosphate	Phosphoamidase
13	6-Bromo-2-naphthyl-α-D-galactopyranoside	α-Galactosidase
14	2-Naphthyl-β-D-galactopyranoside	β-Galactosidase
15	Naphthyl-AS-Bl-β-glucuronide	β-Glucosidase
16	2-Naphthyl-α-D-glucopyranoside	α-Glucosidase
17	6-Bromo-2-naphthyl-β-D-glucopyranoside	β-Glucosidase
18	1-Naphthyl-N-acetyl-β-D-glucosaminide	N-Acetyl-β-glucosaminidase
19	6-Bromo-2-naphthyl-α-D-mannopyranoside	α-Mannosidase
20	2-Naphthyl-α-L-fucopyranoside	α-Fucosidase

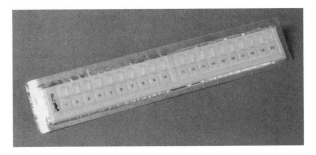

FIGURE 6 API ZYM. Miniaturized cupules are inoculated with a turbid bacterial suspension. Nineteen enzymatic reactions are read and interpreted after a 4-h aerobic incubation.

tion of inocula, inoculation of test strips, and reading and interpretation of results is not significantly different for most of these miniaturized systems. With experience, most require 2 to 4 min to complete these steps. These commercial products are being used widely in diagnostic microbiology laboratories. In different evaluation studies, identification accuracies of the various packaged kits have varied widely, depending upon many variables (97). Some of the more important variables include the numbers and sources of strains, the species included in the study, and the identification system(s) selected for comparison purposes (e.g., the gold standard). A number of the products have not been evaluated in several years since the initial studies were published. Thus, a number of their databases, tables, and schema for identification need updating with respect to current nomenclature and taxonomic classification. Composite data on the performance of four different packaged systems for identification of commonly encountered and/or medically significant anaerobes to genus and species levels is given in Table 13. The identifications were based on data generated by the kit and do not include data based on testing with additional biochemical tests. In general, these products provide accurate identification of the most common indole-negative species of the *B. fragilis* group. The RapID ANA II system has been successful in identification of selected difficult-to-identify species within the pigmented *Prevotella-Porphyromonas* group (139).

An-IDENT

The An-IDENT (Fig. 7) is one of the first commercially available 4-h packaged kits to be marketed with a database for identification (i.e., the An-IDENT Analytical Profile Index). Each An-IDENT test strip has 20 microcupules containing dehydrated substrates to test for 21 different miniaturized biochemical reactions (Table 14). To test for indole production, either Kovac's reagent or acidic *p*-dimethylamino-cinnamaldehyde (i.e., DMAC Indole reagent) can be added back to the first microcupule in the strip. The next nine microcupules in a test strip contain chromognic substrates whose reactions are detected by the liberation of *p*-nitrophenol, *o*-nitrophenol, or indoxyl from the substrates (no. 2 to 10 in Table 14); these are read without the need for the microbiologist to add back indicator. Nine other microcupules (no. 12 to 20 in Table 14) require the addition of cinnamaldehyde reagent to detect free β-naphthylamine. The arginine dihydrolase microcupule contains phenol red as the indicator. To test for catalase, 3% hydrogen peroxide is added back to the third microcupule; this microcupule is also used to test for α-glucosidase. The manufacturer's package insert provides descriptions of color

TABLE 13 Data compiled from published commercial kit evaluations on the identification of commonly encountered anaerobes to genus and species level, correctly without the use of additional tests

Organisms	% Correctly identified to genus and species (no. of isolates tested)			
	An-IDENT[a]	RapID-ANA II[b]	Vitek ANI[c]	MicroScan[d]
B. fragilis group	94 (87)	66 (221)	83 (99)	63 (54)
Pigmented *Prevotella* spp.	ND[f]	84 (32)	57 (14)	71 (17)
Porphyromonas spp.	ND	86 (7)	60 (5)	63 (8)
Other *Prevotella* spp.	83 (24)	85 (60)	61 (18)	84 (31)
B. ureolyticus	ND	100 (14)	100 (2)	100 (4)
Fusobacterium spp.	66 (3)	26 (27)	27 (11)	73 (15)
F. magna-M. micros[e]	ND	98 (66)	74 (27)	ND
P. anaerobius	ND	71 (31)	89 (9)	ND
Other cocci	80 (35)	73 (139)	65 (74)	82 (22)
Propionibacterium spp.	93 (14)	94 (49)	89 (18)	ND
Other gram-positive anaerobic rods	75 (8)	67 (101)	45 (20)	41 (17)
C. perfringens	98 (53)	100 (26)	100 (8)	ND
C. ramosum	ND	100 (25)	0 (1)	ND
C. septicum	ND	100 (5)	50 (2)	ND
C. difficile	100 (20)	55 (11)	64 (11)	ND
Other *Clostridium* spp.	89 (19)	61 (97)	71 (17)	71 (69)

[a]An-INDENT data compiled from reference 113.
[b]RapID ANA II composite data compiled from references 25a and 92.
[c]Vitek ANI data compiled from reference 119.
[d]MicroScan data compiled from reference 131.
[e]*F. magna* and *M. micros* were formerly classified as *P. magnus* and *P. micros* (109)
[f]ND, not determined.

FIGURE 7 An-IDENT. This was one of the first commercially available 4-h packaged kits with a database for the identification of anaerobes.

reactions. After the reactions are recorded, the microorganism can be identified with the aid of differential charts or the Analytical Profile Index provided by the manufacturer.

RapID ANA II

Originally developed by Innovative Diagnostic Systems, Inc. (IDS) (Atlanta, Ga.), the RapID ANA system was upgraded and improved as the RapID ANA II system in 1989 (Fig. 8). Included in the second-generation system was a revised test selection to address newer taxa, especially in problem areas; a redesigned panel sequence; and an updated and expanded "code compendium." Subsequently, IDS was purchased by Remel and Remel now manufactures and distributes the RapID ANA II system. We are not aware of any substantial differences in the product or its performance since this change in manufacturers occurred. The RapID ANA II system tests for 18 preformed enzymes using 10 wells. The dehydrated substrates included in these wells are listed in Table 14.

Preparation of inocula: important considerations. A turbid inoculum is prepared from a pure culture suspension of a fresh isolate (e.g., growing on anaerobe blood agar), to be equivalent to at least a McFarland no. 3 turbidity standard. An important variable addressed (in part) in the package insert is the culture medium used for preparation of inocula. Thus, Remel recommends that inocula be prepared from any of the following nonselective agar media supplemented with 5 to 7% sheep blood: brucella, Columbia, brain heart infusion, Lombard-Dowell, Trypticase soy base, or CDC Anaerobic Blood Agar. Also recommended are any of the following selective or differential agar media: phenylethylalcohol, paromomycin-vancomycin, and kanamycin-vancomycin media and egg yolk agar.

The manufacturer recommends that a number of other media not be used. Schaedler blood agar and other media supplemented with disaccharides or monosaccharides should not be used because they may suppress glycolytic activity and reduce test reactivity. Bacteroides bile esculin agar and kanamycin-bile esculin agars are not recommended because a ferric-esculitin complex can form that interferes with test interpretation. In addition, so-called reducible agars containing reducing agents or palladium chloride should not be used. Reducing agents can interfere with certain enzyme reactions.

A problem that needs to be addressed is the preparation of inocula from media that are not enriched with vitamin K, L-cystine, or yeast extract (all found in CDC Anaerobe Blood agar). For example, trypticase soy blood agar (TSA) may not be adequately enriched to meet the nutritional requirements of certain pigmented *Prevotella-Porphyromonas* species or of certain other anaerobe groups or species. The growth of these organisms may be very poor on TSA, which could lead to incorrect test results or failure to identify certain isolates. In general, at this time, there is little published information on the effect of growth medium used to prepare the inoculum on enzymatic test results obtained using commercially packaged rapid test kits.

In a comparison of the RapID ANA II system against VPI conventional methods in our laboratory, 68% of 566 clinical anaerobe isolates (overall) were identified correctly to the species level by the RapID ANA II system without the use of additional tests (92). Those correctly identified to the species level included 62% of 204 anaerobic gram-negative rods, 72% of 163 anaerobic cocci, 70% of anaerobic non-spore-forming gram-positive rods, and 74% of clostridia. Within the *B. fragilis* group, the kit correctly identified the majority of the indole-negative species including 26 of 28 (93%) *B. fragilis* strains, all 19 strains of *Bacteroides distasonis*, and 19 of 27 (70%) *B. vulgatus* strains. In our hands, the indole-positive species of the *B. fragilis* group were problematic. For example, only 1 of 29 (3%) *Bacteroides ovatus* strains and 4 of 14 (29%) *Bacteroides uniformis* strains were correctly identified to the species level by the RapID ANA II system; they were often confused with each other or with *Bacteroides thetaiotaomicron*. In addition, the database is significantly out of date with respect to nomenclature of the pigmented *Prevotella* and *Porphyromonas* species. Without additional tests, the kit could identify only 1 of 11 (9%) strains of *F. nucleatum* and 1 of 4 strains of *F. necrophorum* correctly to the species level. In our experience, *F. nucleatum* is the most common *Fusobacterium* species encountered. About two-thirds of 24 *P. anaerobius* strains were identified correctly using the system; this is one of the more common anaerobic cocci isolated from properly collected clinical materials. Another area of concern is that the RapID ANA II system does not appear to be able to identify *Actinomyces* species with an acceptable degree of accuracy (97). However, it does correctly identify most *F. magna* (formerly *Peptostreptococcus magnus*) (26 of 27 or 96% correctly identified), *Micromonas micros* (formerly *P. micros*) (20 of 20 or 100% correctly identified), *Propionibacterium acnes* (24 of 25 or 96% correctly identified), *C. perfringens* (19 of 19 correctly identified), and *C. ramosum* (20 of 20 correctly identified). In general, the use of a few simple and practical tests, in addition to the RapID ANA II system, can improve the accuracy of the system. For example, growth in or on 20% bile medium and fermentation of a few key carbohydrates can improve the accuracy of identifying *Bacteroides* species. An SPS disk test aids in differentiating *P. anaerobius* from the other anaerobic cocci. Reactions on egg yolk agar aid in identifying several *Clostridium* spp. (92).

Vitek ANI

Intended for use with the Vitek system for automated identification of medically important anaerobic and microaerophilic bacteria of human origin, the Vitek Anaerobe Identification (ANI) card is a small (~9-by-6-cm) white plastic card with 30 wells, 28 of which contain substrates for the biochemical tests listed in Table 14 (Fig. 9). Of these 28 wells, the first 20 contain chromogenic substrates to test for constitutive enzymes; the remaining 8 contain substrates for modified conventional test determinations. The card was originally developed for Vitek by IDS (now owned by Remel). Although some of the tests in the Vitek ANI and the RapID ANA II systems are based on use of the same substrates, others are different (as can be seen in Table 14). The inoculum for each card, a saline suspension of bacteria equivalent to at least a McFarland no. 3 turbidity

TABLE 14 Summary of biochemical test reactions or substrates available in the 4-h commercial packaged microsystems

	Reaction or substrate available in indicated system				
An-IDENT	RapID ANA II	Vitek ANI	MicroScan	rapid ID 32A	BBL Crystal ANR ID[a]
1. Indole production	1. Urea	1. p-Nitrophenyl phosphate	1. p-Nitrophenyl-β-D-galactopyranoside	1. Urease	1. L-Arginine-7-AMC[a]
2. N-Acetylglucosaminidase	2. p-Nitrophenyl-β,D-lactoside	2. p-Nitrophenyl phosphate choline	2. p-Nitrophenyl-α-D-galactopyranoside	2. Arginine dihydrolase	2. L-Histidine-AMC
3. α-Glucosidase	3. p-Nitrophenyl-α,L-arabinoside	3. p-Nitrophenyl-β,D-galactopyranoside	3. Bis-p-nitrophenylphosphate	3. α-Galactosidase	3. 4-Methylumbelliferon (MU)-α-D-mannoside
4. α-Arabinosidase	4. Leucyl-glycyl-β-naphthylamide	4. p-Nitrophenyl-α,D-galactopyranoside	4. p-Nitrophenyl-N-acetyl-β-D-glucosaminide	4. β-Galactosidase	4. L-Serine-AMC
5. β-Glucosidase	5. o-Nitrophenyl-β,D-galactoside	5. p-Nitrophenyl-β,D-glucopyranoside	5. p-Nitrophenyl-α-D-glucopyranoside	5. β-Galactosidase-6-phosphatase	5. L-Isoleucine-AMC
6. α-Fucosidase	6. Glycyl-β-napthylamide	6. p-Nitrophenyl-α,D-glucopyranoside	6. o-Nitrophenyl-β-D-glucopyranoside	6. Glucosidase	6. 4MU-β-D-mannoside
7. Phosphatase	7. p-Nitrophenyl-α,D-glucoside	7. p-Nitrophenyl-β,D-glucuronide	7. p-Nitrophenyl phosphate	7. β-Glucosidase	7. Glycine-AMC
8. α-Galactosidase	8. Prolyl-β-naphthylamide	8. p-Nitrophenyl-β,D-lactoside	8. p-Nitrophenyl-α-L-fucopyranoside	8. α-Arabinosidase	8. L-Alanine-AMC
9. β-Galactosidase	9. p-Nitrophenyl-β,D-glucoside	9. p-Nitrophenyl-α,D-mannopyranoside	9. p-Nitrophenyl-α-D-mannopyranoside	9. β-Glucuronidase	9. 4MU-N-acetyl-β-D-galactosaminide
10. C-2 esterase	10. Phenylalanyl-β-naphthylamide	10. p-Nitrophenyl-α,L-fucopyranoside	10. L-Leucine-β-naphthylamide	10. β-N-Acetyl-glucosaminidase	10. L-Pyroglutamic acid
11. Arginine dihydrolase	11. p-Nitrophenyl-α,D-galactoside	11. p-Nitrophenyl-β,D-fucopyranoside	11. DL-Methionine-β-naphthylamide	11. Mannose fermentation	11. L-Lysine-AMC
12. Leucine aminopeptidase	12. Arginyl-β-naphthylamide	12. p-Nitrophenyl-β,D-xylopyranoside	12. L-Lysine-β-naphthylamide (alkaline)	12. Raffinose fermentation	12. L-Methionine-AMC
13. Proline aminopeptidase	13. p-Nitrophenyl-α,L-fucoside	13. p-Nitrophenyl-α,L-arabinofuranoside	13. L-Lysine-β-naphthylamide (acid)	13. Glutamic acid decarboxylase	13. 4MU-β-D-cellobiopyranoside
14. Pyroglutamic acid arylamidase	14. Seryl-β-naphthylamide	14. p-Nitrophenyl-N-acetylglucosaminide	14. Glycylglycine-β-naphthylamide	14. α-Fucosidase	14. 4MU-β-D-xyloside
15. Tyrosine aminopeptidase	15. p-Nitrophenyl-N-acetyl-β,D-glucosaminide	15. N-Benzoyl-DL-arginine p-nitroanilide	15. Glycine-β-naphthylamide	15. Nitrate	15. L-Phenylalanine-AMC
16. Arginine aminopeptidase	16. Pyrrolidonyl-β-naphthylamide	16. L-Leucine p-nitroanilide	16. L-Proline-β-naphthylamide	16. Indole	16. L-Leucine-AMC
17. Alanine aminopeptidase	17. p-Nitrophenyl phosphate	17. L-Proline p-nitroanilide	17. L-Arginine-β-naphthylamide	17. Alkaline phosphatase	17. Ecosyl
18. Histidine aminopeptidase	18. Tryptophan	18. L-Alanine p-nitroanilide	18. L-Pyrrolidonyl-β-naphthylamide	18. Arginine arylamidase	18. Disaccharide

(Continued on next page)

TABLE 14 (*Continued*)

Reaction or substrate available in indicated system

An-IDENT	RapID ANA II	Vitek ANI	MicroScan	rapid ID 32A	BBL Crystal ANR ID
19. Phenylalanine aminopeptidase		19. L-Lysine p-nitroanilide	19. L-Tryptophan-β-naphthylamide	19. Proline arylamidase	19. Furanose
20. Glycine aminopeptidase		20. γ-Glutamyl p-nitroanilide	20. 3-Indoxyl-phosphate	20. Leucyl glycine arylamidase	20. Pyranose
21. Catalase		21. Triphenyl tetrazolium	21. Trehalose	21. Phenylalanine arylamidase	21. p-Nitrophenyl-α-D-galactoside
		22. Arginine	22. Urea	22. Leucine arylamidase	22. p-Nitrophenyl-β-D-galactoside
		23. Urea	23. Indole	23. Pyroglutamic acid arylamidase	23. p-Nitrophenyl phosphate
		24. Glucose	24. Nitrate	24. Tyrosine arylamidase	24. p-Nitrophenyl-α-D-glucoside
		25. Trehalose		25. Alanine arylamidase	25. p-Nitrophenyl-N-acetyl-glucosaminide
		26. Arabinose		26. Glycine arylamidase	26. L-Proline-p-nitroanilide
		27. Raffinose		27. Histidine arylamidase	27. p-Nitrophenyl-α-L-fucoside
		28. Xylose		28. Glutamyl glutamic acid arylamidase	28. p-Nitrophenyl-β-D-glucoside
				29. Serine arylamidase	29. L-Alanyl-L-alanine-p-nitroanilide

^aAMC, aminomethylcoumarin.

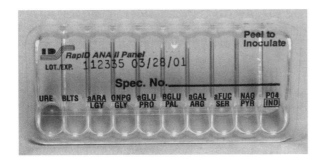

FIGURE 8 RapID ANA II. This 4-h rapid test system for the identification of anaerobes contains 10 wells for testing the performance of 18 preformed enzymatic tests.

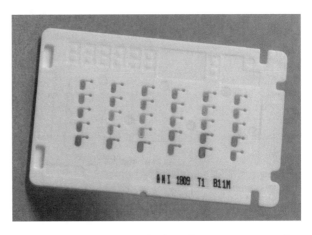

FIGURE 9 Vitek ANI card. This 4-h test system is incubated aerobically and comprises a single plastic card with 30 wells for performing various chromogenic and conventional biochemical tests.

standard, is prepared from fresh (24- to 48-h) colonies from an anaerobe blood agar plate. The ANI cards are inoculated automatically using the Vitek filling module. However, the ANI cards are not incubated in the Vitek reader-incubator module. Instead, the ANI cards are incubated aerobically for 4 h at 35°C in an ambient air incubator. Following incubation, the ANI test reactions are read manually using the Vitek card reader. The ANI test results, along with the results of Gram staining, bacterial morphologic characteristics, and spot indole determination are entered manually into the Vitek computer module. The ANI cards cannot be read in the Vitek reader-incubator module because the ANI test reactions are in the red-yellow portion of the spectrum. The Vitek reader-incubator is capable of reading color reactions in the blue-green part of the spectrum only. The Vitek ANI system was evaluated by Schreckenberger and colleagues with 341 bacterial isolates against conventional methods (120). Those correctly identified to species level included 73% of 149 anaerobic gram-negative rods, 69% of anaerobic cocci, 66% of 38 anaerobic non-spore-forming gram-positive rods, and 64% of 44 clostridia. A more detailed listing of the results is given in Table 13.

MicroScan Rapid Anaerobe Panel

The MicroScan Rapid Anaerobe Identification Panel is a clear plastic 96-well microdilution tray with 24 wells containing dehydrated substrates (Tables 13 and 14; Fig. 10). The system uses chromogenic tests and modified conventional tests to identify anaerobic bacteria following 4 h of incubation aerobically at 35 to 37°C. An inoculum suspension equal to at least a McFarland no. 3 turbidity standard is prepared using a suspension of fresh (24 to 48 h) colonies in autoclaved deionized water. As for the other 4-h rapid identification systems discussed in this section, preformed enzymes are detected and growth of the microorganism in the panel is not necessary. After incubation, changes in pH, utilization of certain substrates, or other biochemical reactions are read visually, and anaerobic bacteria are identified with the aid of a biotype codebook. The MicroScan Rapid Anaerobe Identification Panel must be read manually as it has not been approved by the Food and Drug Administration for use with the MicroScan Instrument systems.

The manufacturer has addressed certain other limitations. With regard to preparation of the inoculum, the product insert indicates that organisms to be tested should be grown only on nonselective media such as TSA or brucella, brain heart infusion, or Columbia agar—all supplemented with 5% sheep blood (only). Furthermore, selective media

that contain antibiotics, including kanamycin-vancomycin, bacteroides bile esculin agar, and phenylethyl alcohol agar, should not be used to prepare inocula for these panels. The problem that some of the media listed above (e.g., TSA) may not be adequately enriched to meet the nutritional requirements of some fastidious anaerobes is probably an additional limiting factor related to this system.

Another limitation listed in the package insert is that, prior to testing with the MicroScan panel, "each organism should be tested for aerotolerance, purity, Gram stain reaction, and colonical morphology. . . . Should more than one species identification be list[ed] for a particular biotype number, supplementary tests may be necessary for the correct identification."

To our knowledge, the accuracy of the MicroScan Rapid Anaerobe Identification system has been addressed in the refereed literature in only a single report (131). Compared to identifications obtained using VPI conventional methods, using this product, correct identifications were reported for 166 of 237 strains (70%) tested. The authors pointed out significant problems with the identification of *B. fragilis*, the pigmented anaerobic gram-negative rods, and anaerobic gram-positive bacilli (Table 13). Stoakes and colleagues emphasized that additional rapid and simple tests—including Gram staining; GLC; growth in the presence of bile;

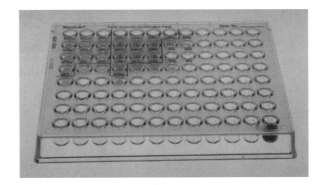

FIGURE 10 MicroScan Rapid Anaerobe Identification Panel. This 96-well microdilution tray contains dehydrated substrates and is incubated for 4 h aerobically.

pitting of agar; production of catalase, indole, lecithinase, lipase, fluorescence, and pigment; and the hydrolysis of esculin—can aid in arriving at correct identifications (131).

rapid ID 32A

In the late 1980s a new prototype of a system for identification of anaerobes, based upon experience with the API ZYM system, began to receive initial evaluations (139). Initially referred to as the API Prototype System and the ATB Anaerobes ID system, it initially contained 32 tests and had a 4-h aerobic incubation period. Subsequently, the prototype was refined to create the ATB 32 A system that first became commercially available in Europe, and later in the United States as the rapid ID 32A. Thus, the current rapid ID 32A is a miniaturized, disposable plastic strip with 32 small cupules (Fig. 11), 29 of which are used as wells containing dehydrated substrates to test for the reactions listed in Table 14. Three of the small cupules are empty. The inoculum suspension, equivalent to a McFarland no. 4 turbidity standard, is prepared from fresh colonies grown 24 to 48 h on a Columbia blood agar plate (supplemented with 5% horse or sheep blood plus vitamin K_3). The package insert indicates that Columbia blood agar is the only medium that can be used.

This system has been the subject of a number of evaluations with widely differing results (67, 76, 84, 111). Without the use of supplemental tests, 73 to 78% of the *B. fragilis* group and 100% of the *C. perfringens* strains tested were identified correctly to the species level. The system had difficulty in identifying *Fusobacterium* species, the less commonly encountered anaerobic gram-negative bacilli (many of those other than the *B. fragilis* group), *Clostridium* species other than *C. perfringens* or *C. ramosum,* and some anaerobic non-spore-forming, gram-positive rods and anaerobic cocci (99). Data compiled from three different evaluations of the system are provided in Table 15.

BBL Crystal ANR ID System

The BBL Crystal Anaerobe (ANR) identification (ID) system, a miniaturized packaged kit with modified conventional, fluorogenic, and chromogenic substrates, was introduced in the mid-1990s. The kit consists of a plastic base with 30 empty reaction wells and a plastic lid with a fluorescence (negative) control and 29 dehydrated biochemical and enzymatic substrates (Table 14) on the tips of plastic prongs (Fig. 12). All 30 wells are filled with a turbid bacterial suspension (prepared in inoculum fluid to be equivalent

to a McFarland no. 4 standard). The test reactions are begun when a plastic base, with its wells filled with bacterial inoculum suspension, and the corresponding lid are aligned and snapped together. Thus, within each reaction well, the bacterial suspension rehydrates the dehydrated substrate on the plastic tip that is inserted into it, and this starts the reaction. According to the package insert, the following BBL media are recommended for preparation of inocula: CDC Anaerobe Blood Agar, Schaedler agar with vitamin K_1 and 5% sheep blood, Columbia agar with 5% sheep blood and brucella blood agar with hemin and vitamin K_1.

Two evaluations of this system, with differences in study design and findings, have appeared in the refereed literature (25, 99). The reported performance of the system is summarized in Table 16. In both of the published reports, the authors indicated that the BBL Crystal ANR ID system was easy to use.

In the most recent evaluation by Cavallaro and colleagues, the results obtained using the ANR system for clinically significant anaerobic bacteria were compared to those obtained using conventional methods of the anaerobe reference laboratory at the CDC (25). Inocula were prepared using CDC Anaerobe Blood Agar. Upon initial testing, the ANR system correctly identified 263 of 322 (81.7%) of the strains tested, and 49 were identified correctly only to the genus level. Upon repeat testing, 23 of the latter 49 strains were identified correctly to the genus and species levels. A total of 26 strains (8.5% of 322) were misidentified at the species level, and an additional 10 strains (3.2%) were not identified.

In the study of the ANR system by Moll and associates, results were obtained for anaerobes grown on BBL Schaedler and BBL Columbia agar plates (99) and compared against those obtained using conventional procedures. In general the ANR system performed better with anaerobes grown on Schaedler agar than it did with anaerobes grown on Columbia agar. The performance of the ANR system, obtained by Moll et al. using Schaedler agar, is summarized in Table 16. Accordingly, for the ANR system, a higher percent of correct identifications to the species level was obtained in the report by Cavallaro and colleagues (25) than in that of Moll and colleagues (99). Explanations for differences in results in these studies probably relate to differences in the media used to prepare the inocula, differences in bacteria included in each of the studies, differences in the conventional procedures and databases used as gold standards for comparisons, and perhaps other variables.

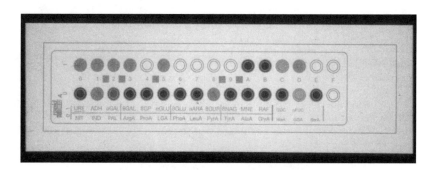

FIGURE 11 rapid ID 32A. A plastic strip with 32 cupules, 29 of which contain dehydrated substrates for rapid biochemical characterization of anaerobes, is incubated for 4 h aerobically.

TABLE 15 Identification of anaerobic bacteria by rapid ID 32 A

Organisms	Reference 99			Reference 84[a]			Reference 76[b]		
	No. tested	No. to species[d]	% Correctly identified	No. tested	No. to species	% Correctly identified	No. tested	No. to species	% Correctly identified
B. fragilis group	52	38	73	88	69	78	73	57	78
Pig. Prevotella	7	4	57	11	10	91	5	2	40
Porphyromonas	3	2	67	3	3	100	ND		
Other Prevotella spp.	2	1	50	11	9	82	18	15	83
B. ureolyticus	ND[e]			4	4	100	ND		
Fusobacterium spp.	10	1	10	22	2	9	10	1	10
Gram-negative rod total	**74**	**46**	**62**	**139**	**97**	**70**	**106**	**75**	**71**
P. magnus-P. micros[c]	5	3	60	10	10	100	11	11	100
P. anaerobius	5	0	0	7	6	86	6	6	100
Other cocci	7	1	14	7	6	86	11	10	91
Anaerobic coccus total	**17**	**4**	**24**	**27**	**24**	**89**	**28**	**27**	**96**
Propionibacterium spp.	15	4	27	11	10	91	11	11	100
Other gram-positive rods	47	6	13	11	6	55	25	11	44
Non-spore-forming gram-positive rod total	**62**	**10**	**16**	**22**	**16**	**73**	**36**	**22**	**61**
C. perfringens	7	7	100	16	16	100	7	7	100
C. ramosum	1	1	100	6	2	33	6	5	100
C. septicum	ND			ND			1	0	0
C. difficile	1	0	0	10	0	0	4	1	25
Other clostridia	5	1	20	39	22	56	25	13	52
Clostridium total	**14**	**9**	**64**	**71**	**40**	**56**	**43**	**26**	**60**
Grand total	167	69	41	259	177	68	214	150	70

[a]Data from Looney et al. do not include one unidentified strain of *Bacteroides splanchnicus* that was listed in their paper.
[b]Data from Kitch and Appelbaum do not include one strain of *Capnocytophaga* sp. that was identified correctly by the rapid ID 32A.
[c]*P. magnus-P. micros* are now called *F. magna* and *M. micros*.
[d]Number identified to species level in agreement with reference system.
[e]ND, not determined.

Interpretation of Identification Results Obtained by Using Commercial Kits

As a note of caution, a species name generated from testing with a commercial packaged kit should be interpreted carefully. For example, if a resulting appellation of an isolate is that of an infrequently encountered species, the microbiologist should not accept this identification without repeat testing, ideally with alternative procedures, or perhaps confirmation by a reference laboratory if warranted clinically. Additional supplemental tests described previously (e.g., the Presumpto system or the Wadsworth approach) are required to identify the less common species with accuracy. Thus, the determination of colonial characteristics and microscopic features of isolates; catalase, lipase, and lecithinase on egg yolk agar; growth in the presence of bile; the reduction of nitrate; inhibition by SPS; inhibition by certain antibiotics (e.g., vancomycin, kanamycin, and others); and metabolic product analysis (GLC) aid in improving the identification accuracies of the kits.

IMMUNODIAGNOSTIC METHODS FOR CDAD

We agree with the view of Johnson and Gerding, that an "optimal test for the diagnosis of [C. *difficile*-associated diarrhea (CDAD)] remains to be developed" (68). Issues and difficulties in diagnosing CDAD have been reviewed elsewhere (4, 40, 48, 49). The state of the art at this writing is that there is no single laboratory test that, by itself, exhibits acceptable specificity and sensitivity for the diagnosis of CDAD. However, some of the newer immunodiagnostic methods show promise.

The cell culture cytotoxin assay, which involves detection and specific neutralization of toxin B, is often considered the gold standard when new detection methods are evaluated and is viewed as the most specific test (Fig. 13). The cytotoxin assay, however, remains relatively slow, requiring 48 h of incubation before a test result can be called negative. However, most positive cytotoxin assay results occur within 18 to 24 h. Stool culture for C. *difficile*,

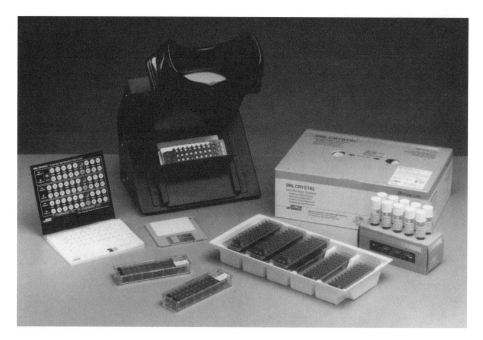

FIGURE 12 BBL Crystal ANR ID. Thirty wells in a plastic base are aligned and snapped together with a lid containing 29 biochemical and enzymatic substrates. Some reactions are visualized with a UV light reader (upper left-center), while others are read and interpreted using ambient light. (Photo courtesy of Becton Dickinson.)

followed by testing of isolates for toxin production, is the most sensitive test and has nearly equivalent specificity compared to the cytotoxicity assay but also has the problem of a relatively long turnaround time (e.g., 2 or more days) (4, 68, 112, 128). Although the diagnosis of CDAD using a combination of clinical criteria and results of cytotoxin testing plus stool culture for toxin-producing C. *difficile* probably represents the ultimate gold standard, the laboratory results cannot be obtained rapidly because of the incubation times involved. Expertise of the bench microbiologist or medical technologists is also an issue for cytotoxin and cell culture testing. Recognizing the need for a rapid, sensitive, specific and simple test for CDAD, manufacturers have brought numerous immunodiagnostic methods to the marketplace (Tables 17 and 18).

Latex Agglutination Tests for Glutamate Dehydrogenase

A rapid latex agglutination test developed in the 1980s was found to react with essentially all noncytotoxic as well as cytotoxic strains of C. *difficile*, and it gave immunologic cross-reactions with *Clostridium sporogenes*, proteolytic *Clostridium botulinum* and P. *anaerobius* (Fig. 14) (95). Lyerly and colleagues found that the protein that reacts in commercial latex tests for C. *difficile* (Table 17) is a glutamate dehydrogenase and not a toxin. Interestingly in clinical trials, the specificity of the tests for glutamate dehydrogenase has actually been very good (i.e., in the 98 to 99% range) (6). As pointed out in some of the more recent studies, the major concern about the latex tests has been their poor sensitivity for diagnosing CDAD (e.g., 47 to 67%) (66, 128).

The ImmunoCard C. *difficile* (Meridian Diagnostics, Cincinnati, Ohio) is an enzyme immunoassay (EIA) that detects glutamate dehydrogenase and is used as a screening test for detecting C. *difficile* in stool specimens from patients suspected of having CDAD (Fig. 15). The test is rapid to

perform (15 to 20 min). The list price is provided in Table 18. As reviewed elsewhere, the sensitivity of the Immuno-Card (e.g., 84 to 92%) was much better than that of latex agglutination, and specificity of the ImmunoCard was in the 95 to 100% range (4, 128).

EIA Systems for C. *difficile* Toxins A and B

Many EIAs have been marketed by various manufacturers for detection of toxin A or toxins A and B in stool (Tables 17 and 18). Although most of these products are in the form of manual test kits, at least one was designed for use with a laboratory instrument that can do other tests (e.g., the VIDAS system [bioMérieux-Vitek] (Fig. 16). In general, a number of these EIAs appear to be very specific but somewhat less sensitive than the cytotoxin assays (68). For this reason, EIAs should not be relied upon as the sole laboratory test for diagnosing CDAD. In addition, many studies have compared EIAs to each other or to cytotoxin assays, but relatively few studies have compared EIAs to toxigenic culture in patients with confirmed CDAD. EIAs for C. *difficile* toxins tend to exhibit even lower sensitivities when compared to toxigenic culture assays than they do in comparisons against cytotoxin assays (13, 49). The laboratory director should evaluate the merits of each EIA kit on an individual basis.

Testing for both toxins A and B in the clinical laboratory is becoming more relevant. Although the incidence of toxin A-negative–toxin B-positive strains of C. *difficile* involved in CDAD was previously thought to be low (85), new cases and outbreaks are being reported with increasing frequency worldwide (1, 28, 72, 83, 100, 117). In these cases, EIAs for toxin A are repeatedly negative. These recent reports suggest the need for an alternative diagnostic test if clinical features suggest CDAD. Thus, in addition to an EIA for toxin A, a cytotoxin assay for toxin B or culture for toxigenic C. *difficile* could be used. Alternatively, an EIA for both toxin A and toxin B should aid in detecting CDAD

TABLE 16 Identification of anaerobic bacteria using the BBL Crystal ANR ID System

Organisms	Reference 25					Reference 99		
	No. tested	No. to species[a]	% Correctly identified initially	No. to species[b]	Total % correctly identified overall (includes retest)	No. tested	No. to species[a]	% correctly identified
B. fragilis group	46	34	74	42	91	52	37	71
Pig. Prevotella	2	1	50	2	100	7	6	86
Porphyromonas	2	1	50	2	100	3	2	67
Other Prevotella spp.	10	10	100	10	100	2	1	50
B. ureolyticus	3	3	100	3	100	ND[d]		
Fusobacterium spp.	26	25	96	26	100	10	9	90
Other gram-negative rods	32	25	78	25	78	3	3	100
Gram-negative rod total	**121**	**99**	**82**	**110**	**91**	**77**	**58**	**75**
P. magnus-P. micros[c]	3	3	100	3	100	5	5	100
P. anaerobius	4	4	100	4	100	5	3	60
Other cocci	12	10	83	12	100	7	2	29
Anaerobic coccus total	**19**	**17**	**90**	**19**	**100**	**17**	**10**	**59**
Propionibacterium spp.	26	26	100	26	100	15	11	73
Other gram-positive rods	53	45	85	52	98	47	19	40
Non-spore-forming gram-positive rod total	**79**	**71**	**90**	**78**	**99**	**62**	**30**	**48**
C. perfringens	8	8	100	8	100	7	5	71
C. ramosum	7	7	100	7	100	1	1	100
C. septicum	8	6	75	8	100	ND		
C. difficile	8	7	88	8	100	1	1	100
Other clostridia	72	48	63	48	63	5	3	60
Clostridium total	**103**	**76**	**74**	**79**	**77**	**14**	**10**	**71**
Grand total	**322**	**263**	**82**	**286**	**89**	**167**	**108**	**65**

[a]Number identified to species level in agreement with reference system initially.
[b]Number identified to species level in agreement with reference system with retest.
[c]P. magnus-P. micros are now called F. magna and M. micros.
[d]ND, not determined.

caused by these toxin A-negative–toxin B-positive strains.

The Tox A/B EIA (TechLab, Inc., Blacksburg, Va.) (Fig. 17) has correlated well and was found to be sensitive and specific compared to the cytotoxicity assay (86). This test takes 1 h to perform and detects toxin A-negative–toxin B-positive strains of C. difficile (100). Recently, the Tox A/B EIA was compared to the Premier Cytoclone CD Toxin A/B system (Meridian Diagnostics) (Fig. 18) and RIDASCREEN C. difficile Toxin A/B test (Rohm Pharma, Darmstadt, Germany). The Tox A/B EIA detected more positives compared to the cytotoxin assay than either the Cytoclone or RIDASCREEN systems and required the least hands-on time (D. M. Lyerly, personal communication; J. S. Thompson, D. E. Smith, and R. Terro, Can. Assoc. Clin. Microbiol. Infect. Dis. Meet., presentation, 1997).

The new Premier Toxins A & B test (Meridian Diagnostics) detects toxin A at levels of ≥1 ng/ml and toxin B at ≥2.4 ng/ml in stool from adult patients. Thus far reported in abstract form only, the test kit was compared to an in-house cytotoxicity assay in a multicenter study of 573 specimens and demonstrated a sensitivity and specificity of 94.7 and 97.3%, respectively; in addition, the Premier Toxins A & B test showed a 96.5% agreement with the Wampole A/B EIA test (C. Gleaves, R. Aldeen, K. Schwartz, M. Campbell, and K. Carroll, Abstr. 100th Gen. Meet. Am. Soc. Microbiol., abstr. C-252, p. 188, 2000).

A critical review by Gröschel in 1996 summarized comparative studies for eight commercial EIAs and reported extremely broad ranges of sensitivity and specificity of 29 to 100% and 88 to 100%, respectively (50a). We reviewed five recent studies which compared six of the newer commercial EIAs to cytotoxin neutralization assays (Table 19) (S. D. Allen and J. A. Siders, Abstr. Anaerobe 2000 Int. Congr. Confed. Anaerobe Soc., abstr. VIII-P1, p. 167, 2000; M.

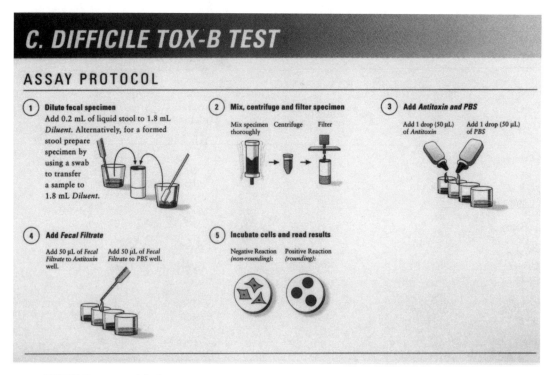

FIGURE 13 C. *difficile* Tox-B cytotoxin assay. Steps involved in the assay protocol are shown. (Photo courtesy of Tech Lab.)

Campion et al., *Abstr. 99th Gen. Meet. Am. Soc. Microbiol.*, abstr., C-2, p. 105, 1999; Gleaves et al., *Abstr. 100th Gen. Meet. Am. Soc. Microbiol.*, 2000; M. R. Smith et al., *Abstr. 100th Gen. Meet. Am. Soc. Microbiol.*, abstr. C-246 [revised], p. 187, 2000; D. Turgeon et al., *Abstr. 100th Gen. Meet. Am. Soc. Microbiol.*, abstr. C-247, p. 187, 2000). Although each study did not compare all EIAs listed in Table 19 to each other or to the same cytotoxin assay, collectively the ranges of sensitivity, specificity, positive predictive values (PPV), and negative predictive values (NPV) for each EIA are as listed in Table 19.

Rapid Membrane EIAs

Rapid membrane-based EIAs have been developed in which one or two antigens specific for *C. difficile* (i.e., toxin A and/or glutamate dehydrogenase) are immobilized on a membrane using specific antibodies. After reagents are added, a positive test is detected visually in about 15 min by the presence or absence of a color bar next to the antigen of interest (Fig. 19). The Triage C. *difficile* Panel (Biosite Diagnostics, San Diego, Calif.) tests for both antigens and performs favorably compared to the cytotoxin assay (J. DeSimone et al., *Abstr. 99th Gen. Meet. Am. Soc. Microbiol.*, abstr. C-5, p. 105, 1999; V. Massey et al., *Abstr. 100th Gen. Meet. Am. Soc. Microbiol.*, abstr. C-250, p. 188, 2000; Y. S. McCarter and I. Ratkiewicz, *Abstr. 99th Gen. Meet. Am. Soc. Microbiol.*, abstr. C-6, p. 106, 1999; Turgeon et al., *Abstr. 100th Gen. Meet. Am. Soc. Microbiol.*; D. Wilson et al., *Abstr. 99th Gen. Meet. Am. Soc. Microbiol.*, abstr. C-7, p. 106, 1999), but it is slightly less sensitive than some (B. L. Bearson et al., *Abstr. 99th Gen. Meet. Am. Soc. Microbiol.*, abstr. C-3, p. 105, 1999; L. D. Slavich et al., *Abstr. 99th Gen. Meet. Am. Soc. Microbiol.*, abstr. C-4, p. 105, 1999) standard EIAs. Most patients with a positive Triage test for glutamate dehydrogenase (common antigen) have CDAD, especially in the presence of toxin A (DeSimone et al., *Abstr. 99th Gen. Meet. Am. Soc. Microbiol.*). However, clinical interpretive guidelines need to be established for glutamate dehydrogenase-positive, toxin-negative test results (Bearson et al., *Abstr. 99th Gen. Meet. Am. Soc. Microbiol.*; Wilson et al., *Abstr. 99th Gen. Meet. Am. Soc. Microbiol.*). The Color-PAC Toxin A test (Becton Dickinson) also performs favorably compared to the cytotoxin assay (C. Kviz et al., *Abstr. 98th Gen. Meet. Am. Soc. Microbiol.*, abstr. C-189, p. 162, 1998) and has even out-performed some standard EIAs (Fig. 20) (W. Greene et al., *Abstr. 98th Gen. Meet. Am. Soc. Microbiol.*, abstr. C-183, p. 161, 1998; M. Tuohy et al., *Abstr. 98th Gen. Meet. Am. Soc. Microbiol.*, abstr. C-190, p. 162, 1998). The Clearview C. DIFF A Test (Wampole Labs, Cranbury, N.J.) is another membrane-based EIA which demonstrated acceptable sensitivity compared to the cytotoxin assay (20).

At this time, a commercial rapid membrane-based EIA for both toxin A toxin B is not available, but probably is under development by at least one manufacturer. In the future, a rapid membrane-based EIA for both toxin A and toxin B could become the optimal method for detection of CDAD, especially if combined with toxigenic culture. For example, specimens testing negative for either toxin by the rapid EIA could be subsequently inoculated for toxigenic culture. Such a protocol could, in fact, offer the most rapid and sensitive means of detecting patients with *C. difficile*-associated disease.

AUTOMATED TESTING PROCEDURES
MIDI Sherlock Microbial Identification System

The MIDI Sherlock Microbial Identification (MIDI, Inc., Newark, Del.) (or, simply, the MIDI) is a fully automated gas-chromatographic (GC) analytical system dedicated to the

TABLE 17 Tests for the detection of *Clostridium difficile* antigens in stool specimens

Method or test name	Manufacturer(s) and location(s)
Tests for glutamate dehydrogenase	
Latex agglutination	
CDT	Becton Dickinson Microbiology Systems, Cockeysville, Md.
Meritec *C. difficile*	Meridian Diagnostics, Inc., Cincinnati, Ohio
EIA	
ImmunoCard *C. difficile*	Meridian Diagnostics, Inc., Cincinnati, Ohio
Triage *C. difficile* panel	Biosite Diagnostics, Inc., San Diego, Calif.
Tests for *C. difficile* toxins	
Toxin B	
C. difficile Tox-B test	TechLab, Inc., Blacksburg, Va., and Wampole Labs, Cranbury, N.J.
Cytotoxicity assay kit	Intracel Corp., Issaquah, Wash.
Toxin A	
C. difficile Toxin A	Oxoid, Inc., Ogdensburg, N.Y.
Premier *C. difficile* Toxin A Microplate	Meridian Diagnostics, Inc., Cincinnati, Ohio
Prospect II *C. difficile* Toxin A microplate	Alexon-Trend, Sunnyvale, Calif.
C. difficile Tox-A Test	TechLab, Inc., Blacksburg, Va., and Wampole Labs, Cranbury, N.J.
Toxin A EIA	Intracel Corp., Issaquah, Wash.
Toxin CD Test	Becton Dickinson Microbiology Systems, Cockeysville, Md.
Triage *C. difficile* Panel	Biosite Diagnostics, Inc., San Diego, Calif.
VIDAS CDA 2	bioMerieux Vitek, Hazelwood, Mo.
ColorPAC	Becton Dickinson Microbiology Systems, Cockeysville, Md.
Immunocard Toxin A	Meridian Diagnostics, Inc., Cincinnati, Ohio
CdTox A OIA	Biostar, Boulder, Colo.
Clearview C. DIFF A	Wampole Labs, Cranbury, N.J.
Toxins A and B	
C. difficile Tox-A/B Test	Techlab, Inc., Blacksburg, Va., and Wampole Labs, Cranbury, N.J.
Cd Toxin A + B (RIDASCREEN)	Rohm Pharma, Darmstadt, Germany
Premier Cytoclone	Meridian Diagnostics, Inc., Cincinnati, Ohio
Premier Toxins A&B	Meridian Diagnostics, Inc., Cincinnati, Ohio

analysis and identification of bacteria (and certain yeasts) by means of cellular fatty acid (CFA) analysis (Fig. 21). Unlike the short-chain fatty acid metabolic products (e.g., volatile fatty acids) of 2 to 8 carbons in length that accumulate in broth cultures during fermentation processes, the CFAs are from 9 to 20 carbons in length and are major constituents of bacterial cell membranes (159). The CFAs differ from the long-chain mycolic acids (24 to 90 carbons in length) or the isoprenoid quinones and menaquinones that also occur in cell membranes. CFA analysis requires a different chromatographic setup than that which can be used for short-chain fatty acids, and it differs from the high-pressure liquid chromatographic procedure that is used for long-chain mycolic acid analyses.

The MIDI system hardware (Fig. 21) consists of a Hewlett-Packard 6890 GC with an automatic sampler and a fused-silica capillary column, a computer, and a printer. An extensive database and procedure for cellular fatty acid analysis of anaerobic bacteria were developed by W. E. C. Moore and colleagues. This has been updated periodically through the years. The current software release (version 3.9) used in our laboratory contains an extensive list of anaerobe genera and species.

For CFA analyses, anaerobic bacteria are grown in a specially formulated PRAS peptone-yeast extract-glucose (PYG) with additional supplements added according to Gram stain reaction. According to the manufacturer's protocol supplied for the VPI broth library, the PYG broth culture is incubated until it reaches maximum turbidity (e.g., usually overnight for clostridia). The cells are then harvested, saponified with the use of NaOH and heated, methylated (using methanol-HCl

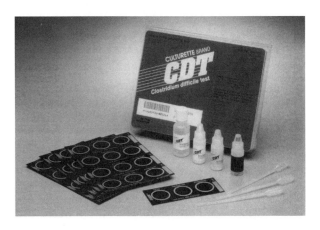

FIGURE 14 Culturette Brand CDT *C. difficile* latex agglutination test for glutamate dehydrogenase. (Photo courtesy of Becton Dickinson, Cockeysville, Md.)

TABLE 18 Commercially available products related to *C. difficile* testing

Product	Manufacturer no.	Package contents	List price ($)
Tests for glutamate dehydrogenase			
Latex agglutination			
BD CULTURETTE Brand CDT Rapid *Clostridium difficile* Test Kit	4364003	100 tests	1,162.00
Meridian *C. difficile* Detection Kit (latex agglutination for common antigen)	204030	Pack of 30	436.00
EIA			
Meridian Immunocard *C. difficile* EIA	706050	Pack of 50	754.00
Biosite Diagnostics Triage *C. difficile* Panel (EIA for antigen and toxin A)	96000	Pack of 20	390.00
Tests for *C. difficile* toxins			
Toxin B			
TechLab *C. difficile* Tox-B	T5003	48 wells	109.40
Intracel/Bartels Cytotoxicity assay kit	B1029-70	48 wells	116.48
Toxin A			
Oxoid *C. difficile* Toxin A	TD0970A	20 tests	328.10
Alexon-Trend Prospect II *C. difficile* Toxin A microplate	725-96	96 tests	637.50
TechLab *C. difficile* Tox-A test	T5001	96 tests	425.00
Intracel/Bartels Toxin A EIA	B1029-69	96 tests	520.00
BD, CULTURETTE BRAND *Toxin* CD (EIA for toxin A)	4954004	96 tests	870.00
Biosite Diagnostics, Triage *C. difficile* Panel	96000	Pack of 20	390.00
BioMerieux VIDAS CDA 2	30193	60 tests	681.00
Becton Dickinson ColorPAC	274030	30 tests	525.00
Meridian, ImmunoCard Toxin A (EIA for Tox A)	711050	Pack of 50	910.00
Biostar CdTox A OIA	CD30	30 tests	450.00
Wampole Clearview C. DIFF A	135050	20 tests	221.70
Toxins A and B			
TechLab *C. difficile* Tox A/B	T5015	96 tests	550.00
Meridian Premier Cytoclone	696-087	96 tests	666.00
Meridian Premier Toxins A&B	616-096	96 tests	680.00

and H_2SO_4-methanol plus heat) and extracted into organic solvent (i.e., hexane and methyl *tert*-butyl ether). The organic layer remaining in the tube is washed with diluted NaOH and then transferred to a GC vial. The sealed vial is loaded onto the automatic sampler of the GC and the analysis of CFAs is performed. In our laboratory, a standard supplied by the manufacturer containing known fatty acids of 9 to 20 carbons in length is chromatographed at the start of each run and

after every 11th sample is analyzed. The MIDI computer compares the CFA results for an unknown isolate with those of known reference strains that have been archived in the library. The identification results are expressed in terms of a similarity index indicating how closely the CFA composition of an unknown isolate matches that of the most closely matched strains in the library.

In an analysis of 216 strains representing 18 species of

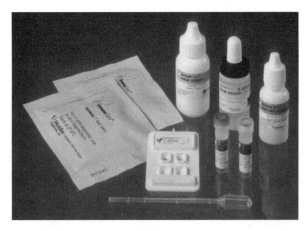

FIGURE 15 ImmunoCard *C. difficile*. This rapid EIA detects glutamate dehydrogenase and is used as a screening test for *C. difficile*.

FIGURE 16 VIDAS immunoassay analyzer (see text).

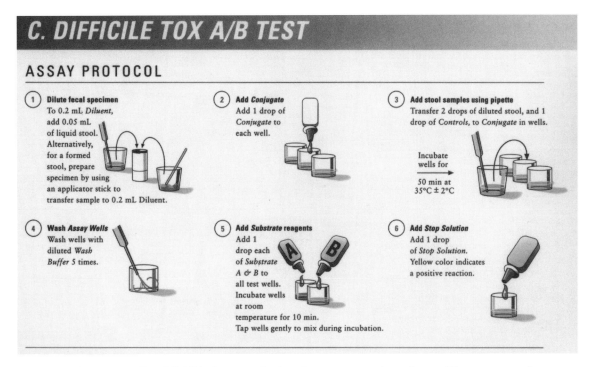

FIGURE 17 Tox A/B EIA. Steps involved in the assay protocol are shown. (Photo courtesy of Tech Lab.)

Clostridium, the MIDI correctly identified 86% of the strains correctly to the species level without the need for supplemental tests, and identified an additional 6% of the strains with the aid of a few supplemental tests. Only 3% of the strains were identified to the genus level incorrectly (8). The CFA patterns of selected medically important species (e.g., *C. perfringens*, *C. difficile*, *C. septicum*, *C. sporogenes*, and *C. botulinum*) obtained using the MIDI were so distinctive that virtually 100% of these species were differentiated correctly. The use of CFA analyses for identification of *Clostridium* species had been investigated previously by Moss and Lewis (106), and by Ghanem et al. (50). CFA profiles of species of the genera *Bacteroides*, *Prevotella*, *Porphyromonas*, *Fusobacterium*, and *Propionibacterium* and *Eubacterium lentum*, as well as several other genera of clinically encountered anaerobes, have been studied extensively (22, 29, 101, 102, 105, 138). Relatively less work has been reported on the use of CFA analysis for identification of the gram-positive anaerobic cocci (107), which are presently undergoing major taxonomic revision (108, 109).

The time required for the chromatography and the automatic computer analysis to take place is about 25 min per sample. Although the initial capital equipment investment is not insignificant, the equipment lasts for many years (one of our instruments has been in service for about 12 years) with relatively low maintenance; the manufacturer's service on the equipment has been commendable. In our laboratory, we prefer to run batches of ~30 samples. The labor involved for a run with this number of samples is about 4 min per isolate. Additional direct costs per isolate include the cost of a purity plate of anaerobe blood agar, the cost of a tube of PYG broth, and a relatively small cost for reagents. We estimate that the direct cost for this procedure is in the range of $5.00 to $8.00 per isolate (not including the initial cost of the equipment).

FUTURE DIRECTIONS IN TESTING
Biolog AN MicroPlate

Described elsewhere in this book, the Biolog Identification System (Biolog, Inc., Hayward, Calif.) is based on the principle of carbon source utilization; the system utilizes a small well microtray format for the characterization of microorganisms. The Biolog AN MicroPlate is a test panel (with 95 tests per MicroPlate) designed specifically for the identification of a wide array of anaerobic bacteria. The anaerobic bacteria database (release 4.01B) contains identification patterns for 363 anaerobic bacterial species or taxa. The list of clinically important anaerobes included in this database is extensive. In addition, a long list of anaerobes of importance in industrial and environmental bacteriology are also included in the database. Although this promising new system has been recently marketed for identification of clinically encountered anaerobes, to our knowledge, no clinical

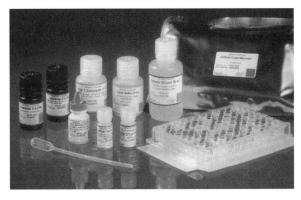

FIGURE 18 Premier Cytoclone CD Toxin A/B. Refer to text for details. (Photo courtesy of Meridian Diagnostics.)

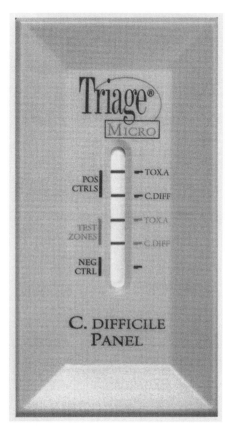

FIGURE 19 Triage C. *difficile* Panel. This rapid membrane EIA detects glutamate dehydrogenase and toxin A. Positive results are indicated by bars appearing next to each corresponding antigen (center of panel). (Photo courtesy of Biosite Diagnostics.)

microbiology laboratory evaluations of it have appeared thus far in the refereed literature.

Molecular Methods

PCR amplification procedures have been used to detect a wide variety of anaerobes including *B. fragilis* and also the enterotoxin gene of *B. fragilis* (121, 146), *Bacteroides distasonis* in river water (78), *Bacteroides forsythus* (51), *Porphyromonas gingivalis* (93), *Prevotella intermedia* and *Prevotella nigrescens* (94), other gram-negative bacteria encountered in the oral cavity (124), and gram-positive non-spore-forming anaerobic bacteria (53). Numerous reports have focused

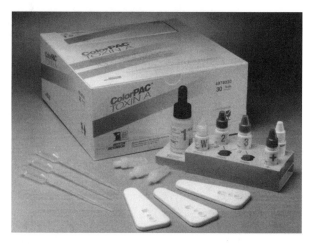

FIGURE 20 ColorPAC Toxin A rapid membrane EIA test kit. (Photo courtesy of Becton Dickinson.)

on the use of PCR methodology to detect enterotoxigenic *C. perfringens* (12, 24, 39, 98, 116, 118, 140, 145), *C. botulinum* or its toxin genes in foods or other materials (10, 38, 41, 45, 46, 57), *Clostridium septicum* (134), and *C. difficile*, particularly to differentiate between toxigenic and nontoxigenic strains (9, 27, 71–74, 100). At present, simple-to-use commercial kits and equipment for specific molecular applications in clinical anaerobic bacteriology have been lacking. Today, rapid sequence analysis of 16S ribosomal DNA (rDNA) or rRNA has found useful applications in polyphasic approaches to bacterial classification (126, 127). It is predictable that kits and equipment that enable 16S rDNA or 16S rRNA gene sequencing of bacteria in general, such as the MicroSeq 16S rRNA Gene Kit (PE Biosystems, Foster City, Calif.), will probably be applied to anaerobe identification at an increasing rate in the next few years.

MALDI-TOF-MS

Mass spectrometry for characterization and identification of bacteria has been studied actively for at least 3 decades (104). The methods have included gas chromatography/mass spectrometry (103), pyrolysis mass spectrometry (87, 88), fast-atom bombardment mass spectrometry (35), and liquid chromatography/microspray/mass spectrometry (79). Matrix-assisted laser desorption/ionization time-of-flight mass spectrometry (MALDI-TOF-MS) of bacterial whole cells is a promising new method for rapid identification of bacteria (11). Recent concern about the potential global threat of bioterrorist activities and biological weapons has been a

TABLE 19 Composite performance data[a] compiled for six immunodiagnostic products marketed for detection of C. *difficile*

Test	Sensitivity (%)	Specificity (%)	PPV (%)	NPV (%)
Alexon-Trend Prospect II CD Toxin A	82–85	98–100	89–100	95–98
BioMerieux VIDAS Toxin A II	65–85	99–100	89–99	74–96
Meridian Premier Cytoclone CD Toxin A/B	77–83	99–100	91–100	96–98
Meridian Premier Toxins A & B	95	97	NA[b]	NA
Meridian Premier CD Toxin A	85–87	99–100	91–100	97–98
TechLab CD Tox A/B	80–85	99–100	95–100	96–97

[a]Data compiled from abstracts by Allen and Siders, Campion et al., Gleaves et al., Smith et al., and Turgeon et al., as cited in the text.
[b]NA, not available.

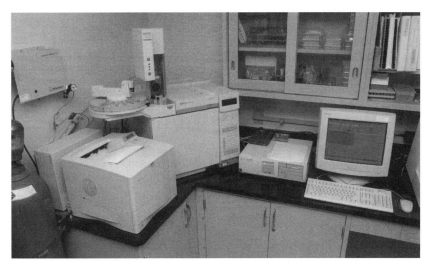

FIGURE 21 MIDI Sherlock Microbial Identification System, an automated GC analytical system. Included in the photo are the following: a Hewlett-Packard 6890 GC with an automatic sampler, a computer, and a printer. The GC contains a fused-silica capillary column and a flame ionization detector.

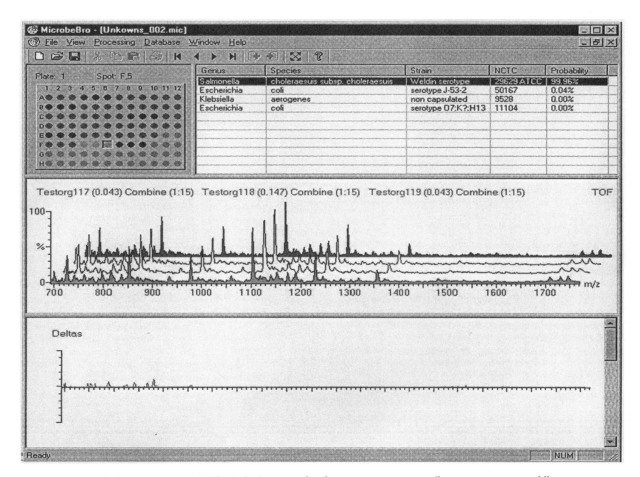

FIGURE 22 MALDI-TOF-MS. An example of a composite spectrum (bottom tracing in middle panel) that was produced from three bacterial mass spectral fingerprints is matched with an 89% probability to the reference spectrum of a strain of *Escherichia coli* in the database. The lower panel shows the difference between the test spectrum and the database entry. (Photograph courtesy of Micromass UK Limited.)

motivating factor to study the use of MALDI-TOF-MS for rapid detection and identification of *Bacillus anthracis*, *Yersinia pestis* and other potential threats (30, 80). Although several species of *Enterobacteriaceae* and other aerobic and facultatively anaerobic bacteria have been actively investigated using this technology (11, 62, 142), little information regarding the use of MALDI-TOF-MS to characterize anaerobic bacteria has been available until recently. During the Anaerobe 2000 Congress of the Confederation of Anaerobe Societies (Manchester, England [July 2000]), a new instrumented system for MALDI-TOF-MS in combination with bio-informatics, called the MicrobeLynx System (Micromass UK Limited, Manchester, England) was presented. With this system, intact bacterial cells from a culture plate are smeared on a stainless steel target plate (or grid) and allowed to cocrystallize with a UV-absorbing matrix. The preparation is dried and then placed into the MALDI-TOF spectrometer. Bacteria in the matrix receive a pulse from a nitrogen laser device, the matrix absorbs energy and surface macromolecules of the bacteria are desorbed and ionized. Mass analysis of the ionized surface macromolecules follows and the results are reported as a mass spectrum with mass plotted on the *x* axis versus abundance on the *y* axis. As illustrated in Fig. 22, a composite spectrum produced from three bacterial mass spectral fingerprints is matched with reference strains in the database and a percent probability to permit rapid identification of the unknown microorganism. Currently, an anaerobe database is being developed for the system.

Thus far reported only in abstract form, MALDI-TOF-MS studies on anaerobic bacteria have focused on a comparison of *Prevotella intermedia* and *Prevotella nigrescens* strains (J. H. Bright et al., *Abstr. Anaerobe 2000 Int. Cong. Confed. Anaerobe Soc.*, abstr. IXa-02, p. 184, 2000), and the ability of this technology to distinguish between representative species of the genera *Bacteroides*, *Prevotella*, *Porphyromonas*, *Treponema*, and *Peptostreptococcus* (C. J. Keys et al., *Abstr. Anaerobe 2000 Int. Congr. Confed. Anaerobe Soc.*, abstr. IXa-01, p. 183, 2000). Patel and colleagues have developed a hybrid neural network (Manchester Metropolitan University Search Engine) which aids in interpreting the large volume and complexity of data generated with these types of analyses by means of pattern recognition and have investigated the discriminating powers of this search engine in an analysis of peptostreptococci (M. Patel et al., *Abstr. Anaerobe 2000 Int. Congr. Confed. Anaerobe Soc.*, abstr. IXa-P3, p. 187, 2000). It appears that MALDI-TOF-MS could be a powerful tool for rapid identification of bacteria that would otherwise be difficult to differentiate with other methods (Keys et al., *Abstr. Anaerobe 2000 Int. Congr. Confed. Anaerobe Soc.*). This new technology holds promise for markedly decreasing the turnaround time for identification of anaerobes and other slow growing microorganisms, thereby enhancing the clinical relevance of this area of clinical microbiology.

REFERENCES

1. **al-Barrak, A., J. Embil, B. Dyck, K. Olekson, D. Nicoll, M. Alfa, and A. Kabani.** 1999. An outbreak of toxin A negative, toxin B positive *Clostridium difficile*-associated diarrhea in a Canadian tertiary-care hospital. *Can. Commun. Dis. Rep.* **25:**65–69.
2. **Allen, S. D.** 1982. Systems for rapid identification of anaerobic bacteria, p. 214–217. *In* R. C. Tilton (ed.), *Rapid Methods and Automation in Microbiology.* American Society for Microbiology, Washington D.C.
3. **Allen, S. D., and B. I. Duerden.** 1998. Infections due to non-sporing anaerobic bacilli and cocci, p. 743–776. *In* L. Collier, A. Balows, and M. Sussman (ed.), *Topley and Wilson's Microbiology and Microbial Infections*, 9th ed., vol. 3. Edward Arnold Publisher, London, United Kingdom.
4. **Allen, S. D., C. L. Emery, and J. A. Siders.** 1999. *Clostridium*, p. 654–671. *In* P. R. Murray, E. J. Baron, M. A. Pfaller, F. C. Tenover, and R. H. Yolken (ed.), *Manual of Clinical Microbiology*, 7th ed. ASM Press, Washington, D.C.
5. **Allen, S. D., and J. A. Siders.** 1982. An approach to the diagnosis of anaerobic pleuropulmonary infections. *Clin. Lab. Med.* **2:**285–303.
6. **Allen, S. D., J. A. Siders, and L. M. Marler.** 1995. Current issues and problems in dealing with anaerobes in the clinical laboratory. *Clin. Lab. Med.* **15:**333–364.
7. **Allen, S. D., J. A. Siders, and L. M. Marler.** 1985. Isolation and examination of anaerobic bacteria, p. 413–443. *In* E. H. Lennette, A. Balows, W. J. Hausler, Jr., and E. J. Shadomy (ed.), *Manual of Clinical Microbiology*, 4th ed. American Society for Microbiology, Washington, D.C.
8. **Allen, S. D., J. A. Siders, M. J. Riddell, J. A. Fill, and W. S. Wegener.** 1995. Cellular fatty acid analysis in the differentiation of *Clostridium* in the clinical microbiology laboratory. *Clin. Infect. Dis.* **20**(Suppl. 2)**:**S198–201.
9. **Alonso, R., C. Munoz, S. Gros, D. Garcia de Viedma, T. Pelaez, and E. Bouza.** 1999. Rapid detection of toxigenic *Clostridium difficile* from stool samples by a nested PCR of toxin B gene. *J. Hosp. Infect.* **41:**145–149.
10. **Aranda, E., M. M. Rodriguez, M. A. Asensio, and J. J. Cordoba.** 1997. Detection of Clostridium botulinum types A, B, E and F in foods by PCR and DNA probe. *Lett. Appl. Microbiol.* **25:**186–190.
11. **Arnold, R. J., and J. P. Reilly.** 1998. Fingerprint matching of *E. coli* strains with matrix-assisted laser desorption/ionization time-of-flight mass spectrometry of whole cells using a modified correlation approach. *Rapid Commun. Mass Spectrom.* **12:**630–636.
12. **Baez, L. A., V. K. Juneja, and S. K. Sackitey.** 1996. Chemiluminescent enzyme immunoassay for detection of PCR-amplified enterotoxin A from *Clostridium perfringens. Int. J. Food Microbiol.* **32:**145–158.
13. **Barbut, F., C. Kajzer, N. Planas, and J.-C. Petit.** 1993. Comparison of three enzyme immunoassays, a cytotoxicity assay, and toxigenic culture for diagnosis of *Clostridium difficile*-associated diarrhea. *J. Clin. Microbiol.* **31:**963–967.
14. **Baron, E. J., and S. D. Allen.** 1993. Should clinical laboratories adopt new taxonomic changes? If so, when? *Clin. Infect. Dis.* **16**(Suppl. 4)**:**S449–450.
15. **Baron, E. J., R. Bennion, J. Thompson, C. Strong, P. Summanen, M. McTeague, and S. M. Finegold.** 1992. A microbiological comparison between acute and complicated appendicitis. *Clin. Infect. Dis.* **14:**227–231.
16. **Baron, E. J., and D. M. Citron.** 1999. Algorithm for identification of anaerobic bacteria, p. 652–653. *In* P. R. Murray, E. J. Baron, M. A. Pfaller, F. C. Tenover, and R. H. Yolken (ed.), *Manual of Clinical Microbiology*, 7th ed. ASM Press, Washington, D.C.
17. **Baron, E. J., P. Summanen, J. Downes, M. C. Roberts, H. Wexler, and S. M. Finegold.** 1989. *Bilophila wadsworthia*, gen. nov. and sp. nov., a unique gram-negative anaerobic rod recovered from appendici-

tis specimens and human faeces. *J. Gen. Microbiol.* **135:**3405–3411.

18. **Bartlett, J. G.** 1998. Anaerobic bacteria, p. 1888–1901. *In* S. L. Gorbach, J. G. Bartlett, and N. R. Blacklow (ed.), *Infectious Diseases,* 2nd ed. W. B. Saunders Company, Philadelphia, Pa.

19. **Bate, G.** 1986. Comparison of Minitek Anaerobe II, API An-Ident, and RapID ANA systems for identification of *Clostridium difficile. Am. J. Clin. Pathol.* **85:**716–718.

20. **Bentley, A. H., N. B. Patel, M. Sidorczuk, P. Loy, J. Fulcher, P. Dexter, J. Richards, S. P. Borriello, K. W. Zak, and E. M. Thorn.** 1998. Multicentre evaluation of a commercial test for the rapid diagnosis of *Clostridium difficile*-mediated antibiotic-associated diarrhoea. *Eur. J. Clin. Microbiol. Infect. Dis.* **17:**788–790.

21. **Brander, M. A., and H. R. Jousimies-Somer.** 1992. Evaluation of the RapID ANA II and API ZYM systems for identification of *Actinomyces* species from clinical specimens. *J. Clin. Microbiol.* **30:**3112–3116.

22. **Brondz, I., and I. Olsen.** 1991. Multivariate analyses of cellular fatty acids in *Bacteroides, Prevotella, Porphyromonas, Wolinella,* and *Campylobacter* spp. *J. Clin. Microbiol.* **29:**183–189.

23. **Buchanon, R. E., and N. E. Gibbons (ed.).** 1974. *Bergey's Manual of Determinative Bacteriology,* 8th ed. Williams & Wilkins, Baltimore, Md.

24. **Buogo, C., S. Capaul, H. Hani, J. Frey, and J. Nicolet.** 1995. Diagnosis of *Clostridium perfringens* type C enteritis in pigs using a DNA amplification technique (PCR). *Zentbl. Veterinarmed. Reihe B* **42:**51–58.

25. **Cavallaro, J. J., L. S. Wiggs, and J. M. Miller.** 1997. Evaluation of the BBL Crystal Anaerobe identification system. *J. Clin. Microbiol.* **35:**3186–3191.

25a. **Celig, D. M., and P. C. Schreckenberger.** 1991. Clinical evaluation of the RapID-ANA II panel for identification of anaerobic bacteria. *J. Clin. Microbiol.* **29:**457–462.

26. **Citron, D. M., E. J. Baron, S. M. Finegold, and E. J. Goldstein.** 1990. Short prereduced anaerobically sterilized (PRAS) biochemical scheme for identification of clinical isolates of bile-resistant *Bacteroides* species. *J. Clin. Microbiol.* **28:**2220–2223.

27. **Citron, D. M., S. H. Gerardo, M. C. Claros, F. Abrahamian, D. Talan, and E. J. Goldstein.** 1996. Frequency of isolation of *Porphyromonas* species from infected dog and cat bite wounds in humans and their characterization by biochemical tests and arbitrarily primed-polymerase chain reaction fingerprinting. *Clin. Infect. Dis.* **23**(Suppl. 1):S78–82.

28. **Cohen, S. H., Y. J. Tang, B. Hansen, and J. Silva, Jr.** 1998. Isolation of a toxin B-deficient mutant strain of *Clostridium difficile* in a case of recurrent *C. difficile*-associated diarrhea. *Clin. Infect. Dis.* **26:**410–412. (Erratum, **26:**1250.)

29. **Cummins, C. S., and C. W. Moss.** 1990. Fatty acid composition of *Propionibacterium propionicum* (*Arachnia propionica*). *Int. J. Syst. Bacteriol.* **40:**307–308.

30. **Demirev, P. A., Y. P. Ho, V. Ryzhov, and C. Fenselau.** 1999. Microorganism identification by mass spectrometry and protein database searches. *Anal. Chem.* **71:**2732–2738.

31. **Dowell, V. R., Jr., and S. D. Allen.** 1981. Anaerobic bacterial infections, p. 171–213. *In* A. Balows and W. J. Hausler (ed.), *Diagnostic Procedures for Bacterial,*

Mycotic, and Parasitic Infections, 6th ed. American Public Health Association, Washington, D.C.

32. **Dowell, V. R., Jr., and T. M. Hawkins.** 1981. *Laboratory Methods in Anaerobic Bacteriology.* CDC Laboratory Manual, HHS publication no. (CDC) 81-8272. Government Printing Office, Washington, D.C.

33. **Dowell, V. R., Jr., and G. L. Lombard.** 1982. Differential agar media for identification of anaerobic bacteria, p. 258–262. *In* R. C. Tilton (ed.), *Rapid Methods and Automation in Microbiology.* American Society for Microbiology, Washington, D.C.

34. **Dowell, V. R., Jr., and G. L. Lombard.** 1984. Procedures for preliminary identification of bacteria. Department of Health, Education, and Welfare, Public Health Service, Centers for Disease Control, Atlanta, Ga.

35. **Drucker, D. B., H. S. Aluyi, V. Boote, J. M. Wilson, and Y. Ling.** 1993. Polar lipids of strains of *Prevotella, Bacteroides* and *Capnocytophaga* analysed by fast atom bombardment mass spectrometry. *Microbios* **75:**45–56.

36. **Durmaz, B., H. R. Jousimies-Somer, and S. M. Finegold.** 1995. Enzymatic profiles of *Prevotella, Porphyromonas,* and *Bacteroides* species obtained with the API ZYM system and Rosco diagnostic tablets. *Clin. Infect. Dis.* **20**(Suppl. 2):S192–194.

37. **Engelkirk, P. G., J. Duben-Engelkirk, and V. R. Dowell, Jr.** 1992. *Principles and Practice of Clinical Anaerobic Bacteriology.* Star Publishing, Belmont, Calif.

38. **Fach, P., M. Gibert, R. Griffais, J. P. Guillou, and M. R. Popoff.** 1995. PCR and gene probe identification of botulinum neurotoxin A-, B-, E-, F-, and G-producing *Clostridium* spp. and evaluation in food samples. *Appl. Environ. Microbiol.* **61:**389–392.

39. **Fach, P., and M. R. Popoff.** 1997. Detection of enterotoxigenic *Clostridium perfringens* in food and fecal samples with a duplex PCR and the slide latex agglutination test. *Appl. Environ. Microbiol.* **63:**4232–4236.

40. **Fekety, R.** 1997. Guidelines for the diagnosis and management of *Clostridium difficile*-associated diarrhea and colitis. American College of Gastroenterology, Practice Parameters Committee. *Am. J. Gastroenterol.* **92:**739–750.

41. **Ferreira, J. L., M. K. Hamdy, S. G. McCay, M. Hemphill, N. Kirma, and B. R. Baumstark.** 1994. Detection of *Clostridium botulinum* type F using the polymerase chain reaction. *Mol. Cell. Probes* **8:**365–373.

42. **Finegold, S. M.** 2000. Anaerobic bacteria: general concepts, p. 2519-2537. *In* G. L. Mandell, J. E. Bennett, and R. Dolin (ed.), *Mandell, Douglas, and Bennett's Principles and Practice of Infectious Diseases,* 5th ed., vol. 2. Churchill Livingstone, Philadelphia, Pa.

43. **Finegold, S. M., and W. L. George (ed.).** 1989. *Anaerobic Infections in Humans.* Academic Press, San Diego, Calif.

44. **Fontana, C., T. Jezzi, G. P. Testore, and B. Dainelli.** 1995. Differentiation of *Clostridium difficile, Clostridium bifermentans, Clostridium sordellii,* and *Clostridium perfringens* from diarrheal stool by API ZYM and API LRA oxidase test. *Microbiol. Immunol.* **39:**231–235.

45. **Franciosa, G., L. Fenicia, C. Caldiani, and P. Aureli.** 1996. PCR for detection of *Clostridium botulinum* type C in avian and environmental samples. *J. Clin. Microbiol.* **34:**882–885.

46. **Franciosa, G., J. L. Ferreira, and C. L. Hatheway.** 1994. Detection of type A, B, and E botulism neurotoxin genes in *Clostridium botulinum* and other *Clostridium* species by PCR: evidence of unexpressed type B toxin

genes in type A toxigenic organisms. *J. Clin. Microbiol.* **32:**1911–1917.

47. **Gargan, R. A., and I. Phillips.** 1979. A comparison of three methods for the transport of clinical specimens containing anaerobes. *Med. Lab. Sci.* **36:**159–169.

48. **Gerding, D. N., and J. S. Brazier.** 1993. Optimal methods for identifying *Clostridium difficile* infections. *Clin. Infect. Dis.* **16**(Suppl. 4)**:**S439–442.

49. **Gerding, D. N., S. Johnson, L. R. Peterson, M. E. Mulligan, and J. Silva, Jr.** 1995. *Clostridium difficile*-associated diarrhea and colitis. *Infect. Control Hosp. Epidemiol.* **16:**459–477.

50. **Ghanem, F. M., A. C. Ridpath, W. E. Moore, and L. V. Moore.** 1991. Identification of *Clostridium botulinum*, *Clostridium argentinense*, and related organisms by cellular fatty acid analysis. *J. Clin. Microbiol.* **29:**1114–1124.

50a.**Gröschel, D. H.** 1996. *Clostridium difficile* infection. *Crit. Rev. Clin. Lab. Sci.* **33:**203–245.

51. **Guillot, E., and C. Mouton.** 1996. A PCR-DNA probe assay specific for *Bacteroides forsythus*. *Mol. Cell. Probes* **10:**413–421.

52. **Hagen, J. C., W. S. Wood, and T. Hashimoto.** 1977. Effect of temperature on survival of *Bacteroides fragilis* subsp. *fragilis* and *Escherichia coli* in pus. *J. Clin. Microbiol.* **6:**567–570.

53. **Hall, V., G. L. O'Neill, J. T. Magee, and B. I. Duerden.** 1999. Development of amplified 16S ribosomal DNA restriction analysis for identification of *Actinomyces* species and comparison with pyrolysis-mass spectrometry and conventional biochemical tests. *J. Clin. Microbiol.* **37:**2255–2261.

54. **Hansen, S. L., and B. J. Stewart.** 1976. Comparison of API and Minitek to Center for Disease Control methods for the biochemical characterization of anaerobes. *J. Clin. Microbiol.* **4:**227–231.

55. **Hanson, C. W., R. Cassorla, and W. J. Martin.** 1979. API and Minitek systems in identification of clinical isolates of anaerobic gram-negative bacilli and *Clostridium* species. *J. Clin. Microbiol.* **10:**14–18.

56. **Head, C. B., and S. Ratnam.** 1988. Comparison of API ZYM system with API AN-Ident, API 20A, Minitek Anaerobe II, and RapID-ANA systems for identification of *Clostridium difficile*. *J. Clin. Microbiol.* **26:**144–146.

57. **Hielm, S., E. Hyytia, J. Ridell, and H. Korkeala.** 1996. Detection of *Clostridium botulinum* in fish and environmental samples using polymerase chain reaction. *Int. J. Food Microbiol.* **31:**357–365.

58. **Hofstad, T.** 1980. Evaluation of the API ZYM system for identification of *Bacteroides* and *Fusobacterium* species. *Med. Microbiol. Immunol.* **168:**173–177.

59. **Holdeman, L. V., E. P. Cato, and W. E. C. Moore.** 1977. *Anaerobe Laboratory Manual.* Virginia Polytechnic Institute and State University, Blacksburg.

60. **Holdeman, L. V., and W. E. Moore.** 1975. Identification of anaerobes in the clinical laboratory. *Am. J. Med. Technol.* **41:**411–416.

61. **Holdeman, L. V., and W. E. C. Moore (ed.).** 1987. *Anaerobe Laboratory Manual (Supplement)*, 4th ed. Virginia Polytechnic Institute and State University, Blacksburg.

62. **Holland, R. D., J. G. Wilkes, F. Rafii, J. B. Sutherland, C. C. Persons, K. J. Voorhees, and J. O. Lay, Jr.** 1996. Rapid identification of intact whole bacteria based on spectral patterns using matrix-assisted laser desorption/ionization with time-of-flight mass spectrometry. *Rapid Commun. Mass Spectrom.* **10:**1227–1232.

63. **Holloway, Y., and J. Dankert.** 1979. Identification of anaerobes on the Minitek System, compared to a conventional system. *Zentbl. Bakteriol. Orig. A* **245:**324–331.

64. **Hudspeth, M. K., S. H. Gerardo, D. M. Citron, and E. J. Goldstein.** 1997. Growth characteristics and a novel method for identification (the WEE-TAB system) of *Porphyromonas* species isolated from infected dog and cat bite wounds in humans. *J. Clin. Microbiol.* **35:**2450–2453.

65. **Hussain, Z., R. Lannigan, B. C. Schieven, L. Stoakes, T. Kelly, and D. Groves.** 1987. Comparison of RapID-ANA and Minitek with a conventional method for biochemical identification of anaerobes. *Diagn. Microbiol. Infect. Dis.* **7:**69–72.

66. **Jacobs, J., B. Rudensky, J. Dresner, A. Berman, M. Sonnenblick, Y. van Dijk, and A. M. Yinnon.** 1996. Comparison of four laboratory tests for diagnosis of *Clostridium difficile*-associated diarrhea. *Eur. J. Clin. Microbiol. Infect. Dis.* **15:**561–566.

67. **Jenkins, S. A., D. B. Drucker, M. G. Keaney, and L. A. Ganguli.** 1991. Evaluation of the RAPID ID 32A system for the identification of *Bacteroides fragilis* and related organisms. *J. Appl. Bacteriol.* **71:**360–365.

68. **Johnson, S., and D. N. Gerding.** 1998. *Clostridium difficile*-associated diarrhea. *Clin. Infect. Dis.* **26:**1027–1036.

69. **Jousimies-Somer, H., and P. Summanen.** 1997. Microbiology terminology update: clinically significant anaerobic gram-positive and gram-negative bacteria (excluding spirochetes). *Clin. Infect. Dis.* **25:**11–14.

70. **Jousimies-Somer, H. R., P. H. Summanen, and S. M. Finegold.** 1999. *Bacteroides, Porphyromonas, Prevotella, Fusobacterium*, and other anaerobic gram-negative rods and cocci, p. 690–711. *In* P. R. Murray, E. J. Baron, M. A. Pfaller, F. C. Tenover, and R. H. Yolken (ed.), *Manual of Clinical Microbiology*, 7th ed. ASM Press, Washington, D.C.

71. **Karasawa, T., T. Nojiri, Y. Hayashi, T. Maegawa, K. Yamakawa, X. M. Wang, and S. Nakamura.** 1999. Laboratory diagnosis of toxigenic *Clostridium difficile* by polymerase chain reaction: presence of toxin genes and their stable expression in toxigenic isolates from Japanese individuals. *J. Gastroenterol.* **34:**41–45.

72. **Kato, H., N. Kato, K. Watanabe, N. Iwai, H. Nakamura, T. Yamamoto, K. Suzuki, S. M. Kim, Y. Chong, and E. B. Wasito.** 1998. Identification of toxin A-negative, toxin B-positive *Clostridium difficile* by PCR. *J. Clin. Microbiol.* **36:**2178–2182.

73. **Kato, N., C. Y. Ou, H. Kato, S. L. Bartley, V. K. Brown, V. R. Dowell, Jr., and K. Ueno.** 1991. Identification of toxigenic *Clostridium difficile* by the polymerase chain reaction. *J. Clin. Microbiol.* **29:**33–37.

74. **Kato, N., C. Y. Ou, H. Kato, S. L. Bartley, C. C. Luo, G. E. Killgore, and K. Ueno.** 1993. Detection of toxigenic *Clostridium difficile* in stool specimens by the polymerase chain reaction. *J. Infect. Dis.* **167:**455–458.

75. **Kelley, R. W.** 1982. Phenotypic differentiation of some of the *Veillonella* species with the API ZYM system. *Can. J. Microbiol.* **28:**703–705.

76. **Kitch, T. T., and P. C. Appelbaum.** 1989. Accuracy and reproducibility of the 4-hour ATB 32A method for anaerobe identification. *J. Clin. Microbiol.* **27:**2509–2513.

77. **Koneman, E. W., S. D. Allen, W. M. Janda, P. C. Schreckenberger, and W. C. Winn, Jr.** 1997. *Color Atlas and Textbook of Diagnostic Microbiology*, 5th ed. Lippincott-Raven Publishers, Philadelphia, Pa.

78. **Kreader, C. A.** 1998. Persistence of PCR-detectable *Bacteroides distasonis* from human feces in river water. *Appl. Environ. Microbiol.* **64:**4103–4105.

79. **Krishnamurthy, T., M. T. Davis, D. C. Stahl, and T. D. Lee.** 1999. Liquid chromatography/microspray mass spectrometry for bacterial investigations. *Rapid Commun. Mass Spectrom.* **13:**39–49.

80. **Krishnamurthy, T., and P. L. Ross.** 1996. Rapid identification of bacteria by direct matrix-assisted laser desorption/ionization mass spectrometric analysis of whole cells. *Rapid Commun. Mass Spectrom.* **10:**1992–1996.

81. **Laughon, B. E., S. A. Syed, and W. J. Loesche.** 1982. API ZYM system for identification of *Bacteroides* spp., *Capnocytophaga* spp., and spirochetes of oral origin. *J. Clin. Microbiol.* **15:**97–102.

82. **Levett, P. N.** 1985. Identification of *Clostridium difficile* using the API ZYM system. *Eur. J. Clin. Microbiol.* **4:**505–507.

83. **Limaye, A. P., D. K. Turgeon, B. T. Cookson, and T. R. Fritsche.** 2000. Pseudomembranous colitis caused by a toxin A⁻ B⁺ strain of *Clostridium difficile. J. Clin. Microbiol.* **38:**1696–1697.

84. **Looney, W. J., A. J. Gallusser, and H. K. Modde.** 1990. Evaluation of the ATB 32 A system for identification of anaerobic bacteria isolated from clinical specimens. *J. Clin. Microbiol.* **28:**1519–1524.

85. **Lyerly, D. M., and S. D. Allen.** 1997. The Clostridia, p. 559–623. *In* A. M. Emmerson, P. Hawkey, and S. Gillespie (ed.), *Principles and Practice of Clinical Bacteriology.* John Wiley & Sons, New York, N.Y.

86. **Lyerly, D. M., L. M. Neville, D. T. Evans, J. Fill, S. Allen, W. Greene, R. Sautter, P. Hnatuck, D. J. Torpey, and R. Schwalbe.** 1998. Multicenter evaluation of the *Clostridium difficile* TOX A/B TEST. *J. Clin. Microbiol.* **36:**184–190.

87. **Magee, J. T., J. M. Hindmarch, K. W. Bennett, B. I. Duerden, and R. E. Aries.** 1989. A pyrolysis mass spectrometry study of fusobacteria. *J. Med. Microbiol.* **28:**227–236.

88. **Magee, J. T., J. M. Hindmarch, B. I. Duerden, and L. Goodwin.** 1992. Classification of oral pigmented anaerobic bacilli by pyrolysis mass spectrometry and biochemical tests. *J. Med. Microbiol.* **37:**56–61.

89. **Mangels, J. I.** 1998. Anaerobic bacteriology, p. 127-167. *In* H. D. Isenberg (ed.), *Essential Procedures for Clinical Microbiology.* ASM Press, Washington, D.C.

90. **Marler, L., S. Allen, and J. Siders.** 1984. Rapid enzymatic characterization of clinically encountered anaerobic bacteria with the API ZYM system. *Eur. J. Clin. Microbiol.* **3:**294–300.

91. **Marler, L. M., J. A. Siders, and S. D. Allen.** 2001. *Direct Smear Atlas: a Monograph of Gram-Stained Preparations of Clinical Specimens.* Lippincott Williams & Wilkins, Philadelphia, Pa.

92. **Marler, L. M., J. A. Siders, L. C. Wolters, Y. Pettigrew, B. L. Skitt, and S. D. Allen.** 1991. Evaluation of the new RapID-ANA II system for the identification of clinical anaerobic isolates. *J. Clin. Microbiol.* **29:**874–878.

93. **Matto, J., M. Saarela, S. Alaluusua, V. Oja, H. Jousimies-Somer, and S. Asikainen.** 1998. Detection of *Porphyromonas gingivalis* from saliva by PCR by using a simple sample-processing method. *J. Clin. Microbiol.* **36:**157–160.

94. **Matto, J., M. Saarela, B. von Troil-Linden, S. Alaluusua, H. Jousimies-Somer, and S. Asikainen.** 1996. Similarity of salivary and subgingival *Prevotella interme-dia* and *Prevotella nigrescens* isolates by arbitrarily primed polymerase chain reaction. *Oral Microbiol. Immunol.* **11:**395–401.

95. **Miles, B. L., J. A. Siders, and S. D. Allen.** 1988. Evaluation of a commercial latex test for *Clostridium difficile* for reactivity with *C. difficile* and cross-reactions with other bacteria. *J. Clin. Microbiol.* **26:**2452–2455.

96. **Miller, J. M., and H. T. Holmes.** 1999. Specimen, collection, transport and storage, p. 33-63. *In* P. R. Murray, E. J. Baron, M. A. Pfaller, F. C. Tenover, and R. H. Yolken (ed.), *Manual of Clinical Microbiology.* ASM Press, Washington, D.C.

97. **Miller, P. H., L. S. Wiggs, and J. M. Miller.** 1995. Evaluation of API An-IDENT and RapID ANA II systems for identification of *Actinomyces* species from clinical specimens. *J. Clin. Microbiol.* **33:**329–330.

98. **Miwa, N., T. Nishina, S. Kubo, and K. Fujikura.** 1996. Nested polymerase chain reaction for detection of low levels of enterotoxigenic *Clostridium perfringens* in animal feces and meat. *J. Vet. Med. Sci.* **58:**197–203.

99. **Moll, W. M., J. Ungerechts, G. Marklein, and K. P. Schaal.** 1996. Comparison of BBL Crystal ANR ID Kit and API rapid ID 32 A for identification of anaerobic bacteria. *Zentbl. Bakteriol.* **284:**329–347.

100. **Moncrief, J. S., L. Zheng, L. M. Neville, and D. M. Lyerly.** 2000. Genetic characterization of toxin A-negative, toxin B-positive *Clostridium difficile* isolates by PCR. *J. Clin. Microbiol.* **38:**3072–3075.

101. **Moore, L. V., D. M. Bourne, and W. E. Moore.** 1994. Comparative distribution and taxonomic value of cellular fatty acids in thirty-three genera of anaerobic gram-negative bacilli. *Int. J. Syst. Bacteriol.* **44:**338–347.

102. **Mosca, A., P. Summanen, S. M. Finegold, G. De Michele, and G. Miragliotta.** 1998. Cellular fatty acid composition, soluble-protein profile, and antimicrobial resistance pattern of *Eubacterium lentum. J. Clin. Microbiol.* **36:**752–755.

103. **Moss, C. W.** 1984. Chemotaxonomic studies of microorganisms using gas chromatography, mass spectrometry, and associated analytical techniques, p. 63–70. *In* A. Sanna and G. Morace (ed.), *New Horizons in Microbiology.* Elsevier Science Publishers, New York, N.Y.

104. **Moss, C. W., and S. B. Dees.** 1975. Identification of microorganisms by gas chromatographic-mass spectrometric analysis of cellular fatty acids. *J. Chromatogr.* **112:**594–604.

105. **Moss, C. W., V. R. Dowell, Jr., D. Farshtchi, L. J. Raines, and W. B. Cherry.** 1969. Cultural characteristics and fatty acid composition of propionibacteria. *J. Bacteriol.* **97:**561–570.

106. **Moss, C. W., and V. J. Lewis.** 1967. Characterization of clostridia by gas chromatography. I. Differentiation of species by cellular fatty acids. *Appl. Microbiol.* **15:**390–397.

107. **Murdoch, D. A.** 1998. Gram-positive anaerobic cocci. *Clin. Microbiol. Rev.* **11:**81–120.

108. **Murdoch, D. A.** 1999. Reclassification of *Peptostreptococcus magnus* (Prevot 1933) Holdeman and Moore 1972 as *Finegoldia magna* comb. nov. and *Peptostreptococcus micros* (Prevot 1933) Smith 1957 as *Micromonas micros* comb. nov. *Anaerobe* **5:**555–559.

109. **Murdoch, D. A., H. N. Shah, S. E. Gharbia, and D. Rajendram.** 2000. Proposal to restrict the genus *Peptostreptococcus* (Kluyver & van Niel 1936) to *Peptostreptococcus anaerobius. Anaerobe* **6:**257–260.

110. **Murray, P. R., E. J. Baron, M. A. Pfaller, F. C. Tenover, and R. H. Yolken (ed.).** 1999. *Manual of Clinical Microbiology,* 7th ed. ASM Press, Washington, D.C.

111. **Pattyn, S. R., M. Ieven, and L. Buffet.** 1993. Comparative evaluation of the Rapid ID 32A kit system, miniaturized standard procedure and a rapid fermentation procedure for the identification of anaerobic bacteria. *Acta Clin. Belg.* **48:**81–85.

112. **Peterson, L. R., P. J. Kelly, and H. A. Nordbrock.** 1996. Role of culture and toxin detection in laboratory testing for diagnosis of *Clostridium difficile*-associated diarrhea. *Eur. J. Clin. Microbiol. Infect. Dis.* **15:**330–336.

113. **Quentin, C., M. A. Desailly-Chanson, and C. Bebear.** 1991. Evaluation of AN-Ident. *J. Clin. Microbiol.* **29:**231–235.

114. **Rodloff, A. C., S. L. Hillier, and B. J. Moncla.** 1999. *Peptostreptococcus, Propionibacterium, Lactobacillus, Actinomyces,* and other non-spore-forming anaerobic gram-positive bacteria, p. 672–689. *In* P. R. Murray, E. J. Baron, M. A. Pfaller, F. C. Tenover, and R. H. Yolken (ed.), *Manual of Clinical Microbiology,* 7th ed. ASM Press, Washington, D.C.

115. **Rosenblatt, J. E.** 1985. Anaerobic identification systems. *Clin. Lab. Med.* **5:**59–65.

116. **Saito, M., M. Matsumoto, and M. Funabashi.** 1992. Detection of *Clostridium perfringens* enterotoxin gene by the polymerase chain reaction amplification procedure. *Int. J. Food Microbiol.* **17:**47–55.

117. **Sambol, S. P., M. M. Merrigan, D. Lyerly, D. N. Gerding, and S. Johnson.** 2000. Toxin gene analysis of a variant strain of *Clostridium difficile* that causes human clinical disease. *Infect. Immun.* **68:**5480–5487.

118. **Schoepe, H., H. Potschka, T. Schlapp, J. Fiedler, H. Schau, and G. Baljer.** 1998. Controlled multiplex PCR of enterotoxigenic *Clostridium perfringens* strains in food samples. *Mol. Cell. Probes* **12:**359–365.

119. **Schreckenberger, P. C., D. M. Celig, and W. M. Janda.** 1988. Clinical evaluation of the Vitek ANI card for identification of anaerobic bacteria. *J. Clin. Microbiol.* **26:**225–230.

120. **Shah, H. N., and S. E. Gharbia.** 1991. *Bacteroides* and *Fusobacterium* classification and relationship to other bacteria, p. 62–84. *In* B. I. Duerden and B. S. Drasar (ed.), *Anaerobes in Human Diseases.* Arnold, London, United Kingdom.

121. **Shetab, R., S. H. Cohen, T. Prindiville, Y. J. Tang, M. Cantrell, D. Rahmani, and J. Silva, Jr.** 1998. Detection of *Bacteroides fragilis* enterotoxin gene by PCR. *J. Clin. Microbiol.* **36:**1729–1732.

122. **Siders, J. A.** 1992. Prereduced anaerobically sterilized biochemicals, p. 2.7. *In* H. D. Isenberg (ed.), *Clinical Microbiology Procedures Handbook.* American Society for Microbiology, Washington, D.C.

123. **Slots, J.** 1981. Enzymatic characterization of some oral and nonoral gram-negative bacteria with the API ZYM system. *J. Clin. Microbiol.* **14:**288–294.

124. **Slots, J., A. Ashimoto, M. J. Flynn, G. Li, and C. Chen.** 1995. Detection of putative periodontal pathogens in subgingival specimens by 16S ribosomal DNA amplification with the polymerase chain reaction. *Clin. Infect. Dis.* **20**(Suppl. 2):S304–307.

125. **Smith, L. D. S., and B. L. Williams.** 1984. *The Pathogenic Anaerobic Bacteria,* 3rd ed. Charles C. Thomas, Springfield, Ill.

126. **Stackebrandt, E., and B. M. Goebel.** 1994. Taxonomic note: a place for DNA-DNA reassociation and 16S

127. **Stackebrandt, E., W. Liesack, and D. Witt.** 1992. Ribosomal RNA and rDNA sequence analyses. *Gene* **115:**255–260.

128. **Staneck, J. L., L. S. Weckbach, S. D. Allen, J. A. Siders, P. H. Gilligan, G. Coppitt, J. A. Kraft, and D. H. Willis.** 1996. Multicenter evaluation of four methods for *Clostridium difficile* detection: ImmunoCard C. *difficile,* cytotoxin assay, culture, and latex agglutination. *J. Clin. Microbiol.* **34:**2718–2721.

129. **Stargel, D., F. S. Thompson, S. E. Phillips, G. L. Lombard, and V. R. Dowell, Jr.** 1976. Modification of the Minitek Miniaturized Differentiation System for characterization of anaerobic bacteria. *J. Clin. Microbiol.* **3:**291–301.

130. **Stargel, M. D., G. L. Lombard, and V. R. Dowell, Jr.** 1978. Alternative procedures for identification of anaerobic bacteria. *Am. J. Med. Technol.* **44:**709–722.

131. **Stoakes, L., T. Kelly, K. Manarin, B. Schieven, R. Lannigan, D. Groves, and Z. Hussain.** 1990. Accuracy and reproducibility of the MicroScan rapid anaerobe identification system with an automated reader. *J. Clin. Microbiol.* **28:**1135–1138.

132. **Summanen, P., E. J. Baron, D. M. Citron, C. A. Strong, H. M. Wexler, and S. M. Finegold.** 1993. *Wadsworth Anaerobic Bacteriology Manual,* 5th ed. Star Publishing, Belmont, Calif.

133. **Summanen, P., and H. Jousimies-Somer.** 1988. Comparative evaluation of RapID ANA and API 20 A for identification of anaerobic bacteria. *Eur. J. Clin. Microbiol. Infect. Dis.* **7:**771–775.

134. **Takeuchi, S., N. Hashizume, T. Kinoshita, T. Kaidoh, and Y. Tamura.** 1997. Detection of *Clostridium septicum* hemolysin gene by polymerase chain reaction. *J. Vet. Med. Sci.* **59:**853–855.

135. **Tanner, A. C., M. N. Strzempko, C. A. Belsky, and G. A. McKinley.** 1985. API ZYM and API An-Ident reactions of fastidious oral gram-negative species. *J. Clin. Microbiol.* **22:**333–335.

136. **Tharagonnet, D., P. R. Sisson, C. M. Roxby, H. R. Ingham, and J. B. Selkon.** 1977. The API ZYM system in the identification of Gram-negative anaerobes. *J. Clin. Pathol.* **30:**505–509.

137. **Traci, P. A., and C. L. Duncan.** 1974. Cold shock lethality and injury in *Clostridium perfringens. Appl. Microbiol.* **28:**815–821.

138. **Tuner, K., E. J. Baron, P. Summanen, and S. M. Finegold.** 1992. Cellular fatty acids in *Fusobacterium* species as a tool for identification. *J. Clin. Microbiol.* **30:**3225–3229.

139. **van Winkelhoff, A. J., M. Clement, and J. de Graaff.** 1988. Rapid characterization of oral and nonoral pigmented *Bacteroides* species with the ATB Anaerobes ID system. *J. Clin. Microbiol.* **26:**1063–1065.

140. **Wang, R. F., W. W. Cao, and C. E. Cerniglia.** 1996. Phylogenetic analysis of *Fusobacterium prausnitzii* based upon the 16S rRNA gene sequence and PCR confirmation. *Int. J. Syst. Bacteriol.* **46:**341–343.

141. **Welch, D. F.** 1991. Applications of cellular fatty acid analysis. *Clin. Microbiol. Rev.* **4:**422–438.

142. **Welham, K. J., M. A. Domin, D. E. Scannell, E. Cohen, and D. S. Ashton.** 1998. The characterization of micro-organisms by matrix-assisted laser desorption/ionization time-of-flight mass spectrometry. *Rapid Commun. Mass Spectrom.* **12:**176–180.

rRNA sequence analysis in the present species definition in bacteriology. *Int. J. Syst. Bacteriol.* **44:**846–849.

143. **Whaley, D. N., L. S. Wiggs, P. H. Miller, P. U. Srivastava, and J. M. Miller.** 1995. Use of Presumpto Plates to identify anaerobic bacteria. *J. Clin. Microbiol.* **33:**1196–1202.

144. **Wimberley, N., L. J. Faling, and J. G. Bartlett.** 1979. A fiberoptic bronchoscopy technique to obtain uncontaminated lower airway secretions for bacterial culture. *Am. Rev. Respir. Dis.* **119:**337–343.

145. **Yamagishi, T., K. Sugitani, K. Tanishima, and S. Nakamura.** 1997. Polymerase chain reaction test for differentiation of five toxin types of *Clostridium perfringens*. *Microbiol. Immunol.* **41:**295–299.

146. **Yamashita, Y., S. Kohno, H. Koga, K. Tomono, and M. Kaku.** 1994. Detection of *Bacteroides fragilis* in clinical specimens by PCR. *J. Clin. Microbiol.* **32:**679–683.

Rapid Systems and Instruments for the Identification of Viruses

MARILYN A. MENEGUS

5

The past 10 to 20 years have brought tremendous changes to the clinical virology laboratory. The laboratory diagnosis of viral diseases has evolved from an almost exclusively tissue culture-based, relatively slow, and somewhat tedious evaluation to a very broadly based potpourri of test types, from immunofluorescence analyses (direct and indirect fluorescent-antibody assays [DFA and IFA] and direct specimen evaluation versus tissue culture confirmation) to enzyme immunoassay (EIA) formats (manual and automated), antigen detection methods, electron microscopy (EM), and molecular amplification methods. Although even in today's clinical virology laboratory routine tissue culture (with the additional use of shell vials) remains the backbone and common "gold standard" to which other evolving methods and kits are compared, most investigators realize that with the introduction of amplification methods, the real gold standard may be undergoing a paradigm shift from tissue culture to more sensitive test formats.

Virus groups, including influenza virus, parainfluenza virus, adenovirus, respiratory syncytial virus (RSV), rotavirus, herpes simplex virus (HSV), varicella-zoster virus (VZV), and cytomegalovirus (CMV), are just some of the important disease-producing agents which are easier to detect and identify due to the many commercial kits and products which are available for the modern clinical virology laboratory. Unfortunately, other clinically important viruses, such as Epstein-Barr virus, rhinoviruses, caliciviruses, astroviruses and other diarrheic viruses, human parvoviruses, and the agents responsible for transmissible spongiform encephalopathies, are still difficult or impossible to isolate and identify in the routine clinical virology laboratory, and commercial methods are either unavailable or in development. Luckily, many antibody-based methods are available either in the hospital laboratory or through large reference laboratories and public health laboratories.

In this chapter, I review the most common commercially available kits and methods for the routine clinical virology laboratory. Due to the rapidly evolving nature of this discipline, I encourage the reader to refer to other authoritative references, such as American Society for Microbiology journals, Cumitechs, and other timely publications, for current and new reviews and evaluations of commercial products, reagents, and instruments. The molecular microbiology section of this manual contains additional information relating to nucleic acid-based products used in clinical virology.

INFLUENZA VIRUSES

Influenza viruses are enveloped, single-stranded, minus-sense RNA viruses that are classified into three groups, A, B, and C, based on the antigenic reactivity of the group-specific nucleocapsid and matrix proteins. The influenza A group is further subtyped based on the antigenic reactivity of the major surface antigens of the virus, hemagglutinin and neuraminidase. The segments (A and B, eight segments; C, seven segments) of the viral genome range in size from approximately 900 to 2,400 nucleotides and predispose the virus to reassortment in vivo (influenza A virus) and in vitro. Reassortment in vivo is important because it results in new virus strains, in particular those responsible for pandemics (89).

Outbreaks of influenza occur annually during the winter. The virus infects the columnar epithelial cells of the upper respiratory tract, and in the majority of those infected, the virus causes a disease characterized by fever, pharyngitis, and myalgias. Less frequently, virus spreads to the lower respiratory tract, causing tracheobronchitis and pneumonia. However, pneumonia associated with influenza virus infection is generally due to secondary bacterial infection.

For years, serology and culture were the only laboratory methods available for establishing the diagnosis of influenza virus infection. The impracticalities associated with obtaining acute- and convalescent-phase sera and the 2- to 3-week delays imposed by employing serology led to greater emphasis on culture as a diagnostic tool. However, traditional culture also has limitations. Depending on the laboratory practices, virus isolation and identification may take days to a week or two (49). The delays associated with both serology and culture prompted exploration of antigen detection strategies for the direct detection of virus in clinical specimens (89).

In the late 1970s and early 1980s, Gardner and McQuillin (23) popularized the application of fluorescent antibodies (FA) to exfoliated respiratory epithelial cells for the diagnosis of virus infections. However, widespread use of immunodiagnostic methods did not occur until the mid-1980s, when commercially developed kits based on the use of monoclonal antibody reagents were developed. Enzyme-linked immunosorbent assay (ELISA) and immunofluorescence kits in a variety of formats were introduced. The latter are discussed below.

TABLE 1. Rapid diagnostic tests for influenza virus

Trade or proprietary name	Manufacturer	Yr cleared by FDA	Virus detected	Method or format
Directigen Flu A	Becton Dickinson & Co., Cockeysville, Md.	1990	A	ELISA
BioStar AB FLU OIA	BioStar, Inc., Boulder, Colo.	1998	A, B[a]	OIA
ZStatFlu Test for Influenza A & B Virus	ZymeTx, Inc., Oklahoma City, Okla.	1998	A, B[a]	Neuraminidase detection
QuickVue Influenza A/B Test	Quidel Corp., San Diego, Calif.	1999	A, B[a]	Lateral ELISA
Roche Diagnostics Influenza A/B Rapid Test[b]	Roche Diagnostics Corp., Indianapolis, Ind.	1999	A, B[a]	Gold-labeled immunoassay
Directigen Flu A+B	Becton Dickinson	2000	A, B[c]	ELISA

[a]Does not distinguish between influenza A and B viruses.
[b]Not currently marketed.
[c]Distinguishes between influenza A and B viruses.

In 1990, Becton Dickinson (Sparks, Md.) introduced the first ELISA for influenza virus. The assay was configured as a single test device contained in a plastic cassette, and later, several similar devices were introduced. Without exception, antigen detection assays for influenza virus are less sensitive than culture, but they do offer something that culture does not: same-day diagnosis. Antigen detection assays are now widely used and are of particular value in clinical situations that require prompt action, such as instituting antiviral therapy and infection control measures.

The commercially available single-test devices for the diagnosis of influenza are listed in Table 1 and illustrated in Fig. 1 to 3. These devices employ different strategies for detection, and the resulting "answer" each provides varies depending on the assay. Some simply score the specimen as positive or negative for influenza virus, while others identify the virus detected as influenza A or B virus.

BD Directigen Flu A and Flu A+B ELISA

The BD Directigen Flu A assay (Becton Dickinson), introduced 10 years ago, detects only influenza A virus, and of all the assays listed, it is the most extensively evaluated. A similarly constructed assay, the Directigen Flu A+B assay (Fig. 2), was cleared by the Food and Drug Administration (FDA) in 2000. The new assay, which separately detects

both influenza A and B virus antigens, was introduced during the 2000–2001 influenza season. The Directigen Flu A+B assay was compared to the old Directigen Flu A assay in a study done in our laboratory that included 216 specimens from mixed sources; 44 (20%) were influenza A virus culture positive. In that study, the sensitivities and specificities of the two assays were equivalent. Therefore, it is reasonable to expect the performance characteristics for influenza A virus determined for the old assay to be similar to those of the new assay (D. W. Newton, W. L. Kuhnert, C. Mayer, and M. A. Menegus, *Abstr. 16th Annu. Meet. Pan Am. Group Rapid Viral Diagn.*, abstr. S17, 2000).

There are eight steps involved in performing the Directigen assays. The steps include specimen extraction, followed by addition of the extracted specimen to the device membrane. If present, the antigen nonspecifically binds to the membrane. Enzyme-conjugated antibody and substrate are then added, with the appropriate wash steps in between. A colored triangle represents a positive finding. The kit also includes positive and negative control samples and a process control in each device that develops only if the test is properly performed.

Sensitivities ranging from 41 to 100% and specificities ranging from 71 to 100% have been reported in independently performed studies (12, 20, 31, 38, 41, 58, 61, 83; D. W. Newton, C. F. Mellen, and M. A. Menegus, *Abstr.*

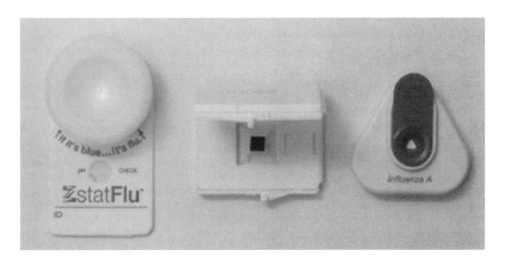

FIGURE 1 ZstatFlu (left), FLU OIA (middle), and Directigen Flu A (right).

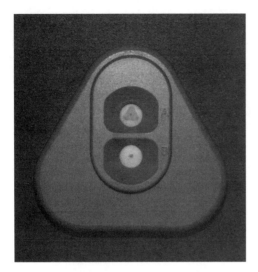

FIGURE 2 Directigen Flu A+B device.

99th Gen. Meet. Am. Soc. Microbiol., abstr. C-319, 1999). The performance variation reported in these studies is not unexpected given the many variables that play an important role in determining the performance of the assay under evaluation and the standard against which it is being measured (32, 37).

The Directigen Flu A+B assay was first evaluated during the 1999-2000 influenza season. The manufacturer-sponsored evaluation was conducted in six different hospital laboratories and included a variety of clinical specimens (n = 1,262); 232 were influenza A virus and 52 were influenza B virus culture positive. The overall sensitivity and specificity of the assay for influenza A virus were 86.2 and 90.7%, and for influenza B virus they were 80.8 and 99.5%, respectively. Variation in performance did occur among the six laboratories, with sensitivities for influenza A virus ranging from 68 to 98% and specificities ranging from 92 to 100%. Only one laboratory reported a significant number of influenza B virus isolates (Newton et al., *Abstr. 16th Annu. Meet. Pan Am. Group Rapid Viral Diagn.*; W. Green and L. McCallister, *Abstr. 16th Annu. Meet. Pan Am. Group Rapid Viral Diagn.*, abstr. S18, 2000; S. J. Todd, G. Thompkins, P. Murphy, and J. L. Waner, *Abstr. 17th Annu. Meet. Pan Am. Soc. Clin. Virol.*, abstr. T17, 2001).

BioStar FLU OIA

Another antigen detection kit, the BioStar, Inc. (Boulder, Colo.) FLU OIA (optical immunoassay) (Fig. 1), employs a unique mechanism for visualizing antigen-antibody complexes. The device detects both influenza A virus and influenza B virus but does not distinguish between them. Viral nucleoprotein in the specimen is extracted and placed on a silicon wafer in the test device. The wafer is an optical surface coated with influenza A and B virus-specific nucleoprotein antibodies. Antigen is captured by antibody, and this binding creates a molecular complex that changes the light path reflected by the optical surface. The alteration in the light path appears to the observer as a purple spot against the unaltered gold background. When no nucleoprotein is present, the optical thickness remains unchanged and the surface retains its original gold color, indicating a negative result. Internal procedural control dots appear if the test steps are performed correctly.

Several investigators have examined the performance of the BioStar FLU OIA. A recently published study by Covalciuc et al. (14) presents the performance characteristics determined by the manufacturer. The study was performed with a selected patient population, restricted to patients seen in an outpatient setting who were ill for <36 h, had a temperature of >100°F, and had at least two influenzalike symptoms. Whenever possible, multiple specimen types were collected from a single patient. Four hundred and four specimens were collected from 182 patients. Influenza virus was recovered from 92 patients (50%) and from 151 specimens (37%). Unfortunately, identification of the recovered viruses is not included in the report, so it is impossible to calculate virus-specific performance characteristics. With 14-day culture as the gold standard, the overall sensitivity and specificity of the assay for influenza virus were found to be 80.1 and 73.1%, respectively. Thirty-two specimens that were OIA positive but culture negative were examined by an influenza virus-specific reverse transcription (RT)-PCR assay, and 21 (66%) were positive. The data suggest that in some cases the OIA detects influenza in culture-negative specimens. In a small study from Japan, reported by Yamazaki et al. (88), the FLU OIA was compared to both culture and PCR. All of the viruses recovered were influenza B. The sensitivities of the FLU OIA were 85.5 and 81.6% compared to culture and RT-PCR, respectively, and the respective specificities were 65.2 and 72.2%.

QuickVue Influenza A/B

Two additional immunoassays, the QuickVue Influenza A/B Test (Quidel Corp., San Diego, Calif.) (Fig. 3) and the Influenza A/B Rapid Test (Roche Diagnostics Corp., Indianapolis, Ind.) were recently cleared by the FDA. Like the FLU OIA, the QuickVue and Roche assays detect both influenza A and B viruses but do not identify the specific virus type detected. The assays are both simple and are designed to be performed by individuals with a wide range of experience and technical skill. Each test employs test strips impregnated with influenza A and B virus antibodies. The strips are placed in tubes containing extracted samples and microparticles charged with influenza A and B virus monoclonal antibodies. If antigen is present in the sample, it forms a complex with the microparticle-conjugated antibodies. The test strip is placed in the sample, and the sample migrates vertically towards the antibody-impregnated area on the test strip. If antigen is present in the sample, microparticle-antigen complex is captured in the antibody-impregnated region and a visible line forms (positive result). A similar strategy is used for the negative control reaction. Unfortunately, these assays have not yet been independently evaluated. The Roche assay is presently being used in the United States for surveillance studies, but it is not available for sale to clinical laboratories.

ZstatFlu Assay

The final influenza virus assay discussed in this section makes use of a different principle for virus detection. As previously noted, an enzyme, neuraminidase, is one of the major surface proteins of both influenza A and B. The ZstatFlu assay (ZymeTx, Inc., Oklahoma City, Okla.) (Fig. 1) makes use of the enzymatic activity of neuraminidase in its detection strategy. The assay is performed by mixing the clinical specimen with a chromogenic neuraminidase substrate. If influenza virus is present in the specimen, the substrate is cleaved to produce free chromogen, an insoluble

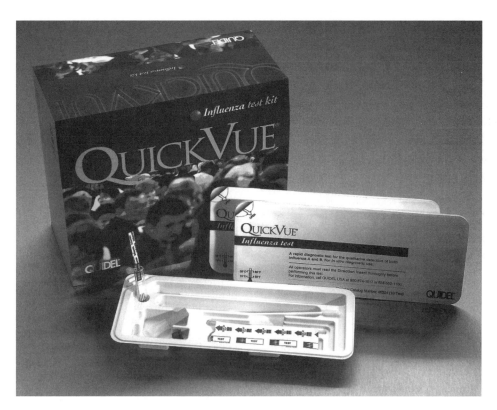

FIGURE 3 QuickVue. Courtesy of Quidel Corp.

blue precipitate. The mixture is added to a device that concentrates the precipitate by filtration, a step that facilitates visualization of the color reaction. The substrate used in the assay is specific and does not react with the neuraminidases produced by the human host, bacteria, or other viruses.

In a recent study, Noyola et al. (52) compared the abilities of the ZstatFlu assay, the Directigen Flu A assay, immunofluorescence, and culture to detect influenza virus in 479 nasal wash specimens. Influenza virus was recovered in cell culture from 124 (61.7%) of the specimens; 102 isolates were influenza A virus, and 22 were influenza B virus. Compared to culture, the sensitivity and specificity of the ZstatFlu assay were 70.1 and 92.4%, respectively. The sensitivity of the ZstatFlu assay was greater for influenza A virus (76.4%) than for influenza B virus (40.9%). Compared to the other two immunoassays examined, the ZstatFlu assay was more sensitive than indirect immunofluorescence for the overall detection of influenza virus (70.1 versus 59.8%) but less sensitive than the Directigen Flu A assay for the detection of influenza A virus (76.4 versus 89.7%). The authors of this study also demonstrated that two variables, specimen quality and the experience of the technologist performing the assay, significantly affected the performance characteristics of the assays being studied. They also speculated that the greater sensitivity (96%) and lower specificity (77%) they observed in an earlier evaluation of the ZstatFlu assay (51) might be due to the use of other specimen types in that study.

The important role that factors other than the specific test device employed play in determining the sensitivity of an assay cannot be overemphasized. This point is illustrated by the results of a study by Kaiser et al. (32), who, using an experimental human infection, examined the influence of specimen type and duration of infection on the performance of culture and the Directigen Flu A assay. Four specimens were collected from 14 volunteers daily for 8 days following infection. The following aggregate specimen-specific sensitivities for the 8-day period were recorded for cell culture: nasopharyngeal-wash specimens, 63%; throat gargles, 46%; nasal swabs, 46%; and throat swabs, 24%. Daily comparison of the Directigen Flu A assay to cell culture in combined nasopharyngeal and throat swabs revealed sensitivities for the Directigen Flu A assay of 78% for specimens collected 3 days postinoculation and 25% for specimens collected 4 days postinoculation. Since extrinsic factors greatly influence sensitivity determinations, they should always be considered when device evaluations are reviewed.

RSV

RSV is a large, pleomorphic, enveloped virus approximately 300 to 350 nm in diameter with a single-stranded, negative-sense RNA genome composed of approximately 15,000 nucleotides. RSV proteins are similar to the proteins of other paramyxoviruses and include a nucleocapsid protein, a phosphoprotein, a large protein believed to have polymerase activity, a matrix protein, and the two surface proteins (G, the attachment glycoprotein, and F, the viral fusion protein). There are two major antigenic groups of RSV, A and B, and additional antigenic variability occurs within the groups. The most extensive antigenic and genetic diversity is found in the attachment glycoprotein, G. During individual epidemic periods, viruses of both antigenic groups may cocirculate or viruses of one group may predominate (71).

RSV is a major cause of viral lower respiratory tract infections among infants and young children worldwide. In the United States alone, approximately 91,000 infants are hospitalized with RSV infections yearly, at an estimated annual cost of at least $300 million. The burden of RSV infections is even greater if outpatient visits for children and adults and RSV morbidity in patients with underlying conditions are included (27).

Unlike influenza virus, RSV is relatively difficult to recover in cell culture. The virus is rather unstable, and as a consequence, viability is often lost during transport to the laboratory. This problem is compounded by the fact that the variables associated with successful virus propagation in cell culture are poorly understood, and reproducible sensitive virus recovery may be difficult to achieve. Further, recovery of RSV in cell culture generally takes 5 to 7 days. The difficulties associated with RSV recovery and its importance as a cause of lower respiratory tract disease in infants and young children made RSV one of the first agents targeted for detection by immunoassay methods.

In the late 1970s, Gardner and McQuillin (23) were the first to explore the use of immunoassays for the direct detection of RSV. They applied indirect immunofluorescence to exfoliated respiratory epithelial cells obtained by nasal aspiration and found the method substantially equivalent to virus recovery in cell culture. However, the technique did not immediately gain widespread acceptance because of the limited availability of quality detector antibody. It was the development of monoclonal antibodies in the early 1980s that led to the expanded use of immunofluorescence as a diagnostic tool and stimulated the commercial development of ELISA in various formats. The use of immunofluorescence for the diagnosis of RSV is discussed in detail in the next section.

Most of the immunoassays for RSV were developed and extensively evaluated in the early 1980s. In general, the outcomes of the many evaluations done during this period were as follows: (i) immunoassays are only slightly less sensitive than optimally performed isolation in cell culture, (ii) IFA and DFA are slightly more sensitive than EIA methods, but overall (iii) the specificity of EIA methods is greater than that of FA methods, largely due to the interpretative nature of the latter approach. In addition, many of the evaluations performed emphasize the importance of specimen quality as a determinant of performance. The greater the number of cells and amount of nasal secretion collected, the greater the sensitivity of the evaluation. Consequently, the sensitivity achieved using nasal aspirates and washes is greater than that achieved using swab specimens. In fact, specimen quality is likely a much more important variable in defining performance than the type of kit or assay strategy used.

VIDAS RSV EIA

VIDAS is the registered trademark for a group of antibody and antigen assays produced by bioMérieux-VITEK, Hazelwood, Mo. The assays are fully automated, with two available instruments, the VIDAS 30 (Fig. 4) and the miniVIDAS, with capacities of 300 and 80 EIAs/8-h shift, respectively. All test reagents are provided in bar-coded plastic multiwell strips (Fig. 5). The sample is processed and placed in the first well of the strip, and the strips and test-specific pipettes are then placed in the VIDAS instrument. Test-specific pipette tips provide the solid-phase receptacle (SPR [Fig. 5]) for the reaction. The inside of the SPR is coated with capture antibody at the time of manufacture. After the strips and SPRs are placed in the instrument, it carries out all of the remaining steps of the assay. Each of the VIDAS instruments is divided into independently functioning sections. Tests with similar protocols may be run at the same time in a single section. Tests with different protocols may also be run at the same time, but in different sections of the instrument.

The VIDAS RSV assay was compared to DFA in a recent study of 231 clinical specimens that reported 92% agreement between the two assays and 82% sensitivity and 94% specificity for the VIDAS assay (26). Miller et al. (48) also reported similar performance characteristics for the VIDAS assay.

FIGURE 4 VIDAS instrument.

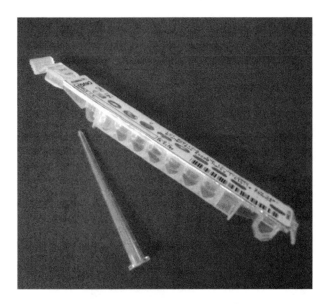

FIGURE 5 VIDAS cuvettes and tip.

Pathfinder RSV EIA

The Pathfinder RSV EIA (Bio-Rad Laboratories, Hercules, Calif.) is a standard ELISA structured in a tube format. Specimen is added to tubes coated with polyclonal capture antibody to RSV (Long), and peroxidase-conjugated monoclonal antibodies specific for epitopes on the group-specific nucleocapsid antigen are used as detector antibodies. The color reaction, which can be read either visually or spectrophotometrically, develops with the addition of substrate. Several investigators (30, 53, 55, 70) examined the performance of the Pathfinder RSV EIA relative to culture and DFA; the sensitivities and specificities reported in these studies ranged from 68.2 to 92% and 80.7 to 99%, respectively.

Abbott TestPack RSV

The FDA first cleared the Abbott Laboratories (North Chicago, Ill.) TestPack RSV assay in 1988. The assay is a single-use ELISA device that makes use of antibody-coated microparticles to capture antigen in the specimen and an enzyme-labeled antibody for detection. The specimen is reacted with the microparticles and conjugated antibody and then added to the device. A membrane in the device retains the antigen-microparticle complex, and chromogenic substrate is then added, resulting in a color reaction. Specimens containing RSV produce a plus sign, while those that do not contain RSV produce a minus sign due to a similarly constructed internal control.

A number of investigators have compared the Abbott TestPack RSV assay to conventional culture, shell vial culture, immunofluorescence methods, or other ELISA methods (20, 28, 46, 48, 54, 60, 69, 72, 75, 87). Almost without exception, the sensitivity of the Abbott TestPack was >90% relative to the gold standards against which it was compared; the specificities were generally also greater than 90%. In four studies (28, 46, 47, 60), the Abbott TestPack RSV assay was directly compared to the RSV Directigen assay, and in three (28, 47, 60) the performance of the Abbott TestPack RSV was superior to that of the RSV

Directigen assay, whereas in one study the two assays were equivalent (46). The sensitivity of the Abbott TestPack RSV assay was also found to be superior to those of the VIDAS and Pathfinder ELISA in two studies that directly compared the tests (48, 87).

BD Directigen RSV Assay

The BD Directigen RSV assay (Becton Dickinson) is a single-test ELISA device similar to the Directigen Flu A device discussed in the previous section and pictured in Fig. 1. The test kit in its entirety is shown in Fig. 6. Lipson et al. (39) recently evaluated the performance of the Directigen RSV assay under "STAT laboratory" conditions. The authors compared the Directigen RSV assay to immunofluorescence and culture. The sensitivity, specificity, and positive and negative predictive values for Directigen, immunofluorescence, and culture were 71, 91, 85, and 80%; 98, 100, 100, and 99%; and 51, 100, 100, and 72%, respectively. As previously noted, in four studies (28, 46, 47, 60) the Abbott TestPack RSV assay was directly compared to the Directigen RSV assay, and in three (28, 47, 60) the performance of the Abbott TestPack RSV assay was superior to that of the Directigen RSV assay. The relative sensitivities of the Directigen and TestPack in these three studies were 83 versus 91%, 76 versus 92%, and 75.8 versus 93.6%, respectively. Overall, the Directigen RSV assay does not appear to be as sensitive as immunofluorescence or the Abbott TestPack.

IMMUNOFLUORESCENCE STAINING OF RESPIRATORY VIRUSES

A number of techniques, including cell culture, antigen detection by ELISA and immunofluorescence, and PCR, can be used to detect respiratory viruses in clinical specimens. However, these techniques differ considerably in a number of variables, including (i) cost, (ii) spectrum of viruses detected, (iii) time to a final result, and (iv) performance characteristics. The most comprehensive and sensitive method for detecting respiratory viruses is cell culture, but it is costly, takes days to weeks to achieve a final result, and requires considerable expertise. Unfortunately, as discussed above, commercial ELISA kits are available for only two viruses, influenza virus and RSV. These assays are of great value in targeted situations, but more than half of the viruses recovered from respiratory tract specimens are not covered by ELISA kits.

Gardner and McQuillan (23) comprehensively examined the value of immunofluorescence staining for the diagnosis of respiratory tract infections in the early 1980s. Their studies were based on the use of polyclonal antibodies that were not available to the clinical laboratory community. Then, in the mid-1980s the development of monoclonal antibody technology led to the production of many high-quality commercial products and consequently to the widespread use of FA staining as a diagnostic tool. Monoclonal antibody reagents that detect influenza A and B viruses, parainfluenza viruses 1, 2, and 3, and adenoviruses are now available through a number of sources, including Chemicon International, Inc., Temecula, Calif.; Trinity Biotech, Dublin, Ireland; Diagnostic Products Corporation, Los Angeles, Calif.; and Dako Diagnostics Ltd., Cambridge, Cambridgeshire, United Kingdom. Fluorescein isothiocyanate (FITC)-conjugated and unconjugated monoclonal antibody reagents are available for DFA and IFA, respectively. In addition to providing monospecific reagents, some

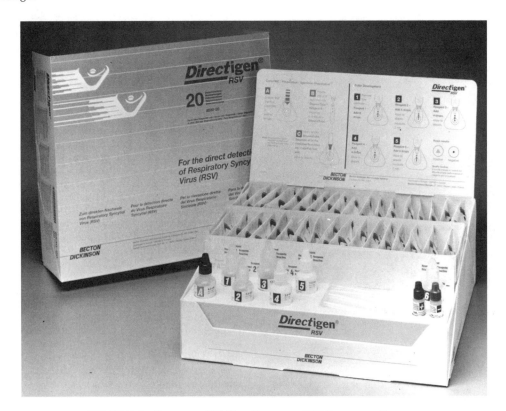

FIGURE 6 Directigen RSV kit. Courtesy of BD Diagnostic Systems.

manufacturers produce pooled monoclonal antibody reagents capable of detecting two or more viruses in a single cell spot.

Monospecific respiratory virus reagents first appeared and were evaluated in the mid-1980s. References to early studies using these reagents and the application of immunofluorescence as a diagnostic tool can be found in the chapters that cover the individual respiratory viruses in both the *Manual of Clinical Laboratory Immunology* (59a) and the *Manual of Clinical Microbiology* (50a), as well as several reviews (23, 33). Therefore, this section will focus on recent studies and newer reagents.

Laboratories interested in providing rapid diagnostic methods for the two most commonly recovered respiratory viruses, influenza virus and RSV, must choose between immunofluorescence and ELISA. Todd et al. (76) compared culture, immunofluorescence, Abbott TestPack RSV, and Directigen Flu A methods for the diagnosis of RSV and influenza A virus infections. In the hands of these investigators, the sensitivities and specificities of the DFAs for both influenza A and RSV were comparable to those of culture, but DFA was more sensitive than both Abbott TestPack for RSV (100 versus 83%) and Directigen Flu A (100 versus 61%). Similar results have been reported by others (83, 84). However, it should be noted that the authors of such studies are often expert in the application of DFA as a diagnostic tool, so the results represent the optimum performance for DFA. In the hands of less-experienced individuals, less sensitivity and specificity might be expected due to the subjective nature of DFA. Todd et al. (76) also compared the cost of DFA and ELISA and found them to be roughly equivalent. While the cost of ELISA kits was high relative to the cost of DFA reagents, the higher reagent cost was off set by the reduced hands-on time required to perform the ELISAs.

SimulFluor RS

Examining a specimen for respiratory viruses with monospecific reagents requires preparation of a cell spot for each virus reagent. With some frequency, specimens do not contain sufficient respiratory epithelial cells to be adequately examined for all viruses. In addition, preparing multiple slides and staining them is time-consuming. A new assay strategy, the SimulFluor Respiratory Screen (RS) (Chemicon International), was developed to overcome these problems. The reagent contains monoclonal antibodies that detect seven respiratory viruses, influenza A and B viruses, parainfluenza viruses 1, 2, and 3, RSV, and adenovirus. The RSV antibodies are rhodamine conjugated, and the antibodies to the remaining viruses are FITC conjugated. Rhodamine-stained cells appear reddish gold, and FITC-stained cells appear apple green when examined with a fluorescence microscope fitted with the appropriate filters. If FITC-stained cells are seen, additional slides must be prepared and stained with monospecific reagents to reveal the identity of the virus. Since influenza antigens are found in the nucleus and cytoplasm and parainfluenza virus antigens are exclusively cytoplasmic, the staining pattern may provide a clue to the identity of the virus and thus direct subsequent staining. The monospecific reagents provided with the kit distinguish between influenza A and B viruses but not among the parainfluenza viruses.

The SimulFluor RS was recently compared to standard single or dual DFA reagents and culture in a large prospective study (36). The authors examined 1,531 respiratory specimens using SimulFluor RS and standard single or dual DFA reagents. A total of 373 specimens were positive and the viruses identified included RSV (62 specimens), influenza A virus (238 specimens), influenza B virus (38 specimens), parainfluenza virus (22 specimens), and adenovirus (13 specimens). The authors found the two immunofluorescence approaches equivalent in sensitivity (98.4 versus 98.7%, respectively). Duplicate testing by SimulFluor RS and culture was performed on 940 specimens; 164 were virus positive. In the author's hands, detecting RSV using the SimulFluor RS DFA was significantly more sensitive than culture (99 versus 58.4%). However, culture and SimulFluor RS DFA detection were equivalent in sensitivity for detecting 30 influenza A virus-positive specimens (86.7 versus 83.3%, respectively), 18 influenza B virus-positive specimens (83.3 versus 83.3%, respectively), and 18 parainfluenza virus-positive specimens (90 versus 95%, respectively). In contrast to RSV, adenovirus detection was best achieved using cell culture rather than immunofluorescence. All 18 adenovirus-positive specimens were detected by culture, whereas only 11 (57.9%) were detected by SimulFluor RS DFA. It should be noted that the authors of this study did not use optimal culture methods for RSV and that others have reported equivalent detection of RSV by culture and DFA (19, 23, 42).

Bartels Viral Respiratory Screening and Identification Kit

The Bartels Viral Respiratory Screening and Identification kit (Trinity Biotech) (Fig. 7), like the SimulFluor RS DFA, makes use of a pool of mouse monoclonal antibodies to detect influenza A and B viruses, RSV, parainfluenza viruses 1, 2, and 3, and adenovirus in patient specimens. However, unlike the SimulFluor RS DFA, the Bartels kit makes use of IFA for detection. The secondary antibody is FITC-conjugated anti-mouse antibody. Specimens identified as positive with pooled reagent require staining with the individual

monoclonal antibodies supplied with the kit for specific virus identification.

The Bartels Viral Respiratory Screening kit individual monoclonal antibody reagents were evaluated for direct detection of respiratory viruses in clinical specimens in a large study reported by Doing et al. (19). Specimens were prepared for staining by using a cytocentrifuge. A total of 946 specimens were examined by culture and immunofluorescence, and 300 were virus positive. Agreement between culture and immunofluorescence detection was 90%. The sensitivity of the IFA relative to culture was 80.4% for 97 influenza A virus-positive specimens, 57.1% for the 28 influenza B virus-positive specimens, 66.7% for 6 parainfluenza virus-positive specimens, 81.9% for 160 RSV-positive specimens, and 55.6% for 9 adenovirus-positive specimens. The overall sensitivity and specificity of IFA as a direct detection tool relative to culture were 76.2 and 96%, respectively. However, the authors argue that the sensitivity of the IFA is likely greater because some of the specimens scored as false positives by IFA probably represent true positives that were incorrectly classified by culture due to the loss of viable virus during transport to the laboratory.

The Bartels Viral Respiratory Screening Kit was evaluated in another study reported by Matthey et al. (42). The authors examined 1,065 respiratory tract specimens submitted over a 3-year period using the pooled reagents as well as the monospecific antisera and compared the results of IFA staining to those of shell vial culture. Washed cell pellets were used to prepare cell spots for staining. A total of 183 (17.2%) specimens could not be interpreted due to inadequate numbers of cells or interfering fluorescence. Use of the pooled-antibody reagent was discontinued early in the study because excessive nonspecific fluorescence interfered with the interpretation of the results. Therefore, like Doing et al. (19), Matthey et al. (42) evaluated only the monospecific reagents. The overall sensitivity of IFA relative to culture was 85.9% for adequate specimens but dropped to 77.3% when uninterpretable specimens were included. The sensitivity of the IFA using monospecific antisera relative to culture was 47.4% for 19 influenza A virus-positive specimens, 78.9% for 28 influenza B virus-positive specimens, 54.5% for 11 parainfluenza virus 1-positive specimens, 100% for 3 parainfluenza virus 2-positive specimens, 83.3% for 11 parainfluenza virus 3-positive specimens, 94.6% for 185 RSV-positive specimens, and 28.6% for 7 adenovirus-positive specimens. IFA detected 80 specimens that were not detected by culture; 72 were RSV.

Imagen Respiratory Virus Screen

The Imagen Respiratory Virus Screen (Dako Diagnostics) is similar in design to the Bartels kit. A pool of mouse monoclonal antibodies that detect influenza A and B viruses, RSV, parainfluenza viruses 1, 2, and 3, and adenovirus, as well as individual antibodies to all seven viruses, are provided with the kit. The indirect staining format is used with FITC-conjugated F(ab')$_2$ fragment of rabbit anti-mouse immunoglobulins serving as the detector antibody. Directly FITC-conjugated monoclonal antibodies for each of the respiratory viruses are also available through the manufacturer as individual kits. The latter reagents were used in a recent study published by Barenfanger et al. (4).

PathoDx Respiratory Virus Panel

Recently, an additional product, the PathoDx Respiratory Virus Panel (Diagnostic Products Corp.) was cleared by the FDA for the qualitative detection of the seven common

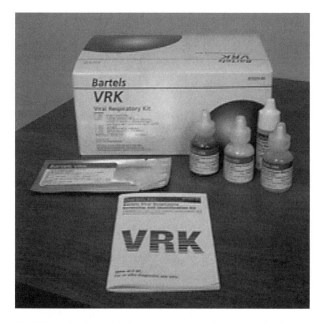

FIGURE 7 Bartels Respiratory kit. Courtesy of Trinity Biotech.

respiratory viruses in clinical specimens. The PathoDx Respiratory Virus kit contains a screening reagent that includes monoclonal antibodies to each of the respiratory viruses and seven virus-specific monoclonal antibody reagents. All of the monoclonal antibody reagents are labeled with FITC.

GASTROINTESTINAL VIRUSES

Many viruses, including astroviruses, caliciviruses (Norwalk and Norwalk-like viruses), coronaviruses (the genus *Torovirus*), adenoviruses, and rotaviruses cause gastrointestinal disease in humans, and unfortunately, all are either difficult or impossible to cultivate in routinely used cell cultures. Initially, each of the agents of gastrointestinal disease was identified in the feces of affected individuals by EM, and for some time EM served as the only diagnostic tool for these agents. While some laboratories continue to offer EM examination diagnostically, immunoassays are now more widely used. However, commercial immunodiagnostic products are available for only two of these agents, rotaviruses and adenoviruses.

Rotavirus

Rotavirus, a member of the family *Reoviridae*, is a major cause of gastroenteritis in infants, young children, and occasionally the elderly. The unenveloped virus is composed of a double-shelled protein capsid that contains a segmented, double-stranded RNA genome. There are five known rotavirus groups (A to E), with the human etiologic agents belonging to groups A, B, and C. Most human infections are caused by group A rotaviruses. Christensen (13) recently reviewed rotaviruses, their clinical significance, and commercially available diagnostic methods in the *Manual of Clinical Microbiology*, 7th ed.

Immunodiagnostic assays for rotavirus are available in several formats, including conventional ELISAs, rapid membrane ELISAs, and latex agglutination assays. Commercial rotavirus kits in a traditional ELISA format include VIDAS Rotavirus (bioMérieux-VITEK), Rotaclone (Meridian Diagnostics, Inc., Cincinnati, Ohio), Kallestad Pathfinder (Bio-Rad Laboratories), and IDEIA Rotavirus (Dako Diagnostics). The rapid membrane ELISAs now available include the ImmunoCard and ImmunoCard STAT! Rotavirus (Meridian Diagnostics, Inc.) (Fig. 8 and 9), the TestPack Rotavirus, and Diarlex Rota-Adeno (Orion Diagnostica, Espoo, Finland). The last test simultaneously detects both rotavirus and adenoviruses in fecal

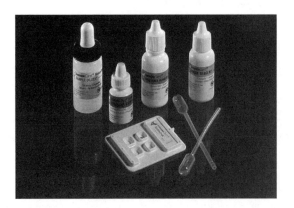

FIGURE 8 ImmunoCard Rotavirus. Courtesy of Meridian Diagnostics.

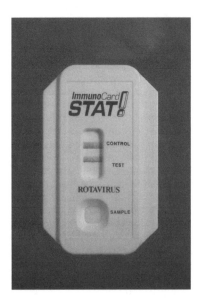

FIGURE 9 ImmunoCard STAT! Rotavirus. Courtesy of Meridian Diagnostics.

specimens. Finally, commercial latex agglutination assays include the Slidex Rota-Kit 2 (bioMérieux-VITEK), Meritec Rota-virus Latex (Meridian Diagnostics), Murex Rotavirus Latex (Murex/Remel, Lenexa, Kans.), Virogen Rotatest (Wampole Laboratories Division, Carter-Wallace, Inc., Cranbury, N.J.), and Rotalex (Orion Diagnostica).

Few recent peer-reviewed reports have critically evaluated commercial rotavirus test kits. However, as stated by Christensen (13), in addition to the 1- to 2-h incubation times required for the standard EIAs and the much shorter incubation times required for the rapid membrane EIA or latex agglutination kits (a range of 2 to 17 min), there appear to be differences in test sensitivity. It is thought that EIAs may be more sensitive than latex agglutination tests, and tests incorporating monoclonal antibodies may be more sensitive than those incorporating polyclonal antibodies (13). Most of the published evaluations for commercially available rotavirus kits are over 10 years old and will not be reviewed in detail in this chapter. However, a recent comparison by Dennehy et al. (17) included the Immunocard STAT! Rotavirus assay, a system utilizing immunogold-based horizontal-flow membrane technology, and two commercially available enzyme immunoassays (TestPack Rotavirus and Rotaclone). Results were confirmed by EM. In this study of 249 stool specimens, and after resolution of discordant results by RT-PCR, these investigators found 125 positive with ImmunoCard STAT!, 127 positive with Rotaclone, and 129 positive with TestPack. The sensitivities, specificities, positive predictive values, and negative predictive values were quite similar in this study and were 94.0, 100, 100, and 93.4% for ImmunoCard STAT!; 95.5, 100, 100, and 95% for Rotaclone; and 97.0, 96.5, 97.0, and 96.5% for TestPack. Thus, these investigators found comparable sensitivities and specificities for detection of group A rotavirus using these three commercial methods.

An earlier study by Dennehy et al. (18) evaluated 464 fecal specimens for the presence of rotavirus using the VIDAS Rotavirus (bioMérieux), Rotaclone (Meridian), and Pathfinder (Bio-Rad) systems. They reported sensitivities, specificities, positive predictive values, and negative predictive values of 98, 99.3, 98.7, and 99%, respectively, for VIDAS Rotavirus; 100, 99, 98.1, and 100% for Rotaclone; and 100, 92.4, 87.3, and 100% for Pathfinder.

Adenoviruses

Adenoviruses are unenveloped, double-stranded DNA viruses with a viral capsid that is composed of 252 capsomeres arranged in the form of an icosahedron. Three proteins, hexons, pentons (fiber plus penton base), and fibers, are included in the capsid. Hexon, the majority protein in the virion (240 capsomeres/virion), carries type-specific, subtype-specific, and group-specific epitopes. Over 45 distinct serotypes of adenovirus are recognized.

The adenoviruses cause a broad spectrum of human disease, including pharyngitis, pneumonia, conjunctivitis, hemorrhagic cystitis, and diarrhea. Disease manifestations vary with the infecting serotype, but it is not uncommon for a single serotype to be associated with several different disease manifestations. Although most adenovirus serotypes can be isolated from fecal specimens, only two, adenovirus types 40 and 41, are clearly associated with diarrheal disease (43).

Adenoviruses have been associated with 3.1 to 13.5% of cases of pediatric diarrhea in studies from Europe, Asia, and North and South America, and types 40 and 41 reportedly account for 37.5 to 100% of these adenoviruses. Because adenoviruses make a minor contribution to the etiology of infantile diarrhea relative to rotaviruses, the number of tests available for these agents is smaller, and many laboratories elect not to offer routine testing for adenoviral gastroenteritis (43). Depending on how they are structured, adenovirus assays are either type specific (i.e., they detect only adenovirus types 40 and 41) or group specific (they detect all adenoviruses regardless of serotype). Three traditional microtiter format ELISAs are available: the adenovirus IDEIA (Dako Diagnostics), the Biotrin International (Dublin, Ireland) Adenovirus EIA, and Adenoclone Type 40/41. The last assay, as its name implies, is type specific, whereas the first two assays are structured to be group specific. In addition, a latex agglutination assay, Adenolex, and a combined membrane immunoassay that detects both rotavirus and adenovirus are available through Orion.

Most of the adenovirus assays listed above have been compared to culture in permissive cells, EM, or both and found roughly comparable to both techniques in sensitivity (25, 43, 81). However, in a recent study performed in South Africa, Moore et al. (50) reported poor sensitivity when a laboratory-developed assay was compared to two commercially produced adenovirus assays, Biotrin and Adenoclone Type 40/41. The insensitivity of the assays was attributed to a viral mutation that changed the antigenic structure of the virus and consequently rendered the monoclonal antibody used in the assays ineffective.

HSV AND VZV

Two herpesviruses, HSV and VZV, commonly cause vesicular exanthems in humans. Less common but important disease manifestations associated with these viruses include encephalitis, disseminated disease in the newborn and immunocompromised host, and ocular disease, including keratitis and acute retinal necrosis. The clinical laboratory is frequently asked to assist in the diagnosis of HSV and VZV infections, in clinically distinct cases to confirm infection with one virus or the other and in more problematic cases to distinguish between the two infections (1, 2).

HSV is an antigenically complex double-stranded DNA virus with two biologically and antigenically distinct serotypes, HSV type 1 (HSV-1) and HSV-2. While distinct, the viruses also share many common antigens. Therefore, distinguishing between the two serotypes is difficult using polyclonal antisera and is most reliably accomplished only by employing type-specific monoclonal antibodies or nucleic acid analyses.

HSV is readily detected in cell cultures inoculated with infected material from dermal lesions and tissues, with most cultures (>90%) exhibiting typical cytopathic effect within 1 to 3 days of specimen inoculation (10). Nevertheless, during the era of rapid antigen detection, several immunodiagnostic tests were developed to more rapidly identify viruses in clinical specimens. The rapid tests include a latex agglutination assay, the Virogen Herpes Slide Test (Wampole Laboratories), membrane ELISA, and the Kodak Surecell Herpes (HSV), as well as two other ELISA-based assays, HerpCheck (Dupont Medical Products, Wilmington, Del.) and VIDAS, an automated ELISA. However, without exception these assays have not withstood the test of time and are no longer marketed, largely due to their insensitivity relative to cell culture and the minimal impact they made in terms of a more rapid diagnosis (68, 90). Immunofluoresence now remains the only immunodiagnostic tool for detecting HSV in clinical specimens.

Direct and indirect immunostaining reagents designed to detect HSV-infected cells found in clinical specimens are available through a number of commercial sources, including Trinity Biotech, Meridian, Diagnostic Products, and Chemicon International. In general, immunostaining, like other immunoassay methods, is not as sensitive as cell culture as a tool for HSV detection. Therefore, culture remains the test of choice in most clinical settings. However, in life-threatening situations that mandate immediate antiviral therapy, such as neonatal HSV infection and disseminated disease in the immunocompromised host, a positive immunostain for HSV can be reassuring (2).

While culture is the test of choice for the diagnosis of HSV in most clinical settings, neither culture nor immunodiagnostic methods can be recommended for cerebrospinal fluid (CSF) specimens from patients thought to have HSV central nervous system (CNS) disease. Until recently, the diagnosis of HSV encephalitis could be reliably accomplished only by brain biopsy followed by immunostaining or culture. However, with the advent of PCR, a new and very valuable tool was added to our diagnostic armamentarium. PCR performed on CSF specimens is now viewed as the method of choice for the diagnosis of both HSV encephalitis and meningitis (59, 73, 86). In addition, several investigators have explored the use of PCR for the diagnosis of HSV infection in more routine clinical settings and found PCR to be more sensitive than the current gold standard, culture (22, 65). It seems likely that PCR will play a greater role in the diagnosis of HSV infections in the future. However, until commercially produced kits become available, this potential is not likely to be realized.

VZV is also an antigenically complex, double-stranded DNA virus, but unlike HSV, only one VZV serotype exists. Also in contrast to HSV, VZV is relatively difficult to recover in cell culture, and successful recovery often takes 5 to 10 days. While shell vial culture improves the time to detection, overall sensitivity does not appear to be improve. Further, VZV is relatively labile, so virus recovery is best achieved by inoculating vesicular fluid directly into the cell cultures (1). The difficulties associated with successful VZV recovery in cell culture likely explain why, in contrast to HSV, VZV is best detected by immunodiagnostic methods (15).

Direct and indirect immunostaining reagents designed to detect VZV-infected cells found in clinical specimens are available through a number of commercial sources, including Trinity Biotech, Meridian, Diagnostic Products, and Light Diagnostics. In addition, a single assay, the Light Diagnostics SimulFluor HSV/VZV Immunofluorescence Assay, is available for the simultaneous detection and identification of HSV-1 and -2 and VZV. The assay uses a single reagent containing fluoresceinated antibodies to the 155-kDa major capsid protein of HSV and to glycoprotein gp1 and the immediate-early antigen of VZV. Illumination of stained cells with UV light employing a FITC filter set results in apple green fluorescence for HSV antigen-antibody complexes and yellow gold for VZV antigen-antibody complexes. Uninfected cells appear dull red due to the presence of Evans blue in the reagent. The HSV-VZV immunofluorescence assay was recently evaluated by Scicchitano et al. (63). Using 167 specimens, 21 VZV positive and 55 HSV positive by conventional methods, they determined the sensitivity and specificity of the SimulFluor HSV/VZV Immunofluorescence Assay to be 100 and 98% for VZV and 62 and 99% for HSV, respectively.

Like the diagnosis of HSV, the diagnosis of VZV is likely to benefit substantially from the addition of PCR to our set of diagnostic tools. Espy et al. (21), in a study of 253 routine clinical specimens, detected VZV in 23 (9.1%) by shell vial cell culture and in 44 (17.4%) by LightCycler PCR. Thus, in this study, PCR was almost twice as sensitive as cell culture for the diagnosis of VZV. In another study, PCR was used to examine whole blood, serum, and plasma in patients with active VZV infections (16). Detection rates of 86% in patients with varicella and 81% in patients with herpes zoster were achieved. In specimens obtained during the first week after the onset of the rash, detection rates were 100 and 89%, respectively. Thus, reliable diagnosis can be achieved using blood, serum, and plasma, all more convenient specimens than lesion swabs and aspirates. PCR has also proved useful for the diagnosis of VZV-associated CNS disease (57) and eye infections, including acute retinal necrosis (34). Multiplex PCRs that detect both HSV and VZV have been developed and successfully applied to CSF (57) and lesion (5) specimens. As with HSV, widespread use of PCR for the diagnosis of VZV infections awaits the development of commercially produced kits and reagents.

CMV

CMV is an antigenically complex, enveloped, double-stranded DNA virus. Only one serotype exists, but strain differences can be demonstrated by a variety of molecular methods. The usefulness of differentiating among strains is limited to establishing links among epidemiologically related isolates.

CMV causes a broad spectrum of disease in humans. In almost all clinical settings the consequences of active infection range from asymptomatic virus shedding to severe disease. Diseases associated with CMV infection include congenital disease in the newborn, a mononucleosislike syndrome in otherwise-healthy adults, and a host of diseases in immunocompromised patients, including retinitis, pneumonia, hepatitis, CNS disease, and gastrointestinal ulcers.

For years, the laboratory diagnosis of CMV infection was accomplished by isolating virus from urine and oral secretions. However, it then became clear that many individuals, particularly immunocompromised hosts, shed virus asymptomatically, and in such individuals, there was poor correlation between positive cultures and disease. Because demonstrating virus in blood was thought to better correlate with disease, blood became the preferred diagnostic specimen for the immunocompromised host. However, even when blood specimens were examined, disease correlation was not perfect, and many now advocate quantitating virus to improve the positive predictive value of testing (8).

CMV is relatively difficult to recover in cell culture. The virus replicates only in human diploid fibroblast cells, and in these cells viral cytopathic effect evolves slowly. While specimens containing high virus titers, such as those from congenitally infected infants, may be positive within the first 7 days of incubation in traditional tube cultures, specimens such as blood that usually contain lower virus titers may take as long as 14 to 21 days to be recognized as positive. Shell vial cultures were introduced in the mid-1980s to improve the time to virus detection, but in the hands of many investigators shell vial cultures are unacceptably labor-intensive and significantly less sensitive than traditional tube cultures.

Early attempts to detect CMV in clinical specimens by ELISA were unsuccessful because beta 2 macroglobulin in the specimen masked the viral surface antigens. However, in 1988, van der Bij et al. (78) found that virus could be detected in circulating leukocytes, specifically neutrophils and macrophages, by immunostaining (Fig. 10). The test strategy, popularly known as the antigenemia assay, is now widely used to detect CMV viremia.

The antigenemia assay is performed by first separating leukocytes from whole blood either by dextran sedimentation or ammonium chloride lysis. The recovered leukocytes are suspended in saline and affixed on slides by using a cytocentrifuge. The leukocytes are then fixed, made permeable, and reacted with a pool of murine monoclonal antibodies to pp65, the lower matrix structural phosphoprotein of the virus that is found in the nuclei of infected cells. FITC-conjugated anti-mouse immunoglobulin secondary antibodies are then added, and the slides are examined with a fluorescence microscope. While the assay is widely used, the need for standardization remains (74, 79).

Many investigators have demonstrated the utility of antigenemia assays for detecting viremia in different patient populations, including solid-organ transplant recipients, bone marrow transplant recipients, and human immunodeficiency virus-positive individuals (3, 45, 62, 64, 67). In the

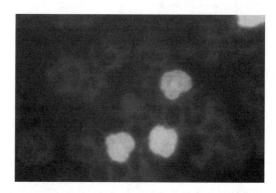

FIGURE 10 Cells stained with Biotest assay. Courtesy of Biotest Diagnostics Corp.

hands of most investigators, the antigenemia assay is two to three times more sensitive than both conventional and shell vial cell culture. Initially, there was some concern that culture-negative, antigenemia-positive findings might represent false positives. However, longitudinal studies of infected patients demonstrated that the antigenemia test identified positive patients sooner than cell culture, suggesting that the culture-negative, antigenemia-positive findings represented greater sensitivity rather than problems with specificity. In addition, many investigators have also compared CMV detection using the antigenemia assay to amplification methods and cell culture. In most studies, the antigenemia and amplification results were similar and significantly more sensitive than those of cell culture. These findings, too, support the validity of the greater sensitivity reported for the antigenemia assay.

Two manufacturers now produce CMV antigenemia kits, Biotest Diagnostics, Denville, N.J., and Trinity Biotech. The FDA cleared two assays produced by these manufacturers, the CMV Brite Antigenemia Test Kit (Biotest) (Fig. 11) and Bartels Cinakit CMV Antigenemia, in 1996 and 1998, respectively. The assays differ in the pp65 monoclonal antibodies used and in the number of cells examined (Bartels, 200,000; Biotest, 150,000). However, the tests are similar in that both recommend dextran sedimentation of blood anticoagulated with EDTA or heparin for leukocyte recovery and both are indirect assays that use FITC-conjugated secondary antibodies. An improved version of the Biotest assay, the CMV Brite Turbo Kit, was cleared by the FDA in 1999. The changes incorporated in the Biotest

assay were designed to reduce the overall processing time and include the use of reduced fixation and permeabilization times and the use of direct ammonium chloride lysis rather than dextran sedimentation followed by lysis for leukocyte recovery. In addition, the new assay recommends examination of 200,000 rather than 150,000 leukocytes. Visser et al. (80) and Landry and Ferguson (35) compared the performance of the CMV Brite Antigenemia Test Kit and the CMV Brite Turbo Kit. Both studies found equivalent performance characteristics and demonstrated a 50% reduction in processing time using the new assay.

Recently, St George et al. (66) reported the results of a large multisite trial designed to compare the performance of the Biotest CMV Brite and Bartels assays. Investigators at four different sites examined a total of 513 specimens from a variety of patient populations. All specimens were tested by culture and by both of the antigenemia assays. A total of 109 specimens were positive for CMV; 97% were detected by the antigenemia assays, and 34% were culture positive. A total of 93 positive specimens (88%) were detected by the Biotest kit, and 86 (81%) were detected by the Bartels kit. When the data from all sites were combined, the performance characteristics of the two assays were equivalent. The finding was somewhat unexpected, since several antecedent studies reported differences in the sensitivities of the individual monoclonal antibodies used in the kits (66).

The benefits of using antigemia assays for detecting CMV—greater sensitivity than cell culture, earlier detection, same-day diagnosis, and quantitative measures of virus load—can also be realized by using molecular detection methods. A number of molecular assays have been developed for both the qualitative and quantitative detection of CMV, including the qualitative and quantitative Amplicor PCR assays (Roche), the Quantiplex bDNA CMV assay, immediate-early mRNA and late pp67 mRNA expression by nucleic acid sequence-based amplification (Organon Teknika), and the CMV hybrid capture assay (Digene in the United States and Murex in Europe). Presently, only the last two assays have been cleared by the FDA. A number of studies have evaluated the performances of individual assays versus the antigenemia assay or culture (7, 9, 11, 29, 40, 44, 62, 82). In addition, several investigators have reported complex comparisons using different assays and specimen types (6, 24, 56, 77, 85). Even though there have been many evaluations of molecular assays for CMV, the tests are still in their infancy and evolving. It will likely be some time before the relative values of these assays in different clinical settings are established. However, there is consensus on several issues: (i) molecular assays, like the antigenemia assay, are all more sensitive than cell culture; (ii) leukocyte assays are more sensitive plasma assays; and (iii) amplification methods are more sensitive than hybridization methods. It also seems clear that molecular assays like the antigenemia assay will play an increasingly important role in the diagnosis and management of CMV infection in the future.

CMV assays are used in a number of different settings: (i) to diagnose CMV disease, (ii) to monitor the efficacy of antiviral therapy, and (iii) to prompt the institution of preemptive therapy. A given assay may be more suitable for one application than another. For example, in the immunocompromised host, qualitative assays with less sensitivity tend to be better predictors of disease than sensitive assays because CMV disease occurs more frequently in those with high virus loads. In contrast, more sensitive

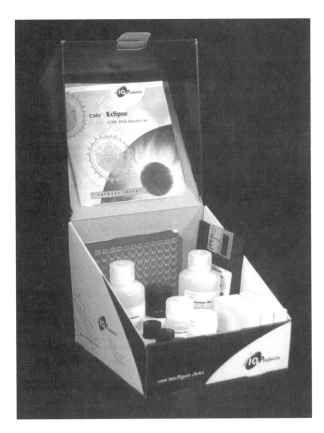

FIGURE 11 Biotest kit in its entirety. Courtesy of Biotest Diagnostics Corp.

assays are most appropriate to prompt the institution of pre-emptive therapy. Strategies for using CMV assays in patient management are evolving in parallel with the assays. To best serve clinicians, laboratories should determine the principal uses of the assays in their institution and select tests or combinations of tests with the most suitable performance characteristics.

REFERENCES

1. **Arvin, A. M.** 1996. Varicella-zoster virus. *Clin. Microbiol. Rev.* **9:**361–381.
2. **Arvin, A. M., and C. G. Prober.** 1999. Herpes simplex viruses, p. 878–887. *In* P. R. Murray, E. J. Baron, M. A. Pfaller, F. C. Tenover, and R. H. Yolken (ed.), *Manual of Clinical Microbiology*, 7th ed. American Society for Microbiology, Washington, D.C.
3. **Baldanti, F., M. G. Revello, E. Percivalle, and G. Gerna.** 1998. Use of the human cytomegalovirus (HCMV) antigenemia assay for diagnosis and monitoring of HCMV infections and detection of antiviral drug resistance in the immunocompromised. *J. Clin. Virol.* **11:**51–60.
4. **Barenfanger, J., C. Drake, N. Leon, T. Mueller, and T. Troutt.** 2000. Clinical and financial benefits of rapid detection of respiratory viruses: an outcomes study. *J. Clin. Microbiol.* **38:**2824–2828.
5. **Beards, G., C. Graham, and D. Pillay.** 1998. Investigation of vesicular rashes for HSV and VZV by PCR. *J. Med. Virol.* **54:**155–157.
6. **Blank, B. S., P. L. Meenhorst, J. W. Mulder, G. J. Weverling, H. Putter, W. Pauw, W. C. van Dijk, P. Smits, S. Lie-A-Ling, P. Reiss, and J. M. Lange.** 2000. Value of different assays for detection of human cytomegalovirus (HCMV) in predicting the development of HCMV disease in human immunodeficiency virus-infected patients. *J. Clin. Microbiol.* **38:**563–569.
7. **Blok, M. J., M. H. Christiaans, V. J. Goossens, J. P. van Hooff, P. Sillekens, J. M. Middeldorp, and C. A. Bruggeman.** 1999. Early detection of human cytomegalovirus infection after kidney transplantation by nucleic acid sequence-based amplification. *Transplantation* **67:**1274–1277.
8. **Boeckh, M., and G. Boivin.** 1998. Quantitation of cytomegalovirus: methodologic aspects and clinical applications. *Clin. Microbiol. Rev.* **11:**533–554.
9. **Caliendo, A. M., K. St George, S. Y. Kao, J. Allega, B. H. Tan, R. LaFontaine, L. Bui, and C. R. Rinaldo.** 2000. Comparison of quantitative cytomegalovirus (CMV) PCR in plasma and CMV antigenemia assay: clinical utility of the prototype AMPLICOR CMV MONITOR test in transplant recipients. *J. Clin. Microbiol.* **38:**2122–2127.
10. **Callihan, D. R., and M. A. Menegus.** 1984. Rapid detection of herpes simplex virus in clinical specimens with human embryonic lung fibroblast and primary rabbit kidney cell cultures. *J. Clin. Microbiol.* **19:**563–565.
11. **Chernoff, D. N., R. C. Miner, B. S. Hoo, L. P. Shen, R. J. Kelso, D. Jekic-McMullen, J. P. Lalezari, S. Chou, W. L. Drew, and J. A. Kolberg.** 1997. Quantification of cytomegalovirus DNA in peripheral blood leukocytes by a branched-DNA signal amplification assay. *J. Clin. Microbiol.* **35:**2740–2744.
12. **Chomel, J. J., M. F. Remilleux, P. Marchand, and M. Aymard.** 1992. Rapid diagnosis of influenza A. Comparison with ELISA immunocapture and culture. *J. Virol. Methods* **37:**337–343.
13. **Christensen, M. L.** 1999. Rotaviruses, p. 999–1004. *In* P. R. Murray, E. J. Baron, M. A. Pfaller, F. C. Tenover, and R. H. Yolken (ed.), *Manual of Clinical Microbiology*, 7th ed. American Society for Microbiology, Washington, D.C.
14. **Covalciuc, K. A., K. H. Webb, and C. A. Carlson.** 1999. Comparison of four clinical specimen types for detection of influenza A and B viruses by optical immunoassay (FLU OIA test) and cell culture methods. *J. Clin. Microbiol.* **37:**3971–3974.
15. **Dahl, H., J. Marcoccia, and A. Linde.** 1997. Antigen detection: the method of choice in comparison with virus isolation and serology for laboratory diagnosis of herpes zoster in human immunodeficiency virus-infected patients. *J. Clin. Microbiol.* **35:**347–349.
16. **de Jong, M. D., J. F. Weel, T. Schuurman, P. M. Wertheim-van Dillen, and R. Boom.** 2000. Quantitation of varicella-zoster virus DNA in whole blood, plasma, and serum by PCR and electrochemiluminescence. *J. Clin. Microbiol.* **38:**2568–2573.
17. **Dennehy, P. H., M. Hartin, S. M. Nelson, and S. F. Reising.** 1999. Evaluation of the ImmunoCardSTAT! rotavirus assay for detection of group A rotavirus in fecal specimens. *J. Clin. Microbiol.* **37:**1977–1979.
18. **Dennehy, P. H., T. E. Schutzbank, and G. M. Thorne.** 1994. Evaluation of an automated immunodiagnostic assay, VIDAS Rotavirus, for detection of rotavirus in fecal specimens. *J. Clin. Microbiol.* **32:**825–827.
19. **Doing, K. M., M. A. Jerkofsky, E. G. Dow, and J. A. Jellison.** 1998. Use of fluorescent-antibody staining of cytocentrifuge-prepared smears in combination with cell culture for direct detection of respiratory viruses. *J. Clin. Microbiol.* **36:**2112–2114.
20. **Dominguez, E. A., L. H. Taber, and R. B. Couch.** 1993. Comparison of rapid diagnostic techniques for respiratory syncytial and influenza A virus respiratory infections in young children. *J. Clin. Microbiol.* **31:**2286–2290.
21. **Espy, M. J., R. Teo, T. K. Ross, K. A. Svien, A. D. Wold, J. R. Uhl, and T. F. Smith.** 2000. Diagnosis of varicella-zoster virus infections in the clinical laboratory by LightCycler PCR. *J. Clin. Microbiol.* **38:**3187–3189.
22. **Espy, M. J., T. K. Ross, R. Teo, K. A. Svien, A. D. Wold, J. R. Uhl, and T. F. Smith.** 2000. Evaluation of LightCycler PCR for implementation of laboratory diagnosis of herpes simplex virus infections. *J. Clin. Microbiol.* **38:**3116–3118.
23. **Gardner, P. S., and J. McQuillin.** 1980. *Rapid Virus Diagnosis: Application of Immunofluorescence*, 2nd ed. Butterworths, London, United Kingdom.
24. **Gerna, G., F. Baldanti, D. Lilleri, M. Parea, E. Alessandrino, A. Pagani, F. Locatelli, J. Middeldorp, and M. G. Revello.** 2000. Human cytomegalovirus immediate-early mRNA detection by nucleic acid sequence-based amplification as a new parameter for preemptive therapy in bone marrow transplant recipients. *J. Clin. Microbiol.* **38:**1845–1853.
25. **Grandien, M., C. A. Pettersson, L. Svensson, and I. Uhnoo.** 1987. Latex agglutination test for adenovirus diagnosis in diarrheal disease. *J. Med. Virol.* **23:**311–316.
26. **Hadziyannis, E., W. Sholtis, S. Schindler, and B. Yen-Lieberman.** 1999. Comparison of VIDAS with direct immunofluorescence for the detection of respira-

tory syncytial virus in clinical specimens. *J. Clin. Virol.* **14:**133–136.

27. **Hall, C. B.** 1999. Respiratory syncytial virus: a continuing culprit and conundrum. *J. Pediatr.* **135:**2–7.

28. **Halstead, D. C., S. Todd, and G. Fritch.** 1990. Evaluation of five methods for respiratory syncytial virus detection. *J. Clin. Microbiol.* **28:**1021–1025.

29. **Hiyoshi, M., S. Tagawa, T. Takubo, K. Tanaka, T. Nakao, Y. Higeno, K. Tamura, M. Shimaoka, A. Fujii, M. Higashihata, Y. Yasui, T. Kim, A. Hiraoka, and N. Tatsumi.** 1997. Evaluation of the AMPLICOR CMV test for direct detection of cytomegalovirus in plasma specimens. *J. Clin. Microbiol.* **35:**2692–2694.

30. **Johnston, S. L., and C. S. Siegel.** 1990. Evaluation of direct immunofluorescence, enzyme immunoassay, centrifugation culture, and conventional culture for the detection of respiratory syncytial virus. *J. Clin. Microbiol.* **28:**2394–2397.

31. **Johnston, S. L., and H. Bloy.** 1993. Evaluation of a rapid enzyme immunoassay for detection of influenza A virus. *J. Clin. Microbiol.* **31:**142–143.

32. **Kaiser, L., M. S. Briones, and F. G. Hayden.** 1999. Performance of virus isolation and Directigen Flu A to detect influenza A virus in experimental human infection. *J. Clin. Virol.* **14:**191–197.

33. **Kellogg, J. A.** 1991. Culture vs direct antigen assays for detection of microbial pathogens from lower respiratory tract specimens suspected of containing the respiratory syncytial virus. *Arch. Pathol. Lab. Med.* **115:**451–458.

34. **Knox, C. M., D. Chandler, G. A. Short, and T. P. Margolis.** 1998. Polymerase chain reaction-based assays of vitreous samples for the diagnosis of viral retinitis. Use in diagnostic dilemmas. *Ophthalmology* **105:**37–44.

35. **Landry, M. L., and D. Ferguson.** 2000. Two-hour cytomegalovirus pp65 antigenemia assay for rapid quantitation of cytomegalovirus in blood samples. *J. Clin. Microbiol.* **38:**427–428.

36. **Landry, M. L., and D. Ferguson.** 2000. SimulFluor respiratory screen for rapid detection of multiple respiratory viruses in clinical specimens by immunofluorescence staining. *J. Clin. Microbiol.* **38:**708–711.

37. **Landry, M. L., S. Cohen, and D. Ferguson.** 2000. Impact of sample type on rapid detection of influenza virus A by cytospin-enhanced immunofluorescence and membrane enzyme-linked immunosorbent assay. *J. Clin. Microbiol.* **38:**429–430.

38. **Leonardi, G. P., H. Leib, G. S. Birkhead, C. Smith, P. Costello, and W. Conron.** 1994. Comparison of rapid detection methods for influenza A virus and their value in health-care management of institutionalized geriatric patients. *J. Clin. Microbiol.* **32:**70–74.

39. **Lipson, S. M., D. Popiolek, Q. Z. Hu, L. H. Falk, M. Bornfreund, and L. R. Krilov.** 1999. Efficacy of Directigen RSV testing in patient management following admission from a paediatric emergency department. *J. Hosp. Infect.* **41:**323–329.

40. **Long, C. M., L. Drew, R. Miner, D. Jekic-McMullen, C. Impraim, and S. Y. Kao.** 1998. Detection of cytomegalovirus in plasma and cerebrospinal fluid specimens from human immunodeficiency virus-infected patients by the AMPLICOR CMV test. *J. Clin. Microbiol.* **36:**2434–2438.

41. **Marcante, R., F. Chiumento, G. Palu, and G. Cavedon.** 1996. Rapid diagnosis of influenza type A infec-

tion: comparison of shell-vial culture, directigen flu-A and enzyme-linked immunosorbent assay. *New Microbiol.* **19:**141–147.

42. **Matthey, S., D. Nicholson, S. Ruhs, B. Alden, M. Knock, K. Schultz, and A. Schmuecker.** 1992. Rapid detection of respiratory viruses by shell vial culture and direct staining by using pooled and individual monoclonal antibodies. *J. Clin. Microbiol.* **30:**540–544.

43. **Mautner, V., V. Steinthorsdottir, and A. Bailey.** 1995. Enteric adenoviruses. *Curr. Top. Microbiol. Immunol.* **199:**229–282.

44. **Mazzulli, T., L. W. Drew, B. Yen-Lieberman, D. Jekic-McMullen, D. J. Kohn, C. Isada, G. Moussa, R. Chua, and S. Walmsley.** 1999. Multicenter comparison of the digene hybrid capture CMV DNA assay (version 2.0), the pp65 antigenemia assay, and cell culture for detection of cytomegalovirus viremia. *J. Clin. Microbiol.* **37:**958–963.

45. **Mazzulli, T., R. H. Rubin, M. J. Ferraro, R. T. D'Aquila, S. A. Doveikis, B. R. Smith, and T. H. The.** 1993. Hirsch MS. Cytomegalovirus antigenemia: clinical correlations in transplant recipients and in persons with AIDS. *J. Clin. Microbiol.* **31:**2824–2827.

46. **Mendoza, J., A. Rojas, J. M. Navarro, C. Plata, and M. de la Rosa.** 1992. Evaluation of three rapid enzyme immunoassays and cell culture for detection of respiratory syncytial virus. *Eur. J. Clin. Microbiol. Infect. Dis.* **11:**452–454.

47. **Michaels, M. G., C. Serdy, K. Barbadora, M. Green, A. Apalsch, and E. R. Wald.** 1992. Respiratory syncytial virus: a comparison of diagnostic modalities. *Pediatr. Infect. Dis. J.* **11:**613–616.

48. **Miller, H., R. Milk, and F. Diaz-Mitoma.** 1993. Comparison of the VIDAS RSV assay and the Abbott Testpack RSV with direct immunofluorescence for detection of respiratory syncytial virus in nasopharyngeal aspirates. *J. Clin. Microbiol.* **31:**1336–1338.

49. **Minnich, L. L., and C. G. Ray.** 1987. Early testing of cell cultures for detection of hemadsorbing viruses. *J. Clin. Microbiol.* **25:**421–422.

50. **Moore, P. L., A. D. Steele, and J. J. Alexander.** 2000. Relevance of commercial diagnostic tests to detection of enteric adenovirus infections in South Africa. *J. Clin. Microbiol.* **38:**1661–1663.

50a. **Murray, P. R., E. J. Baron, M. A. Pfaller, F. C. Tenover, and R. H. Yolken (ed.).** 1999. *Manual of Clinical Microbiology,* 7th ed. American Society for Microbiology, Washington, D.C.

51. **Noyola, D. E., A. E. Paredes, B. Clark, and G. J. Demmler.** 2000. Evaluation of a neuraminidase assay for the rapid detection of influenza A and B infection in children. *Pediatr. Dev. Pathol.* **3:**162–167.

52. **Noyola, D. E., B. Clark, F. T. O'Donnell, R. L. Atmar, J. Greer, and G. J. Demmler.** 2000. Comparison of a new neuraminidase detection assay with an enzyme immunoassay, immunofluorescence, and culture for rapid detection of influenza A and B viruses in nasal wash specimens. *J. Clin. Microbiol.* **38:**1161–1165.

53. **Olsen, M. A., K. M. Shuck, A. R. Sambol, V. A. Bohnert, and M. L. Henery.** 1993. Performance of the Kallestad Pathfinder enzyme immunoassay in the diagnosis of respiratory syncytial virus infections. *Diagn. Microbiol. Infect. Dis.* **16:**325–329.

54. **Olsen, M. A., K. M. Shuck, and A. R. Sambol.** 1993. Evaluation of Abbott TestPack RSV for the diagnosis of

respiratory syncytial virus infections. *Diagn. Microbiol. Infect. Dis.* **16**:105–109.

55. **Pedneault, L., L. Robillard, and J. P. Turgeon.** 1994. Validation of respiratory syncytial virus enzyme immunoassay and shell vial assay results. *J. Clin. Microbiol.* **32**:2861–2864.

56. **Pellegrin, I., I. Garrigue, D. Ekouevi, L. Couzi, P. Merville, P. Merel, G. Chene, M. H. Schrive, P. Trimoulet, M. E. Lafon, and H. Fleury.** 2000. New molecular assays to predict occurrence of cytomegalovirus disease in renal transplant recipients. *J. Infect. Dis.* **182**:36–42.

57. **Read, S. J., and J. B. Kurtz.** 1999. Laboratory diagnosis of common viral infections of the central nervous system by using a single multiplex PCR screening assay. *J. Clin. Microbiol.* **37**:1352–1355.

58. **Reina, J., M. Munar, and I. Blanco.** 1996. Evaluation of a direct immunofluorescence assay, dot blot enzyme immunoassay, and shell vial culture in the diagnosis of lower respiratory tract infections caused by influenza A virus. *Diagn. Microbiol. Infect. Dis.* **25**:143–145.

59. **Revello, M. G., and R. Manservigi.** 1996. Molecular diagnosis of herpes simplex encephalitis. *Intervirology* **39**:185–192.

59a. **Rose, N. R., E. Conway de Macario, J. D. Folds, H. C. Lane, and R. M. Nakamura (ed.).** 1997. *Manual of Clinical Laboratory Immunology*, 5th ed. American Society for Microbiology, Washington, D.C.

60. **Rothbarth, P. H., M. C. Hermus, and P. Schrijnemakers.** 1991. Reliability of two new test kits for rapid diagnosis of respiratory syncytial virus infection. *J. Clin. Microbiol.* **29**:824–826.

61. **Ryan-Poirier, K. A., J. M. Katz, R. G. Webster, and Y. Kawaoka.** 1992. Application of Directigen FLU-A for the detection of influenza A virus in human and nonhuman specimens. *J. Clin. Microbiol.* **30**:1072–1075.

62. **Schirm, J., A. Kooistra, W. J. van Son, W. van der Bij, E. Verschuuren, H. G. Sprenger, P. C. Limburg, and T. H. The.** 1999. Comparison of the Murex Hybrid Capture CMV DNA (v2.0) assay and the pp65 CMV antigenemia test for the detection and quantitation of CMV in blood samples from immunocompromised patients. *J. Clin. Virol.* **14**:153–165.

63. **Scicchitano, L. M., B. Shetterly, and P. P. Bourbeau.** 1999. Evaluation of Light Diagnostics SimulFluor HSV/VZV immunofluorescence assay. *Diagn. Microbiol. Infect. Dis.* **35**:205–208.

64. **Singh, N., D. L. Paterson, T. Gayowski, M. M. Wagener, and I. R. Marino.** 2000. Cytomegalovirus antigenemia directed pre-emptive prophylaxis with oral versus I.V. ganciclovir for the prevention of cytomegalovirus disease in liver transplant recipients: a randomized, controlled trial. *Transplantation* **70**:717–722.

65. **Slomka, M. J., L. Emery, P. E. Munday, M. Moulsdale, and D. W. Brown.** 1998. A comparison of PCR with virus isolation and direct antigen detection for diagnosis and typing of genital herpes. *J. Med. Virol.* **55**:177–183.

66. **St George, K., M. J. Boyd, S. M. Lipson, D. Ferguson, G. F. Cartmell, L. H. Falk, C. R. Rinaldo, and M. L. Landry.** 2000. A multisite trial comparing two cytomegalovirus (CMV) pp65 antigenemia test kits, Biotest CMV Brite and Bartels/Argene CMV antigenemia. *J. Clin. Microbiol.* **38**:1430–1433.

67. **Stocchi, R., K. N. Ward, R. Fanin, M. Baccarani, and J. F. Apperley.** 1999. Management of human cytomegalovirus infection and disease after allogeneic bone marrow transplantation. *Haematologica* **84**:71–79.

68. **Storch, G. A., C. A. Reed, and Z. A. Dalu.** 1988. Evaluation of a latex agglutination test for herpes simplex virus. *J. Clin. Microbiol.* **26**:787–788.

69. **Subbarao, E. K., M. C. Dietrich, T. M. De Sierra, C. J. Black, D. M. Super, F. Thomas, and M. L. Kumar.** 1989. Rapid detection of respiratory syncytial virus by a biotin-enhanced immunoassay: test performance by laboratory technologists and housestaff. *Pediatr. Infect. Dis. J.* **8**:865–869.

70. **Subbarao, E. K., N. J. Whitehurs, and J. L. Waner.** 1987. Comparison of two enzyme-linked immunosorbent assay (EIA) kits with immunofluorescence and isolation in cell culture for detection of respiratory syncytial virus (RSV). *Diagn. Microbiol. Infect. Dis.* **8**:229–234.

71. **Sullender, W. M.** 2000. Respiratory syncytial virus genetic and antigenic diversity. *Clin. Microbiol. Rev.* **13**:1–15.

72. **Swierkosz, E. M., R. Flanders, L. Melvin, J. D. Miller, and M. W. Kline.** 1989. Evaluation of the Abbott TESTPACK RSV enzyme immunoassay for detection of respiratory syncytial virus in nasopharyngeal swab specimens. *J. Clin. Microbiol.* **27**:1151–1154.

73. **Tang, Y.-W., P. S. Mitchell, M. J. Espy, T. F. Smith, and D. H. Persing.** 1999. Molecular diagnosis of herpes simplex virus infections in the central nervous system. *J. Clin. Microbiol.* **37**:2127–2136.

74. **The, T. H., A. P. van den Berg, M. C. Harmsen, W. van der Bij, and W. J. van Son.** 1995. The cytomegalovirus antigenemia assay: a plea for standardization. *Scand. J. Infect. Dis.* **99**(Suppl.):25–29.

75. **Thomas, E. E., and L. E. Book.** 1991. Comparison of two rapid methods for detection of respiratory syncytial virus (RSV) (Testpack RSV and Ortho RSV ELISA) with direct immunofluorescence and virus isolation for the diagnosis of pediatric RSV infection. *J. Clin. Microbiol.* **29**:632–635.

76. **Todd, S. J., L. Minnich, and J. L. Waner.** 1995. Comparison of rapid immunofluorescence procedure with TestPack RSV and Directigen FLU-A for diagnosis of respiratory syncytial virus and influenza A virus. *J. Clin. Microbiol.* **33**:1650–1651.

77. **Tong, C. Y., L. E. Cuevas, H. Williams, and A. Bakran.** 2000. Comparison of two commercial methods for measurement of cytomegalovirus load in blood samples after renal transplantation. *J. Clin. Microbiol.* **38**:1209–1213.

78. **van der Bij, W., R. Torensma, W. J. van Son, J. Anema, J. Schirm, A. M. Tegzess, and T. H. The.** 1988. Rapid immunodiagnosis of active cytomegalovirus infection by monoclonal antibody staining of blood leucocytes. *J. Med. Virol.* **25**:179–188.

79. **Verschuuren, E. A., M. C. Harmsen, P. C. Limburg, W. van Der Bij, A. P. van Den Berg, A. M. Kas-Deelen, B. Meedendorp, W. J. van Son, and T. H. The.** 1999. Towards standardization of the human cytomegalovirus antigenemia assay. *Intervirology* **42**:382–389.

80. **Visser, C. E., C. J. van Zeijl, E. P. de Klerk, B. M. Schillizi, M. F. Beersma, and A. C. Kroes.** 2000. First experiences with an accelerated CMV antigenemia test: CMV Brite Turbo assay. *J. Clin. Virol.* **17**:65–68.

81. **Vizzi, E., D. Ferraro, A. Cascio, R. Di Stefano, and S.**

Arista. 1996. Detection of enteric adenoviruses 40 and 41 in stool specimens by monoclonal antibody-based enzyme immunoassays. *Res. Virol.* **147:**333–339.

82. **Walmsley, S., K. O'Rourke, C. Mortimer, A. Rachlis, I. Fong, and T. Mazzulli.** 1998. Predictive value of cytomegalovirus (CMV) antigenemia and digene hybrid capture DNA assays for CMV disease in human immunodeficiency virus-infected patients. *Clin. Infect. Dis.* **27:**573–581.

83. **Waner, J. L., N. J. Whitehurst, S. J. Todd, H. Shalaby, and L. V. Wall.** 1990. Comparison of Directigen RSV with viral isolation and direct immunofluorescence for the identification of respiratory syncytial virus. *J. Clin. Microbiol.* **28:**480–483.

84. **Waner, J. L., S. J. Todd, H. Shalaby, P. Murphy, and L. V. Wall.** 1991. Comparison of Directigen FLU-A with viral isolation and direct immunofluorescence for the rapid detection and identification of influenza A virus. *J. Clin. Microbiol.* **29:**479–482.

85. **Wattanamano, P., J. L. Clayton, J. J. Kopicko, P. Kissinger, S. Elliot, C. Jarrott, S. Rangan, and M. A. Beilke.** 2000. Comparison of three assays for cytomegalovirus detection in AIDS patients at risk for retinitis. *J. Clin. Microbiol.* **38:**727–732.

86. **Whitley, R. J., and F. Lakeman.** 1995. Herpes simplex virus infections of the central nervous system: therapeutic and diagnostic considerations. *Clin. Infect. Dis.* **20:**414–420.

87. **Wren, C. G., B. J. Bate, H. B. Masters, and B. A. Lauer.** 1990. Detection of respiratory syncytial virus antigen in nasal washings by Abbott TestPack enzyme immunoassay. *J. Clin. Microbiol.* **28:**1395–1397.

88. **Yamazaki, M., K. Kimura, S. Watanabe, O. Komiyama, Y. Mishiku, K. Yamamoto, N. Sugaya, Y. Hashimoto, N. Hagiwara, T. Maezawa, and M. Imai.** 1999. Use of a rapid detection assay for influenza virus, on nasal aspirate specimens. *Kansenshogaku Zasshi* **73:**1064–1068.

89. **Zeigler, T., and N. J. Cox.** 1999. Influenza viruses, p. 928-935. *In* P. R. Murray, E. J. Baron, M. A. Pfaller, F. C. Tenover, and R. H. Yolken (ed.), *Manual of Clinical Microbiology*, 7th ed. American Society for Microbiology, Washington, D.C.

90. **Zimmerman, S. J., E. Moses, N. Sofat, W. R. Bartholomew, and D. Amsterdam.** 1991. Evaluation of a visual, rapid membrane enzyme immunoassay for the detection of herpes simplex virus antigen. *J. Clin. Microbiol.* **29:**842–845.

Human Immunodeficiency Virus

RICHARD L. HODINKA

6

At present, a variety of methods are available for use in the diagnosis and management of patients infected with human immunodeficiency virus (HIV) (Table 1). The selection of tests to perform depends on the patient population and clinical situation and the intended use of the individual assays (Table 2). Serological screening and confirmatory assays designed to detect antibodies to the virus are most commonly used for the laboratory diagnosis of HIV infection. Tests for p24 antigen, molecular amplification assays for viral DNA or RNA, and isolation of the virus in culture allow the detection of acute HIV infection in newborn babies and other individuals at risk and aid in shortening the diagnostic "window period" between infection and seroconversion. The use of quantitative nucleic-acid-based assays provides invaluable information about the prognosis and response to therapy, while phenotypic and genotypic tests can be used to detect antiretroviral-drug resistance. The choice of whether to incorporate these methods into the clinical laboratory depends on the number of specimens to be tested, cost, turnaround time, ease of testing, and the resources and capabilities of the individual laboratory.

The focus of this chapter is on the various technologies that are currently available from commercial companies for the diagnosis and monitoring of HIV infections, and it is not meant to duplicate more extensive reviews of HIV. Many of the important characteristics of the assays are described, and the key advantages and disadvantages are discussed. Because of competition among manufacturers and strict regulation worldwide, many excellent commercial diagnostic products have been developed that are well standardized and of high sensitivity and specificity. Most of the commercial assays have received widespread clinical use in Europe, Asia, or Japan, while a much smaller number have been registered in the United States and have been approved by the U.S. Food and Drug Administration (FDA) for diagnostic use. The pace of development of HIV-related tests, however, has led to an ever-changing market, with the introduction of many new assays and the discontinued use of others. A number of commercial companies are new to the field, while others have merged or left the market place. As a result, products have been transferred from one vendor to another, and many of the kit names have changed. A number of the products are also made by one company but distributed and sold under multiple brand names, making it more difficult to compile a comprehensive list. The information presented about each commercial assay was obtained either from the published literature or from the individual manufacturers through their computer websites, written materials, and/or personal communications with company representatives. While every attempt has been made to fully represent any and all manufacturers of HIV-related test products, readers are advised that the amount of material is great and not every available test may be included. Individuals should contact the manufacturer listed in the appendix for a more comprehensive description and the current list price and availability of a particular kit.

ANTIBODY ASSAYS

Infection with HIV type 1 (HIV-1) results in the induction of a humoral antibody response specific to viral proteins, with the production of immunoglobulin A (IgA), IgM, and IgG. The structural proteins of HIV-1 are the targets for the majority of the circulating antibodies directed against the virus. These include the envelope (*env*) proteins (surface glycoprotein [gp120], transmembrane glycoprotein [gp41], and their precursor glycoprotein [gp160]), polymerase (*pol*) proteins (reverse transcriptase [p65], endonuclease-integrase [p31], and protease [p10]), and core (*gag*) proteins (matrix protein [p18], internal capsid protein [p24], nucleocapsid protein [p7], and their precursor protein [p55]). Antibodies to HIV-1 are detectable in most persons within 4 to 12 weeks after infection and in virtually all patients within 6 to 12 months. Variable levels of IgM appear in the serum first, quickly reaching a peak and declining over the following weeks. About 1 week later, IgG antibody levels rise significantly, reach a plateau within a few months, and remain high for many years. In some individuals, antibody to core proteins (p24 and/or p18) and sometimes polymerase proteins (p65 and/or p31) may drop significantly and even become undetectable with progression to HIV disease. Elevated levels of HIV-specific serum IgA antibody can be found in infants as early as 3 weeks after birth and may persist for an extended time following acute infection.

Tests for the detection of HIV-specific antibodies are at the front line for diagnosis of HIV infection (see reference 45 for an extensive review). They have been effectively used to determine if an individual has been exposed to HIV, to screen blood and plasma donations, and for epidemiological surveillance. The diagnosis of HIV infection in adults and

TABLE 1 Characteristics of available virological methods for the diagnosis and monitoring of HIV infection

Test	Specimen required	Cost	Technical difficulty	Turnaround (days)	Diagnostic value
Antibody EIA or WB[a]	Serum or plasma	Low	Low	1–2	High in patients >15–18 mo of age
Standard p24 antigen	Serum or plasma	Low	Low	1–2	Low in all groups
ICD p24 antigen	Serum or plasma	Low	Low	1–2	High in patients >1 month of age
DNA PCR	PBMC[b]	Moderate	Moderate	1–2	High in all groups
Virus culture	PBMC[c]	High	High	14–28	High in all groups
RNA quantification	Plasma	Moderate	Moderate	1–2	High in all groups
Phenotypic drug susceptibility assays	Patient's virus isolate	High	High	14–42	Most likely high; remains to be fully determined
Genotypic drug susceptibility assays	Plasma	High	High	1–2	High in all groups

[a]WB, Western blot.
[b]DNA PCR also can be performed on whole blood, plasma, other body fluids, and tissues.
[c]Virus culture also can be performed on plasma, other body fluids, and tissues.

children older than 18 months of age is readily made by detecting the presence of antibodies to HIV in serum or plasma. A two-stage testing strategy is universally used. Infection is first identified by using a highly sensitive screening test called an enzyme immunoassay (EIA). If specimens are nonreactive for HIV antibodies when initially tested by EIA, no further testing is performed and the individual is considered to be uninfected. Specimens with initially reactive results by EIA must be retested in duplicate using the same EIA. If one or both of the duplicate samples of the specimen are reactive by EIA, the final EIA result is considered positive. If the two repeat samples of the specimen are nonreactive, the specimen is considered negative for HIV antibodies by EIA. Repeatedly reactive specimens are then confirmed by Western blot or immunofluorescence assay (IFA) to ensure the specificity of the EIA result. A determination of positive reactivity by EIA followed by confirmation using one of the confirmatory tests remains the most convincing laboratory evidence for a diagnosis of HIV infection.

EIAs

A wide selection of traditional EIA screening kits that detect HIV-specific antibodies are available from various commercial sources (Table 3). The assays offer the distinct advantage of using highly standardized and stable immunoreagents that provide accurate and objective results. They require minimal training and equipment and are applicable to large numbers of specimens at a reasonable cost. The different test kits vary in their configurations, the number and type of antigens used to capture HIV-specific antibodies, the specimen sources that can be tested, and whether the assays are designed for detecting antibodies to either HIV-1 or HIV-2 or for the simultaneous detection of antibodies to both viruses.

Since being licensed for diagnostic use in 1985, EIAs for the detection of HIV-specific antibodies have evolved through three generations of change and improvement, with the source of antigen(s) used being a significant factor in determining the overall sensitivity and specificity of the tests. The original first-generation kits use viral proteins prepared from lysates of HIV-infected human T-cell lines as the source of antigens that are bound to a solid phase for capturing HIV-specific antibodies from a patient's specimen. The antigen-antibody complexes that form are then detected with the addition of an enzyme-labeled, anti-human antibody that binds to the complexes and reacts

TABLE 2 Use of laboratory tests for HIV diagnosis and management

Clinical situation	Recommended test(s)
Blood donor screening	Antibody EIA and WB; p24 antigen; plasma RNA detection
Routine diagnosis of HIV infection (excluding infants and acute infection)	Antibody EIA and WB
Acute HIV infection	DNA PCR; p24 antigen; HIV culture; plasma RNA detection[b]
Infant (≤18 mo of age) born to HIV-infected mother	DNA PCR; p24 antigen; HIV culture; plasma RNA detection
Indeterminate HIV-1 WB[a]	Repeat HIV-1 antibody EIA and WB; perform HIV-2-specific antibody EIA and WB; DNA PCR; HIV culture
Prognosis	Plasma RNA quantification
Response to therapy	Plasma RNA quantification
Antiretroviral drug resistance	Phenotypic and/or genotypic resistance assays

[a]WB, Western blot.
[b]Assays for the detection of HIV RNA are reported to be more sensitive than HIV DNA PCR procedures for detection of acute HIV infection in infants (50, 243)

TABLE 3 Screening EIAs for the detection of HIV antibodies[a]

Manufacturer	Product name	Assay type	Sample	Detection	Antigen	Selected reference(s)
Abbott Laboratories	HIVAB HIV-1[c]	BEIA	Serum, plasma, dried blood spot	HIV-1	VL	82, 117, 186
	HIV-1/HIV-2 (rDNA), 3rd Generation[c]	BEIA	Serum, plasma	HIV-1, HIV-2	RP	6, 70, 79, 80, 117, 120, 174, 191, 205, 219, 231, 232, WHO[d]
	HIV-1/HIV-2 3rd Generation Plus	BEIA	Serum, plasma	HIV-1 including group O, HIV-2	RP	228, 230
	AxSYM HIV-1/HIV-2	MPEIA	Serum, plasma	HIV-1 including group O, HIV-2	RP	126, 228
	IMx HIV-1/HIV-2 III Plus	MPEIA	Serum, plasma	HIV-1 including group O, HIV-2	RP	124, 138, 205, WHO
	Murex Wellcozyme HIV Recombinant	MWEIA	Serum, plasma	HIV-1	RP	5, 132, 140, 150, 152, 168, 190, 209, 219, 223, 229, 231
	Murex Wellcozyme HIV 1 + 2	MWEIA	Serum, plasma	HIV-1, HIV-2	RP + SP	34, 120, 121, 139, 140, 168, 205, 218, 219, 228, 232
	Murex ICE HIV-1.O.2	MWEIA	Serum, plasma	HIV-1 including group O, HIV-2	RP + SP	129, 140, 151, 160, 205, WHO
	Murex ICE HIV-2	MWEIA	Serum, plasma	HIV-2	SP	140, 150, 223
BioChem Pharma, Inc.	Detect HIV-1/2	MWEIA	Serum, plasma	HIV-1, HIV-2	SP	70, 106, 139, 219, WHO
Biokit S.A.	Bioelisa HIV 1 + 2 (rec)	MWEIA	Serum, plasma	HIV-1, HIV-2	RP	139, 169, 205
bioMerieux	VIDAS HIV 1 + 2	Automated SPR ELFA	Serum, plasma	HIV-1, HIV-2	SP	10, 139, 168, 229
Bio-Rad	Genetic Systems HIV1 rLAV[c]	MWEIA	Serum, plasma, dried blood spot	HIV-1	RP	82, 117, 219
	Genetic Systems HIV-1/HIV-2 Peptide[c]	MWEIA	Serum, plasma	HIV-1, HIV-2	SP	59, 93, 117, 191
	Genetic Systems HIV-2[c]	MWEIA	Serum, plasma	HIV-2	VL	219
	Sanofi Diagnostics Pasteur Genscreen HIV-1/2	MWEIA	Serum, plasma	HIV-1, HIV-2	SP	6, 75, 126, WHO
	Sanofi Diagnostics Pasteur Genelavia Mixt	MWEIA	Serum, plasma	HIV-1, HIV-2	RP + SP	106, 129, 132, 139, 140, 191, 197, 205, 219, 232
	Sanofi Diagnostics Pasteur ELAVIA I	MWEIA	Serum, plasma	HIV-1	VL	190
	Sanofi Diagnostics Pasteur ELAVIA II	MWEIA	Serum, plasma	HIV-2	VL	17, 154, 190

Company	Test	Method	Specimen	Type	Antigen	References
Biotest AG	Biotest Anti-HIV-1/2 recombinant	MWEIA	Serum, plasma	HIV-1, HIV-2	RP	75, 120, 132, 139, 205, 232, WHO
Calypte Biomedical	Calypte HIV-1 Urine EIA[b,c]	MWEIA	Urine	HIV-1	RP	77, 141, 210–212
Dade Behring	Enzygnost Anti-HIV 1/2	MWEIA	Serum, plasma	HIV-1, HIV-2	RP	5, 6, 174, 205, 221, 229, 231, WHO
	Enzygnost Anti-HIV 1/2 Plus	MWEIA	Serum, plasma	HIV-1 including group O, HIV-2	RP	140, 151, 209, 215, WHO
DiaSorin, Inc.	ETI-AB-HIV-1/2 K	MWEIA	Serum, plasma	HIV-1, HIV-2	SP	139, WHO
Innogenetics	Innotest HIV-1/HIV-2	MWEIA	Serum, plasma	HIV-1 including group O, HIV-2	SP	5, 139, 205, WHO
Organon Teknika	Vironostika HIV-1[c]	MWEIA	Serum, plasma, dried blood spot, oral fluid	HIV-1	VL	79, 117, 161, 219
	Vironostika Uni-Form II	MWEIA	Serum, plasma, oral fluid	HIV-1, HIV-2	VL + SP	120, 140, 197, WHO
	Vironostika Uni-Form II Plus O	MWEIA	Serum, plasma	HIV-1 including group O, HIV-2	VL + SP	75, 150, 205, 213–216, WHO
Roche Diagnostics	Boehringer Mannheim Enzymun Test Anti-HIV 1 + 2, Generation 3	MWEIA	Serum, plasma	HIV-1 including group O, HIV-2	RP + SP	205, 227, 231, 236
	Cobas Core Anti-HIV-1/HIV-2 EIA DAGS	BEIA	Serum, plasma	HIV-1, HIV-2	RP + SP	32, 205
Span Diagnostics, Ltd.	ENZAIDS Fast	MWEIA	Serum, plasma	HIV-1, HIV-2	RP + SP	NF
ThermoLabsystems	HIV EIA	MWEIA	Serum, plasma	HIV-1 including group O, HIV-2	SP	WHO
Trinity Biotech	Recombigen HIV-1/HIV-2	MWEIA	Serum, plasma	HIV-1, HIV-2	RP	120, 232, WHO
	Recombigen HIV-1	MWEIA	Serum, plasma	HIV-1	RP	113, 122, 152, 171
United Biomedical, Inc.	UBI HIV-1/2 EIA[c]	MWEIA	Serum, plasma	HIV-1, HIV-2	SP	5, 76, 139, 186, 205, WHO
	UBI HIV EIA[c]	MWEIA	Serum, plasma	HIV-1	SP	104

[a] Abbreviations: BEIA, bead enzyme immunoassay; MWEIA, microwell enzyme immunoassay; MPEIA, microparticle enzyme immunoassay; SPR, solid-phase receptacle; ELFA, enzyme-linked fluorescent assay; RP, recombinant proteins; SP, synthetic peptides; VL, viral lysate; NF, none found.
[b] The Calypte HIV-1 Urine EIA is also marketed under the name Seradyn Sentinel HIV-1 Urine EIA.
[c] Approved by the FDA for use in the United States.
[d] WHO, http://www.who.int/pht/blood_safety/hivkits.html.

with a chromogenic substrate to produce a color change. The intensity of the color generated is measured in a spectrophotometer and compared with a set of positive and negative controls performed with each batch of specimens. Horseradish peroxidase and alkaline phosphatase are the most common enzyme labels. The surfaces of microwell plates and polystyrene beads are normally used as the solid-phase carrier.

Technological advances over the last decade have led to the incorporation of recombinant antigens and synthetic peptides into second- and third-generation serological assays to improve their sensitivity and specificity over traditional tests based on whole viral lysates. Figure 1 depicts one of the third-generation double-antigen sandwich EIAs (HIV-1/HIV-2 [rDNA], 3rd Generation, Abbott Laboratories, Abbott Park, Ill.) commonly used in the United States for the simultaneous detection of antibodies to HIV-1 and HIV-2. In this third-generation EIA, human serum or plasma is incubated with a polystyrene bead coated with recombinant HIV-1 *env* and *gag* and HIV-2 *env* proteins. HIV-specific antibodies in the specimen react with the antigens on the coated bead to form antigen-antibody complexes. The antigen-antibody complexes on the bead are detected by incubating the bead with a solution containing HIV-1 *env* and *gag* and HIV-2 *env* recombinant proteins labeled with horseradish peroxidase to form an antigen-antibody-antigen sandwich. The sandwich is then detected by adding a colorless substrate (*o*-phenylenediamine) that is cleaved by the enzyme to give off a color reaction. This format has the distinct advantage of efficiently and simultaneously detecting IgG, IgM, and IgA class antibodies to HIV-1 and HIV-2, while first- and second-generation assays primarily detect IgG antibodies. Third-generation assays also have a greater sensitivity for detecting HIV antibodies in the early stages of infection, thereby shortening the time to seroconversion.

In general, the traditional EIAs are considered to be highly sensitive and specific for the detection of antibodies to HIV, regardless of whether they are first-, second-, or third-generation assays (117, 139, 140, 191, 205, 219; http://www.who.int/pht/blood_safety/hivkits.html). The predictive value of a positive result, however, can vary considerably and depends on the prevalence of HIV infection in the population being tested. Few false-positive results are observed when testing high-risk populations, while a high rate of false-positive results may occur in populations with low prevalence of HIV infection, such as blood donors. When using first-generation assays, biological false-positive reactions are mainly due to reactivity of antibodies to

human leukocyte antigen proteins that are expressed by T-cell lines used to prepare the viral lysates. With second- and third-generation assays, the use of recombinant antigens or synthetic peptides significantly decreases the number of false-positive results but does not eliminate them. False positives may be the result of passive immunoglobulin administration, recent exposure to certain vaccine preparations (e.g., influenza vaccine), or the patient having cross-reactive antibodies to contaminating bacterial or yeast proteins used in recombinant antigen-based EIAs. Therefore, the EIA, although an excellent test for screening, should not be used as the sole test for making a diagnosis of HIV infection. A positive confirmation test is also necessary to exclude false-positive results.

A negative antibody EIA result normally means HIV infection is unlikely. However, false-negative results may occur and may be due to immunosuppressive therapy, replacement transfusion, severe hypogammaglobulinemia (B-cell dysfunction and defective antibody synthesis), genetic diversity of the virus itself, and testing too early (before seroconversion) or too late in the course of illness. Although it is possible that levels of HIV-specific antibodies may fall during advanced disease, they rarely drop below the detection limits of current EIAs. A newly identified group of HIV-1 (group O) has been described that has evolved to the extent that antibodies to this group are not detected by many of the EIAs on the market. Group O HIV-1 strains have been seen predominantly in Africa and Europe, although a small number of infections with this group have also been identified in the United States. EIA kits using viral lysates as the antigen source are more efficient at detecting this highly divergent group than kits that use recombinant antigens or synthetic peptides. A number of the current commercial EIAs have recently been or are being reformulated to contain specific antigens to group O in order to recognize infection with this group. Similarly, many commercial HIV-1 and HIV-2 combination tests have been developed and are now being used to effectively screen the blood supply. Separate HIV-2 tests are also commercially available and should be considered if it is suspected that HIV-2 infection is probable. The sensitivity of HIV-1 antibody assays for HIV-2 antibody ranges from 60 to 91%, making it necessary to have tests that are specific for HIV-2. As the prevalence of HIV-2 infection increases worldwide, there will be an obvious need for these combination and HIV-2-specific assays. The Centers for Disease Control and Prevention, however, do not currently recommend screening for HIV-2 outside of the blood bank setting because of the low prevalence of HIV-2 in the United States. HIV-2 antibody testing should be considered for persons from West Africa, where the virus is endemic, individuals who have received blood transfusions from or have had sexual relations with someone from this region, and children of women at risk of infection or known to be infected with HIV-2.

Additional innovations and modifications in the current methods for HIV antibody detection have led to significant advances in the field. Recently, new fourth-generation screening EIAs have been developed and released in Europe for the simultaneous detection of HIV-1 p24 antigen and antibodies to HIV-1 (including group O) and HIV-2 (Table 4). These assays provide an increase in sensitivity over tests that only detect antibody and allow a reduction in the window period between infection with HIV and laboratory diagnosis. The list price of these new assays will be comparable to those of the current third-generation antibody tests, allowing for a cost-

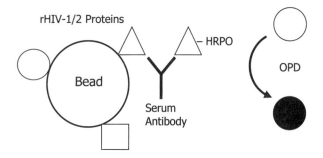

FIGURE 1 Schematic of a third-generation screening enzyme immunoassay for the detection of antibodies to HIV-1 and HIV-2. Abbreviations: rHIV, recombinant HIV; HRPO, horseradish peroxidase; OPD, *o*-phenylenediamine.

TABLE 4 Fourth-generation screening EIAs for the simultaneous detection of HIV antibodies and p24 antigen[a]

Manufacturer	Product name	Test format	Detection	Solid-phase Ag and Ab	Selected references
Abbott	AxSYM HIV Ag/Ab	Automated MEIA	IgG and IgM Ab to HIV-1, including groups O and M, and HIV-2; HIV-1 p24 Ag	Ab, recombinant proteins; Ag, monoclonal anti-p24 Ab	NF
bioMerieux	Vidas HIV Duo	Automated EIA	IgG Ab to HIV-1, including groups O and M, and HIV-2; HIV-1 p24 Ag	Ab, synthetic peptides; Ag, monoclonal anti-p24 Ab	125, 126, 180, 230
Bio-Rad	Sanofi Diagnostics Pasteur Genscreen Plus HIV Ag/Ab	Microplate EIA	IgG Ab to HIV-1, including groups O and M, and HIV-2; HIV-1 p24 Ag	Ab, recombinant proteins; Ag, monoclonal anti-p24 Ab	NF
Roche Diagnostics	Boehringer Mannheim Enzymun-Test HIV Combi	Automated EIA	IgG and IgM Ab to HIV-1, including groups O and M, and HIV-2; HIV-1 p24 Ag	Ab, synthetic peptides and recombinant proteins; Ag, monoclonal anti-p24 Ab	80, 230
	COBAS Core HIV Combi	Automated EIA	IgG and IgM Ab to HIV-1, including groups O and M, and HIV-2; HIV-1 p24 Ag	Ab, synthetic peptides and recombinant proteins; Ag, monoclonal anti-p24 Ab	NF
	Boehringer Mannheim Elecsys HIV Combi	Automated ECA	IgG and IgM Ab to HIV-1, including groups O and M, and HIV-2; HIV-1 p24 Ag	Ab, synthetic peptides and recombinant proteins; Ag, monoclonal anti-p24 Ab	NF
Dade Behring	Enzygnost HIV Integral	Microplate EIA	IgG Ab to HIV-1, including groups O and M, and HIV-2; HIV-1 p24 Ag	Ab, synthetic peptides and recombinant proteins; Ag, rabbit polyclonal anti-p24 Ab	31, 126
Organon Teknika	Vironostika HIV Uniform II Ag/Ab	Microplate EIA	IgG Ab to HIV-1, including groups O and M, and HIV-2; HIV-1 p24 Ag	Ab, synthetic peptides; Ag, monoclonal anti-p24 Ab	126, 214, 216

[a]Abbreviations: MEIA, microparticle enzyme immunoassay; ECA, electrochemiluminescence assay; Ab, antibody; Ag, antigen; NF, none found.

effective means of screening donated blood products while eliminating the current need in the United States and other countries to use a separate p24 antigen assay at a higher cost.

Rapid (10- to 20-min) and less sophisticated EIAs for the detection of HIV antibodies are also being developed by a number of manufacturers (Table 5). The tests offer the distinct advantages of lower costs and same-day results and are packaged as ready-to-use kits with all reagents and materials included. The kits are designed to be performed at the site where the specimen was collected using a single step or a few simple steps and self-contained, disposable devices. The performance of these rapid assays requires no specialized equipment and only limited technical expertise; some of the kits have the added advantage of possessing stabilized biochemicals that guarantee a long shelf life when stored at room temperature (see references 26 and 175, and http://www.who.int/pht/blood_safety/hivkits.html for more extensive reviews).

There are four basic formats for these rapid HIV tests, membrane flow-through immunoconcentration devices, lateral-flow immunochromatographic strips, particle agglutination, and immunodot comb assays. In membrane flow-through devices, HIV antigens immobilized on a membrane capture and concentrate HIV-specific antibodies on the surface of the device as the specimen flows through and is absorbed into an absorbent pad. This is followed by the sequential addition of enzyme-labeled anti-human antibody and a colorless substrate. Enzymatic hydrolysis of the substrate leads to a colorimetric result that is read visually as a dot or line that forms on the membrane. Many of the membrane flow-through kits include procedural controls to verify the satisfactory performance of the assay. Lateral-flow immunochromatographic strips, or so-called dipsticks, are the most recent addition to the development of rapid and simple assays for HIV. The specimen is applied to an absorbent pad and migrates by capillary action along a solid-phase strip, where it combines with HIV antigens and detector reagents to produce a visible line on the strip when HIV-specific antibodies are present. A procedural control is normally included on the strip and is also indicated by a visible line. Immunodot comb assays use a solid plastic comb with teeth that are sensitized at several spots with different HIV antigens and control material. Patient specimens are placed in individual wells that accommodate single teeth of the comb. The comb is then transferred from specimen wells to reagent wells, and the teeth are saturated by the different solutions; positive results are indicated by spots that form at reactive positions on the individual teeth. Lastly, in particle agglutination assays, HIV-specific antibodies in a patient's specimen will visibly agglutinate or aggregate when cross-linked with particles (e.g., latex, red blood cells, or gelatin) that are coated and sensitized with HIV antigens. In one of the commercial agglutination kits (Capillus HIV; Trinity

TABLE 5 Rapid and/or simple EIAs for the detection of HIV antibodies[a]

Manufacturer	Product name	Assay type	Sample	Detection	Antigen	Selected reference(s)
Abbott Laboratories	Determine HIV-1/2	LF	Serum, plasma, whole blood	HIV-1 including group O, HIV-2	RP	3, 7, 26, 116, 124, 157, 175
	TestPack HIV-1/HIV-2 AB	FT	Serum, plasma	HIV-1, HIV-2	RP	28, 114, 120, 128, 176, 197, 198, 208, 219, 247
	SUDS HIV-1[c]	FT	Serum, plasma	HIV-1	SP	26, 34, 48, 110, 117, 123, 163, 172, 176, 186, 198, 208, 226
AccuDx	Retro Cell	PA	Serum, plasma	HIV-1	VL	26, 181, 198, WHO[d]
	AccuSpot HIV-1 & 2	FT	Serum, plasma	HIV-1, HIV-2	RP	WHO
	AccuSpot Plus HIV 1 & 2 (with procedural control)	FT	Serum, plasma	HIV-1, HIV-2	RP	NF
Agen	SimpliRED HIV-1/2	RCA	Serum, plasma	HIV-1, HIV-2	SP	26, 44, 85, 193, 207, WHO
	MicroRED HIV-1/2	PA	Serum, plasma	HIV-1, HIV-2	SP	26, WHO
Biokit S.A.	Biorapid HIV 1 + 2	LF	Serum, plasma, whole blood	HIV-1, HIV-2	SP	NF
Bionor AS	HIV-1 & 2 Test Kit	PEIA	Serum, plasma, whole blood, saliva	HIV-1 including group O, HIV-2	SP	26, 175, WHO
Bio-Rad	Sanofi Diagnostics Pasteur Multispot HIV-1/HIV-2	FT	Serum, plasma	HIV-1, HIV-2	RP + SP	5, 26, 137, 140, 163, 175, 223, 226
	Genetic Systems Genie HIV1/HIV2	FT	Serum, plasma	HIV-1, HIV-2	SP	9, 26, 28, 40, 47, 85, 114, 116, 175, 198, 219
Chembio Diagnostic Systems	HIV 1/2 STAT-PAK	LF	Serum, plasma, whole blood	HIV-1, HIV-2	SP	NF
Fujirebio	Serodia HIV-1/2	PA	Serum, plasma	HIV-1, HIV-2	VL	15, 26, 46, 47, 59, 105, 124, 142, 166, 196–198, 219, 221, 224, WHO
	SFD HIV 1/2 PA	PA	Serum, plasma	HIV-1, HIV-2	RP	NF
Genelabs Diagnostics	HIV Spot	FT	Serum, plasma	HIV-1, HIV-2	RP + SP	3, 26, 47, 109, 116, 140, 145, 175, 176, 218
J Mitra & Co., Ltd.	HIV Tri-Dot	FT	Serum, plasma	HIV-1, HIV-2	RP + SP	26, 109, WHO
MedMira Laboratories	MedMira Rapid HIV 1/2	FT	Serum, plasma, whole blood	HIV-1 including groups O and M, HIV-2	SP	26
	MedMira Rapid HIV/HCV	FT	Serum, plasma, whole blood	HIV-1 including groups O and M, HIV-2, HCV	SP	NF
OraSure Technologies	OraQuick HIV-1/2[c]	LF	Oral fluid	HIV-1, HIV-2	SP	26, 175

Organics	ImmunoComb II HIV 1 & 2 BiSpot	ID	Serum, plasma	HIV-1, HIV-2	SP	5, 47, 107, 178, 197, 219, 223, WHO
	ImmunoComb II HIV 1 & 2 CombFirm (detects and confirms)	ID	Serum, plasma	HIV-1 including group O, HIV-2	SP	NF
	ImmunoComb II HIV 1 & 2 Saliva	ID	Saliva	HIV-1, HIV-2	SP	1, 179, 241
	DoubleCheck HIV 1 & 2	LF	Serum, plasma	HIV-1 including group O, HIV-2	RP + SP	26, 175
Ortho Diagnostics	HIVCHEK 1 + 2	FT	Serum, plasma	HIV-1, HIV-2	RP	15, 26, 47, 105, 128, 142, 161, 163, 175, 196, 198, 200, 219, 224, WHO
PATH (Program for Appropriate Technology in Health)	HIV Dipstick Test[b]	ID	Serum, plasma	HIV-1, HIV-2	RP + SP	164, 174, 219
Saliva Diagnostic System	Sero Strip HIV	LF	Serum, plasma	HIV-1, HIV-2	SP	26, 56, 137, 163, 175, 225, 246, WHO
	Hema Strip HIV	LF	Whole blood	HIV-1, HIV-2	SP	26, 146, 175, 225, 242
	Saliva Strip HIV	LF	Oral fluid	HIV-1, HIV-2	SP	175, 183, 225
Savyon Diagnostics Ltd.	HIV Sav 1 & 2	LF	Serum, plasma	HIV-1, HIV-2	SP	26, 175, WHO
Span Diagnostics Ltd.	CombAIDS-RS	ID	Serum, plasma	HIV-1, HIV-2	RP + SP	26, 140, 208, WHO
Trinity Biotech	SeroCard HIV	FT	Serum, plasma, whole blood	HIV-1, HIV-2	SP	26, 56, 73, 175, WHO
	SalivaCard HIV	FT	Oral fluid	HIV-1, HIV-2	SP	26, 175, 179, 241
	Uni-Gold HIV-Peptide[c]	LF	Serum, plasma, whole blood	HIV-1, HIV-2	SP	26, 73
	Uni-Gold HIV-Recombinant[c]	LF	Serum, plasma, whole blood	HIV-1, HIV-2	RP	175
	Capillus HIV	PA	Serum, plasma, whole blood	HIV-1, HIV-2	RP	5, 26, 56, 111, 116, 120, 124, 163, 164, 175, 176, 240, WHO
Universal Healthwatch	Recombigen HIV RTD	FT	Serum, plasma	HIV-1, HIV-2	RP	5, 139
	Quix HIV-1/2/O	FT	Serum, plasma	HIV-1 including group O, HIV-2	SP	26, 47

[a] Abbreviations: HCV, hepatitis C virus; LF, lateral flow; FT, flow-through; PA, particle agglutination; PEIA, particle EIA; ID, immunodot; RCA, red cell agglutination; RP, recombinant proteins; SP, synthetic peptides; VL, viral lysate; NF, none found.
[b] Manufactured under several names, including DIA HIV-1+2 and Entebe HIV Dipstick.
[c] Approved by the FDA for use in the United States. The FDA has approved an investigational device exemption (IDE) for the Trinity Biotech Uni-Gold and the OraSure Technologies OraQuick rapid HIV tests.
[d] WHO, http://www.who.int/pht/blood_safety/hivkits.html.

Biotech, Bray, County Wicklow, Ireland), a narrow capillary flow channel is used to enhance the binding of specific antibodies to the sensitized particles, thereby promoting aggregation. Most of the described rapid tests use multiple recombinant and/or synthetic peptides to differentiate antibodies to HIV-1 and HIV-2, and some even include specific proteins to efficiently detect HIV-1 group O and the various subtypes of HIV-1 group M. Serum and plasma are the specimens of choice for the majority of the rapid devices, while some of the assays have been adapted to allow the use of whole blood, saliva, or oral fluids. These assays have sensitivities and specificities similar to those of the more traditional EIAs when performed and read by properly trained personnel, although some assays perform better than others. Predictive values comparable to those of the standard combination of EIA and Western blot testing can be obtained using multitest algorithms composed of a combination of two or more rapid tests. Some countries outside of the United States now use these combinations of rapid tests as a less expensive and more rapid alternative to using EIA and Western blotting for blood screening, diagnostic testing, and epidemiological surveillance. Only the SUDS (Single Use Diagnostic System) HIV-1 assay (Abbott Laboratories) is approved by the FDA as a rapid, less sophisticated test for diagnostic use in the United States. The FDA has also approved an investigational device exemption for the Uni-Gold Peptide and Recombinant assays (Trinity Biotech) and the OraQuick HIV-1/2 (OraSure Technologies, Bethlehem, Pa.).

Rapid HIV tests are widely used in developing countries, where resources and facilities may make it impractical or even impossible to perform the more technically demanding conventional EIAs that require time, sophisticated equipment, and completion in a clinical laboratory. The tests are intended for use in emergency departments, hospital clinics, sexually transmitted disease clinics, family planning clinics, and HIV outreach programs. In these settings, rapid testing may provide same-visit results to individuals seeking HIV testing, as many persons, including those infected with HIV, never return to receive their results following the delay in reporting that may occur when using traditional testing. The rapid availability of test results may assist in providing more timely essential medical and prevention services to these individuals. Rapid tests also may be useful in assessing the risk of HIV transmission following exposure to possibly HIV-contaminated materials or in screening pregnant women presenting for delivery with unknown HIV serostatus. It has been shown that antiretroviral therapy reduces occupational transmission of HIV after percutaneous exposures and reduces vertical transmission when used in the intra- or postpartum period.

The approval by the FDA of dried blood spots (16, 19, 81, 220), urine (35, 134, 135, 206, 210–212), and oral-fluid specimens (91, 130, 133, 135, 143, 204) for HIV antibody provides fast, convenient, and relatively noninvasive alternatives in specimen collection. Particular attention has been given to the value of oral fluids and urine specimens for the diagnosis of infection with HIV. Screening assays primarily intended for serum or plasma have been modified for use with oral-fluid and urine specimens, and extremely sensitive assays specifically designed for these specimen types have also been developed and FDA approved. When such changes are made and assay protocols are optimized to accept oral-fluid or urine specimens, the sensitivities and specificities of these assays for the detection of HIV-specific antibodies are equal to those when testing serum or plasma. Several commercial devices have also been developed specifically for the collection of oral mucosal transudate specimens (Table 6), and one of the devices (OraSure; OraSure Technologies) has been approved by the FDA for diagnostic use. The devices provide a homogeneous specimen rich in plasma-derived IgG and IgM that is passively transferred to the mouth across the mucosa and through the gingival crevices (for a detailed description of the devices, see reference 91).

Testing oral fluids or urine for antibody to HIV has wide application in the management of patients and epidemiological surveillance, particularly under circumstances in which collecting serum or plasma is less practical. The collection of these specimens does not require laboratory personnel with special training; patients can easily obtain the sample themselves. HIV antibodies in urine or oral fluids are stable for extended times at room temperature, and specimens can be mailed or shipped without degradation. Collection of oral fluids or urine also increases compliance and alleviates the fear that patients may experience when having their blood drawn. It reduces the potential danger to the health professional through blood exposure and may benefit more challenging populations whose blood may be difficult to obtain, including children, hemophiliacs, obese people, and the elderly and infirm. The use of urine or oral fluids may permit improved access for the surveillance of intravenous drug users, homeless persons, sex industry workers, and persons in developing countries. Finally, collection of these specimens may afford a greater opportunity to screen for HIV antibodies in point-of-care settings, physicians' and dentists' offices, public health institutions, and community outreach programs.

Lastly, EIAs have the greatest potential for automation, and a number of semiautomated and fully automated immunoassay analyzers are now commercially available for the performance of HIV serological assays. The majority of the automated immunoassay analyzers provide walk-away simplicity to perform assays from sample processing through interpretation of results (see chapter 13).

Confirmatory and Supplemental Tests

The Western blot assay is the principal supplemental test used worldwide to confirm the specificity of positive results obtained from HIV EIAs. Western blot assays are essentially solid-phase EIAs that use immobilized viral antigens to detect antibodies to specific proteins. The major advantage of Western blot assays over EIAs is that the specific interaction of antibody and antigen can be directly visualized. These assays have the disadvantages, however, of being technically demanding, relatively expensive, and subject to interpretation. A number of Western blot assays have been commercially developed for detecting antibodies to either HIV-1 or HIV-2 (Table 7). Only a few have been FDA approved as confirmatory tests. Most Western blot tests are similar to first-generation EIAs in that viral lysates are used as the source of HIV antigens. Several commercial manufacturers have recently developed immunoblot assays that utilize recombinant HIV proteins and/or synthetic peptides that are applied in separate lines or bands to nitrocellulose membranes. These assays can differentiate between HIV-1 and HIV-2 infection and are more sensitive and specific than conventional Western blot assays. In the United States, due to the low prevalence of HIV-2, Western blotting is usually performed for the confirmation of antibodies specific to HIV-1. HIV-2-specific Western blotting is not routinely done on all specimens that are positive by HIV EIA but is performed only when warranted by the clinical situation.

TABLE 6 Commercial devices for the collection of oral-fluid specimens for the detection of HIV antibodies

Manufacturer	Product name	Features	Selected references
OraSure Technologies	OraSure[a]	A compressed, absorbent cotton pad treated with a dried salt solution and attached to a plastic handle is placed between the lower cheek and gum, and the subject is instructed to rub the pad back and forth until moist. The pad is then held in place for 2 min, removed from the mouth, and placed in a stoppered transport vial containing buffered preservatives. In the laboratory, the fluid is eluted from the pad, and the eluate is recovered by centrifugation.	71, 79, 91, 156, 194
Saliva Diagnostic Systems	Omni-SAL	A compressed, absorbent cotton pad attached to a plastic stem is placed under the tongue and absorbs fluid from the floor of the mouth. The device incorporates an indicator on the plastic stem that turns from white to blue when an adequate amount of sample has been collected. The collection pad is then inserted into a stoppered transport tube containing buffered preservatives. In the laboratory, the collection pad is compressed and the eluate is filtered with a piston-style filter.	64, 69, 78, 91, 113, 179, 241
Sarstedt, Ltd.	Salivette	A compressed cylinder of cotton is placed in the mouth and is gently chewed for ~1 min to enhance release of oral mucosal transudates. The saturated cotton is then placed inside the provided stoppered inner tube, which has a small hole in its base. The inner tube fits into an outer tube with a conical base, and oral fluids pass through the hole into the outer tube upon centrifugation.	49, 91, 121, 133, 217
Trinity Biotech	Orapette	A small rayon ball is placed in the mouth, and the subject is asked to concentrate on collecting oral fluids from around the teeth and gum area until the rayon is completely saturated. The saturated rayon ball is then placed in the receiving container, and a plunger is screwed into the container, compressing the rayon and releasing drops of oral fluid from a small opening in the container.	91, 179, 242

[a]The OraSure oral-specimen collection device is the only apparatus that has been approved by the FDA for use in the detection of HIV antibodies in oral fluids.

Figure 2 illustrates the steps involved in the Western blot assay technique. In the Western blot assay, whole-virus lysates of inactivated and disrupted viral proteins are separated by electrophoresis according to their molecular weights or relative mobilities as they migrate through a polyacrylamide gel in the presence of sodium dodecyl sulfate. The resolved protein bands are then transblotted (transferred) to a sheet of nitrocellulose paper. The nitrocellulose paper is then cut into strips that are reacted with serum specimens. If HIV-1-specific antibodies are present in the serum, binding of the antibodies occurs in bands corresponding to the presence of the separated viral proteins. The bands are directly visualized by using an enzyme-labeled anti-human antibody followed by a chromogenic substrate (Fig. 3).

If the results of a repeatedly reactive screening test are confirmed by a positive Western blot, a diagnosis of HIV-1 infection can be made. Specificity for detecting HIV-1 antibodies has been reported to approach 100% with the combined use of an HIV-1 EIA and a Western blot assay. A negative Western blot assay result on a serum that is repeatedly reactive by HIV-1 EIA suggests a false-positive EIA result. Specimens producing banding patterns that do not meet the criteria for a positive result are classified as indeterminate. Indeterminate results can be observed in both the lysate-based Western blots and the newly developed immunoblot assays. Indeterminate HIV-1 Western blots may represent evidence of recent infection in high-risk per-

sons exposed to HIV-1, infection with HIV-2, or false reactions in persons at low risk for HIV-1 infection. In this situation, it is recommended that the EIA and Western blot assay be repeated at intervals over time (at least beyond 6 months). Individuals at low risk for HIV-1 infection with persistently indeterminate HIV-1 Western blots are most likely uninfected (96, 100). The most common types of indeterminate Western blots for someone who is uninfected include a p24 band only, a p18 band only, p24 and p55 bands, and a p55 band only (e.g., *gag*-derived proteins) in the absence of antibodies to the envelope glycoproteins gp41, gp120, and gp160 (52). The sensitivity of a Western blot assay is approximately equal to that of the first- and second-generation EIA formats but is less than that of the antigen sandwich EIAs. The Western blot assay should not be used as a screening test, since its specificity is less than that of EIA, with a significant number of indeterminate blots seen in low-risk populations.

IFA is a very useful and inexpensive alternative to performing Western blot assays for confirmation of HIV-specific antibody responses (94, 201). Although several IFA kits have been developed and used in various countries over the years, only one (Fluorgonost HIV-1 IFA; Sanochemia Pharmazeutika, Vienna, Austria) remains commercially available and has received approval from the FDA for use in the United States. The Fluorognost HIV-1 IFA has also been FDA approved as a screening test in hospital laboratories, medical clinics, physicians' offices, emergency care

TABLE 7 Confirmatory and supplemental tests for the detection of HIV antibodies[a]

Manufacturer	Product name	Assay type	Sample	Detection	Antigen	Selected reference(s)
Biokit S.A.	Bioblot HIV 1 Plus	WB	Serum, plasma	HIV-1; indicates suspected HIV-2	VL + SP	135
	Bioblot HIV 2	WB	Serum, plasma	HIV-2	VL	NF
Bio-Rad	Novapath HIV-1 Immunoblot[b]	WB	Serum, plasma	HIV-1	VL	NF
	Genetic Systems HIV-1 Western Blot[b]	WB	Serum, plasma, dried blood spots	HIV-1	VL	148
	Genetic Systems HIV-2 Western Blot	WB	Serum, plasma	HIV-2	VL	NF
	Sanofi Diagnostics Pasteur Pepti-Lav 1-2	LIA	Serum, plasma	HIV-1/HIV-2	SP	5, 25, 27, 72, 140, 150, 151, 154, 219, 223
	Sanofi Diagnostics Pasteur New Lav-Blot 1	WB	Serum, plasma	HIV-1	VL	75, 140, 153, 154, 190, 229, 231
	Sanofi Diagnostics Pasteur New-Lav Blot 2	WB	Serum, plasma	HIV-2	VL	140, 153, 154, 190, 219, 231
Calypte Biomedical	Cambridge Biotech HIV-1 Western Blot[b]	WB	Serum, plasma, urine	HIV-1	VL	98, 141, 211, 212
Chiron Corporation	RIBA HIV-1/HIV-2 SIA	LIA	Serum, plasma	HIV-1/HIV-2	RP + SP	108, 115, 136, 150
Genelabs Diagnostics	HIV-1 Blot 1.3	WB	Serum, plasma	HIV-1	VL	NF
	HIV Blot 2.2	WB/LIA	Serum, plasma	HIV-1/HIV-2	VL + SP	108, 124
	HIV-2 Blot 1.2	WB	Serum, plasma	HIV-2	VL	NF
Innogenetics	Inno-Lia	LIA	Serum, plasma	HIV-1/HIV-2	RP + SP	5, 108, 149, 150, 167, 219
Organon Teknika	LiaTek-1/2	LIA	Serum, plasma	HIV-1/HIV-2	RP + SP	119, 140, 173, 244
OraSure Technologies	OraSure HIV-1 Western Blot[b]	WB	Oral fluids	HIV-1	VL	71
Sanochemia Pharmazeutika (formerly Waldheim Pharmazeutika)	Fluorgonost HIV-1 IFA[b]	IFA	Serum, plasma, dried blood spots	HIV-1	IC	94, 201

[a]Abbreviations: WB, Western blot; LIA, line immunoassay; RIBA, recombinant immunoblot assay; VL, viral lysate; SP, synthetic peptides; RP, recombinant proteins; IC, infected Pall T-cells; NF, none found.

[b]Approved by the FDA for use in the United States.

situations, and blood banks or other settings where EIAs are not practical or available.

The Fluorognost HIV-1 IFA is based on the specific binding of HIV-1 antibodies in a specimen to HIV-1 antigens expressed on the surfaces of immortalized human T- cells fixed to glass slides. Specific antibody-antigen complexes are then detected using an anti-human antibody conjugated with fluorescein isothiocyanate. Uninfected cells of the same type are used as negative controls. The stained slides are carefully examined and compared to infected and uninfected cells reacted with positive and negative control sera. A test is considered positive based on the percentage of fluorescent cells (40 to 60%) and the pattern (cytoplasmic versus peripheral staining) and intensity of fluorescence obtained when read with a fluorescence microscope. The assay offers the advantages of speed and simplicity, and the kit contains all of the necessary reagents, including substrate slides, labeled secondary antibody, wash buffer, mounting fluid, and positive and negative control sera. The major disadvantages of the IFA kit are that it requires a fluorescence microscope and darkroom for examining slides and that extensive training and critical evaluation is needed to accurately read and interpret the test results. In addition, IFA does not distinguish reactivity to specific HIV proteins, as the Western blot assay allows. Therefore, some sera that may nonspecifically react with infected cells may be incorrectly classified as positive.

SEROLOGICAL TESTING OF NEONATES

Current antibody tests are not useful for the diagnosis of HIV in infants less than 18 months of age born to HIV-infected mothers. These newborns possess passively transmitted maternal IgG antibodies that may persist for up to 15 to 18 months (median, 10 to 12 months) after birth. As such, a positive HIV serology in this population only indicates exposure and possible infection. Additional testing is required to determine if such newborns are truly infected. Evaluation of the patient's clinical status and the results of sequential antibody tests performed beyond 18 months of age can be diagnostic, although results from this approach occur too late for decisions regarding specific antiretroviral therapy or therapy for opportunistic infections. Children who remain antibody positive (seroconverter) beyond 18 months are considered to be infected, while children who remain asymptomatic and become antibody negative (seroreverter) are not infected. Rare instances of asymptomatic infants who have seroreverted and have positive

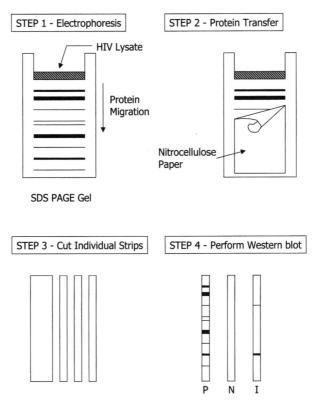

FIGURE 2 Illustration of the steps involved in the Western blotting technique. Abbreviations: SDS, sodium dodecyl sulfate; PAGE, polyacrylamide gel electrophoresis; P, positive; N, negative; I, indeterminate.

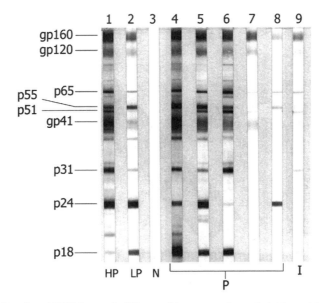

FIGURE 3 Results of HIV-1-specific Western blot assays. Lane 1, high-positive (HP) control; lane 2, low-positive (LP) control; lane 3, negative (N) control; lanes 4 to 8, positive (P) sera; lane 9, indeterminate (I) serum. The numbers at the left refer to approximate molecular weights of HIV-1 antigens. The criteria used for the interpretation of a positive HIV-1 Western blot is that of the Centers for Disease Control and Prevention and the Association of Public Health Laboratories and includes any two bands of p24, gp41, or gp160-gp120 (37, 38). For a specimen to be negative for HIV-1 antibodies by Western blotting, no banding pattern can be observed. Any banding pattern on a Western blot that does not meet the criteria of a positive result is interpreted as indeterminate for antibodies to HIV-1.

results from viral culture or molecular assays for viral RNA or DNA or who subsequently produced antibodies to HIV have been reported. Moreover, some children have hypogammaglobulinemia and remain seronegative despite clinical symptoms suggesting HIV infection. Detection of HIV-specific IgM and IgA antibodies may be useful for diagnosis in the newborn period, but such assays are not readily available and are generally less sensitive and specific than other methods during the first 3 months of life. Although IgM and IgA assays are appealing because of the relatively simple technology, current assays have problems with false-positive results, and IgM is not always produced at detectable levels in HIV-infected newborns. IgA persists for a longer time than IgM in infants but can be difficult to detect before 3 to 6 months of age, making this test less useful for rapid diagnosis in the newborn. Early diagnosis of HIV infection in neonates requires direct detection of the virus using p24 antigen assays, molecular assays for HIV DNA or RNA, and/or viral culture.

p24 Antigen Detection

A number of commercial assays are available for the detection of HIV-1 p24 antigen (Table 8), although only two have been FDA approved for use in the United States. The kits are simple and inexpensive to use and are relatively sensitive and specific. Figure 4 depicts the commercial p24 antigen test developed and marketed by Abbott Laboratories. In this assay, a monoclonal antibody bound to a solid support (bead) is used to capture HIV-1 p24 antigen from the sample. A second antibody specific to p24 antigen is then added and, if p24 antigen is present in the sample, the antibody will bind to the antigen captured on the bead to form an antibody-antigen-antibody sandwich. The antibody-antigen-antibody complexes are then detected by incubating the bead with a third antibody labeled with horseradish peroxidase followed by a colorless substrate (o-phenylenediamine) that is cleaved by the enzyme to give off a yellow-orange color reaction. The intensity of the color formed is proportional to the amount of HIV-1 p24 antigen present in the sample. Many manufacturers directly conjugate the second antibody with an enzyme label and do not use the third antibody as part of the detection system. Similar to the HIV antibody assays described above, patient specimens with initially reactive results for HIV-1 antigen must be retested in duplicate using the same EIA. Repeatedly reactive specimens must be further tested to

TABLE 8 EIAs for the detection of HIV-1 p24 antigen[a]

Manufacturer	Product name	Sample	Test format	Qualitative/ quantitative results	Confirmatory/ ICD reagents	Selected reference(s)
Abbott Laboratories	HIVAG-1 Monoclonal[b]	Serum, plasma	Bead sandwich EIA	Y/N	Y/N	65, 86, 99, 222, 233, 237
Advanced Biotechnologies, Inc.	HIV-1 p24 Antigen Capture	Virus culture supernatants	Microplate sandwich EIA	Y/Y	N/N	NF
Beckman Coulter	HIV-1 p24 Antigen EIA[b]	Serum, plasma, virus culture supernatants	Microplate sandwich EIA	Y/Y	Y/Y	8, 67, 68, 99, 233, 237
bioMerieux	VIDAS HIV p24 II	Serum, plasma	Automated SPR sandwich ELFA	Y/Y	Y/N	67
Bio-Rad	Genetic Systems HIV-1 Ag EIA	Serum, plasma, virus culture supernatants, CSF	Microplate sandwich EIA	Y/Y	Y/N	67
Genelabs Diagnostics	HIV-1 p24 Antigen ELISA	Serum, plasma, virus culture supernatants	Microplate sandwich EIA	Y/Y	Y/Y	NF
Innogenetics	Innotest HIV Antigen mAb	Serum, plasma, virus culture supernatants	Microplate sandwich EIA	Y/Y	Y/Y	67, 68, 222
NEN Life Science Products	HIV-1 p24 ELISA	Serum, plasma, virus culture supernatants	Microplate sandwich EIA	Y/Y	Y/Y	237
Organon Teknika	Vironostika p24 Antigen	Serum, plasma, virus culture supernatants	Microplate sandwich EIA	Y/Y	Y/Y	NF
Roche Diagnostics	Boehringer Mannheim Elecsys HIV Ag	Serum, plasma, virus culture supernatants	Automated EIA	Y/N	Y/N	233
ZeptoMetrix	Retro-Tek HIV-1 p24 Antigen ELISA	Serum, plasma, virus culture supernatants	Microplate sandwich EIA	Y/Y	Y/Y	95

[a]Abbreviations: SPR, solid-phase receptacle; ELFA, enzyme-linked fluorescent assay; CSF, cerebrospinal fluid; Y, yes; N, no; NF, none found.
[b]Approved by the FDA for use in the United States.

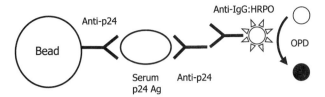

FIGURE 4 Schematic of an EIA for the detection of HIV-1 p24 antigen. Abbreviations: Ag, antigen; HRPO, horseradish peroxidase; OPD, o-phenylenediamine.

confirm the results. This is accomplished by using a neutralization assay in which a neutralizing antibody specific to HIV-1 p24 antigen is incubated with the specimen and binds the p24 antigen in immune complexes, thus preventing its detection upon performance of the test.

Tests that detect HIV-1 p24 antigen may be useful for the early diagnosis of HIV in persons at risk, as p24 antigen levels are high early in infected individuals, usually before antibodies to HIV can be detected. HIV-1 p24 antigen can also be detected during the later stages of illness when increasing failure of the immune system and increasing levels of viral replication are observed. Other uses for p24 antigen assays include screening blood donors for early infection, monitoring the effect of antiviral therapy, assessing the prognosis for disease progression, diagnosis of HIV infection in infants born to HIV-infected mothers, and the detection of HIV growth in viral culture. It has been observed that HIV infection is detectable by p24 antigen tests approximately 6 days earlier than with antibody tests (33). The window period is estimated to be 22 days for the detection of HIV-specific antibody using current EIAs and approximately 16 days for tests that detect p24 antigen. Consequently, the FDA now recommends that both HIV antibody and p24 antigen assays be used to screen all donated blood and blood products in the United States (39). The detection of p24 antigen in newborn infants has been problematic. Only low levels of circulating p24 antigen are present in the sera of newborns during the first months of life, and excess maternal HIV-specific antibodies complex to any free p24 antigen, rendering it undetectable in the antigen assay. Similarly, p24 antigen is only detected in a minority of asymptomatic HIV-infected adults due mostly to immune complex formation. The detection of HIV p24 antigen in infants and asymptomatc HIV-infected individuals has been greatly enhanced by pretreatment of serum with acid, base, or heat to dissociate immune complexes (8, 22, 57, 158, 159, 170). Appropriate reagents are supplied with many of the current commercial p24 antigen assays to allow for immune complex dissociation (ICD) before completion of the testing (Table 8); ICD p24 antigen assays are sensitive and specific for the diagnosis of HIV in infants, although they are often negative in infected infants younger than 1 month of age. Of additional concern for the use of p24 antigen assays is the fact that, even with dissociation of immune complexes, these assays are not sufficiently sensitive for testing asymptomatic individuals at risk of HIV-1 infection or for monitoring those already infected following seroconversion (30, 165). As a result, p24 antigen assays are mainly used to screen donors and test culture supernatants for viral growth and have been largely replaced by more sensitive molecular DNA PCR assays for diagnosis of HIV-1 and quantitative molecular RNA assays for monitoring prognosis and antiviral therapy.

DNA PCR

PCR has proved to be a rapid and effective method for the early detection of HIV infection in virtually all HIV-infected individuals. The method can be used to detect integrated HIV proviral DNA obtained from purified peripheral blood mononuclear cells (PBMCs) or anticoagulated whole blood or HIV RNA in plasma or serum after the RNA is converted to cDNA by reverse transcription (50, 243). The use of qualitative PCR for the detection of HIV DNA from purified PBMCs or whole blood is favored for diagnosis of HIV infection in children under 18 months of age born to HIV-infected mothers. Conserved sequences in the *gag* or *pol* gene are routinely used in DNA PCR assays for detecting proviral HIV-1 DNA. The amplified products are detected using enzyme-labeled or radiolabeled oligonucleotide probes. The assay can be performed on small quantities of blood, and small amounts of target nucleic acid can be detected. The sensitivity and specificity of DNA PCR are in the high 90th percentile and can approach 100% when procedures and conditions for testing are carefully monitored and controlled (54, 97, 189). The sensitivity of PCR in the first 6 months of life depends on the virus load and the timing of infection. Infants infected in utero may be identified at birth or in the first week, while several weeks or months may pass before infection acquired intrapartum can be detected. The American Academy of Pediatrics recommends that infants born to HIV-infected women be tested by HIV DNA PCR during the first 48 h of life. A second test should be performed at 1 to 2 months of age. Specimens obtained as early as 14 days of age, if positive, may facilitate administration of antiretroviral therapy at an earlier age. A third specimen, if necessary, is recommended at 3 to 6 months of age. Other indications for HIV DNA PCR include the resolution of indeterminate serologies and testing high-risk individuals for acute HIV infection before seroconversion.

Despite the importance of using DNA PCR to detect HIV in certain clinical situations, the availability of commercial PCR assays is still limited. Only one commercial HIV DNA PCR kit, the AMPLICOR HIV-1 test (Roche Molecular Diagnostics, Indianapolis, Ind.), has been developed (21, 101, 118, 127, 235), and this assay has not been submitted to the FDA for approval as a diagnostic test. A commercial RNA-based transcription-mediated amplification assay has been jointly developed by Gen-Probe Inc. (San Diego, Calif.) and Chiron Corp. (Emeryville, Calif.) for the early detection of HIV and hepatitis C virus in blood and blood products; the assay is currently being evaluated by the American Red Cross Blood Services under an FDA-sponsored Investigational New Drug protocol. The manufacturers of this assay have also developed a fully automated instrument (TIGRIS) to be used for future testing of single units of donated blood. Abbott Laboratories has recently reported on the development of a qualitative RNA multiplex PCR LCx assay for the detection of HIV-1, including group M subtypes A through G and group O, as well as HIV-2 (2). The assay was reported to have a sensitivity of 20 to 50 RNA copies/ml and a specificity of 99.6% and could detect HIV infection an average of 6 days before p24 antigen and 11 days before antibodies can be detected. Like the Roche AMPLICOR HIV-1 test, the Abbott PCR LCx assay has not been submitted to the FDA for approval to be used as a diagnostic test.

Virus Culture

HIV-1 can be isolated from purified PBMCs of anticoagulated whole blood or from other body fluids or tissues of infected individuals. The conditions for the recovery of HIV-1 in culture have been optimized and standardized (36, 92), but the technology involved does not lend itself to the development of a commercially available culture-based assay. The method involves the cocultivation of patient PBMCs with mitogen-stimulated, metabolically active donor PBMCs that are collected and purified from individuals who are not infected with HIV-1. Growth of the virus in culture is detected by measuring levels of p24 antigen in the culture supernatant over time. While virus culture is highly sensitive and specific and is still considered the most definitive method of detecting HIV, it is mainly used in support of HIV research studies and is seldom needed in a clinical setting for the diagnosis and management of patients with HIV infection. Isolation of HIV in culture can be particularly useful for the diagnosis of HIV infection in neonates born to HIV-infected mothers, detecting acute HIV infection in other age groups, resolving indeterminate serological results, and examining the phenotypic resistance patterns of patient isolates to various antiretroviral drugs. However, the method is expensive, labor-intensive, and, because of the need for special facilities and biosafety precautions, cannot be performed in all laboratories. For the most part, HIV cultures are done in highly specialized independent reference laboratories or large clinical and/or research laboratories associated with academic institutions. It can take up to 4 weeks to complete a culture, although most HIV isolates will be detected within the first 7 to 10 days. It is sometimes difficult to obtain sufficient quantities of blood from newborns (3 to 5 ml is recommended) to perform culture, although microculture assays have been developed for use with smaller specimen volumes. A total of 10 ml of whole blood is recommended for adults.

QUANTIFICATION OF HIV RNA

Our present understanding of HIV-1 infection has been facilitated by using laboratory parameters that correlate with the stage of disease progression and the degree of viral replication. However, methods used for monitoring HIV-1 infection, such as $CD4^+$-cell counts and levels of p24 antigen, neopterin, β_2-microglobulin, and interferon, use surrogate markers that do not always correlate with the disease state or the antiviral activity of therapeutic regimens. Accurate measurement of the virus burden can be accomplished using endpoint dilution cocultures of plasma or PBMCs of infected individuals. Such cultures have yielded useful information, but the procedure is time-consuming, labor-intensive, and expensive and requires special precautions and facilities. The development of molecular assays to quantitate the levels of HIV RNA in infected patients has provided one of the most valuable tools to assess the progression of HIV disease, monitor the impact of antiviral therapy, predict treatment failure and the emergence of drug-resistant viruses, and facilitate our understanding of the natural history and pathogenesis of this virus. The quantitation of high levels of HIV-1 RNA has been related to vertical transmission of the virus from mother to fetus, a more rapid progression to AIDS and death in both adult and pediatric patients, and the emergence of drug resistance during prolonged antiretroviral therapy. Molecular methods, such as quantitative competitive PCR, reverse transcriptase-PCR, nucleic acid sequence-based amplification, branched-DNA (bDNA) technology, and transcription-mediated amplification, have been developed for the accurate quantification of HIV RNA from infected individuals. A comparative summary of the important characteristics of the commercial assays is provided in Table 9. The

AMPLICOR HIV-1 Monitor 1.0 assay (Roche Molecular Diagnostics), the Versant HIV-1 RNA 3.0 assay (bDNA; Bayer Diagnostics, Tarrytown, N.Y.), and the NucliSens HIV-1 QT RNA assay (Organon Teknika, Durham, N.C.) are commercially available in the United States. Only the AMPLICOR HIV-1 Monitor 1.0 assay is approved by the FDA for diagnostic use. The NucliSens HIV-1 QT RNA assay, the Versant HIV-1 RNA 3.0 assay, and version 1.5 of the AMPLICOR HIV-1 Monitor assay are currently being reviewed by the FDA. The Abbott LCx RT-PCR and the Gen-Probe HIV-1 viral load assays are under development and are not available in the United States. A more detailed description of the various commercial assays and their clinical applications can be found in recent reviews (74, 90). Such tests may influence the choice of initial therapy and tailor treatment regimens to reduce the risk of viral mutation and subsequent viral resistance. Sustained decreases in patient virus load during treatment suggests an appropriate antiviral effect, while a return of the virus load to pretreatment levels is indicative of drug failure and the possible need for alternative therapy. Measurement of plasma HIV-1 RNA levels or virus loads is the strongest predictor of clinical outcome and the predominant test used by clinicians to monitor HIV-infected patients on antiretroviral therapy. The monitoring of the HIV-1 virus load is essential for making appropriate patient management decisions regarding the initiation and alteration of therapy.

PHENOTYPIC AND GENOTYPIC ASSAYS FOR DRUG RESISTANCE

A number of antiretroviral drugs have been approved for the treatment of HIV-1 infection. They are divided into three main classes based on their modes of inhibition of viral replication and include the nucleoside and nonnucleoside inhibitors of the viral RNA-dependent DNA polymerase (reverse transcriptase) and the viral protease inhibitors. Therapy with these antiretroviral agents is beneficial to prolong survival, to reduce the incidence and severity of opportunistic infections in patients with advanced HIV-1 disease, and to delay disease progression in asymptomatic HIV-1-infected patients.

The development of HIV resistance to various antiretroviral drugs has been well documented in patients receiving antiretroviral therapy. Currently, there are approximately 175 HIV-1 drug resistance mutations, of which 88 occur in the gene for reverse transcriptase, 52 occur in the protease gene, 34 occur in the envelope gene, and 1 occurs in the gene coding for the integrase. Because HIV-resistant mutants rapidly emerge in patients treated with a single drug, combination therapy of nucleoside reverse transcriptase inhibitors with nonnucleoside reverse transcriptase inhibitors and protease inhibitors is the current standard of care. The single exception is the prophylactic use of zidovudine in pregnant women to reduce the risk of perinatal HIV transmission. The emergence of HIV antiviral-drug resistance is thought to be a function of the length of therapy, stage of disease or virus burden, and high spontaneous mutation rate in the viral genome during active replication of the virus. Prolonged therapy and advancing disease increase the likelihood of developing viral resistance. Both in vitro and in vivo resistance to various antiretroviral drugs is due to the accumulation of specific mutations in the coding regions of either the reverse transcriptase or the protease of the virus, leading to single-amino-acid substitutions in the viral-protein targets. The extent of resistance appears to be correlated with the number of mutations, although

TABLE 9 Comparison of commercial molecular quantitative HIV-1 RNA assays

Characteristic	Manufacturer and information[a]				
	Abbott	Bayer Diagnostics	Gen-Probe	Organon Teknika	Roche
Kit name	LCx	Versant 3.0	To be named	NucliSens	Amplicor Monitor[b]
Amplification method	RT-PCR (target)	bDNA (signal)	TMA (target)	NASBA (target)	RT-PCR (target)
Gene region	*pol*	*pol*	*pol*	*gag*	*gag*
Plasma specimen	EDTA, ACD	EDTA	Any type of anticoagulant	Any type of anticoagulant	EDTA, ACD
Sample size (microliters)	200 and 1,000	1,000	500	200–2,000	200
Kit size (batch size)	100 tests (≥21)	96 tests (≥18)	Not yet established	50 tests (≥10)	24 tests (≥12)
Turnaround time (days)	1 or 2	2	1 or 2	1 or 2	1 or 2
Standards and controls	1 internal QS, 6 external calibrators, 3 external controls	6 external standards, 3 external controls	5 external calibrators, 2 external controls	3 internal calibrators	1 internal QS, 3 external controls
Detection method	Microparticle EIA	Chemiluminescence	Chemiluminescence	Electrochemiluminescence	EIA
Upper limit (copies/milliliter)	5.0×10^6 (200 μl) 1.0×10^6 (1,000 μl)	5×10^5	5.0×10^5	4.0×10^7	7.5×10^5 (version 1.0) 5.0×10^4 (version 1.5)
Lower limit (copies/milliliter)	178 (200 μl) 50 (1,000 μl)	50	50	40	400 (version 1.0) 50 (version 1.5)
Selected references	74, 103, 202, 203, 245	4, 18, 41, 60, 61, 63, 74, 90, 131, 144, 182, 192, 203, 245	20, 61, 74	29, 66, 74, 90, 144, 188, 195, 245	4, 18, 24, 29, 41, 60, 61, 63, 66, 74, 90, 144, 182, 188, 192, 195, 202, 203

[a]Abbreviations: RT-PCR, reverse transcriptase PCR; NASBA, nucleic acid sequence-based amplification; TMA, transcription-mediated amplification; EDTA, ethylenediaminetetraacetic acid; ACD, acid citrate dextrose; QS, quantitation standard.

[b]Approved by the FDA for use in the United States.

individual mutations can be interactive, causing changes in the resistance or susceptibility of HIV isolates.

The described emergence of antiviral-drug resistance has led to a definite need for in vitro antiviral susceptibility testing. Laboratory confirmation of drug resistance is essential for defining the mechanisms of antiviral resistance, for determining the frequency with which drug-resistant viral mutants emerge in clinical practice, to help guide initial antiretroviral therapy, to predict treatment failure and identify cross-resistance to other antiviral agents, for the institution of the most appropriate alternative therapy, and when evaluating new antiviral agents (11, 58, 83, 84, 88, 234, 238, 248). To this end, both phenotypic and genotypic antiviral susceptibility assays have been developed for the determination of HIV-1 drug resistance (Table 10). Recommendations for the use of anti-retroviral-drug resistance testing have recently been published by the International AIDS Society-USA Panel and by the Department of Health and Human Services (89; http://www.hivatis.org).

Phenotypic Assays

Phenotypic assays for susceptibility testing of HIV-1 directly measure the amount of drug needed to inhibit or suppress virus replication. The susceptibility of a particular virus isolate is traditionally measured in cell culture using primary PBMCs from HIV-1-negative donors (102). Virus replica-

tion in the presence or absence of an antiretroviral drug is monitored by quantitating the HIV-1 p24 antigen by EIA. The susceptibility of a virus isolate is expressed as the drug concentration required to inhibit virus replication by 50% (IC_{50}) relative to a no-drug control. The results are compared to the IC_{50}s obtained from resistant and susceptible control strains that are tested with each batch of patient isolates. The disadvantages of this assay are many and include the fact that it is costly, labor-intensive, and highly variable and has a turnaround time of a number of weeks. The procedure also requires that infected PBMCs from the patient be cocultivated with donor PBMCs to produce a stock of the clinical isolate to be tested. This may lead to the selection of subpopulations of the original HIV-1 isolate that are not representative of the total virus population. Comparison of phenotypic in vitro susceptibility data among different laboratories is also valid only if the same assay, host cell, viral inoculum, and range of drug concentrations are used. The definition of a sensitive or resistant isolate may differ for each assay system and for different laboratories performing the same assay. Efforts to develop standardized consensus assays for HIV-1 are in progress. Recently, recombinant-virus phenotypic susceptibility assays have been developed for HIV-1 testing (23, 55, 87, 112, 147). In these assays, total RNA is extracted from a patient's plasma sample, and the reverse transcriptase and protease genes are amplified by PCR. The amplified viral sequences are then inserted into

TABLE 10 Commercial phenotypic and genotypic assays for detection of HIV-1 resistance to reverse transcriptase and protease inhibitors

Assay type	Manufacturer	Product name	Technology	Selected references
Phenotyping	Virco	Antivirogram	Culture-based; recombinant virus assay	83, 87
	ViroLogic	PhenoSense	Culture-based; recombinant virus assay	53, 162
Genotyping	Applied Biosystems	ViroSeq	Electrophoretic; sequencing-based assay	11, 14, 51, 62, 185, 239, 248
	Affymetrix	GeneChip	Oligonucleotide probe microarrays	13, 239
	Innogenetics, Inc.	Line Probe	Non-sequencing-based; reverse-hybridization assay	12, 42, 187, 199, 239
	Pyrosequencing	Pyrosequencing	Nonelectrophoretic sequencing-based assay	155
	Virco	VircoGen	Electrophoretic; sequencing-based assay with virtual phenotype	83
	Visible Genetics	Trugene	Electrophoretic; sequencing-based assay	14, 58, 62, 187

an HIV-1 molecular vector that has these sequences deleted but contains the remainder of the HIV-1 genome. The chimeric virus that is generated can replicate efficiently in CD4+ T lymphocytes, eliminating the need for initial isolation of the virus in primary coculture and the subsequent development of high-titer virus stocks in order to complete the phenotypic testing. Since the newly formed chimeric virus contains the reverse transcriptase and protease gene sequences from the wild-type virus, the recombinant-virus assays can directly measure the susceptibility of the patient's isolate to virtually any antiretroviral drug. Two recombinant-virus assays have been developed commercially: the PhenoSense assay (53, 162) from ViroLogic (San Francisco, Calif.) and the Antivirogram (83, 87) assay from VircoLab (Baltimore, Md.). These assays have been validated by each manufacturer and are now available on a fee-for-service basis in a limited number of reference laboratories. The recombinant-virus assays are also highly automated and improve both the throughput and reproducibility of the traditional phenotypic assay.

Genotypic Assays

Genotypic assays have also been developed to identify specific gene mutations that may lead to resistance when patients are treated with antiretroviral drugs (Table 10) (12–14, 43, 55, 184). These assays normally involve using PCR for amplification of the reverse transcriptase and protease genes of HIV-1 and direct sequencing of the amplified products. Direct sequencing is considered to be the "gold standard" for detecting HIV-1 drug resistance mutations (see reference 14 for a review). The method is based on a modification of the Sanger dideoxynucleotide chain terminator chemistry (177), using specific sequencing primers to produce DNA amplicons labeled with four different fluorescent dyes, one specific for each nucleotide. The DNA is then separated by electrophoresis, and an argon laser excites the fluorescence-labeled nucleotides. The nucleotides are identified one by one as they pass through the laser detector. Specialized software is used to analyze the sequence data and is displayed graphically as an electropherogram or a linear string of one-letter nucleotide bases or codes. To identify the presence of drug resistance mutations, the generated base sequences are then compared to a genomic database containing a consensus sequence of wild-type HIV-1 and sequence variations of well-characterized resistance mutations that correlate with phenotypic drug resistance. Two commercial kits and accompanying instrumentation have been developed and evaluated for PCR

amplification and automated, high-speed sequencing to detect HIV-1 resistance mutations. These include the Applied Biosystems (Foster City, Calif.) ViroSeg HIV-1 kit and genetic analyzers 310, 377, and 3100 (11, 14, 51, 62, 185, 239, 248) and the Visible Genetics (Toronto, Ontario, Canada) TruGene HIV-1 kit and OpenGene System (14, 58, 62, 187). The TrueGene HIV-1 kit has been submitted for FDA approval. A third direct-sequencing kit, the Pyrosequencing HIV-1 kit and PSQ 96 automated luminometric instrument, is currently under development by Pyrosequencing Inc. (Westborough, Mass.), and is a nonelectrophoresis-based assay that utilizes PCR, specific primers, and an enzyme cascade system for sequencing (155). VircoLab has commercialized a genotyping method (VircoGEN) that utilizes direct sequencing and a proprietary mutation expert interpretation system that generates a "virtual" phenotype from the genotype (83). Like the phenotypic assay developed by this company, the genotypic assay has been internally validated and is available on a fee-for-service basis from a limited number of reference laboratories.

A microarray chip-based technology (HIV PRT GeneChip; Affymetrix Inc., Santa Clara, Calif.) (13, 239) and two reverse-hybridization assays (INNO-LiPA HIV-1 RT and INNO-LiPA HIV-1 protease) from Innogenetics (Alpharetta, Ga.) (12, 42, 187, 199, 239) have been developed as alternatives to direct sequencing for detecting selected drug resistance mutations in the reverse transcriptase and protease genes of HIV-1. Mutation identification in these assays is dependent upon the hybridization of viral sequences to oligonucleotide probes that are complementary to the viral reverse transcriptase and protease genes. In the GeneChip assay, the microarray is a silicon-glass chip that contains over 16,000 unique oligonucleotide probes that are available for hybridization with fluorescein-labeled target nucleic acid. Based on the patterns of hybridization of the target sequences to the probes, HIV genomic sequences and specific point mutations are simultaneously identified. The LiPA assays consist of nitrocellulose strips that contain oligonucleotide probes immobilized as parallel bands on the membrane. Biotinylated target sequences hybridize to the probes attached to specific locations on the strip and form colored precipitate bands. The immobilized probes contain both the wild type and single-base changes of the reverse transcriptase and protease genes of HIV-1 known to confer phenotypic resistance to specific antiretroviral drugs. The locations of these colored bands on the strip determine the presence of specific wild-type codons, mixtures of codons, and mutant codons.

TABLE 11 Advantages and disadvantages of HIV-1 phenotypic and genotypic antiviral resistance assays

Assay type	Advantages	Disadvantages
Phenotyping	Direct measure of viral susceptibility to any given drug Identifies drug-resistant viruses with novel mutations to given drugs Measures amount of drug needed to inhibit viral replication Can test samples with virus loads of 500 RNA copies/ml or more and can evaluate non-B clade HIV-1 strains	Assays are complex, costly, highly variable and labor-intensive Tests for the effect of one drug at a time; difficult to test effect of drug combinations Turnaround time, 2 to 6 weeks, depending on assay May not detect minority drug-resistant species at levels of <10–30% of total virus population or minority species with low phenotypic resistance Clinical relevance of testing not well established for each drug Testing likely to remain centralized in large, highly specialized reference laboratories Thresholds for defining drug susceptibility are arbitrary and not standardized
Genotyping	Detects actual viral mutations associated with drug resistance Turnaround time of 1–2 days Less costly; higher throughput and more reproducible Clinical utility is more widely established Less technically demanding and complex May allow earlier detection of emergence of drug resistance Can determine effect of multiple mutations on resistance to one drug or combinations of drugs	Indirect measure of viral susceptibility to drug Only detects known drug-resistant mutations; interpretation requires prior knowledge of mutational changes Results may be confounded by presence of mutations that have no bearing on drug resistance May not detect minority drug-resistant species at levels of <10–30% of total virus population or minority species with low phenotypic resistance Reproducible results are not achieved with samples below 1,000 RNA copies/ml Complexity of assays still makes them less than routine for most clinical laboratories

A summary of the advantages and disadvantages of HIV-1 phenotypic and genotypic antiviral resistance assays is presented in Table 11. Phenotypic assays offer the advantage of being direct measures of viral susceptibility to any antiretroviral drug and can provide data on the concentration of drug needed to inhibit virus replication. Genotypic assays offer the distinct advantages of speed and efficiency in analyzing large numbers of virus isolates and may allow earlier detection of the emergence of drug resistance than phenotypic assays. The clinical relevance of genotypic testing has also been established, while data on the usefulness of phenotypic assays are limited. Current genotypic assays, however, detect only known drug-resistant mutations, and phenotypic assays are still required to identify drug-resistant viruses with novel mutations for antiviral resistance. Neither assay may detect mutants present at low frequency or distinguish minority species with low phenotypic resistance. In general, phenotypic assays are more cumbersome and technically demanding than genotypic assays, and their complexity prohibits their use by most clinical laboratories. The availability of commercial kits and automated instrumentation for genotypic assays makes them more accessible and desirable to use. The performance and accuracy of genotypic assays, however, still depends on the use of highly skilled personnel, expert interpretation of the results obtained, and stringent quality control of the procedures.

REFERENCES

1. **Abrao Ferreira, P. R., R. Gabriel, T. M. Furlan, A. de Camargo Soares, M. E. Myazaky, J. O. Bordim, A. Castelo, and D. S. Lewis.** 1999. Anti-HIV-1/2 antibody detection by dot-ELISA in oral fluid of HIV positive/AIDS patients and voluntary blood donors. *Braz. J. Infect. Dis.* **3:**134–138.
2. **Abravaya, K., C. Esping, R. Hoenle, J. Gorzowski, R. Perry, P. Kroeger, J. Robinson, and R. Flanders.** 2000. Performance of a multiplex qualitative PCR LCx assay for detection of human immunodeficiency virus type 1 (HIV-1) group M subtypes, group O, and HIV-2. *J. Clin. Microbiol.* **38:**716–723.
3. **Aidoo, S., W. K. Ampofo, J. A. M. Brandful, S. V. Nuvor, J. K. Ansah, N. Nii-Trebi, J. S. Barnor, F. Apeagyei, T. Sata, D. Ofori-Adjei, and K. Ishikawa.** 2001. Suitability of a rapid immunochromatographic test for detection of antibodies to human immunodeficiency virus in Ghana, West Africa. *J. Clin. Microbiol.* **39:**2572–2575.
4. **Anastassopoulou, C. G., G. Touloumi, A. Katsoulidou, H. Hatzitheodorou, M. Pappa, D. Paraskevis, M. Lazanas, P. Gargalianos, and A. Hatzakis.** 2001. Comparative evaluation of the QUANTIPLEX HIV-1 RNA 2.0 and 3.0 (bDNA) assays and the AMPLICOR HIV-1 MONITOR v1.5 test for the quantitation of human immunodeficiency virus type 1 RNA in plasma. *J. Virol. Methods* **91:**67–74.
5. **Andersson, S., Z. da Silva, H. Norrgren, F. Dias, and G. Biberfeld.** 1997. Field evaluation of alternative testing strategies for diagnosis and differentiation of HIV-1 and HIV-2 infections in an HIV-1 and HIV-2-prevalent area. *AIDS* **11:**1815–1822.
6. **Apetrei, C., I. Loussert-Ajaka, D. Descamps, F. Damond, S. Saragosti, F. Brun-Vezinet, and F. Simon.**

1996. Lack of screening test sensitivity during HIV-1 non-subtype B seroconversions. *AIDS* **10**:57–60.

7. **Arai, H., B. Petchclai, K. Khupulsup, T. Kurimura, and K. Takeda.** 1999. Evaluation of a rapid immunochromatographic test for detection of antibodies to human immunodeficiency virus. *J. Clin. Microbiol.* **37**:367–370.

8. **Ascher, D. P., C. Roberts, and A. Fowler.** 1992. Acidification modified p24 antigen capture assay in HIV seropositives. *J. Acquir. Immune Defic. Syndr.* **5**:1080–1083.

9. **Asihene, P. J., R. L. Kline, M. W. Moss, A. V. Carella, and T. C. Quinn.** 1994. Evaluation of rapid test for detection of antibody to human immunodeficiency virus type 1 and type 2. *J. Clin. Microbiol.* **32**:1341–1342.

10. **Azevedo-Pereira, J. M., M. H. Lourenco, F. Barin, R. Cisterna, F. Denis, P. Moncharmont, R. Grillo, and M. O. Santos-Ferreira.** 1994. Multicenter evaluation of a fully automated screening test, VIDAS HIV 1 + 2, for antibodies to human immunodeficiency virus types 1 and 2. *J. Clin. Microbiol.* **32**:2559–2563.

11. **Baxter, J. D., D. L. Mayers, D. N. Wentworth, J. D. Neaton, M. L. Hoover, M. A. Winters, S. B. Mannheimer, M. A. Thompson, D. I. Abrams, B. J. Brizz, J. P. A. Ioannidis, T. C. Merigan, and the CPCRA 046 Study Team for the Terry Beirn Community Programs for Clinical Research on AIDS (CPCRA).** 2000. A randomized study of antiretroviral management based on plasma genotypic antiretroviral resistance testing in patients failing therapy. *AIDS* **14**:F83–F93.

12. **Bean, P.** 2000. HIV genotyping by reverse hybridization. *Am. Clin. Lab.* **19**:14–15.

13. **Bean, P., and J. W. Wilson.** 2000. HIV genotyping by chip technology: the GeneChip HIV system. *Am. Clin. Lab.* **19**:16–17.

14. **Bean, P., D. J. Robbins, H. Hamdan, and T. M. Alcorn.** 2000. HIV genotyping by DNA sequencing in the U.S. *Am. Clin. Lab.* **19**:18–20.

15. **Behets, F., K. Bishagara, A. Disasi, S. Likin, R. W. Ryder, C. Brown, and T. C. Quinn.** 1992. Diagnosis of HIV infection with instrument-free assays as an alternative to the ELISA and western blot testing strategy: an evaluation in Central Africa. *J. Acquir. Immune Defic. Syndr.* **5**:878–882.

16. **Behets, F., M. Kashamuka, M. Pappaioanou, T. A. Green, R. W. Ryder, V. Batter, J. R. George, W. H. Hannon, and T. C. Quinn.** 1992. Stability of human immunodeficiency virus type 1 antibodies in whole blood dried on filter paper and stored under various tropical conditions in Kinshasa, Zaire. *J. Clin. Microbiol.* **30**:1179–1182.

17. **Berry, N., J. Pepin, I. Gaye, D. Parker, M. Jarvill, A. Wilkins, H. Whittle, and R. Tedder.** 1993. Competitive EIA for anti-HIV-2 detection in The Gambia: use as a screening assay and to identify possible dual infections. *J. Med. Virol.* **39**:101–108.

18. **Best, S. J., A. P. Gust, E. I. Johnson, C. H. McGavin, and E. M. Dax.** 2000. Quality of human immunodeficiency virus viral load testing in Australia. *J. Clin. Microbiol.* **38**:4015–4020.

19. **Biggar, R. J., W. Miley, P. Miotti, T. E. Taha, A. Butcher, J. Spadoro, and D. Waters.** 1997. Blood collection on filter paper: a practical approach to sample collection for studies of perinatal HIV transmission. *J. Aquir. Immune Defic. Syndr. Hum. Retrovirol.* **14**:368–373.

20. **Bodrug, S., R. Domingo, J. Holloway, M. Sanders, K. Nunomura, C. Sloan, and B. Billyard.** 1997. Gen-Probe single tube quantitative HIV assay. *J. Clin. Microbiol. Infect.* **3**:1050.

21. **Bogh, M., R. Machuca, J. Gerstoft, C. Pedersen, N. Obel, B. Kvinesdal, H. Nielsen, and C. Nielsen.** 2001. Subtype-specific problems with qualitative Amplicor HIV-1 DNA PCR test. *J. Clin. Virol.* **20**:149–153.

22. **Bollinger, R. C., Jr., R. L. Kline, H. L. Francis, M. W. Moss, J. G. Bartlett, and T. C. Quinn.** 1992. Acid dissociation increases the sensitivity of p24 antigen detection for the evaluation of antiviral therapy and disease progression in asymptomatic human immunodeficiency virus-infected persons. *J. Infect. Dis.* **165**:913–916.

23. **Boucher, C. A. B., W. Keulen, T. van Bommel, M. Nijhuis, D. de Jong, M. D. de Jong, P. Schipper, and N. K. T. Back.** 1996. Human immunodeficiency virus type 1 drug susceptibility determination by using recombinant viruses generated from patient sera tested in a cell-killing assay. *Antimicrob. Agents Chemother.* **40**:2404–2409.

24. **Brambilla, D. J., S. Granger, C. Jennings, and J. W. Bremer.** 2001. Multisite comparison of reproducibility and recovery from the standard and ultrasensitive Roche Amplicor HIV-1 Monitor assays. *J. Clin. Microbiol.* **39**:1121–1123.

25. **Brandful, J. A., W. K. Ampofo, F. A. Apeagyei, K. Asare-Bediako, and M. Osei-Kwasi.** 1997. Predominance of HIV-1 among patients with AIDS and AIDS-related complex in Ghana. *East Afr. Med. J.* **74**:17–20.

26. **Branson, B. M.** 2000. Rapid testing for HIV antibody. *AIDS Rev.* **2**:76–83.

27. **Brattegaard, K., D. Soroh, F. Zadi, H. Digbeu, K. M. Vetter, and K. M. De Cock.** 1995. Insensitivity of a synthetic peptide-based test (Pepti-LAV 1-2) for the diagnosis of HIV infection in African children. *AIDS* **9**:656–657.

28. **Brattegaard, K., J. Kouadio, M. L. Adom, R. Doorly, J. R. George, and K. M. De Cock.** 1993. Rapid and simple screening and supplemental testing for HIV-1 and HIV-2 infections in West Africa. *AIDS* **7**:883–885.

29. **Bremer, J., M. Nowicki, S. Beckner, D. Brambilla, M. Cronin, S. Herman, A. Kovacs, and P. Reichelderfer.** 2000. Comparison of two amplification technologies for detection and quantitation of human immunodeficiency virus type 1 RNA in the female genital tract. Division of AIDS Treatment Research Initiative 009 Study Team. *J. Clin. Microbiol.* **38**:2665–2669.

30. **Brown, A. E., J. R. Lane, K. F. Wagner, S. Zhou, R. Chung, K. L. Ray, S. P. Blatt, and D. S. Burke.** 1995. Rates of p24 antigenemia and viral isolation in comparable white and black HIV-infected subjects. Military Medical Consortium for Applied Retroviral Research. *AIDS* **9**:325–328.

31. **Brust, S., H. Duttman, J. Feldner, L. Gurtler, R. Thorstensson, and F. Simon.** 2000. Shortening of the diagnostic window with a new combined HIV p24 antigen and anti-HIV-1/2/O screening test. *J. Virol. Methods* **90**:153–165.

32. **Burgisser, P., F. Simon, M. Wernli, T. Wust, M. F. Beya, and P. C. Frei.** 1996. Multicenter evaluation of new double-antigen sandwich enzyme immunoassay for measurement of anti-human immunodeficiency virus type 1 and type 2 antibodies. *J. Clin. Microbiol.* **34**:634–637.

33. **Busch, M. P., L. L. Lee, G. A. Satten, D. R. Henrard, H. Farzadegan, K. E. Nelson, S. Read, R. Y. Dodd,**

and L. R. Petersen. 1995. Time course of detection of viral and serologic markers preceding human immunodeficiency virus type 1 seroconversion: implications for screening of blood and tissue donors. *Transfusion* **35:**91–97.

34. **Carducci, A., B. Casini, F. Morleo, A. Giuntini, M. Parenti, and R. Moretti.** 1999. Comparison of three assays for HIV antibodies detection in urine to be applied to epidemiological setting. *Eur. J. Epidemiol.* **15:**545–548.

35. **Carpenter, C. L., D. Longshore, K. Annon, J. J. Annon, and M. D. Anglin.** 1999. Prevalence of HIV-1 among recent arrestees in Los Angeles County, California: serial cross-sectional study, 1991–1995. *J. Acquir. Immune Defic. Syndr.* **21:**172–177.

36. **Castro, B. A., C. D. Weiss, L. D. Wiviott, and J. A. Levy.** 1988. Optimal conditions for recovery of the human immunodeficiency virus from peripheral blood mononuclear cells. *J. Clin. Microbiol.* **26:**2371–2376.

37. **Centers for Disease Control and Prevention.** 1989. Interpretation and use of the Western blot assay for serodiagnosis of human immunodeficiency virus type 1 infections. *Morb. Mortal. Wkly. Rep.* **38**(Suppl. 7):1.

38. **Centers for Disease Control and Prevention.** 1991. Interpretation criteria used to report Western blot results for HIV-1 antibody testing: United States. *Morb. Mortal. Wkly. Rep.* **40:**692.

39. **Centers for Disease Control and Prevention.** 1996. U.S. Public Health Service guidelines for testing and counseling blood and plasma donors for human immunodeficiency virus type 1 antigen. *Morb. Mortal. Wkly. Rep.* **45**(RR-2):1.

40. **Chan, E. L., F. Sidaway, and G. B. Horsman.** 1996. A comparison of the Genie and western blot assays in confirmatory testing for HIV-1 antibody. *J. Med. Microbiol.* **44:**223–225.

41. **Clarke, J. R., S. Galpin, R. Braganza, A. Ashraf, R. Russell, D. R. Churchill, J. N. Weber, and M. O. McClure.** 2000. Comparative quantification of diverse serotypes of HIV-1 in plasma from a diverse population of patients. *J. Med. Virol.* **62:**445–449.

42. **Clarke, J. R., S. Kaye, A. G. Babiker, M. H. Hooker, R. Tedder, and J. N. Weber.** 2000. Comparison of a point mutation assay with a line probe assay for the detection of the major mutations in the HIV-1 reverse transcriptase gene associated with reduced susceptibility to nucleoside analogues. *J. Virol. Methods* **88:**117–124.

43. **Cockerill, F. R.** 1999. Genetic methods for assessing antimicrobial resistance. *Antimicrob. Agents Chemother.* **43:**199–212.

44. **Conradie, J. D., and D. R. Tait.** 1992. SimpliRED—a rapid and reliable test for HIV. *S. Afr. Med. J.* **81:**624–625.

45. **Constantine, N. T.** 1993. Serologic tests for the retroviruses: approaching a decade of evolution. *AIDS* **7:**1–13.

46. **Constantine, N. T., E. Fox, E. A. Abbatte, and J. N. Woody.** 1989. Diagnostic usefulness of five screening assays for HIV in an east African city where prevalence of infection is low. *AIDS* **3:**313–317.

47. **Constantine, N. T., L. Zekeng, A. K. Sangare, L. Gurtler, R. Saville, H. Anhary, and C. Wild.** 1997. Diagnostic challenges for rapid human immunodeficiency virus assays. Performance using HIV-1 group O, HIV-1 group M, and HIV-2 samples. *J. Hum. Virol.* **1:**45–51.

48. **Constantine, N. T., X. Zhang, L. Li, J. Bansal, K. C. Hyams, and J. E. Smialek.** 1994. Application of a

rapid assay for detection of antibodies to human immunodeficiency virus in urine. *Am. J. Clin. Pathol.* **101:**157–161.

49. **Covell, R., E. Follett, I. Coote, M. Bloor, A. Finlay, M. Frischer, D. Goldberg, S. Green, S. Haw, and N. McKeganey.** 1993. HIV testing among injecting drug users in Glasgow. *J. Infect.* **26:**27–31.

50. **Cunningham, C. K., T. T. Charbonneau, K. Song, D. Patterson, T. Sullivan, T. Cummins, and B. Poiesz.** 1999. Comparison of human immunodeficiency virus 1 DNA polymerase chain reaction and qualitative and quantitative RNA polymerase chain reaction in human immunodeficiency virus 1-exposed infants. *Pediatr. Infect. Dis. J.* **18:**30–35.

51. **Cunningham, S., B. Ank, D. Lewis, W. Lu, M. Wantman, J. Dileanis, J. B. Jackson, P. Palumbo, P. Krogstad, and S. H. Eshleman.** 2001. Performance of the Applied Biosystems ViroSeq human immunodeficiency virus type 1 (HIV-1) genotyping system for sequence-based analysis of HIV-1 in pediatric plasma samples. *J. Clin. Microbiol.* **39:**1254–1257.

52. **Davey, R. T., Jr., L. R. Deyton, J. A. Metcalf, M. Easter, J. A. Kovacs, M. Vasudevachari, M. Psallidopoulos, L. M. Thompson III, J. Falloon, M. A. Polis, H. Masur, and H. C. Lane.** 1992. Indeterminate western blot patterns in a cohort of individuals at high risk for human immunodeficiency virus (HIV-1) exposure. *J. Clin. Immunol.* **12:**185–192.

53. **Deeks, S. G., N. S. Hellman, R. M. Grant, N. T. Parkin, C. J. Petropoulos, M. Becker, W. Symonds, M. Chesney, and P. A. Volberding.** 1999. Novel four-drug salvage treatment regimens after failure of a human immunodeficiency virus type 1 protease inhibitor-containing regimen: antiviral activity and correlation of baseline phenotypic drug susceptibility with virologic outcome. *J. Infect. Dis.* **179:**1375–1381.

54. **Defer, C., H. Agut, A. Garbarg-Chenon, M. Moncany, F. Morinet, D. Vignon, M. Mariotti, and J.-J. Lefrere.** 1992. Multicentre quality control of polymerase chain reaction for detection of HIV DNA. *AIDS* **6:**659–663.

55. **Demeter, L., and R. Haubrich.** 2001. Phenotypic and genotypic resistance assays: methodology, reliability, and interpretations. *J. Acquir. Immune Defic. Syndr.* **26:**S3–S9.

56. **Downing, R. G., R. A. Otten, E. Marum, B. Biryahwaho, M. G. Alwano-Edyegu, S. D. Sempala, C. A. Fridlund, T. J. Dondero, C. Campbell, and M. A. Rayfield.** 1998. Optimizing the delivery of HIV counseling and testing services: the Uganda experience using rapid HIV antibody test algorithms. *J. Acquir. Immune Defic. Syndr. Hum. Retrovirol.* **18:**384–388.

57. **Duiculescu, D. C., R. B. Geffin, G. B. Scott, and W. A. Scott.** 1994. Clinical and immunological correlates of immune-complex-dissociated HIV-1 p24 antigen in HIV-1-infected children. *J. Acquir. Immune Defic. Syndr.* **8:**807–815.

58. **Durant, J., P. Clevenbergh, P. Halfon, P. Delgiudice, S. Porsin, P. Simonet, N. Montagne, C. A. B. Boucher, J. M Schapiro, and P. Dellamonica.** 1999. Drug-resistance genotyping in HIV-1 therapy: the VIRADAPT randomized controlled trial. *Lancet* **353:**2195–2199.

59. **Elavia, A. J., A. Thomas, J. Nandi, G. D. Coyaji, and V. Bhavalkar-Potdar.** 1995. Performance evaluation of a particle agglutination test for antibody to human immunodeficiency virus 1: comparison with enzyme immunoassay. *Vox Sang.* **69:**23–26.

60. **Elbeik, T., E. Charlebois, P. Nassos, J. Kahn, F. M. Hecht, D. Yajko, V. Ng, and K. Hadley.** 2000. Quantitative and cost comparison of ultrasensitive human immunodeficiency virus type 1 RNA viral load assays: Bayer bDNA Quantiplex versions 3.0 and 2.0 and Roche PCR Amplicor Monitor version 1.5. *J. Clin. Microbiol.* **38:**1113–1120.

61. **Emery, S., S. Bodrug, B. A. Richardson, C. Giachetti, M. A. Bott, D. Panteleeff, L. L. Jagodzinski, N. L. Michael, R. Nduati, J. Bwayo, J. K. Kreiss, and J. Overbaugh.** 2000. Evaluation of performance of the Gen-Probe human immunodeficiency virus type 1 viral load assay using primary subtype A, C, and D isolates from Kenya. *J. Clin. Microbiol.* **38:**2688–2695.

62. **Erali, M., S. Page, L. G. Reimer, and D. R. Hillyard.** 2001. Human immunodeficiency virus type 1 drug resistance testing: a comparison of three sequence-based methods. *J. Clin. Microbiol.* **39:**2157–2165.

63. **Erice, A., D. Brambilla, J. Bremer, J. B. Jackson, R. Kokka, B. Yen-Lieberman, and R. W. Coombs.** 2000. Performance characteristics of the QUANTIPLEX HIV-1 RNA 3.0 assay for detection and quantitation of human immunodeficiency virus type 1 RNA in plasma. *J. Clin. Microbiol.* **38:**2837–2845.

64. **Ettiegne-Traore, V., P. D. Ghys, C. Maurice, Y. M. Hoyi-Adonsou, D. Soroh, M. L. Adom, M. J. Teurquetil, M. O. Diallo, M. Laga, and A. E. Greenberg.** 1998. Evaluation of an HIV saliva test for the detection of HIV-1 and HIV-2 antibodies in high-risk populations in Abidjan, Cote d'Ivoire. *Int. J. STD AIDS* **9:**173–174.

65. **Feorino, P., B. Forrester, C. Schable, D. Warfield, and G. Schochetman.** 1987. Comparison of antigen assay and reverse transcriptase assay for detecting human immunodeficiency virus in culture. *J. Clin. Microbiol.* **25:**2344–2346.

66. **Fiscus, S. A., D. Brambilla, R. W. Combs, B. Yen-Lieberman, J. Bremer, A. Kovacs, S. Rasheed, M. Vahey, T. Schutzbank, P. S. Reichelderfer, and O. A. Group.** 2000. Multicenter evaluation of methods to quantitate human immunodeficiency virus type 1 RNA in seminal plasma. *J. Clin. Microbiol.* **38:**2348–2353.

67. **Fransen, K., G. Beelaert, and G. van der Groen.** 2001. Evaluation of four HIV antigen tests. *J. Virol. Methods* **93:**189–193.

68. **Fransen, K., G. Mertens, D. Stynen, A. Goris, P. Nys, J. Nkengasong, L. Heyndrickx, W. Janssens, and G. van der Groen.** 1997. Evaluation of a newly developed HIV antigen test. *J. Med. Virol.* **53:**31–35.

69. **Frerichs, R. R., N. Silarug, N. Eskes, P. Pagcharoenpol, A. Rodklai, S. Thangsupachai, and C. Wongba.** 1994. Saliva-based HIV-antibody testing in Thailand. *AIDS* **8:**885–894.

70. **Galli, R. A., S. Castriciano, M. Fearon, C. Major, K. W. Choi, J. Mahony, and M. Chernesky.** 1996. Performance characteristics of recombinant enzyme immunoassay to detect antibodies to human immunodeficiency virus type 1 (HIV-1) and HIV-2 and to measure early antibody responses in seroconverting patients. *J. Clin. Microbiol.* **34:**999–1002.

71. **Gallo, D., J. R. George, J. H. Fitchen, A. S. Goldstein, and M. S. Hindahl.** 1997. Evaluation of a system using oral mucosal transudate for HIV-1 antibody screening and confirmatory testing. OraSure HIV Clinical Trials Group. *JAMA* **277:**254–258.

72. **George, J. R., C. Y. Ou, B. Parekh, K. Brattegaard, V. Brown, E. Boateng, and K. M. De Cock.** 1992.

Prevalence of HIV-1 and HIV-2 mixed infections in Cote d'Ivoire. *Lancet* **340:**337–339.

73. **Giles, R. E., K. R. Perry, and J. V. Parry.** 1999. Simple/rapid test devices for anti-HIV screening: do they come up to the mark? *J. Med. Virol.* **59:**104–109.

74. **Ginocchio, C. C.** 2001. HIV-1 viral load testing. *Lab. Med.* **32:**142–152.

75. **Gobin, E., J. M. Desruelle, and J. P. Vigier.** 2001. Evaluation of the analytic performance of blood collection tubes (BD Vacutainer SST) for the screening of anti-HIV, anti-HTLV, anti-HCV, anti-HBc, anti-CMV antibodies, and of HBs p24 HIV antigens, and of alanine aminotransferase. *Transfus. Clin. Biol.* **8:**44–50.

76. **Gonzalez, L., R. W. Boyle, M. Zhang, J. Castillo, S. Whittier, P. Della-Latta, L. M. Clarke, J. R. George, X. Fang, J. G. Wang, B. Hosein, and C. Y. Wang.** 1997. Synthetic-peptide-based enzyme-linked immunosorbent assay for screening human serum or plasma for antibodies to human immunodeficiency virus type 1 and type 2. *Clin. Diagn. Lab. Immunol.* **4:**598–603.

77. **Gottfried, T. D., J. C. Sturge, and H. B. Urnovitz.** 1999. A urine test system for HIV-1 antibodies. *Am. Clin. Lab.* **18:**4.

78. **Granade, T. C., S. K. Phillips, B. Parekh, C. P. Pau, and J. R. George.** 1995. Oral fluid as a specimen for detection and confirmation of antibodies to human immunodeficiency virus type 1. *Clin. Diagn. Lab. Immunol.* **2:**395–399.

79. **Granade, T. C., S. K. Phillips, B. Parekh, P. Gomez, W. Kitson-Piggott, H. Oleander, B. Mahabir, W. Charles, and S. Lee-Thomas.** 1998. Detection of antibodies to human immunodeficiency virus type 1 in oral fluids: a large-scale evaluation of immunoassay performance. *Clin. Diagn. Lab. Immunol.* **5:**171–175.

80. **Gurtler, L., A. Muhlbacher, U. Michl, H. Hofmann, G. G. Paggi, V. Bossi, R. Thorstensson, R. G. Villaescusa, A. Eiras, J. M. Hernandez, W. Melchior, F. Donie, and B. Weber.** 1998. Reduction of the diagnostic window with a new combined p24 antigen and human immunodeficiency virus antibody screening assay. *J. Virol. Methods* **75:**27–38.

81. **Hannon, W. H., D. S. Lewis, W. K. Jones, and M. K. Powell.** 1989. A quality assurance program for human immunodeficiency virus seropositivity screening of dried-blood spot specimens. *Infect. Control Hosp. Epidemiol.* **10:**8–13.

82. **Hardy, C. T., T. A. Damrow, D. B. Villareal, and G. E. Kenny.** 1992. Evaluation of viral-lysate enzyme-linked immunosorbent assay kits for detection of human immunodeficiency virus (type 1) infections using human sera standardized by quantitative western blotting. *J. Virol. Methods* **37:**259–273.

83. **Harrigan, P. R., K. Hertogs, W. Verbiest, R. Pauwels, B. Larder, S. Kemp, S. Bloor, B. Yip, R. Hogg, C. Alexander, and J. S. Montaner.** 1999. Baseline HIV drug resistance profile predicts response to ritonavir-saquinavir protease inhibitor therapy in a community setting. *AIDS* **13:**1863–1871.

84. **Haubrich, R., and L. Demeter.** 2001. Clinical utility of resistance testing: retrospective and prospective data supporting use and current recommendations. *J. Acquir. Immune Defic. Syndr.* **26:**S51–S59.

85. **Healey, D. S., and E. M. Dax.** 1992. An evaluation of two simple synthetic peptide based anti-HIV assays. *J. Virol. Methods* **38:**305–312.

86. **Henrard, D. R., S. Wu, J. Phillips, D. Wiesner, and J. Phair.** 1995. Detection of p24 antigen with and with-

out immune complex dissociation for longitudinal monitoring of human immunodeficiency virus type 1 infection. *J. Clin. Microbiol.* **33:**72–75.

87. **Hertogs, K., M.-P. de Bethune, V. Miller, T. Ivens, P. Schel, A. Van Cauwenberge, C. Van Den Eynde, V. Van Gerwen, H. Azijn, M. Van Houtte, F. Peeters, S. Staszewski, M. Conant, S. Bloor, S. Kemp, B. Larden, and R. Pauwels.** 1998. A rapid method for simultaneous detection of phenotypic resistance to inhibitors of protease and reverse transcriptase in recombinant human immunodeficiency virus type 1 isolates from patients treated with antiretroviral drugs. *Antimicrob. Agents Chemother.* **42:**269–276.

88. **Hirsch, M. S., and D. D. Richman.** 2000. The role of genotypic resistance testing in selecting therapy for HIV. *JAMA* **284:**1649–1650.

89. **Hirsch, M. S., F. Brun-Vezinet, R. T. D'Aquila, S. M. Hammer, V. A. Johnson, D. R. Kuritzkes, C. Loveday, J. W. Mellors, B. Clotet, B. Conway, L. M. Demeter, S. Vella, D. M. Jacobsen, and D. D. Richman.** 2000. Antiretroviral drug resistance testing in adults with HIV infection: recommendations of an International AIDS Society-USA Panel. *JAMA* **283:**2417–2426.

90. **Hodinka, R. L.** 1998. The clinical utility of viral quantitation using molecular methods. *Clin. Diagn. Virol.* **10:**25–47.

91. **Hodinka, R. L., T. Nagashunmugam, and D. Malamud.** 1998. Detection of human immunodeficiency virus antibodies in oral fluids. *Clin. Diagn. Lab. Immunol.* **5:**419–426.

92. **Hollinger, F. B., J. W. Bremer, L. E. Myers, J. W. Gold, and L. McQuay.** 1992. Standardization of sensitive human immunodeficiency virus coculture procedures and establishment of a multicenter quality assurance program for the AIDS Clinical Trials Group. The NIH/NIAID/DAIDS/ACTG Virology Laboratories. *J. Clin. Microbiol.* **30:**1787–1794.

93. **Holloman, D. L., C. P. Pau, B. Parekh, C. Schable, I. Onorato, G. Schochetman, and J. R. George.** 1993. Evaluation of testing algorithms following the use of combination HIV-1/HIV-2 EIA for screening purposes. *AIDS Res. Hum. Retrovir.* **9:**147–151.

94. **Iltis, J. P., N. M. Patel, S. R. Lee, S. L. Barmat, and W. C. Wallen.** 1990. Comparative evaluation of an immunofluorescent antibody test, enzyme immunoassay and western blot for the detection of HIV-1 antibody. *Intervirology* **31:**122–128.

95. **Jackson, B., K. Sannerud, F. Rhame, R. Tsang, and H. H. Balfour, Jr.** 1987. Comparison of reverse transcriptase assay with the Retro-Tek viral capture assay for detection of human immunodeficiency virus. *Diagn. Microbiol. Infect. Dis.* **7:**185–192.

96. **Jackson, J. B.** 1992. Human immunodeficiency virus (HIV)-indeterminate western blots and latent HIV infection. *Transfusion* **32:**497–499.

97. **Jackson, J. B., J. Drew, H. J. Lin, P. Otto, J. W. Bremer, F. B. Hollinger, S. M. Wolinksy, The ACTG PCR Working Group, and The ACTG PCR Virology Laboratories.** 1993. Establishment of a quality assurance program for human immunodeficiency virus type 1 DNA polymerase chain reaction assays by the AIDS Clinical Trials Group. *J. Clin. Microbiol.* **31:**3123–3128.

98. **Jackson, J. B., J. S. Parsons, L. S. Nichols, N. Knoble, S. Kennedy, and E. M. Piwowar.** 1997. Detection of human immunodeficiency virus type 1 (HIV-1) antibody

by western blotting and HIV-1 DNA by PCR in patients with AIDS. *J. Clin. Microbiol.* **35:**1118–1121.

99. **Jackson, J. B., K. J. Sannerud, and H. H. Balfour, Jr.** 1989. Comparison of two serum HIV antigen assays for selection of asymptomatic antigenemic individuals into clinical trials. *J. Acquir. Immune Defic. Syndr.* **2:**394–397.

100. **Jackson, J. B., M. R. Hanson, G. M. Johnson, T. G. Spahlinger, H. F. Polesky, and R. J. Bowman.** 1995. Long-term follow-up of blood donors with indeterminate human immunodeficiency virus type 1 results on Western blot. *Transfusion* **35:**98–102.

101. **Jackson, J. B., E. M. Piwowar, J. Parsons, P. Kataaha, G. Bihibwa, J. Onecan, S. Kabengera, S. D. Kennedy, and A. Butcher.** 1997. Detection of human immunodeficiency virus type 1 (HIV-1) DNA and RNA sequences in HIV-1 antibody-positive blood donors in Uganda by the Roche AMPLICOR assay. *J. Clin. Microbiol.* **35:**873–876.

102. **Japour, A. J., D. L. Mayers, V. A. Johnson, D. R. Kuritzkes, L. A. Becket, J.-M. Arduino, J. Lane, R. J. Black, P. S. Reichelderfer, R. T. D'Aquila, C. S. Crumpacker, The RV-43 Study Group, and The AIDS Clinical Trials Group Virology Committee Resistance Working Group.** 1993. Standardized peripheral blood mononuclear cell culture assay for determination of drug susceptibilities of clinical human immunodeficiency virus type 1 isolates. *Antimicrob. Agents Chemother.* **37:**1095–1101.

103. **Johanson, J., K. Abravaya, W. Caminiti, D. Erickson, R. Flanders, G. Leckie, E. Marshall, C. Mullen, Y. Ohhashi, R. Perry, J. Ricci, J. Salituro, A. Smith, N. Tang, M. Vi, and J. Robinson.** 2001. A new ultrasensitive assay for quantitation of HIV-1 RNA in plasma. *J. Virol. Methods* **95:**81–92.

104. **Johnson, J. E.** 1992. Detection of human immunodeficiency virus type 1 antibody by using commercially available whole-cell viral lysate, synthetic peptide, and recombinant protein enzyme immunoassay systems. *J. Clin. Microbiol.* **30:**216–218.

105. **Jorgensen, A., J. Shao, S. Maselle, E. Yangi, A. Thomsen, S. Matunda, I. Bygbjerg, P. Gotzsche, J. Svendsen, P. Skinhoj, and V. Faber.** 1990. Evaluation of simple tests for detection of HIV antibodies: analysis of interobserver variation in Tanzania. *Scand. J. Infect. Dis.* **22:**283–285.

106. **Kamat, H. A., M. Adhia, G. V. Koppikar, and B. K. Parekh.** 1999. Detection of antibodies to HIV in saliva. *Natl. Med. J. India* **12:**159–161.

107. **Kamat, H. A., D. D. Banker, and G. V. Koppikar.** 2000. Human immunodeficiency viruses types 1 and 2 infection among replacement blood donors in Mumbai (Bombay). *Indian J. Med. Sci.* **54:**43–51.

108. **Kannangai, R., S. Ramalingam, K. J. Prakash, O. C. Abraham, R. George, R. C. Castillo, D. H. Schwartz, M. V. Jesudason, and G. Sridharan.** 2000. Molecular confirmation of human immunodeficiency virus (HIV) type 2 in HIV-seropositive subjects in south India. *Clin. Diagn. Lab. Immunol.* **7:**987–989.

109. **Kannangai, R., S. Ramalingam, S. Pradeepkumar, K. Damodharan, and G. Sridharan.** 2000. Hospital-based evaluation of two rapid human immunodeficiency virus antibody screening tests. *J. Clin. Microbiol.* **38:**3445–3447.

110. **Kassler, W. J., C. Haley, W. K. Jones, A. R. Gerber, E. J. Kennedy, and J. R. George.** 1995. Performance of a rapid, on-site human immunodeficiency virus

antibody assay in a public health setting. *J. Clin. Microbiol.* **33:**2899–2902.

111. **Kassler, W. J., M. G. Alwano-Edyegu, E. Marum, B. Biryahwaho, P. Kataaha, and B. Dillon.** 1998. Rapid HIV testing with same-day results: a field trial in Uganda. *Int. J. STD AIDS* **9:**134–138.

112. **Kellam, P., and B. A. Larder.** 1994. Recombinant virus assay: a rapid, phenotypic assay for assessment of drug susceptibility of human immunodeficiency virus type 1 isolates. *Antimicrob. Agents Chemother.* **38:**23–30.

113. **King, A., S. A. Marion, D. Cook, M. Rekart, P. J. Middleton, M. V. O'Shaughnessy, and J. S. Montaner.** 1995. Accuracy of a saliva test for HIV antibody. *J. Acquir. Immune Defic. Syndr. Hum. Retrovirol.* **9:**172–175.

114. **Kline, R. L., A. Dada, W. Blattner, and T. C. Quinn.** 1994. Diagnosis and differentiation of HIV-1 and HIV-2 infection by two rapid assays in Nigeria. *J. Acquir. Immune Defic. Syndr.* **7:**623–626.

115. **Kline, R. L., D. McNairn, M. Holodniy, L. Mole, D. Margolis, W. Blattner, and T. C. Quinn.** 1996. Evaluation of Chiron HIV-1/HIV-2 recombinant immunoblot assay. *J. Clin. Microbiol.* **34:**2650–2653.

116. **Koblavi-Deme, S., C. Maurice, D. Yavo, T. S. Sibailly, K. N'guessan, Y. Kamelan-Tano, S. Z. Wiktor, T. H. Roels, T. Chorba, and J. N. Nkengasong.** 2001. Sensitivity and specificity of human immunodeficiency virus rapid serological assays and testing algorithms in an antenatal clinic in Abidjan, Ivory Coast. *J. Clin. Microbiol.* **39:**1808–1812.

117. **Koch, W. H., P. S. Sullivan, C. Roberts, K. Francis, R. Downing, T. D. Mastro, J. Nikengasong, D. Hu, S. Masciotra, C. Schable, and R. B. Lal.** 2001. Evaluation of United States-licensed human immunodeficiency virus immunoassays for detection of group M viral variants. *J. Clin. Microbiol.* **39:**1017–1020.

118. **Kovacs, A., J. Xu, S. Rasheed, X. L. Li, T. Kogan, M. Lee, C. Liu, and L. Chan.** 1995. Comparison of a rapid nonisotopic polymerase chain reaction assay with four commonly used methods for the early diagnosis of human immunodeficiency virus type 1 infection in neonates and children. *Pediatr. Infect. Dis.* **14:**948–954.

119. **Kulkarni, S., M. Thakar, J. Rodrigues, and K. Banerjee.** 1992. HIV-2 antibodies in serum samples from Maharashtra state. *Indian J. Med. Res.* **95:**213–215.

120. **Kuun, E., M. Brashaw, and A. D. Heyns.** 1997. Sensitivity and specificity of standard and rapid HIV-antibody tests evaluated by seroconversion and nonseroconversion low-titre panels. *Vox Sang.* **72:**11–15.

121. **Lamey, P. J., A. Nolan, E. A. Follett, I. Coote, T. W. MacFarlane, D. H. Kennedy, A. Connell, and J. V. Parry.** 1996. Anti-HIV antibody in saliva: an assessment of the role of the components of saliva, testing methodologies and collection systems. *J. Oral Pathol. Med.* **25:**104–107.

122. **Lepine, D. G., P. W. Neumann, S. L. Frenette, and M. V. O'Shaughnessy.** 1990. Evaluation of a human immunodeficiency virus test algorithm utilizing a recombinant protein enzyme immunoassay. *J. Clin. Microbiol.* **28:**1169–1171.

123. **Li, L., N. T. Constantine, X. Zhang, and J. E. Smialek.** 1993. Determination of human immunodeficiency virus antibody status in forensic autopsy cases using a rapid and simple FDA-licensed assay. *J. Forensic Sci.* **38:**798–805.

124. **Lien, T. X., N. T. Tien, G. F. Chanpong, C. T. Cuc, V. T. Yen, R. Soderquist, K. Laras, and A. Corwin.** 2000. Evaluation of rapid diagnostic tests for the detection of human immunodeficiency virus types 1 and 2, hepatitis B surface antigen, and syphilis in Ho Chi Minh City, Vietnam. *Am. J. Trop. Med. Hyg.* **62:**301–309.

125. **Ly, T. D., C. Edlinger, A. Vabret, O. Morvan, B. Greuet, and B. Weber.** 2000. Contribution of combined detection assays of p24 antigen and anti-human immunodeficiency virus (HIV) antibodies in diagnosis of primary HIV infection by routine testing. *J. Clin. Microbiol.* **38:**2459–2461.

126. **Ly, T. D., S. Laperche, and A. M. Courouce.** 2001. Early detection of human immunodeficiency virus infection using third- and fourth-generation screening assays. *Eur. J. Clin. Microbiol. Infect. Dis.* **20:**104–110.

127. **Lyamuya, E., E. Olausson-Hansson, J. Albert, F. Mhalu, and G. Biberfeld.** 2000. Evaluation of a prototype Amplicor PCR assay for detection of human immunodeficiency virus type 1 DNA in blood samples from Tanzanian adults infected with HIV-1 subtypes A, C and D. *J. Clin. Virol.* **17:**57–63.

128. **Lyons, S. F.** 1993. Evaluation of rapid enzyme immunobinding assays for the detection of antibodies to HIV-1. *S. Afr. Med. J.* **83:**115–117.

129. **Lyons, S. F., E. T. Bowers, G. M. McGillivray, N. K. Blackburn, and G. E. Gray.** 1997. Evaluation of the MUREX*ICE HIV-1.O.2 capture enzyme immunoassay for early identification of HIV-1 seroreverting infants in a developing country. *Clin. Diagn. Virol.* **8:**1–8.

130. **Malamud, D., and H. M. Friedman.** 1993. HIV in the oral cavity: virus, viral inhibitory activity, and antiviral antibodies: a review. *Crit. Rev. Oral Biol. Med.* **4:**461–466.

131. **Manegold, C., C. Krempe, H. Jablonowski, L. Kajala, M. Dietrich, and O. Adams.** 2000. Comparative evaluation of two branched-DNA human immunodeficiency virus type 1 RNA quantification assays with lower detection limits of 50 and 500 copies per milliliter. *J. Clin. Microbiol.* **38:**914–917.

132. **Martin, D. J., N. K. Blackburn, K. F. O'Connell, E. T. Brant, and E. A. Goetsch.** 1995. Evaluation of the World Health Organisation antibody-testing strategy for the individual patient diagnosis of HIV infection (strategy III). *S. Afr. Med. J.* **85:**877–880.

133. **Martinez, P., R. Ortiz de Lejarazu, J. M. Eiros, E. Perlado, M. Flores, M. A. del Pozo, and A. Rodriguez-Torres.** 1995. Comparison of two assays for detection of HIV antibodies in saliva. *Eur. J. Clin. Microbiol. Infect. Dis.* **14:**330–336.

134. **Martinez, P., R.O. Lejarazu, J. M. Eiros, J. D. Benito, and A. Rodriguez-Torres.** 1996. Urine samples as a possible alternative to serum for human immunodeficiency virus antibody screening. *Eur. J. Clin. Microbiol. Infect. Dis.* **15:**810–813.

135. **Martinez, P. M., A. R. Torres, R. O. de Lejarazu, A. Montoya, J. F. Martin, and J. M. Eiros.** 1999. Human immunodeficiency virus antibody testing by enzyme-linked fluorescent and Western blot assays using serum, gingival-crevicular transudate, and urine samples. *J. Clin. Microbiol.* **37:**1100–1106.

136. **Mas, A., V. Soriano, M. Gutierrez, F. Fumanal, A. Alonso, and J. Gonzalez-Lahoz.** 1997. Reliability of a new recombinant immunoblot assay (RIBA HIV-1/HIV-2 SIA) as a supplemental (confirmatory) test

for HIV-1 and HIV-2 infections. *Transfus. Sci.* **18:** 63–69.

137. **Masciotra, S., D. L. Rudolph, G. van der Groen, C. Yang, and R. B. Lal.** 2000. Serological detection of infection with diverse human and simian immunodeficiency viruses using consensus *env* peptides. *Clin. Diagn. Lab. Immunol.* **7:**706–709.

138. **Matter, L., and D. Germann.** 1995. Detection of human immunodeficiency virus (HIV) type 1 antibodies by new automated microparticle enzyme immunoassay for HIV types 1 and 2. *J. Clin. Microbiol.* **33:**2338–2341.

139. **McAlpine, L., J. Gandhi, J. V. Parry, and P. P. Mortimer.** Thirteen current anti-HIV-1/HIV-2 enzyme immunoassays: how accurate are they? *J. Med. Virol.* **42:**115–118.

140. **Meda, N., L. Gautier-Charpentier, R. B. Soudre, H. Dahourou, R. Ouedraogo-Traore, A. Ouangre, A. Bambara, A. Kpozehouen, H. Sanou, D. Valea, F. Ky, M. Cartoux, F. Barin, and P. Van de Perre.** 1999. Serological diagnosis of human immunodeficiency virus in Burkina Faso: reliable, practical strategies using less expensive commercial test kits. *Bull. W. H. O.* **77:**731–739.

141. **Meehan, M. P., N. K. Sewankambo, M. J. Wawer, D. McNairn, T. C. Quinn, S. Lutalo, S. Kalibbala, C. Li, D. Serwadda, F. Wabwire-Mangen, and R. H. Gray.** 1999. Sensitivity and specificity of HIV-1 testing of urine compared with serum specimens: Rakai, Uganda. The Rakai Project Team. *Sex. Transm. Dis.* **26:**590–592.

142. **Mitchell, S. W., S. Mboup, J. Mingle, D. Sambe, P. Tukei, K. Milenge, J. Nyamongo, O. K. Mubarak, J. L. Sankale, D. S. Hanson, and T. C. Quinn.** 1991. Field evaluation of alternative HIV testing strategy with a rapid immunobinding assay and an agglutination assay. *Lancet* **337:**1328–1331.

143. **Mortimer, P. P., and J. V. Parry.** 1994. Detection of antibody to HIV in saliva: a brief review. *Clin. Diagn. Virol.* **2:**231–243.

144. **Murphy, D. G., L. Cote, M. Fauvel, P. Rene, and J. Vincelette.** 2000. Multicenter comparison of Roche COBAS AMPLICOR MONITOR version 1.5, Organon Teknika NucliSens QT with extractor, and Bayer Quantiplex version 3.0 for quantification of human immunodeficiency virus type 1 RNA in plasma. *J. Clin. Microbiol.* **38:**4034–4041.

145. **Newman, L. M., F. Miguel, B. B. Jemusse, A. C. Macome, and R. D. Newman.** 2001. HIV seroprevalence among military blood donors in Manica Province, Mozambique. *Int. J. STD AIDS* **12:**225–228.

146. **Ng, K. P., T. L. Saw, A. Baki, J. He, N. Singh, and C. M. Lyles.** 1999. Evaluation of a rapid test for the detection of antibodies to human immunodeficiency virus type 1 and 2. *Int. J. STD AIDS* **10:**401–404.

147. **Nijhuis, M., R. Schuurman, and C. A. B. Boucher.** 1997. Homologous recombination for rapid phenotyping of HIV. *Curr. Opin. Infect. Dis.* **10:**475–479.

148. **Nishanian, P., J. M. Taylor, E. Korns, R. Detels, A. Saah, and J. L. Fahey.** 1987. Significance of quantitative enzyme-linked immunosorbent assay (ELISA) results in evaluation of three ELISAs and Western blot tests for detection of antibodies to human immunodeficiency virus in a high-risk population. *J. Clin. Microbiol.* **25:**395–400.

149. **Nkengasong, J., I. Van Kerckhoven, G. Vercauteren, P. Piot, and G. van der Groen.** 1992. Alternative con-firmatory strategy for anti-HIV antibody detection. *J. Virol. Methods* **36:**159–169.

150. **Nkengasong, J. N., C. Maurice, S. Koblavi, M. Kalou, C. Bile, D. Yavo, E. Boateng, S. Z. Wiktor, and A. E. Greenberg.** 1998. Field evaluation of a combination of monospecific enzyme-linked immunosorbent assays for type-specific diagnosis of human immunodeficiency virus type 1 (HIV-1) and HIV-2 infections in HIV-seropositive persons in Abidjan, Ivory Coast. *J. Clin. Microbiol.* **36:**123–127.

151. **Nkengasong, J. N., C. Maurice, S. Koblavi, M. Kalou, D. Yavo, M. Maran, C. Bile, K. N'guessan, J. Kouadio, S. Bony, S. Z. Wiktor, and A. E. Greenberg.** 1999. Evaluation of HIV serial and parallel serologic testing algorithms in Abidjan, Cote d'Ivoire. *AIDS* **13:**109–117.

152. **Nunn, A. J., B. Biryahwaho, R. G. Downing, G. van der Groen, A. Ojwiya, and D. W. Mulder.** 1993. Algorithms for detecting antibodies to HIV-1: results from a rural Ugandan cohort. *AIDS* **7:** 1057–1061.

153. **Obi, C. L., B. A. Ogbonna, E. O. Igumbor, R. N. Ndip, and A. O. Ajayi.** 1993. HIV seropositivity among female prostitutes and nonprostitutes: obstetric and perinatal implications. *Viral Immunol.* **6:**171–174.

154. **Ollero, M., E. Pujol, P. Marquez, A. Gimeno, R. Alcoucer, and D. Mora.** 1991. Spread pattern of HIV-2 in patients at risk. Evaluation of serologic markers. *An. Med. Interna.* **8:**116–121.

155. **O'Meara, D., K. Wilbe, T. Leitner, B. Hejdeman, J. Albert, and J. Lundeberg.** 2001. Monitoring resistance to human immunodeficiency virus type 1 protease inhibitors by pyrosequencing. *J. Clin. Microbiol.* **39:**464–473.

156. **Osmond, D. H., J. Catania, L. Pollack, J. Canchola, D. Jaffe, D. MacKellar, and L. Valleroy.** 2000. Obtaining HIV test results with a home collection test kit in a community telephone sample. *J. Acquir. Immune Defic. Syndr.* **24:**363–368.

157. **Palmer, C. J., J. M. Dubon, E. Koenig, E. Perez, A. Ager, D. Jayaweera, R. R. Cuadrado, A. Rivera, A. Rubido, and D. A. Palmer.** 1999. Field evaluation of the Determine rapid human immunodeficiency virus diagnostic test in Honduras and the Dominican Republic. *J. Clin. Microbiol.* **37:**3698–3700.

158. **Palomba, E., V. Gay, M. de Martino, C. Fundaro, L. Perugini, and P. A. Tovo.** 1992. Early diagnosis of human immunodeficiency virus infection in infants by detection of free and complexed p24 antigen. *J. Infect. Dis.* **165:**394–395.

159. **Panakitsuwan, S., N. Yoshihara, N. Hashimoto, K. Miyamura, and T. Chotpitayasunondh.** 1997. Early diagnosis of vertical HIV infection in infants by rapid detection of immune complex-dissociated HIV p24 antigen. *AIDS Patient Care STDS* **6:**429–433.

160. **Pasquier, C., P. Y. Bello, P. Gourney, J. Puel, and J. Izopet.** 1997. A new generation of serum anti-HIV antibody immunocapture assay for saliva testing. *Clin. Diagn. Virol.* **8:**195–197.

161. **Perriens, J. H., K. Magazani, N. Kapila, M. Konde, U. Selemani, P. Piot, and G. van der Groen.** 1993. Use of a rapid test and an ELISA for HIV antibody screening of pooled serum samples in Lubumbashi, Zaire. *J. Virol. Methods* **41:**213–221.

162. **Petropoulos, C. J., N. T. Parkin, K. L. Limoli, Y. S. Lie, T. Wrin, W. Huang, H. Tian, D. Smith, G. A. Winslow, D. J. Capon, and J. M. Whitcomb.** 2000. A novel phenotypic drug susceptibility assay for human

immunodeficiency virus type 1. *Antimicrob. Agents Chemother.* **44:**920–928.

163. **Phillips, S., T. C. Granade, C. P. Pau, D. Candal, D. J. Hu, and B. S. Parekh.** 2000. Diagnosis of human immunodeficiency virus type 1 infection with different subtypes using rapid tests. *Clin. Diagn. Lab. Immunol.* **7:**698–699.

164. **Plourde, P. J., S. Mphuka, G. K. Muyinda, M. Banda, K. Sichali-Sichinga, D. Chama, and A. R. Ronald.** 1998. Accuracy and costs of rapid human immunodeficiency virus testing technologies in rural hospital in Zambia. *Sex. Transm. Dis.* **25:**254–259.

165. **Pokriefka, R. A., O. Manzor, N. P. Markowitz, L. D. Saravolatz, P. Kvale, and R. M. Donovan.** 1993. Increased detection of human immunodeficiency virus antigenemia after dissociation of immune complexes at low pH. *J. Clin. Microbiol.* **31:**1656–1658.

166. **Poljak, M., N. Zener, K. Semer, and L. Kristancic.** 1997. Particle agglutination test "Serodia HIV-1/2" as a novel anti-HIV-1/2 screening test: comparative study on 3311 serum samples. *Folia Biol.* **43:**171–173.

167. **Pollet, D. E., E. L. Saman, D. C. Peeters, H. M. Warmenbol, L. M. Heyndrickx, C. J. Wouters, G. Beelaert, G. van der Groen, and H. Van Heuverswyn.** 1991. Confirmation and differentiation of antibodies to human immunodeficiency virus 1 and 2 with a strip-based assay including recombinant antigens and synthetic peptides. *Clin. Chem.* **37:**1700–1707.

168. **Portincasa, P., G. Conti, T. Zannino, S. Visalli, and C. Chezzi.** 1994. Radioimmune western blotting in comparison with conventional western blotting, second and third generation ELISA assays for the serodiagnosis of HIV-1 infection. *New Microbiol.* **17:**169–176.

169. **Preiser, W., N. S. Brink, A. Hayman, J. Waite, P. Balfe, and R. S. Tedder.** 2000. False-negative HIV antibody test results. *J. Med. Virol.* **60:**43–47.

170. **Quinn, T. C., R. Kline, M. W. Moss, R. A. Livingston, and N. Hutton.** 1993. Acid dissociation of immune complexes improves diagnostic utility of p24 antigen detection in perinatally acquired human immunodeficiency virus infection. *J. Infect. Dis.* **167:**1193–1196.

171. **Raboud, J. M., C. Sherlock, M. T. Schechter, D. G. Lepine, and M. V. O'Shaughnessy.** 1993. Combining pooling and alternative algorithms in seroprevalence studies. *J. Clin. Microbiol.* **31:**2298–2302.

172. **Rajegowda, B. K., B. B. Das, R. Lala, S. Rao, and D. F. McNeeley.** 2000. Expedited human immunodeficiency virus testing of mothers and newborns with unknown HIV status at time of labor and delivery. *J. Perinat. Med.* **28:**458–463.

173. **Ramirez, E., P. Uribe, D. Escanilla, G. Sanchez, and R. T. Espejo.** 1992. Reactivity patterns and infection status of serum samples with indeterminate Western immunoblot tests for antibody to human immunodeficiency virus type 1. *J. Clin. Microbiol.* **30:**801–805.

174. **Ray, C. S., P. R. Mason, H. Smith, L. Rogers, O. Tobaiwa, and D. A. Katzenstein.** 1997. An evaluation of dipstick-dot immunoassay in the detection of antibodies to HIV-1 and 2 in Zimbabwe. *Trop. Med. Int. Health* **2:**83–88.

175. **Respess, R. A., M. A. Rayfield, and T. J. Dondero.** 2001. Laboratory testing and rapid HIV assays: applications for HIV surveillance in hard-to-reach populations. *AIDS* **15**(Suppl. 3)**:**S49–S59.

176. **Samdal, H. H., B. G. Gutigard, D. Labay, S. I. Wiik, K. Skaug, and A. G. Skar.** 1996. Comparison of the sensitivity of four rapid assays for the detection of antibodies to HIV-1/HIV-2 during seroconversion. *Clin. Diagn. Virol.* **7:**55–61.

177. **Sanger, F., S. Nicklen, and A. R. Coulson.** 1977. DNA sequencing with chain-terminating inhibitors. *Proc. Natl. Acad. Sci. USA* **74:**5463–5467.

178. **Saran, R., and A. K. Gupta.** 1995. HIV-2 and HIV-1/2 seropositivity in Bihar. *Indian J. Public Health* **39:**119–120.

179. **Saville, R. D., N. T. Constantine, C. Holm-Hansen, C. Wisnom, L. DePaola, and W. Falkler, Jr.** 1997. Evaluation of two novel immunoassays designed to detect HIV antibodies in oral fluids. *J. Clin. Lab. Anal.* **11:**63–68.

180. **Saville, R. D., N. T. Constantine, F. R. Cleghorn, N. Jack, C. Bartholomew, J. Edwards, P. Gomez, and W. A. Blattner.** 2001. Fourth-generation enzyme-linked immunosorbent assay for the simultaneous detection of human immunodeficiency virus antigen and antibody. *J. Clin. Microbiol.* **39:**2518–2524.

181. **Scheffel, J. W., D. Wiesner, A. Kapsalis, D. Traylor, and A. Suarez.** 1990. RETROCELL HIV-1 passive hemagglutination assay for HIV-1 antibody screening. *J. Acquir. Immune Defic. Syndr.* **3:**540–545.

182. **Schmid, I., E. Arrer, T. Hawranek, and W. Patsch.** 2000. Evaluation of two commercial procedures for quantification of human immunodeficiency virus type 1 RNA with respect to HIV-1 viral subtype and antiviral treatment. *Clin. Lab.* **46:**355–360.

183. **Schramm, W., G. B. Angulo, P. C. Torres, and A. Burgess-Cassler.** 1999. A simple saliva-based test for detecting antibodies to human immunodeficiency virus. *Clin. Diagn. Lab. Immunol.* **6:**577–580.

184. **Schuurman, R.** 1997. State of the art of genotypic HIV-1 drug resistance. *Curr. Opin. Infect. Dis.* **10:**480–484.

185. **Schuurman, R., L. Demeter, P. Reichelderfer, J. Tijnagel, T. de Groot, and C. Boucher.** 1999. Worldwide evaluation of DNA sequencing approaches for identification of drug resistance mutations in the human immunodeficiency virus type 1 reverse transcriptase. *J. Clin. Microbiol.* **37:**2291–2296.

186. **Schwartz, D. H., A. Mazumdar, S. Winston, and S. Harkonen.** 1995. Utility of various commercially available human immunodeficiency virus (HIV) antibody diagnostic kits for use in conjunction with efficacy trials of HIV-1 vaccines. *Clin. Diagn. Lab. Immunol.* **2:**268–271.

187. **Servais, J., C. Lambert, E. Fontaine, J.-M. Plesseria, I. Robert, V. Arendt, T. Staub, F. Schneider, R. Hammer, G. Burtonboy, and J.-C. Schmit.** 2001. Comparison of DNA sequencing and a line probe assay for detection of human immunodeficiency virus type 1 drug resistance mutations in patients failing highly active antiretroviral therapy. *J. Clin. Microbiol.* **39:**454–459.

188. **Shepard, R. N., J. Schock, K. Robertson, D. C. Shugars, J. Dyer, P. Vernazza, C. Hall, M. S. Cohen, and S. A. Fiscus.** 2000. Quantitation of human immunodeficiency virus type 1 RNA in different biological compartments. *J. Clin. Microbiol.* **38:**1414–1418.

189. **Sheppard, H. W., M. S. Ascher, M. P. Busch, P. R. Sohmer, M. Stanley, M. C. Luce, J. A. Chimera, R. Madej, G. C. Rodgers, C. Lynch, H. Khayam-Bashi, E. L. Murphy, B. Eble, W. Z. Bradford, R. A. Royce, and W. Winkelstein, Jr.** 1991. A multicenter proficien-

cy trial of gene amplification (PCR) for the detection of HIV-1. *J. Acquir. Immune Defic. Syndr.* **4:**277–283.

190. **Shokunbi, W. A., I. Saliu, and E. M. Essien.** 1995. Incidence of dual presence of antibodies to HIV1 and HIV2 in seropositive cases seen in Ibadan, Nigeria. *Afr. J. Med. Sci.* **24:**249–253.

191. **Silvester, C., D. S. Healey, P. Cunningham, and E. M. Dax for the Australian HIV Test Evaluation Group.** 1995. Multisite evaluation of four anti-HIV-1/HIV-2 enzyme immunoassays. *J. Acquir. Immune Defic. Syndr. Hum. Retrovirol.* **8:**411–419.

192. **Si-Mohamed, A., L. Andreoletti, I. Colombet, M.-P. Carreno, G. Lopez, G. Chatelier, M. D. Kazatchkine, and L. Belec.** 2001. Quantitation of human immunodeficiency virus type 1 (HIV-1) RNA in cell-free cervicovaginal secretions: comparison of reverse transcription-PCR amplification (AMPLICOR HIV-1 MONITOR 1.5) with enhanced-sensitivity branched-DNA assay (Quantiplex 3.0). *J. Clin. Microbiol.* **39:**2055–2059.

193. **Sirivichayakul, S., P. Phanuphak, S. Tanprasert, S. Thanomchat, C. Uneklabh, T. Phutiprawan, C. Mungklavirat, and Y. Panjurai.** 1993. Evaluation of a 2-minute anti-human immunodeficiency virus (HIV) test using the autologous erythrocyte agglutination technique with populations differing in HIV prevalence. *J. Clin. Microbiol.* **31:**1373–1375.

194. **Soto-Ramirez, L. E., L. Hernandez-Gomez, J. Sifuentes-Osornio, G. Barriga-Angulo, D. Duarte de Lima, M. Lopez-Portillo, and G. M. Ruiz-Palacios.** 1992. Detection of specific antibodies in gingival crevicular transudate by enzyme-linked immunosorbent assay for diagnosis of human immunodeficiency virus type 1 infection. *J. Clin. Microbiol.* **30:**2780–2783.

195. **Spearman, P., S. A. Fiscus, R. M. Smith, R. Shepard, B. Johnson, J. Nicotera, V. L. Harris, L. A. Clough, J. McKinsey, and D. W. Haas.** 2001. Comparison of Roche MONITOR and Organon Teknika NucliSens assays to quantify human immunodeficiency virus type 1 RNA in cerebrospinal fluid. *J. Clin. Microbiol.* **39:**1612–1614.

196. **Spielberg, F. A., C. M. Kabeya, T. C. Quinn, R. W. Ryder, N. K. Kifuani, J. Harris, T. R. Bender, W. L. Heyward, M. R. Tam, and K. Auditore-Hargreaves.** 1990. Performance and cost-effectiveness of a dual rapid assay system for screening and confirmation of human immunodeficiency virus type 1 seropositivity. *J. Clin. Microbiol.* **28:**303–306.

197. **Sreedharan, A., R. S. Jayshree, A. Desai, H. Sridhar, A. Chandramuki, and V. Ravi.** 1995. Assessment of suitability and ease of performance of seven commercial assay systems for the detection of antibodies to HIV-1. *Indian J. Med. Res.* **101:**179–182.

198. **Stetler, H. C., T. C. Granade, C. A. Nunez, R. Meza, S. Terrell, L. Amador, and J. R. George.** 1997. Field evaluation of rapid HIV serologic tests for screening and confirming HIV-1 infection in Honduras. *AIDS* **3:**369–375.

199. **Stuyver, L., A. Wyseur, A. Rombout, J. Louwagie, T. Scarcez, C. Verhofstede, D. Rimland, R. F. Schinazi, and R. Rossau.** 1997. Line probe assay (LiPA) for rapid detection of drug-selected mutations in the human immunodeficiency virus type 1 reverse transcriptase gene. *Antimicrob. Agents Chemother.* **41:**284–291.

200. **Suarez, M. A., Jr., B. Blanco, L. P. Brion, M. Schulman, T. A. Calvelli, J. Youchah, Y. Devash, A. Rubinstein, and H. Goldstein.** 1993. A rapid test for the detection of human immunodeficiency virus antibodies in cord blood. *J. Pediatr.* **123:**259–261.

201. **Sullivan, M. T., H. Mucke, S. D. Kadey, C. T. Fang, and A. E. Williams.** 1992. Evaluation of an indirect immunofluorescence assay for confirmation of human immunodeficiency virus type 1 antibody in U.S. blood donor sera. *J. Clin. Microbiol.* **30:**2509–2510.

202. **Swanson, P., B. J. Harris, V. Holzmayer, S. G. Devare, G. Schochetman, and J. Hackett, Jr.** 2000. Quantification of HIV-1 group M (subtypes A–G) and group O by the LCx HIV RNA quantitative assay. *J. Virol. Methods* **89:**97–108.

203. **Swanson, P., V. Soriano, S. G. Devare, and J. Hackett, Jr.** 2001. Comparative performance of three viral load assays on human immunodeficiency virus type 1 (HIV-1) isolates representing group M (subtypes A to G) and group O: LCx HIV RNA quantitative, AMPLICOR HIV-1 MONITOR version 1.5, and Quantiplex HIV-1 RNA version 3.0. *J. Clin. Microbiol.* **39:**862–870.

204. **Tamashiro, H., and N. T. Constantine.** 1994. Serological diagnosis of HIV infection using oral fluid samples. *Bull. W. H. O.* **72:**135–143.

205. **Thorstensson, R., S. Andersson, S. Lindback, F. Dias, F. Mhalu, H. Gaines, and G. Biberfeld.** 1998. Evaluation of 14 commercial HIV1/HIV2 antibody assays using serum panels of different geographical origin and clinical stage including a unique seroconversion panel. *J. Virol. Methods* **70:**139–151.

206. **Tiensiwakul, P.** 1998. Urinary HIV-1 antibody patterns by western blot assay. *Clin. Lab. Sci.* **11:**336–338.

207. **Toye, P., and M. S. Riyat.** 1997. Specificity of a novel blood cell agglutination assay ('Simpli-RED') for HIV-1/HIV-2 infection. *East Afr. Med. J.* **74:**237–238.

208. **Tribble, D. R., G. R. Rodier, M. D. Saad, G. Binson, F. Marrot, S. Salah, C. Omar, and R. R. Arthur.** 1997. Comparative field evaluation of HIV rapid diagnostic assays using serum, urine, and oral mucosal transudate specimens. *Clin. Diagn. Virol.* **7:**127–132.

209. **Urassa, W., K. Godoy, J. Killewo, G. Kwesigabo, A. Mbakileki, F. Mhalu, and G. Biberfeld.** 1999. The accuracy of an alternative confirmatory strategy for detection of antibodies to HIV-1: experience from a regional laboratory in Kagera, Tanzania. *J. Clin. Virol.* **14:**25–29.

210. **Urnovitz, H. B., W. H. Murphy, T. D. Gottfried, and A. E. Friedman-Kien.** 1996. Urine-based diagnostic technologies. *Trends Biotechnol.* **14:**361–364.

211. **Urnovitz, H. B., J. C. Sturge, and T. D. Gottfried.** 1997. Increased sensitivity of HIV-1 antibody detection. *Nat. Med.* **11:**1258.

212. **Urnovitz, H. B., J. C. Sturge, T. D. Gottfried, and W. H. Murphy.** 1999. Urine antibody tests: new insights into the dynamics of HIV-1 infection. *Clin. Chem.* **45:**1602–1613.

213. **Van Binsbergen, J., D. de Rijk, H. Peels, C. Dries, J. Scherders, M. Koolen, L. Zekeng, and L. G. Gurtler.** 1996. Evaluation of a new third generation anti-HIV-1/anti-HIV-2 assay with increased sensitivity for HIV-1 group O. *J. Virol. Methods* **60:**131–137.

214. **Van Binsbergen, J., W. Keur, A. Siebelink, M. van de Graaf, A. Jacobs, D. de Rijk, L. Nijholt, J. Toonen, and L. G. Gurtler.** 1998. Strongly enhanced sensitivity of a direct anti-HIV1/2 assay in seroconversion by incorporation of HIV p24 Ag detection: a

new generation Vironostika HIV Uniform II. *J. Virol. Methods* **76:**59–71.

215. **Van Binsbergen, J., W. Keur, M. van de Graaf, A. Siebelink, A. Jacobs, D. de Rijk, J. Toonen, L. Zekeng, E. A. Ze, and L. G. Gurtler.** 1997. Reactivity of a new HIV-1 group O third generation A-HIV-1/-2 assay with an unusual HIV-1 seroconversion panel and HIV-1 group O/group M subtyped samples. *J. Virol. Methods* **69:**29–37.

216. **Van Binsbergen, J., A. Siebelink, A. Jacobs, W. Keur, F. Bruynis, M. van de Graaf, J. van der Heijden, D. Kambel, and J. Toonen.** 1999. Improved performance of seroconversion with a 4th generation HIV antigen/antibody assay. *J. Virol. Methods* **82:**77–84.

217. **Van den Akker, R., J. A. van den Hoek, W. M. van den Akker, H. Kooy, E. Vijge, G. Roosendaal, R. A. Coutinho, and A. M. van Loom.** 1992. Detection of HIV antibodies in saliva as a tool for epidemiological studies. *AIDS* **6:**953–957.

218. **Van Hoogstraten, M. J., E. C. Consten, C. P. Henny, H. A. Heij, and J. J. van Lanschot.** 2000. Are there simple measures to reduce the risk of HIV infection through blood transfusion in a Zambian district hospital? *Trop. Med. Int. Health* **5:**668–673.

219. **Van Kerckhoven, I., G. Vercauteren, P. Piot, and G. van der Groen.** 1991. Comparative evaluation of 36 commercial assays for detecting antibodies to HIV. *Bull. W. H. O.* **69:**753–760.

220. **Varnier, O. E., F. B. Lillo, S. Reina, A. De Maria, A. Terragna, and G. Schito.** 1988. Whole blood collection on filter paper is an effective means of obtaining samples for human immunodeficiency virus antibody assay. *AIDS Res. Hum. Retrovir.* **4:**131–136.

221. **Vercauteren, G., G. Beelaert, and G. van der Groen.** 1995. Evaluation of an agglutination HIV-1 + 2 antibody assay. *J. Virol. Methods* **51:**1–8.

222. **Vercauteren, G., P. Piot, M. Vandenbruaene, and G. van der Groen.** 1989. Evaluation of two enzyme immunoassays for detection of human immunodeficiency virus antigen in African and European sera. *Eur. J. Clin. Microbiol. Infect. Dis.* **8:**892–895.

223. **Walther-Jallow, L., S. Andersson, Z. da Silva, and G. Biberfeld.** 1999. High concordance between polymerase chain reaction and antibody testing of specimens from individuals dually infected with HIV types 1 and 2 in Guinea-Bissau, West Africa. *AIDS Res. Hum. Retrovir.* **15:**957–962.

224. **Walzman, M., D. Natin, and M. Shahmanesh.** 1990. The usefulness of the HIV-CHEK assay as a simple, rapid and sensitive screening test for HIV infection. *Int. J. STD AIDS* **1:**182–183.

225. **Webber, L. M., C. Swanevelder, W. O. Grabow, and P. B. Fourie.** 2000. Evaluation of a rapid test for HIV antibodies in saliva and blood. *S. Afr. Med. J.* **90:**1004–1007.

226. **Webber, M. P., P. Demas, E. Enriquez, R. Shanker, W. Oleszko, S. T. Beatrice, and E. E. Schoenbaum.** 2001. Pilot study of expedited HIV-1 testing of women in labor at an inner-city hospital in New York City. *Am. J. Perinatol.* **18:**49–57.

227. **Weber, B.** 1998. Multicenter evaluation of the new automated Enzymun-Test Anti-HIV 1 + 2 + subtype O. *J. Clin. Microbiol.* **36:**580–584.

228. **Weber, B., N. Behrens, and H. W. Doerr.** 1997. Detection of human immunodeficiency virus type 1 and type 2 antibodies by a new automated microparticle immunoassay AxSYM HIV-1/HIV-2. *J. Virol. Methods* **63:**137–143.

229. **Weber, B., and H. W. Doerr.** 1993. Evaluation of the automated VIDAS system for the detection of anti-HIV-1 and anti-HIV-2 antibodies. *J. Virol. Methods* **42:**63–73.

230. **Weber, B., E. H. M. Fall, A. Berger, and H. W. Doerr.** 1998. Reduction of diagnostic window by new fourth-generation human immunodeficiency virus screening assays. *J. Clin. Microbiol.* **36:**2235–2239.

231. **Weber, B., G. Hess, R. Koberstein, and H. W. Doerr.** 1993. Evaluation of the automated 'Enzymen-Test Anti HIV-1 + 2' and 'Enzymen-Test Anti HIV-1/2 selective' for the combined detection and differentiation of anti-HIV-1 and anti-HIV-2 antibodies. *J. Virol. Methods* **44:**251–260.

232. **Weber, B., M. Moshtaghi-Boronjeni, M. Brunner, W. Preiser, M. Breiner, and H. W. Doerr.** 1995. Evaluation of the reliability of 6 current anti-HIV-1/HIV-2 enzyme immunoassays. *J. Virol. Methods* **55:**97–104.

233. **Weber, B., A. Muhlbacher, U. Michl, G. Paggi, V. Bossi, C. Sargento, R. Camacho, E. H. Fall, A. Berger, U. Schmitt, and W. Melchior.** 1999. Multicenter evaluation of a new rapid automated human immunodeficiency virus antigen detection assay. *J. Virol. Methods* **78:**61–70.

234. **Weinstein, M. C., S. J. Goldie, E. Losina, C. J. Cohen, J. D. Baxter, H. Zhang, A. D. Kimmel, and K. A. Freeburg.** 2001. Use of genotyping resistance testing to guide HIV therapy: clinical impact and cost-effectiveness. *Ann. Intern. Med.* **134:**440–450.

235. **Whetsell, A. J., J. B. Drew, G. Milman, R. Hoff, E. A. Dragon, K. Adler, J. Hui, P. Otto, P. Gupta, H. Farzadegan, and S. M. Wolinsky.** 1992. Comparison of three nonradioisotopic polymerase chain reaction-based methods for detection of human immunodeficiency virus type 1. *J. Clin. Microbiol.* **30:**845–853.

236. **Wienhues, U., E. Faatz, W. Melchior, and H. Bayer.** 1993. Boehringer Mannheim modular test concepts in HIV and hepatitis immunoassays. *Clin. Biochem.* **26:**295–299.

237. **Willoughby, P. B., A. Lisker, and J. D. Folds.** 1989. Evaluation of three enzyme immunoassays for HIV-1 antigen detection. *Diagn. Microbiol. Infect. Dis.* **12:**319–326.

238. **Wilson, J. W., and P. Bean.** 2000. A physician's primer to antiretroviral drug resistance testing. *AIDS Read.* **10:**469–478.

239. **Wilson, J. W., P. Bean, T. Robins, F. Graziano, and D. H. Persing.** 2000. Comparative evaluation of three human immunodeficiency virus genotyping systems: the HIV-genotypeR method, the HIV PRT genechip assay, and the HIV-1 RT line probe assay. *J. Clin. Microbiol.* **38:**3022–3028.

240. **Windsor, I. M., M. L. Gomes dos Santos, L. I. de la Hunt, A. A. Wadee, S. Khumalo, F. Radebe, Y. Dangor, and R. C. Ballard.** 1997. An evaluation of the capillus HIV-1/HIV-2 latex agglutination test using serum and whole blood. *Int. J. STD AIDS* **8:**192–195.

241. **Wisnom, C., L. DePaola, R. D. Saville, N. T. Constantine, and W. Falkler, Jr.** 1997. Clinical applications of two detection systems for HIV using saliva. *Oral Dis.* **3**(Suppl. 1):S85–S87.

242. **Wu, Z., K. Rou, and R. Detels.** 2001. Prevalence of HIV infection among former commercial plasma donors in rural eastern China. *Health Policy Plan.* **16:**41–46.

243. **Young, N. L., N. Shaffer, T. Chaowanachan, T. Chotpitayasunondh, N. Vanparapar, P. A. Mock, N. Waranawat, K. Chokephaibulkit, R. Chuachoowong, P. Wasinrapee, T. D. Mastro, R. J. Simonds, and The Bangkok Collaborative Perinatal HIV Transmission Study Group.** 2000. Early diagnosis of HIV-1-infected infants in Thailand using RNA and DNA PCR assays sensitive to non-B subtypes. *J. Acquir. Immune Defic. Syndr.* **24:**401–407.

244. **Zaaijer, H. L., G. A. van Rixel, J. N. Kromosoeto, D. R. Balgobind-Ramdas, H. T. Cuypers, and P. N. Lelie.** 1998. Validation of a new immunoblot assay (LiaTek HIV III) for confirmation of human immunodeficiency virus infection. *Transfusion* **38:**776–781.

245. **Zanchetta, N., G. Nardi, L. Tocalli, L. Drago, C. Bossi, F. R. Pulvirenti, C. Galli, and M. R. Gismon-**do. 2000. Evaluation of the Abbott LCx HIV-1 RNA Quantitative, a new assay for quantitative determination of human immunodeficiency virus type 1 RNA. *J. Clin. Microbiol.* **38:**3882–3886.

246. **Zaw, M., R. R. Frerichs, K. Y. Oo, and N. Eskes.** 1999. Local evaluation of a rapid HIV assay for use in developing countries. *Trop. Med. Int. Health* **4:**216–221.

247. **Zehner, R., H. Bratzke, and D. Mebs.** 1995. Evaluation of a rapid assay system, HIV-1/HIV-2 TestPack, Abbott, to detect human immunodeficiency virus antibodies in postmortem blood. *J. Forensic Sci.* **40:**113–115.

248. **Zolopa, A. R., R. W. Shafer, A. Warford, J. G. Montoya, P. Hsu, D. Katzenstein, T. C. Merigan, and B. Efron.** 1999. HIV-1 genotypic resistance patterns predict response to saquinavir-ritonavir therapy in patients in whom previous protease inhibitor therapy had failed. *Ann. Intern. Med.* **131:**813–821.

Chlamydia trachomatis

ROBERT C. JERRIS AND CAROLYN M. BLACK

7

This chapter (reprinted in part from reference 17) discusses commercially available methods for identification of Chlamydia trachomatis. Selected molecular assays for Neisseria gonorrhoeae are also included in this chapter, since the evaluation of methods and the performance of tests for C. trachomatis and N. gonorrhoeae are often done concomitantly.

The first portion of the chapter describes the epidemiology and clinical manifestation of infections with C. trachomatis. The second portion details the laboratory aspects of detection of C. trachomatis. The third portion of the chapter details the test performance of the products. While we have attempted to review representative products and the literature on each, we make no claim that we have included every company that might have a product for chlamydia detection or that this is an exhaustive review of a continually expanding wealth of literature. The reader is encouraged to review the limited amount of information included here and to obtain the specific references for a complete review. An important point to consider in review of some of the older citations is that commercial manufacturers continually update their products and may have different "versions" or "generations" that tend to improve test performance based on problems in previously released products. These details require careful review and scrutiny.

The last sets of tables are preceded by indexes in table form (see Tables 15 and 92). Readers should identify the products in Tables 15 and 92 and proceed to the respective table for details.

EPIDEMIOLOGY AND CLINICAL MANIFESTATIONS

Biology

The chlamydiae are nonmotile gram-negative bacterial pathogens that were once mistakenly thought to be viruses. Because of their obligate intracellular life cycle, chlamydiae are metabolically deficient in their ability to synthesize ATP and thus require an exogenous source of this high-energy compound. Chlamydiae undergo a unique biphasic developmental cycle, forming distinctive intracellular inclusions that permit identification by light or fluorescence microscopy. Chlamydiae are susceptible to tetracyclines, macrolides, and quinolones, but because these microorganisms lack a cell wall, beta-lactam antibiotics have no activity against them.

Historically, four species have been recognized in the genus Chlamydia: C. trachomatis, Chlamydia psittaci, Chlamydia pneumoniae, and Chlamydia pecorum. C. trachomatis includes the agents of trachoma, lymphogranuloma venereum (LGV), urogenital tract disease, and inclusion conjunctivitis. Within the species C. trachomatis, there are three biovars, or clusters, based on etiologic potential for disease categories, including the trachoma, LGV, and murine biovars. The murine biovar is much less closely related genetically to the other biovars and is now thought by many investigators to be misclassified within the species C. trachomatis (144). The species is also divided into 15 well-characterized serotypes known as serovars, as well as several additional serovariants that have been identified (80, 145). These serovars are based on differences in the major outer membrane proteins (MOMP) of the organism (24). All of the chlamydia species possess a heat-stable cell membrane lipopolysaccharide (LPS). The MOMP and LPS are major targets in the immunologically based assays for the organisms.

A proposal has been made to modify the taxonomy of the family Chlamydiaceae by dividing the family into two genera, Chlamydia and Chlamydophila, based on analysis of the 16S and 23S rRNA genes. Two new species, Chlamydia muradarum and Chlamydia suis, have been proposed to join C. trachomatis in the genus Chlamydia. The old C. pecorum, C. psittaci, and C. pneumoniae have been proposed to be included in the new genus Chlamydophila (46).

No attempt has been made in this chapter to comprehensively review the biology, pathogenesis, or epidemiology of C. trachomatis infections, since several excellent reviews have been published on these subjects (17, 26, 117, 160). The following information, taken in part from the review article by Black (17), represents highlights of our understanding of disease caused by C. trachomatis.

Epidemiology

It is currently estimated that about 4 million new chlamydial infections occur each year in the United States at an estimated annual cost exceeding $2.4 billion (28, 158, 159). Worldwide, it is estimated that there are more than 50 million new cases of C. trachomatis infection annually (78). C. trachomatis infections are among the sexually transmitted diseases (STDs) known to increase the risk for human immunodeficiency virus (HIV) infection (79); thus, treatment of chlamydial infections could delay the spread of HIV in some groups.

The prevalence of *C. trachomatis* infection in sexually active adolescent women, the population considered most at risk, generally exceeds 10%, and in some adolescent and STD clinic populations of women, the prevalence can reach 40% (8, 28, 128). The prevalence of *C. trachomatis* infection ranges from 4 to 10% in asymptomatic men (94, 124) and from 15 to 20% in men attending STD clinics (140). Chlamydial infections in newborns occur as a result of perinatal exposure; approximately 65% of babies born from infected mothers become infected during vaginal delivery (12, 129).

The biggest challenge to the control of chlamydial disease is that as many as 70 to 80% of women and up to 50% of men who are infected do not experience any symptoms (131, 139, 168). Contributing to this challenge is the fact that immunity following infection is thought to be type specific and only partially protective; therefore, recurrent infections are common (20, 62, 92).

Risk and Demographic Factors for *C. trachomatis* Infections

The most common demographic correlate of infection with *C. trachomatis* in women is youth (<20 years). The biological basis for this association is thought to be anatomic differences in the cervix in younger women wherein the squamocolumnar junction, a primary host target for *C. trachomatis*, is everted and thus more exposed, a condition known as ectopy. Demographic factors associated with older women include unmarried status, nulliparity, and poor socioeconomic conditions (26, 160). Higher numbers of sexual partners, a new sexual partner, lack of use of barrier contraceptive devices, and concurrent gonococcal infection are also consistently associated with infections due to chlamydiae (87).

Clinical Sequelae of *C. trachomatis* Infections in Women

Although most infections caused by *C. trachomatis* in women are asymptomatic, clinical manifestations include cervicitis, urethritis, endometritis, pelvic inflammatory disease (PID), or abscess of the Bartholin glands (19, 28). The urethra and rectum may also be infected (43, 103). PID, which results from ascending infection, is responsible for most of the morbidity and cost resulting from chlamydial infection.

When symptoms do occur, they most commonly consist of vaginal discharge and dysuria. Postcoital bleeding is often reported. Symptoms of chlamydial PID, which may be subtle, include pelvic, uterine, or adnexal pain. Clinical signs associated with chlamydial infection include mucopurulent cervicitis, cervical friability, and culture-negative pyuria.

Asymptomatic chlamydial infections are an important cause of PID and ectopic pregnancy. Silent and untreated salpingitis is now recognized as a major cause of infertility; more than 50% of women with documented tubal occlusion report no history of PID but show serologic evidence of previous *C. trachomatis* infection (160). Similarly, multiple studies have shown associations between previous chlamydial infection, both symptomatic and asymptomatic, and ectopic pregnancy (26, 162).

The prevalence of *C. trachomatis* infection in pregnant women ranges from 2 to 35% (92, 127, 146). Pregnant women with chlamydial infections are at increased risk for adverse outcomes of pregnancy and postpartum PID (59, 60, 92, 151). Diagnosis and treatment of women who are infected with *C. trachomatis* during pregnancy and their sexual partners will prevent these adverse outcomes, as well as postpartum and perinatal disease.

Clinical Sequelae of *C. trachomatis* Infections in Infants

C. trachomatis is the most common cause of neonatal conjunctivitis and one of the most common causes of pneumonia in early infancy (36, 58). Prophylactic treatment of the eyes with silver nitrate does not prevent chlamydial infection; 15 to 25% of treated infants who were exposed at birth develop conjunctivitis, and 3 to 16% develop pneumonia (121). Symptoms of conjunctivitis usually develop within 2 weeks of delivery, and if the infection is untreated, chlamydial pneumonia can develop at 4 to 17 weeks after delivery (36). These conditions are occasionally difficult to treat, and prolonged hospitalization may be necessary (28). Infants with chlamydial pneumonia are at increased risk for later pulmonary dysfunction and possibly for chronic respiratory disease (161).

Clinical Sequelae of *C. trachomatis* Infections in Men

Among heterosexual men, chlamydial infections are usually urethral, and up to 50% are asymptomatic (139, 168). When symptoms do occur, usually 1 to 3 weeks following exposure, they are indistinguishable from those of gonorrhea (urethral discharge and/or pyuria). However, compared with gonococcal urethritis, chlamydial urethritis is more likely to be asymptomatic. Nongonococcal urethritis is the most common clinical syndrome seen in men in the United States; 30 to 50% of cases of nongonococcal urethritis are caused by *C. trachomatis* (64). Chlamydial urethritis is presumptively diagnosed by history, urethral discharge, and the presence of four or more polymorphonuclear leukocytes per oil immersion field of a Gram-stained urethral smear or pyuria noted on urinalysis. Untreated infections may lead to arthritis or Reiter's syndrome.

Epididymitis, or infection of the sperm ducts of the testicles, is most often due to *C. trachomatis* or *N. gonorrhoeae* in sexually active men less than 35 years of age (14). Unilateral scrotal pain is the primary symptom, and common clinical signs of this infection include scrotal swelling and tenderness and fever.

Among homosexual and bisexual men, the prevalence of chlamydial urethritis is about one-third of that reported in heterosexual men (140). In STD clinic populations, 4 to 8% of homosexual men have chlamydial infections of the rectum, most of which are asymptomatic (126).

LGV

LGV is a disease caused by *C. trachomatis* serovars L1 to L3. LGV is uncommon in the United States but prevalent in parts of Africa, Asia, and South America, and it occurs in both men and women (118). The LGV serovars of *C. trachomatis* are more invasive than other genital serovars, resulting in infection of the epithelial layers and underlying soft tissue. The primary symptom is a painless genital ulcer or papule. In homosexual men and women who practice receptive anal intercourse, the ulcer is often accompanied by proctocolitis with symptoms resembling inflammatory bowel disease. The most common manifestation of the secondary stage of LGV in men, and the reason most men seek treatment, is inflammation and swelling of the inguinal lymph nodes (118). Women tend to be less symptomatic at this stage: only 20 to 30% of women present with inguinal lymphadenopathy, and approximately one-third of women without proctocolitis present with lower abdominal and back pain (118). The secondary stage of infection is characterized by systemic symptoms, including fever, malaise, chills, anorexia, myalgia, and arthralgia (117). Untreated infections can lead to late complications, including ulceration and hypertrophy of the genitalia, arthritis, and fistula formation involving the rectum, bladder, vagina, or vulva (118).

LABORATORY TESTING FOR
C. TRACHOMATIS

The traditional approach to laboratory diagnostic testing for C. trachomatis infections has consisted of cell culture of inocula prepared from urogenital specimens. Antigen and nucleic acid detection technologies were developed during the 1980s and have found widespread application in diagnosis due to lesser demands of cost, expertise, preservation of infectivity during transport, and time required to obtain results. Most recently, nucleic acid amplification technologies have been developed, and application of these tests has taught us that culture is not as sensitive as it was thought to be and that the prevalence of C. trachomatis infection is higher in most populations than was previously believed. Modern approaches to laboratory diagnostic testing for C. trachomatis have become complex, sometimes involving the use of combinations of different test technologies for screening and confirmation. Issues such as ensuring the adequacy of specimen collection, the importance of the positive predictive values of tests in low-prevalence populations, the need for an improved "gold standard," and the value of screening asymptomatic populations have evolved as important concerns.

Specimen Collection

Proficiency in specimen collection and transport is paramount for accuracy in diagnostic testing for C. trachomatis. Both the sensitivities and the specificities of diagnostic tests for C. trachomatis have been shown to be directly related to the adequacy of the specimen (67, 72, 98, 120, 135). The lack of specimen adequacy remains a serious shortcoming in many screening programs and research studies, with as many as 30% of specimens being inadequate in some Centers for Disease Control and Prevention (CDC)-sponsored studies, in spite of extensive clinician training. Because chlamydiae are obligate intracellular pathogens, the objective of specimen collection should usually be to include the host cells that harbor the organism. Specimens that contain secretions or exudate but lack the cells that harbor chlamydiae (urethral or endocervical columnar cells) are not satisfactory. The most sensitive technologies, such as DNA amplification, may not require intact chlamydial elementary bodies, since in theory only a few gene copies are needed for a positive result; however, a recent study has shown that even a DNA amplification test can be affected by specimen adequacy measured as the presence of host cells (J. A. Kellog, J. W. Seiple, J. L. Klinedinst, and E. S. Stroll, Abstr. 95th Gen. Meet. Am. Soc. Microbiol. 1995, abstr. C-494, p. 86, 1995). Because the requirements of specimen collection and transport are very different, depending on whether the viability of the chlamydiae must be maintained (for culture versus nonculture), these topics are treated separately.

Collection and Transport of Specimens for Culture

Specimens that are collected in a medium designed to maintain the viability of chlamydiae should be assumed to also carry the potential to contain other infectious agents, such as hepatitis B virus or HIV. Standard precautions should be followed when handling such specimens, and laboratory workers should be vaccinated against hepatitis B virus (27).

The most common anatomic site used to obtain specimens for the isolation of C. trachomatis from women is the endocervix, which is sampled with a swab or cytologic brush. The type of swab used to collect the specimen is an important consideration, since some types have been reported to cause toxicity to cell cultures or to inhibit chlamydial growth within cells (88). Dacron, cotton, rayon, and calcium alginate-tipped swabs may be used, and individual lots should be tested for toxicity to cell culture (88, 91). Swabs with wooden shafts must be avoided. The swab should be inserted into the cervical os past the squamocolumnar junction, about 1 to 2 cm deep, rotated for 15 to 30 s, and removed without touching the vaginal mucosa. Cytologic brushes collect more cells than swabs and thus are thought by some investigators to improve isolation rates (101); however, the brushes are more invasive; induce bleeding, which may inhibit some nonculture tests (1); and cannot be used for pregnant women. In cases where the clinician is well trained in collecting specimens, the use of a cytologic brush is probably not advantageous (73).

Specimens are collected following the removal of secretions and discharge from the cervix, which decreases bacterial contamination and toxicity for culture and improves the appearance of direct fluorescent-antibody (DFA) stains (45, 63). Specimens for C. trachomatis detection should also be obtained following any needed specimens for Gram-stained smears or for the culture of N. gonorrhoeae. Since 1993, the CDC has recommended that specimens for Papanicolaou (Pap) smears be collected before obtaining an endocervical specimen for C. trachomatis culture (28); however, cytologic brushes used for the Pap smear induce bleeding, which may affect the performance of nonculture tests. To avoid this problem, some public health clinics have performed the sampling for the Pap smear after collecting samples for N. gonorrhoeae and C. trachomatis; however, there are insufficient data to address whether this practice diminishes the performance of Pap smears.

The pooling of a urethral-swab specimen with the endocervical-swab specimen increases culture sensitivity by 23% (71). The same swab type that is used for male urethral specimens is suitable for this purpose. The swab should be inserted 1 cm into the female urethra, rotated once prior to removal, and placed in culture transport medium, either together with the endocervical swab or in a separate tube.

The preferred site of sample collection from males is the anterior urethra. A dry swab is placed 3 to 4 cm into the urethra and rotated prior to removal. The subject should not have urinated within the previous hour, since urination will reduce the sensitivity of most diagnostic tests by washing out infected columnar cells.

For conjunctival specimens, as for endocervical specimens, any purulent exudate should be removed before collecting epithelial cells by rubbing a dry swab over the everted palpebral conjunctiva. For suspected LGV infections, bubo pus, saline aspirates of the bubo, swabs of the rectum, or biopsy specimens of the lower gastrointestinal tract aided by anoscopy are best. Optimal sites within the gastrointestinal tract are hypertrophic or ulcerative lesions (48).

The likelihood of isolation is optimized if specimens are refrigerated at 2 to 8°C immediately after collection and kept at this temperature during transport to the laboratory. The time between collection and laboratory processing of specimens for culture should ideally be less than 48 h; however, specimens that cannot be processed within this time may be frozen at −70°C until they are processed. Freezing a specimen is likely to result in at least a 20% loss of viability (88, 123). Freezing specimens or cultures at −20°C has a deleterious effect on the viability and antigens of C. trachomatis and should be avoided (108).

Traditionally, media for the transport of specimens for chlamydial culture have consisted of variations of medium formulations originally developed for the transport of rickettsiae, most commonly 2-sucrose phosphate (2SP) (55, 106) or sucrose-glutamate phosphate (21). The addition of 2 to 5% fetal bovine serum is favored by some investigators and helps to preserve the viability of chlamydiae in specimens that must be frozen. Antimicrobial agents to which chlamydiae are not susceptible are often added to the transport medium to inhibit or prevent the growth of fungi and bacteria present in clinical specimens. Broad-spectrum antibiotics, such as tetracyclines, penicillins, or macrolides, should be avoided. Gentamicin (10 μg/ml) or vancomycin (100 μg/ml) for bacterial contaminants and amphotericin B (2.5 to 4.0 μg/ml) or nystatin (25 to 50 U/ml) for fungal contaminants are most commonly used. Viral transport media are not suitable substitutes. Recently, synthetic transport media for culture and some nonculture tests have been developed and approved for diagnostic use (M4 transport medium [MicroTest, Inc., Lilburn, Ga.] and FlexTrans medium [Bartels Diagnostics]). Limited studies have found that the new M4 synthetic or "universal" medium is comparable to 2SP for culture and was equivalent or superior to other commercial formulations for nonculture tests, such as enzyme immunoassay (EIA) and PCR (V. C. Salmon, B. R. Kenyon, J. C. Overall, Jr., and R. Anderson, *Abstr. 10th Annu. Clearwater Virol. Symp. 1994*, abstr. 57, p. 33, 1994). The lack of standardization of traditional culture transport medium formulations and their unsuitability for nonculture tests make the new universal media attractive. Both 2SP and the commercial universal media are effective for use in the PCR test, making it possible to perform both culture and DNA amplification from a single swab specimen (Salmon et al., *Abstr. 10th Annu. Clearwater Virol. Symp. 1994*; R. P. Verkooven, A. Luijendijk, W. H. F. Goessens, W. M. Huisman, J. A. J. Kluytmans, J. H. van Rijsoort-Vos, and H. A. Verbrugh, *Abstr. 35th Intersci. Conf. Antimicrob. Agents Chemother.*, abstr. D104, p. 84, 1995).

Collection of Specimens for Nonculture Tests

Collection and transport of specimens for commercially licensed nonculture tests should be performed as instructed by the manufacturer and are generally performed as described for culture tests. Commercial diagnostic tests should be used only with the types of specimens approved for the test as defined in the package insert. Specimens for which nonculture methods have not been developed, approved, or adequately evaluated should be tested by the culture method. Examples of specimens that are generally not acceptable for testing by nonculture methods are vaginal, rectal, nasopharyngeal, and female urethral specimens (28). In cases of suspected sexual assault or abuse, only culture tests should be used regardless of the specimen site (28).

Since infection of the urethra occurs commonly during *C. trachomatis* and *N. gonorrhoeae* infections, copies of target DNA remain present in urine and are detectable by nucleic acid amplification tests. Development of these improved urine tests for the detection of *C. trachomatis* has been a major advance and allows access to asymptomatic populations for purposes of noninvasive screening. Urine specimens are not, however, free from concerns about adequacy, since test manufacturers specify that urine specimens should be collected as first catch, should be of appropriate volume, and should be obtained within no less than 1 to 2 h of previous urination. In contrast to nucleic acid amplification tests, culture of urine specimens has historically not been useful, and antigen detection tests for urine have been relatively insensitive compared to DNA amplification tests, performing best with urine specimens from symptomatic males (31, 33). Nucleic acid amplification testing with urine specimens from females can also detect cervical infections (82), presumably as a result of the urine washing over

mucosal surfaces that have been in contact with cervical or vaginal secretions during collection.

Urine specimens for antigen detection and DNA amplification tests, such as PCR or ligase chain reaction (LCR), should be collected as directed by the test manufacturer. Subjects should not have urinated within the previous hour, and females should be instructed not to clean the perineum prior to urinating. The first catch of 10 to 20 ml of urine should be collected in a clean collection cup and refrigerated immediately at 2 to 8°C. Times since the last void of >3 h dramatically decrease the sensitivity of antigen detection tests for urine from women (136) but not from men (136, 147). Ambient-temperature storage of fresh unprocessed urine should be minimized, since the low pH and high urea content rapidly denature DNA present in the specimen, especially at 25°C and above. Several studies of LCR performed on frozen urine specimens have been reported, but none of these include direct comparisons of the sensitivity of LCR on fresh and frozen urine (6, 34, 82). PCR is currently not approved for use on frozen urine specimens; however, freezing and thawing may improve the sensitivity of PCR for urine specimens in which transient inhibitory factors are present (138). Once processed as specified by the manufacturer, urine specimens can be stored at 2 to 8°C for up to 4 days for PCR.

When specimens obtained by both invasive and noninvasive methods are available for testing, the sensitivity of detection may be improved by collection and testing of both types. In a study of endocervical and first-catch urine specimens collected from mostly asymptomatic women, no single test by either culture, DNA testing, PCR, or LCR on a single specimen identified all of the infected women (M. A. Chernesky, H. Lee, D. Jang, K. Luinstra, S. Chong, J. Sellors, L. Pickard, and J. Mahony, *Abstr. 35th Intersci. Conf. Antimicrob. Agents Chemother.*, abstr. D105, p. 85, 1995). Results on whether endocervical specimens detect more or fewer *C. trachomatis* infections than do first-catch urine specimens from the same women when both are tested by DNA amplification methods have been discrepant among laboratories (Chernesky et al., 35th ICAAC; C. Farshy, J. Bullard, C. Woodfill, E. Urdez, C. M. Black, Z. Simmons, B. Jones, D. Blackbourn, E. Ginsburg, and C. Sague, *Abstr. 35th Intersci. Conf. Antimicrob. Agents Chemother.*, abstr. K82, p. 302, 1995; J. Moncada, E. Williams, and J. Schachter, *Abstr. 35th Intersci. Conf. Antimicrob. Agents Chemother.*, abstr. K81, p. 302, 1995). These discrepancies could not be explained by differences in the prevalences or symptomatic states of the populations studied. Differences among study sites in adequacy of endocervical specimens or in the site of infection, whether cervical and/or urethral, could account for the discrepancy in results.

A new and promising approach to noninvasive testing is the collection of vaginal introitus or vulval specimens for testing by nucleic acid amplification. Non-nucleic acid amplification test methods have not proven satisfactory with vaginal or extragenital specimens. However, preliminary reports have shown that in one study, PCR testing of vaginal introitus specimens was as sensitive as culture of cervical specimens (H. C. Wiesenfeld, R. P. Heine, F. M. DiBiasi, C. A. Repp, A. Rideout, I. Macio, and R. L. Sweet, *Abstr. 11th Meet. Int. Soc. STD Res.*, abstr. 040, p. 42, 1995), and in another study, LCR testing of vulval specimens was as sensitive as testing of cervical specimens (A. Stary, B. Chouiere, and H. Lee, *Abstr. 11th Meet. Int. Soc. STD Res.*, abstr. 041, p. 43, 1995). The adequacy of the cervical specimens, if assessed, was not reported in the latter study. Studies comparing the recovery of *C. trachomatis* DNA from different specimen sites could underestimate the performance of cervical specimens if they are not collected adequately. An intriguing fact is that in one of these studies (Wiesenfeld et al., *Abstr. 11th Meet. Int. Soc. STD Res.*), the vaginal introitus specimens were self-collected by the patients. This method might prove to be useful in settings where women do not normally seek clinical care and could collect and mail in their own specimens for testing. Further studies are needed to determine the utility of these unconventional sampling sites for *C. trachomatis* tests.

Quality Assurance of Specimen Collection

Several studies have shown that without quality assurance of specimen adequacy, more than 10% of specimens will be unsatisfactory because they contain exudate and lack urethral or endocervical columnar cells (72, 73, 86). Two methods have been developed for the determination of specimen adequacy by visualization of columnar cells: (i) performing a direct specimen smear and staining with a chlamydia-specific fluorescent antibody followed by a counterstain or by Giemsa stain and (ii) centrifugation of an aliquot of a specimen in non-detergent-based transport medium, such as that used for culture or EIA, onto a microscope slide and staining by a DFA test (29, 52). For determination of specimen adequacy, many laboratorians find a rapid differential hematology stain such as Diff-Quik (catalog no. B4132-1; Dade-Behring, Deerfield, Ill.; http://www.dadebehring.com) easier to read than fluorescent antibody-stained slides due to improved differentiation of cell types. A specimen is considered adequate if it contains at least one columnar or metaplastic cell per slide. A specimen would be determined to be inadequate if any one of the following conditions exist: no cellular components, no columnar or metaplastic cells, or the presence of only squamous epithelial cells or polymorphonuclear leukocytes. Detergent-based transport media, such as those conventionally used for DNA probe or DNA amplification tests, will lyse chlamydial elementary bodies and columnar cells and cannot be used for assessment of specimen adequacy. Routine performance of DFA tests for *C. trachomatis* makes it possible to continuously monitor specimen adequacy. When non-DFA tests are used, periodic cytologic examination of specimen quality is recommended by the CDC to ensure that specimen collection remains adequate over time (28).

Commercially Available Methods for Identification
Stains
Giemsa Stain

Giemsa staining of clinical specimens is a standard method for demonstrating chlamydial inclusions. It is simple to perform, is available in most laboratories, and provides a permanent stain for referral. When performed by experienced personnel, this stain is extremely reliable for diagnosing ocular infections in neonates, using corneal scrapings as specimens. The stain lacks enough sensitivity to be useful with urogenital specimens. This microscopic examination, however, is time-consuming and tedious and requires a skilled observer for interpretation.

Immunofluorescence

Detection of organisms by the DFA test (Table 1) is the staining method of choice for diagnosing chlamydial infections. The assay is as sensitive as Giemsa staining for diagnosis of ocular infections and is more sensitive for urogenital specimens.

In the DFA test, an appropriate specimen is obtained from the patient and applied to a fluorescence microscope-compatible slide. The specimen is then fixed to the slide (with ethanol or by another suitable method as dictated by the manufacturers). The conjugate, containing fluorescein-isothiocyanate-labeled antibodies directed to either chlamydial MOMP or LPS, is added to the specimen. The slide is incubated in a humid chamber to prevent drying of the conjugate. The slide is then washed to remove excess reagent. A drop of mounting fluid (pH balanced to optimize fluorescence) is placed on the specimen. The specimen is then cover slipped and examined at ×400 to 600 with a fluorescence microscope equipped with filter combinations specific for fluorescein isothiocyanate. If elementary bodies are present in the specimen, indicative of the presence of chlamydia, the conjugate specifically binds to

TABLE 1 Commercially available immunofluorescent reagents for detection of *C. trachomatis*

Kit name	MicroTrak *Chlamydia trachomatis* Direct Specimen Test	PathoDx *Chlamydia trachomatis* Direct Specimen
Manufacturer and phone no.	Wampole Laboratories, (800) 257-9525	Diagnostic Products Corp. (DPC), (800) 372-1782
Approved specimen types	Urogenital; rectal; conjunctival (inclusion conjunctivitis, not trachoma); nasopharyngeal (infants)	M urethral; cervical
Reagent	Pooled monoclonal directed against MOMP; Evans' blue counterstain	Pooled monoclonal directed against MOMP; Evans' blue counterstain
Specimen collection	Kit containing single-well glass slide; cytobrush; 2 dacron swabs, 1 lg, 1 small; methanol fixative	PathoDx specimen collection systems for *C. trachomatis*
Controls	Positive and negative per run; not provided; sold separately	Controls include 5 slides (each slide contains a positive and a negative)
No of tests/kit	2-ml volume (60 tests)	4.5-ml volume (100 tests)
Staining time	15 min (RT)	15 min (RT)
Miscellaneous	Cost for control slides, $27.68 for 10 slides (5 positive, 5 negative)	Culture confirmation DFA kit sold separately
Kit cost (varies based on volume of assays performed)	$322.88	$200.00

them and they fluoresce as distinct, "apple green," uniformly circular particles (measuring 350 to 400 nm in diameter) against a black background. The slides in the DFA assay are less fatiguing to read than the Giemsa stain as elementary bodies fluoresce apple green against a dark background. Direct stains allow assessment of specimen adequacy. Host cells appear red when examined under a fluorescence microscope due to the uptake of the Evan's blue counterstain in the reagents. The DFA assay can be used with residual specimens from EIAs for confirmation of positive reactions or to resolve equivocal EIA results. The DFA technique requires technologists to be fully trained in the dynamics and troubleshooting of fluorescent-antibody staining and microscopy and in interpretation by differentiating true positives from nonspecific reactions. The reagents are more expensive than Giemsa stain but moderate in cost compared to other types of assays for the diagnosis of chlamydia. The DFA stain requires more technologist time than batch assays, as slides are read completely and thoroughly and the entire slide is scanned before being reported as negative. The cost of the fluorescence microscope is a capital expenditure. The DFA stain is not to be used as a test-of-cure post-antibiotic therapy.

DFA assays are categorized by the Clinical Laboratory Improvements Act (CLIA) as high complexity and are most suitable for routine clinical and reference laboratories. The Current Procedural Terminology (CPT) code is 87270. The Medicare reimbursement fee schedules vary by state. Profiles for commercially available reagents are described below, and an example of a positive DFA smear is shown in Fig. 1.

EIAs

EIAs, or enzyme-linked immunoassays, consist of antibodies directed against organisms firmly bound to a solid matrix. In this format, the enzyme retains its ability to catalyze a reaction that yields a detectable product, either visual, fluorimetric, spectrophotometric, or radiometric, while attached to the antibodies.

In solid-phase EIA, antibodies are bound inside the wells of microtiter plates, on the outsides of plastic or metal spheres, or on other, similar substances. When antigen is

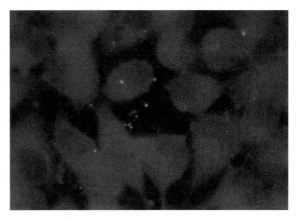

FIGURE 1 Direct immunofluorescence image demonstrating *C. trachomatis* elementary bodies. (Courtesy of Diagnostic Products Corporation.)

present in the test sample, antigen-antibody complexes are formed. Washing removes the excess, unbound reagent, and a second antibody directed against the antigen is added. The second antibody is labeled with an enzyme. If antigen is present, this labeled antibody binds, forming a sandwich with the antigen in the center. Washing removes unbound reagents. Enzyme-specific substrates are added, which, when hydrolyzed by the enzyme, yield a detectable reaction. These assays are well suited to batch processing and automation.

Membrane-bound EIAs take advantage of the large surface area and the flowthrough porous nature of nitrocellulose, nylon, and similar substances. Absorbent material below the surface membrane rapidly removes liquid reagents from the membrane, enhancing the speed and sensitivity of the EIA. All reagents and washing solutions are added to the membrane. If present in the specimen, the antigen is sandwiched as described above, yielding a colored reaction product on the membrane. Several variations of the membrane-bound assays exist in which treated samples

Pathfinder *Chlamydia trachomatis* Direct Specimen	Bartels Chlamydiae Fluorescent Monoclonal Antibody Test	Merifluor Chlamydia
Sanofi Pasteur, (800) 666-5111	Intracel, Bartels Division, (800) 227-8357	Meridian Diagnostics, (800) 503-1980
Urogenital; rectal; pediatric ocular; pediatric nasopharyngeal	Cervical; M urethral	
Pooled monoclonal directed against MOMP; Evans' blue counterstain	Monoclonal to LPS	
Pathfinder *C. trachomatis* specimen collection kit	Kit available (B1029-68) (20 for $35.30): 1 slide; 2 Dacron swabs, 1 lg, 1 small; acetone	
Controls include 5 slides (each slide contains a positive and a negative)	Controls include 5 positive control slides and 5 negative control slides	
2.5-ml volume (50 tests)	5-ml volume (100 tests)	
15 min (RT)	30 min (35–37°C)	
Culture confirmation DFA kit sold separately	Culture confirmation by DFA or immunoperoxidase stains sold separately	Kit approved only for culture confirmation
$351.00	$520.00	

flow by capillary action through different zones of reagents to yield results (see Wampole Clearview in Table 2). Membrane-bound assays are generally for single use and small batches. A novel approach is found in the optical immunoassay (OIA), which detects changes in light reflection on an inert matrix following antigen-antibody complexing (e.g., Biostar in Table 2).

A significant limitation for EIA is the inability to assess specimen adequacy. A screening method to detect the presence of columnar and cuboidal cells is recommended. False-positive reactions in EIAs may result from cross-reacting antigens, and in microwell formats they may result from spillover from a positive sample into a negative sample during improper washing. Therefore, particularly in low-prevalence populations, a blocking antibody assay is recommended to confirm positive reactions. Blocking antibody assays provide supplemental evidence for the presence of chlamydial antigen by binding to the antigen and preventing the binding and capture antibodies from reacting. This results in a decreased or negative reaction compared to the assay performed concurrently with the same specimen in the absence of the blocking agents.

False-negative results may result from specimens contaminated with spermicides and some gynecologic medications. The effect of lubricants is uncertain. The performance of these assays for specimens collected during or immediately after antimicrobial therapy is unknown; however, the tests may identify nonviable organisms.

Commercially available EIAs are described in Tables 2, 3, and 4. Nonamplified nucleic acid detection and nucleic acid amplification assays are reviewed in chapter 12. Tables 5 and 6 describe methods available for the detection of C. trachomatis.

Assay Performance

The following section is divided into two parts detailing assay performance and comparison of different tests. Tables 7 through 14 detail selected references before 1995 (or as indicated) (17). Tables 16 to 91 and 93 to 111 are based on selected references after 1995, while Tables 15 and 92 are indexes detailing comparative assays and their respective literature citations.

Abbreviations Used Only in Tables

Ab, antibody; Ag, antigen; AMP-CT, GenProbe Amplified *Chlamydia trachomatis*; CI$_{95}$, 95% confidence interval; CT, C. *trachomatis*; CT/GC, C. *trachomatis*-N. *gonorrhoeae* dual-screen assay, Digene Hybrid Capture (HC); CT/NG, C. *trachomatis*-N. *gonorrhoeae* dual-screen assay, Roche Amplicor PCR (both products amplified simultaneously); ELISA, enzyme-linked immunosorbent assay; F, female(s); FVU, first-void urine; GC, gonococcus (N. *gonorrhoeae*); HPA, hybridization protection assay; HRP, horseradish peroxidase; IC, internal control; ID, identification; Ig, immunoglobulin; lg, large; M, male(s); MT, Micro Trak II; NA, not available; ND, not done; NG, N. *gonorrhoeae*; NPV, negative predictive value; NT, not tested; OD$_{450}$, optical density at 450 nm; PBS, phosphate-buffered saline; PPV, positive predictive value; Prep, preparation; Prev, prevalence; RT, room temperature; SDA, strand displacement amplification (Becton Dickinson amplification assay); Sens, sensitivity; Spec, specificity; TC, thermal cycler; TMA, transcription-mediated amplification (GenProbe amplified assay); UPP, urine-processing pouch.

We thank the manufacturers' representatives who reviewed and edited the specific tables detailing their respective products for accuracy.

TABLE 2 Single-use EIA systems for detection of C. *trachomatis* antigens

Manufacturer and kit name	Abbott TestPak Chlamydia (Fig. 2)
Phone no.	(800) 323-9100
Assay format	Single-use reaction disk EIA
Target	LPS
No. of tests/kit	20
Controls	2 Internal (procedural) controls are included in each disk. Abbott does not recommend external controls.
Specimen types	Endocervical swab
Collection device	Abbott STD Collection System for TestPak (included in kit)
Kit storage	Kit reagents D, E, and G require refrigeration; other reagents and reaction disks may be stored at 2 to 30°C
Specimen storage	If delay in processing, refrigerate at 2 to 8°C; may be held for 96 h prior to testing
Technical procedure and/or principle	Add extraction reagent to the transport tube; add 2 drops of dithiothreitol to tube and mix. Place filtration device in tube. Remove reaction disk from pouch and pour contents of tube into disk. Remove focus device from reaction disk and add antibody reagent. Wait 5 min. Add anti-rabbit IgG. Wait 5 min. Wash with PBS reagent. If processed according to directions and no deterioration of reagents has occurred, a line should appear, indicating a valid assay. Add chromogen reagent. Wait 5 min. A " + " indicates the presence of C. *trachomatis*.
Cost	$297
CPT code	87320
CLIA class	Moderate

FIGURE 2 Abbott Laboratories TestPack Chlamydia (courtesy of Abbott Laboratories).

TABLE 2   *(Continued)*

Manufacturer and kit name	Wampole Clearview Chlamydia (Fig. 3)
Phone no.	(800) 257-9525
Assay format	Disposable single-use test kit
Target	LPS
No. of tests/kit	20
Controls	An internal control is included in each kit to ensure reagent integrity and proper product performance. A positive control is included in the kit.
Specimen types	Endocervical swab
Collection device	Clearview Chlamydia Female Specimen Collection Kit
Kit storage	2 to 8°C
Specimen storage	Refrigerate at 2 to 8°C; may be tested up to 5 days. Extracted specimen stable for 3 h at RT

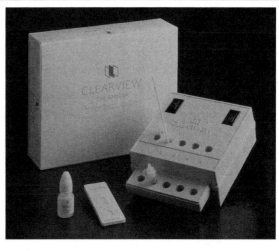

FIGURE 3   Wampole Laboratories Clearview Chlamydia (courtesy of Wampole Laboratories).

Technical procedure and/or principle	Chlamydia antigen is extracted by heating swabs at 80°C in extraction buffer. The sample is added to the sample window. The sample pad of the test unit contains dried colored latex particles coated with mouse monoclonal Abs to the genus-specific chlamydial LPS. The sample pad is in contact with a membrane strip, which contains 2 bands of immobilized Ab. The result window contains immobilized monoclonal Ab directed to chlamydial LPS. The control window contains immobilized polyclonal Ab directed against mouse Ig. When the patient sample is added, the latex-labeled Ab is hydrated and the sample-latex mixture moves by capillary action into the two reaction zones. If chlamydial LPS is present in the extract, it will react with the Ab on the latex particles, forming a complex that becomes immobilized in the reaction window, resulting in a blue line. If negative, no line occurs. The latex continues to move through the strip, and regardless of the result in the reaction window, reacts with the immobilized Ab in the control window to form a blue line, indicating a valid test.
Cost	$263.35
CPT code	87320
CLIA class	Moderate

(Continued on next page)

TABLE 2 Single-use EIA systems for detection of *C. trachomatis* antigen (*Continued*)

Manufacturer and kit name	Biostar Chlamydia Immunoassay (Fig. 4 and 5)
Phone no.	(800) 637-3717
Assay format	Single-use optical test kit (provides permanent results)
Target	LPS
No. of tests/kit	30
Controls	An internal-control dot in the center of the test surface is included to ensure proper procedural technique. A positive control is included in the kit.
Specimen types	Endocervical swab
Collection device	Proprietary (included in kit)
Kit storage	2 to 8°C
Specimen storage	RT for 24 h. Refrigerate at 2 to 8°C; test within 5 days

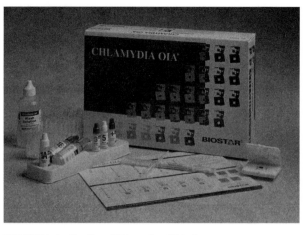

FIGURE 4 BioStar Chlamydia OIA (kit components; courtesy of BioStar).

a. Weak positive

b. Moderate positive

c. Strong positive

FIGURE 5 BioStar Chlamydia OIA (reactions; courtesy of BioStar). (a) Weak positive; (b) moderate positive; (c) strong positive.

Technical procedure and/or principle	Add extraction reagent to extraction tube. Add patient swab and thoroughly mix. Wait for 2 min. Add second extraction reagent to the extraction tube. Wait 2 min. Add neutralization reagent. Use this mixture within 2 h. Add 1 drop to center of gold test surface. Wait 5 min and add 1 drop of conjugate reagent directly to the sample drop. Wait 5 min. Using wash reagent, thoroughly wash reagents from the test surface. Two blotter positions labeled I and II are noted in the kit lid. It is preset to position I. Close the device for 10 s to remove residual moisture. Open the lid, move the blotter to position II, and add 1 drop of substrate reagent. Wait 5 min. Wash. Close lid for 10 s. Open and examine for color change.
	PRINCIPLE: Chlamydia OIA uses OIA technology to detect genus-specific chlamydial LPS. OIA uses a thin molecular film and differs from previously described membrane assays. The film is coated with an optical layer that binds extracted Ag. When labeled Ab is added, the Ag-Ab complex binds on the test surface. Substrate is added, and the complex increases the thickness of the film and optical properties to give a permanent blue color configured as a circle against a negative gold background.
Cost	$450
CPT code	87320 or 87810
CLIA class	Moderate

TABLE 3 Microwell systems for detection of *C. trachomatis* antigens

Manufacturer and kit name	Bio-Rad Laboratories (formerly Sanofi Diagnostics Pasteur) Chlamydia Microplate EIA (Fig. 6)
Phone no.	(800) 224-6723
Assay format	Microwell EIA; disposable wells
Target	LPS
No. of tests/kit	96 or 192
Controls	4/run
Throughput	92 or 188 (max) specimens/run
Assay time	Three 60-min incubations and one 30-min incubation
Specimen types	M urethra; endocervix
Collection device	Proprietary; chlamydia swab collection kit; chlamydia brush (nonpregnant females)
Kit storage	2–8°C
Specimen storage	7 days; RT
Technical procedure and principle	Pipette samples and controls, incubate for 60 min; add polyclonal Ab; incubate for 60 min; add HRP conjugate and incubate for 60 min. Wash 5 times; prepare and pipette substrate into wells; incubate for 30 min; add stop reagent; read absorbance on spectrophotometer; calculate OD cutoff value for positive and negative results.
Maintenance	Spectrophotometer
Other	Blocking assay required for positive reaction in low-prevalence populations and is suggested on all positive specimens that have had delays of >4 days before testing. A shortened assay procedure has been validated that requires a shaker incubator. This procedure decreases all incubation steps to 30 min.
Cost	$1.30–3.75/assay
CPT code	87320
CLIA class	High

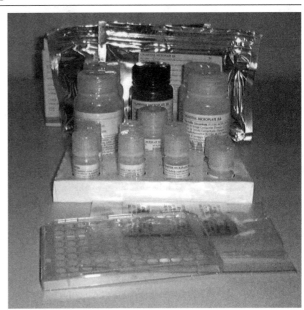

FIGURE 6 Bio-Rad (Sanofi Pasteur) Microwell EIA kit components (courtesy of Bio-Rad).

(Continued on next page)

TABLE 3 Microwell systems for detection of *C. trachomatis* antigens (*Continued*)

	Bio-Rad Laboratories (formerly Sanofi Diagnostics Pasteur) Chlamydia Blocking Assay	Intracel Bartels Chlamydia ELISA Kit
Manufacturer and kit name	Bio-Rad Laboratories (formerly Sanofi Diagnostics Pasteur) Chlamydia Blocking Assay	Intracel Bartels Chlamydia ELISA Kit
Phone no.	(800) 224-6723	(800) 227-8357
Assay format	EIA blocking assay	Microwell EIA; disposable wells
Target	NA	LPS
No. of tests/kit	100	96
Controls	NA; 2 wells required per specimen, one for the blocking assay, the other as a blank	3 (2 negative, 1 positive)
Throughput	NA	93 patient samples/plate
Assay time	Short assay procedure suggested; 4 incubations of 30 min	See "Technical Procedure"
Specimen types	Positive specimen requiring confirmation	M urethra; endocervix; ophthalmic
Collection device	NA	Bartels chlamydia EIA transport medium (included in kit); institution supplies swab
Kit storage	2–8°C	2–8°C
Specimen storage	Positive specimen must be tested within 48 h. Store at 2–8°C.	2–8°C for 8 days; −70°C for 6 mo
Technical procedure and principle	Patient specimen is added to 2 wells; a blank Ab (normal equine IgG) and a blocking Ab (anti-chlamydia) are added to the wells. The shortened assay is run with the shaker. Results are read as previously described, and a true positive yields a result in the blocking Ab well that is <50% of the absorbance of the blank well.	Heat specimens and controls at 100°C for 10 min. Place microwells in plate frame. In order, add conjugate, specimen, and chlamydia-specific Ab. Incubate at 35–37°C for 60 min. Wash 5 times. Prepare substrate reagent 10 min in advance of use, and add substrate. Incubate for 20 min in the dark. Add stop solution. Read at OD_{450}. Calculate cutoff values. Interpret results.
Maintenance	Spectrophotometer	Spectrophotometer; heat block
Other		No blocking assay described
Cost	$1.30–3.75/assay	$624/kit
CPT code	87320	87320
CLIA class	High	High

TABLE 3 *(Continued)*

	Wampole (previously Syva) MicroTrak II	Meridian Diagnostics Premier Chlamydia
Manufacturer and kit name	Wampole (previously Syva) MicroTrak II	Meridian Diagnostics Premier Chlamydia
Phone no.	(800) 257-9525	(800) 343-3858
Assay format	Microwell EIA; disposable wells	Microwell EIA; disposable wells
Target	LPS	LPS
No. of tests/kit	96	96
Controls	4 (3 negative, 1 positive)	4 (required for direct specimen detection); 5 (required for blocking assay)
Throughput	94	92
Assay time	See "Technical Procedure"	See "Technical Procedure"
Specimen types	M urethra; M urine; endocervix; ophthalmic	M urethra; endocervix; conjunctival
Collection device	MicroTrak II specimen collection kit	Proprietary kits for endocervix, M urethra, and conjunctiva
Kit storage	2–8°C	2–8°C
Specimen storage	2–25°C; test within 7 days.	RT for 48 h. For >48 h, hold at 2–8°C. Test within 7 days. May also be frozen at −20°C
Technical procedure and principle	Prepare substrate 5–10 min before use. Prepare working buffer. Heat specimens and controls at 100°C for 10 min. Place microwells in plate frame. In order, add conjugate, specimen, and chlamydia-specific Ab. Incubate at 35–37°C for 60 min. Wash 5 times. Add substrate. Incubate for 20 min in the dark. Add stop solution. Read at OD_{450}. MicroTrak reader will calculate cutoff values, interpret results, and check control validity.	Prepare wash buffer; prepare substrate solution (15 min before use); prepare Ag standard. Add extraction buffer to the specimen transport tubes; vortex; incubate for 15–60 min at RT; vortex. Add sample to microwell; incubate for 60 min at RT; wash 3 times. Add detector Ab; incubate for 30–35 min at RT; wash 3 times. Add freshly prepared conjugate Ab; incubate for 30–35 min at RT; wash 3 times. Add freshly prepared substrate; incubate for 30–35 min at RT; add stop reagent; read within 30 min at OD_{490}. Assess acceptability of controls; calculate cutoff values; interpret.
Maintenance	Spectrophotometer; heat block	Spectrophotometer
Other	Blocking assay available	Blocking antibody included in kit for verification of positive reaction in low-prevalence populations (<5%)
Cost	$435	$535
CPT code	87320	87320
CLIA class	High	High

TABLE 4 Semiautomated systems for detection of *C. trachomatis* antigens

	Beckman Coulter Access CHLAMYDIA (Fig. 7 and 8)	BioMerieux VIDAS Chlamydia (CHL) and Chlamydia Blocking Assay (CHB) (Fig. 9–11)	Abbott Chlamydiazyme (Fig. 12)	Abbott IMx Chlamydia (Fig. 13)	Wampole MicroTrak XL (Fig. 14)
Manufacturer and kit name					
Phone no.	(800) 854-3633	(800) 638-4835	(800) 323-9100	(800) 527-1869	(800) 257-9525
Assay format	Semiautomated EIA	Semiautomated EIA	Semiautomated EIA	Microparticle EIA	Automated EIA microwell plate processing system
Target	LPS	LPS	LPS	LPS	LPS
No. of tests/kit	200 antigen detection; 50 blocking assays	60 (CHL); 30 (CHB)	100 and 500	100 reagent packs	96
Controls	Negative calibration; positive calibration; low positive (all valid for 24 h). Unblocked calibration control; blocked calibration control (valid for 24 h).	Positive; negative. Standard for instrument calibration required per new lot (valid for 14 days).	1 positive; 3 negative	1 positive; 1 negative. Unique calibration per lot number.	1 positive; 3 negative. Calibration test each month.
Instrument throughput	100/h	30/h	Commander system: 8 trays of 60 (less 4 controls) = 476 specimens in 3.5 h	24 samples per carousel	192 samples
Assay time	See "Procedure"	1 h (after sample preparation; see "Procedure")	See "Procedure"	70 min/run	3.5 h
Specimen types	M urethra; M urine; endocervix	M urethra; M urine; endocervix	M urethra; M urine; endocervix; neonate conjunctiva; neonate nasopharynx	M urethra; endocervix	M urethra; M urine; endocervix; ocular
Collection device[a]	Proprietary collection kit	Proprietary	Proprietary; included in kit	Proprietary	Proprietary
Kit storage	2–10°C	2–8°C	2–8°C	2–8°C; extraction buffer and reaction cells, 15–30°C	2–23°C
Specimen storage	Urine, 2–8°C tested within 3 days. Swabs, RT (refrigeration preferred) tested within 7 days. Note: if swab tested is ≥4 days old, blocking assay must be performed to confirm positives.	Swabs in transport, 2–25°C for 48 h in transit, then 2–8°C. Urine, 2–8°C. All specimens must be tested in 7 days.	Swabs, 48 h at RT in transport tube. If further delay, store at 2–8°C. Test within 7 days. Male urine, 2–8°C up to 48 h prior to centrifugation. Centrifuge at 1,800–3,000 × g for 20 min; remove supernatant. Test pellet within 5 days.	2–8°C; test within 7 days	2–23°C
Technical procedure and principle	Add specimen treatment solution to transport tube. Incubate for 10 min and	Add sample treatment reagent to collection tube and incubate for 15–30	Prepare OPD substrate immediately prior to use. Prepare sample by adding	Add extraction buffer to transport tubes; vortex. Place in heat block at 95–100°C for	Place bar code on specimen tube. Add treatment solution to specimen. Heat at 95–100°C in

TABLE 4 (Continued)

Technical procedure and principle					
vortex. Wring swab to remove excess liquid and discard swab. Add treatment solution 2. Vortex and incubate for 10 min. Add solution 3. Vortex. Assay within 48 h. For urine, centrifuge 10 ml of urine at 3,500 × g for 10 min. Decant supernatant and continue as described for swab specimen. For instrument analysis, on the instrument's CRT, access the test request screen from the main menu. For each sample, assign a tray position, enter sample information, and request the chlamydial antigen test. Aliquot treated specimen to the appropriate sample cup. (Note that the volume of aliquot depends on the assay to be performed—screening, repeat, or blocking.) Press the run key. PRINCIPLE: Chlamydia-specific LPS monoclonal Abs are covalently coupled to a paramagnetic particle. Samples containing chlamydia bind to the particle. A mixture of polyclonal rabbit anti-chlamydia Ab and alkaline phosphatase-conjugated polyclonal goat anti-rabbit IgG is added and binds in succession. Separation in a magnetic field allows washing unreacted material not bound to the solid phase. A chemiluminescent	min at RT. Vortex, wring, and remove swab. Place tube in heating block for 15–30 min to remove endogenous alkaline phosphatase activity. Vortex and add sample to reagent strip (see "Other" below for strip description). For male urine, 10–12 ml of urine is centrifuged at 1,400–3,000 × g for 15–30 min. Decant supernatant, add treatment reagent, and proceed as described above. The VIDAS Chlamydia is an enzyme-linked fluorescent immunoassay. All assay steps are performed and controlled by the instrument. A pipette tip-like disposable device known as the Solid Phase Receptacle (SPR) functions as the solid phase as well as the pipetter for the assay. The patient sample is cycled into the SPR, and antigen adheres. The sample is then transferred into a mouse anti-chlamydial Ab solution, and specific binding occurs. Wash steps remove unbound reagents, and anti-mouse IgG conjugated with alkaline phosphatase is cycled through the SPR. Wash steps remove unbound reagents, and substrate (4-methylumbelliferyl phosphate) is cycled	dilution buffer. Vortex for 3 cycles of 15 s. Remove excess liquid by wringing swab; discard swab. Pipette into sample reaction tray. All remaining steps, including addition of reagents, washes, incubations, reading and interpretation, are performed by the instrument. Treated beads are reacted with patient samples. If Ag is present, it adsorbs to the bead. The bead is washed and then incubated with Ab to *C. trachomatis*. Ab complexes with Ag if present. The bead is incubated with Ab-enzyme conjugate (HRP), which reacts with Ag-Ab complex on the bead. The bead is exposed to the enzyme substrate (o-phenylenediamine) and hydrogen peroxide, and a yellow color develops in proportion to the amount of Ag present. Absorbance is read by the spectrophotometer in the Abbott COMMANDER system, cutoff values are determined, and interpretation is given.	10–15 min. Cool. Insert filtration device into tube and filter into separate tube. Pipette 150 µl into sample wells in the carousel. Program load list in computer, add reagent packs, and run assay. The IMx chlamydia assay is based on microparticle EIA technology. The IMx probe-electrode assembly dispenses microparticles to the sample wells containing treated specimen. If present, chlamydial LPS binds, forming a microparticle-LPS complex. An aliquot of the solution is transferred to the glass fiber matrix of the reaction cell. The microparticle-LPS complex binds irreversibly. Rabbit anti-chlamydia LPS Ab is added and complexes with the LPS. Biotinylated goat anti-rabbit Ab is dispensed onto the matrix and binds specifically. After washing, 4-methylumbelliferyl phosphate is added to the matrix, and the instrument reads fluorescence. Cutoff values are determined, interpretation is performed by the instrument, and results are printed.	heat block for 15 min. Cool to RT. Vortex for 1 min. Remove swab. Add specimen to carrier rack. Load in instrument. Place reagents on reagent wheel, and load in the instrument. Start run on instrument. The system runs the assays automatically, including diluting and transferring the patient specimens, adding reagents, washing, incubating, reading the reaction wells, processing the data, and reporting results. The system is controlled through a keyboard, function keys, and a CRT display. The principle is identical to MicroTrakII (as previously described), as it is a semiautomated microwell assay.	

(Continued on next page)

TABLE 4 Semiautomated systems for detection of *C. trachomatis* antigens (*Continued*)

Technical procedure and principle	substrate is added, and the reaction is read in a luminometer. The blocking assay uses equine anti-chlamydial Ab in a competition assay. Results are determined and automatically calculated.	through the SPR. The enzyme, if present, catalyzes conversion of substrate to 4-methylumbelliferone, and the fluorescence intensity is measured by the instrument. Results are analyzed by the computer, and a test value is computed and printed for each patient specimen. The Vidas Chlamydia Blocking Assay with the VIDAS chlamydia reagent strips is used as a dual strip assay to verify positive results in low-prevalence populations.			
Maintenance	Daily; weekly	Daily; weekly	Daily; weekly; monthly	Per shift; daily; weekly; monthly	Daily; weekly; monthly
Other	Universal access substrate and buffer purchased separately	Each CHL strip is a polyprolylene strip of 10 wells covered with a foil seal and paper label. The first well is for the sample; the next 8 wells are for the various reagents, and the last well is an optically clear cuvette for the fluorimetric determination. All assay reagents are resident in the strip. A heating block (95–100°C) is required. MINI-VIDAS holds 12 strips (compact version of VIDAS 30).	For blocking assay, a specific blocking reagent is required. This assay must be performed manually, using the Abbott Quantum spectrophotometer. The assay blocks *C. trachomatis* only and not other chlamydia species.	Blocking assay available; heat block required	Blocking assay available; heat block required
Cost	$2,240 (Ag assay and blocking kit) $129,800 (Access system)	$378 (CHL) $50,300 (Vidas 30 system) $27,300 (mini-VIDAS system)	$806 (100-test kit) $4,030 (500-test kit) $28,000 (COMMANDER system)	$735 (blocking reagent) $21 (4 ml) $50,950 (instrument)	NA
CPT Code	87320	87320	87320	87320	87320
CLIA Class	High	High	High	High	High

Extra costs may be associated with collection device.

FIGURE 7 Beckman Coulter Access system (courtesy of Beckman Coulter).

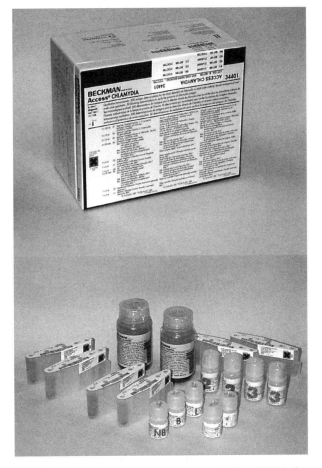

FIGURE 8 Beckman Coulter Access CHLAMYDIA kit components (courtesy of Beckman Coulter).

FIGURE 9 bioMerieux, Inc., VIDAS multiparametric semiautomated immunoanalysis system (courtesy of bioMerieux, Inc.).

FIGURE 10 bioMerieux, Inc., MiniVIDAS multiparametric semiautomated immunoanalysis system (courtesy of bioMerieux, Inc.).

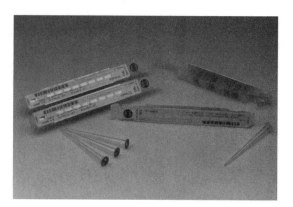

FIGURE 11 bioMerieux, Inc., VIDAS Chlamydia kit cartridges (courtesy of bioMerieux, Inc.).

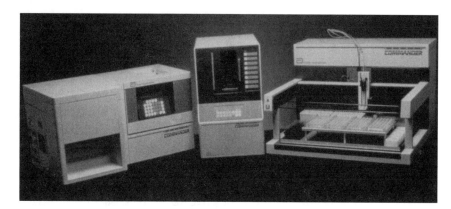

FIGURE 12 Abbott Laboratories Chlamydiazyme system (courtesy of Abbott Laboratories).

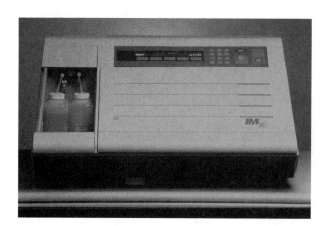

FIGURE 13 Abbott Laboratories IMx system (courtesy of Abbot Laboratories).

FIGURE 14 Wampole (Syva) MicroTrak XL system.

TABLE 5 Direct nucleic acid probe for detection of *C. trachomatis*

Manufacturer and kit name	GenProbe PACE 2[a] (Fig. 15)
Phone no.	(800) 523-5001
Assay format	Direct DNA probe
Target	rRNA for *C. trachomatis* and *N. gonorrhoeae*
No. of tests/kit	100 and 1,000
Controls	3 negative reference samples and a positive control
Instrument throughput	Manual single tube; 250- and 400-sample systems (instrument dependent)
Assay time	2.5 h
Specimen types	Endocervical swab, M urethral swab, and conjunctival swab
Collection device	Proprietary transport
Kit storage	Probe and separation reagent must be stored at 2–8°C. Once reconstituted the probe reagent is stable for 3 weeks. Once the separation suspension is prepared, it remains stable for 6 h at 20–25°C; all other components stable at 2–25°C.
Specimen storage	Transport at 2–25°C; store at 2–25°C; stable for 7 days. If longer storage required, freeze at −20 to −70°C.
Technical procedure/principle	The PACE 2 assays are direct DNA probes that rely on the ability of complementary nucleic acid strands to form a stable complex. The PACE 2 system uses a proprietary single-stranded DNA probe with a chemiluminescent label that is complimentary to an rRNA target. The DNA-RNA hybrids are separated from nonhybridized nucleic acids, and the signal from the labeled hybrids is measured in a luminometer.
Maintenance	Run tritium standard once/wk. Routine periodic maintenance for cleaning every 2 wk.
Other	After specimen transfer, all steps of procedure and detection take place in a single tube. Competition assay meets CDC guidelines in low-risk populations. Assay requires a water bath for thermal-incubation steps.
Cost	Volume dependent
CPT code	87490 (*C. trachomatis*); 87590 (*N. gonorrhoeae*). Use both for combination assay
CLIA class	High complexity

[a]Combination *C. trachomatis* and *N. gonorrhoeae*, individual *C. trachomatis* and *N. gonorrhoeae*, and probe competition assay (for confirmation).

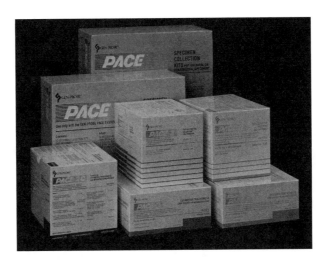

FIGURE 15 GenProbe PACE 2 kits (courtesy of GenProbe).

TABLE 6 Nucleic acid amplification assays for detection of C. trachomatis

Manufacturer and kit name	Abbott Diagnostics Ligase Chain Reaction for Chlamydia trachomatis and Neisseria gonorrhoeae (Fig. 16)	Roche Diagnostics Corp. Amplicor CT/NG/IC (simultaneous amplification of C. trachomatis, N. gonorrhoeae, and internal control) (Fig. 17)	Digene Hybrid Capture II CT/GC DNA Test (dual-screen assay), HC II GC-ID DNA Test, and HC II CT-ID DNA Test[a] (Fig. 18)	Gen-Probe Amplified Chlamydia trachomatis Assay (Fig. 19)	Becton-Dickinson Probe Tec ET; C. trachomatis and N. gonorrhoeae (Fig. 20)
Phone no.	(800) 527-1869	(800) 526-1247	Digene, (800) 344-3631; Abbott, (800) 323-9100	(800) 523-5001	(800) 638-8663
Assay format	LCR	PCR	HC RNA probe with signal amplification	Transcription-mediated amplification	SDA
Target	C. trachomatis, cryptic plasmid DNA; N. gonorrhoeae, 48-bp sequence in the conserved Opa genes	C. trachomatis, 207-nucleotide sequence on the cryptic plasmid DNA; N. gonorrhoeae, 201-nucleotide sequence within M.NgoPII	The C. trachomatis-N. gonorrhoeae probe cocktail contains a probe mixture specifically chosen to minimize cross-reactivity with DNA sequences from human cells, other bacterial species, Chlamydia species other than C. trachomatis or Neisseria species other than N. gonorrhoeae. The C. trachomatis-N. gonorrhoeae probe cocktail supplied with the HC II CT/GC DNA Test is complementary to approximately 39,000 bp (4%) of the C. trachomatis genomic DNA and 7,500 bp (100%) of the cryptic plasmid and 9,700 bp (5%) of the N. gonorrhoeae genomic DNA and 4,200 bp (100%) of the cryptic lasmid. A specimen testing positive with the dual C. trachomatis-N. gonorrhoeae test must be tested by the individual CT-ID assay or the GC-ID assay to specify the organism.	Chlamydia, a specific 23S rRNA sequence; no N. gonorrhoeae at present	C. trachomatis, cryptic plasmid DNA; N. gonorrhoeae, a region within the multicopy pilin gene-inverting protein homologue
No. of tests/kit	Separate C. trachomatis and N. gonorrhoeae amplification kits. Each kit contains 96 vials for assays and a detection reagent pack for 100 assays.	C. trachomatis and N. gonorrhoeae specimen prep, 100 tests; amplification kit, 96 tests; N. gonorrhoeae detection kit, 100 tests; C. trachomatis detection kit,	96-well format	50 tests; test tube format	Kits contain 96 microwell plates (providing for 384 determinations)

TABLE 6 (*Continued*)

Controls	2 negative and 2 calibration controls per run.	Swabs, positive and negative swab control; urine, 1 positive and 1 negative urine controls	100 tests; generic reagents, 100 tests	1 positive, 1 negative; both in triplicate	1 positive and 1 negative control per run. Optional specimen-processing control.	1 positive and 1 negative per plate
Instrument throughput	Specimens are placed on a carousel that allows 24 samples. Four samples are dedicated to controls, leaving 20 samples for patient specimens. Samples for *C. trachomatis* and *N. gonorrhoeae* are placed on separate carousels, as reagents loaded into the instrument are unique to each agent.		Manual Amplicor, 96 tests (less urine and swab controls). COBAS AMPLICOR, fully automated amplification and detection. Manual initial specimen preparation, then placement on instrument. Basic mode is essentially walk-away mode. Two amplification rings (24 samples) are used in this mode. Samples are placed in the ring, the instrument is started, and all procedures are automatically performed (start to finish). Parallel mode (continuous-feed mode) is used to increase throughput. The system has the ability to have 2 runs ongoing concurrently: 2 amplification rings (12 specimens per ring subjected to a 2-h TC process) and 2 detection positions. The user has the ability in parallel mode to intervene once the amplification steps are completed and to manually move these rings to the detection position and load new amplification rings. Simultaneous detection and amplification increases throughput. COBAS AMPLICOR detection results are generated 1 h post-TC with subsequent results	90 patient samples/plate	Single-tube format allows flexible format and performance. A Tecan automated pipetting instrument can be used for high-volume testing.	Simultaneous detection of *C. trachomatis* and *N. gonorrhoeae*. In an 8-h day, *C. trachomatis* and *N. gonorrhoeae* with amplification controls allow for 180 determinations per 8-h shift

(*Continued on next page*)

TABLE 6 Nucleic acid amplification assays for detection of *C. trachomatis* (*Continued*)

Assay time	4 h start to finish; hands on at 2.5–3.0 min/specimen	Prep time, 24 swabs in 45 min; 24 urine specimens in 90 min. Manual detection, 1-h hybridization; 15-min conjugate incubation; 10-min substrate incubation; read within 60 min of stop solution addition. The entire microwell plate is read within 1 min. every 72 s (50 determinations per h post-TC).	4 h (hands-on, 2 h)	3–3.5 h	Swabs, 2.5 h with 1.0 min/specimen hands-on time. Urine, 3 h.
Specimen types	Endocervical swabs; F and M urine; M urethral swabs	*C. trachomatis*, Endocervical swabs, F and M urine, M urethral swabs. *N. gonorrhoeae*, M urine (from asymptomatic and symptomatic patients) and M urethral swabs (symptomatic patients); F endocervical swab only (symptomatic and asymptomatic patients)	Endocervical swabs; M urethral swabs; M urine	Endocervical swabs; F and M urine; M urethral swabs	Endocervical swabs; F and M urine; M urethral swabs
Collection device	Sterile cup for FVU; proprietary transport system for swab specimens	Sterile cup for 10–50 ml of FVU. Approved swab transport media: 2SP, Bartels Chlam trans, SPG, M-4. Note: there is no longer a proprietary specimen transport system. The approved transport medium allows for culture of *C. trachomatis* as necessary.	Proprietary endocervical brush (Digene cervical sampler) or proprietary swab (Digene F swab collection kit). Note: a Pap transport medium compatible with the assay is under development.	FVU (30–50 ml) in sterile container; proprietary swab	FVU (15–60 ml) in sterile cup; proprietary swab
Kit storage	LCR reaction cells; system diluent and inactivation diluent (keep this away from sunlight) can be stored at 15–30°C; all other kit reagents must be stored at 2–8°C.	Kit components, 2–8°C.	2–8°C	2–8°C; however, some components do not require refrigeration	All at RT

TABLE 6 (Continued)

Specimen storage	FVU, transported and stored at 2–8°C; must be tested in 4 days or frozen at −20°C or lower and tested within 60 days. Swabs, transported and stored at 2–30°C; tested within 4 days or frozen at −20°C or lower and tested within 60 days.	Urine, transport at 18–30°C within 24 h. If delay, 2–8°C tested within 7 days or frozen at −20°C or below and test within 30 days. Swabs, RT if within 1 h; if delay of >1 h, send at 2–8°C and process within 7 days, or freeze at −20°C or below and test within 30 days.	2 wk at RT and one additional wk at 2–8°C; frozen at −20°C up to 3 mo	Urine, test within 24 h if stored at 15–30°C; test within 7 days if stored at 2–8°C. Swabs, may be held at 2–25°C for 7 days, frozen at −70°C for 60 days.	Urine, add the UPP directly to the urine at the time of voiding. Urine must be in contact with the UPP for a minimum of 2 h prior to processing. The UPP removes many interfering substances that may inhibit the reaction. The UPP containing urine may be held at RT for 2 days or refrigerated for 4–6 days. Alternatively, the urine may be refrigerated immediately postvoiding and held for 4–6 days at 2–8°C prior to testing. Swab specimen, may be held at RT for 4–6 days.
Technical procedure and principle	The Abbott assay uses the LCR to detect and amplify specific target gene sequences. Four oligonucleotide probes in the assay recognize and hybridize to their specific target sequences. These oligonucleotides were designed in pairs to be complimentary to the target so that in the presence of target, the probes will join adjacent to each other. They are then enzymatically joined to form the amplification product, which then becomes the target sequence for additional amplification. Each end of the joined oligonucleotide probes contains haptens. One hapten (capture hapten) is recognized by an antibody on a microparticle, while the other hapten (detection hapten) is recognized by an antibody labeled with	The Amplicor assay is based on 4 major processes: specimen preparation (releasing DNA), PCR amplification of the DNA target using specific complementary primers (making millions of copies of the target), hybridization of the amplified DNA to specific probes (capturing the product on microparticles or coated microwell plates), and detection of the probe-bound amplified DNA by colorimetric determination.	Samples are denatured. DNA is hybridized with RNA probes. Hybrids are captured on microplates coated with Abs directed against the RNA-DNA complex. Alkaline phosphatase anti-RNA-DNA Abs are added for signal amplification. The microplate is washed. Signal is generated by cleavage of the chemiluminescent substrate, and the microplate is read in a DML 2000 luminometer.	This assay utilizes the TMA and HPA procedures to qualitatively detect *C. trachomatis* rRNA. The GenProbe TMA system used in this test amplifies a specific 23S rRNA target via DNA intermediates with detection of the amplified rRNA sequences (amplicon) using the HPA nucleic acid hybridization method. The chemiluminescent single-stranded DNA probe is complementary to the amplicon and combines with the amplicon to form a stable RNA-DNA hybrid. The selection reagent differentiates hybridized from nonhybridized probe, and during the detection step, light emitted from the labeled RNA-DNA hybrids is measured as photon signals in a luminometer. The results are interpreted based on an	This assay is based on the simultaneous amplification and detection of amplified DNA using amplification primers and fluorescently labeled detector probes. The SDA reagents are dried in a disposable microwell format. The treated specimen is added to the priming microwell containing the amplification primers, fluorescently labeled detector probe, and other reagents necessary for amplification. After incubation, the sample is transferred to an amplification microwell containing DNA polymerase and a restriction endonuclease necessary for the SDA. The microwells are sealed and incubated in a thermally controlled incubator-reader, which monitors and reads

(Continued on next page)

TABLE 6 Nucleic acid amplification assays for detection of *C. trachomatis* (*Continued*)

	alkaline phosphatase. During processing and instrument analysis, if the amplification product is present, the patented Microparticle Enzyme Immunoassay used in many Abbott instruments detects it.			assigned equivocal zone and assigned assay cutoff.	fluorescence. Each sample is tested in 3 distinct wells to detect *C. trachomatis*, *N. gonorrhoeae*, and substances that may inhibit the SDA.
Maintenance	Moderate in time and complexity. Daily (20 min), weekly (30 min), and monthly (30 min) to monitor all aspects of amplification, thermal cycling, contamination, and instrument performance.	COBAS AMPLICOR, 5 min daily maintenance. Weekly, rinse wash reservoir. Every 2 mo, clean surfaces, change transfer probe and tubing Manual Amplicor, clean washer. TC, weekly heater-chiller test; monthly system performance test.	DML 2000 luminometer, no maintenance	Depends on type of luminometer. Weekly, run tritium standards and clean carrying cassettes (large luminometer). Every 2 weeks (or as needed), warm water flush of lines. Monthly (or as needed), clean chamber of single-tube luminometer.	Daily, check paper in printer, record displayed temperature of instrument, and record heat block temperature. Monthly, clean instrument air filter and perform precision check of instrument temperature (internal).
Other	As with PCR, two dedicated areas are recommended for performing LCR. Area 1 is for processing and addition of samples to amplification vials. Equipment (pipettes, dry bath, and microcentrifuge) should be a permanent part of area 1. Area 2 is dedicated to amplification and detection of the amplified product. The TC, LCR analyzer, and all accessories for the analyzer should be kept in this area. Note: postdetection, to reduce the risk of contamination, the amplification product is automatically inactivated with a 2-reagent chemical inactivation system.	Unidirectional work flow or two dedicated areas are recommended for prevention of contamination. In addition, the proprietary Amperase system is used in the assays to decrease the contamination from amplified products. Extended gray zone implemented to compensate for cross-reaction with *Neisseria subflava* and *Neisseria cinerea*.	Signal amplification increases sensitivity over direct DNA probe.	Single room with unidirectional flow; isothermal amplification	Single room to perform all functions. The assay is isothermal and does not require a TC. There is the option of running an amplification control with all samples to detect interfering substances that may inhibit the reaction.
Cost	Volume dependent	Volume dependent	Volume dependent	Volume dependent	Volume dependent
CPT code	*C. trachomatis*, 87491; *N. gonorrhoeae*, 87591	*C. trachomatis*, 87491; *N. gonorrhoeae*, 87591	87491	87491	*C. trachomatis*, 87491; *N. gonorrhoeae*, 87591
CLIA class	High complexity	High complexity	High complexity	High complexity	High complexity

[a]Marketed and sold throughout the United States, Europe, the Middle East, and Africa through Abbott Laboratories.

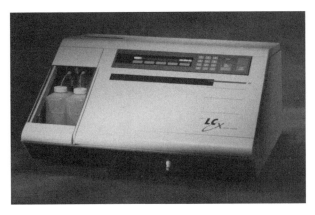

FIGURE 16 Abbott Laboratories LCx analyzer (courtesy of Abbott Laboratories).

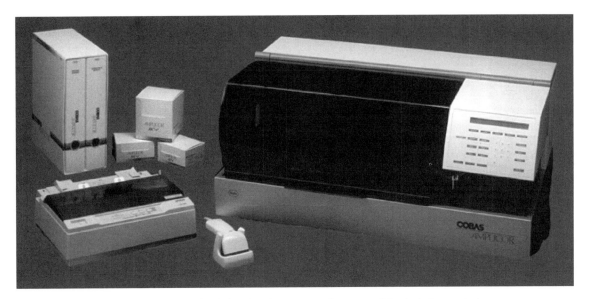

FIGURE 17 Roche Cobas Amplicor system (courtesy of Roche Laboratories).

FIGURE 18 Digene Hybrid Capture system and reagents (courtesy of Digene Corporation).

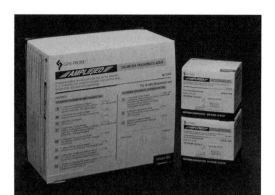

FIGURE 19 GenProbe Amplified *Chlamydia trachomatis* assay kit (courtesy of GenProbe).

FIGURE 20 Becton Dickinson ProbeTec ET system (courtesy of Becton Dickinson).

TABLE 7 Selected studies before 1995 evaluating EIA (Chlamydiazyme) detection of endocervical or male urethral *C. trachomatis* infection by using a culture standard or an expanded culture standard

Authors, yr (reference)	No. of subjects (sex)	Prev (%)	Sens (%)	Spec (%)	Expanded standard method[a]	PPV (%)	NPV (%)
Biro et al., 1994 (16)	479 (F)	16	65	100[c]	DFA, consensus[d]	100	95
Chernesky et al., 1994 (32)[b]	447 (F)	6	78	100[c]	LCR	100	98
Weisenfeld et al., 1994 (164)	474 (F)	13	43	100[c]	PCR[e]	100	92
Clarke et al., 1993 (37)[b]	217 (F)	22	78	98.8	DFA	93	94
Warren et al., 1993 (157)[b]	1,037 (F)	5	80	99.3	DFA	86	99
Altaie et al., 1992 (2)[b]	1,240 (F)	9	80	100	DFA	100	97
Mills et al., 1992 (95)	502 (F)	11	91	98	None	86	99
Van Dyck et al., 1992 (153)	670 (F)	15	54	96[c]	None	70	92
Genc et al., 1991 (52)	245 (F)	12	85	99[c]	DFA	92	97
	196 (M)	11	62	96[c]	DFA	70	94
Gaydos et al., 1990 (50)[b]	307 (F + M)	15	77	98	DFA	95	83
Kluytmans et al., 1990 (76)	611 (F)	8	68	96	None	58	97
	280 (M)	14	92	92	None	66	98.6
Moncada et al., 1990 (98)[f]	2,891 (F)	9	81	99	DFA	99	97.8

[a]Method added to culture for expanded definition of true positive; if none, culture alone was used as the standard.
[b]Also compared with other nonculture tests.
[c]With use of blocking antibody reagent.
[d]Two consensus nonculture test positive results included in definition of true positive.
[e]PCR used as standard; no culture performed.
[f]Multicenter study.

TABLE 8 Selected studies prior to 1995 evaluating EIA (Chlamydiazyme) detection of C. *trachomatis* in urine from men by using culture with and without nonculture tests of urethral specimens as a standard

Authors, yr (reference)	No. of subjects	Prev (%)	Sens (%)	Spec (%)	Expanded standard method[a]	PPV (%)	NPV (%)
Chernesky et al., 1994 (32)	305	18	68	100[c]	LCR	100	93
Talbot and Romanowski, 1994 (147)[b]	318	15	76	100[c]	DFA	100	96
Moncada et al., 1994 (100)[b]	1,341	15	76	97	DFA	80	96
Backman et al., 1993 (4)	167	34	55	98	None	94	81
Ehret et al., 1993 (44)	540	14	83	96	None	76	97
Genc et al., 1991 (52)	196	11	77	94[c]	DFA	67	96
Mumtaz et al., 1991 (105)	312	11	78	100[c]	DFA	100	97

[a]Method added to culture for expanded definition of true positive. If none, culture alone was used as the standard.
[b]Also compared with other nonculture tests.
[c]With use of blocking antibody reagent.

TABLE 9 Select studies prior to 1995 evaluating EIA (MicroTrak) detection of endocervical or male urethral *C. trachomatis* infection by using an expanded culture standard

Authors, yr (reference)	No. of subjects (sex)	Prev (%)	Sens (%)	Spec (%)	Expanded standard method[a]	PPV (%)	NPV (%)
Biro et al., 1994 (16)[b]	479 (F)	16	80	98	DFA consensus[c]	90	96
Chan et al., 1994 (29)	32,495 (F)	7	96	99[c]	DFA	92	99
	5,113 (M)	15	98	97	DFA	84	99
Skulnick et al., 1994 (138)[b]	993 (F)	2	61	100	PCR, MOMP	92	99
Thomas et al., 1994 (149)[b]	151 (F)	24	74	100	DFA	100	92
Clarke et al., 1993 (37)[b]	217 (F)	22	80	99	DFA	93	94
Altaie et al., 1992 (2)[b]	1,240 (F)	9	94	100	DFA	100	99
Moncada et al., 1992 (99)[b]	1,254 (F)	10	93	99	DFA	94	99
Gaydos et al., 1990 (50)[b]	792 (F)	13	89	96	DFA	80	98
	240 (M)	10	97	98	DFA	91	99

[a]Method added to culture for expanded definition of true positive.
[b]Also compared with other nonculture tests.
[c]Two consensus nonculture test positive results included in definition of true positive.

TABLE 10 Selected studies prior to 1995 evaluating EIA (MicroTrak) detection of *C. trachomatis* in urine from males by using a culture with or without nonculture tests of urethral specimens as a standard

Authors, yr (reference)	No. of subjects	Prev (%)	Sens (%)	Spec (%)	Expanded standard method[a]	PPV (%)	NPV (%)
Domeika et al., 1994 (42)	184	18	85	100	PCR, MOMP[b]	100	77.5
Moncada et al., 1994 (100)[c]	1,341	15	86	99	DFA	94	97
Talbot et al., 1994 (147)[b]	318	15	78	99	DFA	97	97
Matthews et al., 1993 (93)	304	33	82	98	None	95	92

[a]Method added to culture for expanded definition of true positive. If none, culture alone was used as the standard.
[b]PCR used as standard; no culture performed. A second PCR was performed with alternate target primers specific for the MOMP.
[c]Also compared with other nonculture tests.

TABLE 11 Selected studies prior to 1995 evaluating DNA probe (PACE 2) detection of endocervical or male urethral *C. trachomatis* infection by using a culture standard or an expanded culture standard

Authors, yr (reference)	No. of subjects (sex)	Prev (%)	Sens (%)	Spec (%)	Expanded standard method[a]	PPV (%)	NPV (%)
Altwegg et al., 1994 (3)	222 (F)	10	77	100	PCR	100	97
Biro et al., 1994 (16)	479 (F)	16	67	96	DFA, consensus[c]	75	94
Blanding et al., 1993 (18)[b]	940 (F)	4	76		Consensus[c]	52	98
Warren et al., 1993 (157)[b]	1,037 (F)	5	96	99	DFA	74	99
Hosein et al., 1992 (66)	246 (F)	13	94	99	None	94	99
Limberger et al., 1992 (84)	398 (F)	5	95	100	None	100	99
Kluytmans et al., 1991 (75)	482 (F)	9	93	98	PCR	83	99
	260 (M)	13	77	99	PCR	96	96
Iwen et al., 1991 (69)	318 (F)	9	93	98	None	85	99
Lees et al., 1991 (83)	909 (F)	3	86	99	None	96	99
Yang et al., 1991 (167)	426 (F)	10	86	99[d]	None	91	98

[a]Method added to culture for expanded definition of true positive. If none, culture alone was used as the standard.
[b]Also compared with other nonculture tests.
[c]Two consensus nonculture test positive results were included in the definition of true positive.
[d]With use of competition probe assay.

TABLE 12 Selected studies prior to 1995 evaluating EIA (Chlamydiazyme) detection of *C. trachomatis* in urine from men by using culture with and without nonculture tests of urethral specimens as a standard

Authors, yr (reference)	No. of subjects	Prev (%)	Sens (%)	Spec (%)	Expanded standard method[a]	PPV (%)	NPV (%)
Altwegg et al., 1994 (3)[b]	222	10	95	96	MOMP,[c] consensus[d]	73	99
Bianchi et al., 1994 (15)	290	4	92	100	MOMP	92	99
Mahony et al., 1994 (89)	770	4	91	100	MOMP	100	99
Skulnick et al., 1994 (138)[b]	993	2	89	99	MOMP	84	99
Bass et al., 1993 (5)	945	8	98	100	MOMP	100	99
Bauwens et al., 1993 (10)	587[e]	10	88	100	MOMP	99	99
	362	8	64	99	MOMP	95	99

[a]Method added to culture for expanded definition of true positive.
[b]Also compared with other nonculture tests.
[c]A second PCR was performed with alternate target primers specific for MOMP.
[d]Two consensus nonculture test positive results are included in the definition of true positive; no culture was performed.
[e]Two separate studies were performed as preclinical and clinical evaluation, respectively.

TABLE 13 Selected studies prior to 1995 evaluating PCR (Amplicor) detection of *C. trachomatis* in urine from men and women by using culture with nonculture tests of urethral or endocervical specimens as a standard

Authors, yr (reference)	No. of subjects (sex)	Prev (%)	Sens (%)	Spec (%)	Expanded standard method[a]	PPV (%)	NPV (%)
Bianchi et al., 1994 (15)	466 (M)	14	93	99	MOMP[b]	73	99
Skulnick et al., 1994 (138)	394 (F)	3	99	99	MOMP	85	99
Bauwens et al., 1993 (9)	365 (M)	9	96	98	MOMP	82	99
Jaschek et al., 1993 (70)	530 (M)	9	95	99	DFA, MOMP	99	99

[a]Method added to culture for expanded definition of true positive. If none, culture alone was used as the standard.
[b]A second PCR was performed with alternate target primers specific for the MOMP.

TABLE 14 Selected studies prior to 1996 evaluating LCR (Abbott) detection of *C. trachomatis* infection with endocervical or urine specimens by using culture with nonculture tests of endocervical or urethral specimens as a standard

Authors, yr (reference)	No. of subjects (sex, sample)	Prev (%)	Sens (%)	Spec (%)	Expanded standard method[a]	PPV (%)	NPV (%)
Bassiri et al., 1995 (6)	447 (F, cx[b])	3	87	100	MOMP[c]	100	99
Lee et al., 1995 (82)[d]	1,937 (F, urine)	8	94	99	DFA, MOMP	99	99
Chernesky et al., 1994 (32)	447 (F, urine)	6	96	100	DFA, MOMP	100	99
	305 (M, urine)	18	96	100	DFA, MOMP	100	99
Chernesky et al., 1994 (34)[d]	542 (M, urine[e])	18	98	100	DFA, MOMP	100	99
	1,043 (M, urine)	14	63	66	DFA, MOMP	98	98
Schachter et al., 1994 (130)[d]	2,132 (F, cx[b])	11	94	100	DFA, MOMP	100	99

[a]Method added to culture for expanded definition of true positive.
[b]Endocervical-swab specimens.
[c]Second PCR performed with alternate target primers specific for the MOMP.
[d]Multicenter study.
[e]Urethral-swab specimen.

The next series of tables reviews selections of literature from 1995 to 2000. Table 15, which precedes the comparative studies, is the index organized by reference number, specimen type, and comparative assays. Note that some of the articles may not compare methods but rather provide information on analytical aspects of the assays or deal with other aspects of test selection for chlamydia assays. To use the index, locate the desired reference and turn to the indicated table that follows the index.

TABLE 15 Index to specific references comparing *C. trachomatis* assays

Reference[a]	Test assay(s)	Comparative assay(s)	Specimens and notes	Table(s)
25	LCR[b]	Gen Probe PACE 2 (direct DNA probe assay)	Endocervical and M urethral swabs; M and F urine	16, 17
35	LCR	Culture; DFA; PCR (Roche Amplicor [manual PCR assay])	M urine for amplification; M urethral swab for culture and DFA	18, 19
51	LCR	Culture	F urine; endocervical swab	20
65	LCR	Culture	Endocervical swab collected by physician; vaginal swab collected by patient	21, 22
40	LCR	PCR	Endocervical swab	23
96	LCR	PCR	Flow cytometry assay used to assess the sensitivities of LCR and PCR	24
148	LCR	EIA (Syva)	Endocervical swabs and F urine; impact of menstrual cycle on test performance	25, 26
150	LCR	EIA (Syva); DFA (urine)	LCR assay of vaginal swab compared to DFA and LCR assay of cervical and F urine	27, 28
141	LCR and COBAS AMPLICOR	EIA (Syva)	Comparison of assays in asymptomatic M	50
11	EIA (Syva)	None	Data justifying the need for confirmatory tests for EIAs	29
30	EIA (Syva)	EIA (Sanofi)	F endocervical and M urethral swabs	30, 31
107	EIA (Syva), DFA (Syva), EIA (Abbott Chlamydiazyme), GenProbe PACE 2	Culture	F endocervical swabs	32, 33, 34
110	GenProbe PACE 2	None	Review of the impact of shipping endocervical-swab specimens to a state health laboratory for testing	35
13	GenProbe PACE2	Culture	Urethral swabs from asymptomatic M patients	36
39, 49	GenProbe AMP-CT (TMA)	Culture	Endocervical and M urethral swabs; M and F urine	37, 38, 39, 40, 41
90			Delineation of inhibitors of amplification assays using urine as specimens with methods used to remove inhibitors	43
54	GenProbe AMP-CT	LCR; PCR (COBAS AMPLICOR); culture	Endocervical and M urethral swabs; M and F urine	44, 45
111	GenProbe AMP-CT	PCR (COBAS AMPLICOR)	M and F urine	46
142	GenProbe AMP-CT	LCR; culture	Endocervical swab; vulval swab; M urethral swab; M and F urine	47, 48
143	PCR (COBAS AMPLICOR)	LCR; EIA (Syva MicroTrak)	M urine	49
134	PCR (COBAS AMPLICOR)	LCR; culture; EIA (Syva MicroTrak)	Urine for amplification assays; endocervical and M urethral swabs for culture and EIA	51
137	PCR (Amplicor)	PCR (COBAS AMPLICOR); LCR; culture; in-house PCR assay using purified DNA	Endocervical and M urethral swab	53
154	PCR (COBAS AMPLICOR)	Culture	Endocervical and M urethral swabs; M and F urine	54, 55

(Continued on next page)

TABLE 15 Index to specific references comparing C. *trachomatis* assays (*Continued*)

Reference	Test assay(s)	Comparative assay(s)	Specimens and notes	Table(s)
85	ProbeTec ET-CT (Becton-Dickinson; amplification by SDA for C. *trachomatis*)	LCR	Endocervical and M urethral swabs; M and F urine	56
30	EIA (Sanofi-Pasteur)	EIA (Syva MicroTrak II); culture	Endocervical and M urethral swabs; M and F urine	57
133	Vidas (bioMerieux; EIA)	Culture	Endocervical and M urethral swabs; M and F urine	58
57	Vidas	PCR (Amplicor); EIA (Abbott Chlamydiazyme)	Endocervical and M urethral swabs; M and F urine	59
122	PCR (COBAS AMPLICOR)	LCR; culture	Endocervical and M urethral swabs; M and F urine	60, 61
113	PCR (Amplicor CT)	PCR (COBAS AMPLICOR); LCR; culture	F urine	62
116	PCR (Amplicor CT)	EIA (Syva MicroTrak)	F urine (Amplicor); endocervical swab (MicroTrak)	63
40	PCR (Amplicor CT)	LCR	Endocervical swab	64
109	PCR (Amplicor CT/NG)	EIA (Syva MicroTrak); DFA	M urine (Amplicor); M urethral swab (DFA; MicroTrak)	65
114	PCR (Amplicor CT)	Gen Probe PACE 2; culture	F urine (Amplicor); endocervical swabs (all assays)	66
119	PCR (Amplicor CT)		Reproducibility problems with the assay (early version of the Amplicor assay)	52
156	EIA (Beckman Access [C. *trachomatis*])	LCR	Endocervical swab	69
7	PCR (Amplicor CT/NG)	None	Performance study of assay using F urine	70
38	PCR (Amplicor CT/NG)	Culture	Endocervical and M urethral swabs; M and F urine	71
74	PCR (Amplicor CT)	Culture	Conjunctival, pharyngeal, and urethral swabs	72
61	PCR (COBAS AMPLICOR)	Radioimmune dot blot for C. *trachomatis*	M urine and M urethral swab	73
165	PCR (Amplicor CT)	GenProbe PACE2; EIA (Abbott Chlamydiazyme)	Endocervical swab	75
41	GenProbe PACE 2	LCR	Endocervical swab	74
104	GenProbe AMP-CT	Culture	Endocervical and M urethral swabs; M and F urine	76
112	GenProbe AMP-CT	PCR (Amplicor CT); culture	Urine for TMA and PCR; Endocervical swabs for culture	77
22	Abbott IMx (EIA)	Culture	Endocervical swab	78
53	DiGene-Hybrid Capture II CT-ID DNA	Culture; PCR (Amplicor CT)	Endocervical swab	79
97	DiGene-Hybrid Capture II CT-ID DNA	GenProbe PACE2	Endocervical and M urethral swabs	80
115	BIOSTAR (OIA)	PCR (Amplicor CT); DFA; culture	Endocervical swab	81
125	BIOSTAR (OIA)	Culture	Ocular swabs (infants)	82
163	BIOSTAR (OIA)	LCR	Endocervical swabs	83
23	LCR	Culture	Endocervical and M urethral swabs; M and F urine	84, 85
152	PCR (Amplicor); PCR (COBAS AMPLICOR)	Culture	Endocervical and M urethral swabs; M and F urine	67, 68
81	Clearview (EIA); GenProbe PACE 2	LCR; GenProbe AMP-CT	Endocervical swabs for all assays; F urine for amplification assays	88, 89

TABLE 15 (*Continued*)

Reference	Test assay(s)	Comparative assay(s)	Specimens and notes	Table(s)
J. C. Lovchick, L. M. Peralta, C. M. Brown, and D. M. Hilligoss, *Abstr. 99th Gen. Meet. Am. Soc. Microbiol.*, abstr. C-119, 1999	BD ProbeTec ET-CT	LCR; culture, DFA	Endocervical swabs; F urine	86
D. V. Ferrero, L. Buck-Barrington, H. Meyers, G. S. Hall, M. Tuohy, D. Wilson, J. Schacter, J. Moncada, and F. Pang, *Abstr. 99th Gen. Meet. Am. Soc. Microbiol.*, abstr. C-120, 1999				87
68			Computer-based decision analysis justifying the use of nucleic acid amplification assays for C. *trachomatis* in prevention of PID	42
102			Mailed, home-obtained urine samples for detection of C. *trachomatis* by amplification assays	90
56			Reproducibility problems with the Abbott LCR assay	91

[a]As test methods are grouped together, the references are not in sequential order.
[b]Amplification by LCR, Abbott Diagnostics.

TABLE 16 LCR and GenProbe PACE 2[a]

Test assay	Comparison assay	Specimen type	n	Prev (%)	Sens	Spec	PPV	NPV	Expanded standard	Other
LCR	GenProbe PACE 2	F, endocervical swab, urine; M, urethral swab, urine	562	16.2	See Table 17	See Table 17	NA	NA	MOMP-based PCR	Positive defined as positive reaction in all 3, or 2 of 3, assays. Patient population, STD clinic.

[a]From reference 25 with permission.

TABLE 17 Summary of sensitivity and specificity of LCR and GenProbe PACE 2[a]

Pathogen	Total no. of samples tested	Total no. of positive samples	Assay	Sens (%)						Spec (%)					
				Before supplemental data analysis			After supplemental data analysis			Before supplemental data analysis			After supplemental data analysis		
				Overall	M	F	Overall	M	F	Overall	M	F	Overall	M	F
C. trachomatis	562	91	Gen-Probe PACE 2	75.9	70.2	84.4	65.9	62.3	71.1	100	100	100	100	100	100
			LCR (swab specimen)	100	100	100	96.7	96.2	97.4	98.1	98.3	97.9	100	100	100
			LCR (urine specimen)	100	100	100	90.1	92.5	86.6	99.4	99.1	99.6	100	100	100

[a]From reference 25 with permission.

TABLE 18 LCR, culture, DFA (Syva), and PCR (Amplicor)[a]

Test assay	Comparison assay	Specimen type	n	Prev	Sens	Spec	PPV	NPV	Expanded standard	Other
LCR	Culture, DFA (Syva), PCR (Amplicor)	M urine	287	See Table 19	See Table 19	See Table 19	See Table 19	See Table 19	MOMP-based PCR or LCR	Patient population, men in STD clinic. Inhibitors detected in 1 of 2 LCR specimens, both of which were positive by PCR, 1 false positive by LCR; hands-on time calculated as 60–90 min for LCR and 90–120 min for PCR.

[a]From reference 35 with permission.

TABLE 19 Prevalence, sensitivity, and specificity of laboratory assays for C. trachomatis[a]

Specimen	Test	% Prev	% Sens[b]	% Spec	PPV	NPV
Urethral swab	Culture	4.5 (13/287)	37.1 (13/35)	100 (252/252)		
	DFA	5.2 (15/287)	42.9 (15/35)	100 (252/252)		
FVU	LCR[c]	11.5 (33/287)	94.3 (33/35)	99.6 (251/252)	97.1 (33/34)	99.2 (251/313)
	PCR[d]	12.2 (35/287)	100 (35/35)	98.0 (247/252)	87.5 (35/40)	100 (247/247)

[a]Values in parentheses are the numbers of samples with positive test results divided by the numbers of samples tested. In all, samples were collected from 287 men. Data from reference 35 with permission.
[b]Calculations are based on 35 men determined by culture or confirmed nonculture assay to be infected with C. trachomatis.
[c]LCR kit (Abbott Laboratories).
[d]Amplicor Chlamydia test (Roche Molecular Systems).

TABLE 20 LCR and culture with female urine[a]

Test assay	Comparison assay	Specimen type	n	Prev (%)	Sens (%)	Spec (%)	PPV (%)	NPV (%)	Expanded standard	Other
LCR	Culture	F urine	465	7.3	88.6	99.7	96.9	99.0	DFA of culture sediment; PCR and MOMP-based LCR	Patient population, asymptomatic military women; inhibitors detected in 3 specimens

[a]Data from reference 51.

TABLE 21 LCR and culture with vaginal and cervical specimens and expanded standard[a]

Test assay	Comparison assay	Specimen type	n	Sens	Spec	PPV	NPV	Expanded standard	Other
LCR	Culture	Endocervical swabs (clinician obtained) Vaginal swabs (collected by patients)	309	See Table 22	See Table 22	See Table 22	See Table 22	DFA of culture transport sediment; MOMP-based LCR	Patient population, F from STD clinic

[a]Data from reference 65.

TABLE 22 Performance of cell culture and LCR for detection of C. trachomatis in patient-obtained vaginal and clinician-obtained cervical specimens[a]

Assay and specimen source	Sens (%)	Spec (%)	PPV (%)	NPV (%)
Culture				
Patient-obtained vaginal swabs	32.7	100	100	88.7
Clinician-obtained endocervical swabs	79.6	100	100	96.3
LCR				
Patient-obtained vaginal swabs	91.8	99.6	97.8	98.5
Clinician-obtained endocervical swabs	98.8	100	100	98.1

[a]Data from reference 65 with permission.

TABLE 23 LCR and PCR with swab specimens and expanded standard[a]

Test assay	Comparison assay	Specimen type	n	Prev (%)	Sens (%)	Spec (%)	PPV (%)	NPV (%)	Expanded standard	Other
LCR	PCR (Amplicor CT)	F endocervical swab	486	8.6	97.6 (LCR) 90 (PCR)	100 (LCR) 99.5 (PCR)	100 (LCR) 95 (PCR)	99.8 (LCR) 99.1 (PCR)	Repeat testing; MOMP-based amplification infection	Patients, ob-gyn population at risk for C. trachomatis

[a]Data from reference 40.

TABLE 24 LCR and PCR with flow cytometry[a]

Test assay	Comparison assay	Comments
LCR	PCR (Amplicor CT)	Using particular counting analysis by flow cytometry, both LCR and PCR had detection limits of 2 C. trachomatis elementary bodies.

[a]Data from reference 96.

TABLE 25 LCR and EIA with urine and swab specimens and expanded standard[a]

Test assay	Comparison assay	Specimen type	n	Prev	Sens	Spec	PPV	NPV	Expanded standard	Other
LCR	EIA (Syva MicoTrak)	F urine and endocervical swab	388	See Table 26	NA	NA	NA	NA	NA	Patients, F attending genitourinary clinic; assessment of impact of menstrual cycle on detection of *C. trachomatis*

[a]Data from reference 148.

TABLE 26 Prevalence of *C. trachomatis* and sensitivity of tests by stage of menstrual cycle[a]

Test	Site	Proportion (%) of all women with positive specimens		Sens (%) of test			
				Overall population		Low-risk population	
		Wks 1–3	Wk 4	Wks 1–3	Wk 4	Wks 1–3	Wk 4
EIA	Cervix	18/273 (6.6)	10/114 (8.8)	64	66	56	44
DFA	Cervix	26/273 (9.5)	14/114 (12)	93	100	94	100
LCR	Cervix	27/273 (9.9)	14/114 (12)	96	100	94	100
DFA	Urine	26/273 (9.5)	13/114 (11.4)	93	93	100	88
LCR	Urine	24/271 (8.4)	13/114 (11.4)	89	100	86	100

[a]Table reprinted from reference 148 with permission of BMJ Publishing.

TABLE 27 LCR and EIA with urine and swab specimens[a]

Test assay	Comparison assay	Specimen type	n	Prev (%)	Sens	Spec	PPV	NPV	Other
LCR	EIA (Syva Microtrak) and DFA	F urine and endocervical swabs	413	11.6	See Table 28	See Table 28	See Table 28	See Table 28	Patients, F from genitourinary clinic; assessment of value of vaginal swabs in amplification assays for diagnosis of *C. trachomatis* infection

[a]Data from reference 150.

TABLE 28 Comparison of results of LCR assay of vaginal-swab samples with those of DFA staining and the LCR assay of cervical samples and urine samples and an EIA of cervical samples from 46 C. trachomatis-positive women[a]

LCR assay results for vaginal swabs	No. of samples with indicated result[b]									
	DFA test of cervical deposit (n = 45)		LCR assay of cervical sample (n = 45)		DFA test of urine deposit (n = 43)		LCR assay of urine deposit (n = 43)		EIA of cervical sample (n = 44)	
	+	−	+	−	+	−	+	−	+	−
+	40	1	41	0	36	3	36	3	28	13
−	3	1	3	1	4	0	4	0	2	1
Sens	40/43 (93%)		41/44 (93%)		36/40 (90%)		36/40 (90%)		28/30 (93%)	
Adjusted sens	41/44 (93%)		41/44 (93%)		39/43 (91%)		39/43 (91%)		41/43 (95%)	

[a]Data from reference 150.
[b]Results are the totals in each vertical column (e.g., DFA result for cervical deposit, 43 positive and 2 negative). LCR vaginal-sample results are shown in the horizontal rows (e.g., for the 45 women tested by DFA with cervical deposits, 41 were positive and 4 were negative).

TABLE 29 EIA[a]

Test assay	Comments
EIA (Syva MicroTrak)	Confirmatory tests of blocking antibody and PCR for all initial positives brought the rate of initial positives down by 8%. There were 14.7% false positives on initial testing. This report supports the CDC recommendation to verify all positives in low-prevalence populations.

[a]Data from reference 11.

TABLE 30 EIA with swab samples[a]

Test assay	Comparison assay	Specimen type	n	Sens	Spec	PPV	NPV	Other
EIA (Syva MicroTrak)	EIA (Sanofi Microwell [short assay]) and culture	Endocervical swab M urethral swab	See Table 31	See Table 31	See Table 31	See Table 31	See Table 31	Cytocentrifugation and DFA for all EIA-positive specimens and for specimens that fell in the range of 30% below the cutoff; blocking assays for both EIA procedures

[a]Data from reference 30 with permission.

TABLE 31 Sensitivity, specificity, and predictive values of EIAs in comparison with the resolved standard[a]

Sex	Test	Group	No. of specimens	Sens (%)	Spec (%)	PPV (%)	NPV (%)
F	Syva EIA	Overall	732	94.5	99.6	94.5	99.6
		High risk	242	100.0	99.1	92.6	100.0
		Low risk	490	90.0	99.8	96.4	99.4
	Sanofi EIA	Overall	732	94.9	100.0	100.0	99.6
		High risk	242	96.2	100.0	100.0	99.5
		Low risk	490	93.9	100.0	100.0	99.6
M	Sanofi EIA	Overall	121	100.0	100.0	100.0	100.0
		Asymptomatic	96	100.0	100.0	100.0	100.0
		Symptomatic	25	100.0	100.0	100.0	100.0

[a]Data from reference 30 with permission.

TABLE 32 Nonculture and culture tests with expanded standard[a]

Test assay	Comparison assay	n	Prev	Sens	Spec	PPV	Expanded standard	Other
Head-to-head evaluation of EIA (Syva MicroTrak), DFA (Syva), EIA (Abbott Chlamydiazyme), and GenProbe PACE 2	Culture	4,980	See Table 33	See Table 33	See Table 33	See Table 33	Blocking assays for EIA and DFA assays; probe competition assay for GenProbe PACE 2	Population, F attending family-planning clinic

[a]Data from reference 107 with permission.

TABLE 33 Performance characteristics of nonculture tests relative to culture before confirmation of positive results

Test	Sens[a] (%)	95% CI (%) for sens	Spec[a] (%)	95% CI (%) for spec	No. of false positives	PPV[b] (%)
GenProbe	75.3 (146/194)	68.6–81.2	99.6 (4,694/4,712)	99.4–99.8	18	89
Syva DFA	74.5 (140/188)	67.6–80.5	99.8 (4,429/4,438)	99.6–99.9	9	94
Syva EIA	71.7 (139/194)	64.8–77.9	99.6 (4,696/4,713)	99.4–99.8	17	89.1
Sanofi EIA	66.8 (129/193)	59.7–73.4	99.6 (4,703/4,720)	99.4–99.8	17	88.4
Abbott EIA	61.9 (120/194)	55.6–68.7	99.6 (4,692/4,713)	99.3–99.7	21	85.1

[a]True positive equals culture positive. The data were generated by using the manufacturer's recommended cutoffs and results prior to confirmation. Values are reported as percents (number of specimens positive/total number of specimens). Data from reference 107 with permission. [b]Prevalence of 3.9% based on cell culture.

TABLE 34 Performance characteristics of culture and nonculture tests after confirmation of positive results[a]

Test (confirmatory test)	Sens[b]	95% CI (%) for sens	Spec[b] (%)	95% CI (%) for spec	No. of false positives	PPV[c] (%)
GenProbe (Sanofi DFA)	75.3 (146/194)	68.6–81.2	99.92 (4,708/4,712)	99.78–99.98	4	97.3
Syva EIA (blocking antibody)	70.6 (137/194)	63.7–76.9	99.96 (4,710/4,712)[d]	99.85–99.99	2	98.6
Abbott EIA (Syva DFA)	61.9 (120/194)	53.9–68.1	99.94 (4,712/4,715)[e]	99.81–99.99	3	97.5

[a]Data from reference 107 with permission.
[b]True positive equals culture positive. The data were generated by using the manufacturer's recommended cutoffs and results prior to confirmation. Values are reported as percents (numbers of specimens positive/total numbers of specimens).
[c]Prevalence of 3.9% based on cell culture.
[d]No confirmatory test was performed on one specimen that was dropped from the analysis.
[e]No confirmatory test was performed on five specimens that were dropped from the analysis.

TABLE 35 PACE 2 with different specimen delivery methods[a]

Test assay	Comparison assay	Specimen type	n (Prev [%])	Comments	Expanded standard	Other
GenProbe PACE 2	Duplicate specimens transported by courier and U.S. mail	Endocervical swab	990 (8.9)	Shipping by mail vs. courier did not change results in 99% of specimens, even with temperatures exceeding the manufacturer's specifications in about 50% of samples. Time to receipt and temperature of >31°C were monitored. Median time for courier delivery of specimens to accessioning was 1 day; range, 1–6 days Median time for specimens by U.S. mail was 2 days; range, 1–29 days	PCR	State Health Laboratory specimens

[a]Data from reference 110.

TABLE 36 PACE 2 and culture[a]

Test assay	Comparison assay	Specimen type	n	Prev (%)	Sens (%)	Spec (%)	PPV (%)	NPV (%)	Other
GenProbe PACE 2	Culture	M urethral swabs	196	16	84	99	93	97	Patient population, asymptomatic M

[a]Data from reference 13.

TABLE 37 AMP-CT and culture[a]

Test assay	Comparison assay	Specimen type	n	Sens	Spec	PPV	NPV	Expanded standard	Other
GenProbe (AMP CT)	Culture	Endocervical swabs	479	See Table 38	See Table 38	See Table 38	See Table 38	DFA; alternate target-based TMA (16S rRNA)	Patient population, M and F attending STD clinic; discussion of advantages of TMA; 28 specimens were TMA positive and culture negative.
		F urine	480						
		M urine	464						

[a]Data from reference 39 with permission.

TABLE 38 Resolved performance characteristics for detection of C. *trachomatis* by AMP-CT for gender and specimen type[a]

Sex	Test	Sample type	No. of samples	Resolved performance			
				Sens (%)	Spec (%)	PPV (%)	NPV (%)
F	AMP CT	Endocervical swab	479	100.0	99.5	97.0	100.0
	AMP CT	Urine	480	93.8	100.0	100.0	99.0
	AMP CT	Matched urine and endocervical swab	474	95.7	99.8	98.5	99.3
	Culture	Endocervical swab	474	54.3	100.0	100.0	92.7
M	AMP CT	Urine	464	95.6	98.7	92.9	99.2
	Culture	Urethral swab	464	58.8	100.0	100.0	93.4

[a]Data from reference 39 with permission.

TABLE 39 AMP-CT and culture[a]

Test assay	Comparison assay	Specimen type (n)	Prev	Sens	Spec	PPV	NPV	Expanded standard	Other
GenProbe AMP-CT	Culture	Endocervical swab, (717); F urine (607); M urine (193); M urethral swab (209)	See Table 40	See Table 40	See Table 40	See Table 40	See Table 40	Reculture; DFA; alternate-target-based TMA	Population, STD and family-planning clinic

[a]Data from reference 49 with permission.

TABLE 40 Performance characteristics for AMP-CT before resolution by discrepant analysis (initial results)[a]

Sample type	Sens (%)	Spec (%)	Prev (%)	PPV (%)	NPV (%)
Urine specimens					
M	83.3	95.4	15.3	76.6	99.9
F	92.3	98.6	5.4	72.8	99.9
Swab specimens					
M	100.0	94.1	15.3	75.4	100.0
F	100.0	98.7	5.4	81.5	100.0

[a]Data from reference 49 with permission.

TABLE 41 Performance characteristics for AMP-CT after resolution by discrepant analysis[a]

Sample type	Sens (%)	Spec (%)	Prev (%)	PPV (%)	NPV (%)
Urine specimens					
M	88.5	100.0	15.3	100.0	99.9
F	93.5	99.5	5.4	88.4	99.9
Combined results	91.2	99.6	7.7	95.0	99.9
Swab specimens					
M	100.0	100.0	15.3	100..0	100.0
F	100.0	100.0	5.4	100.0	100.0
Combined results	100.0	100.0	7.7	100.0	100.0

[a]From reference 49 with permission.

TABLE 42 Amplification[a]

Test assay	Comments
Amplification	A computer-based decision analysis model was developed to evaluate and compare the cost and disease outcomes associated with seven screening strategies. The authors present data to conclusively demonstrate that compared to EIA testing, the strategy using a combination of DNA amplification for women receiving pelvic examinations and DNA amplification of urine for women with no medical indications requiring a pelvic examination prevents more cases of PID and gives higher cost savings.

[a]Data from reference 68.

TABLE 43 Delineation of inhibitors of amplification using urine specimens[a]

Assay	Inhibitory substance	Removal of inhibitors
TMA	Hemoglobin; nitrites; crystals	Dilute 1:1; perform phenol-chloroform extraction. Dilute 1:4 or 1:10; incubate overnight at 4°C or −70°C.[b]
LCR	Nitrites	Dilute 1:1; perform phenol-chloroform extraction. Dilute 1:4; incubate overnight at 4°C. Dilute 1:10; incubate overnight at −70°C.
PCR	Beta human chorionic gonadotropin; crystals	Dilute 1:1; perform phenol-chloroform extraction. Dilute 1:4 or 1:10; incubate overnight at 4°C or −70°C.[b]

[a]Data from reference 90.
[b]Not all specimens were rendered noninhibitory by this method.

TABLE 44 AMP-CT, PCR, LCR, and culture[a]

Test assay	Comparison assay	Specimen type	n	Sens	Spec	PPV	NPV	Expanded standard	Other
GenProbe AMP-CT	PCR, LCR, and culture	Endocervical swab; F urine; M urethral swab; M urine	544 (M), 456 (F)	See Table 45	See Table 45	See Table 45	See Table 45	In-house PCR	Population, STD clinic

[a]Data from reference 54 with permission.

TABLE 45 Detection of true *C. trachomatis* infections by cell culture and amplification assays[a]

Method	No. (%) of patients positive[b]		% Sens (CI95)[c]	% Spec
	F (n = 57)	M (n = 66)		
Cell culture	37 (64.9)	34 (51.5)	55.7 (48.4–66.6)	99.3
LCR	45 (78.9)	58 (87.9)	83.7 (75.9–89.9)	99.9
AMP CT	44 (77.2)	61 (92.4)	85.4 (77.8–91.1)	99.1
COBAS AMPLICOR	50 (87.7)	64 (97)	92.7 (86.6–96.6)	99.4

[a]Truly infected patients (n = 23) are defined as patients whose urine specimens tested positive by two or more techniques or whose endocervical-swab or urethral-swab specimens were confirmed by the in-house PCR. Data from reference 54 with permission.
[b]No statistical significance was obtained for COBAS AMPLICOR versus LCR or AMP CT after analysis of the male or female results separately.
[c]Statistical significance was obtained for COBAS versus LCR ($P = 0.047$). No statistical significance was obtained for COBAS versus AMP CT ($P = 0.101$).

TABLE 46 AMP-CT (TMA) and COBAS AMPLICOR (PCR)[a, b]

Test assay	Comparison assay	Specimen type	n	Prev (%)	Expanded standard	Other
GenProbe AMP CT (TMA)	COBAS AMPLICOR (PCR)	Urine	320 (M) 338 (F)	11.2	Retest of both the original specimen and a 1:10 dilution of the specimen; alternate-target TMA; MOMP-based PCR	PCR gave 2 false-negative and 6 false-positive results in the initial testing; 9 (1.5%) specimens had inhibitors that were resolved on retesting or by diluting urine 1:10. Patient population, outpatient clinics.

[a]Data from reference 111.
[b]Results after confirmatory testing were as follows: positive by PCR and TMA, 72; positive by PCR and negative by TMA, 0; negative by PCR and positive by TMA, 2; negative by both tests, 584.

TABLE 47 AMP-CT (TMA), LCR, and culture[a]

Test assay	Comparison assay	Specimen type	n	Prev (%)	Results	Expanded standard	Other
GenProbe AMP CT (TMA)	LCR; culture	F, endocervical swab; urine; vulva; M, urethra, urine	308 (F) 240 (M)	8.1 (F) 18.3 (M)	See Table 48	DFA; alternate-target TMA	Population, STD clinic

[a]Data from reference 142 with permission.

TABLE 48 Performance characteristics of TMA assay, LCR, and culture for the detection of C. trachomatis in urogenital specimens[a]

Specimen type	Sens (%)			Spec (%)			PPV (%)			NPV (%)		
	TMA	LCR	Culture	TMA	LCR	Culture	TMA	LCR	Culture	TMA	LCR	Culture
Endocervical	88.0	92.0	52.0	99.6	99.6	100.0	95.7	95.8	100.0	99.9	99.3	95.9
Vulvar	92.0	92.0	8.0	99.6	100.0	100.0	95.8	100.0	100.0	99.3	99.3	92.5
Urine (F)	76.0	96.0[b]	<ND>	99.3	100.0	ND	90.5	100.0	ND	97.9	99.6	ND
Urethral (M)	93.2	93.2	50.0	99.0	99.5	100.0	95.3	97.6	100.0	98.5	98.5	89.9
Urine (M)	88.6	86.4	<ND>	99.0	99.5	ND	95.1	97.4	ND	97.5	97.0	ND

[a]Data from reference 142 with permission.
[b]Significant (P = 0.0253). All other differences of sensitivity between TMA and LCR are not significant.

TABLE 49 COBAS AMPLICOR and LCR[a]

Test assay	Comparison assay	Specimen type	n	Prev (%)	Sens (%)	Spec (%)	PPV (%)	NPV (%)	Expanded standard	Other
COBAS AMPLICOR	LCR	Endocervical swab; M urine	302 (98 F, 204 M)	15.3 (F) 13.2 (M)	100 (COBAS) 83.3 (97.6)[b] (LCR)	99.6 (COBAS) 100 (LCR)	97.6 (COBAS) 100 (LCR)	100 (COBAS) 97.3 (99.6)[b] (LCR)	MOMP-based PCR	High-risk patients from STD clinic. Inhibitors to amplification detected in 2% of urine specimens and 20% of endocervical swabs. Specimens demonstrating inhibitors were diluted 1:9 and retested; 100% agreement with 15 positive endocervical swabs in all methods. Of 8 PCR-positive, LCR-negative specimens, freeze-thaw and dilution 1:9 resolved 7 of 8 discrepancies.

[a]Data from reference 143.
[b]Revised value after freeze-thaw and dilution of LCR urine specimens.

TABLE 50 LCR, COBAS AMPLICOR, and EIA[a]

Test assay	Comparison assay	Specimen type	n	Prev (%)	Sens (%)	Spec (%)	PPV (%)	NPV (%)	Expanded standard	Other
LCR and COBAS AMPLICOR (PCR)	EIA (Syva Microtrak)	M urine (asymptomatic patients)	705	4.1	93.1 (LCR) 62.1 (PCR) 37.9 (EIA)	100 (LCR) 99.6 (PCR) 98.1 (EIA)	100 (LCR) 85.7 (PCR) 45.8 (EIA)	99.9 (LCR) 98.4 (PCR) 97.4 (EIA)	Retesting of original sample DFA confirmation of EIA results DFA and MOMP-based LCR	Military hospital (United Nations) 29 positive specimens

[a]Data from reference 141.

TABLE 51 LCR, COBAS AMPLICOR (PCR), culture, and EIA[a]

Test assay	Comparison assay	Specimen type	n	Prev (%)	Sens (%)	Spec (%)	PPV (%)	NPV (%)	Expanded standard	Other
LCR and COBAS AMPLICOR (PCR)	Culture and EIA (Syva MicroTrak)	Urine for amplification assay; endo-cervical swabs and urethral swabs for culture and EIA	1,005 (418 F; 587 M)	3.8 (F) 4.1 (M)	95 (PCR)[b] 75 (LCR)[b] 58 (culture) 45 (EIA)	99.9 (PCR)[b] 100 (LCR)[b] 100 (culture) 100 (EIA)	97.4 (PCR)[b] 100 (LCR)[b] 100 (culture) 100 (EIA)	99.8 (PCR)[b] 99 (LCR)[b] 98 (culture) 98 (EIA)	MOMP-based PCR	Positive defined as culture positive or positive in any 2 independent tests. Patient population, STD clinic. All EIAs confirmed with a blocking assay. In specimens seeded with *C. trachomatis*, inhibitors detected in 16 and 26% of PCR and LCR samples, respectively.

[a]Data from reference 134.
[b]Urine.

TABLE 52 Amplicor CT (PCR)[a]

Test assay	Comment
Amplicor CT (PCR)	In an evaluation for an Ag detection test for *C. trachomatis*, 591 sets of cervical samples were studied. Thirty-five specimens were retested due to equivocal PCR results (n = 19) or discrepancy between the results of culture, PCR, and an Ag assay. For 13 of these specimens, the PCR assay value as measured by OD varied dramatically. When tested in triplicate with primers to the chlamydia plasmid and in duplicate with primers to the outer membrane, only 3 of 13 samples gave the same interpretation with these 5 replicates. The Amplicor assay used in this citation was an early version (generation) of the assay and has been modified. The inability to detect inhibitors in the sample with this version of the assay may have influenced the results.

[a]Data from reference 119.

TABLE 53 Amplicor (PCR), COBAS AMPLICOR (PCR), LCR, culture, and nested PCR[a]

Tests	Comparison assay	Specimen type	n	Prev (%)	Sens (%)	Spec (%)	PPV (%)	NPV (%)	Expanded standard	Other
Amplicor (PCR), COBAS AMPLICOR (PCR), LCR	Culture and an in-house nested PCR using purified DNA	Endocervical swab; M urethral swab	245	9.3	87.5 (COBAS) 82 (Amplicor) 82 (LCR) 78 (culture)	99.5 (COBAS) 99 (Amplicor) 100 (LCR) 100 (culture)	91.3 (COBAS) 82.6 (Amplicor) 82.6 (LCR) 78.2 (culture)	99.1 (COBAS) 98.2 (Amplicor) 98.2 (LCR) 97.8 (culture)	DFA; MOMP-based PCR. The comparative assay using purified DNA is novel and demonstrated an 8% increase in sens over commercial PCR.	Patient population, genitourinary clinic and a local general practice; 23 positives. Positive defined as (i) culture positive, (ii) culture negative but DFA positive, or (iii) positive in 2 or more of the other assays. Titration of elementary bodies demonstrated manual Amplicor, COBAS AMPLICOR, and LCR assays all had detection limits of ≥4 elementary bodies.

[a]Data from reference 157.

TABLE 54 COBAS AMPLICOR and culture[a]

Test assay	Comparison assay	No. of specimens (type)	Prev	Sens	Spec	PPV	NPV	Expanded standard	Other
COBAS AMPLICOR	Culture	2,014 (endocervical swab) 1,278 (F urine) 373 (M urethra) 254 (M urine)	See Table 55	See Table 55	See Table 55	See Table 55	See Table 55	MOMP-based PCR	A total of 2.9% of specimens demonstrated inhibitors to PCR on initial testing. Patients were enlisted from family-planning clinics, general practitioners, dermatologists, ob-gyn, and STD clinics. Multicenter clinical trial.

[a]Data from reference 154 with permission.

TABLE 55 Test results by specimen type[a]

Specimen source and type	Sens (%)[b]	Spec (%)[c]	PPV (%)[d]	NPV (%)[e]
F				
Swab	96.5 (83/86)	99.4 (1,912/1,924)	87.4 (83/95)	99.8 (1,912/1,915)
Urine	95.1 (39/41)	99.8 (1,204/1,207)	92.9 (39/42)	99.8 (1,204/1,206)
M				
Swab	100.0 (41/41)	98.5 (325/330)	89.1 (41/46)	100.0 (325/325)
Urine	94.4 (17/18)	100.0 (236/236)	100.0 (17/17)	99.6 (236/237)

[a]Sensitivity and specificity were calculated by taking the IC results into account. Results for all specimens that were initially COBAS AMPLICOR negative and IC negative for C. trachomatis were interpreted as positive, negative, or inhibitory on the basis of the result of the repeat COBAS AMPLICOR test. Data from reference 154 with permission.
[b]Values in parentheses are true positive/(true positive + false negative).
[c]Values in parentheses are true negative/(true negative + false positive).
[d]Values in parentheses are true positive/(true positive + false positive).
[e]Values in parentheses are true negative/(true negative + false negative).

TABLE 56 SDA and LCR[a]

Test assay	Comparison assay	Specimen type	n	Prev (%)	Sens (%)	Spec (%)	PPV (%)	NPV (%)	Other
SDA (ProbeTec)	LCR	Endocervical swab and urine (F) Urethral swab and urine (M)	122	8.1	100 (LCR) 100 (SDA)	100 (LCR) 100 (SDA)	100 (LCR) 100 (SDA)	100 (LCR) 100 (SDA)	Patient population, family practice clinic. Two urine specimens and 2 swab specimens were "equivocal" by SDA and not included in the tabulated data.

[a]Data from reference 85.

TABLE 57 EIA and culture[a]

Test assay	Comparison assay	Specimen type	Sens (%)	Spec (%)	PPV (%)	NPV (%)	Expanded standard	Other
Sanofi Pasteur (Microwell EIA) and MT (MicroTrak II EIA)	Culture	F endocervical swab (n = 732); M urethral swab (n = 121)	94.5 (MT, F) 94.9 (Sanofi, F) 100 (Sanofi, M)	99.6 (MT, F) 100 (Sanofi, F) 100 (Sanofi, M)	94.5 (MT, F) 100 (Sanofi, F) 100 (Sanofi, M)	99.6 (MT, F) 99.6 (Sanofi, F) 100 (Sanofi, M)	EIA blocking assay; DFA	Shortened protocol for Sanofi Pasteur validated in this article. Patient population, family clinic and STD clinic.

[a]Data from reference 30.

TABLE 58 Vidas and culture (TC)[a]

Test assay	Comparison assay	Specimen type	Prev (%)	Sens (%)	Spec (%)	PPV (%)	NPV (%)	Expanded standard
Vidas	Culture (TC)	F endocervical swabs (n = 2,453)	5.5	89.6 (Vidas) 94.1 (TC)	99.9	99.2	99.9	DFA or PCR
		M urethral swabs	10.3	90.9 (Vidas) 86.4 (TC)	99.9	98.8	99.9	
		M urine (n = 850)[b]	10.7[b]	81.2 (Vidas) 77.7 (TC)	100	100	97.5	

[a]Data from reference 133.
[b]Male urethral swabs plus urine.

TABLE 59 Vidas, Amplicor, and Chlamydiazyme[a]

Test assay	Comparison assay	Specimen type (n)	Assay	Sens (%)	Spec (%)	Prev (%)	PPV (%)	NPV (%)	Expanded standard	Other
Vidas	Amplicor (PCR) and Chlamydiazyme	Endocervical swab (330)	Vidas	63.2	100	5.7	100	97.8	MOMP-based PCR; EIA blocking assay	Patient population, STD clinic
			Chlamydiazyme	73.7	100		100	98.4		
			PCR	100	NA		NA	NA		
		M urine (100)	Vidas	68.8	100	19	100	94.4		
			Chlamydiazyme	87.5	100		100	97.7		
			PCR	93.8	NA		NA	NA		

[a]Data from reference 57.

TABLE 60 COBAS AMPLICOR (PCR), LCR, and culture[a]

Test assay	Comparison assay	Specimen	n	Prev	Sens	Spec	PPV	NPV	Expanded standard	Other
COBAS AMPLICOR (PCR) and LCR	Culture	F endocervical swab and urine; M urethral swab	1,015	See Table 61	See Table 61	See Table 61	See Table 61	See Table 61	LCR retested with urine diluted 1:10	

[a]Data from reference 122 with permission.

TABLE 61 Diagnostic performance of different samples in detection of urogenital C. trachomatis infection[a]

Patient sex	Test (specimen)[b]	No. of specimens positive/no. of specimens tested (%)	Sens (%)[c]	Spec (%)[c]	PPV (%)[c]	NPV (%)[c]	Other
F	Culture (swab)	19/449 (4.2)	67.9 (19/28)[d]	100 (421/421)	100 (19/19)	97.9 (421/430)	Population, STD clinic. Inhibitors in urine detected in 3% of samples for PCR; 2 false-negative urine results by LCR (representing 4% of the positive urine specimens)
	PCR (swab)	27/447 (6.0)	78.6 (22/28)	98.8 (414/419)	81.5 (22/27)	98.6 (414/420)	
	LCR (swab)	22/448 (4.9)	81.5 (22/27)	100 (421/421)	100 (22/22)	98.8 (421/426)	
	PCR (FVU)	28/442 (5.3)	96.4 (27/28)	99.8 (413/414)	96.4 (27/27)	99.8 (413/414)	
	LCR (FVU)	25/443 (5.6)	92.6 (25/27)	100 (406/406)	100 (25/25)	99.5 (406/408)	
M	Culture (swab)	42/565 (7.4)	85.7 (42/49)	100 (516/516)	100 (42/42)	98.7 (516/523)	
	PCR (swab)	52/565 (9.2)	100 (49/49)	99.4 (513/516)	94.2 (49/52)	100 (513/513)	
	LCR (swab)	45/557 (3.1)	100 (45/45)	100 (512/512)	100 (45/45)	100 (512/512)	
	PCR (FVU)	50/565 (8.8)	95.9 (47/49)	99.4 (513/516)	94.0 (47/50)	99.6 (513/515)	
	LCR (FVU)	45/560 (8.0)	93.8 (45/48)	100 (512/512)	100 (45/45)	99.4 (512/515)	
Combined	Culture (swab)	61/1,014 (6.0)	79.2 (61/77)	100 (937/937)	100 (61/61)	98.3 (937/953)	
	PCR (swab)	70/1,012 (7.8)	92.2 (71/77)	99.1 (927/935)	89.9 (71/79)	99.4 (927/933)	
	LCR (swab)	67/1,005 (6.7)	93.1 (67/72)	100 (933/933)	100 (67/67)	99.7 (926/929)	
	PCR (FVU)	78/1,007 (7.7)	96.1 (74/77)	99.6 (926/930)	94.9 (74/78)	99.7 (926/929)	
	LCR (FVU)	70/993 (7.0)	93.3 (70/75)	100 (918/918)	100 (70/70)	99.5 (918/923)	

[a]Data from reference 122 with permission.
[b]Swab, endocervical swab from females and urethral swab from males.
[c]Values in parentheses are numbers used to calculate sensitivity, specificity, PPV, and NPV.

TABLE 62 Amplicor CT (PCR), COBAS AMPLICOR, LCR, and culture[a]

Test assay	Comparison assay	Specimen	n	Prev (%)	Sens (%)	Spec (%)	PPV (%)	NPV (%)	Expanded standard	Other
Amplicor CT (PCR)	COBAS AMPLICOR; LCR; culture	F urine; endocervical swab (for culture only; serves as reference method)	442	11.3	100 (Amplicor) 94 (COBAS) 94 (LCR) 88 (culture)	99.7 (Amplicor) 99.2 (COBAS) 100 (LCR) 100 (culture)	98.0 (Amplicor) 92.2 (COBAS) 100 (LCR) 100 (culture)	100 (Amplicor) 99.2 (COBAS) 99.2 (LCR) 98.5 (culture)	MOMP-based amplification on specimens with results of cell culture negative and either PCR or LCR positive	

[a]Data from reference 113.

TABLE 63 Amplicor CT (PCR) and EIA[a]

Test assay	Comparison assay	Specimen	n	Prev (%)	Sens (%)	Spec (%)	PPV (%)	NPV (%)	Expanded standard	Other
Amplicor CT (PCR)	EIA (Syva MicroTrak)	F urine (Amplicor) and endocervical swab (Micro Trak)	1,090	5.6	85.2 (Amplicor) 90.2 (EIA and DFA)	99.9 (Amplicor) 100 (EIA and DFA)	98.1 (Amplicor) 100 (EIA and DFA)	99.9 (Amplicor) 99.4 (EIA and DFA)	DFA from EIA tubes or MOMP-based PCR	Population, asymptomatic females from family-planning clinic and university student health clinic; DFA performed on EIA specimens when ≥0.35 times cutoff. No specific data given on inhibitors of PCR in urine.

[a]Data from reference 116.

TABLE 64 LCR and Amplicor CT (PCR)[a]

Test assay	Comparison assay	Specimen type	n	Prev (%)	Sens (%)	Spec (%)	PPV (%)	NPV (%)	Expanded standard	Other
LCR	Amplicor CT (PCR)	Endocervical swab	486	8.6	97.6 (LCR) 90 (PCR)	100 (LCR) 99.5 (PCR)	100 (LCR) 95 (PCR)	99.8 (LCR) 99.1 (PCR)	Repeat testing; MOMP-based PCR and LCR	Patient population; ob-gyn clinic; 42 positive patients. Differences not statistically significant.

[a]Data from reference 40.

TABLE 65 Amplicor CT/NG, EIA, and DFA[a]

Test assay	Comparison assay	Specimen type	n	Prev (%)	Sens (%)	Spec (%)	PPV (%)	NPV (%)	Expanded standard	Other
Amplicor CT/NG	EIA (Syva MicroTrak); DFA (urethral swab)	M urine (Amplicor); M urethral swab (MicroTrak)	73	13.7	80 (Amplicor) 60.0 (EIA and DFA)	95.2 (Amplicor) 100 (EIA and DFA)	72.7 (Amplicor) 100 (EIA and DFA)	96.8 (Amplicor) 94.0 (EIA and DFA)	In-house PCR (target different from Amplicor's)	

[a]Data from reference 109.

TABLE 66 Amplicor CT, PACE 2, and culture[a]

Test assay	Comparison assay	Specimen type	n	Prev (%)	Sens (%)	Spec (%)	PPV (%)	NPV (%)	Expanded standard	Other
Amplicor CT	GenProbe PACE 2 and culture	F urine (Amplicor); cervical swabs (all assays)	666	5.9	82.0 (Amplicor, urine), 82.0 (Amplicor, swab), 79.5 (PACE 2), 84.6 (culture)	99.7 (Amplicor, urine), 99.8 (Amplicor, swab), 100 (PACE 2), 100 (culture)	94.1 (Amplicor, urine), 96.9 (Amplicor, swab), 100 (PACE 2), 100 (culture)	93 (Amplicor, urine), 98.8 (Amplicor, swab), 99.0 (PACE 2), 98.7 (culture)	Positive defined as (i) cell culture positive, (ii) PACE 2 and PCR positive confirmed with probe competition assay, or (iii) PCR positive only confirmed with MOMP-based PCR.	By eliminating inhibitors of PCR in urine by dilution, the sensitivity for PCR increases to 97.4%.

[a]Data from reference 114.

TABLE 67 Amplicor, COBAS AMPLICOR, and culture[a]

Test assay	Comparison assay	Specimen type	n	Prev (%)	Sens (%)	Spec (%)	PPV (%)	NPV	Expanded standard	Other
Amplicor and COBAS AMPLICOR	Culture	Endocervical and M urethral swab; M and F urine	1,134 (asymptomatic F), 1,173 (symptomatic F), 723 (asymptomatic M), 1,298 (symptomatic M)	2.4 (F), 7.2 (M)	See Table 68	See Table 68	See Table 68	See Table 68	Specimens positive by PCR and negative by culture were resolved by DFA testing of the pellet from centrifuged transport medium; if DFA was negative, a MOMP-based PCR was performed; If MOMP-based PCR was negative, the other specimen type was tested by MOMP-based PCR.	Detailed data given for COBAS AMPLICOR in this multicenter study with patients from STD clinics and family-planning clinics. Data calculated on a patient basis (including results from culture and all 4 PCR assays [urine and swab by both Amplicor and COBAS AMPLICOR]), and on a sample basis (demonstrating performance if only one specimen type was tested in one assay format). The table below represents the per sample results, which demonstrate higher sensitivity than the per patient basis. The Amplicor and COBAS AMPLICOR tests yielded concordant results for 98.1% of the specimens. One hundred and eight specimens were repeatedly inhibitory in either the Amplicor or the COBAS AMPLICOR assay.

[a]Data from reference 152 with permission.

TABLE 68 Sensitivity, specificity, and positive and negative predictive values for detection of *C. trachomatis* by COBAS AMPLICOR calculated on a sample basis[a]

Sex	Symptom	Specimen	n	Prev (%)	Sens (%)	Spec (%)	PPV (%)	NPV (%)
F	Asymptomatic	Endocervical	1,098	7.7	97.6	99.5	94.3	99.8
		Urine		8.1	93.3	99.6	95.4	99.4
	Symptomatic	Endocervical	1,138	9.1	96.1	99.2	92.5	99.6
		Urine		9.2	93.3	98.5	86.0	99.3
M	Asymptomatic	Urethral	710	11.7	98.8	98.7	91.1	99.8
		Urine		12.5	92.1	98.4	89.1	98.9
	Symptomatic	Urethral	1,230	19.0	97.4	98.7	94.6	99.4
		Urine		20.6	92.5	98.5	94.0	98.1

[a]Data from reference 152 with permission.

TABLE 69 ACCESS *Chlamydia trachomatis* and LCR[a]

Test assay	Comparison assay	Specimen type	n	Prev (%)	Sens (%)	Spec (%)	PPV (%)	NPV (%)	Expanded standard	Other
Beckman ACCESS *Chlamydia trachomatis*	LCR	Endocervical swab	356	8.7	83.9	99.7	96.3	98.5	MOMP-based LCR	Population, university emergency department. Good cost comparison data (total cost for ACCESS, $20.08; LCR, $17.83).

[a]Data from reference 156.

TABLE 70 Amplicor CT/NG and performance study of CT, PCR, and DFA[a]

Test assay	Comparison assay	Specimen type	Test results for 3,340 urine samples from women[b]					Expanded standard	Other
			No. of samples	PCR results		DFA result	No. of samples regarded as positive after resolution		
				Plasmid-based PCR[c]	MOMP-based PCR				
Amplicor CT/NG	Performance study with CT initial positives repeated by the plasmid-based PCR and MOMP-based PCR. Discrepancies resolved with DFA.	F urine	3,258	−	NT	NT	0	MOMP-based PCR, then DFA	Inhibitors detected in 1.8% of urine samples
			74	+,+	+	NT	74		
			6	+,+	−	+	6		
			1	+,−,−	+	+	1		
			1	+,−,−	+	−	0		

[a]Data from reference 7 with permission.
[b]The women were attending clinics for pregnancy termination or contraceptive advice and were tested for *C. trachomatis* by a plasmid-based PCR (Multiplex AMPLICOR CT/NG kit), and the results were resolved by an MOMP-based PCR and DFA.
[c]Multiple results are the results of consecutive runs.

TABLE 71 Amplicor CT/GC (PCR) and culture[a]

Test assay	Comparison assay	Specimen type	Resolved performance characteristics for detection of C. trachomatis by PCR for gender and specimen type					Expanded standard	Other
			Test sample	Per specimen		Per patient			
				Sens (%)	Spec (%)	Sens (%)	Spec (%)		
Amplicor CT/GC (PCR)	Culture	F urine and endocervical swab; M urethral swab and urine	M (n = 344) PCR					MOMP-based PCR; DFA	Patient population, STD clinic
			Urine	88.2	98.6	80.4	98.6		
			Urethral	96.2	99.3	91.1	99.3		
			Urethral culture	51.8	100	51.8	100		
			F (n = 192) PCR						
			Urine	100	99.4	82.4	99.4		
			Endocervical	100	100	94.1	100		
			Endocervical culture	55.9	100	55.9	100		

[a]Data from reference 38.

TABLE 72 Amplicor CT (PCR) and culture[a]

Test assay	Comparison assay	Specimen type	n	Prev (%)	Sens (%)	Spec (%)	PPV (%)	NPV (%)	Expanded standard	Other
Amplicor CT (PCR)	Culture	Conjunctiva (n = 258) Pharyngeal (n = 182) Urethral swabs (n = 75)	515	6.6	85.3 (culture) 100 (PCR) 82.4 (DFA)	100 (culture) 99.8 (PCR) 99.6 (DFA)	NA	NA	DFA	Population, ophthalmologic and venereologic clinics; 34 positive patients using DFA as expanded standard; 4 of 13 conjunctival specimens were positive only in PCR, not culture (3 were also DFA positive). One positive amplified conjunctival specimen was considered false positive.

[a]Data from reference 74.

TABLE 73 COBAS AMPLICOR CT/GC and radioimmune dot blot[a]

Test assay	Comparison assay	Specimen type	n	Prev (%)	Sens (%)	Spec (%)	PPV (%)	NPV (%)	Expanded standard	Other
COBAS AMPLICOR CT/GC	Radioimmune dot blot for *C. trachomatis*	M urine and urethral swab	390	15.4	90 (M urethra) 86.7 (urine)	100 (M urethra) 100 (urine)	NA	NA	MOMP-based PCR and DFA	Population, genitourinary clinic. Inhibitors detected in 0.77 and 0.25% of urethral and urine specimens, respectively.

[a]Data from reference 61.

TABLE 74 PACE 2 and LCR[a]

Test assay	Comparison assay	Specimen type	n	Prev (%)	Sens (%)	Spec (%)	PPV (%)	NPV (%)	Expanded standard	Other
GenProbe PACE 2	LCR	Endocervical swab	493	3.1	100 (PACE 2) 100 (LCR)	100 (PACE 2) 100 (LCR)	100 (PACE 2) 100 (LCR)	100 (PACE 2) 100 (LCR)	Probe competition assay (GenProbe)	Population, university health clinic, STD clinic, community health clinic, and women's health clinic

[a]Data from reference 41.

TABLE 75 PACE 2, EIA, and Amplicor CT[a]

Test assay	Comparison assay	Specimen type	n	Prev (%)	Comparison to Amplicor CT					Expanded standard	Other
					Assay	Sens (%)	Spec (%)	PPV (%)	NPV (%)		
GenProbe PACE 2 and EIA (Abbott Chlamydiazyme)	Amplicor CT	Endocervical swab	787	10.4	PACE 2 Chlamydiazyme	79.3 63.4	100 100	100 100	100 95.9	Retest; blocking Ab assay for Chlamydiazyme; probe competition assay for PACE 2	Population, rural diagnostic clinics. True positive defined as any 2 of the independent assays positive.

[a]Data from reference 165.

TABLE 76 AMP-CT (TMA) and culture[a]

Test assay	Comparison assay	Specimen type	n	Prev (%)	Sens (%)	Spec (%)	PPV (%)	NPV (%)	Expanded standard	Other
GenProbe AMP-CT (TMA)	Culture	F urine and endocervical swab; M urine and urethral swab	544 (M) 456 (F)	13	84.3 (AMP-CT, M) 100 (AMP-CT, F) 72.5 (culture)	98.8 (AMP-CT, M) 99.2 (AMP-CT, F) 99.2 (culture)	89.6 (AMP-CT, M) 93.1 (AMP-CT, F) 92.5 (culture)	98 (AMP-CT, M) 100 (AMP-CT, F) 98 (culture)	In-house PCR	Population, STD clinic

[a]Data from reference 104.

TABLE 77 AMP-CT (TMA), Amplicor (PCR), and culture[a]

Test assay	Comparison assay	Specimen	n	Prev (%)	Sens (%)	Spec (%)	PPV (%)	NPV (%)	Expanded standard	Other
GenProbe AMP-CT (TMA)	Roche Amplicor (PCR) and culture	Urine for TMA and PCR Endocervical swabs for culture	561 (F)	12.3	91.4, 98.6[b] (AMP-CT) 97.1, 100[b] (PCR) 85.7 (culture)	99.6 (AMP-CT) 99.8 (PCR) 100 (culture)	96.8 (AMP-CT) 98.5 (PCR) 100 (culture)	98.8 (AMP-CT) 99.6 (PCR) 98.0 (culture)	Retesting	Population, STD clinic

[a]Data from reference 112.
[b]On reanalysis.

TABLE 78 Abbott IMX and culture[a]

Test assay	Comparison assay	Specimen type	n	Prev (%)	Sens (%)	Spec (%)	PPV (%)	NPV (%)	Expanded standard	Other
Abbott IMX	Cell culture	Endocervical swab	622	5.1	84.4 (IMX) 87.5 (culture)	100 (IMX) 100 (culture)	100 (IMX) 99.4 (culture)	99.2 (IMX) 99.5 (culture)	EIA blocking assay; DFA on discrepant results	Population, STD clinics. Culture of female urethral swab yielded nine patients that were positive for CT but negative by both culture and EIA from endocervical specimens. These nine results were not included in the calculations for this table.

[a]Data from reference 22.

TABLE 79 HC II, CT-ID, culture, and Amplicor CT (PCR)[a]

Test assay	Comparison assay	Specimen type	n	Prev (%)	Sens (%)	Spec (%)	PPV (%)	NPV (%)	Expanded standard	Other
HC II, CT-ID DNA Test	Culture and Amplicor CT (PCR)	F endocervical swab	587	11.1	95.4 (HC) 90.8 (PCR) 81.5 (culture)	99.0 (HC) 99.6 (PCR) 100 (culture)	92.5 (HC) 96.7 (PCR) 100 (culture)	99.4 (HC) 98.9 (PCR) 97.8 (culture)	DFA	Population, STD clinic. Cytological-brush specimen used for some nonpregnant female patients.

[a]Data from reference 53.

TABLE 80 HC II, CT-ID, and PACE 2[a]

Test assay	Comparison assay	Specimen type	n	Sens (%)	Spec (%)	PPV (%)	NPV (%)	Expanded standard	Other
HC II, CT-ID DNA Test (Digene)	GenProbe PACE 2	F endocervical swab; M urethral swab	1,746	100 (74 of 74) (Digene) 86.5 (64 of 74) (PACE 2)	99.8 (1,683 of 1,687) (Digene) 99.9 (1,686 of 1,687) (PACE 2)	94.9 (Digene) 98.5 (PACE 2)	100 (Digene) 99.4 (PACE 2)	PCR on extracted DNA from transport tube	Population, STD, family practice clinics, and private practice offices; 24 discrepant results resolved by PCR indicated 9 false-positive results using Digene assay; the GenProbe transport medium was used in all assays.

[a]Data from reference 97.

TABLE 81 BIOSTAR, culture, DFA, and Amplicor CT (PCR)[a]

Test assay	Comparison assay	Specimen type	n	Prev (%)	Sens (%)	Spec (%)	PPV (%)	NPV (%)	Expanded standard	Other
BIOSTAR	Culture, DFA, Amplicor CT (PCR)	F endocervical swab	306	13.7	92.9 (culture) 59.5 (DFA) 92.9 (PCR) 73.8 (BIOSTAR)	100 (culture) 99.6 (DFA) 96.2 (PCR) 100 (BIOSTAR)	100 (culture) 96.2 (DFA) 79.6 (PCR) 100 (BIOSTAR)	98.9 (culture) 93.9 (DFA) 98.8 (PCR) 96 (BIOSTAR)	DFA for specimens yielding only a single positive test result	Population, F from STD clinic. For DFA to be interpreted as positive, >10 elementary bodies were required to be present; if this level were reduced to 2–7 elementary bodies, the sensitivity of DFA would be 88%.

[a]Data from reference 115.

TABLE 82 BIOSTAR and culture[a]

Test assay	Comparison assay	Specimen type	n	Prev (%)	Sens (%)	Spec (%)	PPV (%)	NPV (%)	Expanded standard	Other
BIOSTAR	Cell culture	Swabs from infants with ocular infections	152 (retrospective) 37 (prospective)	32.2 (retrospective) 27 (prospective)	94.2 (retrospective) 100 (prospective)	97 (retrospective) 92.6 (prospective)	94.2 (retrospective) 83.3 (prospective)	97 (retrospective) 100 (prospective)	DFA	Population, infants from a large metropolitan hospital

[a]Data from reference 125.

TABLE 83 BIOSTAR and LCR[a]

Test assay	Comparison assay	Specimen type	n	Prev (%)	Sens (%)	Spec (%)	PPV (%)	NPV (%)	Other
BIOSTAR	LCR	F endocervical swabs	415	9.2	31.6	98.9	75	93.5	Population, F from outpatient clinics at three institutions. Range in sens and spec varied dramatically among institutions, from 20 to 50% and from 98.1 to 100%, respectively.

[a]Data from reference 163.

TABLE 84 LCR and culture[a]

Test assay	Comparison assay	Specimen type	Prev	Sens	Spec	PPV	NPV	Expanded standard	Other
LCR	Cell culture	F endocervical swab and urine; M urethral swab and urine	See Table 85	See Table 85	See Table 85	See Table 85	See Table 85	Repeat testing and MOMP-based LCR assay	Patients from STD clinic

[a]Data from reference 23 with permission.

TABLE 85 Evaluation of LCR and cell culture assays for *C. trachomatis* relative to the number of confirmed cases of infection among 614 men and 602 women[a]

Sex	Assay	Specimen	Sens (%)[b]	Spec (%)[c]	PPV (%)[b]	NPV (%)[c]
M	Culture	Urethral swab	57.3 (43/75)	100 (539/539)	100 (43/43)	94.4 (539/571)
	LCR	Urethral swab	93.3 (70/75)	100 (539/539)	100 (70/70)	99.1 (539/544)
	LCR	Urine	77.3 (58/75)	99.4 (536/539)	95.1 (58/61)	96.9 (536/553)
F	Culture	Cervical swab	45.5 (30/66)	100 (536/536)	100 (30/30)	93.7 (536/572)
	LCR	Cervical swab	87.9 (58/66)	100 (536/536)	100 (58/58)	98.5 (536/544)
	LCR	Urine	78.8 (52/66)	99.4 (533/536)	94.5 (52/55)	97.4 (533/547)

[a]Data from reference 23 with permission.
[b]Values in parentheses are numbers of specimens with test results of assay in question of specimens with results according to "gold standard" (positive culture or at least one specimen confirmed to be positive by PCR).
[c]Values in parentheses are numbers of specimens with test results confirmed according to gold standard/total number of specimens with results of assay in question.

TABLE 86 BD ProbeTec ET (Chlamydia), LCR, and culture[a]

Test assay	Comparison assay	Specimen type	n	Prev (%)	Sens (%)	Spec (%)	Expanded standard	Other
BD ProbeTec ET (Chlamydia)	LCR and culture-DFA	Endocervical swabs and F urine	294	10.9	96.9 (31 of 32) (ProbeTec ET swab) 90.6 (29 of 32) (ProbeTec ET urine)	96.6 (253 of 262) (ProbeTec ET swab) 98.4 (239 of 245) (ProbeTec ET urine)	ProbeTec ET results compared to LCR and culture-DFA where any 2 of the assays yielding positive results were taken to indicate an infected (positive) patient.	F patients from adolescent, family-planning and obstetrics clinics. A total of 13 specimens contained inhibitors of amplification.

[a]Data from Lovchick et al., *Abstr. 99th Gen. Meet. Am. Soc. Microbiol.*, abstr. C-119.

TABLE 87 BD ProbeTec ET (Chlamydia), culture, DFA, and LCR[a]

Test assay	Comparison assay	Specimen type	n	Prev (%)	Sens (%)	Spec (%)	PPV (%)	NPV (%)	Expanded standard	Other
BD Probe Tec ET (Chlamydia)	Culture; DFA; LCR	Endocervical swab; F urine	657	5.0	96.9 (31 of 32) (ProbeTec ET swab) 100 (30 of 30) (ProbeTec ET urine)	99.0 (608 of 614) (ProbeTec ET swab) 98.3 (513 of 522) (ProbeTec ET urine)	83.8 (31 of 37) (ProbeTec ET swab) 76.9 (30 of 39) (urine)	99.8 (608 of 609) (ProbeTec ET swab) 100 (513 of 513) (urine)	A combination of culture, DFA, and LCR	Multicenter study with females from family practice, teen, ob/gyn, and STD clinics.

[a]Data from Ferrero et al., Abstr. 99th Gen. Meet. Am. Soc. Microbiol., abstr. C-120.

TABLE 88 PACE 2, Clearview EIA, LCR, and AMP-CT (TMA)[a]

Test assay	Comparison assay	Specimen type	n	Prev (%)	Sens	Spec	PPV	NPV	Expanded standard	Other
GenProbe PACE 2; Clearview EIA	Abbott LCR; GenProbe AMP-CT (TMA)	Endocervical swabs for all assays; F urine for amplification assays	787	8.4	See Table 89	See Table 89	See Table 89	See Table 89	Amplification assay discrepancies resolved by amplification of alternative target	Patient population; F from ob/gyn clinics. Patients considered positive if 3 or more assays were positive (swab and/or urine).

[a]Data from reference 81 with permission.

TABLE 89 Comparison of tests using endocervical swabs and urine[a]

Specimen and assay	Sens (%)[b]	Spec (%)[b]	PPV (%)[b]	NPV (%)[b]
Endocervical swabs				
Clearview EIA	50 (5/10)	100 (55/55)	100 (5/5)	91.7 (55/60)
GenProbe PACE 2	80.6 (50/62)	99.9 (675/676)	98.0 (50/51)	98.3 (675/687)
Abbott LCR	96.8 (60/62)	99.9 (675/676)	98.4 (60/61)	99.7 (675/677)
Urine				
GenProbe AMP CT	100 (62/62)	99.9 (675/676)	98.4 (62/63)	100 (677/677)
Abbott LCR	98.4 (61/62)	100 (685/685)	100 (61/61)	99.9 (685/686)
GenProbe AMP-CT	80.6 (50/62)	100 (685/685)	100 (50/50)	98.2 (685/697)

[a]Data from reference 81 with permission.
[b]Values in parentheses are number of specimens positive/total number of specimens tested.

TABLE 90 Amplicor (PCR) and LCR[a]

Test assay	Comments
Roche Amplicor (PCR) and Abbott LCR	Mailed, home-obtained urine specimens were submitted for detection of C. trachomatis. C. trachomatis was detectable in samples by both PCR and LCR after 1 wk of storage and/or transport at RT.

[a]Data from reference 102.

TABLE 91 Summary of improved reproducibility following procedural changes in the Abbott LCR C. trachomatis and N. gonorrhoeae tests[a,b]

Parameter	Results before changes	Results after changes
No. of samples	14,420	7,728
No. of repeats	181	514[c]
No. of discrepant results/no. of repeats (%)	30/181 (17)	13/514 (2.5)
No. of discrepant results detected by Abbott "equivocal range"/total no. of discrepant results (%)	5/30 (17)	6/13 (46)
Mean difference (sample/cutoff ratio [n]) between pairs of results		
Urine (%)	2.0 (14)	1.9 (2)
Swabs (%)	0.88 (16)	0.49 (11)[d]

[a]Data from reference 56.
[b]Reproducibility problems that have been observed for other amplification procedures are described for Abbott's LCR assay. The authors implemented a series of guidelines to reduce and prevent reproducibility problems that included procedural issues, calibration issues, and quality assurance parameters. The table summarizes their experience before and after implementing these changes.
[c]Includes 475 urine specimens that were tested in duplicate.
[d]Differs significantly ($P < 0.05$) from values before changes.

TABLE 92 Index to specific references comparing *N. gonorrhoeae* molecular assays[a]

Reference(s)	Test assay(s)	Comparative assay(s)	Specimen and notes	Table(s)
77	GenProbe PACE 2	LCR, culture	Endocervical and M urethral swabs; M and F urine	93
13, 155	GenProbe PACE 2	Culture	M urethral swabs	94, 95, 96
132	DiGene HC II–GC-ID DNA	Culture	Endocervical swab; brush	97
85; C. J. Lenderman, K. R. Smith, E. W. Hook III, S. Leister, C. Gaydos, T. Quinn, R. B. Jones, and B. Vanderpol, *Abstr. 99th Gen. Meet. Am. Soc. Microbiol.*, abstr. C-118, 1999	BD ProbeTec ET-GC	LCR	Endocervical and M urethral swabs; M and F urine	98, 99
25	LCR	Culture	Endocervical and M urethral swabs	100, 101
166	LCR	Culture	Endocervical swab (LCR and culture); F urine (LCR)	102
J. C. Lovchick, L. M. Peralta, C. M. Brown, and D. M. Hilligoss, *Abstr. 99th Gen. Meet. Am. Soc. Microbiol.*, abstr. C-119, 1999; D. V. Ferrero, L. Buck-Barrington, H. Meyers, G. S. Hall, M. Tuohy, D. Wilson, J. Schacter, J. Moncada, and F. Pang, *Abstr. 99th Gen. Meet. Am. Soc. Microbiol.*, abstr. C-120, 1999	BD ProbeTec ET-GC	LCR, culture-DFA	Endocervical swabs and F urine	103, 104
97	DiGene HC II–GC-ID DNA	GenProbe PACE 2	Endocervical and M urethral swab; M and F urine	106
109	PCR (Amplicor CT/NG)	Culture and microscopy	M urethral swabs (culture) and M urine (PCR)	107
38	PCR (Amplicor CT/NG)	Culture	Endocervical and M urethral swab; M and F urine	108
23	LCR	Culture	Endocervical and M urethral swabs; M and F urine	109, 110
M. Tuohy, G. Hall, N. Maldeis, D. Shank, D. Farrell, and G. W. Procop, *Abstr. 100th Gen. Meet. Am. Soc. Microbiol.*, abstr. C-378, 2000	BD ProbeTec ET	Culture	Analysis of nongonococcal neisseria isolates for cross-reactivity	105
47	Amplicor		Limitations of the assay compared to a nested PCR and 16S rRNA PCR	111

[a]As test methods are grouped together, the references are not in sequential order.

TABLE 93 *N. gonorrhoeae* PACE 2, LCR, and culture[a]

Test assay	Comparison assay	Specimen type	No. positive	Sens (%)	Spec (%)[b]	Other
GenProbe PACE 2	Culture	Endocervical swab	329 (from 10 studies)	92.1	99.7 (99.1)	Assays for *N. gonorrhoeae*. Excellent review article detailing performance of direct nucleic acid (Gen-Probe PACE 2) detection and LCR amplification and detection methods for diagnosis of *N. gonorrhoeae* infection. Summary of different studies with different patient populations, clinicians, and laboratories; evaluation of the quality of each study based on sample size, test quality, blinding of samples, and assembly of data (including demographics, risk factors, source, and clinical symptoms), and whether the reference test was applied independently of the test under evaluation.
LCR	Culture	Endocervical swab	121 (from 3 studies)	96.7	98.3 (99.7)	
GenProbe PACE 2	Culture	M urethra	500 (from 5 studies)	96.4	97.9 (98.8)	
LCR	Culture	M urethra	311 (from 3 studies)	98.6	97.8 (99.9)	
LCR	Culture	F urine	80 (from 4 studies)	96.2	99.2 (100)	
LCR	Culture	M urine	60 (from 4 studies)	98.3	99.2 (100)	

[a]Data from reference 77.
[b]Numbers in parentheses indicate specificity after discrepant analysis.

TABLE 94 PACE 2 (GC) and culture[a]

Test assay	Comparison assay	Specimen type	n	Prev (%)	Sens (%)	Spec (%)	PPV (%)	NPV (%)	Other
GenProbe PACE 2 (*N. gonorrhoeae*)	Culture (*N. gonorrhoeae*)	M urethral swabs	444	2–3	54	99.5	78	99	Patient population, asymptomatic M. A urine assay or leukocyte esterase assay was obtained prior to the study swabs and may be responsible for lower sensitivities than in some studies.

[a]Data from reference 13.

TABLE 95 PACE 2 and culture[a]

Test assay	Comparison assay	Specimen type	Prev (%)	Other
GenProbe PACE 2	Culture	Endocervical swab, M urethra, pharynx, rectum	14.9 (M) 8.7 (F)	Population, STD clinic

[a]Data from reference 155.

TABLE 96 Test results by sample type[a]

Sample Type	n	Sens (%)	Spec (%)	PPV (%)	NPV (%)
F	1,069	94.2	99.3	87.5	99.6
M	681	98.8	98.8	92.5	99.8
Genital	1,567	97.0	99.1	90.8	99.7
Nongenital	183	100	99.4	87.5	100
Total	1,750	97.1	99.1	90.6	99.8

[a]Data from reference 155.

TABLE 97 HC II GC-ID and culture[a]

Test assay	Comparison assay	Specimen type	n	Prev (%)	Sens (%)[b]	Spec (%)[b]	PPV (%)[b]	NPV (%)[b]	Expanded standard	Other
HC II GC-ID DNA Test (Digene)	Culture	Endocervical swab; brush	1,370	6.9	92.6 (93.0)	98.5 (99.0)	82.1 (87.7)	99.4 (99.4)	PCR	Multicenter clinical trial from family-planning and STD clinics; 19 positive Digene assays were culture negative, hence the lower spec and PPV.

[a]Data from reference 132.
[b]Numbers in parentheses indicate values after discrepant analysis.

TABLE 98 ProbeTec ET GC and LCR[a]

Test assay	Comparison assay	Specimen type	n	Prev	Other
ProbeTec ET GC	LCR	Endocervical swab and F urine; M urethral swab and M urine	58	Only 1 positive detected and 57 negative detected with total agreement	Patient population, family practice clinic. SDA for N. gonorrhoeae; limited data due to low prevalence in the specific patient population.

[a]Data from reference 85.

TABLE 99 ProbeTec ET GC and LCR[a]

Test assay	Comparison assay	Specimen type	Sens (%)	Spec (%)	PPV (%)	NPV (%)	Other
ProbeTec ET GC	LCR	Endocervical swab (n = 471)	98.2	99.0	93.3	99.8	Population; STD clinic
		F urine (n = 463)	85.3	98.2	86.3	98.2	
		M urine (n = 663)	96.9	96.7	92.7	95.6	

[a]Data from Lenderman et al., Abstr. 99th Gen. Meet. Am. Soc. Microbiol., abstr. C-118.

TABLE 100 LCR and culture[a]

Test assay	Comparison assay	Specimen type	n	Prev (%)	Sens	Spec	PPV	NPV	Expanded standard	Other
LCR	Culture	Endocervical swab; M urethral swab	562	5.5	See Table 101	See Table 101	NA	NA	PCR of the pilin gene of N. gonorrhoeae	Patient population, STD clinic

[a]Data from reference 25 with permission.

TABLE 101 Summary of assay sensitivity and specificity for *N. gonorrhoeae*[a]

No. of samples	Total positive	Assay	Sens (%) Before supplemental data analysis			Sens (%) After supplemental data analysis			Spec (%) Before supplemental data analysis			Spec (%) After supplemental data analysis		
			Total	M	F	Total	M	F	Total	M	F	Total	M	F
562	31	Culture	86.2	100	63.6	80.6	94.7	58.3	100	100	100	100	100	100
		LCR (swab)	100	99.6	100	96.8	100	91.7	99.6	99.6	100	100	100	100
		LCR (urine)	96.6	100	90.1	93.5	94.7	91.7	99.6	99.6	99.6	99.8	99.6	100

[a]Data from reference 25 with permission.

TABLE 102 LCR and culture (for swabs)[a]

Test assay	Comparison assay	Specimen type	n	Prev (%)	Sens (%) Swab	Sens (%) Urine	Spec (%) Swab	Spec (%) Urine	PPV (%) Swab	PPV (%) Urine	NPV (%) Swab	NPV (%) Urine	Expanded standard	Other
LCR	Culture (for swabs)	Endocervical swabs and F urine	330	10.3 (34 of 330)	82.3	88.2	98.9	100	90.3	100	98	98.7	PCR (developed to amplify *N. gonorrhoeae* DNA from urine samples)	Patient population, high prevalence; adolescent clinics. LCR sens was 6% higher than culture.

[a]Data from reference 166.

TABLE 103 ProbeTec ET GC, LCR, and culture-DFA[a]

Test assay	Comparison assay	Specimen type	ProbeTec ET Sens (%) Swab	ProbeTec ET Sens (%) Urine	ProbeTec ET Spec (%) Swab	ProbeTec ET Spec (%) Urine	Other
ProbeTec ET GC	LCR and culture-DFA	Endocervical swabs and F urine	100 (21 of 21)	84.2 (16 of 19)	99.6 (273 of 274)	98.8 (255 of 258)	F patients from adolescent, family-planning, and obstetrics clinics. A total of 13 specimens contained inhibitors of amplification.

[a]Data from Lovchick et al., Abstr. 99th Gen. Meet. Am. Soc. Microbiol., abstr. C-119.

TABLE 104 ProbeTec ET GC, culture, DFA, and LCR[a]

Test assay	Comparison assay	Specimen type	n	Prev (%)	Sens (%) Swab	Sens (%) Urine	Spec (%) Swab	Spec (%) Urine	PPV (%) Swab	PPV (%) Urine	NPV (%) Swab	NPV (%) Urine	Expanded standard	Other
ProbeTec ET GC	Culture, DFA, and LCR	Endocervical swab; F urine	657	1.6	100 (10 of 10)	88.9 (8 of 9)	99.7 (629 of 631)	99.4 (528 of 531)	83.3 (10 of 12)	72.7 (8 of 11)	100 (629 of 629)	99.8 (528 of 529)	Culture, DFA, and LCR	Multicenter study with F from family practice, teen, ob/gyn, and STD clinics.

[a]Data from Ferrero et al., Abst. 99th Gen. Meet. Am. Soc. Microbiol., abstr. C-120.

TABLE 105 BD ProbeTec ET GC and LCR[a]

Test assay	Comments
BD ProbeTec ET GC and LCR	Because certain strains of *Neisseria subflava* and *Neisseria cinerea* have been shown to contain nucleic acid sequences that amplify with the Amplicor *N. gonorrhoeae* PCR assay, strains of nongonococcal isolates were evaluated for amplification using the BD ProbeTec ET GC assay and the Abbott LCR assay. At concentrations of 10^8 CFU/ml, 2 isolates of *N. cinerea* were false positive in the ProbeTec assay. Dilution of the original inoculum revealed that 1 strain remained reactive at 10^7 CFU/ml (not below this threshold) while another strain remained reactive at 10^2 CFU/ml. No amplification was detected with the LCR assay. The clinical implications of these findings are uncertain.

[a]Data from Tuohy et al., Abstr. 100th Gen. Meet. Am. Soc. Microbiol., abstr. C-378.

TABLE 106 HC II GC-ID and PACE 2[a]

Test assay	Comparison assay	Specimen type	n	Prev (%)	Sens (%)	Spec (%)	PPV (%)	NPV (%)	Expanded standard	Other
HC II GC-ID DNA Test (Digene)	GenProbe PACE 2	F endocervical swab M urethral swab (n = 19)	1,750	1.8	100 (31 of 31) (Digene) 87.1 (27 of 31) (PACE 2)	99.7 (1,714 of 1,719) (Digene) 100 (1,719 of 1,719) (PACE 2)	86.1 (Digene) 100 (PACE 2)	100 (Digene) 99.8 (PACE 2)	PCR on extracted DNA from transport tube	Population, STD and family practice clinics, and private practice offices; 24 discrepant results resolved by PCR indicated 9 false-positive results using Digene assay. The GenProbe transport medium was used in all assays.

[a]Data from reference 97.

TABLE 107 Amplicor CT/NG and culture and/or microscopy[a]

Test assay	Comparison assay	Specimen type	n	Prev (%)	Sens (%)	Spec (%)	PPV (%)	NPV (%)	Expanded standard	Other
Amplicor CT/NG	Culture and/or microscopy for *N. gonorrhoeae*	M urethral swab for culture and/or microscopy; M urine for Amplicor assay	73	52.1	100 (38 of 38) (urine Amplicor) 86.8 (33 of 38) (urethral swab)	100 (urine Amplicor) 100 (urethral swab)	100 (urine Amplicor) 100 (urethral swab)	100 (urine Amplicor) 87.5 (urethral swab)	Supplemental PCR (developed in house)	Population, M seeking treatment for STD and M identified as contacts of sexual partners who had infections with *C. trachomatis* or *N. gonorrhoeae*.

[a]Data from reference 109.

TABLE 108 Amplicor CT/NG and culture[a]

Test assay	Comparison assay	Specimen type	n	Sens (%)	Spec (%)	PPV (%)	NPV (%)	Expanded standard	Other
Amplicor CT/NG	Culture for N. gonorrhoeae	F urine and endocervical swab; M urine and urethral swab	344 (M) 192 (F)	94.4 (M Amplicor urine) 97.3 (urethral) 76.6 (urethral culture) 90.0 (F Amplicor urine) 100 (endocervical swab) 65.2 (endocervical culture)	98.5 (M Amplicor urine) 97 (urethral) 100 (urethral culture) 95.9 (F Amplicor urine) 99.4 (endocervical swab) 100 (endocervical culture)	94.4 (M Amplicor urine) 90.1 (urethral) 72.0 (F Amplicor urine) 95.8 (endocervical swab)	98.5 (M Amplicor urine) 99.2 (urethral) 97.0 (F Amplicor urine) 100 (endocervical swab)	16S RNA PCR from swab transport or urine	Patient population, STD clinic; data are resolved performance characteristics for detection of N. gonorrhoeae by specimen.

[a]Data from reference 38.

TABLE 109 LCR and culture[a]

Test assay	Comparison assay	Specimen type	n	Prev	Sens	Spec	PPV	NPV	Expanded standard	Other
LCR	Cell culture	F endocervical swab and urine; M urethral swab and urine	See Table 110	See Table 110	See Table 110	See Table 110	See Table 110	See Table 110	Repeat testing and an alternative-target N. gonorrhoeae-based LCR assay	Patients from STD clinic

[a]Data from reference 23.

TABLE 110 Evaluation of LCR and cell culture assays for *N. gonorrhoeae* relative to the number of confirmed cases of infection among 220 men and 383 women[a]

Sex	Assay	Specimen	Sens (%)[b]	Spec (%)[b]	PPV (%)[c]	NPV (%)[a]
M	Culture	Urethral swab	72.2 (13/18)	100 (202/202)	100 (13/13)	97.6 (202/207)
	LCR	Urethral swab	100 (18/18)	100 (202/202)	100 (18/18)	100 (202/202)
	LCR	Urine	88.9 (16/18)	100 (202/202)	100 (16/16)	99.0 (202/204)
F	Culture	Cervical swab	50.0 (11/22)	100 (361/361)	100 (11/11)	97.0 (361/362)
	LCR	Cervical swab	95.4 (21/22)	100 (361/361)	100 (21/21)	99.7 (361/362)
	LCR	Urine	50 (11/22)	100 (361/361)	100 (11/11)	97.0 (361/372)

[a]Data from reference 23 with permission.
[b]Values in parentheses are number of specimens with test results of assay in question of specimens with results according to "gold standard" (expanded standard).
[c]Values in parentheses are number of specimens with test results confirmed according to gold standard/total number of specimens with results of assay in question.

TABLE 111 Roche Amplicor *Neisseria gonorrhoeae*[a]

Assay	Comments
Roche Amplicor *Neisseria gonorrhoeae*	The author describes key limitations of the assay. Six of 14 clinical strains of *Neisseria subflava* produced false-positive results in the Amplicor assay and were negative in the other two PCR assays. For 207 clinical specimens selected from a patient population with a prevalence rate of ~9%, 15 of 96 (15.6%) Amplicor-positive specimens and 14 of 17 (82.3%) Amplicor-equivocal specimens were not confirmed by the other two PCR assays. These data suggest that the assay is acceptable as a screening tool, but in this population positive results should be confirmed by a more specific assay.

[a]Data from reference 47.

REFERENCES

1. **Akane, A., K. Matsubara, H. Nakamura, S. Takahashi, and K. Kimura.** 1994. Identification of the heme compound copurified with deoxyribonucleic acid (DNA) from bloodstains, a major inhibitor of polymerase chain reaction amplification. *J. Forensic Sci.* **39**:362–372.

2. **Altaie, S. S., F. A. Meier, R. M. Centor, M. Wakabongo, D. Toksoz, K. M. Harvey, E. Basinger, B. A. Johnson, R. R. Brookman, and H. P. Dalton.** 1992. Evaluation of two ELISA's for detecting *Chlamydia trachomatis* from endocervical swabs. *Diagn. Microbiol. Infect. Dis.* **15**:579–586.

3. **Altwegg, M., D. Burger, U. Lauper, and G. Schar.** 1994. Comparison of Gen-Probe PACE 2, Amplicor Roche, and a conventional PCR for the detection of *Chlamydia trachomatis* in genital specimens. *Med. Microbiol. Lett.* **3**:181–187.

4. **Backman, M., A. K. Ruden, O. Ringertz, and E. G. Sandstrom.** 1993. Detection of *Chlamydia trachomatis* in urine from men with urethritis. *Eur. J. Clin. Microbiol. Infect. Dis.* **12**:447–449.

5. **Bass, C. A., D. L. Jungkind, N. S. Silverman, and J. M. Bondi.** 1993. Clinical evaluation of a new polymerase chain reaction assay for detection of *Chlamydia trachomatis* in endocervical specimens. *J. Clin. Microbiol.* **31**:2648–2653.

6. **Bassiri, M., H. Y. Hu, M. A. Domeika, J. Burczak, L.-O. Svensson, H. H. Lee, and P.-A. Mårdh.** 1995. Detection of *Chlamydia trachomatis* in urine specimens from women by ligase chain reaction. *J. Clin. Microbiol.* **33**:898–900.

7. **Bassiri, M., P.-A. Mårdh, M. Domeika, and The European Chlamydia Epidemiology Group.** 1997. Multiplex AMPLICOR PCR screening for *Chlamydia trachomatis* and *Neisseria gonorrhoeae* in women attending non-sexually transmitted disease clinics. *J. Clin. Microbiol.* **35**:2556–2560.

8. **Batteiger, B. E., and R. B. Jones.** 1987. Chlamydial infections. *Infect. Dis. Clin. N. Am.* **1**:55–81.

9. **Bauwens, J. E., A. M. Clark, M. J. Loeffelholz, S. A. Herman, and W. E. Stamm.** 1993. Diagnosis of *Chlamydia trachomatis* urethritis in men by polymerase chain reaction assay of first-catch urine. *J. Clin. Microbiol.* **31**:3013–3016.

10. **Bauwens, J. E., A. M. Clark, and W. E. Stamm.** 1993. Diagnosis of *Chlamydia trachomatis* endocervical infections by a commercial polymerase chain reaction assay. *J. Clin. Microbiol.* **31**:3023–3027.

11. **Beebe, J. L., H. Masters, D. Jungkind, D. M. Heltzel, and A. Weinberg.** 1996. Confirmation of the Syva MicroTrak enzyme immunoassay for *Chlamydia trachomatis* by Syva Direct Fluorescent Antibody Test. *Sex. Transm. Dis.* **23**:465–470

12. **Bell, T. A., W. E. Stamm, C. C. Kuo, S. P. Wang, K. K. Holmes, and J. T. Grayston.** 1987. Delayed appearance *Chlamydia trachomatis* infections acquired at birth. *Pediatr. Infect. Dis.* **6**:928–931.

13. **Beltrami, J. F., T. A. Farley, J. T. Hamrick, D. A. Cohen, and D. H. Martin.** 1998. Evaluation of the Gen-Probe PACE 2 assay for the detection of asymptomatic *Chlamydia trachomatis* and *Neisseria gonorrhoeae* infections in male arrestees. *Sex. Transm. Dis.* **25**:501–504.

14. **Berger, R. E., E. R. Alexander, J. P. Harnisch, C. A. Paulsen, G. D. Monda, J. Ansell, and K. K. Holmes.** 1979. Etiology and therapy of acute epididymitis: prospective study of 50 cases. *J. Urol.* **121**:750–754.

15. **Bianchi, A., C. Scieux, N. Brunat, D. Vexiau, M. Kermanach, P. Pezin, N. Janier, P. Morel, and P. H. Lagrange.** 1994. An evaluation of the polymerase chain reaction Amplicor *Chlamydia trachomatis* in male urine and female urogenital specimens. *Sex. Transm. Dis.* **21**:196–200.

16. **Biro, F. M., S. F. Reising, J. A. Doughman, L. M. Kollar, and S. L. Rosenthal.** 1994. A comparison of diagnostic methods in adolescent girls with and without symptoms of *Chlamydia* urogenital infection. *Pediatrics* **93**:476–480.

17. **Black C. M.** 1997. Current methods of laboratory diagnosis of *Chlamydia trachomatis* infections. *Clin. Microbiol. Rev.* **10**:160–184.

18. **Blanding, J., L. Hirsch, N. Stranton, T. Wright, S. Aarnaes, L. M. de la Maza, and E. M. Peterson.** 1993. Comparison of the Clearview Chlamydia, the PACE 2 assay, and culture for detection of *Chlamydia trachomatis* from cervical specimens in a low-prevalence population. *J. Clin. Microbiol.* **31**:1622–1625.

19. **Bleker, O. P., D. J. Smalbraak, and M. F. Schutte.** 1990. Bartholin's abscess: the role of *Chlamydia trachomatis*. *Genitourin. Med.* **66**:24–25.

20. **Blythe, M. J., B. P. Katz, B. E. Batteiger, J. A. Ganser, and R. B. Jones.** 1992. Recurrent genitourinary chlamydial infections in sexually active female adolescents. *J. Pediatr.* **121**:487–493.

21. **Bovarnick, M. R., J. C. Miller, and J. C. Snyder.** 1950. The influence of certain salts, amino acids, sugars, and proteins on the stability of rickettsiae. *J. Bacteriol.* **59**:509–522.

22. **Brokenshire, M. K., P. J. Say, A. H. van Vonno, and C. Wong.** 1997. Evaluation of the microparticle enzyme immunoassay Abbott IMx Select Chlamydia and the importance of urethral site sampling to detect *Chlamydia trachomatis* in women. *Genitourin. Med.* **73**:498–502.

23. **Buimer, M., G. J. van Doornum, S. Ching, P. G. Peerbooms, P. K. Plier, D. Ram, and H. H. Lee.** 1996. Detection of *Chlamydia trachomatis* and *Neisseria gonorrhoeae* by ligase chain reaction-based assays with clinical specimens from various sites: implications for diagnostic testing and screening. *J. Clin. Microbiol.* **34**:2395–2400.

24. **Caldwell, H. D., and J. Schacter.** 1982. Antigenic analysis of the outer membrane protein of *Chlamydia* spp. *Infect. Immun.* **35**:1024–1031.

25. **Carroll, K. C., W. E. Aldeen, M. Morrison, R. Anderson, D. Lee, and S. Mottice.** 1998. Evaluation of the Abbott LCx ligase chain reaction assay for detection of *Chlamydia trachomatis* and *Neisseria gonorrhoeae* in urine and genital swab specimens from a sexually transmitted disease clinic population. *J. Clin. Microbiol.* **36**:1630–1633.

26. **Cates, W., Jr., and J. N. Wasserheit.** 1991. Genital chlamydial infections: epidemiology and reproductive sequelae. *Am. J. Obstet. Gynecol.* **164**:1771–1781.

27. **Centers for Disease Control and Prevention.** 1989. Guidelines for prevention of transmission of human immunodeficiency virus and hepatitis B virus to health care and public safety workers. *Morb. Mortal. Wkly. Rep.* **38**:1–37.

28. **Centers for Disease Control and Prevention.** 1993. Recommendations for the prevention and management of *Chlamydia trachomatis* infections. *Morb. Mortal. Wkly. Rep.* **42**(RR-12):1–39.

29. **Chan, E. L., K. Brandt, and G. B. Horsman.** 1994. A 1-year evaluation of Syva Microtrak Chlamydia enzyme immunoassay with selective confirmation by direct fluorescent-antibody assay in a high-volume laboratory. *J. Clin. Microbiol.* **32:**2208–2211.

30. **Chan, E. L., K. Brandt, and G. Horsman.** 1995. Evaluation of Sanofi Diagnostics Pasteur Chlamydia Microplate EIA shortened assay and comparison with cell culture and Syva Chlamydia MicroTrak II EIA in high- and low-risk populations. *J. Clin. Microbiol.* **33:**2839–2841.

31. **Chernesky, M., S. Castriciano, J. Sellors, I. Stewart, I. Cunningham, S. Landis, W. Seidelman, W. Grant, L. Devlin, and J. Mahoney.** 1990. Detection of *Chlamydia trachomatis* antigens in urine as an alternative to swabs and cultures. *J. Infect. Dis.* **161:**124–126.

32. **Chernesky, M. A., D. Jang, H. Lee, J. D. Burczak, H. Hu, J. Sellors, S. J. Tomazic-Allen, and J. B. Mahony.** 1994. Diagnosis of *Chlamydia trachomatis* infections in men and women by testing first-void urine by ligase chain reaction. *J. Clin. Microbiol.* **32:**2682–2685.

33. **Chernesky, M. A., D. Jang, J. Sellors, P. Coleman, J. Bodner, I. Hrusovsky, S. Chong, and J. B. Mahony.** 1995. Detection of *Chlamydia trachomatis* antigens in male urethral swabs and urines with a microparticle enzyme immunoassay. *Sex. Transm. Dis.* **22:**55–59.

34. **Chernesky, M. A., H. Lee, J. Schachter, J. D. Burczak. W. E. Stamm, W. M. McCormack, and T. C. Quinn.** 1994. Diagnosis of *Chlamydia trachomatis* urethral infection in symptomatic and asymptomatic men by testing first void urine in a ligase chain reaction assay. *J. Infect. Dis.* **170:**1308–1311.

35. **Chernesky, M. A., S. Chong, D. Jang, K. Luinstra, J. Sellors, and J. B. Mahoney.** 1997. Ability of commercial ligase chain reaction and PCR assays to diagnose *Chlamydia trachomatis* infections in men by testing first-void urine. *J. Clin. Microbiol.* **35:**982–984.

36. **Claesson, B. A., B. Trollfors, I. Brolin, M. Granstrom, J. Henrichsen, U. Jodal, P. Juto, I. Kallings, K. Kanclerski, T. Lagergard, et al.** 1989. Etiology of community-acquired pneumonia in children based on antibody responses to bacterial and viral antigens. *Pediatr. Infect. Dis. J.* **8:**856–862.

37. **Clarke, L. M., M. F. Sierra, B. J. Daidone, N. Lopez, J. M. Covino, and W. M. McCormack.** 1993. Comparison of the Syva Microtrak enzyme immunoassay and Gen-Probe PACE 2 with cell culture for diagnosis of cervical *Chlamydia trachomatis* infection in a high-prevalence female population. *J. Clin. Microbiol.* **31:**968–971.

38. **Crotchfelt, K. A., L. E. Welsh, D. DeBonville, M. Rosenstraus, and T. C. Quinn.** 1997. Detection of *Neisseria gonorrhoeae* and *Chlamydia trachomatis* in genitourinary specimens from men and women by a coamplification PCR assay. *J. Clin. Microbiol.* **35:**1536–1540.

39. **Crotchfelt, K. A., B. Pare, C. Gaydos, and T. C. Quinn.** 1998. Detection of *Chlamydia trachomatis* by the Gen-Probe AMPLIFIED *Chlamydia trachomatis* Assay (AMP CT) in urine specimens from men and women and endocervical specimens from women. *J. Clin. Microbiol.* **36:**391–394.

40. **Davis, J. D., P. K. Riley, C. W. Peters, and K. H. Rand.** 1998. A comparison of ligase chain reaction to polymerase chain reaction in the detection of *Chlamydia trachomatis* endocervical infections. *Infect. Dis. Obstet. Gynecol.* **6:**57–60.

41. **Doing, K. M., K. Curtis, J. W. Long, and M. L. Volock.** 1999. Prospective comparison of the Gen-Probe PACE 2 assay and the Abbott ligase chain reaction for the direct detection of *Chlamydia trachomatis* in a low prevalence population. *J. Med. Microbiol.* **48:**507–510.

42. **Domeika, M., M. Bassiri, and P. A. Mardh.** 1994. Diagnosis of genital *Chlamydia trachomatis* infections in asymptomatic males by testing urine by PCR. *J. Clin. Microbiol.* **32:**2350–2352.

43. **Dunlop, E. M. C., B. T. Goh, S. Darougar, and R. Woodland.** 1985. Triple culture tests for the diagnosis of Chlamydial infection of the female genital tract. *Sex. Transm. Dis.* **12:**68–71.

44. **Ehret, J. M., J. C. Leszcynski, J. M. Douglas, S. L. Genova, M. A. Chernesky, J. Moncada, and J. Schachter.** 1993. Evaluation of chlamydiazyme enzyme immunoassay for detection of *Chlamydia trachomatis* in urine specimens from men. *J. Clin. Microbiol.* **31:**2702–2705.

45. **Embil, J. A., H. J. Thiebaux, F. R. Manuel, L. H. Pereira, and S. W. MacDonald.** 1983. Sequential cervical specimens and the isolation of *Chlamydia trachomatis*: factors affecting detection. *Sex. Transm. Dis.* **10:**62–66.

46. **Everett, K. D., R. M. Bush, and A. A. Andersen.** 1999. Emended description of the order Chlamydiales, proposal of *Parachlamydiaceae* fam. nov. and *Simkaniaceae* fam. nov., each containing one monotypic genus, revised taxonomy of the family *Chlamydiaceae*, including a new genus and five new species, and standards for the identification of organisms. *Int. J. Syst. Bacteriol.* **49:**415–440.

47. **Farrell, D. J.** 1999. Evaluation of Amplicor *Neisseria gonorrhoeae* PCR using cppB nested PCR and 16S rRNA PCR. *J. Clin. Microbiol.* **37:**386–390.

48. **Fedorko, D. P., and T. F. Smith.** 1991. Chlamydial infections, p. 95–125. *In* B. B. Wentworth, F. N. Judson, and M. J. R. Gilchrist (ed.), *Laboratory Methods for the Diagnosis of Sexually Transmitted Diseases*, 2nd ed. American Public Health Association, Washington, D.C.

49. **Ferrero, D. V., H. N. Meyers, D. E. Schultz, and S. A. Willis.** 1998. Performance of the Gen-Probe AMPLIFIED *Chlamydia trachomatis* assay in detecting *Chlamydia trachomatis* in endocervical and urine specimens from women and urethral and urine specimens from men attending sexually transmitted disease and family planning clinics. *J. Clin. Microbiol.* **36:**3230–3233.

50. **Gaydos, C. A., C. A. Reichart, J. M. Long, L. E. Welsh, T. M. Neumann, E. W. Hook III, and T. C. Quinn.** 1990. Evaluation of Syva enzyme immunoassay for detection of *Chlamydia trachomatis* in genital specimens. *J. Clin. Microbiol.* **28:**1541–1544.

51. **Gaydos, C. A., M. R. Howell, T. C. Quinn, J. C. Gaydos, and K. T. McKee, Jr.** 1998. Use of ligase chain reaction with urine versus cervical culture for detection of *Chlamydia trachomatis* in an asymptomatic military population of pregnant and nonpregnant females attending Papanicolaou smear clinics. *J. Clin. Microbiol.* **36:**1300–1304.

52. **Genc, M., A. Stary, S. Bergman, and P. A. Mardh.** 1991. Detection of *Chlamydia trachomatis* in first-void urine collected from men and women attending a venereal clinic. *APMIS* **99:**455–459.

53. **Girdner, J. L., A. P. Cullen, T. G. Salama, L. He, A. Lorincz, and T. C. Quinn.** 1999. Evaluation of the Digene Hybrid Capture II CT-ID test for detection of *Chlamydia trachomatis* in endocervical specimens. *J. Clin. Microbiol.* **37:**1579–1581.

54. Goessens, W. H., J. W. Mouton, W. I. van der Meijden, S. Deelen, T. H. van Rijsoort-Vos, N. Lemmensden Toom, H. A. Verbrugh, and R. P. Verkooyen. 1997. Comparison of three commercially available amplification assays, AMP CT, LCx, and COBAS AMPLICOR, for detection of Chlamydia trachomatis in first-void urine. J. Clin. Microbiol. 35:2628–2633.

55. Gordon, F. B., I. A. Harper, A. L. Quan, J. D. Treharne, R. S. Dwyer, and J. A. Garland. 1969. Detection of Chlamydia (Bedsonia) in certain infections of man. 1. Laboratory procedures: comparisons of yolk sac and cell culture for detection and isolation. J. Infect. Dis. 120:451–462.

56. Gronowski, A. M., S. Copper, D. Baorto, and P. R. Murray. 2000. Reproducibility problems with the Abbott LCx assay for Chlamydia trachomatis and Neisseria gonorrhoeae. J. Clin. Microbiol. 38:2416–2418.

57. Gun-Munro, J., J. Mahony, P. Lyn, K. Luinstra, F. Smaill, and H. Richardson. 1996. Detection of Chlamydia trachomatis in genitourinary tract specimens using an automated enzyme-linked fluorescent immunoassay. Sex. Transm. Dis. 23:115–119.

58. Hammerschlag, M. R., C. Cummings, P. M. Roblin, T. H. Williams, and I. Delke. 1989. Efficacy of neonatal ocular prophylaxis for the prevention of chlamydial and gonococcal conjunctivitis. N. Engl. J. Med. 320:769–772.

59. Hardy, P. H., J. B. Hardy, E. E. Nell, D. A. Graham, M. R. Spencer, and R. C. Rosenbaum. 1984. Prevalence of six sexually transmitted disease agents among pregnant inner city adolescents and pregnancy outcome. Lancet ii:333–337.

60. Harrison, H. R., E. R. Alexander, L. Weinstein, M. Lewis, M. Nash, and D. A. Sim. 1983. Cervical Chlamydia trachomatis and mycoplasmal infections in pregnancy: epidemiology and outcomes. JAMA 250:1721–1727.

61. Higgins, S. P., P. E. Klapper, J. K. Struthers, A. S. Bailey, A. P. Gough, R. Moore, G. Corbitt, M. N. Bhattacharyya. 1998. Detection of male genital infection with Chlamydia trachomatis and Neisseria gonorrhoeae using an automated multiplex PCR system (Cobas Amplicor). Int. J. STD AIDS 9:21–24.

62. Hillis, S. D., A. Nakashima, P. A. Marchbanks, D. G. Addiss, and J. P. Davis. 1994. Risk factors for recurrent C. trachomatis infections in women. Am. J. Obstet. Gynecol. 170:801–806.

63. Hobson, D., P. Karayiannis, R. E. Byng, I. Rees, I. A. Tait, and J. A. Davies. 1980. Quantitative aspects of chlamydial infection of the cervix. Br. J. Vener. Dis. 56:156–162.

64. Holmes, K. K., H. H. Handsfield, S. P. Wang, B. B. Wentworth, M. Turck, J. B. Anderson, and E. R. Alexander. 1975. Etiology of nongonococcal urethritis. N. Engl. J. Med. 292:1199–1205.

65. Hook, E. W., III, K. Smith, C. Mullen, J. Stephens, L. Rinehardt, M. S. Pate, and H. H. Lee. 1997. Diagnosis of genitourinary Chlamydia infections by using the ligase chain reaction on patient-obtained vaginal swabs. J. Clin. Microbiol. 35:2133–2135.

66. Hosein, I. K., A. M. Kaunitz, and S. J. Craft. 1992. Detection of cervical Chlamydia trachomatis and Neisseria gonorrhoeae with deoxyribonucleic acid probe assays in obstetric patients. Am. J. Obstet. Gynecol. 167:588–591.

67. Howard, C., D. L. Friedman, J. K. Leete, and N. L. Christensen. 1991. Correlation of the percent of positive Chlamydia trachomatis direct fluorescent antibody detection tests with the adequacy of specimen collection. Diagn. Microbiol. Infect. Dis. 14:233–237.

68. Howell, M. R., T. C. Quinn, W. Brathwaite, C. A. Gaydos. 1998. Screening women for Chlamydia trachomatis in family planning clinics: the cost-effectiveness of DNA amplification assays. Sex. Transm. Dis. 25:108–117.

69. Iwen, P. C., T. M. Blair, and G. L. Woods. 1991. Comparison of the Gen-Probe PACE 2 system, direct fluorescent antibody, and cell culture for detecting Chlamydia trachomatis in cervical specimens. Am. J. Clin. Pathol. 95:578–582.

70. Jaschek, G., C. A. Gaydos, L. E. Welsh, and T. C. Quinn. 1993. Direct detection of Chlamydia trachomatis in urine specimens from symptomatic and asymptomatic men by using a rapid polymerase chain reaction assay. J. Clin. Microbiol. 31:1209–1212.

71. Jones, R. B., B. P. Katz, B. VanderPol, V. A. Caine, B. E. Batteiger, and W. J. Newhall. 1986. Effect of blind passage and multiple sampling on recovery of Chlamydia trachomatis from urogenital specimens. J. Clin. Microbiol. 24:1029–1033.

72. Kellogg, J. A., J. W. Seiple, J. L. Klinedinst, and J. S. Levisky. 1991. Impact of endocervical specimen quality on apparent prevalence of Chlamydia trachomatis infections diagnosed using an enzyme-linked immunosorbent assay method. Arch. Pathol. Lab. Med. 115:1223–1227.

73. Kellogg, J. A., J. W. Seiple. J. L. Klinedinst, and J. S. Levisky. 1992. Comparison of cytobrushes with swabs for recovery of endocervical cells and for Chlamydiazyme detection of Chlamydia trachomatis. J. Clin. Microbiol. 30:2988–2990.

74. Kessler, H. H., K. Pierer, D. Stuenzner, P. Auer-Grumbach, E. M. Haller, and E. Marth. 1994. Rapid detection of Chlamydia trachomatis in conjunctival, pharyngeal, and urethral specimens with a new polymerase chain reaction assay. Sex. Transm. Dis. 21:191–195.

75. Kluytmans, J. A., H. G. Niesters, J. W. Mouton, W. G. Quint, J. A. Ijpelaar, J. H. van Rijsoort, L. Habbema, E. Stolz, M. F. Michel, and J. H. Wagenvoort. 1991. Performance of a nonisotopic DNA probe for detection of Chlamydia trachomatis in urogenital specimens. J. Clin. Microbiol. 29:2685–2689.

76. Kluytmans, J. A., A. H. van der Willigen, B. Y. van Heyst, W. I. van der Meyden, E. Stolz. and J. H. Wagenvoort. 1990. Evaluation of an enzyme immunoassay for detection of Chlamydia trachomatis in urogenital specimens. Int. J. STD AIDS 1:49–52.

77. Koumans, E. H., R. E. Johnson, J. S. Knapp, and M. E. St. Louis. 1998. Laboratory testing for Neisseria gonorrhoeae by recently introduced nonculture tests: a performance review with clinical and public health considerations. Clin. Infect. Dis. 27:1171–1180.

78. Krul, K. G. 1995. Closing in on Chlamydia. CAP Today 9:1–20.

79. Laga, M., A. Manoka, M. Kivuvu, B. Malele, M. Tuliza, N. Nzila, J. Goeman, et al. 1993. Nonulcerative sexually transmitted diseases as risk factors for HIV-1 transmission in women: results from a cohort study. AIDS 7:95–102.

80. Lampe, M. F., R. J. Suchland, and W. E. Stamm. 1993. Nucleotide sequence of the variable domains within the major outer membrane protein gene from serovariants of Chlamydia trachomatis. Infect. Immun. 61:213–219.

81. **Lauderdale, T., L. Landers, I. Thorneycraft, and K. Chapin.** 1999. Comparison of the PACE 2 assay, two amplification assays, and Clearview EIA for detection of *Chlamydia trachomatis* in female endocervical and urine specimens. *J. Clin. Microbiol.* **37:**2223–2229.

82. **Lee, H. H., M. A. Chernesky, J. Schachter, J. D. Burczak, W. W. Andrews, S. Muldoon, G. Leckie, and W. E. Stamm.** 1995. Diagnosis of *Chlamydia trachomatis* genitourinary infection in women by ligase chain reaction assay of urine. *Lancet* **345:**213–216.

83. **Lees, M. I., D. M. Newnan, and S. M. Garland.** 1991. Comparison of a DNA probe assay with culture for the detection of *Chlamydia trachomatis. J. Med. Microbiol.* **35:**159–161.

84. **Limberger, R. J., R. Biega, A. Evancoe, L. McCarthy, L. Slivienski, and M. Kirkwood.** 1992. Evaluation of culture and the Gen-Probe PACE 2 assay for detection of *Neisseria gonorrhoeae* and *Chlamydia trachomatis* in endocervical specimens transported to a state health laboratory. *J. Clin. Microbiol.* **30:**1162–1166.

85. **Little, M. C., J. Andrews, R. Moore, S. Bustos, L. Jones, C. Embres, G. Durmowicz, J. Harris, D. Berger, K. Yanson, C. Rostkowski, D. Yursis, J. Price, T. Fort, A. Walters, M. Collis, O. Llorin, J. Wood, F. Failing, C. O'Keefe, B. Scrivens, B. Pope, T. Hansen, K. Marino, K. Williams, et al.** 1999. Strand displacement amplification and homogeneous real-time detection incorporated in a second-generation DNA probe system, BDProbeTecET. *Clin. Chem.* **45:**777–784.

86. **Lossick, J., S. DeLisle, D. Fine, D. Mosure, V. Lee, and C. Smith.** 1990. Regional program for widespread screening for *Chlamydia trachomatis* in family planning clinics, p. 575–579. *In* W. R. Bowie, H. D. Caldwell, R. P. Jones, et al. (ed.), *Chlamydial Infections. Proceedings of the Seventh International Symposium on Human Chlamydial Infections.* Cambridge University Press, Cambridge, England.

87. **Magder, L. S., H. R. Harrison, J. M. Ehret, T. S. Anderson, and F. N. Judson.** 1988. Factors related to genital *Chlamydia trachomatis* and its diagnosis by culture in a sexually transmitted disease clinic. *Am. J. Epidemiol.* **128:**298–308.

88. **Mahony, J. B., and M. A. Chernesky.** 1985. Effect of swab type and storage temperature on the isolation of *Chlamydia trachomatis* from clinical specimens. *J. Clin. Microbiol.* **22:**865–867.

89. **Mahony J. B., K. E. Luinstra, J. W. Sellors, L. Pickard, S. Chong, D. Jang, and M. A. Chernesky.** 1994. Role of confirmatory PCRs in determining performance of Chlamydia Amplicor PCR with endocervical specimens from women with a low prevalence of infection. *J. Clin. Microbiol.* **32:**2490–2493.

90. **Mahony, J. B., S. D. Chong, D. Jang, K. Luinstra, M. Faught, D. Dalby, J. Sellors, and M. Chernesky.** 1998. Urine specimens from pregnant and nonpregnant women inhibitory to amplification of *Chlamydia trachomatis* nucleic acid by PCR, ligase chain reaction, and transcription-mediated amplification: identification of urinary substances associated with inhibition and removal of inhibitory activity. *J. Clin. Microbiol.* **36:**3122–3126.

91. **Mardh, P. A., and B. Zeeberg.** 1981. Toxic effect of sampling swabs and transportation test tubes on the formation of intracytoplasmic inclusions of *Chlamydia trachomatis* in McCoy's cell cultures. *Br. J. Vener. Dis.* **57:**268–272.

92. **Martin, D. H., L. Koutsky, D. A. Eschenbach, J. R. Daling, E. R. Alexander, J. K. Benedetti, and K. K. Holmes.** 1982. Prematurity and perinatal mortality in pregnancies complicated by maternal *Chlamydia trachomatis* infections. *JAMA* **247:**1585–1588.

93. **Matthews, R. S., P. G. Pandit, S. D. Bonigal, R. Wise, and K. W. Radcliffe.** 1993. Evaluation of an enzyme-linked immunoassay and confirmatory test for the detection of *Chlamydia trachomatis* in male urine samples. *Genitourin. Med.* **69:**47–50.

94. **McNagny, S. E., R. M. Parker, J. M. Zenilman, and J. S. Lewis.** 1992. Urinary leukocyte esterase test; A screening method for the detection of asymptomatic chlamydial and gonococcal infections in men. *J. Infect. Dis.* **165:**573–576.

95. **Mills, R. D., A. Young, K. Cain, T. M. Blair, M. A. Sitorius, and G. L. Woods.** 1992. Chlamydiazyme plus blocking assay to detect *Chlamydia trachomatis* in endocervical specimens. *Am. J. Clin. Pathol.* **97:**209–212.

96. **Miyashita, N., A. Matsumoto, Y. Niki, and T. Matsushima.** 1996. Evaluation of the sensitivity and specificity of a ligase chain reaction test kit for the detection of *Chlamydia trachomatis. J. Clin. Pathol.* **49:**515–517.

97. **Modarress, K. J., A. P. Cullen, W. J. Jaffurs, Sr, G. L. Troutman, N. Mousavi, R. A. Hubbard, S. Henderson, and A. T. Lorincz.** 1999. Detection of *Chlamydia trachomatis* and *Neisseria gonorrhoeae* in swab specimens by the Hybrid Capture II and PACE 2 nucleic acid probe tests. *Sex. Transm. Dis.* **26:**303–308.

98. **Moncada, J., J. Schachter, G. Bolan, J. Engelman, L. Howard, I. Mushahwar, G. Ridgway, G. Mumtaz, W. Stamm, and A. Clark.** 1990. Confirmatory assay increases specificity of the Chlamydiazyme test for *Chlamydia trachomatis* infection of the cervix. *J. Clin. Microbiol.* **28:**1770–1773.

99. **Moncada, J., J. Schachter, G. Bolan, J. Nathan, M. A. Shafer, A. Clark, J. Schwebke, W. Stamm, T. Mroczkowski, and D. Martin.** 1992. Evaluation of Syva's enzyme immunoassay for the detection of *Chlamydia trachomatis* in urogenital specimens. *Diagn. Microbiol. Infect. Dis.* **15:**663–668.

100. **Moncada, J., J. Schachter, M. A. Shafer, E. Williams, L. Gourlay, B. Lavin, and G. Bolan.** 1994. Detection of *Chlamydia trachomatis* in first catch urine samples from symptomatic and asymptomatic males. *Sex. Transm. Dis.* **21:**8–12.

101. **Moncada, J., J. Schachter, M. Shipp, G. Bolan, and J. Wilber.** 1989. Cytobrush in collection of cervical specimens for detection of *Chlamydia trachomatis. J. Clin. Microbiol.* **27:**1863–1866.

102. **Morre, S. A., I. G. M. van Valkengoed, A. deJong, A. J. P. Boeke, J. T. M. van Eijk, C. J. L. M. Meijer, and A. J. C. van den Brule.** 1999. Mailed, home-obtained urine specimens: a reliable screening approach for detecting asymptomatic *Chlamydia trachomatis* infections. *J. Clin. Microbiol.* **37:**976–980.

103. **Morris, R., J. Legault, and C. Baker.** 1993. Prevalence of isolated urethral asymptomatic *Chlamydia trachomatis* infection in the absence of cervical infection of incarcerated adolescent girls. *Sex. Transm. Dis.* **20:**198–200.

104. **Mouton, J. W., R. Verkooyen, W. I. van der Meijden, T. H. van Rijsoort-Vos, W. H. Goessens, J. A. Kluytmans, S. D. Deelen, A. Luijendijk, and H. A. Verbrugh.** 1997. Detection of *Chlamydia trachomatis*

in male and female urine specimens by using the amplified *Chlamydia trachomatis* test. *J. Clin. Microbiol.* **35:**1369–1372.

105. **Mumtaz, G., G. L. Ridgway, S. Clark, and E. Allason-Jones.** 1991. Evaluation of an enzyme immunoassay (Chlamydiazyme) with confirmatory test for the detection of chlamydial antigen in urine from men. *J. STD AIDS* **2:**359–361.

106. **Nash, P., and M. M. Krenz.** 1991. Culture media, p. 1226–1288. *In* A. Balows, W. J. Hausler, Jr., K. L. Herrmann, H. D. Isenberg, and H. J. Shadomy (ed.), *Manual of Clinical Microbiology*, 5th ed. American Society for Microbiology, Washington, D.C.

107. **Newhall, W. J., R. E. Johnson, S. DeLisle, D. Fine, A. Hadgu, B. Matsuda, D. Osmond, J. Campbell, and W. E. Stamm.** 1999. Head-to-head evaluation of five chlamydia tests relative to a quality-assured culture standard. *J. Clin. Microbiol.* **37:**681–685.

108. **Ossewaarde, J. M., and M. Rieffe.** 1989. Storage conditions of *Chlamydia trachomatis* antigens. *Eur. J. Clin. Microbiol. Infect. Dis.* **8:**658–660.

109. **Palladino, S., J. W. Pearman, I. D. Kay, D. W. Smith, G. B. Harnett, M. Woods, L. Marshall, and J. McCloskey.** 1999. Diagnosis of *Chlamydia trachomatis* and *Neisseria gonorrhoeae*. Genitourinary infections in males by the Amplicor PCR assay of urine. *Diagn. Microbiol. Infect. Dis.* **33:**141–146.

110. **Parker, E. K., A. Wozniak, S. D. White, C. Beckham, and D. Roberts.** 1999. Stability study on specimens mailed to a state laboratory and tested with the Gen-Probe PACE 2 assay for chlamydia. *Sex. Transm. Dis.* **26:**213–215.

111. **Pasternack, R., P. Vuorinen, and A. Miettinen.** 1999. Comparison of a transcription mediated amplification assay and polymerase chain reaction for detection of *Chlamydia trachomatis* in first void urine. *Eur. J. Clin. Microbiol. Infect. Dis.* **18:**142–144.

112. **Pasternack, R., P. Vuorinen, and A. Miettinen.** 1997. Evaluation of the Gen-Probe *Chlamydia trachomatis* transcription-mediated amplification assay with urine specimens from women. *J. Clin. Microbiol.* **35:**676–678.

113. **Pasternack, R., P. Vuorinen, T. Pitkajarvi, M. Koskela, and A. Miettinen.** 1997. Comparison of manual Amplicor PCR, COBAS AMPLICOR PCR, and LCx assays for detection of *Chlamydia trachomatis* infection in women by using urine specimens. *J. Clin. Microbiol.* **35:**402–405.

114. **Pasternack, R., P. Vuorinen, A. Kuukankorpi, T. Pitkajarvi, and A. Miettinen.** 1996. Detection of *Chlamydia trachomatis* infections in women by Amplicor PCR: comparison of diagnostic performance with urine and cervical specimens. *J. Clin. Microbiol.* **34:**995–998.

115. **Pate, M. S., P. B. Dixon, K. Hardy, M. Crosby, and E. W. Hook III.** 1998. Evaluation of the Biostar Chlamydia OIA assay with specimens from women attending a sexually transmitted disease clinic. *J. Clin. Microbiol.* **36:**2183–2186.

116. **Paukku, M., M. Puolakkainen, D. Apter, S. Hirvonen, and J. Paavonen.** 1997. First-void urine testing for *Chlamydia trachomatis* by polymerase chain reaction in asymptomatic women. *Sex. Transm. Dis.* **24:**343–346.

117. **Pearlman, M. D., and S. G. McNeeley.** 1992. A review of the microbiology, immunology, and clinical implications of *Chlamydia trachomatis* infections. *Obstet. Gynecol. Surv.* **47:**448–461.

118. **Perine, P. L., and A. O. Osoba.** 1990. Lymphogranuloma venereum, p. 195–204. *In* K. K. Holmes, P. A. Mardh, P. F. Sparling, and P. J. Wiesner (ed.), *Sexually Transmitted Diseases*. McGraw Hill Book Co., New York, N.Y.

119. **Peterson, E. M., V. Darrow, J. Blanding, S. Aarnaes, and L. M. de la Maza.** 1997. Reproducibility problems with the Amplicor PCR *Chlamydia trachomatis* test. *J. Clin. Microbiol.* **35:**957–959.

120. **Phillips, R. S., P. A. Hanff, R. S. Kauffman, and M. D. Aronson.** 1987. Use of a direct fluorescent antibody test for detecting *Chlamydia trachomatis* cervical infection in women seeking routine gynecologic care. *J. Infect. Dis.* **156:**575–581.

121. **Preece, P. M., J. M. Anderson, and R. G. Thompson.** 1989. *Chlamydia trachomatis* infection in infants: a prospective study. *Arch. Dis. Child* **64:**525–529.

122. **Puolakkainen, M., E. Hiltunen-Back, T. Reunala, S. Suhonen, P. Lahteenmaki, M. Lehtinen, and J. Paavonen.** 1997. Comparison of two commercially available tests, a PCR assay and a ligase chain reaction test, in detection of urogenital *Chlamydia trachomatis* infection. *J. Clin. Microbiol.* **36:**1489–1493.

123. **Reeve, P., J. Owen, and J. D. Oriel.** 1975. Laboratory procedures for the isolation of *Chlamydia trachomatis* from the human genital tract. *J. Clin. Pathol.* **28:**910–914.

124. **Rietmeijer, C. A. M., F. N. Judson, M. B. van Hensbroek, J. M. Ehret, and J. M. Douglas, Jr.** 1991. Unsuspected *Chlamydia trachomatis* infection in heterosexual men attending a sexually transmitted diseases clinic: evaluation of risk factors and screening methods. *Sex. Transm. Dis.* **18:**28–35.

125. **Roblin, P. M., M. Gelling, A. Kutlin, N. Tsumura, and M. R. Hammerschlag.** 1997. Evaluation of a new optical immunoassay for diagnosis of neonatal chlamydial conjunctivitis. *J. Clin. Microbiol.* **35:**515–516.

126. **Rompalo, A. M., P. Roberts, K. Johnson, and W. E. Stamm.** 1988. Empirical therapy for the management of acute proctitis in homosexual men. *JAMA* **260:**348–353.

127. **Ryan, G., T. N. Abdella, S. G. McNeeley, V. S. Baselski, and D. E. Drummond.** 1990. *Chlamydia trachomatis* infection in pregnancy and effect of treatment on pregnancy outcome. *Am. J. Obstet. Gynecol.* **162:**34–39.

128. **Schachter, J.** 1989. Why we need a program for the control of *Chlamydia trachomatis*. *N. Engl. J. Med.* **320:**802–804.

129. **Schachter, J., M. Grossman, R. L. Sweet, J. Holt, C. Jordan, and E. Bishop.** 1986. Prospective study of perinatal transmission of *Chlamydia trachomatis*. *JAMA* **255:**3374–3377.

130. **Schachter, J., W. E. Stamm, T. C. Quinn, W. W. Andrews, J. D. Burczak, and H. H. Lee.** 1994. Ligase chain reaction to detect *Chlamydia trachomatis* infection of the cervix. *J. Clin. Microbiol.* **32:**2540–2543.

131. **Schachter, J., E. Stoner, and J. Moncada.** 1983. Screening for chlamydial infections in women attending family planning clinics. *West. J. Med.* **138:**375–379.

132. **Schachter, J., E. W. Hook III, W. M. McCormack, T. C. Quinn, M. Chernesky, S. Chong, J. I. Girdner, P. B. Dixon, L. DeMeo, E. Williams, A. Cullen, and A. Lorincz.** 1999. Ability of the Digene Hybrid Capture II test to identify *Chlamydia trachomatis* and *Neis-*

seria gonorrhoeae in cervical specimens. *J. Clin. Microbiol.* **37:**3668–3671.

133. **Schachter, J., R. B. Jones, R. C. Butler, B. Rice, D. Brooks, B. Van der Pol, M. Gray, and J. Moncada.** 1997. Evaluation of the Vidas Chlamydia test to detect and verify *Chlamydia trachomatis* in urogenital specimens. *J. Clin. Microbiol.* **35:**2102–2106.

134. **Schepetiuk, S., T. Kok, L. Martin, R. Waddell, and G. Higgins.** 1997. Detection of *Chlamydia trachomatis* in urine samples by nucleic acid tests: comparison with culture and enzyme immunoassay of genital swab specimens. *J. Clin. Microbiol.* **35:**3355–3357.

135. **Schwebke, J. R., W. E. Stamm, and H. H. Handsfield.** 1990. Use of sequential enzyme immunoassay and direct fluorescent antibody tests for detection *of Chlamydia trachomatis* infections in women. *J. Clin. Microbiol.* **28:**2473–2476.

136. **Sellors, J., M. Chernesky, L. Pickard, D. Jang, S. Walter, J. Krepel, and J. Mahony.** 1993. Effect of time elapsed since previous voiding on the detection of *Chlamydia trachomatis* antigens in urine. *Eur. J. Clin. Microbiol. Infect. Dis.* **12:**285–289.

137. **Shattock, R. M., C. Patrizio, P. Simmonds, and S. Sutherland.** 1998. Detection of *Chlamydia trachomatis* in genital swabs: comparison of commercial and in house amplification methods with culture. *Sex. Transm. Infect.* **74:**289–293.

138. **Skulnick, M., R. Chua, A. E. Simor, D. E. Low, H. E. Khosid, S. Fraser, E. Lyons, E. A. Legere, and D. A. Kitching.** 1994. Use of the polymerase chain reaction for the detection of *Chlamydia trachomatis* from endocervical and urine specimens in an asymptomatic low-prevalence population of women. *Diagn. Microbiol. Infect. Dis.* **20:**195–201.

139. **Stamm, W. E., and B. Cole.** 1986. Asymptomatic *Chlamydia trachomatis* urethritis in men. *Sex. Transm. Dis.* **13:**163–165.

140. **Stamm, W. E., L. A. Koutsky, J. K. Benedetti, J. L. Jourden, R. C. Brunham, and K. K. Holmes.** 1984. *Chlamydia trachomatis* urethral infections in men. Prevalence, risk factors, and clinical manifestations. *Ann. Intern. Med.* **100:**47–51.

141. **Stary, A., S. Tomazic-Allen, B. Choueiri, J. Burczak, K. Steyrer, and H. Lee.** 1996. Comparison of DNA amplification methods for the detection of *Chlamydia trachomatis* in first-void urine from asymptomatic military recruits. *Sex. Transm. Dis.* **23:**97–102.

142. **Stary, A., E. Schuh, M. Kerschbaumer, B. Götz, and H. Lee.** 1998. Performance of transcription-mediated amplification and ligase chain reaction assays for detection of chlamydial infection in urogenital samples obtained by invasive and noninvasive methods. *J. Clin. Microbiol.* **36:**2666–2670.

143. **Steingrimsson, O., K. Jonsdottir, J. H. Olafsson, S. M. Karlsson, R. Palsdottir, and S. Davidsson.** 1998. Comparison of Roche Cobas Amplicor and Abbott LCx for the rapid detection of *Chlamydia trachomatis* in specimens from high-risk patients. *Sex. Transm. Dis.* **25:**44–48.

144. **Stephens, R. S.** 1990. Molecular genetics of Chlamydia, p. 63–72. *In* W. R. Bowie III, H. D. Caldwell, R. P. Jones, P. A. Mardh, G. L. Ridgway, J. Schachter, W. E. Stamm, and M. E. Ward (ed.), *Chlamydial infections. Proceedings of the 7th International Symposium on Human Chlamydial Infections.* Cambridge University Press, Cambridge, England.

145. **Suchland, R. J., and W. E. Stamm.** 1991. Simplified microtiter well cell culture method for rapid immunotyping of *Chlamydia trachomatis. J. Clin. Microbiol.* **29:**1333–1338.

146. **Sweet, R. L., D. V. Landers, C. Walker, and J. Schachter.** 1987. *Chlamydia trachomatis* infection and pregnancy outcome. *Am. J. Obstet. Gynecol.* **156:**824–833.

147. **Talbot, H., and B. Romanowski.** 1994. Factors affecting urine EIA sensitivity in the detection of *Chlamydia trachomatis* in men. *Genitourin. Med.* **70:**101–104.

148. **Taylor-Robinson, D., B. Thomas, T. Pierpoint, and A. Renton.** 1998. Ligase chain reaction assay for *Chlamydia trachomatis* during the menstrual cycle. *Lancet* **351:**1290.

149. **Thomas, B. J., E. J. MacLeod, P. E. Hay, P. J. Horner, and D. Taylor-Robinson.** 1994. Limited value of two widely used enzyme immunoassays for detection of *Chlamydia trachomatis* in women. *Eur. J. Clin. Microbiol. Infect. Dis.* **13:**651–655.

150. **Thomas, B. J., T. Pierpoint, D. Taylor-Robinson, and A. M. Renton.** 1998. Sensitivity of the ligase chain reaction assay for detecting *Chlamydia trachomatis* in vaginal swabs from women who are infected at other sites. *Sex. Transm. Infect.* **74:**140–141.

151. **Thompson, S., B. Lopez, K. H. Wong, et al.** 1982. A prospective study of chlamydial and mycoplasmal infections during pregnancy, p. 155–159. *In* P. A. Mardh, K. K. Holmes, J. D. Oriel, J. Schachter, and P. Piot (ed.), *Chlamydial Infections.* Fernstrom Foundation Series. Elsevier Biomedical Press, Amsterdam, The Netherlands.

152. **Van der Pol, B., T. C. Quinn, C. A. Gaydos, K. Crotchfelt, J. Schacter, J. Moncada, D. Jungkind, D. H. Martin, B. Turner, C. Peyton, and R. B. Jones.** 2000. Multicenter evaluation of the AMPLICOR and COBAS AMPLICOR CT/GC tests for detection of *Chlamydia trachomatis. J. Clin Microbiol.* **38:**1105–1112.

153. **Van Dyck, E., N. Samb, A. D. Sarr, L. Van de Velden, J. Moran, S. Mboup, I. Ndoye, J. L. Lamboray, A. Meheus, and P. Piot.** 1992. Accuracy of two enzyme immunoassays and cell culture in the detection of *Chlamydia trachomatis* in low and high risk populations in Senegal. *Eur. J. Clin. Microbiol. Infect. Dis.* **11:**527–534.

154. **Vincelette, J., J. Schirm, M. Bogard, A. M. Bourgault, D. S. Luijt, A. Bianchi, P. C. van Voorst Vader, A. Butcher, and M. Rosenstraus.** 1999. Multicenter evaluation of the fully automated COBAS AMPLICOR PCR test for detection of *Chlamydia trachomatis* in urogenital specimens. *J. Clin. Microbiol.* **37:**74–80.

155. **Vlaspolder, F., J. A. Mutsaers, F. Blog, and A. Notowicz.** 1993. Value of a DNA probe assay (GenProbe) compared with that of culture for diagnosis of gonococcal infection. *J. Clin. Microbiol.* **31:**107–110.

156. **Waites, K. B., K. R. Smith, M. A. Crum, R. D. Hockett, A. H. Wells, and E. W. Hook III.** 1999. Detection of *Chlamydia trachomatis* endocervical infections by ligase chain reaction versus ACCESS Chlamydia antigen assay. *J. Clin. Microbiol.* **37:**3072–3073.

157. **Warren, R., B. Dwyer, M. Plackett, K. Pettit, N. Rizvi, and A. M. Baker.** 1993. Comparative evaluation of detection assays for *Chlamydia trachomatis. J. Clin. Microbiol.* **31:**1663–1666.

158. **Washington, A. E., R. E. Johnson, and L. L. Sanders.** 1987. *Chlamydia trachomatis* infections in the United States. What are they costing us? *JAMA* **257:**2070–2072.

159. **Washington, A. E., R. E. Johnson, L. L. Sanders, R. C. Barnes, and E. R. Alexander.** 1986. Incidence of *Chlamydia trachomatis* infections in the United States using reported *Neisseria gonorrhoeae* as a surrogate, p. 487–490. *In* D. Oriel, G. Ridgway, J. Schachter, et al. (ed.), *Chlamydia Infections. Proceedings of the Sixth International Symposium on Human Chlamydial Infections.* Cambridge University Press, Cambridge, England.

160. **Weinstock, H., D. Dean, and G. Bolan.** 1994. *Chlamydia trachomatis* infections. Sexually transmitted diseases in the AIDS era. Part II. *Infect. Dis. Clin. N. Am.* **8:**797–819.

161. **Weiss, S. G., R. W. Newcomb, and M. O. Beem.** 1986. Pulmonary assessment of children after chlamydial pneumonia of infancy. *J. Pediatr.* **108:**659–664.

162. **Westrom, L., L. P. H. Bengtsson, and P. A. Mardh.** 1981. Incidence, trends and risks of ectopic pregnancy in a population of women. *Br. Med. J.* **282:**15–18.

163. **Widjaja, S., S. Cohen, W. E. Brady, K. O'Reilly, Susanto, A. Wibowo, Cahyono, R. R. Graham, and K. R. Porter.** 1999. Evaluation of a rapid assay for detection of *Chlamydia trachomatis* infections in outpatient clinics in South Kalimantan, Indonesia. *J. Clin. Microbiol.* **37:**4183–4185.

164. **Wiesenfeld, H. C., M. Uhrin, B. W. Dixon, and R. L. Sweet.** 1994. Diagnosis of male *Chlamydia trachomatis* urethritis by polymerase chain reaction. *Sex. Transm. Dis.* **21:**268–271.

165. **Wylie, J. L., S. Moses, R. Babcock, A. Jolly, S. Giercke, and G. Hammond.** 1998. Comparative evaluation of Chlamydiazyme, PACE 2, and AMP-CT assays for detection of *Chlamydia trachomatis* in endocervical specimens. *J. Clin. Microbiol.* **36:**3488–3491.

166. **Xu, K., V. Glanton, S. R. Johnson, C. Beck-Sague, V. Bhullar, D. H. Candal, K. S. Pettus, C. E. Farshy, and C. M. Black.** 1998. Detection of *Neisseria gonorrhoeae* infection by ligase chain reaction testing of urine among adolescent women with and without *Chlamydia trachomatis* infection. *Sex. Transm. Dis.* **25:**533–538.

167. **Yang, L. I., E. S. Panke, P. A. Leist, R. J. Fry, and R. F. Lee.** 1991. Detection of *Chlamydia trachomatis* endocervical infection in asymptomatic and symptomatic women: comparison of deoxyribonucleic acid probe test with tissue culture. *Am. J. Obstet. Gynecol.* **165:**1444–1453.

168. **Zelin, J. M., A. J. Robinson, G. L. Ridgway, E. Allason-Jones, and P. Williams.** 1995. Chlamydial urethritis in heterosexual men attending a genitourinary medicine clinic: prevalence, symptoms, condom usage and partner change. *Int. J. STD AIDS* **6:**27–30.

Mycoplasmas

KEN B. WAITES, DEBORAH F. TALKINGTON, AND CECILE M. BÉBÉAR

8

INTRODUCTION AND CLINICAL CONSIDERATIONS

Mycoplasmas and ureaplasmas represent a complex and unique group of microorganisms that has previously been ignored by most diagnostic laboratories. This situation has changed somewhat in recent years because of greater appreciation for their clinical importance, improved methods for detection, and availability of commercially manufactured growth media, serological assays, and products sold as complete diagnostic kits in some countries. However, the number and types of products available for the detection and characterization of mycoplasmas is still very limited in comparison to products marketed for other types of bacteria, and not all products sold commercially have proven adequate when compared to reference methods established by researchers in mycoplasmology. A number of products that can be purchased for use in mycoplasma detection have never been subjected to rigorous external evaluation, so recommendations for their use cannot be made based on evidence of performance.

It is beyond the scope of this chapter to review in depth the taxonomy, disease associations, and biological characteristics of mycoplasmas and ureaplasmas of humans. The reader is referred to chapter 56 of the seventh edition of the *Manual of Clinical Microbiology* (39) or other reference texts for more detailed information on these aspects of human mycoplasmas, and only a very brief synopsis of this information is included here.

Bacteria commonly referred to as mycoplasmas (fungus form) are included within the class *Mollicutes* (soft skin), which comprises four orders, five families, eight genera, and over 150 known species. Sixteen species have been isolated from humans, excluding occasional animal mycoplasmas that have been detected in humans from time to time but are generally considered transient colonizers. Mollicutes are the smallest life forms capable of an independent existence. They permanently lack cell walls and peptidoglycan precursors and require a complex enriched medium containing serum, amino acid precursors, nucleotides, yeast extract, and other supplements.

Among mollicutes isolated from humans, three organisms are of major concern. *Mycoplasma pneumoniae* is a well-established pathogen, whereas *Mycoplasma hominis* and *Ureaplasma* spp. are generally considered opportunists. The two biovars of *Ureaplasma urealyticum* have recently been pro-

posed for division into separate species: *U. urealyticum* and *Ureaplasma parvum*. Separation of these species is not possible except by PCR. Therefore, they will be considered together as *Ureaplasma* spp. Descriptions of growth media and diagnostic products still refer to all ureaplasmas as *U. urealyticum*.

M. pneumoniae causes tracheobronchitis and pneumonia in persons of all age groups. Some people may experience extrapulmonary complications, including skin rashes, pericarditis, hemolytic anemia, arthritis, meningoencephalitis, peripheral neuropathy, and pericarditis. *M. pneumoniae* has been isolated from extrapulmonary sites, such as synovial fluid, cerebrospinal fluid, pericardial fluid, and skin lesions. Clinical manifestations are not sufficiently unique to allow differentiation from infections caused by other common microorganisms.

Ureaplasma spp. and *M. hominis* can be isolated from the lower genital tract in many sexually active adults, leading to difficulty in accepting these organisms as causes of disease. Nevertheless, there is evidence that the organisms play etiologic roles in some conditions. *Ureaplasma* spp. and *Mycoplasma genitalium* are causes of nongonococcal urethritis in men. *M. hominis* causes a small portion of cases of pyelonephritis and has been isolated from the endometrium and fallopian tubes of women with salpingitis accompanied by specific antibody response. *M. genitalium* may also play a role in salpingitis. Ureaplasmas can cause placental inflammation and may invade the amniotic sac early in pregnancy in the presence of intact fetal membranes, causing persistent infection and an adverse pregnancy outcome, including premature birth. *M. hominis* and *Ureaplasma* spp. have been isolated from the blood of women with postpartum or postabortal fever, but not from afebrile women who have had abortions or from healthy pregnant women. Congenital pneumonia, bacteremia, progression to chronic lung disease of prematurity, and even death occurring in very low birthweight infants have been attributed to ureaplasmal infection of the lower respiratory tract. Both *M. hominis* and *Ureaplasma* spp. have been isolated from maternal and umbilical cord blood, as well as the blood of neonates. Both organisms can also invade the cerebrospinal fluid.

Mycoplasmas and ureaplasmas can cause invasive disease of the joints and respiratory tract with bacteremic dissemination, especially in individuals with hypogammaglobulinemia. M. hominis bacteremia has been demonstrated after renal transplantation, trauma, and genitourinary manipulations and has also been found in wound infections. Mycoplasma fermentans, Ureaplasma spp., and M. hominis have been detected in the synovial fluid of persons with rheumatoid arthritis and other inflammatory arthritides, although their precise contribution to these diseases is uncertain.

Recent evidence from PCR assays has shown that more fastidious and/or slow-growing mycoplasmal species, such as M. genitalium, may be of etiologic significance in various genital tract diseases. M. fermentans, Mycoplasma penetrans, Mycoplasma pirum, and others may be associated with disseminated diseases in immunocompromised states, including AIDS. M. fermentans has also been reported to cause an acute respiratory syndrome in adults, although the frequency of its occurrence is unknown.

MANUAL TESTING PROCEDURES
Culture-Based Tests
Specimens Appropriate for Culture

Culture is considered relatively insensitive for the detection of M. pneumoniae, and alternative techniques, such as PCR and serology, should be considered, even if culture is attempted. Respiratory tract specimens, including nasopharyngeal and throat swabs, sputum, pleural fluid, broncheoalveolar lavage fluid, endotracheal aspirates, and lung tissue, are all acceptable specimens for detection of M. pneumoniae by culture. Urethral swabs in men are preferred over urine for detection of genital mycoplasmas. Prostatic secretions, semen, and urinary calculi can be cultured. For females, urine and cervical or vaginal swabs are acceptable. Specimens that are contaminated by lubricants or antiseptics should be avoided. Urine samples from females are most meaningful when obtained by catheter or suprapubic aspiration and if numbers of organisms are quantitated. Endometrial tissue, tubal samples, or pouch of Douglas fluid can be obtained to confirm the mycoplasmal etiology of pelvic inflammatory disease or postpartum fever. For women with clinical amnionitis, the amniotic fluid, blood, and placenta should be cultured. Culture of nasopharyngeal, throat, and endotracheal secretions of neonates is appropriate, especially if the birthweight is <1,500 g and there is clinical, radiographic, laboratory, or other evidence of pneumonia.

Extragenital or extrapulmonary specimens submitted for culture should reflect the site of infection and the disease process. Ureaplasmas and mycoplasmas should always be sought from synovial fluid in the setting of acute arthritis in hypogammaglobulinemia. Other sterile fluids, including peritoneal fluid, pericardial fluid, cerebrospinal fluid, and blood, are suitable for culture. Bone chips from patients with chronic osteomyelitis without a proven bacterial etiology are also appropriate for culture, as are wound aspirates and tissue collected by biopsy or autopsy. Mollicutes are inhibited by sodium polyanethol sulfonate, present in most commercial blood culture media, but the effect can be overcome by adding 1% gelatin (24). Commercial blood culture media designed for use in automated instruments may support the growth of M. hominis, but the instruments usually do not flag the bottles containing this organism as positive (39a). Successful isolation of M. hominis and Ureaplasma spp. from blood can be achieved by inoculating ≥10 ml directly into liquid mycoplasmal growth medium in at least a 1:10 ratio. Smaller volumes can be used for neonates or children.

There is sometimes difficulty in detecting a color change in liquid media in the presence of large amounts of blood due to hemolysis, and there may be a slight color change immediately after introducing blood into liquid media. Serial dilution of the original specimen and subcultures to agar will help distinguish such nonspecific color changes.

Specimen Collection and Transport

Mycoplasmas are extremely sensitive to adverse environmental conditions, particularly to drying and heat. Specimens should be inoculated at bedside whenever possible, using appropriate transport or mycoplasma culture medium. Transport media include liquid mycoplasma culture media, such as SP4 (35) for Mycoplasma spp. or 10B broths (28) for M. hominis or Ureaplasma spp. These media are manufactured and sold in the United States by Remel Inc. 2 SP (10% [vol/vol] heat-inactivated fetal calf serum with 0.2 M sucrose in 0.02 M phosphate buffer, pH 7.2) and tryptic soy broth with 0.5% (vol/vol) bovine serum albumin (Remel Inc.) are also acceptable transport media. Other media available commercially in the United States for transport and storage of specimens include Mycotrans (Irvine Scientific), A3B, and arginine broth (Remel Inc.). Specialized liquid transport media, such as A3B, have been specially designed by deletion of some of the growth supplements present in other growth media so that the increase in pH caused by urea hydrolysis by ureaplasmas will be delayed and result in less toxicity to the organisms during transport. Specialized transport media for genital mycoplasmas are also available from various suppliers in France and several other countries in Europe and elsewhere. These include R1 medium (bioMérieux), A3 (International Microbio and PBS Orgenics), and UMMt (International Microbio).

Regardless of which transport medium is used, if specimens are obtained from body sites with indigenous bacterial flora, such as the lower urogenital tracts of women or the throat, incorporation of antibiotics or other inhibitors is essential to prevent bacterial overgrowth. The same consideration must be applied to growth media. Some commercial products already contain inhibitors, whereas others provide them with the product in the form of disks or tablets to be added at the time of culture inoculation. Beta lactam antibiotics, such as ampicillin, penicillin, or cefoperazone, and an antifungal agent, such as amphotericin B or nystatin, should be added. Some products incorporate thallous acetate. However, this substance has some drawbacks in that it is inhibitory to some mycoplasmal species and ureaplasmas and can also be toxic to humans.

Liquid specimens or tissues do not require special transport media if cultures can be inoculated within 1 h, provided the specimens are protected from drying. Tissues can be placed in a sterile container, which can be delivered to the laboratory immediately. Otherwise, tissue specimens should be placed in transport medium if delay in culture inoculation is anticipated. When swabs are obtained, care must be taken to sample the desired site vigorously to obtain as many cells as possible, since mycoplasmas are cell associated. Calcium alginate, dacron, and polyester swabs with aluminum or plastic shafts are preferred. Cotton swabs with wooden shafts should be avoided because of potential inhibitory effects. Specimens should be refrigerated if immediate transportation to the laboratory is not possible. If specimens must be shipped and/or if the storage time is likely to exceed 24 h prior to processing, the specimen in transport medium should be frozen to prevent loss of viability. Mollicutes can be stored for long periods in appropriate growth or transport media at −70°C or in liquid nitrogen. Storage at −20°C for even short periods will result in loss of viability. Frozen specimens may be shipped with dry ice to a reference laboratory if necessary. When specimens are to be examined, they should be thawed rapidly in a 37°C water bath. In order to maximize the yield of mycoplasmas from clinical specimens, serial 10-fold dilution in liquid medium to 10^{-3} to remove inhibitors and inoculation of broth dilutions onto appropriate agar have been suggested (35, 39). Tissues should be minced with scissors prior to dilution.

Growth Media and Diagnostic Kits

Many different liquid and solid media have been developed by researchers over the years for cultivation of mycoplasmas that occur in humans. No single formulation is ideal for all pertinent species due to their different properties, optimum pHs, and substrate requirements. Mycoplasmas of importance for human infections utilize glucose and/or arginine as metabolic substrates, whereas ureaplasmas utilize urea. The various medium formulations designed for their cultivation must provide these substances along with other essential ingredients, including serum as a cholesterol source, yeast extract, and a pH indicator for detection of growth when the substrate is metabolized. SP4 broth and agar (35) can be used for both M. pneumoniae and M. hominis. These media can also be used to cultivate other fastidious and slow-growing species. Shepard's formulations of U9 or 10B broth (28) have been used successfully for cultivation of Ureaplasma spp., with A7 or A8 agars (29) as the corresponding solid media. Other media, such as bromthymol blue (B broth), Hayflick's medium, Boston broth, PPLO medium, and several others have also been used for detection of mycoplasmas and/or ureaplasmas in clinical specimens. Detailed description of essential components of growth medium, formulations, instructions for the preparation of various nonproprietary types, and incubation conditions for different mycoplasmas can be found in reference texts (39). Positive and negative controls are recommended to ensure adequate detection of growth in all media.

Biphasic media have been used successfully for many years for cultivation of M. pneumoniae. An agar slant is prepared in a small screw-cap bottle to which broth is added to fill one-half to two-thirds the height of the agar. The rationale for this approach is that it supplies a wide range of atmospheric conditions, and inhibitor metabolites present in the specimen may be absorbed by the agar, further promoting the possibility of growth. This idea has been adapted by Irvine Scientific in its Mycotrim medium formulations.

For self-prepared media, quality control is crucial for each of the main components. New lots or batches of broth are considered satisfactory if the numbers of organisms that grow in them are within 10-fold of the number that grow in the reference batch. Agar plates should support the growth of at least 90% of the colonies that are supported by the reference media. Remel Inc. sells lyophilized cultures of American Type Culture Collection (ATCC) type strains in the form of BACTI disks that can be used for laboratory quality control procedures for mycoplasma media. ATCC type strains may also be purchased from Irvine Scientific. Low-passage clinical isolates of Mycoplasma spp. and Ureaplasma spp. should be included in quality control testing. For further information on this topic, see reference 39. Almost any formulation of mycoplasma growth medium that is prepared correctly can be expected to grow prototype

strains. The real challenge is to isolate mycoplasmas and ureaplasmas from clinical specimens, hence the need for inclusion of known positive specimens or recent clinical isolates. Testing inhibitory properties of media against the growth of various other organisms likely present in specimens from nonsterile sites may also be worthwhile to prevent loss of mycoplasmas due to overgrowth of contaminating organisms. If commercially prepared media or kits are to be utilized, it is advisable that laboratories perform internal quality control tests, and users should be aware of the potential limitations of existing products.

Several companies sell growth media patterned after the original formulations developed by researchers in mycoplasmology, and some have developed kits for the detection and preliminary identification of the most common clinically significant species. The products available vary from one location to another, with a wider selection produced and sold in Europe and certain other countries than in North America.

Mycoscreen GU (Irvine Scientific) is available in the United States as a broth culture to screen for M. hominis and Ureaplasma spp. Growth of M. hominis in Mycoscreen GU is suggested by a change in the color of the broth from orange to orange-red, whereas Ureaplasma spp. cause a color change from orange to red. Vials showing color change must be subcultured to the Mycotrim GU triphasic flask system or other solid medium for isolation and identification. The triphasic flask system incorporates an agar growth surface and enrichment broth containing arginine, urea, and phenol red separated by a humid air phase in a single flask. Mycotrim GU agar is similar to A7 agar in basic composition but differs in some of the supplements included and uses CaCl₂ as the urease indicator, in the same manner as A8 agar. A pipette is used to add liquid specimens and to inoculate the agar surface. Swab specimens are streaked directly across the agar surface. The broth-specimen mixture is washed over the agar phase by rotating the flask. Using this product, growth in the broth medium, indicated by a color change, can be verified and further characterized by the observation of colonies. Ureaplasmas and M. hominis may produce an initial color change in Mycotrim GU broth in as little as 24 h. The flask can be placed directly onto a microscope stage, and the surface of the agar can be examined for the presence of colonies. A comparable system adapted to M. pneumoniae detection, Mycotrim RS, using enriched glucose agar and broth, is also available. M. pneumoniae will require ≥72 h in Mycotrim RS to produce a color change from red to yellow-orange in the broth. Prior to incubating these systems, antibiotic disks that are supplied with the product are added to the liquid to prevent bacterial overgrowth. Gross turbidity indicates bacterial contamination, since the small cell size of mycoplasmas and ureaplasmas does not typically produce turbid growth in liquid media. Bacterial overgrowth, if present, can completely mask the presence of mycoplasmas and ureaplasmas in clinical specimens. Mycoscreen GU broths are sold in boxes of 96 with a list price of $70.65 ($0.74 each). Mycotrim GC triphasic flasks are sold in boxes of 12 for $144.40 ($12.03 each), and Mycotrim RS triphasic flasks are sold for $186.00 ($15.50 each).

Remel Inc. markets several formulations of growth medium, including 10B broth and A7 and A8 agars for cultivation of M. hominis and Ureaplasma spp., 10B broth with arginine and arginine broth for M. hominis, and SP4 broth and agar for M. pneumoniae, M. hominis, and other Mycoplasma spp. PPLO broth and agar are also available for general-purpose cultivation of mycoplasma species, but since the broth does not contain a pH indicator, it is not possible to detect growth by direct observation of the liquid medium. The manufacturer will produce customized orders of various sizes for most of the media, with or without inhibitors or special additives as desired by the purchaser. Most Remel broths and transport media are available in lyophilized form to prolong shelf life.

Mycoplasma Experience, a British company, is primarily involved in the sale and manufacture of diagnostic products for use in detecting mycoplasmas of veterinary importance, but they also produce their own line of liquid and solid media for human mycoplasmas and ureaplasmas. Mycoplasma Experience manufactures broths containing specific biochemical substrates, e.g., for glucose fermentation or arginine hydrolysis to aid in preliminary characterization of species. Agar plates are available in different sizes, and broths can be purchased in a range of volumes, with individual units of up to 100 ml. Solid media can be puchased from Mycoplasma Experience as three components: agar, freeze-dried supplement, and diluent. Liquid media can be purchased as two components: freeze-dried medium and diluent. Freeze-dried media are stable for at least 12 months prior to reconstitution. At this time, Mycoplasma Experience does not market its products in the United States.

Several French companies produce a variety of liquid and solid media for urogenital-mycoplasma detection. Some of the products are the companies' versions of the traditional media initially described by mycoplasma researchers, while others are proprietary products of their own design. Some transport systems and liquid and solid growth media are now available as components of complete diagnostic kits for human mycoplasmas and are marketed primarily in Europe and also in other regions, but not in the United States at present. Diagnostic kits for the detection and preliminary characterization of mycoplasmas are generally similar, consisting of strips with wells containing specific dried or lyophilized substrates and inhibitors. Specimens are placed in a suspension medium that is used to inoculate the wells. The detection and identification of organisms is based on the color change of specific wells containing biochemical substrates and inhibitors. Some products also contain antimicrobial agents so that in vitro susceptibilities can be determined simultaneously.

Descriptions of some of the most widely used commercial mycoplasma growth media, corresponding diagnostic kits, and their manufacturers are provided in Table 1. The kits described in Table 1 include the products that have undergone external validation in clinical trials that have been published in peer-reviewed journals. Citations of published studies, when available, are also indicated in the table. Other commercial products exist, some marketed by these same companies, but they have not been subjected to rigorous comparison with standard methods of testing. The principles and test procedures of the other media and kits not shown are generally similar to those which have been described in detail and mainly involve detection and/or susceptibility testing of urogenital mycoplasmas. International Microbio also produces the Pneumofast M. pneumoniae culture for isolation, quantitation, and susceptibility testing of M. pneumoniae, but the product has not been independently evaluated.

Commercial products developed for use in the detection and characterization of urogenital infections caused by M. hominis and Ureaplasma spp. have been designed for use with specimens from the genital tract. No information or recommendations are available regarding whether these products are suitable for specimens from other sterile body sites or for respiratory tract specimens in neonates.

A few studies have been published within the past several years examining the capabilities of commercial media sold in the United States to detect M. hominis and ureaplasmas in clinical specimens. Only studies describing currently available products have been included in this discussion. Results indicate that certain products may be comparable to self-prepared media in some settings but that problems continue to exist with recovery of both M. hominis and ureaplasmas. Problems identified in the past with commercial media are likely related to contamination of serum with mycoplasmas of animal origin, unsatisfactory ability of other individual medium components to sustain the growth of mycoplasmas or ureaplasmas, inadequate quality control using fastidious strains, or possibly inattention to important procedures, such as serial dilutions and broth-to-agar culture techniques for satisfactory recovery of these organisms. The importance of the broth-to-agar technique and serial dilutions for recovery of ureaplasmas from respiratory tract specimens from neonates has been demonstrated previously (36).

Wood et al. (40) reported that the Mycotrim GU broth-agar system worked as well as arginine and urea broths, glucose agar, and A7B agar they prepared themselves in an evaluation of 100 clinical specimens, 18 of which were positive for Mycoplasma spp. and 33 of which were positive for Ureaplasma spp. These investigators were able to detect ureaplasma colonies slightly more often with the Mycotrim system than with conventional nonproprietary media. In contrast to the findings of Wood and colleagues (40), Phillips et al. (23) found that the Mycotrim GU system detected only 84% of Ureaplasma strains and 73% of M. hominis strains in comparison to 89 and 93%, respectively, using A7 agar. Broitman et al. (6) compared Remel's A7 and A8 agars in combination with Mycotrim GU broth alone versus the Mycotrim triphasic flask system. The triphasic flask system detected Ureaplasma spp. in only 25% of 64 positive specimens versus 98% with GU broth and A7 agar and 100% with GU broth and A8 agar. The Mycotrim triphasic flask system detected 94% of 18 positive M. hominis cultures, all of which were detected using GU broth and A7 or A8 agar. They concluded that Mycotrim GU broth inoculated simultaneously with A7 or A8 agar was more sensitive and cost-effective than the Mycotrim triphasic flask system. Similar results were obtained by Cavicchini et al. (7). More recently, Welborn et al. (W. Welborn, L. Skodack-Jones, and K. Carroll, Abstr. 97th Gen. Meet. Am. Soc. Microbiol., abstr. G-8, p. 281, 1997) compared the Mycotrim triphasic flask sytem with Remel's 10B broth and A7 agar in 195 clinical specimens. Seventy-nine (40.5%) positive cultures were detected with the Remel media versus 69 (35.4%) with Mycotrim. Mycotrim had a 13% bacterial contamination rate versus <1% for the Remel media. No published studies are available for Mycotrim RS.

Several types of broths, agars, and diagnostic kits sold outside the United States for the detection and preliminary identification of Ureaplasma spp. and M. hominis have been evaluated in comparison to one another and to other nonproprietary products. Sillis (31) performed a limited evaluation and reported that Mycoplasma-LYO, an arginine-urea broth and agar system (bioMérieux), provides qualitative and quantitative results suitable for diagnostic use. Abele-Horn and colleagues (1) conducted the most comprehensive study to date. They performed cultures of 298 clinical specimens using Mycoscreen broth and agar (International

TABLE 1 Mycoplasma detection and susceptibility testing kits produced in France and sold in various countries[a]

				Product and information				
Parameter	Mycoplasma IST	Mycoplasma-Lyo	Mycoplasma "All-In"	Mycofast Evolution 2	Mycokit-NUM	Mycokit-ATB	Mycoplasma DUO	SIR Mycoplasma
Manufacturer	bioMérieux	bioMérieux	International Microbio	International Microbio	PBS Orgenics	PBS Orgenics	Bio-Rad (formerly Sanofi Diagnostics Pasteur)	Bio-Rad (formerly Sanofi Diagnostics Pasteur)
Kit description	R1 medium flasks R2 urea-arginine broth flasks Microplates with 16 wells	R1 urea-arginine broth flasks R2 sterile water flasks R3 agar flasks R4 additive-component flasks R5 sterile-water flasks	Microplates with 4 rows of 2 wells A3 medium flasks A7 agar plates Mh supplement dropper bottle	Racks composed of 2 rows of 10 wells UMMt transport medium flasks UMMlyo growing medium flasks Mh supplement dropper bottle	A3 medium vials U9 urea medium flask M42 arginine medium flask Racks composed of 2 rows of 8 wells Adhesive sheets	U9 urea medium flask M42 arginine medium flask Racks composed of 2 rows of 8 wells Adhesive sheets	Microplates with 6 wells Suspension medium vials Diluent dropper bottle Micropipettes Adhesive sheets	SIR microplates with 2 rows of 8 wells Adhesive sheets
Specimen types	Vaginal, endocervical, urethral swabs; semen; urines	Vaginal, endocervical, urethral swabs; semen; urine specimens	Vaginal, endocervical, urethral swabs; semen; urines	Vaginal, endocervical, urethral swabs; semen; urines	Vaginal, endocervical, urethral swabs	Broth culture of M. hominis or Ureaplasma spp. from A3 broth inoculated with original sample	Vaginal, endocervical, urethral swabs; semen; urines	Broth culture of M. hominis or Ureaplasma spp. from well X of Mycoplasma DUO or from A3 broth inoculated with the sample and enriched with mycoplasmas
Principle and technical procedure	Principle: detection, identification, and quantitation of M. hominis and Ureaplasma spp. based on differential utilization of biochemical substrates contained in broth media and determination of antimicrobial	Principle: detection, identification, and quantitation of M. hominis and Ureaplasma spp. in broth and agar media. Procedure: To prepare the urea-arginine broth, dissolve R1 in 2 ml of R2. For agar plates, boil	Principle: detection, identification, and quantitation of M. hominis and Ureaplasma spp. based on differential utilization of biochemical substrates contained in broth media. Procedure: Soak the swab or put	Principle: detection, identification, and quantitation of M. hominis and Ureaplasma spp. based on differential utilization of biochemical substrates contained in broth media and determination of antimicrobial	Principle: detection, identification, and quantitation of urogenital mycoplasmas, M. hominis and Ureaplasma spp. Procedure: Soak the swab in the A3 tranport medium vial. Distribute 180 µl of U9 medium in	Principle: Liquid-medium susceptibility testing of urogenital mycoplasmas, M. hominis and Ureaplasma spp. Procedure: According to the mycoplasma titer obtained by preculture in A3 broth from the Mycokit-NUM,	Principle: detection, identification, and quantitation of urogenital mycoplasmas, M. hominis and Ureaplasma spp. based on differential utilization of biochemical substrates contained in broth media. Identification.	Principle: liquid medium susceptibility testing of urogenital mycoplasmas, M. hominis and Ureaplasma spp.; 8 antibiotics are tested, 6 at 2 different concentrations (doxycycline, minocycline, tetracycline,

(Continued on next page)

TABLE 1 Mycoplasma detection and susceptibility testing kits produced in France and sold in various countries[a] (Continued)

Parameter	Mycoplasma IST	Mycoplasma-Lyo	Mycoplasma "All-In"	Mycofast Evolution 2	Mycokit-NUM	Mycokit-ATB	Mycoplasma DUO	SIR Mycoplasma
	susceptibilities by broth microdilution. Procedure: Soak the swab or put 200 μl of urine or semen in the R1 transport medium; then put 3 ml of the seeded R1 medium in the R2 broth flask. Distribute 55 μl of seeded R2 and then 2 drops of paraffin oil in each well of the microplate. Cover the microplate with its top. Incubate the microplate and the remaining R2 seeded broth for 48 h at 37°C. Reading of results consists of identifying the color changes obtained in the seeded flask or in microplate wells. Check the color change in the R2 seeded flask after 24–48 h. Color change (yellow to red-orange for Ureaplasma spp., yellow to red-	flask R3 for 15–20 min, and then cool it to 45–50°C. Dissolve R4 in 10 ml of R5, and then put it in the dissolved agar (R3). After homogenization, pour it in plates (6 ml per plate). Soak the swab or put 200 μl of urine or semen in the urea-arginine broth. Inoculate 3 drops of the seeded broth onto an agar plate and incubate both broth and agar plate for 24–48 h at 37°C under anaerobic or microaerophilic conditions. Reading of results consists of identifying the color changes obtained in the broth flask and the colony morphology on agar. Color change of yellow to red-pink and fried-egg colonies indicate M.	100 μl of urine or semen in the A3 medium, and then distribute successively 150 μl of A3 in well U (urea), then 100 μl in well A (arginine) with a drop of M. hominis supplement. Add 2 drops of paraffin oil to both wells. Inoculate 100 μl or 3 drops of A3 medium seeded with the specimen in the urea-arginine broth. Incubate both microplate and agar plate for 24–48 h at 37°C under anaerobic conditions. Reading of results consists of identifying the color changes obtained in the wells. Color change (yellow to red-orange) in well U means ≥10^4 CCU of Ureaplasma spp./ml. Color change in well A (yellow to red-	susceptibilities using broth microdilution. Procedure: Soak the swab or put 200 μl of urine or semen in the transport medium UMMt, and then put all the seeded UMMt medium in the UMMlyo flask. Distribute 100 μl of UMMlyo seeded with the specimen in wells 1 to 10, then 2 drops of Mh supplement in wells 9 and 10, then 2 drops of paraffin oil in wells 1 to 10. Cover the rack with an adhesive sheet, and then incubate it for 24–48 h at 37°C. Reading of results consists of identifying the color changes obtained. Reading is identical for both mycoplasmas: yellow well, no Ureaplasma spp. or M. hominis; red well, pres-	all wells of the first row and 180 μl of M42 medium in wells of the second row. Put 20 μl of the A3 seeded medium in the first well (well A) of each row, and then make doubling dilutions until the last well (well H) from each row. Cover the rack with an adhesive sheet and incubate it for 48 h at 37°C. Reading of results consists of identifying the color changes obtained in wells. Yellow well, no Ureaplasma spp. or M. hominis; green-blue well, presence of Ureaplasma spp. or M. hominis. The rack with U9 medium indicates growth of Ureaplasma spp., while the second rack indicates M. hominis growth. Quantitation is deter-	dilute the A3 broth culture with medium U9 for Ureaplasma spp. or medium M42 for M. hominis to yield a final inoculum at 10^3 CCU/ml; 4 ml is necessary for the susceptibility tests. Distribute 200 μl of the inoculum into each well of the rack. Cover it with an adhesive sheet and incubate it for 48 h at 37°C. Seven antibiotics, each at 2 different concentrations, are tested: fucidic acid, doxycycline, erythromycin, lincomycin, minocycline, pristinamycin, and josamycin. Interpret as soon as control wells have changed color from yellow to green-blue. Two yellow wells mean a lack of growth (susceptible), while 2 green-	cation of Ureaplasma spp. and M. hominis is based on specific hydrolysis of urea by Ureaplasma spp. or arginine by M. hominis, which is indicated by a change in color of the well containing the relevant substrate, well U containing urea or well H containing arginine. Procedure: Transfer 4 drops (200 μl) from the diluent dropper bottle to each of the 3 wells of the lower row of the microplate (U ≥ 10^4, D, and H ≥ 10^5). Soak the swab, or put 200 μl of urine or semen in the suspension medium. Distribute 100 μl of the suspension medium seeded with the specimen in each of the 3 wells of the upper row of the microplate (U, X, and H) and 25 μl in well D.	josamycin, erythromycin, and ofloxacin) and 2 at 1 concentration (clindamycin and pristinamycin). To inoculate the susceptibility test medium with an inoculum of 10^3 to 10^5 CCU/ml, perform a preculture of the medium inoculated with the sample using the Mycoplasma DUO kit, leading to a growth titer of 10^6 to 10^7 CCU/ml. A calibrated inoculum is obtained by dilution of this preculture to 1/100 in U9 broth or arginine broth, depending on the isolated species. From the Mycoplasma DUO kit, the contents of well X after 24–48 h of incubation for M. hominis or 24 h for Ureaplasma spp. correspond to the preculture and is diluted as

Product and information

TABLE 1 (Continued)

Parameter	Mycoplasma IST	Mycoplasma-Lyo	Mycoplasma "All-In"	Mycofast Evolution 2	Mycokit-NUM	Mycokit-ATB	Mycoplasma DUO	SIR Mycoplasma
				Product and information				
	pink for M. hominis) indicates growth. Microplate reading has to be done after 16–24 h for well 4 (Ureaplasma spp. quantitation) and after 24–48 h for the others. A yellow-to-red color change indicates a positive reaction. The microplate is divided into 3 parts. Wells 1 to 3 correspond to the identification part (well 2 for Ureaplasma spp., well 3 for M. hominis). Wells 4 and 5 correspond to the quantitation component; a color change in well 4 means $\geq 10^4$ CCU of Ureaplasma spp./ml, and a color change in well 5 means $\geq 10^4$ CCU of M. hominis/ml. Wells 6 to 16 correspond to	hominis growth. Color change of yellow to red-orange and small brown granular colonies indicate growth of Ureaplasma spp. In case of association of both Ureaplasma spp. and M. hominis in the same specimen, color change in broth will be yellow to red-pink, and the association of both colony types on agar will make the diagnosis. Quantitation is realized directly by numbering colonies on agar (>1 colony, 10^3 CFU; 1–5 colonies, 10^4 CFU; 5–15 colonies, 10^5 CFU; >15 colonies, 10^6 CFU).	pink) means $\geq 10^4$ CCU of M. hominis/ml. On A7 agar, more than 1 colony in a microscopic field means $\geq 10^4$ CFU/ml.	ence of Ureaplasma spp. or M. hominis. Wells 1 to 3 contain lincomycin, which inhibits M. hominis growth, and are used for Ureaplasma spp. identification and quantitation. Wells 4 to 6 are used for testing susceptibility to 3 antibiotics, doxycycline, roxithromycin, and ofloxacin, at a single concentration. Wells 7 to 9, containing lincomycin, trimethoprim-sulfamethoxazole (SXT), and erythromycin, are used for Ureaplasma spp. or M. hominis identification using their natural antibiotic susceptibility profiles. Ureaplasma spp. are naturally susceptible to erythromycin and resistant to	mined from serial dilutions, from 10^{-2} (well A) to 10^{-9} (well H); 10^{-1} corresponds to the seeded A3 vial. The titer in CCU/ml corresponds to the last dilution with a well color change.	blue wells mean growth in the presence of antibiotic (resistant). An intermediate strain will give a green-blue low-antibiotic-concentration well and a yellow high-antibiotic-concentration well.	Homogenize the contents of well D, and transfer 25 µl to well D, and transfer 25 µl to well U $\geq 10^4$ and 25 µl to well H $\geq 10^4$. Cover the microplate with an adhesive sheet, and incubate it for 24–48 h at 37°C. Reading of the results is identical for both mycoplasmas: well U or H yellow, no Ureaplasma spp. or M. hominis; well U or H red, presence of Ureaplasma spp. or M. hominis. Read wells U and U $\geq 10^4$ to evaluate Ureaplasma spp., and read wells H and H $\geq 10^4$ to evaluate M. hominis.	described above to obtain the standardized inoculum (20 µl of the contents in 2 ml of the appropriate broth). A3 broth is inoculated with the sample and incubated for 16 h at 37°C to enrich it with mycoplasmas, and 20 µl is then suspended in 2 ml of the appropriate broth. Distribute 100 µl of the standardized inoculum into each well of the SIR microplate. Cover the microplate with an adhesive sheet, and incubate it at 37°C for 24–48 h. Read the results as soon as the control wells have changed color from yellow to red. Two yellow wells mean a lack of growth (susceptible), while 2 red

(Continued on next page)

207

TABLE 1 Mycoplasma detection and susceptibility testing kits produced in France and sold in various countries[a] (Continued)

Parameter	Product and information							
	Mycoplasma IST	Mycoplasma-Lyo	Mycoplasma "All-In"	Mycofast Evolution 2	Mycokit-NUM	Mycokit-ATB	Mycoplasma DUO	SIR Mycoplasma
	the susceptibility testing component, with 6 antibiotics tested (5 with 2 different concentrations, 1 with a single concentration). For antibiotics with 2 concentrations (doxycycline, josamycin, ofloxacin, erythromycin, and tetracycline), 2 yellow wells indicate a lack of growth (susceptible) while 2 red wells mean growth in the presence of antibiotic (resistant). An intermediate strain will give a red low-antibiotic-concentration well and a yellow high-antibiotic-concentration well. For the antibiotic with a single concentration (pristinamycin), only 2 results are possible, susceptible or resistant.			SXT and lincomycin, while M. hominis is naturally susceptible to lincomycin and resistant to SXT and erythromycin. Well 10, containing erythromycin, which inhibits growth of Ureaplasma spp., is used for M. hominis identification and quantitation.				wells mean growth in the presence of antibiotic (resistant). An intermediate strain will give a red low-antibiotic-concentration well and a yellow high-antibiotic-concentration well. For the 2 antibiotics with a single concentration, only 2 results are possible, susceptible or resistant.

TABLE 1 (Continued)

| | Product and information | | | | | | | |
Parameter	Mycoplasma IST	Mycoplasma-Lyo	Mycoplasma "All-In"	Mycofast Evolution 2	Mycokit-NUM	Mycokit-ATB	Mycoplasma DUO	SIR Mycoplasma
No. of tests/kit	25	40	15	12 or 30	24	6	20	10
Kit storage temp (°C)	2–8	2–8	2–8	2–8	2–8	2–8	2–8	2–8
Specimen storage	If delay in processing is anticipated, storage at −20°C (up to 6 months) or −70°C (several years) is recommended.	If delay in processing is anticipated, storage at −20°C (up to 6 months) or −70°C (several years) is recommended.	If delay in processing is anticipated, storage at −20°C (6 months) or −70°C (several years) is recommended.	If delay in processing is anticipated, storage at −20°C (6 months) or −70°C (several years) is recommended.	If delay in processing is anticipated, storage at −20°C (6 months) or −70°C (several years) is recommended.	If delay in processing is anticipated, storage at −20°C (up to 6 months) or −70°C (several years) is recommended.	If delay in processing is anticipated, storage at −20°C (6 months) or −70°C (several years) is recommended.	If delay in processing is anticipated, storage at −20°C (6 months) or −70°C (several years) is recommended.
Controls	Check that all wells are clear after incubating. A cloudy medium indicates bacterial contamination.	Check that all wells are clear after incubating. A cloudy medium indicates bacterial contamination.	Check that all wells are clear after incubating. A cloudy medium indicates bacterial contamination.	Check that all wells are clear after incubating. A cloudy medium indicates bacterial contamination.	Check that all wells are clear after incubating. A cloudy medium indicates bacterial contamination.	Check that all wells are clear after incubating. A cloudy medium indicates bacterial contamination.	Check that all wells are clear after incubating. A cloudy medium indicates bacterial contamination.	Check that all wells are clear after incubating. A cloudy medium indicates bacterial contamination.
Cost (euros)	154	211	157	124 (12 tests) 296 (30 tests)	95	33	124	28
Reference(s)	1, 10	31	1	Poulin and Kundsin, Abstr. 97th Gen. Meet. Am. Soc. Microbiol.			1, 10	1, 26

[a]Countries where these products are sold vary. None of these products is currently approved for use in the United States. Specific information about product availability should be obtained directly from the manufacturer. CCU, color-changing unit.

Microbio), arginine broth and A7 agar (bioMérieux), U9 broth, arginine broth, and A7 agar (Bio-Rad [formerly Sanofi Diagnostics Pasteur]), and mycoplasma broth and A7 agar (Biotest AG). Isolation of M. hominis and Ureaplasma spp. with the commercial media was compared to recovery of these organisms with standard media that were not specifically identified or described by the authors. A total of 73 isolates of Ureaplasma spp. and 33 M. hominis isolates were detected. They reported that the performance of the various commercial media was similar to that of standard media for identification of Ureaplasma spp. and M. hominis, with specificities of about 100% for all cultures and sensitivities of 94 to 100% for Ureaplasma spp. and 8 to 10% for M. hominis. Poulin and Kundsin (Abstr. 97th Gen. Meet. Am. Soc. Microbiol., abstr. G-9, p. 281, 1997) reported that the Mycofast Evolution 2 and the Mycofast US kits (International Microbio) correctly identified all cultures positive for Ureaplasma spp. and/or M. hominis as determined by standard methods using Boston broth and A7 agar, with no false-positive results. Abele-Horn et al. (1) evaluated the Mycoplasma IST (bioMérieux), the Mycoplasma DUO (Bio-Rad [formerly Sanofi Diagnostics Pasteur]), and the quantitative Mycoplasma "All-In" (International Microbio) products with 298 clinical specimens and found the results were comparable to those with standard media and procedures. The specificities were 100% for all cultures, and the sensitivities were 85 to 97% for Ureaplasma spp. and 79 to 83% for M. hominis. However, the standard methods were superior to the commercial kits for quantitative identification, with a 71% correlation for M. hominis and a 71 to 80% correlation for Ureaplasma spp. The authors concluded that the commercial media were generally satisfactory for organism detection and identification but that when quantitation is desired, standard methods are still recommended. Clegg et al. (10) compared the Mycoplasma IST kit (bioMérieux) with their own arginine broth, 10C broth, and A7 agar to determine the prevalence of M. hominis and Ureaplasma spp. in 100 vaginal specimens. The Mycoplasma IST kit gave a sensitivity and specificity of 92.9 and 86.7%, respectively, for M. hominis versus 97.4 and 72.7% respectively, for Ureaplasma spp. They noted that false-positive reactions may occur with the Mycoplasma IST kit if contaminating bacteria are present. Therefore, organism identification and purity should be verified by colony morphology on agar plates. Lower organism numbers detected by commercial kits in comparison to standard methods have been attributed to the fact that standard methods, unlike the kits, include recommendations for serial dilution of the original specimen to remove inhibitors that may be present. Some of the kits and media, such as those described above, have shelf lives of several months, making them attractive for use in laboratories that have only an occasional need for performing mycoplasma cultures, but they are considerably more expensive than media and reagents prepared in house.

For completeness, it is relevant to mention that M. hominis is unique among the mycoplasmal species that occur in humans in that it may occasionally be detected growing in clinical specimens inoculated onto bacteriologic media, such as Columbia CNA, chocolate, or even tryptic soy agar with sheep blood after 3 to 5 days of incubation, usually under 5 to 10% CO_2 or anaerobic conditions. Many accidental detections of M. hominis infections of the urogenital tract or systemic sites have been reported because of the organism's ability to survive and grow under conditions not specifically designed for its cultivation. Laboratories that encounter pinpoint colonies that do not Gram stain should consider the possibility of M. hominis and attempt to establish the identity of the organism as a mycoplasma by subculturing it to appropriate mycoplasma medium and proceeding accordingly or placing the organism in suitable transport medium and referring it to a reference laboratory, if clinically indicated. The availability of frozen or freeze-dried mycoplasma media from commercial sources facilitates such transport or subculture as may be needed on an infrequent basis for low-volume laboratories that do not offer complete diagnostic service for mycoplasmas.

Diagnostic Approach for Detection and Characterization of Mycoplasmas in Clinical Laboratories Using Culture

Detection of mycoplasmas and ureaplasmas in clinical specimens by culture involves careful consideration of the type of specimen to be cultured, the type of patient suspected of having infection, and the organisms sought. For respiratory specimens, if M. pneumoniae is the only organism of interest, it is sufficient to inoculate a single general-purpose medium type such as SP4 broth and agar, using either a broth-to-agar technique or a biphasic medium system. Likewise, if only M. hominis or Ureaplasma spp. are of interest, it is sufficient to set up cultures using a single broth and agar formulation, such as 10B broth and A8 agar. Mycoplasmas of human origin grow best in an atmosphere supplemented with 5 to 10% CO_2, necessitating the use of a CO_2 incubator or candle jar. If the specimen to be cultured is from an extragenital or extrapulmonary site, is a normally sterile body fluid or tissue, and/or the patient is immunosuppressed, any mycoplasma of human origin or accidental infection with a mycoplasma of animal origin should be considered. Many of these species are not reliably detected in culture by the culture protocols described for the more common species, and alternative non-culture-based tests should also be performed. However, isolation by culture may occasionally be successful, usually after prolonged incubation. To maximize the potential yield, SP4 broth and agar, containing both glucose and arginine, should also be inoculated and incubated anaerobically.

Even though the Gram stain is not useful for observation of mycoplasmas due to their exremely small size and lack of a cell wall, this technique can assist in excluding the presence of other bacteria from a clinical specimen. To stain mycoplasmas in clinical specimens or cultures, it is sometimes helpful to employ a DNA fluorochrome stain, such as Hoechst 33258 (21). This stain is commercially available from ICN Biomedicals, Inc.

For detailed procedures concerning the performance and interpretation of mycoplasma cultures and identification of species, the reader is referred to reference 39. Therefore, only brief comments will be presented here. Ureaplasma spp. can easily be identified by the characteristic brown granular colony morphology that occurs due to urease activity on an agar, such as A8, containing a $CaCl_2$ indicator. In order to identify an unknown large-colony mycoplasma definitively to species level, a number of different techniques are available. Although none is practical for a diagnostic laboratory that may encounter such organisms very rarely, there are a variety of procedures, using a combination of common biochemical reactions, colony morphologies, and growth rates, that can assist in the preliminary characterization of the most frequently encountered isolates and that will be sufficient in most clinical settings. These are described in detail elsewhere (39).

A slow-growing glycolytic organism that produces spherical colonies on SP4 medium and exhibits hemolytic activity when colonies are overlaid with guinea pig erythrocytes is most likely to be M. pneumoniae. Oral commensal mycoplasmas can sometimes cause diagnostic confusion with M. pneumoniae in respiratory specimens but should not hemadsorb in this manner. Isolation of M. pneumoniae from respiratory tract specimens is clinically significant in most instances and should be correlated with the presence of clinical respiratory disease, since a small proportion of asymptomatic carriers may exist. Isolation of ureaplasmas in any quantity from normally sterile body fluids or tissues is significantly associated with disease. Fewer than 10^4 ureaplasmas in the male urethra is unlikely to be significant; thus, dilution and quantitation of organisms in clinical specimens can be valuable. For genital specimens yielding an arginine-hydrolyzing, urease-negative organism that produces fried-egg colonies after 3 to 4 days of incubation, a report of presumptive M. hominis is appropriate, and in most instances, no further workup is required. Isolation of M. hominis in any quantity from normally sterile body fluids or tissues is significantly associated with disease, but reports quantitating the numbers of organisms present may be of value in other circumstances.

Nonculture Methods of Diagnosis

Culture is well adapted to species, such as M. hominis and Ureaplasma spp., which can be isolated easily and rapidly from clinical specimens, and it has the advantage of being able to provide quantitative results and an isolate for antimicrobial susceptibility testing. However, it is not satisfactory for the detection of fastidious and/or extremely slow-growing organisms, such as M. genitalium and, to some degree, M. pneumoniae. Therefore, alternative rapid methods have been devised based on knowledge concerning the characterization of epitopes and DNA sequences.

Several types of rapid methods have been developed for antigenic detection of M. pneumoniae. DNA hybridization techniques for the diagnosis of M. pneumoniae infection were developed in the early 1980s. The utility and general acceptance of DNA probes have been limited by low sensitivity, typically requiring 10^3 to 10^6 CFU for detection; low specificity, due to the antigenic similarity between M. pneumoniae and M. genitalium; and nonspecific reactions with other bacteria. Commercial kits were previously sold in the United States by Gen-Probe but are no longer available. International Microbio produces an antigen detection kit, the Pneumofast Ag, in France, and Institut Virion markets the Virion ELISA antigen test for M. pneumoniae in some European countries. Because of the relative insensitivity of nonamplified detection sytems, amplified techniques, such as PCR, have supplanted them. Little attention is being paid to further development of nonamplified tests. The success of commercially manufactured products using PCR technology for detection of another fastidious microorganism, Chlamydia trachomatis, in urogenital infections justifies interest in development of similar products for M. pneumoniae and perhaps even M. genitalium and M. fermentans, even though the disease associations and clinical significance of the last two organisms are somewhat less appreciated by clinicians. As of early 2001, however, there were no complete kits available for detection or characterization of mycoplasmas, even though a few reference laboratories in the United States offer PCR-based testing for M. pneumoniae using their own in-house reagents based on published primer sequences that are available for purchase from vari-

ous sources. Detailed information concerning the PCR assay methodologies and controls and their applications for mycoplasmas has been recently reviewed by Razin (25).

Serological Diagnosis

Despite its drawbacks in immunosuppressed persons who are unable to mount an antibody response, serological diagnosis of M. pneumoniae respiratory infections has been extremely important because of the relative lack of sensitivity and time-consuming nature of culture, as well as the carrier state that may occur in an unknown percentage of persons in the absence of acute infection.

M. pneumoniae possesses both protein and glycolipid antigens that elicit antibody responses in infected individuals. Following an initial infection, the normal immune system responds by rapidly producing antibodies that can be detected after about 1 week of illness, peaking at 3 to 6 weeks, followed by a gradual decline over months to years. Elevation of M. pneumoniae-specific immunoglobulin M (IgM) alone is often interpreted as evidence of acute infection, since this antibody typically appears approximately 2 weeks before IgG antibody (22, 30). This approach has the theoretical advantage that only a single specimen, taken approximately 7 to 10 days after infection, is required. However, the presence of IgM is considered most significant in pediatric populations, in which there have been fewer opportunities for repeated exposures. Adults who have been infected repeatedly over a period of years may not respond to mycoplasma antigens with a brisk IgM response. In these cases, reinfection leads directly to an IgG response; therefore, the absence of a positive IgM test result does not rule out an acute infection. When it does occur, the IgM response may persist for months following infection (41), and in these cases, a positive IgM test result may not reflect a current or recent infection. In view of these considerations, it is advisable to test simultaneously for both IgM and IgG in paired specimens collected 2 to 3 weeks apart for the most accurate diagnosis of recent or current M. pneumoniae infection, especially in adults (34). A fourfold or greater rise in antibody titer indicates a current or recent infection.

IgA, while often overlooked as a diagnostic antibody class, may actually be a better indicator of recent infections in all age groups (30). IgA antibodies are produced early in the course of disease, rise quickly to peak levels, and decrease earlier than IgM or IgG levels (15). In theory, a single early specimen could allow detection of an acute infection even after multiple reinfections. Research into IgA responses in adult and pediatric populations and assessment of antibody levels in other body fluids is warranted.

Cold Agglutinins

Cold agglutinins are IgM antibodies to the I antigen of erythrocytes and are detected by agglutination of type O Rh-negative erythrocytes at 4°C. They are produced 1 to 2 weeks after initial infection and persist for several weeks. Cold agglutinins are found in only 30 to 50% of M. pneumoniae infections. Therefore, a negative result does not exclude mycoplasma infection. Because other bacteria, viruses, or even collagen vascular diseases can induce cold agglutinins, a positive test result is not specific for mycoplasmal infection. Better tests are available that have eliminated the need for this assay, and cold-agglutinin testing is no longer recommended for diagnosis of M. pneumoniae infections.

CF

The complement fixation (CF) test using a chloroform methanol lipid extract of M. pneumoniae has long been used for serodiagnosis of M. pneumoniae pneumonia. CF tests with whole-organism antigen give results similar to those with the lipid antigen. The CF test is a two-step procedure which is based on the ability of an antigen-antibody complex to bind added complement. If added complement is bound by specific antibody-mycoplasma complexes, it cannot later initiate lysis of the sheep erythrocyte indicator system. Absence of erythrocyte lysis correlates with the presence of anti-mycoplasma antibodies in serum. The titer is usually defined as the greatest dilution of serum that shows 0 to 30% hemolysis in a test using a specific amount of antigen and complement and serial twofold dilutions of serum. CF tests measure IgM and, to a lesser extent, IgG. Therefore, this technique cannot differentiate among antibody classes, and recent infection in adults who have minimal IgM response may be missed.

Historically, the CF test gained early popularity among laboratories that routinely ran CF tests for viral agents. However, mycoplasmas are much more antigenically complex than viruses, leading to nonspecific reactions. The glycolipids of M. genitalium are highly cross-reactive with M. pneumoniae due to shared lipid antigens (19) and cause problems in CF tests. Sera from patients with bacterial meningitis also tend to have high CF titers. Cross-reactions of CF antigens with other organ-specific antigens unrelated to microorganisms may also occur.

Kenny et al. (17) reported that among M. pneumoniae culture-positive, X-ray-proven pneumonia patients, 53% showed a fourfold titer increase and 36% showed antibody titers of ≥32. Using both high titers and high stationary titers as criteria, the sensitivity of the CF test was 90% and the specificity was 88%. Single titers of >32 are sometimes considered to be indicative of recent infection. However, this endpoint varies greatly among laboratories, and antibodies to glycolipid antigens may persist for long periods. Confirmation of CF test results with Western immunoblots can aid in interpretation but adds greatly to the time and expense of testing.

In view of the many limitations of CF tests and their time-consuming, labor-intensive nature, they have largely been replaced by improved methods using alternative technologies. The target antigens vary among tests; some are proteins, while others are glycolipids. Whether there is a significant cross-reactivity with M. genitalium in these M. pneumoniae antibody test systems has not been established. A great many publications have appeared since the early 1980s describing a number of different techniques for measuring antibody response to M. pneumoniae infection, but many of the assays have never been developed for commercial distribution. In consideration of the primary purpose of this book, only the tests that can be purchased as kits will be discussed in detail. Several of the existing commercial kits have been evaluated in clinical studies, using CF as the reference method. However, in most instances it is not as accurate as some of the newer assay formats, and considering its lack of ability to distinguish antibody classes and tendency for cross-reactivity with other microorganisms, it is not really suitable as a reference standard (4, 27). This fact makes the interpretation of the specificity and sensitivity results of comparative studies using CF as the reference standard somewhat problematic. Recent studies evaluating commercial kits have merely compared one method or product with another without consideration of CF data, but

since results and conclusions from some studies are based upon assay of a single acute-phase serum sample while others have used paired specimens, direct comparisons and extrapolations from multiple studies become rather complex and are not always feasible.

Assay formats adapted for commercial distribution include indirect immunofluorescence assays (IFA), particle agglutination (PA), and enzyme immunoassays (EIA). A German company, Genzyme Virotech GmbH, has developed a Western blot assay for detection of IgG and IgA, but this test is not sold in the United States. Some of the serology kits sold commercially may be marketed under different names in various countries and even by different companies, and some that have been marketed in the past are no longer sold, making it quite difficult to keep track of all of the different products available worldwide. Therefore, discussion will be limited primarily to those tests currently sold in the United States. A summary of the major commercial serological test kits and their various assay formats is provided in Table 2.

IFA

M. pneumoniae antigen affixed to glass slides and specific antibody is detected after staining with anti-human IgM or IgG fluorochrome conjugate. Zeus Scientific distributes separate IFA kits for IgM and IgG. These kits have been studied by several investigators in comparison to other methods (2, 3, 13) and have been shown to provide accurate, quantitative serological data. The Zeus IFA is relatively simple and requires about 2 h to perform, but its interpretation is subjective and it requires a fluorescence microscope. The results can be affected by the presence of rheumatoid factor and high M. pneumoniae-specific IgG antibody levels, and additional procedures are required to validate IgM results in these settings (2).

Particulate Antigen-Antibody Assays

PA tests make use of particles, such as latex, gelatin, or erythrocytes, coated with M. pneumoniae-specific antigen. When erythrocytes are used, the test is referred to as the indirect hemagglutination assay. Erythrocytes are treated with tannic acid or chromium chloride to facilitate adherence of the antigen. The antigen-coated particles are incubated with test serum, and if the serum contains specific antibodies, the particles agglutinate, resulting in a visible reaction. PA tests use a mixture of M. pneumoniae antigens to detect IgG and IgM simultaneously (3, 4, 12, 16, 18, 22). A potential disadvantage of using erythrocytes as the carrier particles in such assays is the presence of other antigens that can interfere with the assay. To circumvent this problem, newer assays using either latex particles or gelatin have been developed. Commercial kits utilizing PA technology (Serodia Myco II) are available in Canada, Europe, and Japan through Fujirebio in a microtiter plate format that provides a quantitative titer result. Lieberman et al. (18) evaluated the Serodia Myco II in comparison to the Sero Mycoplasma pneumoniae antibody-capture EIA kit and concluded that the antibody-capture EIA was better than PA for the diagnosis of current M. pneumoniae infection when testing a single serum sample, but they stressed that testing of acute- and convalescent-phase sera is required for precise diagnosis even with the EIA. Barker et al. (4) also compared the Serodia Myco II PA kit with IgM-specific antibody-capture EIA and IFA. They determined that this PA test was not as specific for detection of IgM as the other tests. Aubert and colleagues (3) felt that PA tests performed about as well as EIA and IFA, but Matas et al. (22) reported that the Meridian ImmunoCard EIA was able to detect lower levels of IgM antibodies than either the Serodia Myco II or CF tests. International Microbio sells their latex agglutination test, the Serofast, in Europe and other locations, but not in the United States at present. This product was evaluated by Karppelin et al. (16) and shown to be less sensitive than CF and EIA and was not recommended as a replacement for either test. In another study (12), the Serofast kit was compared to an antibody capture EIA (MpTest; Diatech Diagnostica) using sera that were shown to be either positive or negative by CF. Both tests were shown to be sensitive and specific, but the performance of the antibody capture EIA for detection of IgM diminished in older persons, as might be expected, because of the likelihood of reinfection without a significant IgM response. Based on evidence presented to date, PA assays do not offer any advantages over other techniques, such as EIA or IFA, except possibly their ease and simplicity of performance.

EIA

EIAs, first developed in the 1970s, have been widely adopted in laboratories, achieving the largest market share of commercial mycoplasma serology tests in the United States. They are amenable to a variety of assay conditions, detect very small amounts of antibody, and can be made isotype specific. Crude multiantigen preparations, purified proteins, μ-capture approaches, purified glycolipids, and synthetic peptides have all been used as targets (4, 9, 12, 13, 14, 27, 41). The basic format is to immobilize an antigen to a solid phase. Patient sera are incubated with the solid phase, and bound antibodies are visualized using substrate and enzyme-labeled conjugates directed against the primary immunoglobulin. The amount of conjugate reacting is proportional to the levels of antibody present in the patient's serum, measured quantitatively with a spectrophotometer. Most EIAs are sold in 96-well microtiter plate formats, although some can be obtained as breakaway microwell strips, which allow smaller numbers of sera to be tested economically. Overall, EIAs have been shown to be more sensitive than CF according to several published studies. Specific examples of commercial EIAs using the microtiter plate format, along with references, are shown in Table 2.

Two EIAs are packaged as qualitative membrane-based procedures for the detection of single test specimens. These are truly rapid EIAs (10 min or less) and are simple to perform. The Meridian ImmunoCard (Fig. 1) is an IgM-only assay that is simple to read and is especially useful for testing pediatric samples (22, 34). Typical of IgM-only assays based on M. pneumoniae protein antigens, the specificity may be somewhat compromised in patients with autoimmune disease, and there are limitations in interpreting the results of IgM-only assays when testing sera from adults, as previously discussed. Other studies using sera from patients with confirmed cases of M. pneumoniae infection have also shown that the ImmunoCard performed better than the CF test (2, 33, 34; W. L. Thacker and D. F. Talkington, Abstr. 97th Gen. Meet. Am. Soc. Microbiol., abstr. G-5, p. 280, 1997).

The Remel EIA (Fig. 2) is a membrane-based assay that detects IgM and IgG simultaneously, and it has shown good sensitivity and specificity compared to other tests, including IFA, the GenBio ImmunoWell EIA, the Diasorin EIA, the Meridian ImmunoCard, and CF (13, 33, 34; Thacker and Talkington, Abstr. 97th Gen. Meet. Am. Soc. Microbiol.).

Although the manufacturers and some investigators (13) have endorsed the use of a single assay such as the Remel IgG and IgM EIA for diagnosis of acute M. pneumoniae infections in young persons, others (33) contend that acute- and convalescent-phase sera are necessary for the greatest accuracy but acknowledge the fact that the value of the single point-of-care tests offered by both the Remel and Meridian kits is lost if paired specimens are required. Other EIA kits are sold in Europe and other locations but not in the United States at present. These include the Platelia Mycoplasma pneumoniae IgM and IgG kit (Bio-Rad) (4), the Virion ELISA M. pneumoniae kit (Institut Virion), and the Genzyme Virotech GmbH Mycoplasma pneumoniae ELISA kit.

Genital Mycoplasma Assays

The ubiquity of most genital mycoplasmas in humans makes interpretation of antibody titers difficult, and the mere existence of antibodies alone cannot be considered significant. However, if invasive extragenital disease occurs, elevation of antibody titers is often apparent. Serological methods for detecting M. hominis, M. genitalium, and/or Ureaplasma spp. include IFA, EIA, microimmunofluorescence, and metabolic inhibition. No single serological test has proven satisfactory for the identification of genital mycoplasmal infections, and most assays for M. hominis and Ureaplasma spp. have not been evaluated for cross-reactions that could occur with sera containing antibodies to M. genitalium. PCR-based studies done at the Centers for Disease Control and Prevention (CDC) showed that respiratory specimens rarely contain M. genitalium. However, this does not preclude the potential for cross-reacting antibodies arising from a genital infection with this organism. No commercial serological kits for detection of antibodies to urogenital mycoplasmas are produced and sold commercially in the United States. However, two French companies, International Microbio and PBS Orgenics, sell kits based on the metabolic-inhibition technique to measure total antibodies. No external clinical evaluations have been reported for these products.

Antimicrobial Susceptibility Testing

Recent publications have provided step-by-step procedures for performing antimicrobial susceptibility tests on human mycoplasmas and should be consulted for detailed indications for testing of mycoplasmas, development of specific laboratory protocols, expected susceptibility patterns of specific organisms, and interpretation and reporting of results (5, 39). Testing in vitro susceptibilities of M. pneumoniae is generally not indicated except for the evaluation of new antimicrobial agents. However, in view of the possibility of tetracycline resistance among M. hominis and Ureaplasma spp., reports of resistant organisms, and poor therapeutic response in some systemic infections in immunosuppressed patients, it may be worthwhile to perform in vitro susceptibility tests to guide patient management in some clinical settings. The activities of different classes of antimicrobials vary to some extent among mycoplasmas and ureaplasmas. An example of this variation is the fact that most macrolides are poorly active against M. hominis whereas lincosamides are active, while the reverse is true for Ureaplasma spp. Depending on the species being tested, it may be useful to determine MICs for tetracycline or doxycycline, erythromycin, clindamycin, streptogramins, and fluoroquinolones, such as ofloxacin, levofloxacin, moxifloxacin, and gatifloxacin.

TABLE 2 Major test kits for detection of serum antibodies to *M. pneumoniae* sold in the United States

Parameter	Information						
Product name	*M. pneumoniae* Antibody (MP) Test System	Chromalex *Mycoplasma pneumoniae* Antibody Latex Test System	ETI-MP IgM or IgG	GenBio ImmunoWELL *Mycoplasma pneumoniae* antibody IgM or IgG	ImmunoCard	*M. pneumoniae* IgG and IgM Antibody Test System	Mycoplasma IgG and IgM ELISA Test System
Manufacturer or distributor	Zeus Scientific, distributed by Wampole Laboratories	Shared Systems, Inc.	Savyon; manufactured for Diasorin	GenBio; distributed by Remel Inc.	Meridian Diagnostics	Remel Inc.	Zeus Scientific, Inc.
Antibodies measured	IgM and IgG separately	All classes simultaneously	IgM and IgG separately	IgM and IgG separately	IgM only	IgM and IgG simultaneously	IgM and IgG separately
Assay format	IFA	Latex PA	EIA	EIA	Qualitative, membrane-based EIA	Qualitative, membrane-based enzyme-linked immunobinding membrane assay	EIA
Kit description and target antigen	Available as a slurry of *M. pneumoniae* antigenic substrate or a Crown Titre of *M. pneumoniae* colonies affixed to microscope slides	Qualitative or semiquantitative agglutination test kit using colored latex particles coated with *M. pneumoniae* protein antigen that are reacted with sera on slides	96-well microtiter plate format coated with a membrane preparation containing the P1 protein of *M. pneumoniae*	96-well microtiter plate format coated with purified glycoplid mycoplasma antigen (*M. pneumoniae* strain FH, ATCC 15531)	Single-sample cards consisting of a test port containing *M. pneumoniae* antigen and a control port containing immobilized human IgM	Qualitative membrane-based single-sample test containing inactivated *M. pneumoniae* protein antigen (primarily cytadhesin protein)	Qualitative system for determination of IgG and IgM antibodies to *M. pneumoniae* using multiwell breakaway strips coated with partially purified inactivated *M. pneumoniae*
No. of tests/kit	100	100	192	96	30	10	96
Specimen throughput	Each slide contains 10 wells	Each slide contains 8 wells	Strips of 8 wells	Strips of 8 wells	1 specimen per card	1 specimen per card	Breakaway strips with 8 wells
No. of controls/run	1 positive, 1 negative, and 1 buffer control are tested with each assay	1 positive and 1 negative control each day assay is used	1 negative and 1 low, 1 medium, and 1 high positive per batch	Positive and negative controls in duplicate	Positive and negative controls are assayed upon receipt of kit. A procedural control is	Positive and negative controls run with each test	For each assay, 1 low-positive control is run in triplicate, along with a high-positive,

TABLE 2 *(Continued)*

Parameter	Information					
				included with every specimen for proper flow and reagent performance.		a negative, and a reagent blank.
Assay time (start to finish)	2.5 h	2.5 h	2.35 h IgG 2.75 h IgM	12 min	10 min	50 min
Assay time (hands on) (min)	30	15–20	15–20	10	2–3	5–10
Kit storage temp (°C)	Freeze at −20	2–8	2–8	2–8	2–8	2–8
Specimen storage	2–8°C for up to 5 days or −20°C. Avoid multiple freeze-thaw cycles.	2–8°C for up to 7 days (adding 0.1% sodium azide is highly recommended). For longer storage, freeze at −20°C. Avoid multiple freeze-thaw cycles.	2–8°C for up to 5 days; −20°C for extended periods	2–8°C for up to 72 h; otherwise, freeze at −20°C	2–8°C for up to 5 days; −10°C for longer periods. Filter sera prior to testing.	Room temperature for up to 8 h or 2–8°C for up to 48 h. If further delay is anticipated, store at −20°C. Avoid multiple freeze-thaw cycles.
Limitations	Avoid lipemic specimens. Rheumatoid factor may cause false-positive IgM tests. All positive IgM tests should be tested separately for rheumatoid factor, and if positive, it must be removed prior to retesting. High IgG can	Avoid lipemic or contaminated specimens. Test may not always distinguish prior from repeat infections. If negative results are obtained and a clinical suspicion still exists, a second specimen should be collected 2 weeks later and	False-negative IgM tests may result from samples being collected too early after disease onset. Avoid grossly lipemic or contaminated samples.	Whether the purified glycolipid antigen used in this assay will cross-react with other substances or microorganisms is unknown.	False-negative tests may result from samples being collected too early after disease onset. Avoid grossly lipemic, contaminated, or hemolyzed samples. Presence of IgM is usually detected in persons with recent primary infection with M.	False-negative tests may result from samples being collected too early after disease onset. A single positive result indicates only previous exposure. Avoid grossly lipemic, hemolyzed, or contaminated samples. Performance has

(Continued on next page)

TABLE 2 Major test kits for detection of serum antibodies to *M. pneumoniae* sold in the United States (*Continued*)

Parameter	Information			
	cause false-negative IgM test unless it is removed prior to testing. If background fluorescence prevents interpretation, result is equivocal. False-negative tests may result from samples being collected too early after disease onset. End-point may vary according to variables associated with microscopy.	retested concurrently with the first specimen to look for seroconversion (four-fold change in titer), which is indicative of a primary infection. Single titers of 1:8 are also suggestive of recent or current infection.	*pneumoniae.* However, IgM may be detected in persons with no evidence of recent infection and can persist 2–12 months.	not been established on neonates and immunocompromised patients.
Technical procedure and equipment required	Dilute sera in phosphate-buffered saline (PBS) 1:8 for IgM and 1:64 for IgG. Dispense 1 drop of patient serum and positive, negative, and buffer controls into appropriate wells on slides that are prewarmed to room temperature. Incubate slides at room	For the Quick Screen Assay, place 10 μl of positive and negative control sera in the appropriately labeled ovals of a test slide. Place 10 μl of test serum into a vacant oval of a test slide using a clean pipette tip for each. Add 1 drop of latex to each appropriate well and	Add 50 μl of negative control, 3 calibrators, and diluted sera to separate microwells. For determination of IgM, prepare wash buffer from concentrate. Dilute each control and specimen 1:100 in specimen diluent. Add 100 μl of 1:100-diluted controls and specimens to 20 μl of absorbent. Add 200 μl of prediluted calibrator to	For determination of IgM, the serum diluent contains anti-human IgG to remove IgG antibodies that might interfere with the reaction. Cover the
			Label test card, and use a single card for each control or specimen. Pipette 2 drops of serum to both lower sample ports. Incubate for 2 min at room temperature. Add 3 drops of enzyme conjugate to both lower ports. Incubate for 2 min at room temperature.	Patient serum (10 μl) is diluted in 200 μl of diluent, and 100 μl of the diluted sample is incubated in microwells coated with antigen for 20 min. The plate is washed 5 times to remove unbound antibody and other serum components. Pipette 1 drop of specimen into cup. Dispense 6 drops of reagent A into the same cup. Dispense 1 drop of the diluted specimen into the test well of the test module. Dispense 1 drop of undiluted negative control into the (−) well. Dispense 1 drop of undi-

TABLE 2 (Continued)

Parameter	Information
	temperature for 30 min for IgG or 37°C for 1 h for IgM. Rinse the slides with PBS and wash for two 5-min intervals with PBS. Blot dry. Add 1 drop of fluorescein isothiocyanate-labeled anti-human IgM or IgG, depending on assay, to each well. Incubate at room temperature for 30 min for IgG or 37°C for 30 min for IgM. Rinse slides with PBS and wash for two 5-min intervals with PBS. Blot dry. Apply 3–5 drops of mounting medium to each slide and coverslip. Examine slides immediately with fluorescent microscope for
	mix. Rock slide gently for 2 min, and read results. If antibodies to M. pneumoniae are present, they will react with antigen-coated particles and agglutination will occur. Undiluted samples exhibiting agglutination must then be tested qualitatively by diluting the serum 1:8 (50 μl in 350 μl of diluent) and retesting using 50 μl of test and control sera and rotating at 100–150 rpm on a mechanical rotator for 10 min. Only those samples remaining positive at 1:8 or greater are considered positive. For semiquantitative testing, 50 μl of specimen diluent is
	plate and incubate it at 37°C for 1 h to allow M. pneumoniae-specific antibodies to bind to the immobilized antigens. A water bath or moisture chamber in an incubator is needed. Wash 3 times with wash buffer, and add 50 μl of diluted horseradish peroxidase conjugate. Incubate mixture at 37°C for 1 h. Wash 3 times, and then add 100 μl of a TMB substrate. Incubate for 15 min at room temperature. Add 100 μl of stop solution. A positive reaction will turn from blue to yellow. Read absorbance at 450 nm. The absorbance is
	40 μl of absorbent. Mix and incubate for 30 min at room temperature. This step is necessary to prevent IgG and rheumatoid factor from interfering with the reaction. Add 100 μl of specimen diluent to the first well as a substrate blank. Pipette 100 μl of the treated calibrator, controls, and specimens into coated microwells, and incubate them for 60 min at room temperature. Aspirate the microwells, and wash them 3 times with wash buffer. Pipette 100 μl of conjugate into the microwells, and incubate it for 30 min at room tem-
	Add 3 drops of wash buffer to both upper ports. Add 2 drops of substrate reagent to both upper ports. Incubate for 5 min at room temperature. Visually read results immediately. Visually detectable blue color in both reaction ports indicates a positive test and the presence of IgM to M. pneumoniae. Blue color in control well alone indicates a negative test. If no blue color develops in the control port, test is invalid. Uncontaminated serum may be centrifuged and retested.
	luted positive control into the (+) well. Allow it to react for 1 min. Dispense 1 drop of reagent A into each well, and allow it to completely wick into the membrane. Dispense 1 drop of reagent B into each well, and allow it to completely wick into the membrane. Dispense 1 drop of reagent C into each well, and allow it to completely wick into the membrane. Repeat above step. Dispense 1 drop of reagent D into each well. Read results at 2–5 min. With the addition of the substrate or chromogen, a visible purple dot of varying inten-
	Peroxidase-conjugated goat anti-human IgG or IgM (100 μl) is added to the wells, and the plate is incubated for 20 min at room temperature. The conjugate will react with IgG or IgM immobilized on the solid phase. The wells are washed 5 times to remove unreacted conjugate. The wells containing immobilized peroxidase conjugate are incubated with 100 μl of peroxidase substrate solution for 10 min. Hydrolysis of the substrate by peroxidase produces a color change. The reaction is stopped by addition of 50 μl of stop

(Continued on next page)

217

TABLE 2 Major test kits for detection of serum antibodies to *M. pneumoniae* sold in the United States (*Continued*)

Parameter	Information
apple-green fluorescence. Presence of IgM in a single serum indicates current infection. Presence of IgG at ≥1:128 indicates active or past infection with *M. pneumoniae*. Presence of IgM at ≥1:16 indicates active or recent infection. Paired specimens demonstrating a fourfold rise in titer are recommended for confirmation of acute infection.	placed in wells 4–8 of the slide. Place 50 µl of diluted positive and negative controls into wells 1 and 2, respectively. Add 50 µl of diluted test serum to wells 3 and 4, and mix with diluent. Transfer 50 µl to the next well, and continue for the desired number of doubling dilutions, discarding the 50-µl excess from the last well. Add 1 drop of latex reagent to each control and test well and mix. Rotate the plate for 10 min at 100–150 rpm on a mechanical rotator, and read results. A test is considered positive if any sign of fine agglutination is visible. The antibody titer is proportional to the level of antibody present. perature. Aspirate the microwells, and wash them 3 times with wash buffer. Prepare fresh color developer. Pipette 100 µl of color developer into the microwells, and incubate it for 30 min at room temperature. Pipette 100 µl of stop solution into the microwells, and read results at 405 nm. The procedure is the same for IgG except that the absorbent step is omitted. Equipment required includes the Microwell washer and Microwell spectrophotometer capable of measurement to 2.00 absorbance units. The absorbance is proportional solution, and the color intensity of the solution is measured spectrophotometrically at 450 nm using a microtiter plate reader. A positive reaction is indicated by a color change from blue to yellow. The color intensity depends on the antibody concentration in the test sample. sity, greater than the negative control, is produced, indicating the presence of IgG and/or IgM. In a negative reaction, the color intensity of the test well is less than that of the negative control. Equivocal reactions indicate retesting after filtering the sera.

TABLE 2 (Continued)

Continued from previous page (under **Information**):

…is expressed as the inverse of the highest dilution that displays agglutination. The test is invalid if either control gives a result inconsistent with its label.

…to the level of antibody present.

Parameter	Information						
List price ($)	373.45 (IgG slurry) 394.74 (IgM slurry) 517.74 (IgG Crown Titre) 546.71 (IgM Crown Titre)	395	555 (IgM) 540 (IgG)	304.85 each for IgM and IgG	289	125.85 (10-test kit) 476.85 (40-test kit)	188 (IgG) 188 (IgM)
Current Procedural Terminology code	86738	86738	86738	86738	86738		86738
Clinical Laboratory Improvement Act complexity classification	High	Moderate	High	High	Moderate		High
References	2, 13	2; Thacker and Talkington, Abstr. 97th Gen. Meet. Am. Soc. Microbiol.	9, 33; Thacker and Talkington, Abstr. 97th Gen. Meet. Am. Soc. Microbiol.	2, 22, 34; Thacker and Talkington, Abstr. 97th Gen. Meet. Am. Soc. Microbiol.	13, 33, 34; Thacker and Talkington, Abstr. 97th Gen. Meet. Am. Soc. Microbiol.		

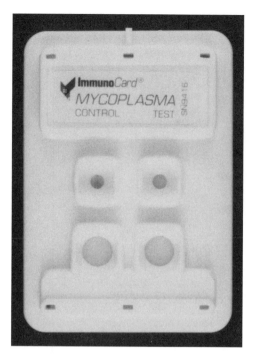

FIGURE 1 Immuno*Card* (Meridian Diagnostics). The Meridian Immuno*Card* is a qualitative, membrane-based EIA that can be performed in 10 min or less to detect the presence of IgM antibody directed against *M. pneumoniae*. A positive test is indicated by the development of a blue color in both of the upper reaction ports. Photograph courtesy of Lanier Thacker.

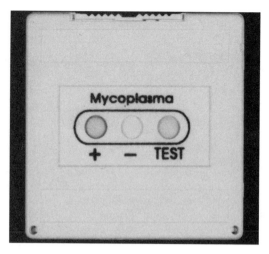

FIGURE 2 *M. pneumoniae* IgG and IgM Antibody Test System (Remel Laboratories). The Remel *M. pneumoniae* IgG and IgM Antibody Test System is a qualitative, membrane-based EIA that can be performed in 10 min or less to detect IgG and IgM antibodies against *M. pneumoniae*. A positive test is indicated by the presence of a visible purple dot in the test well. Photograph courtesy of Lanier Thacker.

Agar disk diffusion has not been recommended in testing mycoplasmas for antimicrobial susceptibilities because the relatively slow growth of many mycoplasmal species on agar will allow widespread diffusion of the antimicrobial from the disk throughout the agar plate before visible growth can be detected and because there are no data to correlate the zone diameters with MICs. Broth and agar dilution techniques and, more recently, the agar gradient (Etest) technique have been adapted for use with these organisms. When properly performed, with appropriate controls, these techniques can provide reliable susceptibility data to aid in the treatment of infections caused by or associated with mycoplasmas and ureaplasmas.

The culture medium used should be appropriate for the organism being tested, providing for the most rapid and sustained growth, and must contain an indicator. It is not possible to standardize on a single medium or pH, since growth requirements differ among species. 10B broth and A8 agar work well for ureaplasmas, whereas SP4 and Hayflick modified broth and agar work equally well for *M. hominis*, *M. pneumoniae*, and other species. Bromthymol blue broth has also been used sucessfully for ureaplasmas. Its theoretical advantages are less serum binding of antibiotics due to a reduced serum component and a lower concentration of urea, which protects against the death of ureaplasmas and makes mycoplasmacidal testing easier. Ureaplasmas require media with acidic pHs (6.0 to 6.5) for good growth, so testing at neutral pH is inappropriate. In contrast, *M. hominis* and *M. pneumoniae* can be tested at pH 7.3 to 7.4. Even though incubation in CO_2 can affect pH and ultimately the MICs obtained, ureaplasmas will grow poorly on agar in the absence of supplemental CO_2 unless the medium is buffered at pH 6.1 to 6.4. Broth-based tests can be incubated under atmospheric conditions. The length of incubation will depend on the growth rates of the species being tested and the technique, e.g., growth in broth will typically be evident before colonies develop on agar, so the MIC may be determined after shorter periods in broth-based tests.

Broth microdilution is the most widely used method for determining the antimicrobial susceptibilities of human mycoplasmas and ureaplasmas. It is economical and allows several antimicrobials to be tested in the same plate, and mycoplasmacidal testing can be performed in the same system. The test provides a quantitative MIC for each antimicrobial agent tested, with susceptibility or resistance determined based on the ability of the organism to metabolize substrates when grown in the presence of the antimicrobial agent and indicator of growth. Any of the commercial sources of 10B (for *Ureaplasma* spp.), SP4 (for *Mycoplasma* spp.), or other liquid media suitable for cultivation of the species of interest that can be shown to sustain adequate growth can be used for broth microdilution MIC testing, since there are no universally approved or standardized media for these organisms.

Susceptibility-testing kits utilizing the broth microdilution technique, such as Mycoplasma IST (bioMérieux); Mycoplasma SIR (Bio-Rad); Mycofast "All-In," Mycofast Evolution 2, and Mycofast ABG (International Microbio); and Mycokit-ATB (PBS Orgenics) are produced in France. Some of these products are described further in Table 1. They consist of microwells containing dried antimicrobials, generally in two or more concentrations corresponding to the thresholds proposed for conventional bacteria to classify a strain as susceptible, intermediate, or resistant. Some of these kits combine organism growth and identification with susceptibility tests in the same product. Abele-Horn et al. (1) evaluated the Mycofast "All In" and Mycoplasma IST kits in comparison to standard methods and found that the results for tetracycline correlated better than those for other agents, which showed wider differences, but they did not endorse the use of these products for clinical purposes. Inoculum size can influence MICs, and direct inoculation with the clinical specimen without a defined inoculum or preculture, as is done with the Mycofast "All In" and Mycoplasma IST kits, can contribute to error. Observations of gynecological specimens indicated that high numbers of mycoplasmas, exceeding 10^5 per ml, caused MICs to be elevated twofold or more with these products. Renaudin and Bébéar (26) also evaluated the Mycoplasma SIR kit and reported that, using a defined inoculum, this product gave results comparable to those obtained by established MIC determination. They considered it a reasonable choice for diagnostic laboratories in countries where the product is available. This product is adapted for use with *Ureaplasma* spp. and *M. hominis* after a primary culture so that a standardized inoculum can be incorporated into the test system. The Mycokit-ATB also includes procedures for standardization of the inoculum prior to determining antimicrobial susceptibilities.

Dilution of antimicrobials in agar has been adapted for use with several mycoplasmal species and ureaplasmas for reference work evaluating antimicrobial activities. It has the advantage of a relatively stable endpoint over time, allows the ready detection of mixed cultures, and is suitable for testing large numbers of organisms simultaneously. However, this technique is not practical for testing small numbers of strains or occasional isolates which may be encountered in diagnostic laboratories. Media must be used when fresh, and there are no commercially available sources for agar plates containing the antibiotics.

Preliminary studies using the agar gradient diffusion, or Etest (AB BIODISK, Piscataway, N.J.), technique for detection of tetracycline and fluoroquinolone susceptibilities in *M. hominis* yielded results comparable to those of both broth microdilution and agar dilution (37, 39). The Etest has also been used successfully for performing in vitro susceptibility tests of *Ureaplasma* spp. (11). Performing susceptibility tests using the Etest can be done by adapting methods used for other techniques (39). Two Etests can be placed on each standard-size (100-mm) plate. Agar plates are incubated until colonies are apparent in the periphery of the

plate and an ellipse appears. The MIC is read under the microscope as the number on the Etest strip corresponding to the intersection of mycoplasmal growth with the strip (Fig. 3). Staining plates with Dienes stain helps to visualize the ellipse. The Etest has the advantage of the simplicity of agar-based testing, has an endpoint that does not shift over time, does not have a large inoculum effect, and can be easily adapted for testing single isolates. This technique is readily adaptable to laboratories that do not specialize in mycoplasma diagnosis but that may encounter isolates needing susceptibility tests only on an occasional basis. Etests are commercially available, can be maintained frozen for 3 to 5 years, and have been shown to work well with Remel SP4 broth and agar (38). The strips can be expensive if a large number of different drugs are evaluated, but many of the drugs used for testing mycoplasmas may also be appropriate for use with other bacteria. For laboratories testing occasional clinical mycoplasma or ureaplasma isolates, the Etest is a very practical technique, since all of the materials needed can be purchased commercially and have long shelf lives.

AUTOMATED TEST PROCEDURES

There are no automated instruments or procedures available to detect mycoplasmas in clinical specimens. The use of automated blood culture instruments to detect *M. hominis* bacteremia has been mentioned earlier. Recent data obtained using the Bact/ALERT instrument (Organon Teknika Corp.) have again demonstrated the inability of automated blood culture systems to flag blood cultures containing mycoplasmas (39a).

UTILIZATION OF DIAGNOSTIC TESTS FOR MYCOPLASMAS

The culture-based procedures and products described in this chapter are suitable for performance in a high-complexity hospital-based general microbiology laboratory or reference laboratory. The availability of prepared media, and now

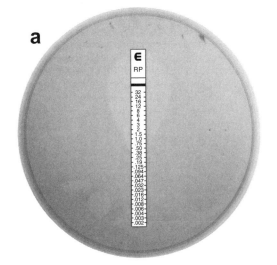

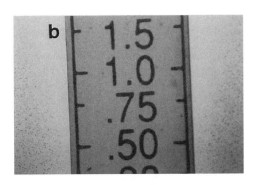

FIGURE 3 Etest technique for determining in vitro susceptibilities for *M. hominis*. Shown is a Remel SP4 agar plate and quinupristin-dalfopristin Etest demonstrating in vitro susceptibility testing of *M. hominis* at low power (a) and high power (b). MIC = 0.5 μg/ml.

diagnostic kits in some countries, makes it easier than before for laboratories to offer diagnostic service for mycoplasmal and ureaplasmal infections. It should be noted, however, that many products have not been thoroughly evaluated in clinical settings, some of those being sold have not performed as well as nonproprietary materials, and little or no external clinical experience or published information is available for validation of commercial media or kits except for use in the detection of M. hominis or Ureaplasma spp. (as opposed to any other mycoplasmal species) in urogenital specimens.

Decisions as to whether to undertake in-house culture-based testing should be based primarily on the volume of tests ordered and the availability of qualified personnel rather than on trepidation because of unfamiliarity with mycoplasmas and ureaplasmas and perceived complexities and difficulties in their detection in clinical specimens. Commonly encountered organisms, such as M. hominis and Ureaplasma spp., are really no more difficult to detect and identify than other fastidious bacteria regularly sought, provided appropriate attention is paid to their specific culture requirements and quality control is practiced. Mycoplasmas that require more definitive identification than that provided by the rudimentary tests described above can be submitted to a reference laboratory in the somewhat-rare event that it is clinically indicated. This may be relevant when the organism is present in a clinically significant infection of a normally sterile site, particularly in an immunosuppressed person. If a laboratory chooses to offer mycoplasmal and ureaplasmal cultures at remote sites, it is recommended that one of the numerous transport media available be kept on the clinical units for direct inoculation of specimens at bedside. This should pose no difficulty, since most of the commercially sold products can be stored freeze-dried or frozen for several weeks to months and used only as needed.

One of the most significant advances in recent years in the diagnosis of M. pneumoniae respiratory infections has been the development of rapid single-specimen EIA and PA tests that are readily adaptable to the primary care physician's office laboratory. This is important because the clinical features of these infections are not sufficiently distinctive to allow differentiation from other respiratory infections, including those caused by various viruses that should not be treated with antibiotics, and because of the time-consuming nature of culture. The Remel IgG and IgM Antibody Test System and the Meridian ImmunoCard for IgM are unique among the EIA-based tests sold in the United States in that both can be performed without special expertise or equipment and can be interpreted in about 10 min. Both have moderate complexity classifications under the Clinical Laboratory Improvement Act (CLIA), making it possible for many physicians to offer serological assays for M. pneumoniae antibodies as point-of-care tests, so they can be used to direct patient management. However, such single-specimen assays do have limitations, as described earlier, with perhaps the most practical utilization of the IgM ImmunoCard being for children and young adults in whom an acute infection with M. pneumoniae is strongly suspected. The Meridian MERISTAR latex agglutination test is also adaptable to use in an ambulatory care setting and carries a moderate complexity classification under the CLIA, but it has not been thoroughly evaluated in a clinical setting. Physicians who desire the most accurate identification and characterization of M. pneumoniae infection in adults over 20 years of age should obtain determinations of both IgM and IgG in paired sera.

FUTURE DIRECTIONS

Interest in mycoplasmas as bonafide agents of clinically significant human infection has increased over the past decade as a result of better documentation of the occurrence of these organisms in human disease, particularly opportunistic infections caused by M. hominis and Ureaplasma spp. in neonates and immunosuppressed persons. This has been made possible partly through the use of powerful molecular techniques, such as PCR, a tool that has also facilitated the discovery and preliminary characterization of other potential human pathogens, most notably M. genitalium. PCR systems have now been reported for almost all mycoplasmas known to occur in humans. Practically, PCR technology appears to be less valuable for routine diagnostic purposes in the case of the more rapidly growing and easily cultivable organisms, such as M. hominis and Ureaplasma spp., except in some specific cases where isolation by culture could be difficult. For instance, the PCR technique may detect urogenital mycoplasmas more easily in nonurogenital specimens, such as synovial fluids and tissue biopsies.

The importance of the detection of M. pneumoniae for diagnostic and management purposes has become better appreciated by some clinicians as a result of the mounting evidence that the organism is a serious cause of pneumonia requiring hospitalization and that it can affect persons in all age groups (20). For M. pneumoniae, and especially for slow-growing, extremely fastidious species for which optimum cultivation techniques are not established, such as M. genitalium and M. fermentans, the use of PCR assays is the most practical means of detecting their presence in clinical material. The PCR assay can also detect the organisms even if tissues have already been fixed for histologic examination (32).

The detection limit of PCR is very high, corresponding to a single organism, or a single copy of the gene, when purified DNA is used. In contrast, culture requires at least 1 to 10 viable organisms; larger numbers are probably needed when clinical specimens are considered. Comparison of the PCR technique with culture and/or serology, in the case of M. pneumoniae, has yielded varied results, and large-scale experience with this procedure is still limited. Positive PCR results for M. pneumoniae in culture-negative persons without evidence of respiratory disease suggests inadequate specificity, persistence of the organism after infection, or its existence in asymptomatic carriers, making interpretation of such results difficult. Thus, even though the many benefits of PCR for detection and speciation of mycoplasmas in humans are readily apparent, a positive result must still be correlated with clinical events. At present, reliable serology, including both acute- and convalescent-phase sera, is still the preferred method for accurate diagnosis of M. pneumoniae respiratory disease. Presently, PCR detection of mycoplasmas is still too labor-intensive, expensive, and complex to be carried out routinely in most clinical microbiology laboratories. Some drawbacks must still be corrected, such as the presence of inhibitors in the specimens and laboratory contamination. The possible development of commercial PCR kits in the future should bring about better standardization of the technique, and if they are available at a reasonable cost, PCR could become a major method for the diagnosis of mycoplasmal infections.

Thus, the future of mycoplasma diagnostics should ideally include the development and refinement of commercial PCR kits for detection of M. pneumoniae that could be employed as screening devices for the presence of respiratory infection by this organism, other mycoplasmas, and unrelated bacteria that cause pneumonias simultaneously. Preliminary studies (8) indicate that such multiplex PCR systems may be feasible but may result in decreased sensitivity. Use of the PCR assay to detect and identify unknown or uncommon mycoplasmas is also likely to become important, but the utilization of such technology would be limited primarily to mycoplasma research or reference laboratories due to the infrequent occurrence of such organisms in clinically significant human infections.

Further improvements and simplification of commercial serological tests that can accurately and rapidly measure IgM, IgA, and IgG antibodies, ideally in an acute-care clinic setting, will also be important goals to achieve. Another attractive approach may be further investigation aimed at producing species-specific antigen targets for serological assays based on the use of synthetic peptides.

Given that culture is likely to remain the preferred diagnostic method for detection of M. hominis and Ureaplasma spp. for the foreseeable future, the most obvious need will be for reliable commercial products to be sold more widely and for validation of media capable of detecting these organisms in all types of clinical specimens. The availability of complete diagnostic kits, similar to those now sold in Europe and some other areas, may further increase the likelihood that hospital-based microbiology laboratories will begin offering diagnostic services for mycoplasmas, provided the kits can be proven to be accurate when compared directly with standard methods.

COMMERCIAL SUPPLIERS OF KITS, MEDIA, AND REAGENTS

Companies that supply kits, media, reagents, and other products tend to change rapidly, and some products are only available in Europe. Thus, the products from the suppliers listed in this chapter may not always be available as they are described. Contact information for products mentioned in this manual is provided in the appendix.

REFERENCES

1. **Abele-Horn, M., C. Blendinger, C. Becher, P. Emmerling, and G. Ruckdeschel.** 1996. Evaluation of commercial kits for quantitative identification and tests on antibiotic susceptibility of genital mycoplasmas. *Zentbl. Bakteriol.* **284:**540–549.
2. **Alexander, T. S., L. D. Gray, J. A. Kraft, D. S. Leland, M. T. Nikaido, and D. H. Willis.** 1996. Performance of Meridian Immuno*Card Mycoplasma* test in a multicenter clinical trial. *J. Clin. Microbiol.* **34:**1180–1183.
3. **Aubert, G., B. Pozzetto, O. G. Gaudin, J. Hafid, A. D. Mbida, and A. Ros.** 1992. Evaluation of five commercial tests: complement fixation, microparticle agglutination, indirect immunofluorescence, enzyme-linked immunosorbent assay and latex agglutination, in comparison to immunoblotting for *Mycoplasma pneumoniae* serology. *Ann. Biol. Clin.* **50:**593–597.
4. **Barker, C. E., M. Sillis, and T. G. Wreghitt.** 1990. Evaluation of Serodia Myco II particle agglutination test for detecting *Mycoplasma pneumoniae* antibody: comparison with μ-capture ELISA and indirect immunofluorescence. *J. Clin. Pathol.* **43:**163–165.
5. **Bébéar, C., and J. A. Robertson.** 1996. Determination of minimal inhibitory concentration, p. 189–197. *In* J. G. Tully and S. Razin (ed.), *Molecular and Diagnostic Procedures in Mycoplasmology.* Academic Press, New York, N.Y.
6. **Broitman, N. L., C. M. Floyd, C. A. Johnson, L. M. de la Maza, and E. Peterson.** 1992. Comparison of commercially available media for detection and isolation of *Ureaplasma urealyticum* and *Mycoplasma hominis*. *J. Clin. Microbiol.* **30:**1335–1337.
7. **Cavicchini, S., M. R. Biffi, A. Brezzi, and E. Alessi.** 1989. *Mycoplasma hominis* and *Ureaplasma urealyticum*: comparison of media for isolation and prevalence among patients with nongonococcal nonchlamydial genital infections. *G. Ital. Dermatol. Venereol.* **124:**321–324.
8. **Choppa, P. C., A. Vodjani, C. Tagle, R. Andrin, and L. Magtoto.** 1998. Multiplex PCR for the detection of *Mycoplasma fermentans*, *M. hominis* and *M. penetrans* in cell cultures and blood samples of patients with chronic fatigue syndrome. *Mol. Cell. Probes* **12:**301–308.
9. **Cimolai, N., and A. C. H. Cheong.** 1996. An assessment of a new diagnostic indirect enzyme immunoassay for the detection of anti-*Mycoplasma pneumoniae* IgM. *Clin. Microbiol. Infect. Dis.* **105:**205–209.
10. **Clegg, A., M. Passey, M. Yoannes, and A. Michael.** 1997. High rates of genital mycoplasma infections in the highlands of Papua New Guinea determined both by culture and by a commercial detection kit. *J. Clin. Microbiol.* **35:**197–200.
11. **Dosa, E., E. Nagy, W. Falk, I. Szoke, and U. Ballies.** 1999. Evaluation of the Etest for susceptibility testing of *Mycoplasma hominis* and *Ureaplasma urealyticum*. *J. Antimicrob. Chemother.* **43:**575–578.
12. **Echevarria, J. M., P. Leon, P. Balfagon, J. A. Lopez, and M. V. Fernandez.** 1990. Diagnosis of *Mycoplasma pneumoniae* infection by microparticle agglutination and antibody-capture enzyme-immunoassay. *Eur. J. Clin. Microbiol. Infect. Dis.* **9:**217–220.
13. **Fedorko, D. P., D. D. Emery, S. M. Franklin, and D. D. Congdon.** 1995. Evaluation of a rapid enzyme immunoassay system for serologic diagnosis of *Mycoplasma pneumoniae* infection. *Diagn. Microbiol. Infect. Dis.* **23:**85–88.
14. **Gerstenecker, B., and E. Jacobs.** 1993. Development of a capture-ELISA for the specific detection of *Mycoplasma pneumoniae* in patient's material, p. 195–205. *In* I. Kahane and A. Adoni (ed.), *Rapid Diagnosis of Mycoplasmas.* Plenum Press, New York, N.Y.
15. **Granström, M., T. Holme, A. M. Sjögren, A. Örtqvist, and M. Kalin.** 1994. The role of IgA determination by ELISA in the early serodiagnosis of *Mycoplasma pneumoniae* infection, in relation to IgG and μ-capture IgM methods. *J. Med. Microbiol.* **40:**288–292.
16. **Karppelin, M., K. Hakkarainen, M. Kleemola, and A. Miettinen.** 1993. Comparison of three serological methods for diagnosing *Mycoplasma pneumoniae* infection. *J. Clin. Pathol.* **46:**1120–1123.
17. **Kenny, G. E., G. G. Kaiser, M. K. Cooney, and H. M. Foy.** 1990. Diagnosis of *Mycoplasma pneumoniae* pneumonia: sensitivities and specificities of serology with lipid antigen and isolation of the organism on soy peptone medium for identification of infections. *J. Clin. Microbiol.* **28:**2087–2093.
18. **Lieberman, D., D. Lieberman, S. Horowitz, O. Horovitz, F. Schlaeffer, and A. Porath.** 1995. Microparticle agglutination versus antibody-capture enzyme immunoassay for diagnosis of community-acquired *Mycoplasma pneumoniae* pneumonia. *Eur. J. Clin. Microbiol. Infect. Dis.* **14:**577–584.
19. **Lind, K.** 1982. Serological cross-reaction between *Mycoplasma genitalium* and *M. pneumoniae*. *Lancet* **ii:**1158–1159.

20. **Marston, B. J., J. F. Plouffe, T. M. File, B. A. Hackman, S.-J. Salstrom, H. B. Lipman, M. S. Kolczak, R. F. Breiman, and the Community Based Pneumonia Study Group.** 1997. Incidence of community-acquired pneumonia requiring hospitalization. *Arch. Intern. Med.* **157:**1709–1718.

21. **Masover, G. K., and F. A. Becker.** 1996. Detection of mycoplasmas by DNA staining and fluorescent antibody methodology, p. 419–429. *In* J. G. Tully and S. Razin (ed.), *Molecular and Diagnostic Procedures in Mycoplasmology.* Academic Press, New York, N.Y.

22. **Matas, L., J. Domínguez, F. De Ory, N. García, N. Gali, P. J. Cardona, A. Hernández, C. Rodrigo, and V. Ausina.** 1998. Evaluation of Meridian ImmunoCard mycoplasma test for the detection of *Mycoplasma pneumoniae*-specific IgM in pediatric patients. *Scand. J. Infect. Dis.* **30:**289–293.

23. **Phillips, L. E., K. H. Goodrich, R. M. Turner, and S. Faro.** 1986. Isolation of *Mycoplasma* species and *Ureaplasma urealyticum* from obstetrical and gynecological patients by using commercially available medium formulations. *J. Clin. Microbiol.* **24:**377–379.

24. **Pratt, B.** 1990. Automatic blood culture systems: detection of *Mycoplasma hominis* in SPS-containing media, p. 778–781. *In* G. Staneck, G. H. Cassell, J. G. Tully, and R. F. Whitcomb (ed.), *Recent Advances in Mycoplasmology.* Gustav Fischer Verlag, Stuttgart, Germany.

25. **Razin, S.** 1994. DNA probes and PCR in diagnosis of mycoplasma infections. *Mol. Cell. Probes* **8:**497–511.

26. **Renaudin, H., and C. Bébéar.** 1990. Evaluation des systemes *Mycoplasma* PLUS et SIR *Mycoplasma* pour la detection quantitative et l'etude de la sensibilite aux antibiotiques des mycoplasmes genitaux. *Pathol. Biol.* **38:**431–435.

27. **Samra, Z., and R. Gadba.** 1993. Diagnosis of *Mycoplasma pneumoniae* infection by specific IgM antibodies using a new capture-enzyme immunoassay. *Eur. J. Epidemiol.* **9:**97–99.

28. **Shepard, M. C.** 1983. Culture media for ureaplasmas, p. 137–146. *In* S. Razin and J. G. Tully (ed.), *Methods in Mycoplasmology,* vol. 1. Academic Press, Inc., New York, N.Y.

29. **Shepard, M. C., and C. D. Lunceford.** 1978. Serological typing of *Ureaplasma urealyticum* isolates from urethritis patients by an agar growth inhibition method. *J. Clin. Microbiol.* **8:**566–574.

30. **Sillis, M.** 1990. The limitation of IgM assays in the serological diagnosis of *Mycoplasma pneumoniae* infection. *J. Med. Microbiol.* **33:**253–258.

31. **Sillis, M.** 1993. Genital mycoplasmas revisited—an evaluation of a new culture medium. *Br. J. Biomed. Sci.* **50:**89–91.

32. **Talkington, D. F., W. L. Thacker, D. W. Keller, and J. S. Jensen.** 1998. Diagnosis of *Mycoplasma pneumoniae* infection in autopsy and open-lung biopsy tissues by nested PCR. *J. Clin. Microbiol.* **36:**1151–1153.

33. **Thacker, W. L., and D. F. Talkington.** 1995. Comparison of two rapid commercial tests with complement fixation for serologic diagnosis of *Mycoplasma pneumoniae* infections. *J. Clin. Microbiol.* **33:**1212–1214.

34. **Thacker, W. L., and D. F. Talkington.** 2000. Analysis of complement fixation and commercial enzyme immunoassays for detection of antibodies to *Mycoplasma pneumoniae* in human serum. *Clin. Diagn. Lab. Immunol.* **7:**778–780.

35. **Tully, J. G., D. L. Rose, R. F. Whitcomb, and R. P. Wenzel.** 1979. Enhanced isolation of *Mycoplasma pneumoniae* from throat washings with a newly modified culture medium. *J. Infect. Dis.* **139:**478–482.

36. **Waites, K. B., G. H. Cassell, L. B. Duffy, and K. B. Searcey.** 1995. Isolation of *Ureaplasma urealyticum* from low birthweight infants. *J. Pediatr.* **126:**502.

37. **Waites, K. B., D. M. Crabb, L. B. Duffy, and G. H. Cassell.** 1997. Evaluation of the Etest for detection of tetracycline resistance in *Mycoplasma hominis. Diagn. Microbiol. Infect. Dis.* **27:**117–122.

38. **Waites, K. B., K. C. Canupp, and G. E. Kenny.** 1999. In vitro susceptibilities of *Mycoplasma hominis* to moxifloxacin determined by Etest and agar dilution. *Drugs* **58**(Suppl. 2):406–407.

39. **Waites, K. B., and D. Taylor-Robinson.** 1999. *Mycoplasma* and *Ureaplasma,* p. 782–794. *In* P. R. Murray, E. J. Baron, M. A. Pfaller, F. C. Tenover, and R. H. Yolken (ed.), *Manual of Clinical Microbiology,* 7th ed. American Society for Microbiology, Washington, D.C.

39a. **Waites, K. B., and K. C. Canupp.** Evaluation of the BacT/ALERT system for detection of *Mycoplasma hominis* in simulated blood cultures. *J. Clin. Microbiol,* in press.

40. **Wood, J. C., R. M. Lu, E. M. Peterson, and L. M. de la Maza.** 1985. Evaluation of Mycotrim-GU for isolation of *Mycoplasma* species and *Ureaplasma urealyticum. J. Clin. Microbiol.* **22:**789–792.

41. **Wreghitt, T. G., and M. Sillis.** 1985. A μ-capture ELISA for detecting *Mycoplasma pneumoniae* IgM: comparison with indirect immunofluorescence and indirect ELISAs. *J. Hyg.* **94:**217–227.

Commercial Methods for Identification and Susceptibility Testing of Fungi

DONNA M. WOLK AND GLENN D. ROBERTS

9

The use of commercial fungal detection and identification systems has increased, largely in an effort to replace the often lengthy and resource-intensive methods used for traditional testing. Rapid diagnosis of fungal pathogens is crucial for immunocompromised patients, for whom opportunistic infections can often be fatal. The advance of human immunodeficiency virus disease, the widespread use of antibiotics, and the growing number of transplants performed each year all contribute to the increased incidence of fungal infections (6). Most currently available commercial systems are relatively rapid (2 to 72 h) and widely used; however, in many cases, these methods still lack sensitivity and require substantial laboratory expertise to interpret supplementary tests, identify microscopic features of organisms, or correctly choose isolates for appropriate testing. Clearly, more efforts will be necessary to advance more rapid and sensitive test methods to diagnose fungal disease.

The methods described in this chapter are meant to represent those that are commercially available, Food and Drug Administration-approved, and published in peer-reviewed journals. For the sake of brevity and applicability to the current state of clinical mycology, publications were reviewed, in most cases, only for the last 10 years. Interest in commercially available solutions for accurate, more rapid, and less costly diagnosis has forced a change in the algorithms used for fungal diagnosis and increased the numbers of published comparisons and evaluations of commercial methods and systems. Certainly, publications of this nature will continue to appear at a relatively rapid pace, as the health care and clinical-diagnostics industries strive to improve patient care and reduce costs.

DIRECT EXAMINATION AND DETECTION METHODS

Accurate and sensitive detection methods for fungal pathogens in clinical specimens are essential for a number of reasons. Many patients infected with fungi are immunosuppressed, and disease in these patients can be severe and progress rapidly. Results from fungal cultures and serological testing often come too late to affect patient management. Direct microscopic examination has traditionally been performed by analysis of KOH preparations with and without Calcofluor white staining, but smears may lack sensitivity (168).

Despite these limitations, the direct recognition of fungi in clinical specimens remains the most rapid means of fungal diagnosis. Unfortunately, only a few pathogens have commercial products specifically designed to enhance their direct detection in clinical specimens. Most available methods target fungal cell wall antigens or fungal enzymes. Molecular testing for fungal DNA may show promise, but no commercial test methods are yet available.

Cryptococcus

Due to its high specificity and increased sensitivity, the *Cryptococcus neoformans* antigen detection test (Table 1) has, for the most part, replaced the more historic India ink test and has become the "gold standard" for *Cryptococcus* testing (79, 107). Antigen tests are more specific than either the immunofluorescence assay or tube agglutinin method for cryptococcal antibody testing, but both methods may be considered when patients have symptoms of pulmonary or meningeal infection (79). The cryptococcal polysaccharide capsular antigen can be detected via latex agglutination or enzyme immunoassay (EIA). Due to the soluble nature of the target cryptococcal antigen, most methods can detect antigen in serum or cerebrospinal fluid (CSF) even before viable organisms can be cultivated. Different cryptococcal serotypes are known to be cross-reactive, so all known serotypes can be detected by these assays.

The amount of antigen detected is generally accepted to correlate with disease severity; however, titers may persist for some time after therapy is completed, especially for AIDS patients, whose titers may decline slowly even after clinical improvement has occurred. Results may be difficult to interpret when patients are immunocompromised, since some patients will never clear the antigen. Antigen titers of 1:8 usually indicate active cryptococcosis, but lower titers can also be indicative of cryptococcal disease. Titers can be helpful for monitoring disease, but care must be taken to compare titers only when and if the serum specimens have been tested with the same test methods or kits during the same batch testing. For titers to have diagnostic meaning, there must be careful adherence to protocols and samples must be collected and frozen over time for simultaneous batch testing (168).

TABLE 1 Comparison of commercial methods for cryptococcal antigen testing

Assay	Company	Method	Target	Comparison	Samples tested	No. of positives tested	Sensitivity (%)	Specificity (%)	Comments[b]	Reference
Crypto-LA	Wampole; International Biological Laboratories	Latex PAb[a]	Polysaccharide antigen	Culture	Serum	30	83	98	No pronase	34
					CSF	19	100	98		
Myco-Immune	Microscan	Latex		Culture	Serum	30	83	100	No pronase	
					CSF	19	100	97		
Immy and Latex-Crypto	Immunomycologics	Latex PAb		Culture	Serum	30	97	93	Pronase	
					CSF	19	93	93		
CALAS	Meridian Diagnostics	Latex PAb	Cryptococcal capsular antigen	Culture	Serum	30	97	95	Pronase	34
					CSF	19	100	96		
				Chart review to establish outcomes	Serum	103	99	100		45
					CSF	18	100	100	No cross-reactions with serum factors but cross-reactions with syneresis fluid	34
					Serum and CSF	120[c]	100	96–100	Multicenter evaluation	54
Crypto-Lex	Trinity Laboratories	MAb[d] to capsular polysaccharide			Serum	44 of 327	97	100	Comparable to CALAS	85
					CSF	23 of 253				
Premier	Meridian Diagnostics	EIA PAb	Cryptococcal capsular antigen	Culture	Serum	30	97	93–100		158
					CSF	19	93–100	93–98		
				CALAS and chart review	Serum	103	12 sera discrepant		EIA titer higher than latex tests	45
				CALAS and chart review	CSF	18	1 CSF discrepant			
					Serum and CSF	120	99	97	Cross-reactions are rare; wide variety of organisms tested; false positive with T. beigelii	54
									No cross-reactions with RF, syneresis fluid, or serum macroglobulins from SLE patients	34

[a] PAb, polyclonal antibody.
[b] Do not compare titers from different kits interchangeably. RF, rheumatoid factor; SLE, systemic lupus erythematosis.
[c] Including 10 RA and 20 ANA-positive sera.
[d] MAb, monoclonal antibody.

Latex Agglutination

CALAS (Meridian Diagnostics, Cincinnati, Ohio), Crypto-LA (Wampole and International Biological Laboratories, Cranbury, N.J.), Immy Latex-Crypto Antigen (Immunomycologics, Norman, Okla.), Myco-Immune (Dade Microscan, West Sacramento, Calif.), and CRYPTO-LEX (Trinity Laboratories, Raleigh, N.C.)

Latex agglutination methods provide an immunochemical technique for detecting the antigen of *C. neoformans* in serum or CSF. Most test kits contains latex particles sensitized with anti-cryptococcal globulins. Normal rabbit globulins are used as the control to test for nonspecific latex agglutination. When the anti-cryptococcal globulin-sensitized latex particle comes in contact with the polysaccharide antigen of *C. neoformans* in the patient specimen, the two form a complex that causes agglutination. The reaction occurs on a slide included in the kit. There are both qualitative and quantitative protocols for this assay. In the quantitative procedure, the titer is the reciprocal of the last dilution to demonstrate agglutination (158).

Despite the sensitive and accurate nature of cryptococcal antigen testing, there are a few pitfalls related to the use of latex test kits. As with any assay, low concentrations of antigen can cause false-negative reactions to occur. False-negative reactions can also be caused by very high antigen concentrations that cause a prozone effect or in cases of immune complex formation. The latter can be avoided by the dissociation of such complexes with pronase pretreatment of sera (60, 154). In addition, false negatives have also been reported in specimens from patients infected with the nonencapsulated variants of *C. neoformans* (79). Alternatively, false-positive reactions are caused by the presence of *Trichosporon beigelii* and *Leptotrichia buccalis* (a gramnegative rod formerly known as *Capnocytophaga* or Centers for Disease Control and Prevention group DF-1) (10, 168). Rheumatoid factor may cause nonspecific agglutination (34, 168) as will the presence of some soaps and disinfectants (12), povidone-iodine from skin preparations (27), and syneresis fluid (69).

Crypto-Lex, a mouse immunoglobin M (IgM) monoclonal antibody latex agglutination reagent, reacts with four serogroups of *C. neoformans* in the Crypto-Lex test. In a comparison with the CALAS assay for 327 serum and 253 CSF samples, the two latex reagents agreed for all except one serum sample and one CSF sample. The correlation coefficient was 0.886, with a sensitivity and specificity of 97 and 100%, respectively, for Crypto-Lex, a 99.6% correlation with CALAS (85).

EIA (Premier; Meridian Diagnostics)

The Premier Cryptococcal Antigen EIA kit (Meridian) uses anti-cryptococcal polyclonal antibodies adsorbed to plastic microwells, which can be used for batch assays or snapped off and used for smaller sample runs. The detection system is based on a sandwich EIA using a monoclonal peroxidase conjugate. If cryptococcal antigen is present in the sample, a complex is formed between the adsorbed antibody, the antigen, and the enzyme conjugate. After the wells are washed to remove unbound conjugate, a substrate solution is added, and color will develop in the presence of the bound enzyme and in proportion to the amount of cryptococcal antigen present. With a spectrophotometer, semiquantitative results can be generated, producing titers that can be used to monitor drug therapy or disease severity. Pro-

zone effects, like those that interfere with latex-based testing, are generally not a problem. Cross-reactions can occur, causing false-positive reactions with the yeast *T. beigelii* (168) and with syneresis fluid, which can contaminate the CSF from debris that accumulates on platinum wire loops. Test endpoints are objective but are more costly to achieve than endpoint titers generated by the latex-based methods (34, 69).

Comparative Studies

In an early commercially performed comparison of the Premier EIA with the CALAS latex agglutination assay, 475 specimens were screened, including 120 positive specimens from serum and CSF. Serum samples that contained rheumatoid factor were tested to determine which methods would be affected and thus generate false-positive results. When the results were compared, the EIA was found to be virtually equivalent to the latex method, with 99% sensitivity and 97% specificity (54).

In another report, the abilities of two Meridian Diagnostics products, the Premier EIA and the CALAS latex agglutination test, to detect cryptococcal capsular antigen in serum ($n = 471$, with 103 positives) and CSF ($n = 123$, with 18 positives) were compared. The methods were evaluated for the dual purposes of sample screening and quantitation. Good concordance (97.8%) was reported between the tests. However, 12 serum samples and 1 CSF sample had discrepant screening results. When used for semiquantitative assays for titer determination, the titers obtained with *Cryptococcus* antigen assays performed by EIA were generally higher than those obtained with latex (45).

In 1994, the Premier Cryptococcal Antigen EIA kit was compared to several other kits by using CSF and serum specimens. Of the 182 CSF specimens, 19 had positive cultures, and of the 90 serum specimens, 30 cultures were positive. The latex agglutination kits used for comparison were CALAS (Meridian Diagnostics), Crypto-LA from International Biological Laboratories (marketed by Wampole Laboratories), Myco-Immune (Microscan), and Immy (Immunomycologics). Routine culture of the specimens was also performed. For CSF testing, the sensitivities for all of the kits were comparable, ranging from 93 to 100%. The specificities were also comparable, ranging from 93 to 98%. The specificities of the kits for serum testing were similar, ranging from 93 to 100%; however, two tests (Crypto-LA and Myco-Immune) were less sensitive (83%) than Immy and Premier, which both had a sensitivity of 97%. Interestingly, the two kits with lower sensitivities for serum testing were also the two kits that do not pretreat serum with pronase prior to testing (158).

Pronase treatment is known to be beneficial for enhancing the sensitivities of polyclonal antibody-based latex agglutination methods (60, 66, 154); similar results were not observed with monoclonal antibody-based methods (162). While it was beneficial in reducing false-positive serum test results, recent studies have shown that pronase pretreatment of CSF is not efficacious, and it is not recommended (66, 162). The choice to pretreat specimens may very well depend on the type of patient population to be tested and the needs and expectations of the particular diagnostic services.

C. albicans

Direct antigen testing for *Candida albicans* (Table 2) is not nearly as efficacious as those methods available for *C. neoformans* but is nonetheless important. The reproducibility of

TABLE 2 Comparison of commercial methods for C. albicans antigen testing

Assay	Company	Method	Target	No. of samples tested	n	Sensitivity (%)	Specificity (%)	Comments	Reference
Cand-tec	Ramco Laboratories	Latex	Unidentified Candida antigen	64 sera	64 patients	29	97	Compared with various standard methods	14
					911 patients	46–95	50–80	Reported early diagnostic role despite low sensitivity and specificity; 71% PPV when titers are increased	16
					139 patients (60 negative; 79 positive)	47.4	98.3		72
Enolase	Becton Dickinson	Immunoassay	48-kDa protein	149		76.9			100
						71.8	87.5		100
					50 patients	54	16		167
						65	97.1		63
Mannan Candida antigen	Sanofi	Immunoassay	Galactomannan			25.6	100		100
					139 patients (60 negative; 79 positive)	52.6	100	Multiple samples increased sensitivity to 76%	72
β-Glucan Limulus test	Sanofi	Immunoassay				84.4			100
Cell wall mannan	Hybritech	Immunoassay			314 patients	86	92	PPV, 60; NPV, 98%	122

the method is reported to be adequate; however, test sensitivity continues to suffer. Few methods are available for use.

Cand-tec Latex Agglutination (Ramco Laboratories, Houston, Tex.) and Pastorex Candida (Sanofi Diagnostics Pasteur, Marnes-la-Coquette, France)

The presence of an unidentified heat-labile candidal antigen can be detected in serum by the Cand-tec latex agglutination test. In an early study of 911 samples, the reproducibility of titer results was reported to be 95 to 100% (16), but the sensitivity for detection of disseminated candidiasis was limited (19 to 100% and 46 to 95%). The specificity for disseminated candidiasis ranged from 22 to 99% (16, 72, 110).

The Cand-tec method may be best used as a measure of the severity of disease, since it has been suggested that the antigens detected by this assay accumulate in sera from patients with renal failure. In a study that measured creatinine (Cr) levels of 22 patients with proven disseminated candidiasis, the Cand-tec assay yielded positive results for a significantly larger number of patients having disseminated candidiasis than those with simple peripheral colonization. In addition, Cr levels exhibited an increase parallel with that of Cand-tec titers. Despite this association, results based on disease prevalence and a positive reaction cutoff of a 1:8 titer showed that a positive Cand-tec assay result had only a 79% probability of being associated with disseminated candidiasis (14).

In an analysis of the Pastorex Candida test with the Cand-tec antigenemia test, the specificities were 100 and 98.3%, respectively. The overall sensitivities of both tests were low, but with multiple testing, a positive antigen test result preceded other indications of candidemia for 6 of 10 Pastorex-positive patients and 5 of 9 Cand-tec-positive patients. Multiple sampling increased the sensitivity of the Pastorex Candida test to a high of 76.9%; however, this was not the case with the Cand-tec assay. Future refinements of these tests may add to their diagnostic value (72).

In addition, the Cand-tec test was compared to methods for the detection of enolase antigen and mannan antigen (Pastorex Candida assay) and to the detection of β-glucan by the *Limulus* test. In 39 patients with candidemia, 76.9% were positive for the heat-labile antigen in the Cand-tec assay. Other test methods were positive for 71.8, 25.6, and 84.4% of the patients. Of 10 patients with superficial *Candida* colonization, 10 patients with deep infections, and 20 healthy subjects, it was found that 2 patients with superficial infections and 1 patient with invasive pulmonary aspergillosis were also positive with the Cand-tec assay. None of the tests in this study appeared satisfactory for the diagnosis of candidemia; however, the tests may be helpful in certain diagnostic situations (100).

Other *Candida* Assays

The *Candida* enolase tests (Becton Dickinson Microbiology Systems, Sparks, Md.), are reported to have sensitivity and specificity values of 65 to 85% and 91 to 96%, respectively (63, 167). The membrane dot immunoassay for cell wall mannan (CWM and Hybritech) is another commercially available technology that uses a polyclonal capture antibody with monoclonal antibodies that act as indicators. The sensitivity and specificity of this assay to diagnose invasive candidiasis were reported to be 86 and 92%, respectively (122). Another commercially available assay

is the L-arabinitol (LA)-*Candida* antigen test (Immunomycologics).

For the semiquantitation of yeast in vaginal candidiasis, Orion Diagnostic (Espoo, Finland) markets the Oricult-N system, which has been compared to Orion's Vagicult system and to wet-mount microscopy. Further testing must be performed to determine the diagnostic value of these methods (19).

DA

Although small amounts of the D enantiomer of arabinitol (DA) and larger amounts of LA are produced normally by humans, *Candida* species, with the exception of *Candida krusei* (production by *Candida [Torulopsis] glabrata* is uncertain), produce larger amounts of DA during infection. Consequently, the additional DA produced by *Candida* species has to be normalized with respect to clearance from the kidney as the DA/Cr ratio or the DA/LA ratio.

Testing indicates that DA/LA ratios are significantly elevated (above the normal concentration of 10 μg/ml) in the urine of acute leukemic patients with candidiasis (20). An enzymatic-fluorometric method is also used to measure the presence of DA fluorometrically in a Cobas Fara II (Roche) centrifugal analyzer. The Cobas Fara II can also measure Cr, thereby facilitating calculation of the DA/Cr ratio. Healthy control subjects have DA/Cr ratios of 1.1 μM/mg/dl, whereas in patients with candidemia, a mean of 2.74 μM has been reported (range, 1.6 to 19.1 μM) (173). Further definitive comparisons and evaluations of these methods are necessary, but elevated DA/Cr ratios appear to precede positive blood cultures, and testing is prognostic in monitoring responses to therapy (135).

β-D-1-3-Glucan

The horseshoe crab (*Limulus*) factor G test measures β-D-1-3-glucan in the plasma of patients infected with pathogenic fungi, such as *Candida* and *Aspergillus* (108), and forms the basis for the WB003 test (Wako Pure Chemical Industries, Osaka, Japan), which uses kinetic turbidimetry to detect the presence of β-D-1-3-glucan. The detection limit of the WB003 test is ~100 pg/ml. The normal concentration in healthy subjects is 0.57 ± 0.1 μg/ml (103). Difficulties have been reported in interpreting the results of this assay, since plasma glucan activity was found to be increased in 36 of 45 patients who were undergoing chemotherapy for hematologic malignancies but who had no signs of invasive fungal disease (78). Other false-positive reactions have also been observed in patients after surgery or hemodialysis, perhaps due to the presence of β-D-1-3-glucan in surgical gauze or cellulose dialyzer tubing (109, 135). A multicenter study to assess the utility of the WB003 test found a mean concentration of plasma glucan of 19.63 ± 73.3 μg/ml in 12 candidemia patients (103).

The Fungitec G Test MK (Seikagaku Kogyo Co. Ltd., Tokyo, Japan) is another test that is specific for β-D-1-3-glucan. Once activated by β-D-1-3-glucan, enzyme cleaves a chromogen, releasing *p*-nitroanilide, which is quantitated by its absorbance at 405 nm. The Fungitec G test has a reported sensitivity limit of 1 pg/ml, exceeding the sensitivity of the WB003 test (135). A comparison of the WB003 and Fungitec G tests was performed with patients with candidiasis and aspergillosis and indicated that β-D-1-3-glucan was detected in 39 of 43 (90.7%) patients with deep mycoses by the Fungitec G test and in 29 of 43 (60.7%) such patients by the Wako WB003 test (75, 135).

Aspergillus

Invasive aspergillosis is a serious problem for all immuno-compromised patients, especially those undergoing bone marrow transplantation or chemotherapy. The disease is severe, progresses rapidly, and is difficult to diagnose. Fatality rates can approach 100% but can be reduced with early diagnosis and proper therapy. Unfortunately, blood and respiratory cultures may be negative during the early stages of disease (74, 168).

Pastorex *Aspergillus* Test (Bio-Rad Laboratories; Previously Sanofi Diagnostics)

The Pastorex *Aspergillus* test was designed for early detection of *Aspergillus* spp., more specifically, to detect circulating galactomannan antigen in the peripheral blood by latex agglutination. The test has been evaluated in several different studies, and the results generated have lacked sensitivity and specificity compared to those of traditional enzyme-linked immunosorbent assays (ELISAs) (96, 134).

In addition, false-positive results can be observed in about 6% of patients in whom there is no evidence of invasive disease (74, 150). In an effort to uncover a possible explanation for this phenomenon, healthy human subjects were tested and found to have galactomannan in their fecal material. It was postulated that fecal galactomannan, previously identified in food and in some antibiotics, may reach the circulation in patients with dysfunction in the intestinal mucosal barrier, precipitating the false-positive antigenemia results (3).

The Pastorex *Aspergillus* test has also been adapted for use with bronchoalveolar lavage (BAL) fluids; however, when BAL fluids from bone marrow transplant patients were tested, the Pastorex assay was found to have little diagnostic value. Still, there have been occasions when the assay has been useful for a limited number of patients (134).

Overall, the utility of this assay for use with BAL and blood specimens is questionable. Even with testing of multiple specimen types, the results are disappointing, and the use of this method should be considered only in conjunction with other laboratory findings (74, 111).

In contrast, in a study that compared the Pastorex method to antibody assays based on immunodiffusion tests of agar gel sensitivity for *Aspergillus fumigatus*, the antigen test was less sensitive but specific. The sensitivity and specificity of the antigen test were 38.8 and 95%, respectively. For the antibody test, the sensitivity and specificity were 55.5 and 100%, respectively (93). Other reports have shown a sensitivity range from 36 to 95% (134).

Sanofi EIA

Though originally devised for serum testing, the EIA method was adapted in one study for use with urine samples. Urine testing by this method was shown to be more reliable than serum testing, and the antigen detection limit in urine was 20 ng/ml. Moreover, antigenuria preceded antigenemia and was more persistent during infection. The urine assay had better sensitivity, specificity, positive predictive value (PPV), and negative predictive value (NPV) (57, 53, 31, and 71%, respectively) than the antigenemia test for autopsy-proven or clinically suspected cases of aspergillosis but was still limited in its clinical usefulness due to low sensitivity and other correlative data. Additionally, the galactomannan assays cannot make a distinction between *Aspergillus* infection and exposure (2). In yet another comparative study, the Pastorex *Aspergillus* antigen latex agglutination test for detection of galactomannan was shown to be too insensitive to be used as a diagnostic test but may

supply supplementary information for diagnostic purposes. In that work, it was shown that serum used for the assay must be stored at $-70°C$ (163). In selected cases, this assay has been helpful in the diagnosis of aspergillosis (68, 74).

For a select group of patients, such as transplant patients, the benefits of early diagnosis are clear, but alternative test methods may be necessary. In a 1996 study, serum from 215 bone marrow transplant patients was tested before and after transplant by the Pastorex latex agglutination test and the Sanofi Diagnostics sandwich immunocapture EIA kit. The immunocapture EIA uses rat galactomannan monoclonal antibodies as the capture and detector antibodies. No positive reactions were obtained prior to transplant. After transplant, the EIA and the latex test were positive in 19 of 25 and 4 of 25 patients, respectively, with confirmed aspergillosis and 14 of 15 and 7 of 15 of those with probable aspergillosis. The EIA was generally more sensitive and detected infection earlier than the latex text. These results favor the use of the EIA for monitoring aspergillosis in bone marrow transplant patients (156).

β-D-1-3-Glucan

The *Limulus* factor G test (see above) forms the basis for the Wako WB003 kinetic turbidimetric test (135). In a multicenter clinical study to assess test utility, plasma from four cases of invasive pulmonary aspergillosis was tested, and a mean glucan concentration of 12.21 ± 31.3 µg/ml was reported (103). In a comparison of the WB003 and Fungitec G tests, 60.7 and 90.7% of patients, respectively, tested positive for the presence of β-D-1-3-glucan when there was evidence of deep mycoses due to either *Candida* or *Aspergillus* ($n = 43$) (75). Plasma glucan activity can be increased in 80% ($n = 45$) of patients with hematologic malignancies who are undergoing chemotherapy but have no signs of invasive fungal disease (78). False-positive reactions can also occur with patients after surgery or hemodialysis, possibly due to glucan present in surgical gauze or cellulose dialyzer tubing (109). Therefore, interpretation of the plasma β-D-1-3-glucan tests remains problematic.

CULTURE AND DETECTION
Fungal Blood Cultures

Increases in cases of fungemia have resulted from greater numbers of immunocompromised patients, the widespread use of invasive medical devices, and the use of broad-spectrum antibiotics. Mortality rates range from 38 to 100%, and early diagnosis is critical. While *Candida* spp. account for a large percentage of isolates, the range of different fungal species isolated is extensive (57). Many publications have compared fungemia detection systems prior to the 1990s, and the reader may refer to reviews of those studies elsewhere (57). Clearly, no system is 100% sensitive or amenable to use in every situation or health care facility. A combination of different methodologies may provide the most sensitive culture, but this is not usually feasible. The shifting prevalence of commonly identified species and emerging pathogens may force the use of algorithms to provide guidelines for the most effective use of resources. Many of the blood culture systems described here are also described and depicted elsewhere in this text in the sections describing bacterial blood culture systems.

Isolator (Wampole)

The lysis-centrifugation blood culture method has long been regarded as one of the most sensitive systems for the detection of fungemia. Blood collection devices are available as 10-ml draws and as 1.5-ml draws for children (ISO 1.5) (119). The Isolator system (Table 3) uses saponin to lyse erythrocytes and leukocytes, releasing microorganisms from white blood cells and inactivating inhibitory serum factors and antimicrobial agents. In this system, the blood is lysed before organisms are concentrated by centrifugation. The sediment is subsequently inoculated onto solid media (159). Improvements to and automation of broth-based systems have made them an attractive possibility for the future, but continued studies performed with the Isolator system reconfirm its merit for the detection of fungemia, primarily for the detection of filamentous fungi.

The Isolator system has been proven to be more rapid and sensitive than detecting fungi in vented biphasic or broth media, such as Septi-Check and Bactec, especially for the recovery of *Histoplasma capsulatum* (57, 106, 159). It was also found to be superior to the ESP system and thiol bottle (Trek Diagnostic Systems, Westlake, Ohio) (13). In one retrospective study, 23,586 matched pairs of fungal blood cultures were compared, and the Isolator system enhanced the detection of fungi by as much as 30% compared to the

TABLE 3 Comparison of Isolator method (Wampole) with other blood culture methods and systems

Assay	Method	Comparison	n	Sensitivity (%)		Comments	Reference
				Isolator	Compared system		
Isolator	Lysis centrifugation	Bactec HBV	93			Isolator provides better detection of *H. capsulatum*; HBV better for *C. glabrata*	170
		Bactec 26B	86			Isolator superior for *H. capsulatum*, *C. glabrata*, others	170
		Bactec 9240 with MYCO/F lytic bottle	24	79.2	70.8	Isolator provides advantage for *C. neoformans*	166
		Organon Teknika	22	100		Improved TTD[a] with Isolator; *H. capsulatum* and *C. neoformans* from Isolator only	92
		BacT/Alert	22		68.2		92
		BacT/Alert for *C. neoformans*	53	88.7	43.4	Shorter TTD with Isolator; both systems costly	11
		BacT/Alert for *H. capsulatum*	18	100	22.2	Shorter TTD with Isolator; both systems costly	11
		ESP				Isolator superior; detected significantly more *Candida* spp.	13, 21
		Septi-Chek				Isolator superior; detected significantly more *Candida* spp.	21
		Conventional broth methods	20 (true positive)	70	80	False negatives in 30% (6 of 20) with Isolator vs. 20% with broth method	104
			42 (false positive)	97.6	2.4	Routine use of Isolator was not advocated	104
		Routine bacterial blood cultures				Isolator was not found to significantly enhance fungal recovery and was not found to be cost-effective	105
Isolator 1.5		Bactec NR660 Aerobic 6A	89	85.4	82	Comparable systems for *Candida* spp. in pediatric population; similar TTDs	119

[a]TTD, time to detection.

Septi-Chek system (62). Fungal recovery can be enhanced by as much as 15 to 50% in comparison to broth cultures, and time to recovery is generally decreased (57); however, newer automated systems have compared more favorably than the traditional broth-based fungal cultures.

In one study, the Isolator method had the highest recovery rate and shortest recovery time compared to both the Bactec and Septi-Chek systems (159). When the Isolator and Bactec BP26 systems were compared to the Bactec high-blood-volume fungal medium (HBV) for 93 test isolates, HBV proved to be comparable, except for the recovery of *H. capsulatum*, where the Isolator system was superior. In contrast, *C. glabrata* was detected earlier in HBV than in the Isolator system. The Isolator system proved to be superior to the Bactec BP26 system for recovery of *H. capsulatum*, *C. glabrata*, and other fungi when 86 cultures positive for fungi were tested (170). The Isolator system also detected significantly more *Candida* spp. than the ESP 80A or the Septi-Check system (21). A significantly improved recovery rate was also documented for 171 isolates of *C. albicans* and *Candida parapsilosis* (84).

The Isolator system was also found to be superior, especially for *Candida* spp., *H. capsulatum*, and *C. neoformans*, to the Organon Teknika Corp. (Raleigh, N.C.) BacT/Alert system. In 22 cases of candidemia, the Isolator system recovered 100% of the isolates, with a mean time to detection of 2.9 days (range, 0 to 9 days) while the BacT/Alert system recovered only 68.2% of the isolates, with a comparable mean time to detection of 2.1 days (range, 0 to 5 days). *H. capsulatum* and *C. neoformans* were recovered only by the Isolator system (92).

When the small-volume Isolator 1.5 was compared to the Bactec NR660 system aerobic NR6A blood culture bottle for detection of fungal infections in pediatric patients, NR6A was found to be comparable to ISO 1.5 in both yield and recovery time (119).

In another study with 53 isolates of *C. neoformans*, the Isolator system detected 47 while the BacT/Alert system detected only 23. Of 18 isolates of *H. capsulatum*, the Isolator system detected 18 and the BacT/Alert system detected 4. Time to detection was shorter for the Isolator system, 14 days for *H. capsulatum* and 6 days for *C. neoformans*, compared to the BacT/Alert with 18 and 11 days, respectively (11).

In a 1998 comparison, the Bactec 9240 system with MYCO/F lytic medium was compared to the Isolator system. Of 24 blood cultures positive for fungi, 12 were positive in both systems, 7 were positive only in the Isolator system, and 5 were positive only with the MYCO/F lytic bottle. Of 14 positive blood cultures containing *H. capsulatum*, 7 were positive in both systems and 7 were positive only in the Isolator system. For fungal isolates in this study, the mean times to detection were similar at 8 to 9 days, but the organisms were more quickly identifiable from the Isolator plates. For *C. neoformans*, the MYCO/F bottle may have an advantage. Of 10 blood cultures positive for *C. neoformans*, 5 were positive in both systems and 5 were positive in only the MYCO/F system. The mean time to detection was 4 days in the MYCO/F system and 7 days with the Isolator system (166).

Despite its widespread use, the Isolator system can produce false-positive results due to contamination events related to the multiple-step processing of IsoSTAT tubes; this is observed mostly with bacterial isolates and not with fungi. It is important to monitor false-positive rates in individual laboratories, to identify the case definition of fungemia, and to determine the types of organisms isolated in each patient population. In cases where a substantial number of false positives are observed, the use of the Iso-STAT tube may not be the most efficacious procedure (104, 105); however, for most situations, it still provides a gold standard for the detection of filamentous fungi.

Not all publications find the use of the Isolator system beneficial. In one retrospective study, the utility of the Isolator system for the detection of fungal recovery was compared to that of culture in the Bactec NR660 system. Strict definitions for both positive and false-positive blood cultures were given, specific to the study. Of 66 positive fungal blood cultures, 42 were considered to be due to contaminants, and all except one derived from the Isolator system. The use of these definitions produced summary results with 16 different species of fungal contaminants (24). Similarly, relatively high contamination rates were reported in other studies and further detract from the cost-effectiveness of a system which is very labor- and cost-intensive (46, 83, 105, 159). Clearly, the utility and cost-effectiveness of the Isolator system for fungal blood cultures will vary with institutional blood culture collections, patient population, processing, and contamination rates.

BacT/Alert (Organon Teknika)

The BacT/Alert system is a continuously monitored closed blood culture system for which no special fungal medium is available. The system is based on the photochromatic detection of CO_2 as organisms grow in the broth. Although the BacT/Alert system has slightly diminished capacity for the detection of some fungi (92), use of the FAN bottle may enhance the recovery of some yeasts for users of this system. The FAN aerobic media isolated significantly more yeasts than standard aerobic medium, with the same mean time to detection (169). When compared to the standard anaerobic bottle, vented for aerobic use, the FAN bottle recovered significantly more *Candida* species (23). In contrast, when the FAN anaerobe bottle was compared to standard aerobic medium, more *C. glabrata* isolates were recovered from the aerobic medium (57, 171). When the BacT/Alert system was compared to the Bactec NR660 or 730 system, the BacT/Alert system had a noticeable but not statistically significant advantage in fungus recovery; however, terminal subcultures of both systems added many of the fungi that otherwise would not have been recovered (172).

Bactec (Becton Dickinson)

Various models of the Bactec blood cultures systems are used in clinical laboratories worldwide. Older models, based on the radiometric, interval-based detection of CO_2, have been replaced by nonradiometric (NR) systems and, most recently, by the Bactec 9000 systems, which use fluorescent sensor technology to continuously monitor CO_2 produced by the organism present in the Bactec MYCO/F line of blood culture broth media. Bar-code-scanning and data management systems are also included with these systems.

In a study of the Bactec NR660 and Oxoid Signal systems, the Bactec system proved to have the advantage in detection time. The NR660 system isolated 18 yeasts, with median times to detection of 39 h for the Bactec system and 168 h for the Signal system (145). In 1992, Bactec PLUS high-volume resin (BP-HBV) and aerobic BP26 and anaerobic BP27 vials were compared to standard Bactec aerobic NR6A and anaerobic NR7A vials. Bactec PLUS vials enhanced the recovery rate for *C. albicans* for patients receiving antibiotics (94).

The Bactec system has a slightly diminished capacity for detection of some fungi compared to the Isolator system.

When Bactec HBV was compared to the Isolator system and Bactec BP26 with 93 isolates tested, HBV proved to be comparable, except for the recovery of *H. capsulatum*, where the Isolator system was superior. In contrast, *C. glabrata* was detected earlier in HBV than in the Isolator system. For HBV versus BP26, 68 positive fungal isolates were identified, and HBV was found to be superior for *C. glabrata* and all fungi combined. No difference was noted in the mean time to detection (170).

When the Bactec 9240 system was evaluated in comparison with the conventional Bactec NR730 system, the two systems were compared using two media with the same formulation: NR26 for the Bactec 730 system and Aerobic Plus F for the Bactec 9240 system. Simulated blood cultures were prepared for 41 strains belonging to 18 different fungal species, meant to represent fungi that are responsible for hematogenous dissemination. In a second phase of the study, two different media, HBV and NR26, were both compared on the Bactec NR730 system. The differences between the two systems and the two media (NR26 and HBV) were not statistically significant, but use of HBV did improve the detection of fungi on the Bactec NR730 system. In addition, the Bactec 9240 system had a mean detection time that was significantly shorter than that of the Bactec NR730 system (50).

In one of two 1998 studies, simulated blood cultures were used to compare the Bactec 9240 medium (Mycosis IF/C) with the Bactec standard medium, the Aerobic Plus F. In this study, 43 strains and 10 species of yeast were tested. The median time to detection was 29.03 ± 13.99 h for Mycosis IF/C compared to 73.92 ± 56.74 h for Aerobic Plus F medium. Mycosis IF/C reduced the time to detection (49). In another study, specifically targeting *C. neoformans*, the MYCO/F medium appears to have an advantage over the Isolator system. Of 10 blood cultures positive for *C. neoformans*, 5 were positive in both systems and 5 were positive in only the MYCO/F medium. The mean time to detection was 4 days in MYCO/F and 7 days with the Isolator system (166).

ESP System (Trek Diagnostics)

The ESP blood culture system detects changes in oxygen consumption and the CO_2 generated by microbes. In 1994, the ESP system with ESP 80A and 80N bottles was compared to the Bactec NR660 system using Bactec 6A/7A, 16A/17A, or pediatric bottles. In this study, ESP 80N anaerobic bottles detected significantly more *Candida* spp. (102). In a recent assessment of neonatal sepsis, the ESP blood culture system isolated 41 yeasts. By 48 h of incubation, 88% (36 of 41) of those isolates were identified as positive cultures. There was no difference in the time to positivity in pretherapy and posttherapy blood cultures (56).

In a comparison of the ESP and the BacT/Alert systems, no significant difference was documented for 31 yeast isolates (174).

IDENTIFICATION SYSTEMS
Media for Isolation and Identification
CHROMagar Candida (CHROMagar, Paris, France, and Becton Dickinson)

CHROMagar is a chromogenic differential and selective medium for isolation and presumptive identification of yeasts directly from clinical specimens. The suggested incubation temperature is 30°C. Colony color and morphology are used for identification purposes. *C. albicans* grows as green colonies with slight green halos due to the presence of the enzyme β-*N*-acetylgalactosaminidase (1). *Candida tropicalis* grows as blue-grey or metallic grey colonies with dark brown or purple halos. *T. beigelii* also exhibits a blue-green color and becomes rough and crenated, distinguishing it from *C. albicans* and *C. tropicalis*. *C. krusei*, the isolation and identification of which are important because of its resistance to the common antifungal fluconazole, grows as large pink or rose-colored crenated colonies with pale edges. CHROMagar supports the growth of yeasts and most fungi while inhibiting most bacteria, paralleling the performance of Sabouraud dextrose agar.

In a 1994 study of CHROMagar, the presumptive identification of *C. albicans*, *C. krusei*, and *C. tropicalis* exceeded 99% for all three species when blind reading tests were performed by four different personnel. Several occurrences of potential misinterpretation are noteworthy, although such instances are rare. For example, rare isolates of *Candida norvegensis* produced a colony morphology similar to that of *C. krusei*. In addition, some *Geotrichum* spp. and *Pichia* spp. formed colonies with a grey, blue, or green color or a dark halo in the agar (112). In addition, the recently described pathogen *Candida dubliniensis*, has been reported to grow as dark-green-colored colonies upon primary isolation (143); however, the presence of this phenotype is variable. When compared to identification by DNA sequencing, only 56% of all *C. dubliniensis* isolates tested ($n = 53$) displayed the green phenotype (160).

In one study, in which 1,537 yeast isolates were tested after 48 h of incubation, sensitivity and specificity were 99 and 100% for *C. albicans*, 93.8 and 99.1% for *C. tropicalis*, and 100 and 100% for *C. krusei*. A large sample size makes this comparison noteworthy, with 970 *C. albicans*, 165 *C. parapsilosis*, 131 *C. glabrata*, 62 *Candida guilliermondii*, 35 *C. krusei*, 32 *C. tropicalis*, 31 *Rhodotorula rubra*, and 23 *Trichosporon* sp. samples and a variety of other species. The accuracy of CHROMagar is reported to be similar to those of germ-tube tests and chlamydospore development for the identification of *C. albicans* and to that of the ATB ID32C (API Biomerieux, Marcy l'Etoile, France) for identification of *C. tropicalis* and *C. krusei* (142).

Additionally, CHROMagar was evaluated for 618 yeast isolates and 128 direct specimen inoculations. After 2 days at 37°C, the sensitivities for identifying species were as follows: *C. albicans*, 99.4% ($n = 341$); *C. glabrata*, 98.9% ($n = 99$); and *C. krusei* ($n = 35$) and *C. tropicalis* ($n = 73$), 100%. The specificities were 100, 99.8, 99.6, and 99.6%, respectively. The authors also report CHROMagar to be useful for the identification and determination of mixed cultures (9). Detection rates for *C. albicans* were reported to be as much as 20% higher than those reported with Sabouraud-chloramphenicol plates ($n = 951$), with 92.2% correct identification after 72 h. The specificity was reported to be 100%. Similar results were reported using the Albicans ID agar (Biomerieux) (7).

Another recent study compared CHROMagar to two other commercial media for identification of *Candida* spp., Albicans ID and Candiselect (Sanofi Diagnostics Pasteur), both of which contain fluorogenic substrates in the agar. All three were compared with isolation on traditional Sabouraud-chloramphenicol agar medium followed by standard yeast identification. A total of 192 clinical specimens were tested on all four agar media. For all yeast species, CHROMagar yielded the largest colonies. The identification rates for *C. albicans* were nearly equivalent at 86.59, 83.79, 83.24, and 84.91%, respectively. At 72 h, Albicans ID, Candiselect, and CHROMagar identified 92.75, 91.3, and 88.57% of the specimens, respectively, but at 24 h the sensitivities for growth and specific pigment were substantially lower at 56.6, 37.68, and 11.59%, respectively. Although CHROMagar did not identify *C. albicans* as quickly as the other two commercial identification media, the agar allowed correct identification of *C. tropicalis*, the main source of false positives on most other chromogenic media. In addition, the authors report that morphology is easier to read with CHRO-Magar (47).

In 1998, two studies were reported. In one, 21 yeast isolates from neutropenic and AIDS patients were tested, both on CHROMagar and Sabouraud dextrose agar. The overall sensitivities for detecting *C. albicans* were the same for both media. The primary isolation on CHROMagar was found to be 100% sensitive and 100% specific for *C. albicans*. For identification purposes, CHROMagar proved to be the most economical and the least time-consuming method. The authors of the study suggest its use for blood culture plating when yeasts are seen in microscopy or when early therapy is imperative (1). In another study, CHROMagar Candida (CHROMagar) was tested with a total of 262 yeast isolates: *C. albicans*, 173; *C. tropicalis*, 21; *C. krusei*, 8; *C. glabrata*, 49; and other yeasts, 12. Compared to conventional identification, *C. albicans* was correctly identified 98% of the time. In addition, CHROMagar helped identify 37 of 46 mixed cultures that were otherwise not detected. The authors suggest that it is reasonable to include CHROMagar for identification of *C. albicans* and as an aid in identification of mixed cultures (127).

Other Comparisons

In a recently published study, differential agars were compared to the Murex Diagnostics (Norcross, Ga. [formerly Carr-Scarborough, Stone Mountain, Ga.]) *C. albicans* kit (also known as *Candida albicans* CA50) and traditional germ tube testing. Albicans ID, Candiselect, CHROMagar Candida, Fluoroplate Candida (Merck, Darmstadt, Germany), Fongiscreen 4H (Sanofi Diagnostics Pasteur), and Murex *C. albicans* were compared to the germ tube test for presumptive identification of *C. albicans*. For 350 isolates of *C. albicans* and 135 isolates of nonalbicans yeast, the sensitivities and specificities for all kits for *C. albicans* identification were >97%. The two presumptive identification systems, CHROMagar and Fongiscreen 4H, did not identify all isolates of *C. glabrata* and *C. tropicalis*. In addition, CHRO-Magar Candida incorrectly identified some nonalbicans yeasts as *C. glabrata* (73). Fongiscreen 4H was also evaluated, and the results were compared to those from another commercial test, the Rapidec albicans (API Biomerieux), for 13 species of yeast. Correct identification was obtained for 100 and 97%, respectively (130). One study tested 1,006 clinical samples containing 723 yeast strains on Albicans ID and found 93.8% sensitivity and 98.6% specificity, including several false positives with *C. tropicalis* (138). In another study, 100% sensitivity and 90.3% specificity were noted (89).

In another recent comparison, yeasts grown on Candiselect agar grew less well than those on Sabouraud dextrose agar, but the sensitivity and specificity for Candiselect were 100 and 96.8%, respectively. Of concern were false-positive reactions occurring with 4 of 19 *C. tropicalis* spp. as well as *Trichosporon* and *Saccharomyces cerevisiae* (18).

Methyl Blue Sabouraud Agar (Becton Dickinson)

Methyl blue Sabouraud agar is prepared from Sabouraud agar but contains 0.01% methyl blue as a supplement (58). This medium has been reported to distinguish between *C. albicans* and *C. dubliniensis* based on the yellow fluorescence of *C. albicans* under a Wood lamp (143).

EMB

In testing with 92 isolates, all *C. glabrata* isolates grew better on eosin-methylene blue (EMB) agar plates than on blood agar plates at 24 h. The criterion of better growth on EMB agar is used as an identification method for *C. glabrata* and is combined with the germ tube test as a cost-effective method for identification of *C. glabrata* and *C. albicans* (5).

Rapid Identification and Screens

C. albicans

Bichro-latex Albicans (Fumoze Diagnostics, Asnières, France)

The Bichro-latex Albicans test consists of red latex particles coated with monoclonal antibody with specificity for the protein moiety of a *C. albicans* cell wall glycoprotein of >200 kDa. The latex particles are suspended in a green dye, which gives the entire mixture a brown color. Yeast colonies are emulsified in a dissociating agent that contains enzymes to expose the antigen, which is recognized by the monoclonal antibody. Tests are reported as quick to perform and easy to read, with positive results appearing as red agglutination on a green background. Tests can be performed directly from a number of primary isolation media. Nonspecific agglutination was easily distinguished, since the background remained brown and did not turn green as for a positive result (48, 132).

In two recent studies, the Bichro-latex Albicans test was compared to identification by traditional methods. Of 4,643 yeast isolates tested, 6 were noted as false positives and 3 as false negatives, for a sensitivity and specificity of 99.74 and 99.87%, respectively. The method proved to be more sensitive than the germ tube test (132) and can be performed directly on blood culture bottles without the need for subculture onto agar media (48). In another report, 322 yeasts and yeastlike organisms, such as *Candida*, *Cryptococcus*, *Geotrichum*, *Saccharomyces*, and *Trichosporon*, were tested. The sensitivity and specificity were reported to be 100% for identification of *C. albicans* (48).

Identification of Unique Enzymes

Traditionally, *C. albicans* has been identified by the germ tube test. In 1987, Perry and Miller evaluated the use of 4-methylumbelliferyl-*N*-acetyl-β-galactosamine (NAG) to differentiate between *C. albicans* and *C. tropicalis* (117). Later, the enzymes L-proline aminopeptidase and β-galactosaminidase further increased the specificity of the test (118).

Three kits (Murex *C. albicans*, Albicans-Sure [JRGA Diagnostics and Clinical Standards Laboratories, Rancho Domingo, Calif.] and BactiCard Candida [Remel Laboratories, Lenexa, Kans.] (Fig. 1) detect the enzymes L-proline aminopeptidase and β-galactosaminidase. *C. albicans* produces both enzymes, while other yeasts produce only one or neither.

The Murex *C. albicans* test uses the enzyme substrates *p*-nitrophenyl-*N*-acetyl-β-D-galactosamide and L-proline-β-naphthylamide, incorporating them onto a paper disk in a kit-provided test tube. The disk is moistened with distilled

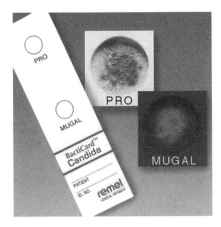

FIGURE 1 BactiCard Candida. Courtesy of Remel.

or deionized water, and a heavy inoculum of yeast is placed on it with an inoculating loop. The tube is capped and incubated at 35 to 37°C for 30 min, after which 0.03% sodium hydroxide is added. If one or both of the enzymes are present, the substrates will be hydrolyzed and a colored reaction will be noted. C. albicans is positive for both enzymes; other yeasts may be positive for one or the other.

With the Albicans-Sure test, two circles on paper cards contain different substrates, NAG and proline-p-nitroanilide. The substrates are dehydrated with distilled or deionized water, and each circle is inoculated with two or three colonies of an organism. The test cards are incubated for 5 min at room temperature, after which a drop of p-dimethylaminocinnamaldehyde reagent is added. If NAG is cleaved by the specific enzyme, a highly fluorescent compound is released. If proline-p-nitroanilide is cleaved, a blue precipitate is formed.

The BactiCard Candida contains 4-methylumbelliferyl-N-acetyl-β-D-galactosaminide and L-proline-β-naphthylamide; the latter is the same substrate found in the Murex system. These substrates are also dehydrated on a packaged test card, which is rehydrated prior to testing and inoculated with a thick paste of yeast cells. If the substrates are cleaved after 5 min of incubation at room temperature, one fluorescent reaction product, 4-methylumbelliferone, and one colorimetric reaction product will be produced. The fluorescent product is viewed under long-wave UV light. The test can be performed on pure cultures that are 18 to 72 h old.

These three tests were compared to the traditional germ tube test for presumptive identification of C. albicans. The API 20 C test (Biomerieux Vitek, Hazlewood, Mo.) and conventional methods were used as the gold standards. A variety of yeasts were tested, including C. albicans (n = 303), C. glabrata (n = 153), C. tropicalis (n = 70), C. parapsilosis (n = 36), Cryptococcus sp. (n = 5), S. cerevisiae (n = 3), and other yeasts (n = 13). Murex C. albicans, Albicans-Sure, BactiCard, and the germ tube test identified 98.7, 99, 99.3, and 94.7% of C. albicans isolates, respectively. One false positive each was observed using the Murex and Bacti-Card tests, while two false positives were seen with the traditional germ tube test. All of the enzymatic methods were noted to be rapid and accurate alternatives to germ tube testing of C. albicans (25). All three tests were also used to test 133 yeast isolates in comparison with the germ tube test and the API 20 C test. With the exception of Albicans-Sure, all were found to correlate 100% (70). The Albicans-

Sure assay had a sensitivity, specificity, NPV, and PPV of 100, 97, 100, and 95%, respectively. All of the assays generated some false-positive results for the noncandida yeasts. The assays were compared for cost and other parameters (70).

The Murex C. albicans test was also compared with standard 2-h germ tube tests (n = 502). In this study, the number of C. albicans isolates tested was 316; the number of nonalbicans yeasts tested was 186. Discrepant reactions were confirmed with the API 20 C test and by chlamydospore production. Two yeast isolates, C. albicans and Candida lusitaniae, gave incorrect results with the Murex test. Compared to the germ tube test in fetal bovine serum and fetal clone II, the sensitivity and specificity of the Murex test were 94.6 and 97.8%, respectively (44).

Methods for Identification of *C. glabrata*

Rapid Trehalose

The rapid screening method for the identification of C. glabrata is based on the principle that C. glabrata can rapidly utilize trehalose in the presence of cycloheximide. The Rapid Assimilation Trehalose Broth, used at the Mayo Clinic, is prepared in house with yeast nitrogen base, trehalose (40%), bromcresol green (0.02%), and cycloheximide (10,000 μg/ml). Three drops of the reagent are added to each well of a microtiter plate. A heavy inoculum (three to five colonies from Sabouraud agar) is emulsified into each well. The microtiter plate is incubated for 1 h at 37°C before being read. A change in color from blue to yellow is considered positive. The microtiter format allows for the screening of large numbers of germ tube-negative yeasts which have morphologic features suggestive of C. glabrata, i.e., small cell size (L. Stockman and G. D. Roberts, Abstr. 85th Annu. Meet. Am. Soc. Microbiol. 1985, p. 377). Several commercial systems are also available to test for trehalose production.

In a 1996 study, a screening protocol for C. glabrata based on trehalose fermentation at 42°C for 24 h (Trehalose Fermentation Broth; Hardy Diagnostics) was compared to identifications by the API 20 C Aux and Dade Microscan Rapid Yeast Identification Panel. Trehalose Fermentation Broth was found to have 97.8% sensitivity and 95.8% specificity when used for presumptive identification of C. glabrata (87). More recently, four methods for the rapid identification of C. glabrata were compared. Included were the Rapid Assimilation Trehalose Broth method described at Mayo Clinic, the 24-h Trehalose Fermentation Broth, the 3-h Remel Rapid Trehalose Assimilation Broth, and the 24-h Remel Yeast Fermentation Broth. All of the methods were compared for accuracy, cost, and turnaround time. Remel Rapid Trehalose Assimilation Broth was recommended as the method of choice based on its rapid results, reasonable sensitivity, and low number of false-positive reactions (42). However, the size of the cells was not considered before testing.

Methods for Identification of *C. krusei*

The Krusei color test (Fumoze Diagnostics) is performed with red latex particles coated with a monoclonal antibody that specifically reacts with a C. krusei antigen located on the cell surface. The appearance of large clumps of red agglutination is considered a positive reaction, while no agglutination or white agglutination is interpreted as negative. Unfortunately, at the time of this study, the background of the test was not colored as is that of its counterpart for C. albicans, the Bichro-latex Albicans kit. The sensitivity was reported to be 100%, and the specificity was

95%. The practical use of this test is limited, since, unlike many C. *albicans* infections, treatment of any nonalbicans yeasts isolated from blood or tissue is not done on the basis of yeast identification alone; rather, isolates are generally tested for susceptibility to antifungal agents (48).

Identification of C. dubliniensis

Commercial Systems

Identification of C. *dubliniensis* is a relatively new diagnostic dilemma, and few commercial systems have been tested. In one such study, 66 isolates of C. *dubliniensis* and 100 isolates of C. *albicans* were tested with the API 20 C Aux and the Vitek Yeast Biochemical Card (YBC) systems. The results were specifically focused on the abilities of the respective systems to identify xylose and α-*methyl*-D-glucoside utilization by C. *albicans* and lack of the same utilization by C. *dubliniensis*. Growth at 45°C was also examined. The results suggested that clinical laboratories could use the lack of growth at 45°C and negative xylose utilization with either the API 20 C Aux or Vitek system to presumptively identify C. *dubliniensis*. Less correlation existed with a negative MDG test on either system, and misclassification may occur if this result is used, especially with the API 20 C Aux system (55).

Molds

Exoantigen

In the only published results of exoantigen testing, *Aspergillus* exoantigen tests from Greer Laboratories (Lenoir, N.C.), Immunomycologics, and Scott Laboratories (Fiskville, R.I.) were tested. The Immunomycologics test correctly grouped all isolates (n = 87) of A. *fumigatus*, *Aspergillus flavus*, *Aspergillus nidulans*, *Aspergillus niger*, and *Aspergillus terreus*, but reactions with A. *nidulans* and A. *terreus* were weak. The other two kits do not supply antibodies to these two species but correctly grouped all other isolates. All commercial serodiagnostic reagents can be effectively used to identify the medically important *Aspergillus* spp., with the kit from Immunomycologics identifying more species than its competitors (148).

Commercial Kits for Identification of Yeasts and Yeastlike Organisms

The advantages of the nonautomated and automated identification systems discussed here (summarized in Table 4) include the fact that identification is based on databases that include a variety of substrate utilization patterns and that contain many yeast biotypes (96).

Uni-Yeast-Tek (Remel)

The Uni-Yeast-Tek system (Fig. 2) was the first widely used commercial system for yeast identification. It consists of a sealed multiwell plate, with each well containing medium for testing carbohydrate utilization, nitrate utilization, urease production, and morphology on cornmeal agar. The kit also includes supplementary tests, such as the C-N screen, glucose beef extract, and sucrose assimilation medium tubes. Evaluations of this product have shown it to be generally satisfactory for the identification of commonly isolated yeasts (128). One disadvantage is that the system may require up to 7 days for complete identification. The system is in the process of redevelopment (128).

API 20 C (Biomerieux Vitek)

The API 20 C test is a micromethod for carbohydrate assimilation of clinically important yeasts. Yeasts are inoculated into basal medium in test strips containing 19 dehy-

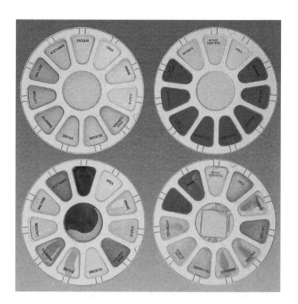

FIGURE 2 Uni-N/F Tek and Uni-Yeast-Tek plates (uninoculated and inoculated). Courtesy of Remel.

drated carbohydrate substrates for assimilation testing. The API 20 C system consists of a strip that contains 20 microcupules; all but 2 contain dehydrated substrates for determining substrate utilization profiles (Fig. 3A). As the yeasts utilize various substrates, the microcupules will appear cloudy (Fig. 3B). The reactions are compared to the first reaction cupule, which does not contain a carbohydrate substrate. The test strips are incubated at 30°C for 24 to 72 h. The results of assimilation reactions are read and converted to a seven-digit biotype profile number. In addition, the system requires that the microscopic morphology be documented on cornmeal agar containing 1% Tween 80 and trypan blue and that this information be used in conjunction with the substrate utilization data for yeast identification (128). The database consists of at least 42 taxa (100). This yeast identification system is perhaps the most widely evaluated and well published of all the current commercial systems. It is frequently used as the gold standard against which comparisons are made.

In early studies performed to compare the API 20 C test with conventional methods, 98% correlation was documented (n = 45). The system was considered to be an acceptable substitute for conventional carbohydrate fermentation and assimilation tests (175).

In a more recent study, 178 yeasts were tested. After 72 h, correct identification occurred for 86.5% of those isolates tested by the API 20 C system compared to 86% with Auxacolor (Sanofi Diagnostics Pasteur), 68% with Mycotube (Roche Diagnostics), and 51.1% with Candifast (International Microbio, Milan, Italy) (144). When 100 common yeasts were tested and reported, the correct identification was obtained for 100% of the isolates (64). Recently, 123 common and 120 rare clinical yeast isolates were tested, and the API 20 C test correctly identified 97 and 88%, respectively (133). Also recently, 202 yeasts belonging to 19 species were tested, and correct results were obtained in 83% compared to Fungichrom 1 and Fungifast Twin from International Microbio, which produced results of 81 and 91%, respectively (116). In another recent report, 90.8% of 120 yeast and yeastlike isolates were correctly identified by the API 20 C test (67).

TABLE 4 Comparison of commercial fungal identification methods

Assay	Company	No. of species tested	n	Correct ID[a] (%)	Correct ID with supplemental tests (%)	Incorrect ID (%)	No ID (%)	Limits	Reference
RapID Yeast Plus	Inovative Diagnostic Systems	9	201	96	6	2	0	No conclusion for C. lusitaniae, Rhodotorula sp., or S. cerevisiae; poor ID for C. glabrata	165
			52	76.9		11.5	11.5		164
		22	156	63.5	14.7	16	5.8		15
		20	160	76					
API 20 C Aux	Biomerieux	9	201	83	34	0	0	No conclusion for C. lusitaniae, Rhodotorula sp., or S. cerevisiae; poor ID with C. krusei	165
			52	59.6	36.5		3.9		164
		25	185	86.5			12.4		144
			200	90.8					67
		19	202	83					116
			123 (common)	97					133
			120 (rare)	88					
		20	100	100					64
			80	71					
API ID32C	Biomerieux		123 (common)	92					133
			120 (rare)	85					
			100	97				Incorrect IDs for C. tropicalis and S. cerevisiae	64
		22	156	80.7	19.2	0	0		51
		20	80	100					
			52	63.5	34.6		1.9	Performed well with atypical strains	98
API Candida	Biomerieux		52	78.8	13.5		7.7		164
		22	156	66	16.7	12.8	4.5	Compared to ID32C	15
		21	619	74	21.8	2.9	1.3	Not suitable for C. lusitaniae; supplemental tests required for many species	52
			52						98
Auxacolor	Sanofi Diagnostics Pasteur	15	100	61	33		6	User friendly, rapid, easy to read; may need subculture to CMA[b]; need experience for microscopic examination	137
		Germ tube negative and common	110	82.7				Compared to API 20 C, quicker, easier to read, similar cost; most errors were due to lack of chlamydospore production;	29
		Germ tube negative and uncommon	65	55.4				limited database of 26 species	

(Continued on next page)

TABLE 4 Comparison of commercial fungal identification methods (*Continued*)

Assay	Company	No. of species tested	n	Correct ID[a] (%)	Correct ID with supplemental tests (%)	Incorrect ID (%)	No ID (%)	Limits	Reference
C. albicans			40	57.5					165
			52	80.8			11.5		164
		22	156	76.9	7.7	3.8	10.3		15
		19	202	61	9				116
		12	97	79.4		5.2	15.5		99
			120	98.3					67
		17	169	95.2		6 strains	7 strains		51
		25	185	86			12.4	Extra tests needed for *C. glabrata*, *C. krusei*, *C. guilliermondii*, and others	144
YeastStar	CLARC Laboratories (Heerlen, The Netherlands)		52	59.6	40.4		0		164
Fungichrom I	International Microbio	22	156	70.5	14.1	4.5	10.9		15
Fungifast Twin	International Microbio	22	156	62.2	15.4	2.6	19.9		15
Mycotube	Roche	25	185	68			29.7		144
Candifast	International Microbio	25	185	51.1			45.4		144

[a]ID, identification.
[b]CMA, cornmeal agar.

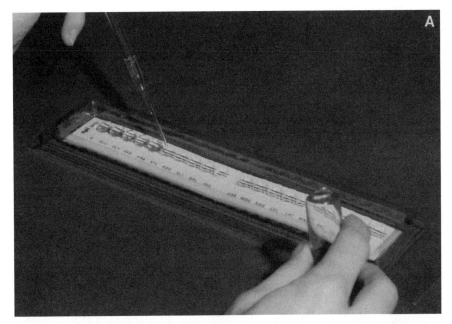

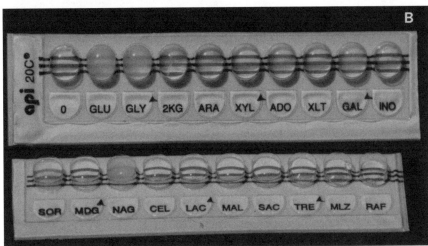

FIGURE 3 (A) Inoculation of the API 20 C (Biomerieux Vitek) strip for identification of yeast isolates. Each strip contains microcupules containing dehydrated substrates, which are rehydrated and examined to determine their ability to support the growth of yeast isolates. (B) Close-up view of substrate utilization by a yeast isolate. Positive reactions appear cloudy in comparison to the negative (0) control.

API Candida System (Biomerieux Vitek)

The API Candida system is one of the newer systems developed for yeast identification and can provide results in 18 to 36 h at 35°C. It consists of a strip of 20 API cupules in which 19 colorimetric tests can be performed. A yeast suspension is adjusted to a turbidity standard, and a test strip is inoculated to hydrate the dried substrates. Reactions can be read visually without the need for the addition of reagents, and the results are converted to a numerical profile.

When the API Candida system was compared to the ID32C system for the identification of 619 yeast isolates, the correlations of the systems for identification of the 15 species that can reportedly be identified with and without additional tests were 97.4 and 75.2%, respectively. Discrepancies were resolved by additional testing by conventional methods (52).

When compared to the API 20 C test, the API Candida system correctly identified 91.4% of 198 yeast isolates; however, only 71.7% were confirmed without additional tests. Identification was better for common yeasts (96.3% identified) than for the more uncommon species (66.6% identified). The system was documented to be easy to use and has applications in the clinical laboratory (8). In a recent multikit comparison, the API Candida system exhibited a total correct identification rate of 62% (15).

In another study, correct identification was reported for 146 of 159 isolates (91.8%), but 23 of the 146 required extra testing (17). In the same study, the API Candida system was compared to the Auxacolor kit (Sanofi Diagnostics Pasteur). The Auxacolor kit gave correct identifications for 145 (91.2%) isolates; however, no identification could be made in 8.2% of the cases compared to the

API Candida system, which gave no identification in only one case. Overall, incorrect identifications were more common with the API Candida system (12 isolates [7.5%]) than with the Auxacolor kit (1 isolate [0.6%]) (17). Most recently, the API Candida system was evaluated for its ability to detect *C. lusitaniae*. The API Candida system was not found to be suitable for the identification of *C. lusitaniae* and was reported to require supplemental testing for the discrimination of several other *Candida* species. The Auxacolor system received a more favorable review; however, the study noted the need for use of the ID32C system for identification of atypical *Candida* strains (98).

ATB ID32C (API Biomerieux [France] or Innovative Diagnostic Systems [United States])

The ID32C system consists of a single-use disposable plastic strip with 32 wells and can perform 29 assimilation tests for carbohydrates, organic acids, and amino acids. It also includes one negative control for the assimilation tests, one colorimetric test for esculin, and a test for susceptibility to cycloheximide. Its database includes 63 species. The strips are incubated at 30°C for 48 h. Descriptions of macroscopic and microscopic morphologies are sometimes required for identification.

In a comparison of 100 common yeast isolates, this product correctly identified 97% of all isolates but had difficulty with identification of *C. tropicalis*, *C. krusei*, and *S. cerevisiae* (64). Recently, 120 rare and 123 common clinical yeast isolates were tested by the ID32C system, with correct identification of 85 and 92%, respectively. For a study that included 156 yeast isolates and comparisons with several kits, including the RapID Yeast Plus system, the Auxacolor system, the Fungichrom I system, the Fungifast I Twin system, and the API Candida system, the ID32C system (with 72 h of incubation) had the highest total identification rate of all systems tested and was recommended as a reference identification system (15, 98).

Auxacolor System (Sanofi Diagnostics Pasteur)

The Auxacolor system consists of disposable plastic microplates containing 16 wells. There is one negative and one positive glucose control. Each of the remaining wells contains a different dehydrated sugar for assimilation by the test organism in the presence of a basic solution and a pH indicator, bromcresol purple. After inoculation with a yeast suspension, growth in the wells is indicated by turbidity and by a color change from blue to yellow as acid is produced. The strip also contains tests for cycloheximide (actidione) resistance and for the detection of phenoloxidase activity of *C. neoformans*.

Several published studies have compared the Auxacolor system with a variety of other methodologies. In one such study, the Auxacolor system was compared to the API 20 C test with a total of 215 yeasts, including 16 species. Yeast identification was confirmed by assimilation and fermentation tests. The Auxacolor system identified 85.7% of all yeasts tested; the API 20 C system identified 88.6%. Incorrect identification was more common with the API 20 C system (7.4%) than with the Auxacolor system (3.7%). In addition, the Auxacolor system performed better than the API 20 C system for the four most common pathogens, *C. glabrata*, *C. parapsilosis*, *C. tropicalis*, and *C. neoformans*. Among these isolates, 91 of 100 were identified with the Auxacolor system compared to 82 of 100 with the API 20 C system; however, the API 20 C system performed better

with the less common germ tube-negative isolates. From a practical perspective, the Auxacolor system was easier to set up and interpret and is comparable in cost to the API 20 C system (29). In another 1995 report, the Auxacolor system was compared to the ID32C system. For 17 common species, correct identification was obtained for 95.2%; 47.1% were identified within 24 h. For 13 more uncommon species, the Auxacolor system did not provide identification (51).

In a 1998 study, 178 yeasts were tested. After 72 h, correct identification was made for 86% of yeasts tested by the Auxacolor system (144), which also provided more rapid results than the API 20 C Aux system. Of all isolates tested, 67.9% were identified in 24 h compared to only 14.6% with the API 20 C system. At 48 h, the Auxacolor system correctly identified 80.9% of the isolates compared to 64% for the API 20 C system. The total identification rate for yeasts was determined to be 65% for the Auxacolor system. Both the Auxacolor system and the Fungichrom system, also tested in this study, were noted by the authors as being most appropriate for use in a clinical laboratory, due to the ease of use, rapidity, cost per test, and relatively high identification rates. However, for reference identification, the ID32C system was still recommended (15).

Most recently, 159 yeasts were tested with the Auxacolor and API Candida systems. The Auxacolor system correctly identified 145 (91.2%) of all isolates tested. No identification was derived in 8.2% of the cases compared with the results of the API Candida test, which gave no identification in only one case. However, incorrect identifications were more common with the API Candida system (12 isolates [7.5%]) than with the Auxacolor system (1 isolate [0.6%]) (17). When the Auxacolor system was compared to conventional biochemical testing, 97 isolates from 12 species produced profiles, for correct identification of 79.4% of the isolates (99); however, correct identification was reported to be as low as 61% in another study testing 19 different species (116).

In a comparison of seven commercial kits (Vitek, API 32 C, API 20 C Aux, YeastStar, Auxacolor, RapID Yeast Plus system, and API Candida) with *n* being 52 and 19 species, the best performance was obtained with the API Candida (78.8% correct identification) and Auxacolor (80.8%) kits. Among germ tube-negative yeasts, the API Candida and Auxacolor kits both identified 93.1% correctly. All of the systems failed to identify the following germ tube-negative yeasts: *C. norvegensis*, *Candida catenulata*, *Candida haemolonii*, and *C. dubliniensis*. For this study, the API 20 C system was noted to be less expensive than the Auxacolor system and required less bench time (164). In the most recent studies, the Auxacolor system identified 100% of 94 clinical yeast isolates after 48 h of incubation compared to conventional identification methods (137). In another recent report, 98.3% of 120 isolates were correctly identified (67).

RapID Yeast Plus (Innovative Diagnostic Systems), Sold as IDS YeastPlus Panel from Remel

The RapID Yeast Plus system is a qualitative method that uses conventional and chromogenic substrates to identify carbohydrate utilization, hydrolysis of fatty acid esters, and constitutive enzymes of yeast and yeastlike organisms. It combines conventional tests and single-substrate chromogenic tests with a 4-h incubation. A suspension of pure yeast culture is made in the RapID inoculation fluid, turbidity is standardized, and the suspension is used to rehydrate the biochemicals in the RapID plastic reaction panel (Fig. 4). With a one-step inoculation, the suspension will disperse in the panel by way of the plastic inoculating trough. After a 4-h incubation at 30°C, the RapID Yeast Plus system is quick and easy to interpret. Some reactions require additional reagents to be added to the reaction panel. The resulting pattern of positive and negative reactions is used to compile a profile number based on the organism's individual reactions in the panel. The RapID Yeast Plus Code Compendium supplies species identification for the biochemical profile numbers, which are generated based on the results of biochemical reactions.

The RapID Yeast Plus method correctly identified 94.1% of 304 clinical yeast isolates within 5 h (86).

In a study of 201 yeasts, when no further tests were performed, the RapID Yeast Plus system was significantly better and easier to use for obtaining results to the species level than the API 20 C Aux system (193 versus 167 correct identifications) or the Vitek YBC (193 versus 173 correct identifications). The API 20 C Aux system did not correctly identify any *C. krusei* isolate without supplemental testing, and this organism accounted for most of the differences seen between the API 20 C system and the RapID Yeast Plus system (165).

In other comparisons, the RapID system exhibited good correlation. When 133 yeasts, specifically, 57 *C. albicans* isolates, 26 *C. tropicalis* isolates, 23 *C. glabrata* isolates, and 27 isolates from other species, were tested by the RapID Card API 20 C systems, the RapID system correctly identified 125 (94% overall accuracy) of all yeasts tested and 99% (105 of 106) of the isolates of the three most common species, *C. albicans*, *C. tropicalis*, and *C. glabrata*. Discrepancies were resolved with Automated Yeast ID cards (Biomerieux Vitek), germ tube production, and microscopic morphology and confirmed the fact that the RapID system compared favorably with the API 20 C system but had a

simpler, more rapid identification process (71). In another recent study, the identification rate for yeasts tested with this system was only 78 to 84% (15).

In a similar study, good correlation was observed between the two systems, with 95.7% agreement with the API 20 C system for *Candida* and *Cryptococcus*. Lower agreement was noted with emerging *Candida* spp. and with other, yeastlike pathogens (79.1 and 75.2%) (40). Most recently, the RapID Yeast Plus system was compared with the Uni-Yeast-Tek system (Remel). A total of 117 fresh and frozen yeast isolates were tested. The Biomerieux Vitek system and morphology were used to resolve discrepant results. The RapID system correctly identified 96.6% of the isolates in 4 h (101).

Other Manual Systems

Other commercially available kits for identification of yeasts and yeastlike organisms include the API Yeast-Ident kit (Biomerieux) (Fig. 5) and the Identicult-Albicans kit (PML Microbiologicals, Wilsonville, Oreg.) (Fig. 6).

Automated Systems

Many of the automated yeast identification systems described here can also be used for identification of bacterial isolates and are depicted elsewhere in this text.

Rapid Yeast Identification Panel (Dade Microscan)

The Dade Microscan Rapid Yeast Identification panel is composed of chromogenic and modified conventional tests for identification of yeasts and yeastlike organisms, such as *Prototheca* spp., from clinical isolates. The 96-well plate format contains 27 dehydrated chromogenic substrates, which are designed to identify yeasts without the need for ancillary tests (Fig. 7). A heavy suspension of a fresh yeast isolate, prepared in water and standardized to a known turbidity standard, is subsequently used to hydrate the plates. The plates are incubated for 4 h at 35 to 37°C before the

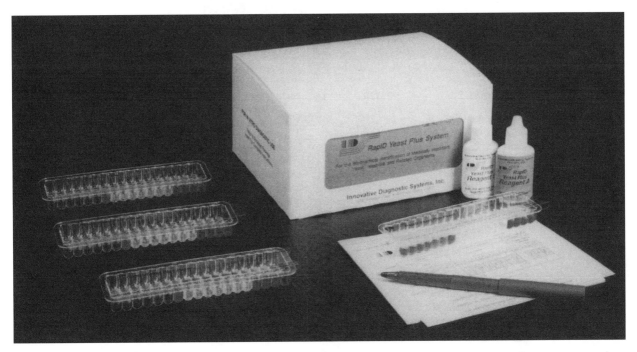

FIGURE 4 The RapID Yeast Plus System (Innovative Diagnostic Systems and Remel).

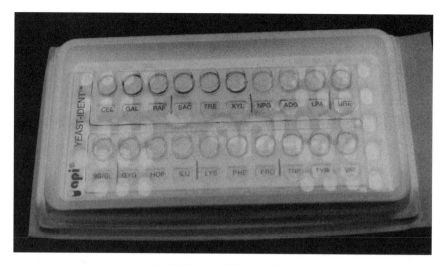

FIGURE 5 API Yeast-Ident.

FIGURE 6 Identicult-Albicans.

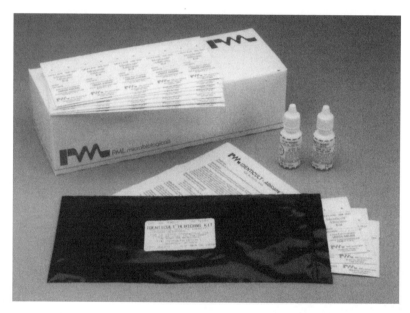

FIGURE 7 The Microscan Rapid Yeast Identification Panel. Courtesy of Dade Microscan.

reactions can be interpreted. The chromogenic readings can be made manually, read by eye, or automated with the Autoscan instrument. Interpretation of the biochemical profile occurs as the results are transformed into a nine-digit biotype number, which is compared with the Microscan cumulative database via the Autoscan instrument or manually, using the Microscan Rapid Yeast Biotype Codebook. The database identifies at least 42 yeasts and yeastlike organisms. In each case, profile identifications are assigned a cumulative relative probability of identification. When the probability of identification is not sufficient to confirm identification, supplemental testing of morphological and phenotypic characteristics is recommended.

When the Rapid Yeast Identification Panel was compared to the API 20 C system, early studies reported 94.1% correlation (208 of 306 isolates) in identifying species of *Candida, Hansenula, Pichia, Rhodotorula, Saccharomyces,* and *Torulopsis* when results were determined after 4 h. A lower correlation (65%) was observed for slower-growing species, such as *Blastoschizomyces* spp., *Cryptococcus* spp., *Trichosporon* spp., and yeastlike organisms, such as *Prototheca* spp. The overall correlation, using the first limited database, was 85% or less (88, 152).

The same system was also tested with results obtained after 72 h. The results were again compared to those with the Microscan Rapid Yeast Identification Panel at 4 h and the API 20 C system with cornmeal-Tween 80 agar at 72 h. Both visual and Autoscan readings of the plates correctly identified 78% of the 357 yeast isolates tested with no supplementary tests required. With supplementary tests, Autoscan identified 96.6% (99.5% of common strains; 92.1% of less common strains, like *C. glabrata, C. fumata, C. lusitaniae, Candida lambica, Candida rugosa, Candida stellatoidea, Cryptococcus albidus, Cryptococcus laurentii,* and *Cryptococcus uniguttulatus*), and the API 20 C system identified 98.9%. Autoscan was shown to be reliable for the identification of common yeasts when supplementary tests were performed (152).

More recently, when the results were compared to those obtained with the API 20 C system and from morphology on Tween-80 agar, the Microscan Walk-Away-96 still correctly identified only 67 to 82% of the 150 yeasts tested. The most commonly misidentified yeasts were *C. tropicalis, C. glabrata,* and *C. parapsilosis* (136).

YBC (Vitek Biomerieux)

The Vitek YBC system is an automated system for yeast identification. The system uses the Vitek test card, which is composed of microwells containing various identification substrates. The YBC is used for the yeasts that are most commonly encountered in hospitals. Yeast suspensions are prepared in sterile water. Tubes are placed in the Vitek filler-sealer, which uses a vacuum to load the liquid into the Vitek test card through a sterile straw. The sealer module removes the straw and seals the test card. The card contains dehydrated medium for a variety of biochemical tests. The reader-incubator unit maintains a controlled environment while the cards incubate. It rotates the cards so that readings can be measured by the laser unit and photodetector. The yeast database contains at least 36 species of commonly encountered yeasts. Results are maintained so that cumulative reports can be generated and can be formatted according to the user's needs. Add-on software is available for purchase to maintain workflow and quality control data. The maximum capacity for the Vitek Junior model is 30 cards per run. The Vitek Senior comes in models that can analyze 60, 120, and 240 cards simultaneously.

In an early study, the Vitek yeast database was compared to the API 20 C Aux system. The Vitek system compared favorably for the identification of common yeasts but less favorably for identifying others. For all yeasts identified, most of the identifications were made within 24 h of incubation (33).

Several studies were published in 1994. In one study, the Vitek YBC correctly identified 93% of common yeasts but only 55% of uncommon yeasts, such as *C. guilliermondii, C. krusei, C. lambica, C. lusitaniae, C. rugosa,* non-*neoformans Cryptococcus* spp., *Geotrichum candidum, Rhodotorula, Saccharomyces,* and *Trichosporon* (96). In another report, the YBC was compared to the API 20 C system with additional biochemicals for the identification of 222 germ tube-negative yeast isolates. The YBC's identifications of commonly isolated yeasts compared favorably; however, the 48-h identifications of less common yeasts were not as successful. Only 55% were correctly identified, and the remaining isolates were either misidentified or not identified (31). The system was not able to identify an isolate of *S. cerevisiae* or *C. tropicalis* but provided correct identification for 98% of the common yeasts (n = 100) (64).

Similarly, the Vitek YBC, with 24- to 48-h results, was compared with the API 20 C system and morphology on Tween-80 agar. Of 150 yeast isolates, 85 to 95% were correctly identified with the Vitek YBC. The most common misidentified yeasts were *C. tropicalis* and *C. glabrata* (136). In another 1994 study, 409 germ tube-negative yeasts and *Geotrichum* spp. were also tested. The results were compared with identification using the API 20 C system and yeast morphology agars. Both systems identified most yeast isolates within 24 h. The YBC correctly identified 89.7% of 409 germ tube-negative yeasts. The API 20 C system correctly identified 99.3%. Of the remaining isolates, some profiles provided no identification of the organism, some profiles converted to misidentifications (the YBC incorrectly identified 7.3% of the isolates tested; the API 20 C system misidentified only 0.7%). Results were comparable for *C. glabrata, C. parapsilosis,* and *C. neoformans.* The Vitek YBC had difficulty with *C. tropicalis, C. krusei, Trichosporon,* and some *Cryptococcus* spp. (43).

A 1999 study compared the Vitek YBC with the API 20 C Aux system without supplementary tests to aid in identification and found no significant difference between them. The API 20 C system correctly identified 167 of 193 isolates, whereas the Vitek YBC correctly identified 173 of 193 isolates. The API 20 C Aux system did not correctly identify any *C. krusei* isolates without supplemental testing, and that species accounted for most of the differences observed between the identification rates of the two systems (165).

A new addition, the Vitek 2 system, allows identification of medically important yeasts and yeastlike organisms in 15 h by using a compendium of 47 biochemical reactions that are monitored by a sensitive fluorescence-based chemistry. The new database with 51 taxa was evaluated using the Vitek ID-YST card. Compared to the ID32C strip (Biomerieux) for 241 strains from 21 species, the Vitek 2 system correctly identified 92.1% of the isolates (59).

Biolog YT Microplate

In the YT Microplate test, yeast cells are suspended in sterile water. The YT Microplate is configured with both of its carbon source oxidation tests using tetrazolium violet as the colorimetric indicator. Carbon-based assimilation tests are scored turbidimetrically and can only be read with the automated Biolog MicroStation. A few wells contain two carbon sources and test for coutilization of various carbon ■

sources with D-xylose. Software support exists for the Biolog Yeast Database. In addition, the SF-N and SF-P Microplates can be used for metabolic testing of sporulating and filamentous microorganisms, such as the actinomycetes and fungi. Special culture medium is used to enhance sporulation, and a gel-forming colloid, which keeps the organisms more evenly suspended, is used to prepare the organism suspension. The SF Microplates are identical to the GN and GP microplates used for aerobic bacteria except they do not contain the tetrazolium redox dye, which is toxic to many of these species. The plates are incubated at 25 to 27°C for several days and read with the automated MicroStation.

Compared to the YBC and API 20 C Aux system with 171 germ tube-negative yeasts, the Biolog system had poor clinical utility, identifying only 48% of the isolates. The YBC and the API 20 C Aux system correctly identified 97% of the 171 strains tested. Both systems required repeat testing less than 10% of the time and required supplemental testing with 28% of the strains. With the Biolog system, there is a need for considerable expertise and a battery of supplemental reagents for correct identification of germ tube-negative yeasts (141).

MIS (Microbial ID, Inc. Newark, Del.)

The Microbial Identification System (MIS) relies on fatty acid analysis to identify yeast isolates. Cellular fatty acid chromatograms are analyzed and compared to the fatty acid profiles in the system's database. Prior to analysis, yeast isolates must be subcultured, with subsequent saponification, methylation, and extraction of fatty acids. In one study using MIS (with the yeast clinical library version 3.8), 477 isolates and 23 different species of yeast were tested. The MIS generated results with a performance of only 75% correct identifications compared to a combination of identification methods, including the Vitek YBC, morphology on cornmeal agar, and the API 20 C Aux system. Predictive values for different species ranged from 47 to 100%, depending on the species (81).

In another similar study, the MIS was compared to the API 20 C system and conventional methods. The results indicated that 374 (68%; n = 550) yeasts were correctly identified to the species level. After repeat testing, 18 more were identified, bringing the overall correct identification to 71.3%. Of all yeasts tested, 15.8% were incorrectly identified and 16.2% were given no identification. The isolates tested represented a variety of species: C. albicans, 294; C. glabrata, 145; C. tropicalis, 58; C. parapsilosis, 33; and other yeasts, 20. C. glabrata was the most commonly misidentified yeast, and it was misidentified as S. cerevisiae 32.4% of the time. Based on these results, version 3.8 was not considered suitable for routine use with clinically important yeasts (26); however, another study performed with the MIS provided evidence that the results can be dependent on the supplier of the agar on which the yeast isolates are grown (82).

MOLECULAR TESTING

Accuprobe (Gen-Probe, San Diego, Calif.)

Because of the slow growth of dimorphic fungi and the potential difficulty and safety risks involved in their confirmatory identification, they seem to be a good choice for molecular identification techniques. Nucleic acid hybridization tests are based on the complementary alignment and binding of nucleic acid strands to form stable double-stranded complexes. The Accuprobe System uses a single-stranded DNA probe with a chemiluminescent acri-

dinium label as the complement of the rRNA of the target organism. Lysates of an organism are prepared by sonication with glass beads and heat treatment. The rRNA is released from the organism, and the labeled DNA subsequently binds with the target organism rRNA to form a stable labeled DNA-RNA hybrid, the presence of which can be detected in the Gen-Probe luminometer. The selection reagent, provided in the kit, provides for differentiation of hybridized and nonhybridized probe during readings taken by the luminometer. Positive reactions are based on cutoff values of relative light units derived from the chemiluminescence. Kits are available for Coccidioides immitis, Histoplasma capsulatum, and Blastomyces dermatitidis. Molds can be tested from growth in both liquid and solid media; however, the assay is not reliable for formalin-killed cultures of C. immitis (61).

The B. dermatitidis culture confirmation kit has been shown to cross-react with all Paracoccidioides brasiliensis isolates tested, so when this organism is suspected, results should be interpreted in conjunction with other microscopic and morphological laboratory data.

The Accuprobe DNA probe was originally compared to the traditional exoantigen test. For 105 isolates of various subspecies of H. capsulatum, all were identified with the Accuprobe with a more rapid result. A total of 103 isolates were identified in 2 h (114). In another 1992 study, 53 of 54 isolates were correctly identified as H. capsulatum by the Accuprobe. One isolate of A. niger was incorrectly identified as H. capsulatum. In this study, the age of the culture, the isolation medium, and the morphological state of the organisms did not adversely affect the results (65).

In 1993, Gen-Probe DNA probes for the identification of H. capsulatum var. capsulatum and C. neoformans were tested. A total of 95 fungal isolates were tested with the Histoplasma kit, 41 of which were H. capsulatum var. capsulatum. A total of 98 yeasts, 42 of which were C. neoformans isolates, were tested with the Cryptococcus probes. The probes exhibited 100% sensitivity and 100% specificity and significantly reduced the time necessary for identification (76, 153).

SUSCEPTIBILITY TESTING

The technical parameters for antifungal susceptibility testing of yeasts used to design the NCCLS M27-P and M27-A guidelines have been reviewed and discussed elsewhere (53, 113, 126, 139, 155). The broth macro- and microdilutions are not themselves commercially available in a packaged format, but individual components are available commercially. Lack of reproducibility and untested or uncertain clinical relevance has partially limited the useful application of antifungal susceptibility testing (126) but has not greatly deterred the requests for testing and interpretation of results. The standardization of methods in recent years has improved the cost-benefit ratio of the testing. Standardized methods and multicenter evaluations have provided some correlative evidence of MIC results and have demonstrated that there are still more evaluations and comparisons to be performed before the utility of these methods reaches that of bacterial susceptibility testing. Although many variables have been evaluated, many more remain to be defined. The NCCLS standards have not yet included specifications for the use of commercial systems; however, standard methods have been adapted, and several commercial methods for yeast susceptibility testing are available for purchase (Table 5).

TABLE 5 Agreement of methods for susceptibility testing of yeasts

Assay	Method	Comparison	Agreement (%)					n	Comments	Reference
			Amphotericin B	Fluconazole	Flucytosine	Itraconazole	Ketoconazole			
Sensititre Alamar, Trek	Microdilution with redox indicator	Macrodilution	85.3	77.9	86.2			125 (Candida spp. and C. neoformans)	Difficulties with C. glabrata and C. tropicalis	161
			100	98	90	98	93	40		129
			>95 (at 24 h)	>95 (at 24 h)	>95 (at 24 h)					121
			>98 (at 48 h)	>98 (at 48 h)	84 (at 48 h)					
				97.7				45 (C. albicans)		39
					93 (5FC)	83		100 (C. albicans)		97
		Microdilution[a]	92–99 (at 48 h)	92–100	92–100	92–100	92–100	1,776		38
			73.2	80	88	70.7	95	42 (C. neoformans)	Discordant for fluconazole and 5-FC (Alamar had lower MICs)	90
E test (AB Biodisk)	Diffusion	Microdilution[a]	71–93	81–100		71–93		32 (C. glabrata)		41
Pasco (Becton Dickinson)	Broth microdilution	Microdilution[a]	89	91	80	85	89	74 (Candida spp. and C. neoformans)	Terconazole, miconazole, and clotrimazole tested	4
ATB Fungus	Micromethod (test strips)	Microdilution[a]	*[b]		*				*	131
PASCO	Microdilution	Microdilution[a]	89	91	80	85	89	74 (Candida spp. and C. neoformans)	Terconazole, miconazole, and clotrimazole also tested	4

[a] NCCLS microdilution.
[b] *, 91.7% correlation with reference strains using amphotericin B and flucytosine; miconazole and econazole also tested.

Sensititre YeastOne (Trek Diagnostics)

The Sensititre system combines a dried microdilution panel and a colorimetric method with an oxidation-reduction indicator to signal the growth of yeasts for susceptibility testing. This system has been compared with both the NCCLS M27-A document for macrodilution susceptibility and the M27-P microdilution method.

The results show good comparison with standard broth macrodilution testing. The method was compared to the NCCLS broth macrodilution method for testing with amphotericin B, fluconazole, and flucytosine. The isolates tested were as follows: C. albicans, 28; C. tropicalis, 17; C. parapsilosis, 15; C. krusei, 12; C. lusitaniae, 10; C. guilliermondii, 9; C. glabrata, 18; and C. neoformans, 25. Overall agreement rates for Candida spp. and C. glabrata were 85.3, 77.9, and 86.2%, respectively, for amphotericin B, fluconazole, and flucytosine. The greatest disagreement was seen with C. glabrata and C. tropicalis. Better correlation for flucytosine was observed after 48 h for Candida spp. and C. glabrata (161). In another more limited but recent comparison with macrodilution, 40 yeast isolates were tested. Agreement was 100% for amphotericin B, 98% for fluconazole, 98% for itraconazole, 93% for ketoconazole, and 90% for flucytosine (129).

The Sensititre system was also compared to broth microdilution susceptibility testing. When 600 isolates of Candida spp. and C. neoformans were tested against amphotericin B, fluconazole, and flucytosine, equivalent results were observed (>95% agreement) for all three drugs with the reference and the colorimetric indicator methods at 24 h, and >98% agreement was observed at 48 h, except for fluconazole, which dropped to 84% agreement (121).

When the microdilution test, read after 24 h at 35°C, was compared with the M27-P method for fluconazole susceptibility testing, excellent agreement (97.7%) was demonstrated for 45 susceptible and resistant isolates of C. albicans. There was also in vivo correlation with fluconazole resistance. Preliminary data indicate that the microdilution methods may serve as less subjective alternatives to the M27-P method for the determination of fluconazole MIC endpoints (39).

The Sensititre system was compared to the current standard microdilution method to compare 100 isolates of C. albicans and susceptibility to 5-fluorocytosine (5FC), fluconazole, itraconazole, and D0870. MICs were determined after 48 h of incubation. Discrepancies of no more than 2 dilutions were used to calculate the percent agreement. With this relatively wide agreement range, the agreement rate was 83% for itraconazole and 93% for 5FC. Both were determined to be acceptable rates for those drugs (97).

In 1998, the alamar blue indicator was specifically tested for its utility in combination with the broth microdilution MIC method. It was shown that when used with RPMI media, alamar blue did not improve the ability of RPMI to detect strains of C. albicans resistant to amphotericin B (91).

In 1999, 1,176 isolates of yeasts and yeastlike organisms were tested with both the NCCLS reference method for broth microdilution and the Sensititre YeastOne Panel. This study is noteworthy for the large number and the variety of isolates tested. Included were isolates of Blastoschizomyces capitus, a Cryptococcus sp., common and emerging Candida spp., a Hansenula sp., a Rhodotorula sp., S. cerevisiae, Sporobolomyces salminocolor, and T. beigelii. The results for the YeastOne Panel were most comparable with those for the 24-h tests of the azoles and flucytosine, with 92 to 100% agreement for most species tested, except for C.

albicans. The best agreement for amphotericin B was at 48 h for C. albicans and most other species tested (92 to 99%), with the exception of C. neoformans at only 76% agreement (38). In addition, the results were comparable when the methods were compared at 48 h for amphotericin B and at 24 h for ketoconazole and itraconazole. The colorimetric test tended to influence the results by only 1 dilution and was not deemed to represent a substantial difference in overall results (140). Good correlation was also observed between the Alamar blue and the microtiter tray (MTT) tests for Candida susceptibilities, and the MTT method also generated interpretable data for Aspergillus spp. (77).

Overall, the YeastOne system has much potential value for clinical mycology laboratories. The test is generally easy to read and is likely to gain more practical utility as further in vivo comparisons are performed. Limitations to the method exist in that discordances with MIC occur for fluconazole and 5-flucytosine. One recent report documented resistance by the MIC reference method but sensitive results for the YeastOne method (90). Clearly, more testing needs to be performed to determine the implications of these discrepancies.

E Test (AB Biodisk, Remel)

The E test has been used to determine MIC for bacteria and has also recently been tested for use with yeast isolates. The E test is an agar-based method which uses strips that are impregnated with antifungal agents and placed on the surfaces of the agar plates containing the inoculum. As the organism grows, the antifungal will inhibit the growth of susceptible organisms around the position of the E test strip. The amount of inhibition correlates with the MIC for the organism. The results are reported as MIC.

Reproducibility was determined as results were compared in an interlaboratory study with broth macrodilution as the reference standard. In this report, the E test quality control limits were 1 dilution greater (4 dilutions overall) than those of the NCCLS broth tube macrodilution method for the following combinations: ketoconazole with C. krusei and amphotericin B and ketoconazole with C. parapsilosis (124). In another study, MICs of fluconazole, itraconazole, and ketoconazole were tested for 100 clinical isolates (C. albicans [25], C. neoformans [25], C. glabrata [20], C. tropicalis [15], and C. parapsilosis [15]). Agreement was observed to be 71% for ketoconazole, 80% for fluconazole, and 84% for itraconazole (22).

Both broth macrodilution and microdilution tests for fluconazole susceptibility were compared to the E test for 238 isolates of C. albicans and C. glabrata. While a 93% correlation was observed within 2 doubling dilutions of the broth methods at 24 h, only 37% was observed by the E test for C. glabrata, 56% for C. tropicalis 93% for C. albicans, and 90% for other Candida spp. The results were slightly better at 48 h (149).

A variety of agars have been used to evaluate E test results, and the type of agar has been found to play a role in the performance of the E test method. When amphotericin B, fluconazole, and itraconazole were tested, the results indicated that Casitone agar may be preferable for use in testing (28, 41). Good correlation was shown between the results with fluconazole and the clinical outcome of the disease. Clinical failure resulted when MICs were >48 μg/ml (28).

In 1996, an interlaboratory evaluation of the E test was performed with Casitone agar and solidified RPMI 1640 medium with 2% glucose. The MICs of amphotericin B, fluconazole, flucytosine, itraconazole, and ketoconazole were tested with 83 isolates of *Candida* spp., *C. neoformans*, and *C. glabrata*. Analysis of 3,420 MICs demonstrated a higher agreement between laboratories when Casitone medium was used to test amphotericin B, itraconazole, and ketoconazole. In contrast, flucytosine and fluconazole results were more reproducible when RPMI medium was used (37). Recently, the performance of the E test was compared to the fluconazole results from the NCCLS microdilution method for 402 yeast isolates. Mueller-Hinton agar, RPMI agar with 2% glucose, and Casitone agar were tested with isolates of various species: *C. albicans*, 161 isolates; *C. glabrata*, 41; *C. tropicalis*, 35; *C. parapsilosis*, 29; *C. krusei*, 32; *C. lusitaniae*, 31; other *Candida* spp., 19; *C. neoformans*, 40; and miscellaneous species, 14. The overall agreement was 97% for Casitone medium and 94% for RPMI. These media may allow the E test to become a viable alternative to broth microdilution for testing the fluconazole susceptibilities of yeasts (125).

Several comparisons with broth microdilution have also been reported. When 169 yeasts were tested, the E test results generally correlated with those of the reference method. There was at least 80% agreement within 2 dilutions of the reference method except for *C. neoformans* isolates, in which only 8% of the results correlated (151). Another study, in which a variety of agar supplements were used to increase the percent agreement, documented 77 to 100% agreement with broth microdilution; however, problems were noted for the determination of azole susceptibility using the E test (41).

The E test and the NCCLS broth microdilution test were compared using 602 yeast isolates for amphotericin B testing. The clinical species tested were *C. albicans*, *C. glabrata*, *C. tropicalis*, *C. parapsilosis*, *C. lusitaniae*, *C. krusei*, *Candida* spp., *C. neoformans*, and *S. cerevisiae*. The overall agreement between the E test with RPMI medium and 2% glucose and the reference method was 98.3%. All E test methods identified two resistant strains for which the MICs were 4.0 to 16 µg/ml, while the reference method failed to distinguish them from 18 other isolates for which the MICs were 2 µg/ml. The significance of this report is clear: the E test may identify subpopulations of yeast isolates for which the amphotericin B MICs are high (123).

Susceptibility testing for mold isolates is more ambiguous than yeast susceptibility testing. When the E test was compared to broth microdilution for mold isolates, 75% agreement was observed. Both amphotericin B and itraconazole were tested by both methods. Among the molds tested were *Absidia corymbifera*, *A. flavus*, *A. fumigatus*, *A. niger*, *A. terreus*, *Exophiala dermatitidis*, *Fusarium solani*, *Scedosporium apiospermum*, *Scedosporium prolificans*, and *Scopulariopsis brevicaulis*. The best agreements were seen for *A. fumigatus* and *A. terreus* (100% of the results for other agents). In every instance, where lack of agreement was observed, it was because of low MICs for molds in broth microdilution and much higher MICs in the E test. The E test appears to be a reasonable approach to susceptibility testing of *Aspergillus* spp. and a few other molds (157). Correlative studies that compare in vivo response to treatment may shed some light on discrepant results observed in this study.

If further testing provides more information relevant to the media of choice and the in vivo correlation data, the E test can be used routinely in clinical laboratories. Results have shown this method to be reproducible and relatively simple to perform. More optimization studies must be done, but improvements to this method may make it a viable alternative for MIC testing of yeast and molds.

Screening with CHROMAgar

When CHROMAgar Candida containing 8 or 16 µg of fluconazole/ml was used as a screen for fluconazole resistance, the results compared favorably with those of the standard broth macrodilution methods when the following criteria were used for interpretation. Susceptible *C. albicans* colonies are smaller on fluconazole medium than on fluconazole-free medium. The correlation with this observation was 96% for MICs with a 16-µg/ml breakpoint and 95% for MICs with an 8-µg/ml breakpoint. The results indicate that this method may be useful as a screen for fluconazole resistance (115); however, a substantial amount of technical expertise will be required for proper interpretation of the results.

FUNGITEST (Bio-Rad Sanofi Diagnostics Pasteur, Paris, France)

The FUNGITEST method, a microplate breakpoint procedure, was compared to the broth macrodilution test for amphotericin B, flucytosine, fluconazole, itraconazole, ketoconazole, and micronazole. Fifty *C. albicans*, 50 *C. glabrata*, 10 *Candida kefyr*, 20 *C. krusei*, 10 *C. lusitaniae*, 20 *C. parapsilosis*, 20 *C. tropicalis*, and 20 *C. neoformans* isolates were tested. Overall, the following agreements were found: 100% for amphotericin B, 95% for flucytosine, 84% for micronazole, 83% for itraconazole, 77% for ketoconazole, and 76% for fluconazole (30).

ASTY (Kyokkuta Pharmaceutical Industrial Co., Ltd.)

The ASTY method is a colorimetric microdilution panel in which the growth of an organism causes the indicator to change from red to purple or blue, indicating growth inhibition or no growth, respectively. The ASTY method was compared in an intralaboratory evaluation with the NCCLS reference broth microdilution method. For the comparison, 802 isolates of various *Candida* spp. were tested, including *C. albicans*, *C. glabrata*, *C. tropicalis*, *C. parapsilosis*, *C. krusei*, *C. lusitaniae*, *C. guilliermondii*, *C. lipolytica*, *C. rugosa*, and *C. zeylanoides*. Reference MIC endpoints for amphotericin B, 5FC, fluconazole, and itraconazole were determined at 48 h and the same endpoints were determined for the ASTY system at 24 and 48 h of incubation. Overall, agreement was 93% at 24 h and 96% at 48 h. At 48 h, the lowest agreement was observed with itraconazole at 92%, ranging to 99% with amphotericin B and 5FC (120).

PASCO (Becton Dickinson, Wheatridge, Colo.)

The PASCO antifungal susceptibility test system is composed of premade fungal panels designed for one-step incoulation of multiwell plates filled with serial dilutions of up to eight antifungal agents on a single plate. The plates are shipped and stored for up to 1 year at −70°C. After inoculation and incubation for 48 h, reaction endpoints are read visually. Compared to the standard NCCLS broth microdilution MIC determination method for 74 yeast isolates, the PASCO method performed reasonably well. Agreement was determined to be at least 80% for medically important *Candida* and *Cryptococcus* spp., with the exception of results for miconazole and clotrimazole and an additional exception for terconazole and *C. krusei* isolates (4).

Miscellaneous Methods

In a 1993 study, the Mycototal, Mycostandard, Candifast, ATB Fungus, and Difftest systems were compared for reproducibility and repeatability, as well as for the concordance of results with MICs. A total of 630 yeast-antifungal agent results for Mycototal and Mycostandard, 540 for Candifast, and 450 for ATB Fungus and Difftest were obtained, with a repeatability of >95% for all kits. Reproducibility was >95% for all tests except Candifast, for which it was 80.1%. The highest concordance with MICs was observed with the Mycototal, Mycostandard, and Difftest systems at 90.3, 90.2, and 80.9%, respectively. Concordance was more limited for the ATB Fungus and Candifast systems, at 75.3 to 91.7 and 51.6%, respectively (32, 35, 36, 131).

IMMUNODIAGNOSTICS

While serological methods are helpful for epidemiological purposes, their diagnostic effect may be limited. Results are often available too late to have an impact on the course of rapidly progressive fungal diseases in what is largely an immunocompromised patient population. Still, the diagnosis of fungal infection can be difficult and can yield false-negative results even with repeated efforts to isolate the pathogen. Some commercial methods are available and may have utility in certain circumstances. Serological diagnosis of fungal disease is not extensively reviewed in this chapter. Refer to the *Manual of Clinical Laboratory Immunology* for detailed information regarding fungal antibody detection methods (79). The tests described here are included because of recent publications comparing these commercially available kits with traditional diagnostic techniques.

Dimorphic Molds

C. immitis ELISA (Premier; Meridian Diagnostics)

The Premier *Coccidioides* enzyme immunoassay (Meridian Diagnostics) is a commercially available test used for the qualitative detection of IgM and IgG antibodies against the tube precipitin (TP) antigen, primarily an IgM response. The complement fixation (CF) antigens of *C. immitis* are also detected by this method. Antigens are adsorbed to microwells so that when a specimen is added, antibodies to TP and CF antigens, if present, will bind to the antigens in the microwells. After the wells are washed to remove unbound antibodies, an anti-IgM enzyme conjugate is added to one microwell and an anti-IgG enzyme conjugate is added to another well. If patient antibodies are bound, a sandwich forms between adsorbed antigens, patient antibodies, and one or both conjugates. After the wells are washed, a substrate solution is added, and color develops in the presence of the bound enzyme conjugate. Positive results are determined by absorbance values. Clinical specimens which can be tested in this assay include serum and CSF. It is the manufacturer's recommendation that the *Coccidioides* EIA be used in conjunction with other laboratory tests and clinical evidence and that positive results be confirmed by traditional immunodiffusion assays, such as those from Scott Laboratories Inc.

In one report, the IgM component of the ELISA was compared to latex agglutination for IgM. The results indicated an agreement of 81.8%, a specificity of 75%, and a sensitivity of 92.6%. The IgG component was compared to the combined results of the CF and immunodiffusion tests, with an agreement of 96.7%, a specificity of 98.5%, and a sensitivity of 94.8%.

Another report documents the fact that ELISA is a reliable assay to determine antibodies against the TP and CF antigens and that the assay did not suffer from the difficulty in interpretation and anticomplement interference associated with traditional assays (95). In another 1995 study, 409 serum and CSF specimens were tested with the Premier EIA tests for IgG and IgM. The results were compared to those of immunodiffusion and CF tests for *C. immitis* antibodies corresponding to those detected by TP or CF tests. The EIA was shown to be sensitive but not as specific as the other methods. Some false positives were also observed (80).

H. capsulatum (Premier EIA; Meridian Diagnostics)

When the Premier EIA method was compared to the microimmunodiffusion (MID) test and the laboratory CF test for a total of 168 sera, including 68 from proven cases of histoplasmosis, the sensitivity of the EIA for IgG was 97% and that of the MID was 100%, and the specificities were 84 and 100%, respectively. In this study, three sera from histoplasmosis patients which were positive for the histoplasmin antigens tested negative for IgG, specific precipitins, and complement-fixing antibodies (146).

B. dermatitidis (Premier EIA, Meridian Diagnostics)

The EIA and the MID test for the A antibody, specific to *B. dermatitidis*, were used to test 103 serum samples. Both tests produced positive results with 20 proven cases of blastomycosis. The sensitivity and specificity of the EIA were 100 and 85.6%, respectively. For the MID test, the sensitivity and specificity were both 100%. Because of low specificity, it is recommended that a positive EIA be confirmed with an MID test (147).

EPILOGUE

Because of the interest in rapid, cost-effective testing for fungi, the pace and number of publications devoted to diagnostic practices in clinical mycology laboratories are growing. This review is meant to represent the state of the art for commercial systems developed for clinical mycology. Molecular methods may well be on the horizon, but at present, no such commercial method is available. Clearly, this review must be supplemented with more current studies as they become available. Still, the overall benefit of the comparisons and contrasts of methods discussed here should be of benefit to those who are managing clinical laboratories and making decisions about test methodology. Each method has its own advantages and limitations and lends itself more easily to some patient populations than to others. The amount of information available is substantial, and review of the literature provided thousands of references. Still, the decisions revolving about test selection and implementation continue to be very difficult. For practical purposes, not all of the references could be included in this format, nor could all of the published literature be retrieved for inclusion. Therefore, we solicit any authors of peer-reviewed publications which are not included here to forward the publications for review and potential citation in the second edition of this work. Likewise, we encourage manufacturers to submit updated information about their test methods.

We thank Errol Reiss (Centers for Disease Control and Prevention) for his contributions to the information related to nonculture methods for systemic opportunistic mycoses.

REFERENCES

1. **Ainscough, S., and C. C. Kibbler.** 1998. An evaluation of the cost-effectiveness of using CHROMagar for yeast identification in a routine microbiology laboratory. *J. Med. Microbiol.* **47:**623–628.

2. **Ansorg, R., E. Heintschel von Heinegg, and P. M. Rath.** 1994. Aspergillus antigenuria compared to antigenemia in bone marrow transplant recipients. *Eur. J. Clin. Microbiol. Infect. Dis.* **13:**582–589.

3. **Ansorg, R., R. van den Boom, and P. M. Rath.** 1997. Detection of Aspergillus galactomannan antigen in foods and antibiotics. *Mycoses* **40:**353–357.

4. **Arthington-Skaggs, B. A., M. Motley, D. W. Warnock, and C. J. Morrison.** 2000. Comparative evaluation of PASCO and national committee for clinical laboratory standards M27-A broth microdilution methods for antifungal drug susceptibility testing of yeasts. *J. Clin. Microbiol.* **38:**2254–2260.

5. **Bale, M. J., C. Yang, and M. A. Pfaller.** 1997. Evaluation of growth characteristics on blood agar and eosin methylene blue agar for the identification of Candida (Torulopsis) glabrata. *Diagn. Microbiol. Infect. Dis.* **28:**65–67.

6. **Banerjee, S. N., T. G. Emori, D. H. Culver, R. P. Gaynes, W. R. Jarvis, T. Horan, J. R. Edwards, J. Tolson, T. Henderson, and W. J. Martone.** 1991. Secular trends in nosocomial primary bloodstream infections in the United States, 1980–1989. National Nosocomial Infections Surveillance System. *Am. J. Med.* **91:**86S–89S.

7. **Baumgartner, C., A. M. Freydiere, and Y. Gille.** 1996. Direct identification and recognition of yeast species from clinical material by using albicans ID and CHROMagar Candida plates. *J. Clin. Microbiol.* **34:**454–456.

8. **Bernal, S., M. E. Martin, M. Chavez, J. Coronilla, and A. Valverde.** 1998. Evaluation of the new API Candida system for identification of the most clinically important yeast species. *Diagn. Microbiol. Infect. Dis.* **32:**217–221.

9. **Bernal, S., M. E. Martin, M. Garcia, A. I. Aller, M. A. Martinez, and M. J. Gutierrez.** 1996. Evaluation of CHROMagar Candida medium for the isolation and presumptive identification of species of Candida of clinical importance. *Diagn. Microbiol. Infect. Dis.* **24:**201–204.

10. **Bernard, K., C. Cooper, S. Tessier, and E. P. Ewan.** 1991. Use of chemotaxonomy as an aid to differentiate among Capnocytophaga species, CDC group DF-3, and aerotolerant strains of Leptotrichia buccalis. *J. Clin. Microbiol.* **29:**2263–2265.

11. **Bianchi, M., R. Negroni, A. Robles, A. Arechavala, S. Helou, and M. Corti.** 1997. Comparative study of three blood culture systems for the diagnosis of systemic mycoses associated with AIDS. *J. Mycol. Med.* **7:**134–136.

12. **Blevins, L. B., J. Fenn, H. Segal, P. Newcomb-Gayman, and K. C. Carroll.** 1995. False-positive cryptococcal antigen latex agglutination caused by disinfectants and soaps. *J. Clin. Microbiol.* **33:**1674–1675.

13. **Boschman, C. R., L. J. Tucker, D. C. Dressel, C. C. Novak, R. T. Hayden, and L. R. Peterson.** 1995. Optimizing detection of microbial sepsis: a comparison of culture systems using packaged sets with directions for blood collection. *Diagn. Microbiol. Infect. Dis.* **23:**1–9.

14. **Bougnoux, M. E., C. Hill, D. Moissenet, D. C. Feuilhade, M. Bonnay, I. Vicens-Sprauel, F. Pietri, M. McNeil, L. Kaufman, and J. Dupouy-Camet.** 1990. Comparison of antibody, antigen, and metabolite assays for hospitalized patients with disseminated or peripheral candidiasis. *J. Clin. Microbiol.* **28:**905–909.

15. **Buchaille, L., A. M. Freydiere, R. Guinet, and Y. Gille.** 1998. Evaluation of six commercial systems for identification of medically important yeasts. *Eur. J. Clin. Microbiol. Infect. Dis.* **17:**479–488.

16. **Cabezudo, I., M. Pfaller, T. Gerarden, F. Koontz, R. Wenzel, R. Gingrich, K. Heckman, and C. P. Burns.** 1989. Value of the Cand-Tec Candida antigen assay in the diagnosis and therapy of systemic candidiasis in high-risk patients. *Eur. J. Clin. Microbiol. Infect. Dis.* **8:**770–777.

17. **Campbell, C. K., K. G. Davey, A. D. Holmes, A. Szekely, and D. W. Warnock.** 1999. Comparison of the API Candida system with the AUXACOLOR system for identification of common yeast pathogens. *J. Clin. Microbiol.* **37:**821–823.

18. **Campbell, C. K., A. D. Holmes, K. G. Davey, A. Szekely, and D. W. Warnock.** 1998. Comparison of a new chromogenic agar with the germ tube method for presumptive identification of Candida albicans. *Eur. J. Clin. Microbiol. Infect. Dis.* **17:**367–368.

19. **Carlson, P., M. Richardson, and J. Paavonen.** 2000. Evaluation of the Oricult-N dipslide for laboratory diagnosis of vaginal candidiasis. *J. Clin. Microbiol.* **38:**1063–1065.

20. **Christensson, B., G. Sigmundsdottir, and L. Larsson.** 1999. D-Arabinitol—a marker for invasive candidiasis. *Med. Mycol.* **37:**391–396.

21. **Cockerill, F. R., III, C. A. Torgerson, G. S. Reed, E. A. Vetter, A. L. Weaver, J. C. Dale, G. D. Roberts, N. K. Henry, D. M. Ilstrup, and J. E. Rosenblatt.** 1996. Clinical comparison of Difco ESP, Wampole Isolator, and Becton Dickinson Septi-Chek aerobic blood culturing systems. *J. Clin. Microbiol.* **34:**20–24.

22. **Colombo, A. L., F. Barchiesi, D. A. McGough, and M. G. Rinaldi.** 1995. Comparison of Etest and National Committee for Clinical Laboratory Standards broth macrodilution method for azole antifungal susceptibility testing. *J. Clin. Microbiol.* **33:**535–540.

23. **Cornish, N., B. A. Kirkley, K. A. Easley, and J. A. Washington.** 1998. Reassessment of the incubation time in a controlled clinical comparison of the BacT/Alert aerobic FAN bottle and standard anaerobic bottle used aerobically for the detection of bloodstream infections. *Diagn. Microbiol. Infect. Dis.* **32:**1–7.

24. **Creger, R. J., K. E. Weeman, M. R. Jacobs, A. Morrissey, P. Parker, R. M. Fox, and H. M. Lazarus.** 1998. Lack of utility of the lysis-centrifugation blood culture method for detection of fungemia in immunocompromised cancer patients. *J. Clin. Microbiol.* **36:**290–293.

25. **Crist, A. E., Jr., T. J. Dietz, and K. Kampschroer.** 1996. Comparison of the MUREX C. albicans, Albicans-Sure, and BactiCard Candida test kits with the germ tube test for presumptive identification of Candida albicans. *J. Clin. Microbiol.* **34:**2616–2618.

26. **Crist, A. E., Jr., L. M. Johnson, and P. J. Burke.** 1996. Evaluation of the Microbial Identification System for identification of clinically isolated yeasts. *J. Clin. Microbiol.* **34:**2408–2410.

27. **D'Amato, R. F., L. Hochstein, and E. A. Fay.** 1990. False-positive latex agglutination test for Neisseria meningitidis groups A and Y caused by povidone-iodine antiseptic contamination of cerebrospinal fluid. *J. Clin. Microbiol.* **28:**2134–2135.

28. **Dannaoui, E., S. Colin, J. Pichot, and M. A. Piens.** 1997. Evaluation of the E test for fluconazole

susceptibility testing of Candida albicans isolates from oropharyngeal candidiasis. *Eur. J. Clin. Microbiol. Infect. Dis.* **16:**228–232.

29. Davey, K. G., P. M. Chant, C. S. Downer, C. K. Campbell, and D. W. Warnock. 1995. Evaluation of the AUXACOLOR system, a new method of clinical yeast identification. *J. Clin. Pathol.* **48:**807–809.

30. Davey, K. G., A. D. Holmes, E. M. Johnson, A. Szekely, and D. W. Warnock. 1998. Comparative evaluation of FUNGITEST and broth microdilution methods for antifungal drug susceptibility testing of *Candida* species and *Cryptococcus neoformans*. *J. Clin. Microbiol.* **36:**926–930.

31. Dooley, D. P., M. L. Beckius, and B. S. Jeffrey. 1994. Misidentification of clinical yeast isolates by using the updated Vitek Yeast Biochemical Card. *J. Clin. Microbiol.* **32:**2889–2892.

32. Druetta, A., A. Freydiere, R. Guinet, and Y. Gille. 1993. Evaluation of five commercial antifungal susceptibility testing systems. *Eur. J. Clin. Microbiol. Infect. Dis.* **12:**336–342.

33. el Zaatari, M., L. Pasarell, M. R. McGinnis, J. Buckner, G. A. Land, and I. F. Salkin. 1990. Evaluation of the updated Vitek yeast identification data base. *J. Clin. Microbiol.* **28:**1938–1941.

34. Engler, H. D., and Y. R. Shea. 1994. Effect of potential interference factors on performance of enzyme immunoassay and latex agglutination assay for cryptococcal antigen. *J. Clin. Microbiol.* **32:**2307–2308.

35. Espinel-Ingroff, A., T. M. Kerkering, P. R. Goldson, and S. Shadomy. 1991. Comparison study of broth macrodilution and microdilution antifungal susceptibility tests. *J. Clin. Microbiol.* **29:**1089–1094.

36. Espinel-Ingroff, A., C. W. Kish, Jr., T. M. Kerkering, R. A. Fromtling, K. Bartizal, J. N. Galgiani, K. Villareal, M. A. Pfaller, T. Gerarden, M. G. Rinaldi, and A. Fothergill. 1992. Collaborative comparison of broth macrodilution and microdilution antifungal susceptibility tests. *J. Clin. Microbiol.* **30:**3138–3145.

37. Espinel-Ingroff, A., M. Pfaller, M. E. Erwin, and R. N. Jones. 1996. Interlaboratory evaluation of Etest method for testing antifungal susceptibilities of pathogenic yeasts to five antifungal agents by using Casitone agar and solidified RPMI 1640 medium with 2% glucose. *J. Clin. Microbiol.* **34:**848–852.

38. Espinel-Ingroff, A., M. Pfaller, S. A. Messer, C. C. Knapp, S. Killian, H. A. Norris, and M. A. Ghannoum. 1999. Multicenter comparison of the Sensititre YeastOne Colorimetric Antifungal Panel with the National Committee for Clinical Laboratory Standards M27-A reference method for testing clinical isolates of common and emerging *Candida* spp., *Cryptococcus* spp., and other yeasts and yeastlike organisms. *J. Clin. Microbiol.* **37:**591–595.

39. Espinel-Ingroff, A., J. L. Rodriguez-Tudela, and J. V. Martinez-Suarez. 1995. Comparison of two alternative microdilution procedures with the National Committee for Clinical Laboratory Standards reference macrodilution method M27-P for in vitro testing of fluconazole-resistant and -susceptible isolates of *Candida albicans*. *J. Clin. Microbiol.* **33:**3154–3158.

40. Espinel-Ingroff, A., L. Stockman, G. Roberts, D. Pincus, J. Pollack, and J. Marler. 1998. Comparison of RapID Yeast Plus system with API 20C system for identification of common, new, and emerging yeast pathogens. *J. Clin. Microbiol.* **36:**883–886.

41. Favel, A., C. Chastin, A. L. Thomet, P. Regli, A. Michel-Nguyen, and A. Penaud. 2000. Evaluation of the E test for antifungal susceptibility testing of *Candida glabrata*. *Eur. J. Clin. Microbiol. Infect. Dis.* **19:**146–148.

42. Fenn, J. P., E. Billetdeaux, H. Segal, L. Skodack-Jones, P. E. Padilla, M. Bale, and K. Carroll. 1999. Comparison of four methodologies for rapid and cost-effective identification of *Candida glabrata*. *J. Clin. Microbiol.* **37:**3387–3389.

43. Fenn, J. P., H. Segal, B. Barland, D. Denton, J. Whisenant, H. Chun, K. Christofferson, L. Hamilton, and K. Carroll. 1994. Comparison of updated Vitek Yeast Biochemical Card and API 20C yeast identification systems. *J. Clin. Microbiol.* **32:**1184–1187.

44. Fenn, J. P., H. Segal, L. Blevins, S. Fawson, P. Newcomb-Gayman, and K. C. Carroll. 1996. Comparison of the Murex Candida albicans CA50 test with germ tube production for identification of C. albicans. *Diagn. Microbiol. Infect. Dis.* **24:**31–35.

45. Frank, U. K., S. L. Nishimura, N. C. Li, K. Sugai, D. M. Yajko, W. K. Hadley, and V. L. Ng. 1993. Evaluation of an enzyme immunoassay for detection of cryptococcal capsular polysaccharide antigen in serum and cerebrospinal fluid. *J. Clin. Microbiol.* **31:**97–101.

46. Fraser, V. J., M. Jones, J. Dunkel, S. Storfer, G. Medoff, and W. C. Dunagan. 1992. Candidemia in a tertiary care hospital: epidemiology, risk factors, and predictors of mortality. *Clin. Infect. Dis.* **15:**414–421.

47. Freydiere, A. M., L. Buchaille, and Y. Gille. 1997. Comparison of three commercial media for direct identification and discrimination of Candida species in clinical specimens. *Eur. J. Clin. Microbiol. Infect. Dis.* **16:**464–467.

48. Freydiere, A. M., L. Buchaille, R. Guinet, and Y. Gille. 1997. Evaluation of latex reagents for rapid identification of Candida albicans and C. krusei colonies. *J. Clin. Microbiol.* **35:**877–880.

49. Fricker-Hidalgo, H., F. Chazot, B. Lebeau, H. Pelloux, P. Ambroise-Thomas, and R. Grillot. 1998. Use of simulated blood cultures to compare a specific fungal medium with a standard microorganism medium for yeast detection. *Eur. J. Clin. Microbiol. Infect. Dis.* **17:**113–116.

50. Fricker-Hidalgo, H., A. Chirpaz-Cerbat, B. Lebeau, P. Ambroise-Thomas, and R. Grillot. 1997. Evaluation of Bactec 9240 blood culture system by using the aerobic 9240 medium for fungemia detection. *J. Mycol. Med.* **7:**128–133.

51. Fricker-Hidalgo, H., B. Lebeau, P. Kervroedan, O. Faure, P. Ambroise-Thomas, and R. Grillot. 1995. Auxacolor, a new commercial system for yeast identification: evaluation of 182 strains comparatively with ID 32C. *Ann. Biol. Clin.* (Paris) **53:**221–225.

52. Fricker-Hidalgo, H., O. Vandapel, M. A. Duchesne, M. A. Mazoyer, D. Monget, B. Lardy, B. Lebeau, J. Freney, P. Ambroise-Thomas, and R. Grillot. 1996. Comparison of the new API Candida system to the ID 32C system for identification of clinically important yeast species. *J. Clin. Microbiol.* **34:**1846–1848.

53. Fromtling, R. A., J. N. Galgiani, M. A. Pfaller, A. Espinel-Ingroff, K. F. Bartizal, M. S. Bartlett, B. A. Body, C. Frey, G. Hall, and G. D. Roberts. 1993. Multicenter evaluation of a broth macrodilution antifungal susceptibility test for yeasts. *Antimicrob. Agents Chemother.* **37:**39–45.

54. Gade, W., S. W. Hinnefeld, L. S. Babcock, P. Gilligan, W. Kelly, K. Wait, D. Greer, M. Pinilla, and

R. L. Kaplan. 1991. Comparison of the PREMIER cryptococcal antigen enzyme immunoassay and the latex agglutination assay for detection of cryptococcal antigens. *J. Clin. Microbiol.* **29:**1616–1619.

55. Gales, A. C., M. A. Pfaller, A. K. Houston, S. Joly, D. J. Sullivan, D. C. Coleman, and D. R. Soll. 1999. Identification of *Candida dubliniensis* based on temperature and utilization of xylose and alpha-*methyl*-D-glucoside as determined with the API 20C AUX and Vitek YBC systems. *J. Clin. Microbiol.* **37:**3804–3808.

56. Garcia-Prats, J. A., T. R. Cooper, V. F. Schneider, C. E. Stager, and T. N. Hansen. 2000. Rapid detection of microorganisms in blood cultures of newborn infants utilizing an automated blood culture system. *Pediatrics* **105:**523–527.

57. Geha, D. J., and G. D. Roberts. 1994. Laboratory detection of fungemia. *Clin. Lab. Med.* **14:**83–97.

58. Goldschmidt, M. C., D. Y. Fung, R. Grant, J. White, and T. Brown. 1991. New aniline blue dye medium for rapid identification and isolation of *Candida albicans*. *J. Clin. Microbiol.* **29:**1095–1099.

59. Graf, B., T. Adam, E. Zill, and U. B. Gobel. 2000. Evaluation of the VITEK 2 system for rapid identification of yeasts and yeast-like organisms. *J. Clin. Microbiol.* **38:**1782–1785.

60. Gray, L. D., and G. D. Roberts. 1988. Experience with the use of pronase to eliminate interference factors in the latex agglutination test for cryptococcal antigen. *J. Clin. Microbiol.* **26:**2450–2451.

61. Gromadzki, S. G., and V. Chaturvedi. 2000. Limitation of the AccuProbe *Coccidioides immitis* culture identification test: false-negative results with formaldehyde-killed cultures. *J. Clin. Microbiol.* **38:**2427–2428.

62. Guerra-Romero, L., R. S. Edson, F. R. Cockerill III, C. D. Horstmeier, and G. D. Roberts. 1987. Comparison of Du Pont Isolator and Roche Septi-Chek for detection of fungemia. *J. Clin. Microbiol.* **25:**1623–1625.

63. Gutierrez, J., C. Maroto, G. Piedrola, E. Martin, and J. A. Perez. 1993. Circulating *Candida* antigens and antibodies: useful markers of candidemia. *J. Clin. Microbiol.* **31:**2550–2552.

64. Gutierrez, J., E. Martin, C. Lozano, J. Coronilla, and C. Nogales. 1994. Evaluation of the ATB 32C automicrobic system and API 20C using clinical yeast isolates. *Ann. Biol. Clin.* (Paris) **52:**443–446.

65. Hall, G. S., K. Pratt-Rippin, and J. A. Washington. 1992. Evaluation of a chemiluminescent probe assay for identification of *Histoplasma capsulatum* isolates. *J. Clin. Microbiol.* **30:**3003–3004.

66. Hamilton, J. R., A. Noble, D. W. Denning, and D. A. Stevens. 1991. Performance of cryptococcus antigen latex agglutination kits on serum and cerebrospinal fluid specimens of AIDS patients before and after pronase treatment. *J. Clin. Microbiol.* **29:**333–339.

67. Hantschke, D. 2000. Differentiation of yeast-like fungi using the commercial Auxacolor system. *Mycoses* **39:**135–140.

68. Hashiguchi, K., H. Wada, O. Yamada, Y. Yawata, K. Yoshida, J. Okimoto, S. Umeki, Y. Niki, and R. Soejima. 1992. A case of chronic myelogenous leukemia with pulmonary aspergillosis diagnosed by the detection of circulating Aspergillus antigen. *Kansenshogaku Zasshi* **66:**1592–1596. (In Japanese.)

69. Heelan, J. S., L. Corpus, and N. Kessimian. 1991. False-positive reactions in the latex agglutination test for *Cryptococcus neoformans* antigen. *J. Clin. Microbiol.* **29:**1260–1261. (Erratum, **29:**2091.)

70. Heelan, J. S., D. Siliezar, and K. Coon. 1996. Comparison of rapid testing methods for enzyme production with the germ tube method for presumptive identification of *Candida albicans*. *J. Clin. Microbiol.* **34:**2847–2849.

71. Heelan, J. S., E. Sotomayor, K. Coon, and J. B. D'Arezzo. 1998. Comparison of the rapid yeast plus panel with the API20C yeast system for identification of clinically significant isolates of *Candida* species. *J. Clin. Microbiol.* **36:**1443–1445.

72. Herent, P., D. Stynen, F. Hernando, J. Fruit, and D. Poulain. 1992. Retrospective evaluation of two latex agglutination tests for detection of circulating antigens during invasive candidosis. *J. Clin. Microbiol.* **30:**2158–2164.

73. Hoppe, J. E., and P. Frey. 1999. Evaluation of six commercial tests and the germ-tube test for presumptive identification of Candida albicans. *Eur. J. Clin. Microbiol. Infect. Dis.* **18:**188–191.

74. Hopwood, V., E. M. Johnson, J. M. Cornish, A. B. Foot, E. G. Evans, and D. W. Warnock. 1995. Use of the Pastorex aspergillus antigen latex agglutination test for the diagnosis of invasive aspergillosis. *J. Clin. Pathol.* **48:**210–213.

75. Hossain, M. A., T. Miyazaki, K. Mitsutake, H. Kakeya, Y. Yamamoto, K. Yanagihara, S. Kawamura, T. Otsubo, Y. Hirakata, T. Tashiro, and S. Kohno. 1997. Comparison between Wako-WB003 and Fungitec G tests for detection of $(1{\rightarrow}3)$-beta-D-glucan in systemic mycosis. *J. Clin. Lab. Anal.* **11:**73–77.

76. Huffnagle, K. E., and R. M. Gander. 1993. Evaluation of Gen-Probe's *Histoplasma capsulatum* and *Cryptococcus neoformans* AccuProbes. *J. Clin. Microbiol.* **31:**419–421.

77. Jahn, B., A. Stuben, and S. Bhakdi. 1996. Colorimetric susceptibility testing for *Aspergillus fumigatus*: comparison of menadione-augmented 3-(4,5-dimethyl-2-thiazolyl)-2,5-diphenyl-2H-tetrazolium bromide and Alamar blue tests. *J. Clin. Microbiol.* **34:**2039–2041.

78. Kami, M., Y. Kanda, S. Ogawa, S. Mori, Y. Tanaka, H. Honda, S. Chiba, K. Mitani, Y. Yazaki, and H. Hirai. 1999. Frequent false-positive results of Aspergillus latex agglutination test: transient Aspergillus antigenemia during neutropenia. *Cancer* **86:**274–281.

79. Kaufman, L., J. A. Kovacs, and E. Reiss. 1997. Clinical immunomycology, p. 585–604. *In* N. R. Rose, E. C. de Macario, J. D. Folds, H. C. Lane, and R. M. Nakamura (ed.), *Manual of Clinical Laboratory Immunology*. ASM Press, Washington, D.C.

80. Kaufman, L., A. S. Sekhon, N. Moledina, M. Jalbert, and D. Pappagianis. 1995. Comparative evaluation of commercial Premier EIA and microimmunodiffusion and complement fixation tests for *Coccidioides immitis* antibodies. *J. Clin. Microbiol.* **33:**618–619.

81. Kellogg, J. A., D. A. Bankert, and V. Chaturvedi. 1998. Limitations of the current microbial identification system for identification of clinical yeast isolates. *J. Clin. Microbiol.* **36:**1197–1200.

82. Kellogg, J. A., D. A. Bankert, and V. Chaturvedi. 1999. Variation in Microbial Identification System accuracy for yeast identification depending on commercial source of Sabouraud dextrose agar. *J. Clin. Microbiol.* **37:**2080–2083.

83. Kellogg, J. A., D. A. Bankert, J. P. Manzella, K. S. Parsey, S. L. Scott, and S. H. Cavanaugh. 1994. Clinical comparison of Isolator and thiol broth with ESP

aerobic and anaerobic bottles for recovery of pathogens from blood. *J. Clin. Microbiol.* **32**:2050–2055.

84. **Kirkley, B. A., K. A. Easley, and J. A. Washington.** 1994. Controlled clinical evaluation of Isolator and ESP aerobic blood culture systems for detection of bloodstream infections. *J. Clin. Microbiol.* **32**:1547–1549.

85. **Kiska, D. L., D. R. Orkiszewski, D. Howell, and P. H. Gilligan.** 1994. Evaluation of new monoclonal antibody-based latex agglutination test for detection of cryptococcal polysaccharide antigen in serum and cerebrospinal fluid. *J. Clin. Microbiol.* **32**:2309–2311.

86. **Kitch, T. T., M. R. Jacobs, M. R. McGinnis, and P. C. Appelbaum.** 1996. Ability of RapID Yeast Plus System to identify 304 clinically significant yeasts within 5 hours. *J. Clin. Microbiol.* **34**:1069–1071.

87. **Land, G., J. Burke, C. Shelby, J. Rhodes, J. Collett, I. Bennett, and J. Johnson.** 1996. Screening protocol for *Torulopsis* (*Candida*) *glabrata*. *J. Clin. Microbiol.* **34**:2300–2303.

88. **Land, G. A., I. F. Salkin, M. el Zaatari, M. R. McGinnis, and G. Hashem.** 1991. Evaluation of the Baxter-MicroScan 4-hour enzyme-based yeast identification system. *J. Clin. Microbiol.* **29**:718–722.

89. **Lipperheide, V., L. Andraka, J. Ponton, and G. Quindos.** 1993. Evaluation of the albicans IDR plate method for the rapid identification of Candida albicans. *Mycoses* **36**:417–420.

90. **Lopez-Jodra, O., J. M. Torres-Rodriguez, R. Mendez-Vasquez, E. Ribas-Forcadell, Y. Morera-Lopez, T. Baro-Tomas, and C. Alia-Aponte.** 2000. In vitro susceptibility of cryptococcus neoformans isolates to five antifungal drugs using a colorimetric system and the reference microbroth method. *J. Antimicrob. Chemother.* **45**:645–649.

91. **Lozano-Chiu, M., M. V. Lancaster, and J. H. Rex.** 1998. Evaluation of a colorimetric method for detecting amphotericin B-resistant Candida isolates. *Diagn. Microbiol. Infect. Dis.* **31**:417–424.

92. **Lyon, R., and G. Woods.** 1995. Comparison of the BacT/Alert and Isolator blood culture systems for recovery of fungi. *Am. J. Clin. Pathol.* **103**:660–662.

93. **Manso, E., M. Montillo, G. De Sio, S. D'Amico, G. Discepoli, and P. Leoni.** 1994. Value of antigen and antibody detection in the serological diagnosis of invasive aspergillosis in patients with hematological malignancies. *Eur. J. Clin. Microbiol. Infect. Dis.* **13**:756–760.

94. **Marcelis, L., J. Verhaegen, J. Vandeven, A. Bosmans, and L. Verbist.** 1992. Evaluation of Bactec high blood volume resin media. *Diagn. Microbiol. Infect. Dis.* **15**:385–391.

95. **Martins, T. B., T. D. Jaskowski, C. L. Mouritsen, and H. R. Hill.** 1995. Comparison of commercially available enzyme immunoassay with traditional serological tests for detection of antibodies to *Coccidioides immitis*. *J. Clin. Microbiol.* **33**:940–943.

96. **Merz, W. G., and G. D. Roberts.** 1999. Algorithms for detection and identification of fungi, p. 1167–1183. *In* P. R. Murray, E. J. Baron, M. A. Pfaller, F. C. Tenover, and R. H. Yolken (ed.), *Manual of Clinical Microbiology*. ASM Press, Washington, D.C.

97. **Messer, S. A., and M. A. Pfaller.** 1996. Clinical evaluation of a dried commercially-prepared microdilution panel for antifungal susceptibility testing. *Diagn. Microbiol. Infect. Dis.* **25**:77–81.

98. **Michel-Nguyen, A., A. Favel, C. Chastin, M. Selva, and P. Regli.** 2000. Comparative evaluation of a com

99. **Milan, E. P., E. S. Malheiros, O. Fischman, and A. L. Colombo.** 1997. Evaluation of the AUXACOLOR system for the identification of clinical yeast isolates. *Mycopathologia* **137**:153–157.

100. **Mitsutake, K., T. Miyazaki, T. Tashiro, Y. Yamamoto, H. Kakeya, T. Otsubo, S. Kawamura, M. A. Hossain, Y. Hirakata, and S. Kohno.** 1996. Enolase antigen, mannan antigen, Cand-Tec antigen, and beta-glucan in patients with candidemia. *J. Clin. Microbiol.* **34**:1918–1921.

101. **Moghaddas, J., A. L. Truant, C. Jordan, and H. R. Buckley.** 1999. Evaluation of the RapID Yeast Plus System for the identification of yeast. *Diagn. Microbiol. Infect. Dis.* **35**:271–273.

102. **Morello, J. A., C. Leitch, S. Nitz, J. W. Dyke, M. Andruszewski, G. Maier, W. Landau, and M. A. Beard.** 1994. Detection of bacteremia by Difco ESP blood culture system. *J. Clin. Microbiol.* **32**:811–818.

103. **Mori, T., H. Ikemoto, M. Matsumura, M. Yoshida, K. Inada, S. Endo, A. Ito, S. Watanabe, H. Yamaguchi, M. Mitsuya, M. Kodama, T. Tani, T. Yokota, T. Kobayashi, J. Kambayashi, T. Nakamura, T. Masaoka, H. Teshima, T. Yoshinaga, S. Kohno, K. Hara, and S. Miyazaki.** 1997. Evaluation of plasma (1→3)-beta-D-glucan measurement by the kinetic turbidimetric Limulus test, for the clinical diagnosis of mycotic infections. *Eur. J. Clin. Chem. Clin. Biochem.* **35**:553–560.

104. **Morrell, R. M., Jr., and B. L. Wasilauskas.** 1992. Tracking laboratory contamination by using a *Bacillus cereus* pseudoepidemic as an example. *J. Clin. Microbiol.* **30**:1469–1473.

105. **Morrell, R. M., Jr., B. L. Wasilauskas, and C. H. Steffee.** 1996. Performance of fungal blood cultures by using the Isolator collection system: is it cost-effective? *J. Clin. Microbiol.* **34**:3040–3043.

106. **Murray, P. R.** 1991. Comparison of the lysis-centrifugation and agitated biphasic blood culture systems for detection of fungemia. *J. Clin. Microbiol.* **29**:96–98.

107. **Musial, C. E., F. R. Cockerill III, and G. D. Roberts.** 1988. Fungal infections of the immunocompromised host: clinical and laboratory aspects. *Clin. Microbiol. Rev.* **1**:349–364.

108. **Muta, T., and S. Iwanaga.** 1996. Clotting and immune defense in Limulidae. *Prog. Mol. Subcell. Biol.* **15**:154–189.

109. **Nakao, A., M. Yasui, T. Kawagoe, H. Tamura, S. Tanaka, and H. Takagi.** 1997. False-positive endotoxemia derives from gauze glucan after hepatectomy for hepatocellular carcinoma with cirrhosis. *Hepato-Gastroenterology* **44**:1413–1418.

110. **Ness, M. J., W. P. Vaughan, and G. L. Woods.** 1989. Candida antigen latex test for detection of invasive candidiasis in immunocompromised patients. *J. Infect. Dis.* **159**:495–502.

111. **Niki, Y.** 1996. Sero-diagnosis for pulmonary aspergillosis—its utility in early diagnosis. *Rinsho Byori* **44**:518–523. (In Japanese.)

112. **Odds, F. C., and R. Bernaerts.** 1994. CHROMagar Candida, a new differential isolation medium for presumptive identification of clinically important *Candida* species. *J. Clin. Microbiol.* **32**:1923–1929.

113. **Odds, F. C., L. Vranckx, and F. Woestenborghs.** 1995. Antifungal susceptibility testing of yeasts: eval-

uation of technical variables for test automation. *Antimicrob. Agents Chemother.* **39:**2051–2060.

114. **Padhye, A. A., G. Smith, D. McLaughlin, P. G. Standard, and L. Kaufman.** 1992. Comparative evaluation of a chemiluminescent DNA probe and an exoantigen test for rapid identification of *Histoplasma capsulatum*. *J. Clin. Microbiol.* **30:**3108–3111.

115. **Patterson, T. F., W. R. Kirkpatrick, S. G. Revankar, R. K. McAtee, A. W. Fothergill, D. I. McCarthy, and M. G. Rinaldi.** 1996. Comparative evaluation of macrodilution and chromogenic agar screening for determining fluconazole susceptibility of *Candida albicans*. *J. Clin. Microbiol.* **34:**3237–3239.

116. **Paugam, A., M. Benchetrit, A. Fiacre, C. Tourte-Schaefer, and J. Dupouy-Camet.** 1999. Comparison of four commercialized biochemical systems for clinical yeast identification by colour-producing reactions. *Med. Mycol.* **37:**11–17.

117. **Perry, J. L., and G. R. Miller.** 1987. Umbelliferyl-labeled galactosaminide as an aid in identification of *Candida albicans*. *J. Clin. Microbiol.* **25:**2424–2425.

118. **Perry, J. L., G. R. Miller, and D. L. Carr.** 1990. Rapid, colorimetric identification of *Candida albicans*. *J. Clin. Microbiol.* **28:**614–615.

119. **Petti, C. A., A. K. Zaidi, S. Mirrett, and L. B. Reller.** 1996. Comparison of Isolator 1.5 and BACTEC NR660 aerobic 6A blood culture systems for detection of fungemia in children. *J. Clin. Microbiol.* **34:**1877–1879.

120. **Pfaller, M. A., S. Arikan, M. Lozano-Chiu, Y. Chen, S. Coffman, S. A. Messer, R. Rennie, C. Sand, T. Heffner, J. H. Rex, J. Wang, and N. Yamane.** 1998. Clinical evaluation of the ASTY colorimetric microdilution panel for antifungal susceptibility testing. *J. Clin. Microbiol.* **36:**2609–2612.

121. **Pfaller, M. A., and A. L. Barry.** 1994. Evaluation of a novel colorimetric broth microdilution method for antifungal susceptibility testing of yeast isolates. *J. Clin. Microbiol.* **32:**1992–1996.

122. **Pfaller, M. A., I. Cabezudo, B. Buschelman, M. Bale, T. Howe, M. Vitug, H. J. Linton, and M. Densel.** 1993. Value of the Hybritech ICON Candida Assay in the diagnosis of invasive candidiasis in high-risk patients. *Diagn. Microbiol. Infect. Dis.* **16:**53–60.

123. **Pfaller, M. A., S. A. Messer, and A. Bolmstrom.** 1998. Evaluation of E test for determining *in vitro* susceptibility of yeast isolates to amphotericin B. *Diagn. Microbiol. Infect. Dis.* **32:**223–227.

124. **Pfaller, M. A., S. A. Messer, A. Bolmström, F. C. Odds, and J. H. Rex.** 1996. Multisite reproducibility of the Etest MIC method for antifungal susceptibility testing of yeast isolates. *J. Clin. Microbiol.* **34:**1691–1693.

125. **Pfaller, M. A., S. A. Messer, Å. Karlsson, and A. Bolmström.** 1998. Evaluation of the Etest method for determining fluconazole susceptibilities of 402 clinical yeast isolates by using three different agar media. *J. Clin. Microbiol.* **36:**2586–2589.

126. **Pfaller, M. A., J. H. Rex, and M. G. Rinaldi.** 1997. Antifungal susceptibility testing: technical advances and potential clinical applications. *Clin. Infect. Dis.* **24:**776–784.

127. **Powell, H. L., C. A. Sand, and R. P. Rennie.** 1998. Evaluation of CHROMagar Candida for presumptive identification of clinically important Candida species. *Diagn. Microbiol. Infect. Dis.* **32:**201–204.

128. **Procop, G. W., and G. D. Roberts.** 1998. Laboratory methods in basic mycology, p. 871–951. *In* B. A. Forbes, D. F. Sahm, and A. S. Weissfeld (ed.), *Bailey & Scott's Diagnostic Microbiology.* Mosby, Inc., St. Louis, Mo.

129. **Qian, Q., I. Sinkeldam, and G. D. Roberts.** 1999. Comparison of a colorimetric microdilution method and the NNCLS macrodilution method for antifungal susceptibility testing of yeasts. *J. Mycol. Med.* **9:**181–184.

130. **Quindos, G., V. Lipperheide, and J. Ponton.** 1993. Evaluation of two commercialized systems for the rapid identification of medically important yeasts. *Mycoses* **36:**299–303.

131. **Quindos, G., R. Salesa, A. J. Carrillo-Munoz, V. Lipperheide, L. Jaudenes, R. San Millan, J. M. Torres-Rodriguez, and J. Ponton.** 1994. Multicenter evaluation of ATB fungus: a standardized micromethod for yeast susceptibility testing. *Chemotherapy* **40:**245–251.

132. **Quindos, G., R. San Millan, R. Robert, C. Bernard, and J. Ponton.** 1997. Evaluation of Bichro-latex Albicans, a new method for rapid identification of *Candida albicans*. *J. Clin. Microbiol.* **35:**1263–1265.

133. **Ramani, R., S. Gromadzki, D. H. Pincus, I. F. Salkin, and V. Chaturvedi.** 1998. Efficacy of API 20C and ID 32C systems for identification of common and rare clinical yeast isolates. *J. Clin. Microbiol.* **36:**3396–3398.

134. **Rath, P. M., R. Oeffelke, K. D. Muller, and R. Ansorg.** 1996. Non-value of Aspergillus antigen detection in bronchoalveolar lavage fluids of patients undergoing bone marrow transplantation. *Mycoses* **39:**367–370.

135. **Reiss, E., T. Obayashi, K. Orle, M. Yashida, and R. M. Zancopé-Oliveira.** 2000. Non-culture based diagnostic tests for mycotic infections. *Med. Mycol.* **38**(Suppl. 1):147–159.

136. **Riddle, D. L., O. Giger, L. Miller, G. S. Hall, and G. L. Woods.** 1994. Clinical comparison of the Baxter MicroScan Yeast Identification Panel and the Vitek Yeast Biochemical Card. *Am. J. Clin. Pathol.* **101:**438–442.

137. **Romney, M. G., E. A. Bryce, R. P. Rennie, and C. A. Sand.** 2000. Rapid identification of clinical yeast isolates using the colorimetric AUXACOLOR system. *Diagn. Microbiol. Infect. Dis.* **36:**137–138.

138. **Rousselle, P., A. M. Freydiere, P. J. Couillerot, H. de Montclos, and Y. Gille.** 1994. Rapid identification of *Candida albicans* by using Albicans ID and fluoroplate agar plates. *J. Clin. Microbiol.* **32:**3034–3036.

139. **Sanati, H., S. A. Messer, M. Pfaller, M. Witt, R. Larsen, A. Espinel-Ingroff, and M. Ghannoum.** 1996. Multicenter evaluation of broth microdilution method for susceptibility testing of *Cryptococcus neoformans* against fluconazole. *J. Clin. Microbiol.* **34:**1280–1282.

140. **Sanchez-Sousa, A., M. E. Alvarez, L. Matz, H. Escobar, and F. Baquero.** 1999. Aspergillus fumigatus susceptibility testing by colorimetric microdilution. *J. Mycol. Med.* **9:**103–106.

141. **Sand, C., and R. P. Rennie.** 1999. Comparison of three commercial systems for the identification of germ-tube negative yeast species isolated from clinical specimens. *Diagn. Microbiol. Infect. Dis.* **33:**223–229.

142. **San Millan, R., L. Ribacoba, J. Ponton, and G. Quindos.** 1996. Evaluation of a commercial medium for identification of Candida species. *Eur. J. Clin. Microbiol. Infect. Dis.* **15:**153–158.

143. **Schoofs, A., F. C. Odds, R. Colebunders, M. Ieven, and H. Goossens.** 1997. Use of specialised isolation media for recognition and identification of Candida dubliniensis isolates from HIV-infected patients. *Eur. J. Clin. Microbiol. Infect. Dis.* **16:**296–300.

144. **Schuffenecker, I., A. Freydiere, H. de Montclos, and Y. Gille.** 1993. Evaluation of four commercial systems for identification of medically important yeasts. *Eur. J. Clin. Microbiol. Infect. Dis.* **12:**255–260.

145. **Schwabe, L. D., E. L. Randall, R. Miller-Catchpole, C. I. Squires, and R. L. Gottschall.** 1990. A comparison of oxoid signal with nonradiometric BACTEC NR-660 for detection of bacteremia. *Diagn. Microbiol. Infect. Dis.* **13:**3–8.

146. **Sekhon, A. S., L. Kaufman, G. S. Kobayashi, N. Moledina, M. Jalbert, and R. H. Notenboom.** 1994. Comparative evaluation of the Premier enzyme immunoassay, micro- immunodiffusion and complement fixation tests for the detection of Histoplasma capsulatum var. capsulatum antibodies. *Mycoses* **37:**313–316.

147. **Sekhon, A. S., L. Kaufman, G. S. Kobayashi, N. H. Moledina, and M. Jalbert.** 1995. The value of the Premier enzyme immunoassay for diagnosing Blastomyces dermatitidis infections. *J. Med. Vet. Mycol.* **33:**123–125.

148. **Sekhon, A. S., P. G. Standard, L. Kaufman, and A. K. Garg.** 1987. Evaluation of commercial serologic test reagents for immunoidentification of medically important aspergilli. *Diagn. Microbiol. Infect. Dis.* **8:**183–187.

149. **Sewell, D. L., M. A. Pfaller, and A. L. Barry.** 1994. Comparison of broth macrodilution, broth microdilution, and E test antifungal susceptibility tests for fluconazole. *J. Clin. Microbiol.* **32:**2099–2102.

150. **Sigler, L., and M. J. Kennedy.** 1999. *Aspergillus, Fusarium,* and other opportunistic moniliaceous fungi, p. 1212–1241. *In* P. R. Murray, E. J. Baron, M. A. Pfaller, F. C. Tenover, and R. H. Yolken (ed.), *Manual of Clinical Microbiology.* ASM Press, Washington, D.C.

151. **Simor, A. E., G. Goswell, L. Louie, M. Lee, and M. Louie.** 1997. Antifungal susceptibility testing of yeast isolates from blood cultures by microbroth dilution and the E test. *Eur. J. Clin. Microbiol. Infect. Dis.* **16:**693–697.

152. **St. Germain, G., and D. Beauchesne.** 1991. Evaluation of the MicroScan Rapid Yeast Identification panel. *J. Clin. Microbiol.* **29:**2296–2299.

153. **Stockman, L., K. A. Clark, J. M. Hunt, and G. D. Roberts.** 1993. Evaluation of commercially available acridinium ester-labeled chemiluminescent DNA probes for culture identification of *Blastomyces dermatitidis, Coccidioides immitis, Cryptococcus neoformans,* and *Histoplasma capsulatum. J. Clin. Microbiol.* **31:**845–850.

154. **Stockman, L., and G. D. Roberts.** 1983. Specificity of the latex test for cryptococcal antigen: a rapid, simple method for eliminating interference factors. *J. Clin. Microbiol.* **17:**945–947.

155. **Sugar, A. M., and X. Liu.** 1995. Comparison of three methods of antifungal susceptibility testing with the proposed NCCLS standard broth macrodilution assay: lack of effect of phenol red. National Committee for Clinical Laboratory Standards. *Diagn. Microbiol. Infect. Dis.* **21:**129–133.

156. **Sulahian, A., M. Tabouret, P. Ribaud, J. Sarfati, E. Gluckman, J. P. Latge, and F. Derouin.** 1996. Comparison of an enzyme immunoassay and latex agglutination test for detection of galactomannan in the diagnosis of invasive aspergillosis. *Eur. J. Clin. Microbiol. Infect. Dis.* **15:**139–145.

157. **Szekely, A., E. M. Johnson, and D. W. Warnock.** 1999. Comparison of E-test and broth microdilution methods for antifungal drug susceptibility testing of molds. *J. Clin. Microbiol.* **37:**1480–1483.

158. **Tanner, D. C., M. P. Weinstein, B. Fedorciw, K. L. Joho, J. J. Thorpe, and L. Reller.** 1994. Comparison of commercial kits for detection of cryptococcal antigen. *J. Clin. Microbiol.* **32:**1680–1684.

159. **Telenti, A., and G. D. Roberts.** 1989. Fungal blood cultures. *Eur. J. Clin. Microbiol. Infect. Dis.* **8:**825–831.

160. **Tintelnot, K., G. Haase, M. Seibold, F. Bergmann, M. Staemmler, T. Franz, and D. Naumann.** 2000. Evaluation of phenotypic markers for selection and identification of *Candida dubliniensis. J. Clin. Microbiol.* **38:**1599–1608.

161. **To, W. K., A. W. Fothergill, and M. G. Rinaldi.** 1995. Comparative evaluation of macrodilution and alamar colorimetric microdilution broth methods for antifungal susceptibility testing of yeast isolates. *J. Clin. Microbiol.* **33:**2660–2664.

162. **Valasco-Martinez, J. J., P. Marti-Belda, A. Guerrero-Espejo, A. Sanchez-Sousa, and F. Baquero.** 1995. Monoclonal/polyclonal antibodies and pronase in detection of Cryptococcal antigen in serum and cerebrospinal fluid. *J. Mycol. Med.* **5:**230–234.

163. **Verweij, P. E., A. J. Rijs, B. E. De Pauw, A. M. Horrevorts, J. A. Hoogkamp-Korstanje, and J. F. Meis.** 1995. Clinical evaluation and reproducibility of the Pastorex Aspergillus antigen latex agglutination test for diagnosing invasive aspergillosis. *J. Clin. Pathol.* **48:**474–476.

164. **Verwij, P. E., I. M. Breuker, A. J. M. M. Rijs, and J. F. G. M. Meis.** 1999. Comparative study of seven commercial yeast identification systems. *J. Clin. Pathol.* **52:**271–273.

165. **Wadlin, J. K., G. Hanko, R. Stewart, J. Pape, and I. Nachamkin.** 1999. Comparison of three commercial systems for identification of yeasts commonly isolated in the clinical microbiology laboratory. *J. Clin. Microbiol.* **37:**1967–1970.

166. **Waite, R. T., and G. L. Woods.** 1998. Evaluation of BACTEC MYCO/F lytic medium for recovery of mycobacteria and fungi from blood. *J. Clin. Microbiol.* **36:**1176–1179.

167. **Walsh, T. J., J. W. Hathorn, J. D. Sobel, W. G. Merz, V. Sanchez, S. M. Maret, H. R. Buckley, M. A. Pfaller, R. Schaufele, and C. Sliva.** 1991. Detection of circulating candida enolase by immunoassay in patients with cancer and invasive candidiasis. *N. Engl. J. Med.* **324:**1026–1031.

168. **Warren, N. G., and K. C. Hazen.** 1999. *Candida, Cryptococcus,* and other yeasts of medical importance, p. 1184–1199. *In* P. R. Murray, E. J. Baron, M. A. Pfaller, F. C. Tenover, and R. H. Yolken (ed.), *Manual of Clinical Microbiology,* 7th ed. ASM Press, Washington, D.C.

169. **Weinstein, M. P., S. Mirrett, L. G. Reimer, M. L. Wilson, S. Smith-Elekes, C. R. Chuard, K. L. Joho, and L. B. Reller.** 1995. Controlled evaluation of BacT/Alert standard aerobic and FAN aerobic blood culture bottles for detection of bacteremia and fungemia. *J. Clin. Microbiol.* **33:**978–981.

170. **Wilson, M. L., T. E. Davis, S. Mirrett, J. Reynolds, D. Fuller, S. D. Allen, K. K. Flint, F. Koontz, and L. B. Reller.** 1993. Controlled comparison of

the BACTEC high-blood-volume fungal medium, BACTEC Plus 26 aerobic blood culture bottle, and 10-milliliter Isolator blood culture system for detection of fungemia and bacteremia. *J. Clin. Microbiol.* **31:**865–871.

171. **Wilson, M. L., M. P. Weinstein, S. Mirrett, L. G. Reimer, R. J. Feldman, C. R. Chuard, and L. B. Reller.** 1995. Controlled evaluation of BacT/Alert standard anaerobic and FAN anaerobic blood culture bottles for the detection of bacteremia and fungemia. *J. Clin. Microbiol.* **33:**2265–2270.

172. **Wilson, M. L., M. P. Weinstein, L. G. Reimer, S. Mirrett, and L. B. Reller.** 1992. Controlled comparison of the BacT/Alert and BACTEC 660/730 nonra-diometric blood culture systems. *J. Clin. Microbiol.* **30:**323–329.

173. **Yeo, S. F., Y. Zhang, D. Schafer, S. Campbell, and B. Wong.** 2000. A rapid, automated enzymatic fluorometric assay for determination of D-arabinitol in serum. *J. Clin. Microbiol.* **38:**1439–1443.

174. **Zwadyk, P., Jr., C. L. Pierson, and C. Young.** 1994. Comparison of Difco ESP and Organon Teknika BacT/Alert continuous-monitoring blood culture systems. *J. Clin. Microbiol.* **32:**1273–1279.

175. **Zwadyk, P., Jr., R. A. Tarlton, and A. Proctor.** 1977. Evaluation of the API 20C for identification of yeasts. *Am. J. Clin. Pathol.* **67:**269–271.

Mycobacteria

GLENN D. ROBERTS, LESLIE HALL, AND DONNA M. WOLK

10

The clinical mycobacteriology laboratory has undergone a dramatic evolution during the past twenty years, particularly with regard to where testing is performed. Today, a limited number of laboratories offer full mycobacteriologic services; others send work to reference laboratories that offer more rapid and advanced methods for the detection and identification of clinically important mycobacteria.

Currently a number of automated commercially available rapid detection systems that can also perform antimycobacterial susceptibility testing are available to clinical laboratories. Further, several commercially available identification systems that utilize molecular methods are used by reference laboratories for the rapid identification of clinically important species of mycobacteria. These systems may be costly, however, and only large laboratories generally are able to use them.

Not all of the methods presented within this chapter are approved by the Federal Food and Drug Administration (FDA). However, it is important for the reader to at least become familiar with those that will be used in the future. The future of mycobacteriology is progressing toward the molecular detection and identification of clinically important species directly from clinical specimens. In addition, the possibility of determining the antimicrobial susceptibility or resistance of an organism is possible without ever culturing it; however, much more work needs to be done before this will be a routine method.

The methods described herein represent those that are commercially available and have been evaluated with data published in peer-reviewed journals. There are many other "home brew" methods that have appeared in the world literature; however, they are not commercially available for all laboratories to purchase.

SPECIMEN PROCESSING

Most of the clinical specimens sent to the mycobacteriology laboratory for culture are contaminated by rapidly growing, normal, bacterial flora. To maximize mycobacterial yield, contaminated specimens require treatment with a liquefying and decontamination procedure. N-Acetyl-L-cysteine-sodium hydroxide solution (NALC-NaOH) is generally recommended as a mild but effective reagent. Many laboratories choose to prepare their own reagents; however, commercially available sources offer some standardization to the process. Commercially available reagents include the following.

I. BBL Myco Prep
   Manufacturer: Becton Dickinson Biosciences Microbiology Products, Sparks, Md.
   Description: The MycoPrep Specimen Digestion/Decontamination kits consist of 10 75- or 100-ml bottles of NALC-NaOH solution and 5 or 10 packages of phosphate buffer, pH 6.

II. NAC-PAC, NAC-PAC2, and AFB Adjusting Reagents
   Manufacturer: Alpha-TEC Systems Incorporated, Vancouver, Wash.
   Description: The kit contains ampoules of premeasured NALC powder and five vials of acid-fast bacillus-based digestant. Buffer and digestion reagents may be purchased separately so that the user may custom conditions for digestion and decontamination.

III. ATSXPR-Plus AFB Neutralizing Buffers, pH 6.0
   Manufacturer: Alpha-Tec Systems Incorporated
   Description: This buffer is used to neutralize the reagents used in the NALC digestion and decontamination procedure of clinical specimens for the increased recovery of acid-fast bacilli. The buffer is supplied as a filtered (0.22-μm pore size) high-efficiency buffer in distilled water. It has been shown that this buffer causes inhibition of the Gen-Probe (San Diego, Calif.) amplified *Mycobacterium tuberculosis* direct (MTD2) test (P. Della-Latta and V. Jonas, Letter, *J. Clin. Microbiol.* **37:**1234–1235, 1999).

STAINING REAGENTS FOR ACID-FAST STAINING OF MYCOBACTERIA

One of the most important components of the laboratory diagnosis of mycobacterial infections is the use of the acid-fast stain; either it is the auramine-rhodamine fluorescent stain or conventional carbol fuchsin-based stain. Many laboratories choose to prepare their own reagents; however, such reagents are available through several commercial sources and include the following:

I. TB Fluorescent Stain Kit, TB Quick Stain Kit, TB Stain Kit K, and TB Stain Kit ZN
   Manufacturer: Becton Dickinson Biosciences Microbiology Products
   Description: Becton Dickinson offers a full spectrum of stains ranging from auramine-rhodamine to the Kiny-

oun stain to the Ziehl-Neelsen stain. All are useful for the detection of mycobacteria in clinical specimens and in cultures. Further, individual reagents may be purchased through Becton Dickinson for laboratories which choose to customize their staining process.

II. TB Auramine-Rhodamine Fluorescent Stain, TB Ziehl-Neelsen Carbol Fuchsin, and TB Kinyoun Carbol Fuchsin Stains
Manufacturer: Remel, Lenexa, Kans.
Description: Remel provides representation of the auramine-rhodamine, Kinyoun, and Ziehl-Neelsen acid-fast stains in kit format along with individual reagents that allow laboratories to customize their own staining method.

III. Alpha-Tech AFB Stains and Controls
Manufacturer: Alpha-Tech Systems Incorporated
Description: Alpha-Tech offers acid-fast bacillus staining control slides. Each slide contains an acid-fast positive staining control of *Mycobacterium scrofulaceum* in simulated sputum and an acid-fast negative staining control consisting of *Escherichia coli* in simulated sputum. In addition, other products include a phenyl acridine orange fluorescence stain and decolorizer used for staining mycobacteria, as well as a modified Kinyoun stain.

IV. Aerospray Slide Stainer 7320 Microbiology Acid-Fast Staining
Manufacturer: Wescor, Inc., Logan, Utah
Description: The Aerospray automated acid-fast stainer (Fig. 1) will perform either the carbol fuchsin or auramine-rhodamine procedure, depending on the choice of the laboratory. Wescor applies reagents that have been modified for instrument compatibility; how-

ever, results are the same as those seen when using traditional staining methods. Primary stain is applied to heat-fixed smears. The excess stain is removed by centrifugation and a water wash. Smears are then decolorized, washed, counterstained, and given a final water wash. The slides are spun dry by 45 s of centrifugation and are ready for examination under the microscope. The entire procedure requires 6 min from beginning to end. An evaluation by Mork-Lewis et al. (K. J. Mork-Lewis, L. Stockman, and G. D. Roberts, *Abstr. 98th Gen. Meet. Am. Soc. Microbiol.*, abstr. C-294, p. 180, 1998) showed that the Wescor Aerospray acid-fast stainer performed in a satisfactory manner and results were equivalent to those obtained by the traditional manual staining methods.

RECOVERY SYSTEMS
Blood Cultures

The clinical significance of mycobacteremia was not realized until it was shown that patients having human immunodeficiency virus infection might also be coinfected with mycobacteria, particularly *Mycobacterium avium-Mycobacterium intracellulare* complex. There has been some emphasis over the years on enhancing blood culture systems so that the mycobacteria are more completely recovered. Perhaps the first system used for the detection of mycobacteremia was the Isolator blood culture system described in the mid-1980s. In 1988, Shanson and Dryden (51) showed that the BACTEC 460 TB system gave the most rapid detection; however, the Isolator gave the best and most rapid isolation

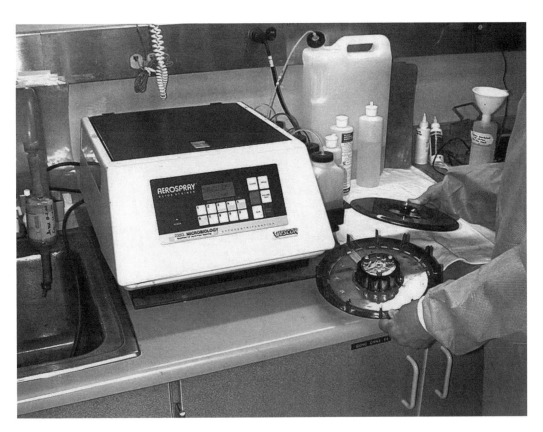

FIGURE 1 Wescor Aerospray stainer.

of individual colonies for identification. Strand et al. suggested that a combination of both methods be used for optimal recovery of mycobacteria from blood (54). In 1989 an evaluation compared BACTEC 13A bottles to BACTEC 12B bottles inoculated with the sediment from the Isolator tube and 7H12 agar also inoculated with the sediment from the Isolator tube. The 13A bottle was shown to recover the greatest number of mycobacteria; however, the 12B bottle inoculated with the sediment from the Isolator tube yielded the shortest recovery time (1). Strand et al. (54) showed that the recovery rates and recovery times were equivalent for both the BACTEC 13A bottle and the Isolator.

In 1997, Tholcken et al. (55) compared the efficiency of the ESP culture system, a continuously monitoring, nonradiometric mycobacterial culture system, for the recovery of mycobacteria using the sediment of blood collected in an Isolator tube. This was compared to the same parameters using Middlebrook 7H11 and 7H11 selective biplates. The ESP system showed a greater overall recovery rate and provided faster detection.

In 1995, Hanna et al. (27) compared blood sediments that were processed in the Isolator system and subsequently inoculated into a mycobacterial growth indicator tube (MGIT) and onto Löwenstein-Jensen slants. Overall, the MGIT and Löwenstein-Jensen slants showed equivalent recovery rates; however, the time for detection was approximately 2 weeks shorter for the MGIT system.

MYCOBACTERIAL RECOVERY SYSTEMS

Several years ago, the Centers for Disease Control and Prevention emphasized the need for rapid turnaround times, not only for acid-fast smear results and recovery of mycobacteria from clinical specimens but also for subsequent antimicrobial susceptibility testing. The guidelines include reporting of acid-fast smears within 24 h of receipt of the specimen and 14 days for the recovery of mycobacteria from clinical specimens. Consequently, the emphasis was directed toward automating culture systems. The following section briefly describes each of the methods, and figures show the details of each.

I. MB Redox
   Manufacturer: Biotest AG, Dreieich, Germany (or Biotest Diagnostics Corporation, Danville, N.J.)
   Description: The method consists of tubes that have Kirchner's medium containing colorless tetrazolium salt, which is reduced by the Mycobacterial Redox system to a formazan colored pink to red to violet. The formazan is accumulated on the cell surface in a granular form, and the growing microcolonies become visible as colored particles. Acid-fast bacilli are easily detected in the medium as pink-to-purple pinhead-sized particles. This system is not automated.

   Table 1 presents a summary of selected reports that evaluated the MB Redox system. The evaluations show its comparison to the MGIT, BACTEC 460 TB system, and solid media. The MB Redox system appears to be comparable to the previously mentioned systems; however, it not available within the United States.

II. MGIT
   Manufacturer: Becton Dickinson Microbiology Systems.
   Description: The system (Fig. 2) uses Middlebrook 7H9 broth (7 ml in plastic tubes) and an oxygen-quenching

TABLE 1 Evaluations of MB Redox culturing system[a]

System	No. of specimens tested	Recovery rate (%)	Recovery time (days)	Contamination rate (%)	Reference	Gold standard(s) for comparison	Conclusion(s)
MB Redox	135	80.5	12.0	NR[f]	28	BACTEC 460	Also compared to MGIT; both comparable to BACTEC 460
MGIT		63.6	17.4				
BACTEC 460		92.2	16.0				
MB Redox	486	72.38 (88.4[b])	6.9[c] 15.5[d]	3.8	53	LJ and 7H11 Middlebrook agar	Compared to MGIT
MGIT		81.3 (94.6[b])	7.2[c] 19.1[d]	3.8	53		
MB Redox	790	87	19.1	NR	46	BACTEC 460, LJ MGIT, and LJ	Also compared to MGIT
MGIT		83	18.2	1.9			
LJ[e]		76	25.9	3.6			
MB Redox	1,580	81	16.3	2.1	46	BACTEC 460 and LJ	MGIT, BACTEC 460, and MB Redox similar
BACTEC 460		84	13.2	2.0			
LJ		77	22.4	NR			

[a]Biotest, AG, or Biotest Diagnostics Corporation.
[b]Plus Löwenstein-Jensen and 7H11 Middlebrook agars.
[c]Smear positive.
[d]Smear negative.
[e]LJ, Löwenstein-Jensen agar.
[f]NR, not reported.

FIGURE 2   Becton Dickinson BACTEC 960 MGIT system.

fluorescence technology. Each automated instrument can accommodate 960 tubes. Tubes have silicon bottoms that are impregnated with an oxygen-sensitive fluorescent indicator. The large amount of oxygen initially present in the broth quenches fluorescence, but as rapid growth of mycobacteria or other mycoorganisms occurs, the oxygen becomes consumed and the indicator fluoresces brightly under a 365-nanometer UV lamp. The MGIT tubes are supplemented prior to growth with oleic acid-albumin dextrose to enhance the growth of mycobacteria and with an antimicrobial agent solution (PANTA) that reduces or inhibits the growth of bacterial contaminants.

Table 2 presents a summary of selected evaluations of the MGIT system. Most studies compared the MGIT to the BACTEC 460 TB system that was developed in the early 1980s and is considered to be the "gold standard." In addition, most investigators included solid culture media in the evaluation simply for traditional reasons. Overall, the MGIT system appeared equivalent to the BACTEC 460 TB system. Recovery rates were essentially the same; however, recovery times were slightly shorter with the BACTEC 460 TB system. A major advantage of the MGIT system is that a manual or automated (BACTEC 960) system is available. Other advantages of the MGIT system include the ease of detection of growth, the fact that there is no need for needles to be used for the inoculation of the

tubes, and the option for smaller laboratories to use a nonautomated system.

III. Septi-Chek System
Manufacturer: Becton Dickinson Microbiology Systems
Description: This system has been available since the early 1990s and consists of a bottle that contains 20 ml of culture media (7H9 broth) under an atmosphere of carbon dioxide. The interior has a paddle containing three different culture media, including Löwenstein-Jensen, Middlebrook 7H11, and chocolate agars, within a plastic tube. In general, bacterial contaminants will grow on a chocolate agar, and the mycobacteria will grow on the remaining two agars. The broth contains several antimicrobials to suppress or inhibit bacterial contamination. Previous evaluations have shown that the overall recovery rate for the Septi-Chek and BACTEC 460 TB systems were essentially equivalent. However, recovery times were much longer for the Septi-Chek system. Current evaluations comparing the Septi-Chek to the MGIT system (30) showed that the MGIT system had a slightly higher recovery rate and that recovery times for M. *tuberculosis* and M. *avium-M. intracellulare* were shorter.

IV. MB/BacT System
Manufacturer: Organon Teknika, Durham, N.C.
Description: The system consists of a bottle that contains a colorimetric sensor at its base. As microorganisms grow and produce carbon dioxide, the sensor

TABLE 2 Evaluations of MGIT[a]

System	No. of specimens tested	Recovery rate (%)	Recovery time (days)	Contamination rate (%)	Reference(s)	Gold standard(s) for comparison	Conclusions
MGIT BACTEC 460 LJ[i]	603	93 95 87	22 14 27	3–4 3–4 3–4	17	BACTEC 460 LJ	MGIT and BACTEC equivalent for recovery; MGIT somewhat slower than BACTEC
MGIT 7H11 LJ	2,832[b]	89.4 50.8 60.7	12.7[c] 13 22.8	NR[h]	33	LJ and 7H11 Middlebrook agars	MGIT superior to solid media
MGIT BACTEC 460 LJ	2,567		13.3 14.8 25.6	10 3.7 17	58	BACTEC 460 and LJ	
MB Redox MGIT LJ	790	87 83 76	19.1 18.2 25.9	NR 1.9 3.6	46, 50	BACTEC 460 and LJ MGIT and LJ	Also compared to MB Redox
MB Redox BACTEC 460 LJ	1,580	81 84 77	16.3 13.2 22.4	2.1 2.0	46, 50		Also compared to MB Redox; MGIT, BACTEC 460, and MB Redox similar
MGIT LJ	2,105	91 84	12[c] 8[d] 20[c] 8[d]	8.7 8.9	33	BACTEC 460 and LJ, 7H11, and 7H11S agars	MGIT system superior
MGIT BACTEC 460 LJ 7H11 agar	3,330	80 75 69	14.4[c] 15.2[c] 24.1[c]	8.1 4.9 NR	26	LJ	Systems comparable; MGIT preferable
MGIT BACTEC 460	1,500	86.7 93.3	9.7[c] 11.9[d] 9.7[c] 13.0[d]	2.0 13.8	44	BACTEC 460 and solid media	Systems comparable
MB Redox MGIT BACTEC 460	135	80.5 63.6 92.2	12.0 17.4 16.0	NR	28	BACTEC 460	Also compared to MGIT; both comparable to BACTEC 460
MB Redox MGIT	486	72.38 (88.4[g]) 81.3 (94.6[g])	6.9[e] 15.5[f] 7.2[e] 19.1[f]	3.8 3.8	53	LJ, 7H11 Middlebrook agar	Compared to MGIT
MB/BacT MGIT BACTEC 460	1,068	83.3 82.5 80	12.6 15.9 11.8	NR	2	BACTEC 460 and LJ	Compared to MB/BacT; both systems comparable

[a]Becton Dickinson, Franklin Lakes, N.J.
[b]And microscopic morphology.
[c]M. tuberculosis.
[d]Nontuberculous mycobacteria.
[e]Smear positive.
[f]Smear negative.
[g]Plus Löwenstein-Jensen and 7H11 Middlebrook agars.
[h]NR, not reported.
[i]LJ, Löwenstein-Jensen agar.

changes from dark green to yellow. The change is continuously monitored by a sensitive reflective monitor in the detection unit. A slight change in the sensor's color is immediately recognized and reported by the instrument.

Table 3 presents selected references that relate to studies comparing the MB/BactT to the MGIT or BACTEC 460C TB systems. Overall, the recovery rates for all systems were essentially comparable. However, two of the evaluations showed that a few isolates of M. tuberculosis were not detected by the MB/BacT system as opposed to the BACTEC 460C system. The time to detection for mycobacteria in the MB/BacT system is consistently 2 to 3 days longer than that seen in the BACTEC 460C TB system.

V. ESP Culture System II

Manufacturer: Trek Diagnostic Systems, West Lake, Ohio
Description: The ESP Culture System II is an automated method that was originally devised for blood cultures and subsequently was adapted for the recovery of mycobacteria in clinical specimens. The technology consists of continuous monitoring of pressure changes due to the consumption or production of gas resulting from active growth of microorganisms in the liquid medium. Metabolism, characterized by utilization of oxygen, makes the organisms detectable by a reduction of pressure in the headspace of the ESP Culture System II. The ESP Culture System II can accommodate both mycobacterial and blood cultures within the same instrument.

Table 4 presents two selected comparisons of the ESP Culture System II and the BACTEC 460C TB system. Overall, recovery rates and recovery times were equivalent for both systems.

VI. BACTEC 9000 MB

Manufacturer: Becton Dickinson Microbiology Systems
Description: The BACTEC 9000 MB is a continuously monitored, automated nonradiometric mycobacterial liquid culture system that employs Middlebrook 7H9 broth with an oxygen-sensitive fluorescence sensor to detect microbial growth. This is the same indicator used by the MGIT culture system. Currently, this system is not advocated by the manufacturer as a mycobacterial recovery system. Zanetti (64) et al. compared the BACTEC 9000 MB to the BACTEC 460C TB system in two separate studies. The results showed

TABLE 3 Evaluations of MB/BacT[a]

System	No. of specimens tested	Recovery rate (%)	Recovery time (days)	Contamination rate (%)	Reference	Gold standard(s) for comparison	Conclusions
MB/BacT MGIT BACTEC 460	1,068	83.3 82.5 80	12.6 15.9 11.8	NR[d]	2	BACTEC 460 and LJ[e]	Compared to MGIT; both systems comparable
MB/BacT BACTEC 460	3,700	73 91	17.2[b] 15.4[b]	6.6 3.2	47	BACTEC 460 and LJ	Sensitivity lower for MB/BacT; a few tuberculosis patients not detected
MB/BacT BACTEC 460	849	94.2 80.7	9.9[b] 7.3[c] 6.9[b] 4.9[c]	NR	36	LJ Gruft	MB/BacT recovery rates same for both
MB/BacT BACTEC 460	1,830	95.6 99.4	11.8[b] 17.1[c] 8.0[b] 12.7[c]	NR	11	BACTEC 460 and LJ	MGIT 960 alternative to BACTEC; recovery time for MB/BacT longer
MB/BacT BACTEC 460	488	89 91	13.7[b] 11.6[c] 11.9[b] 9.4[c]	4.1 7.0	4	BACTEC 460	MB/BACT 960 missed one M. tuberculosis isolate
MB/BacT BACTEC 460	1,078	86.3 91.8	17.5 14.3	4.3 2.9	48	BACTEC 460 and LJ	Systems comparable
MB/BacT BACTEC 460	600	94.3 90.2	14.2 11.7	5.0 1.8	38	LJ and BACTEC 460	Systems comparable

[a]Organon Teknika.
[b]M. tuberculosis.
[c]Nontuberculous mycobacteria.
[d]NR, not reported.
[e]LJ, Löwenstein-Jensen agar.

TABLE 4 Evaluations of ESP Culture System II[a]

System	No. of specimens tested	Recovery rate (%)	Recovery time (days)	Contamination rate (%)	Reference	Gold standard(s) comparison	Conclusions
ESP Culture System II	2,283	87	13.1	8.6	62	BACTEC 460 and 7H11 and 7H11S Middlebrook agars	Systems comparable
BACTEC 460		81	14.4	4.0			
ESP Culture System II	>2,500	79	18.0	7.9	59	BACTEC 460 and LJ[b]	Systems comparable
BACTEC 460		89	18.0	4.0			

[a]Trek Diagnostics.
[b]LJ, Löwenstein-Jensen agar.

that both systems had similar recovery rates and detection times. The BACTEC 9000 MB nonradiometric system proved to be as efficient as the BACTEC 460 TB system.

In general, the automated, continuously monitored systems that were designed for mycobacterial detection are being used by most major laboratories. The advantages of such systems include (i) they are less labor-intensive, (ii) data management systems are available to use, (iii) there is no need for needles to be used for inoculation, (iv) no radioactive isotopes are used, and (v) the problem of cross-contamination as seen with the BACTEC 460C system is eliminated.

MOLECULAR METHODS FOR THE DIRECT DETECTION OF MYCOBACTERIA IN CLINICAL SPECIMENS

The future of mycobacteriology is headed toward the development of molecular method-based tests that not only will detect and identify mycobacteria in clinical specimens but also will detect point mutations responsible for antimicrobial drug resistance. Currently, a few commercially available methods have the capacity to detect the presence of M. tuberculosis in acid-fast smear-positive or smear-negative specimens. The full clinical utility of these commercially available methods is still being realized, and several studies have shown that they are very useful in making a rapid diagnosis of tuberculosis. These methods are rapid, specific, and expensive. However, the cost is overshadowed by the clinical usefulness in making the diagnosis in a patient with tuberculosis. They may be placed in isolation earlier so that they may no longer transmit the organism to others. Currently, the FDA still requires that the culture be considered the gold standard, and it must be done in conjunction with the performance of each molecular method-based test.

I. Amplified *Mycobacterium tuberculosis* Direct (MTD2) Test
Manufacturer: Gen-Probe, Inc.
Description: The MTD2 test is an isothermal transcription-mediated amplification system based on specific mycobacterial rRNA targets using DNA intermediates. The RNA amplicons produced are identified by hybridization protection assay using an acridinium

ester-labeled M. tuberculosis complex-specific DNA probe. The Amplified MTD2 test is the only commercially available method approved by the FDA for use with both acid-fast smear-positive and smear-negative specimens.

Table 5 presents selected studies for the MTD2 test compared to either another commercially available system or culture smear and clinical diagnosis. The overall consensus was that the MTD2 test is very useful for the rapid diagnosis of tuberculosis in patients having both acid-fast smear-positive and smear-negative specimens. The test is rapid, specific, and cost-effective when considering the potential implications of an undiagnosed case of tuberculosis.

II. AMPLICOR *Mycobacterium tuberculosis* (MTB) Test
Manufacturer: Roche Diagnostics Corp., Indianapolis, Ind.
Description: The AMPLICOR MTB test (Fig. 3) is a target-amplified in vitro diagnostic test for the qualitative detection of the M. tuberculosis complex. DNA is concentrated from the sediments prepared from patient specimens and is performed in four parts. These include specimen preparation, target amplification by PCR, hybridization of amplified products to oligonucleotide probes, and detection of the probe-bound amplified product. The COBAS portion of the assay is an automated version of the same product. The 16S ribosomal gene using biotinylated probes is amplified in probes that are hybridized to the amplicons, and the product is detected using an avidin-biotin conjugate. The format for this method is the PCR microwell hybridization method. At the present time, this assay has not been approved for use with acid-fast smear-negative specimens.

Overall, the COBAS AMPLICOR system appears to be useful for making a diagnosis of tuberculosis in acid-fast smear-positive specimens (Table 6). Sensitivity is much lower for acid-fast smear-negative specimens, and the method has not been approved by the FDA for use with the latter. There is a tendency toward lower detection rates with nonrespiratory specimens compared to respiratory tract specimens. The COBAS AMPLICOR system is satisfactory for making a primary diagnosis of tuberculosis; however, it must always be supplemented with traditional culture and acid-fast smear results.

TABLE 5 Evaluations of MTD2 Assay[a]

Method	No. of samples	Sensitivity (%)	Specificity (%)	Reference	Gold standard(s) for comparison	Comments
TMA[a] PCR[b]	486	85.7 94.2	100 100	50	Direct smear, culture, and clinical correlation	Compared to COBAS AMPLICOR; both systems comparable; cultures recommended for confirmation
TMA	823	100	99.6	16	BACTEC 460 TB, clinical correlation	Sensitive, rapid, and cost-effective
TMA	262	94.1	100	8	ESP Culture System II	
TMA	338	59 (low[f]) 100 (moderate[f]) 100 (high[f])	97 100 100	15	Clinical correlation	Clinical assessment strengthens use
TMA	995	90.9	99.1	11	Smear, BACTEC 460 TB, 7H11 Middlebrook agar, clinical correlation	83.3% positive in smear-negative specimens
TMA LCx[c]	457 183[e] 373[d]	78.6[a,d,e] 53.6[c,d] 75.7[c,d,e] 53.6	99.3 99.3 92.8 99.3	45	Culture, clinical correlation	Also compared to LCx; MTD2 was more sensitive
TMA	682 272[d] 410[e]	 94.7 86.8	 100 100	22	Clinical correlation	Sensitive and specific for M. tuberculosis
PCR LCx TMA	230	96.1 100 98.6	100 99.3 99.4	60	Direct smear, BACTEC 460	Compared to LCx and TMA; all three comparable
PCR TMA LCx	63	75[d] 50[e] 65.2[d] 66.7[e] 79.2[d] 60.0[d]	100 100 100 75 100 100	10	Direct smear Culture (LJ[g])	Both respiratory and nonrespiratory specimens used; sensitivity lower in nonrespiratory specimens; interpret data with caution

[a]MTD2, Gen-Probe, Inc.
[b]AMPLICOR MTB, Roche Diagnostics.
[c]LCx MTB, Abbott Laboratories.
[d]Respiratory specimens.
[e]Nonrespiratory specimens.
[f]Clinical suspicion.
[g]LJ, Löwenstein-Jensen agar.

III. LCx MTB Assay, Abbott LCx Probe System
Manufacturer: Abbott Laboratories, Abbott Park, Ill.
Description: The ligase chain reaction is a DNA amplification method similar to PCR; however, four primers are used with an enzyme called ligase to ligate two segments of DNA. In this assay, the probe molecule is amplified, not the target. Basically, probes are added to the sample containing the target sequence. The mixture is heated and cooled to allow the probes to bind to the target DNA. DNA ligase links the adjoining probes. Detection is performed by microparticle enzyme immunoassay using the LCx instrument (Fig. 4). This method is not approved by the FDA for use in the United States. The target is PAB.347, a single-copy chromosomal gene encoding protein antigen B (PAB).

Overall, the ligase chain reaction used for detecting M. tuberculosis in clinical specimens appears to have promise, particularly in detecting M. tuberculosis in nonrespiratory tract specimens (Table 7). Yet, like all of the other molecular assays, this method requires that culture and acid-fast smear also be performed before the final diagnosis of tuberculosis can be made.

IV. BD ProbeTec ET System
Manufacturer: Becton Dickinson Biosciences Microbiology Products

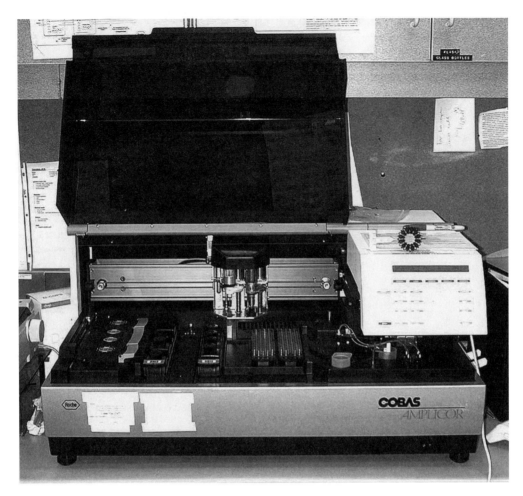

FIGURE 3 Roche COBAS AMPLICOR.

Description: The BD ProbeTec ET system (Fig. 5) utilizes strand displacement and fluorescent energy transfer and amplifies and detects in a closed homogenous format. The system uses the isothermal amplification technology based on restriction enzyme and a polymerase. Target sequences of IS6110 and 16S rRNA gene are simultaneously amplified so that the identification of M. *tuberculosis* and other mycobacteria may be obtained. Detection occurs by a chemiluminescence microwell assay that employs the simultaneous hybridization and capture of strand displacement products within a biotinylated capture probe and alkaline phosphatase detector probe (Table 8).

To date, a limited number of studies have been performed using the BD ProbeTec ET system for the rapid detection of mycobacteria in clinical specimens. However, the existing studies have shown that the system certainly offers great promise and is useful for mycobacterial detection. Further, the system is automated, and this is an advantage in the large laboratory setting.

IDENTIFICATION METHODS

Traditional identification of clinically important mycobacteria has included macroscopic morphologic features and biochemical testing for the production of metabolic byproducts or substrate utilization and/or degradation. Time

has brought significant change to those laboratories fortunate enough to have the finances to invest in new technology. These laboratories have utilized high-performance liquid chromatography (HPLC), nucleic acid probes, nucleic acid sequencing, and the line probe assay.

Molecular technology has made a dramatic impact on shortening the turnaround time for the identification of certain species of mycobacteria. It seems apparent that molecular analysis will further simplify and expedite our ability to detect and identify mycobacteria; perhaps use of the conventional culture will disappear. However, only those laboratories with resources will be able to utilize them fully until testing becomes reasonably priced. The technology, most likely, will be used primarily by reference laboratories; however, it will be available, though, to those offering testing in smaller laboratories.

Biochemical Testing

I. BD Biosciences
   Manufacturer: Becton Dickinson Biosciences Microbiology Products
   Description: Media and reagents for the identification of mycobacteria include the following:
   A. Aryl sulfate broth (product numbers 4395654 and 4397156)
   B. Chloride for salt tolerance (product number 4321896)

TABLE 6 Evaluations of Roche AMPLICOR MTB test[a]

Method	No. of samples	Sensitivity (%)	Specificity (%)	Reference	Gold standard(s) for comparison	Comments
PCR[a]	1,681	66.3[d] 96.4[e] 98.5[f]	99.7 75.0 98.5	20	Culture	Comparison included culture, in-house PCR, and AMPLICOR
PCR LCx[b] TMA[c]	230	96.1 100 98.6	100 99.3 99.4	60	Direct smear, BACTEC 460	Compared to LCx and TMA; all three comparable
PCR	428	50[e] 93[g]	89 100 80	18	Direct smear Culture Clinical correlation	Compared to in house IS6110; nucleic acid amplification tests less sensitive than culture
PCR TMA	422	97.8 100	98.9 99.3	31	Septi-Chek	Compared to MTD2; both systems comparable
PCR	784	76.7	97.7	19	Direct smear Culture Clinical correlation	Must be used with smear and culture results
PCR	956	97.6[e] 40.0[f] 78.7[d]	100 99.5 99.3	6	Culture Clinical correlation	Useful for smear-positive specimens
PCR	7,194	82.0[d] 91.4[e] 60.9[f]	96.1	5	Culture Clinical correlation	More sensitive in smear-positive specimens
PCR	2,073	86[d] 94.5[e] 74[f]	98	14	Direct smear Culture (solid and liquid media)	95% of patients with tuberculosis diagnosed with AMPLICOR
PCR MTD2[c] LCx	63	75[h] 50[i] 65.2[h] 66.7[i] 79.2[h] 60.0[i]	100 100 100 75 100 100	10	Direct smear Culture (LJ[j])	Both respiratory and nonrespiratory specimens used; sensitivity less in nonrespiratory specimens; interpret data with caution
MTD2 PCR	486	85.7 94.2	100 100	50 50	Direct smear, culture, and clinical correlation	Compared to COBAS AMPLICOR; both systems comparable; cultures recommended for confirmation

[a]AMPLICOR MTB, Roche Diagnostics.
[b]LCx, a ligase chain reaction.
[c]MTD2, Gen-Probe.
[d]Overall.
[e]Smear positive.
[f]Smear negative.
[g]Culture positive.
[h]Respiratory specimens.
[i]Nonrespiratory specimens.
[j]LJ, Löwenstein-Jensen agar.

C. Löwenstein-Jensen medium deeps for semiquantitative catalase (product number 4321257)
D. Löwenstein-Jensen medium with 5% sodium
E. Taxo TB Niacin test strips (product number 4331741)
F. Taxo Urea for detection of urease production (product number 4331738)
G. Taxo X Factor strips for detection of *Mycobacterium haemophilum* (product numbers 4331730, 4331107, and 4331106)

II. Remel
Manufacturer: Remel
Description: Media and reagents for the identification of mycobacteria include the following:
A. Löwenstein-Jensen medium with 5% NaCl for NaCl tolerance (product number 08-518)
B. Middlebrook 7H9 broth with Poly 80 for tellurite reduction (product number 09-556)
C. Niacin reagent strip to detect niacin production (product numbers 21-133, 21090, and 77-31820)

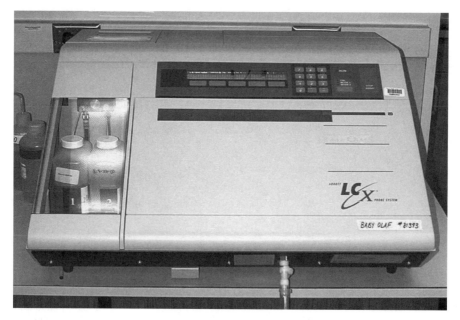

FIGURE 4 Abbott LCx system.

TABLE 7 Evaluations of the LCx MTB assay[a]

Method	No. of samples	Sensitivity (%)	Specificity (%)	Reference	Conclusions	Gold standard(s) for comparison	Comments
LCx[a]	457	75.7[b] 53.6[c]	99.3[b] 99.3[c]	45	Assay satisfactory for detecting M. tuberculosis in clinical specimens	Culture, clinical correlation	Positive rate best for respiratory specimens
LCx	205	90.4	98.4	42	Assay useful for detecting M. tuberculosis in nonrespiratory specimens	Septi-Chek AFB, LJ[f]	
LCx	235	76.5	95.8	3	LCx should be used as a supplement to culture	Direct smear and cultures	
PCR[d] LCx TMA[e]	230	96.1 100 98.6	100 99.3 99.4	60	Compared to PCR and TMA; all three comparable	Direct smear, BACTEC 460	
PCR TMA LCx	63	75[b] 50[c] 65.2[b] 66.7[c] 79.2[b] 60[c]	100 100 100 75 100 100	10	Both respiratory and nonrespiratory specimens used; sensitivity lower in nonrespiratory specimens; interpret data with caution; compared to PCR and TMA	Direct smear Culture (LJ)	

[a]LCx MTB, Abbott Laboratories.
[b]Respiratory specimens.
[c]Nonrespiratory specimens.
[d]AMPLICOR MTB, Roche Diagnostics.
[e]MTD2, Gen-Probe.
[f]LJ, Löwenstein-Jensen agar.

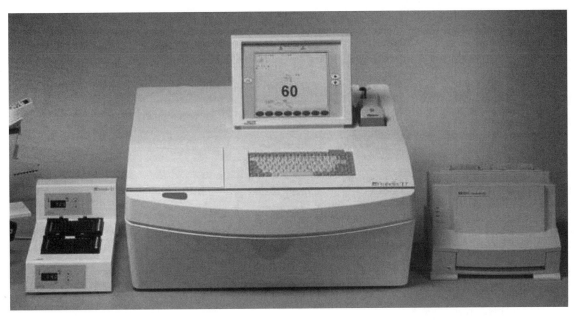

FIGURE 5 Becton Dickinson ProbeTec ET system.

D. Nitrate strips for detection of nitrate reduction (product number 77-31830)
E. Nitrate substrate broth for nitrate utilization (product number 61548-A)
F. Pyrazinamidase agar for pyrazinamide determination (product numbers 07136 and 07138)
G. TCH agar for TCH susceptibility for identification of *Mycobacterium bovis* (product numbers 08-812 and 08-816)
H. Tween 80 hydrolysis substrate concentrate (product number 21-276)
I. Urea broth for acid-fast bacilli (product number 65222)

III. Nucleic Acid Probes (Accuprobe Culture Identification Tests)

Manufacturer: Gen-Probe, Inc.

Description: Nucleic acid hybridization tests are based on the complementary alignment and binding of nucleic acid strands to form stable double-stranded complexes.

The Accuprobe System uses a single-stranded DNA probe with a chemiluminescent acridinium label as the complement of the rRNA of the targent organism. Lysates of an organism are prepared by sonication with glass beads and heat treatment. The rRNA is released from the organism, and the labeled DNA subsequently binds with the target organism rRNA to form a stable labeled DNA-RNA hybrid, the presence of which can be detected in the Gen-Probe luminometer. The selection reagent, provided in the kit, provides for differentiation of hybridized and nonhybridized probe during readings taken by the luminometer. Positive reactions are based on cutoff values of relative light units derived from the chemiluminescence. Kits are available for *M. tuberculosis* complex, *M. avium*, *M. intracellulare*, *M. avium* complex, *M. kansasii*, and *M. gordonae*.

One of the single most important advancements in clinical mycobacteriology was the development of nucleic acid probes for the identification of the aforementioned species. Combined with rapid detection systems, the Accuprobes have provided clinical laboratories with the capability of identifying isolates of *M. tuberculosis* on the same day they are recovered. Accuprobes have become the standard for identification of *M. tuberculosis*, *M. avium* complex, *M. gordonae*, and *M. kansasii*.

Diagnostic algorithms have been developed so that Accuprobes may be used in a more cost-effective

TABLE 8 Evaluations of BD ProbeTec ET System strand displacement assay

No. of samples	Sensitivity (%)	Specificity (%)	Reference	Gold standard for comparison	Conclusion
600	93.8	99.8	9	BACTEC 460 and 7H11 Middlebrook	Performance is good for smear-positive specimens
530	89.5[a] 82.8[b]	99.8	32	Septi-Chek, clinical correlation	System is useful for detection of mycobacteria in clinical specimens

[a]*M. tuberculosis* complex.
[b]*M. avium-M. intracellulare* complex.

manner (25, 39, 41). Studies have shown that Accuprobes for the identification of mycobacteria have a sensitivity and specificity of 98.9 to 100 and 100; 98.4 and 95.2 to 100; 100 and 100; and 100 and 98.8% for *M. tuberculosis*, *M. avium* complex, *M. gordonae*, and *M. kansasii*, respectively (25, 37, 57) Further, they may be used for identification of organisms growing on solid media or in BACTEC 12B bottles, MGIT tubes, or ESP Culture System II Myco bottles (35).

HPLC

HPLC has been exclusively used by the Centers for Disease Control and Prevention since 1984 for the identification of mycobacteria. When tuberculosis was noted to increase in the United States, state laboratories were granted funding to adopt the technology. Others have used HPLC as a means of determining the species of mycobacteria on the basis of their mycolic acid profiles (12, 24, 56). Further, HPLC may be use to identify mycobacteria in BACTEC 12B bottles (13).

I. Nucleic Acid Sequencing
   Manufacturer: Applied Biosystems, Foster City, Calif.
   Description: The Applied Biosystems MicroSeq identification system utilizes DNA extracted from mycobacteria. A 500-bp 16S ribosomal DNA fragment is amplified from the 5′ end of the gene. The components of the PCR mixture are supplied in the kit. Amplified products are purified, and forward and reverse sequencing reactions are performed for each amplified product. Sequencing reaction mixtures are purified, and sequence analysis is performed on a sequencing analyzer. The sequence data (electropherograms) are analyzed with the MicroSeq computer program. The program allows the assembly of the forward and reverse sequences into a consensus sequence, editing of the consensus sequence to resolve discrepancies between the two strands by evaluation of the electropherograms, and comparison to the MicroSeq database containing 16S ribosomal DNA sequences.

   A single study (43) showed that the MicroSeq program correctly identified 82% of isolates tested. This technology promises to be the most accurate identification method and will probably be the standard for mycobacterial identification in the future.

II. Line Probe Assay
   Manufacturer: InnoGenetics, Temse, Belgium
   Description: The LiPA MYCOBACTERIA assay allows for the simultaneous detection and identification of *Mycobacterium* species. The LiPA method involves reverse hybridization of a biotinylated mycobacterial fragment, a 16 to 235 rRNA spacer region, to oligonucleotide probes immobilized as parallel lines on a membrane strip. Visualization of bound DNA occurs using a streptavidin alkaline phosphatase conjugate and a nitroblue tetrazolium–5-bromo-4-chloro-3-indolylphosphate chromogen. The time from processing until a result is generated is up to 6 h. Species of mycobacteria that may be identified by the LiPA method include *M. tuberculosis* complex, *M. kansasii*, *M. xenopi*, *M. gordonae*, *M. avium*, *M. intracellulare*, *M. scrofulaceum*, *M. avium-M. intracellulare-M. scrofulaceum* complex, and *M. chelonae*.

   A single study by Miller et al. (40) evaluated the use of the LiPA method for identification of mycobacterial

isolates using culture fluid from positive BACTEC 12B bottles. The assay detected 60 isolates from 59 patients. Complete agreement was shown between the LiPA MYCOBACTERIA assay and conventional laboratory test results used for identification. The assay was completed within 3 h. This method may be a suitable alternative to nucleic acid sequencing; however, a much smaller number of species can be identified with the LiPA method.

Miscellaneous Methods for the Identification of *M. tuberculosis* Complex

I. Novocastra Laboratories
   Manufacturer: Novocastra Laboratories Limited, Newcastle upon Tyne, United Kingdom
   Description: The following described products are for research use only.
   A. *M. tuberculosis* identification and typing primer sets
      These are reagents that are available to type strains of *M. tuberculosis* during outbreaks, for identification of the source, and for the development of procedures to prevent the transmission of infection. Contact tracing is also facilitated. Within-laboratory cross-contamination can also be investigated. This assay is based on the polymorphism of the IS61210 insertion sequence, but in contrast to the conventional restriction fragment length polymorphism method, the heminested inverse PCR method is relatively quick and more discriminatory.
   B. *M. bovis* BCG primer set
      These reagents are available to be used in PCR to differentiate *M. bovis* BCG from other mycobacteria. In this PCR, three primers permit the detection of a characteristic deletion in the RD1 region of a genome of BCG, which differentiates it from all other members of the *M. tuberculosis* complex. Testing should be restricted to organisms already shown to belong to the *M. tuberculosis* complex.
   C. *M. tuberculosis* complex primer set
      These reagents allow the user to perform a PCR for the direct detection of the organism from sputum. Somewhat low-level infections will be missed, but will later be detected by culture. However, the PCR will identify 60 to 80% of the *M. tuberculosis* isolates detected by conventional microscopy. It is based on the detection of the IS6110 insertion sequence found in all organisms of the *M. tuberculosis* complex. These reagents also have an internal control present.
   D. *M. tuberculosis* antibodies
      This reagent was developed to confirm the mycobacterial etiology of tissue granulomas. These antibodies have to be made to the 36-kDa antigen of the *M. tuberculosis* complex. Presently, there are three antibodies available; two react with several mycobacterial species, including members of the *M. tuberculosis* complex. One antibody is thought to be specific for *M. tuberculosis* complex. It is recommended that all three tests be performed using the three antibodies for the greatest specificity and sensitivity. Further, it is recommended that these antibodies be used with traditional Ziehl-Neelsen staining to be able to locate the acid-fast bacilli to be tested.

ANTIMICROBIAL SUSCEPTIBILITY TESTING

I. Traditional Assays

Traditional antimicrobial susceptibility testing for mycobacteria generally refers to the 1% proportion method established decades ago. The basis is that at least 1% of the population of cells of a susceptible isolate of M. *tuberculosis* will exhibit innate resistance to the primary antituberculous drugs used today, including isoniazid, ethambutol, streptomycin, rifampin, and pyrazinamide. The breakpoint of ≥1% is used to define resistance of an isolate cultured on Middlebrook 7H10 or 7H11 agar.

II. Conventional Media

Manufacturer: Remel

Description:

A. TB Susceptibility Quad 1-IWP: isoniazid, 0.2 mg/ml; isoniazid, 1.0 mg/ml; and ethambutol, 7.5 mg/ml (product number 03-501)

B. TB Susceptibility Quad 2-IWP: streptomycin, 2 mg/ml; streptomycin, 10 mg/ml; rifampin, 1 mg/ml; and rifampin, 5 mg/ml (Product number 03-502)

III. Antimicrobial Disks for Use in Culture Media and Antimicrobial Agents for Use with BACTEC 460C TB System.

Manufacturer: Becton Dickinson Biosciences Microbiology Products

Description: Sensi-Disk Antimycobacterial Disks for use in culture media for susceptibility

A. Antimicrobial agents

1. Ethambutol, 25 mg/ml (product number 4331575)
2. Ethambutol, 50 mg/ml (product number 4331576)
3. Ethionamide, 25 mg/ml (product number 433157)
4. Isoniazid, 1 mg/ml (product number 4331571)
5. Isoniazid, 5.0 mg/ml (product number 4331572)
6. p-Aminosalicylic acid, 10 mg/ml (product number 31573)
7. p-Aminosalicylic acid, 50 mg/ml (product number 31574)
8. Rifampin, 25 mg/ml (product number 4331578)
9. Streptomycin, 50 mg/ml (product number 4331570)

B. BACTEC 460 TB system differentiation kit and reagents

1. Anti-TB drugs, including streptomycin, isoniazid, rifampin, and ethambutol (product number 4402102)
2. Diluting fluid for antimicrobial susceptibility testing (product number 4402104)
3. Isoniazid for susceptibility testing (product number 4402146)

IV. ESP Myco Susceptibility Kit

Manufacturer: Trek Diagnostic Systems

Description: The ESP Myco Susceptibility Kit includes two bottles each of rifampin and isoniazid and three bottles of ethambutol in lyophilized form. The complete test system also includes a liquid culture medium (ESP Myco Broth) and a growth supplement (Myco GS) supplied in specific lots qualified for use with each kit.

The ESP Culture System II contains the monitors and culture bottles inoculated with isolates of M. *tuberculosis* recovered from clinical specimens. The ESP Myco Susceptibility Kit provides rapid results that allow initiation of the appropriate drug therapy and institution of measures to prevent the spread of this infectious disease. Laboratory safety is affected by the ESP's closed system design and the elimination of the need for radioactive material disposal. To date, this is the only antimycobacterial susceptibility testing method approved by the FDA for use with continuous-monitoring systems.

V. E-Test

Manufacturer: AB Bio Disk NA, Inc., Piscataway, N.J.

Description: E-Test is a quantitative technique for the determination of antimicrobial susceptibility of bacteria, including mycobacteria. The system is comprised of a predefined antibiotic gradient, which is used to determine the MIC (in micrograms per milliliter) of individual antimicrobials for mycobacteria when tested in agar media using extended incubation. A continuous and exponential gradient of antibiotic concentrations is created directly underneath the carrier. After incubation, whereby bacterial growth becomes visible, a symmetrical inhibition ellipse centered along the carrier is seen. The zone edge intersects the strip at the MIC (given in micrograms per milliliter).

The following antimycobacterial drugs are available in the E-Test format:

A. Cefoxitin (0.016 to 256 μg/ml), product number 51000658
B. Ciprofloxacin (0.002 to 32 μg/ml), product number 51000868
C. Imipenem (0.002 to 32 μg/ml), product number 51001368
D. Levofloxacin (0.002 to 32 μg/ml), product number 51002748
E. Ofloxacin (0.002 to 32 μg/ml), product number 51001968
F. Clarithromycin (0.016 to 256 μg/ml), product number 51000878
G. Doxycycline (0.016 to 256 μg/ml), product number 51000978
H. Fusidic acid (0.016 to 256 μg/ml), product number 51001158
I. Streptomycin (0.064 to 1024 μg/ml), product number 51002688
J. Streptomycin (0.016 to 256 μg/ml), product number 51002188
K. Ethambutol (0.016 to 256 μg/ml), product number 51002778
L. Ethionamide (0.016 to 256 μg/ml), product number 51002758
M. Isoniazid (0.016 to 256 μg/ml), product number 51002798

A thorough evaluation of the E-Test for antimicrobial susceptibility testing of mycobacteria has not been done. However, several studies have shown that there is good correlation with the standard 1% proportion method. Fabry et al. (21) showed that the E-Test was suitable for testing slowly growing mycobacteria other than M. *tuberculosis* and had the potential to replace the most laborious dilution methods, particularly when testing M. *kansasii*. Subsequently, the same authors (21) evaluated the E-Test for antimicrobial susceptibility testing of M. *avium*-M. *intracellulare* and found good correlation between the E-Test and 1% proportion methods using amikacin, streptomycin, fusidic acid, rifampin, ciprofloxacin, ofloxacin, and fleroxacin. However, clarithromycin yielded discrepant results. Wanger and Mills (61) compared the E-Test to the BACTEC broth method for performing antimicrobial susceptibility testing of M. *tuberculosis* to the primary antituberculous agents. E-Test results were easily read within 5 to 10 days after inoculation and compared favorably with those of the BACTEC broth method. The authors concluded that the E-Test appeared to be a suitable alternative for the susceptibility testing of M. tuberculosis isolates to the primary antituberculous drugs. The E-Test has been most extensively evaluated using the rapidly growing mycobacteria, including the following species: M. fortuitum, M. abscessus, and M. chelonae. Hoffner et al. (29) compared the E-Test to a reference agar dilution method and showed good agreement between MICs obtained by the E-Test compared to the referenced method for 90% of all tests performed. The conclusion of their study suggests that the E-Test is well suited for studies of drug resistance using the rapidly growing mycobacteria. Most recently, Woods et al. (63), in a multicenter study with four laboratories, tested the inter- and intralaboratory reproducibility of the E-Test using rapidly growing mycobacteria compared to a broth microdilution method for susceptibility testing of rapidly growing mycobacteria. In general, the E-Test did not perform as well as the broth microdilution method for the susceptibility testing of rapidly growing mycobacteria. It appears that further studies are necessary to validate the use of the E-Test for antimicrobial susceptibility testing of the mycobacteria. To date, the method is not currently accepted as being one that would be used in the routine laboratory setting.

Molecular Methods for Determining Antimicrobial Resistance of Mycobacteria

I. INNO-LiPA RIF.TB Amplification Kits
   Manufacturer: InnoGenetics
   Description: These PCR-based hybridization assays were developed to detect and identify M. tuberculosis complex isolates and simultaneously provide information relating to the susceptibility resistance of the organism to rifampin. The basic principle is that each nucleotide change should impede the hybridization of the target and the corresponding wild-type probes.
   A. RIF.TB Amplification Kit
      This method utilizes PCR to amplify the rpoB gene found in M. tuberculosis isolates. Biotinylated outer primers complimentary to the conserved regions flanking the target sequence are used to generate a biotinylated exact copy of the template sequence, thus yielding numerous copies of an amplified biotinylated target sequence.
   B. INNO-LiPA RIF.TB Kit
      This method is based on reverse hybridization. Amplified biotinylated DNA is hybridized with specific oligonucleotide probes immobilized as parallel lines on a membrane-based strip. After hybridization, strepavidin labeled with alkaline phosphatase is added and binds to any biotinylated hybrid previously formed. Incubation with chromogen results in a purple-brown precipitate. The probes used for testing were designed to hybridize with the most commonly observed point mutation sites that code for resistance to rifampin.
      A limited number of studies (23, 49) have shown that the INNO-LiPA RIF.TB kit yielded concordant results compared to in vitro susceptibility testing of isolates of M. tuberculosis for rifampin resistance. Further, Gamboa et al. (23) also showed that the INNO-

LiPA RIF.TB kit was a useful tool for the detection of M. tuberculosis directly in clinical specimens.

CONCLUSION

As previously mentioned, the future of clinical mycobacteriology appears to be headed toward molecular detection, identification, and drug resistance detection using molecular methods. This approach is just beginning and will require extensive evaluation before any of the methods may be placed into routine use within the clinical laboratory. Currently the gold standard still consists of traditional culture and antimicrobial susceptibility testing procedures. However, we feel that molecular methods will allow laboratories to detect, identify, and determine drug resistance of mycobacteria without culturing the organism. The future of clinical mycobacteriology appears bright, and all of us anxiously await these developments and subsequently the use of these methods in a routine clinical laboratory setting in the future. However, before this will be possible, clinical correlative studies must show that a strong in vitro-to-in vivo correlation exists.

REFERENCES

1. **Agy, M. B., C. K. Wallis, J. J. Plorde, L. C. Carlson, and M. B. Coyle.** 1989. Evaluation of four mycobacterial blood culture media: BACTEC 13A, Isolator/BACTEC 12B, Isolator/Middlebrook agar, and a biphasic medium. *Diagn. Microbiol. Infect. Dis.* **12:**303–308.
2. **Alcaide, F., M. A. Benítez, J. M. Escribà, and R. Martin.** 2000. Evaluation of the BACTEC MGIT 960 and the MB/BacT systems for recovery of mycobacteria from clinical specimens and for species identification by DNA AccuProbe. *J. Clin. Microbiol.* **38:**398–401.
3. **Alonso, P., A. Orduna, M. A. Bratos, A. San Miguel, and T. A. Rodriguez.** 1998. Clinical evaluation of a commercial ligase-based gene amplification method for detection of *Mycobacterium tuberculosis*. *Eur. J. Clin. Microbiol. Infect. Dis.* **17:**371–376.
4. **Benjamin, W. H., Jr., K. B. Waites, A. Beverly, L. Gibbs, M. Waller, S. Nix, S. A. Moser, and M. Willert.** 1998. Comparison of the MB/BacT system with a revised antibiotic supplement kit to the BACTEC 460 system for detection of mycobacteria in clinical specimens. *J. Clin. Microbiol.* **36:**3234–3238.
5. **Bennedsen, J., V. O. Thomsen, G. E. Pfyffer, G. Funke, K. Feldmann, A. Beneke, P. A. Jenkins, M. Hegginbothom, A. Fahr, M. Hengstler, G. Cleator, P. Klapper, and E. G. Wilkins.** 1996. Utility of PCR in diagnosing pulmonary tuberculosis. *J. Clin. Microbiol.* **34:**1407–1411.
6. **Bergmann, J. S., and G. L. Woods.** 1996. Clinical evaluation of the Roche AMPLICOR PCR *Mycobacterium tuberculosis* test for detection of M. tuberculosis in respiratory specimens. *J. Clin. Microbiol.* **34:**1083–1085.
7. **Bergmann, J. S., and G. L. Woods.** 1999. Enhanced Amplified Mycobacterium Tuberculosis Direct Test for detection of *Mycobacterium tuberculosis* complex in positive BACTEC 12B broth cultures of respiratory specimens. *J. Clin. Microbiol.* **37:**2099–2101.
8. **Bergmann, J. S., and G. L. Woods.** 1999. Enhanced *Mycobacterium tuberculosis* direct test for detection of M. tuberculosis complex in positive ESP II broth cultures of nonrespiratory specimens. *Diagn. Microbiol. Infect. Dis.* **35:**245–248.

9. Bergmann, J. S., W. E. Keating, and G. L. Woods. 2000. Clinical evaluation of the BDProbeTec ET system for rapid detection of Mycobacterium tuberculosis. J. Clin. Microbiol. 38:863–865.

10. Brown, T. J., E. G. Power, and G. L. French. 1999. Evaluation of three commercial detection systems for Mycobacterium tuberculosis where clinical diagnosis is difficult. J. Clin. Pathol. 52:193–197.

11. Brunello, F., F. Favari, and R. Fontana. 1999. Comparison of the MB/BacT and BACTEC 460 TB systems for recovery of mycobacteria from various clinical specimens. J. Clin. Microbiol. 37:1206–1209.

12. Butler, W. R., K. C. Jost, Jr., and J. O. Kilburn. 1991. Identification of mycobacteria by high-performance liquid chromatography. J. Clin. Microbiol. 29:2468–2472.

13. Cage, G. D. 1994. Direct identification of Mycobacterium species in BACTEC 7H12B medium by high-performance liquid chromatography. J. Clin. Microbiol. 32:521–524.

14. Carpentier, E., B. Drouillard, M. Dailloux, D. Moinard, E. Vallee, B. Dutilh, J. Maugein, E. Bergogne-Berezin, and B. Carbonnelle. 1995. Diagnosis of tuberculosis by Amplicor Mycobacterium tuberculosis test: a multicenter study. J. Clin. Microbiol. 33:3106–3110.

15. Catanzaro, A., S. Perry, J. E. Clarridge, S. Dunbar, S. Goodnight-White, P. A. LoBue, C. Peter, G. E. Pfyffer, M. F. Sierra, R. Weber, G. Woods, G. Mathews, V. Jonas, K. Smith, and P. Della-Latta. 2000. The role of clinical suspicion in evaluating a new diagnostic test for active tuberculosis: results of a multicenter prospective trial. JAMA 283:639–645.

16. Chedore, P., and F. B. Jamieson. 1999. Routine use of the Gen-Probe MTD2 amplification test for detection of Mycobacterium tuberculosis in clinical specimens in a large public health mycobacteriology laboratory. Diagn. Microbiol. Infect. Dis. 35:185–191.

17. Chew, W. K., R. M. Lasaitis, F. A. Schio, and G. L. Gilbert. 1998. Clinical evaluation of the Mycobacteria Growth Indicator Tube (MGIT) compared with radiometric (Bactec) and solid media for isolation of Mycobacterium species. J. Med. Microbiol. 47:821–827.

18. Dalovisio, J. R., S. Montenegro-James, S. A. Kemmerly, C. F. Genre, R. Chambers, D. Greer, G. A. Pankey, D. M. Failla, K. G. Haydel, L. Hutchinson, M. F. Lindley, B. M. Nunez, A. Praba, K. D. Eisenach, and E. S. Cooper. 1996. Comparison of the amplified Mycobacterium tuberculosis (MTB) direct test, Amplicor MTB PCR, and IS6110-PCR for detection of MTB in respiratory specimens. Clin. Infect. Dis. 23:1099–1106.

19. Devallois, A., E. Legrand, and N. Rastogi. 1996. Evaluation of Amplicor MTB test as adjunct to smears and culture for direct detection of Mycobacterium tuberculosis in the French Caribbean. J. Clin. Microbiol. 34:1065–1068.

20. Eing, B. R., A. Becker, A. Sohns, and R. Ringelmann. 1998. Comparison of Roche Cobas Amplicor Mycobacterium tuberculosis assay with in-house PCR and culture for detection of M. tuberculosis. J. Clin. Microbiol. 36:2023–2029.

21. Fabry, W., E. N. Schmid, and R. Ansorg. 1996. Comparison of the E test and a proportion dilution method for susceptibility testing of Mycobacterium avium complex. J. Med. Microbiol. 44:227–230.

22. Gamboa, F., G. Fernandez, E. Padilla, J. M. Manterola, J. Lonca, P. J. Cardona, L. Matas, and V. Ausina. 1998. Comparative evaluation of initial and new versions of the Gen-Probe Amplified Mycobacterium tuberculosis Direct Test for direct detection of Mycobacterium tuberculosis in respiratory and nonrespiratory specimens. J. Clin. Microbiol. 36:684–689.

23. Gamboa, F., P. J. Cardona, J. M. Manterola, J. Lonca, L. Matas, E. Padilla, J. R. Manzano, and V. Ausina. 1998. Evaluation of a commercial probe assay for detection of rifampin resistance in Mycobacterium tuberculosis directly from respiratory and nonrespiratory clinical samples. Eur. J. Clin. Microbiol. Infect. Dis. 17:189–192.

24. Glickman, S. E., J. O. Kilburn, W. R. Butler, and L. S. Ramos. 1994. Rapid identification of mycolic acid patterns of mycobacteria by high-performance liquid chromatography using pattern recognition software and a mycobacterium library. J. Clin. Microbiol. 32:740–745.

25. Gonzalez, R., and B. A. Hanna. 1987. Evaluation of Gen-Probe DNA hybridization systems for the identification of Mycobacterium tuberculosis and Mycobacterium avium-intracellulare. Diagn. Microbiol. Infect. Dis. 8:69–77.

26. Hanna, B. A., A. Ebrahimzadeh, L. B. Elliott, M. A. Morgan, S. M. Novak, S. Rusch-Gerdes, M. Acio, D. F. Dunbar, T. M. Holmes, C. H. Rexer, C. Savthyakumar, and A. M. Vannier. 1999. Multicenter evaluation of the BACTEC MGIT 960 system for recovery of mycobacteria. J. Clin. Microbiol. 37:748–752.

27. Hanna, B. A., S. B. Walters, S. J. Bonk, and L. J. Tick. 1995. Recovery of mycobacteria from blood in Mycobacteria Growth Indicator Tube and Lowenstein-Jensen slant after lysis-centrifugation. J. Clin. Microbiol. 33:3315–3316.

28. Heifets, L., T. Linder, T. Sanchez, D. Spencer, and J. Brennan. 2000. Two liquid medium systems, Mycobacteria Growth Indicator Tube and MB redox tube, for Mycobacterium tuberculosis isolation from sputum specimens. J. Clin. Microbiol. 38:1227–1230.

29. Hoffner, S. E., L. Klintz, B. Olsson-Liljequist, and A. Bolmstrom. 1994. Evaluation of Etest for rapid susceptibility testing of Mycobacterium chelonae and M. fortuitum. J. Clin. Microbiol. 32:1846–1849.

30. Ichiyama, S., Y. Iinuma, S. Yamori, Y. Hasegawa, K. Shimokata, and N. Nakashima. 1997. Mycobacterium Growth Indicator Tube testing in conjunction with the AccuProbe or the AMPLICOR-PCR assay for detecting and identifying mycobacteria from sputum samples. J. Clin. Microbiol. 35:2022–2025.

31. Ichiyama, S., Y. Iinuma, Y. Tawada, S. Yamori, Y. Hasegawa, K. Shimokata, and N. Nakashima. 1996. Evaluation of Gen-Probe Amplified Mycobacterium tuberculosis Direct Test and Roche PCR-microwell plate hybridization method (AMPLICOR MYCOBACTERIUM) for direct detection of mycobacteria. J. Clin. Microbiol. 34:130–133.

32. Ichiyama, S., Y. Ito, F. Sugiura, Y. Iinuma, S. Yamori, M. Shimojima, Y. Hasegawa, K. Shimokata, and N. Nakashima. 1997. Diagnostic value of the strand displacement amplification method compared to those of Roche Amplicor PCR and culture for detecting mycobacteria in sputum samples. J. Clin. Microbiol. 35:3082–3085.

33. Idigoras, P., X. Beristain, A. Iturzaeta, D. Vicente, and E. Perez-Trallero. 2000. Comparison of the automated nonradiometric Bactec MGIT 960 system with Lowenstein-Jensen, Coletsos, and Middlebrook 7H11 solid media for recovery of mycobacteria. Eur. J. Clin. Microbiol. Infect. Dis. 19:350–354.

34. Katila, M. L., P. Katila, and R. Erkinjuntti-Pekkanen. 2000. Accelerated detection and identification of mycobacteria with MGIT 960 and COBAS AMPLICOR systems. *J. Clin. Microbiol.* **38:**960–964.

35. Labombardi, V. J., L. Carter, and S. Massarella. 1997. Use of nucleic acid probes to identify mycobacteria directly from Difco ESP-Myco bottles. *J. Clin. Microbiol.* **35:**1002–1004.

36. Laverdiere, M., L. Poirier, K. Weiss, C. Beliveau, L. Bedard, and D. Desnoyers. 2000. Comparative evaluation of the MB/BacT and BACTEC 460 TB systems for the detection of mycobacteria from clinical specimens: clinical relevance of higher recovery rates from broth-based detection systems. *Diagn. Microbiol. Infect. Dis.* **36:**1–5.

37. Lebrun, L., F. Espinasse, J. D. Poveda, and V. Vincent-Levy-Frebault. 1992. Evaluation of nonradioactive DNA probes for identificaiton of mycobacteria. *J. Clin. Microbiol.* **30:**2476–2478.

38. Manterola, J. M., F. Gamboa, E. Padilla, J. Lonca, L. Matas, A. Hernandez, M. Gimenez, P. J. Cardona, B. Vinado, and V. Ausina. 1998. Comparison of a nonradiometric system with Bactec 12B and culture on egg-based media for recovery of mycobacteria from clinical specimens. *Eur. J. Clin. Microbiol. Infect. Dis.* **17:**773–777.

39. Metchock, B., and L. Diem. 1995. Algorithm for use of nucleic acid probes for identifying *Mycobacterium tuberculosis* from BACTEC 12B bottles. *J. Clin. Microbiol.* **33:**1934–1937.

40. Miller, N., S. Infante, and T. Cleary. 2000. Evaluation of the LiPA MYCOBACTERIA assay for identification of mycobacterial species from BACTEC 12B bottles. *J. Clin. Microbiol.* **38:**1915–1919.

41. Nelson, S. M., and C. P. Cartwright. 1998. Comparison of algorithms for selective use of nucleic-acid probes for identification of *Mycobacterium tuberculosis* from BACTEC 12B bottles. *Diagn. Microbiol. Infect. Dis.* **31:**537–541.

42. Palacios, J. J., J. Ferro, P. N. Ruiz, S. G. Roces, H. Villar, J. Rodriguez, and P. Prendes. 1998. Comparison of the ligase chain reaction with solid and liquid culture media for routine detection of *Mycobacterium tuberculosis* in nonrespiratory specimens. *Eur. J. Clin. Microbiol. Infect. Dis.* **17:**767–772.

43. Patel, J. B., D. G. Leonard, X. Pan, J. M. Musser, R. E. Berman, and I. Nachamkin. 2000. Sequence-based identification of *Mycobacterium* species using the MicroSeq 500 16S rDNA bacterial identification system. *J. Clin. Microbiol.* **38:**246–251.

44. Pfyffer, G. E., H. M. Welscher, P. Kissling, C. Cieslak, M. J. Casal, J. Gutierrez, and S. Rusch-Gerdes. 1997. Comparison of the Mycobacteria Growth Indicator Tube (MGIT) with radiometric and solid culture for recovery of acid-fast bacilli. *J. Clin. Microbiol.* **35:**364–368.

45. Piersimoni, C., A. Callegaro, C. Scarparo, V. Penati, D. Nista, S. Bornigia, C. Lacchini, M. Scagnelli, G. Santini, and G. De Sio. 1998. Comparative evaluation of the new Gen-Probe *Mycobacterium tuberculosis* Amplified Direct test and the semiautomated Abbott LCx *Mycobacterium tuberculosis* assay for direct detection of *Mycobacterium tuberculosis* complex in respiratory and extrapulmonary specimens. *J. Clin. Microbiol.* **36:**3601–3604.

46. Piersimoni, C., C. Scarparo, P. Cichero, M. De Pezzo, I. Covelli, G. Gesu, D. Nista, M. Scagnelli, and F. Mandler. 1999. Multicenter evaluation of the MB-Redox medium compared with radiometric BACTEC system, Mycobacteria Growth Indicator Tube (MGIT), and Lowenstein-Jensen medium for detection and recovery of acid-fast bacilli. *Diagn. Microbiol. Infect. Dis.* **34:**293–299.

47. Roggenkamp, A., M. W. Hornef, A. Masch, B. Aigner, I. B. Autenrieth, and J. Heesemann. 1999. Comparison of MB/BacT and BACTEC 460 TB systems for recovery of mycobacteria in a routine diagnostic laboratory. *J. Clin. Microbiol.* **37:**3711–3712.

48. Rohner, P., B. Ninet, C. Metral, S. Emler, and R. Auckenthaler. 1997. Evaluation of the MB/BacT system and comparison to the BACTEC 460 system and solid media for isolation of mycobacteria from clinical specimens. *J. Clin. Microbiol.* **35:**3127–3131.

49. Rossau, R., H. Traore, H. De Beenhouwer, W. Mijs, G. Jannes, P. de Rijk, and F. Portaels. 1997. Evaluation of the INNO-LiPA Rif. TB assay, a reverse hybridization assay for the simultaneous detection of *Mycobacterium tuberculosis* complex and its resistance to rifampin. *Antimicrob. Agents Chemother.* **41:**2093–2098.

50. Scarparo, C., P. Piccoli, A. Rigon, G. Ruggiero, M. Scagnelli, and C. Piersimoni. 2000. Comparison of enhanced *Mycobacterium tuberculosis* amplified direct test with COBAS AMPLICOR *Mycobacterium tuberculosis* assay for direct detection of *Mycobacterium tuberculosis* complex in respiratory and extrapulmonary specimens. *J. Clin. Microbiol.* **38:**1559–1562.

51. Shanson, D. C., and M. S. Dryden. 1988. Comparison of methods for isolating *Mycobacterium avium-intracellulare* from blood of patients with AIDS. *J. Clin. Pathol.* **41:**687–690.

52. Sharp, S. E., C. A. Suarez, M. Lemes, and R. J. Poppiti, Jr. 1996. Evaluation of the Mycobacteria Growth Indicator Tube compared to Septi-Chek AFB for the detection of mycobacteria. *Diagn. Microbiol. Infect. Dis.* **25:**71–75.

53. Somoskövi, Á., and P. Magyar. 1999. Comparison of the Mycobacteria Growth Indicator Tube with MB Redox, Löwenstein-Jensen, and Middlebrook 7H11 media for recovery of mycobacteria in clinical specimens. *J. Clin. Microbiol.* **37:**1366–1369.

54. Strand, C. L., C. Epstein, S. Verzosa, E. Effatt, P. Hormozi, and S. H. Siddiqi. 1989. Evaluation of a new blood culture medium for mycobacteria. *Am. J. Clin. Pathol.* **91:**316–318.

55. Tholcken, C. A., S. Huang, and G. L. Woods. 1997. Evaluation of the ESP Culture System II for recovery of mycobacteria from blood specimens collected in isolator tubes. *J. Clin. Microbiol.* **35:**2681–2682.

56. Tortoli, E., A. Bartoloni, C. Burrini, A. Mantella, and M. T. Simonetti. 1995. Utility of high-performance liquid chromatography for identification of mycobacterial species rarely encountered in clinical laboratories. *Eur. J. Clin. Microbiol. Infect. Dis.* **14:**240–243.

57. Tortoli, E., M. T. Simonetti, and F. Lavinia. 1996. Evaluation of reformulated chemiluminescent DNA probe (AccuProbe) for culture identification of *Mycobacterium kansasii*. *J. Clin. Microbiol.* **34:**2838–2840.

58. Tortoli, E., P. Cichero, C. Piersimoni, M. T. Simonetti, G. Gesu, and D. Nista. 1999. Use of BACTEC MGIT 960 for recovery of mycobacteria from clinical specimens: multicenter study. *J. Clin. Microbiol.* **37:**3578–3582.

59. Tortoli, E., P. Cichero, M. G. Chirillo, M. R. Gismondo, L. Bono, G. Gesu, M. T. Simonetti, G. Volpe, G. Nardi, and P. Marone. 1998. Multicenter

comparison of ESP Culture System II with BACTEC 460TB and with Lowenstein-Jensen medium for recovery of mycobacteria from different clinical specimens, including blood. *J. Clin. Microbiol.* **36**:1378–1381.

60. **Wang, S. X., and L. Tay.** 1999. Evaluation of three nucleic acid amplification methods for direct detection of *Mycobacterium tuberculosis* complex in respiratory specimens. *J. Clin. Microbiol.* **37**:1932–1934.

61. **Wanger, A., and K. Mills.** 1996. Testing of *Mycobacterium tuberculosis* susceptibility to ethambutol, isoniazid, rifampin, and streptomycin by using Etest. *J. Clin. Microbiol.* **34**:1672–1676.

62. **Woods, G. L., G. Fish, M. Plaunt, and T. Murphy.** 1997. Clinical evaluation of Difco ESP Culture System II for growth and detection of mycobacteria. *J. Clin. Microbiol.* **35**:121–124.

63. **Woods, G. L., J. S. Bergmann, F. G. Witebsky, G. A. Fahle, B. Boulet, M. Plaunt, B. A. Brown, R. J. Wallace, Jr., and A. Wanger.** 2000. Multisite reproducibility of Etest for susceptibility testing of *Mycobacterium abscessus, Mycobacterium chelonae,* and *Mycobacterium fortuitum. J. Clin. Microbiol.* **38**:656–661.

64. **Zanetti, S., F. Ardito, L. Sechi, M. Sanguinetti, P. Molicotti, G. Delogu, M. P. Pinna, A. Nacci, and G. Fadda.** 1997. Evaluation of a nonradiometric system (BACTEC 9000 MB) for detection of mycobacteria in human clinical samples. *J. Clin. Microbiol.* **35**:2072–2075.

Diagnostic Medical Parasitology

LYNNE S. GARCIA

11

With the increase in world travel and access to many different populations and geographic areas, it is very likely that we will see more tropical diseases and infections due to the rapidity with which people and organisms can be transmitted from one place to another. The transportation of infectious agents, as well as human travelers, has been clearly demonstrated during the last few years, particularly via air travel. Travel has become available and more affordable for many people throughout the world, including those who are in some way compromised in terms of their overall health status. With the increase in the number of patients whose immune systems are compromised, through either underlying illness, chemotherapy, transplantation, AIDS, or age, we are much more likely to see increasing numbers of opportunistic infections, including those caused by parasites. Also, we continue to discover and document organisms that were thought to be nonpathogenic that, when found in the compromised host, can cause serious disease (8, 19, 20).

Control of parasitic infections depends on a number of different factors, including geographic location, public health infrastructure, political stability, available funding, social and behavioral customs and beliefs, trained laboratory personnel, health care support teams, environmental constraints, poor understanding of organism life cycles, and opportunities for control and overall commitment. Vectors and other carriers of infectious agents do not recognize political or control boundaries, both of which become meaningless. When newer infectious agents and/or diseases are recognized, there is often very little information available regarding the organism life cycle, potential reservoir hosts, and environmental requirements for survival. Priorities may change, and in areas of the world where disease epidemiology was considered important in the past, these important issues may have been moved lower on the priority list. Unfortunately, funding often plays a major role in decisions that impact disease control measures.

As new etiologic agents are discovered and the need for new therapeutics increases, more sensitive and specific diagnostic methods will become mandatory in order to assess the efficacy of newer drugs and alternative therapies.

DIAGNOSTIC PARASITOLOGY TESTING
General Comments

Diagnostic procedures in the field of medical parasitology require a great deal of judgment and interpretation and are generally classified by the Clinical Laboratory Improvement Act of 1988 (CLIA '88) as high-complexity procedures. Very few procedures can be automated, and organism identification relies on morphologic characteristics that can be very difficult to differentiate. Although morphology can be learned at the microscope, knowledge about the life cycle, epidemiology, infectivity, geographic range, clinical symptoms, range of illness, disease presentation depending on immune status, and recommended therapy are critical to the operation of any laboratory providing diagnostic services in medical parasitology. As laboratories continue to downsize and reduce staff, crosstraining will become more common and critical to financial success. Maintaining expertise in fields such as diagnostic parasitology will become more difficult, particularly when standard manual methods are used. Also, the lower the positive rate for parasitic infections is, the more likely we are to generate both false-positive and false-negative laboratory reports. It is important for members of the health care team to thoroughly recognize those areas of the clinical laboratory that require experienced personnel and why various procedures are recommended above others. Health care delivery settings where physicians provide parasitology diagnostic testing may occasionally provide "simple" test results (CLIA '88 waived tests) based on wet-mount examinations. However, in spite of the CLIA classification of these diagnostic methods, wet-mount examinations are often very difficult to perform. The key to performance of diagnostic medical parasitology procedures is formal training and experience. As the laboratory setting changes during the early part of the 21st century, it is important to recognize that these changes will continue to require a thorough understanding of the skills required to perform diagnostic parasitology procedures and the pros and cons of available diagnostic methods. Laboratories will continue to have a number of diagnostic options; whatever approach is selected by an individual laboratory, it is important that the clinical relevance of the approach be thoroughly understood and conveyed to the client user of the laboratory services.

The majority of diagnostic parasitology procedures can be performed either within the hospital setting or in an offsite location. There are very few procedures within this discipline that must be performed and reported on a STAT basis. Two procedures fall into the STAT category:

request for examination of blood films for the diagnosis of malaria and examination of cerebrospinal fluid for the presence of free-living amebae, primarily *Naegleria fowleri*. Any laboratory providing diagnostic parasitology procedures must be prepared to examine these specimens on a STAT basis 7 days a week, 24 hours a day (19, 27, 42). Unfortunately, these two procedures can be very difficult to perform and interpret; crosstrained individuals with little microbiology training or experience will find this work difficult and subject to error. It has been well documented that automated hematology instrumentation lacks the sensitivity to diagnose malaria infections, particularly since most patients seen in an emergency room have a very low parasitemia (27). However, even a low parasitemia can be life threatening in an infection with *Plasmodium falciparum*.

Diagnostic laboratories outside of the hospital setting are also appropriate settings for this type of diagnostic testing; the test requests, for the most part, are routine and are batch tested, rather than tested singly. With very few exceptions, STAT requests are not relevant and/or sent to such laboratory locations and do not require immediate testing and reporting. Point-of-care testing within the hospital (ward laboratories, intensive care units, emergency rooms, and bedside) are usually not considered appropriate sites for diagnostic parasitology testing; one exception might be the emergency room, where patients with malaria may first present with fever and general malaise. Alternative testing sites (outpatient clinics, shopping malls, senior citizen groups, and others) are generally not considered appropriate settings for diagnostic parasitology testing, although relevance might be dictated by geographic location, particularly outside of the United States, and by the development of newer, less subjective methods. The majority of physician office laboratories are not involved in diagnostic parasitology testing; however, as more molecular (nonmicroscopic) methods are developed, they may become more widely used in this setting.

Specific Options

Diagnostic Tests

The selection and use of diagnostic procedures will often depend on a number of factors, including geographic area, population served, overall positivity rate, client preference, number of test orders, staffing, personnel experience, turnaround time requirements, epidemiology considerations, clinical relevance of test results, and cost. Diagnostic tests generally have a wide range in both sensitivity and specificity. As an example, the ova and parasite (O&P) examination (direct wet mount, concentration, and permanent-stained smear) could be considered a screening method for the detection of a number of different intestinal protozoa and helminth infections; this procedure is moderately sensitive but relatively nonspecific. Molecular test methods tend to be very specific (generally for a single organism such as *Giardia lamblia*) and often more sensitive than the routine O&P examination. However, the test results are limited in scope; either the organism is present or it is not, and none of the other possible etiologic agents can be ruled in or out!

In certain situations a laboratory may offer tests on request; an example would be testing for the presence of *Cryptosporidium parvum*. If a potential waterborne outbreak were suspected, the laboratory might change its approach and begin testing all stool specimens submitted for an O&P examination rather than testing only those specimens accompanied by a specific test request. These decisions require close communication with other laboratories, water companies, public health personnel, pharmacies (e.g., communication of an increase in the purchase of antidiarrheal over-the-counter medications), and physicians.

Routine Methods

Routine methods may also be screening methods; however, this is not always the case. "Routine" can imply a widely used, well-understood laboratory test; it can also imply a low or moderately complex method, rather than a high-complexity procedure. Routine diagnostic parasitology procedures could include the O&P exam, preparation and examination of blood films and pinworm tapes and/or paddles, occult-blood tests, and examination of specimens from other body sites (urine, sputum, duodenal aspirates, urogenital sites, etc.).

Diagnostic parasitology includes laboratory procedures that are designed to detect organisms within clinical specimens using morphologic criteria and visual identification. Some clinical specimens, such as those from the intestinal tract, contain numerous artifacts that complicate the differentiation of parasites from surrounding debris. Final identification is usually based on light microscopic examination of stained preparations, often using high-magnification techniques such as oil immersion ($\times$1,000) (19, 36, 37, 39).

Specimen preparation often requires some type of concentration, all of which are designed to increase the chances of finding the organism(s). Microscopic examination requires review of the prepared clinical specimen using multiple magnifications and different examination times; organism identification also depends to a great degree on the skill of the microscopist.

Protozoa range from 1.5 μm (microsporidia) to ~80 μm (*Balantidium coli* [ciliate]) in size. Some parasites are intracellular, and multiple isolation and staining methods may be required for identification. Helminth infections are usually diagnosed by finding eggs, larvae, and/or adult worms in various clinical specimens, primarily those from the intestinal tract. Identification to the species level for many human parasites may require microscopic examination of the specimen. The recovery and identification of blood parasites can require concentration, culture, and microscopy. Confirmation of suspected parasitic infections depends on the proper collection, processing, and examination of clinical specimens, and multiple specimens may be required in order to confirm the presence of organisms (12, 19, 21, 23, 26, 28, 33, 35, 39, 41, 53, 57).

Special Testing

Special procedures such as parasite culture are often performed in limited numbers of laboratories. These procedures require the maintenance of positive-control cultures used for quality control checks on all patient specimens and special expertise and time often not available in many clinical laboratories. Although some standardized reagents are now commercially available, many clinical laboratories choose to send their request for serologic testing for parasitic disease to other laboratories. Often, the Centers for Disease Control and Prevention (CDC) perform serologic testing on specimens submitted to each state's Department of Public Health. Generally, specimens for parasitic serologies are submitted not directly to the CDC but through state public health laboratories. In an emergency situation, consultation with your county or state public health laboratory may allow shipment of a specimen directly to the CDC.

Diagnostic Medical Parasitology

Equipment

Equipment required for diagnostic parasitology work is very minimal; however, the one expense that should not be trimmed would be for one or more microscopes with good optics. Each microscope should be equipped with high-quality (flat-field) objectives (10×, 40×, 50 or 60× oil immersion, and 100× oil immersion objectives). The oculars should be a minimum of 10×. Depending on the range of immunoassay testing available, a fluorescent microscope or enzyme immunoassay (EIA) reader might be desirable. The availability of this equipment will vary tremendously from one laboratory to another and may be shared with other groups within the laboratory. Another option would be a fume hood, in which the staining could be performed; this is not required but is recommended, particularly if the laboratory is still using xylene for dehydration of permanent-stained fecal smears. The rest of the equipment is quite common and can be shared with other areas within the laboratory. Equipment would include refrigerators, freezers, pipette systems, and other common items normally found within a clinical laboratory.

SOLICITATION OF PRODUCT INFORMATION

Commercial companies that supply collection systems, stool concentration devices, reagents, test kits, and other relevant products used in the practice of diagnostic medical parasitology were contacted for inclusion in this book on laboratory medicine diagnostic products. Those companies that replied have been included and are presented below. Not all companies provide the same products, and you will see differences as you review the available products used in this field. In some cases, peer-reviewed manuscripts on the products have not been published. Based on company responses, the data are current.

SPECIMEN COLLECTION SYSTEMS

Various collection systems are available for clinical specimens suspected of containing parasites or parasitic elements (Table 1). The selection of collection options should be based on a thorough understanding of the value and limitations of each. The final laboratory results are based on parasite recovery and identification and will depend on the initial handling of the organisms. Unless the appropriate specimens are properly collected and processed, the examination may lead to false-negative results. Specimen rejection criteria have become much more important for all diagnostic parasitology procedures. Laboratory results based on improperly collected specimens may require inappropriate expenditures of time and supplies and may also mislead the physician.

FRESH STOOL SPECIMEN COLLECTION

Fecal specimens should be collected in clean, wide-mouth containers; often a waxed cardboard or plastic container with a tight-fitting lid is selected for this purpose. These collection containers are available from most of the microbiology product distributors and are not difficult to locate; there are very minor differences, none of which are critical. Stool specimen containers should be placed in plastic bags when transported to the laboratory for testing. Specimens should be identified with the following information: patient's name and identification number, physician's name, the date and time the specimen was collected (if the laboratory is computerized, the date and time may reflect arrival in the laboratory, not the actual collection time), and the laboratory procedures requested. Although it may be helpful to have information concerning the presumptive diagnosis or relevant travel history, this information is rarely available.

Fresh specimens are required for the recovery of motile trophozoites (amebae, flagellates, or ciliates). The protozoan trophozoite stage is normally found in cases of diarrhea; the gastrointestinal tract contents are moving through the system too rapidly for cyst formation to occur. Once the stool specimen is passed from the body, trophozoites do not encyst but may disintegrate if not examined or preserved within a short time after passage. However, most helminth eggs and larvae, coccidian oocysts, and microsporidian spores will survive for extended periods. Liquid specimens should be examined within 30 min of passage, not 30 min from the time they reach the laboratory. If this general time recommendation of 30 min is not possible, then the specimen should be placed in one of the available fixatives. Soft (semiformed) specimens may have a mixture of protozoan trophozoites and cysts and should be examined within 1 h of passage; again, if this time frame is not possible, then preservatives should be used. Immediate examination of formed specimens is not as critical; in fact, if the specimen is examined any time within 24 h after passage, the protozoan cysts should still be intact (Table 2) (19, 20, 21, 39, 41).

PRESERVATION OF STOOL SPECIMENS

If there are likely to be delays in specimen delivery to the laboratory, fecal preservatives should be used (Table 3). To preserve protozoan morphology and to prevent the continued development of some helminth eggs and larvae, the stool specimens can be placed in preservative either immediately after passage (by the patient using a collection kit) or once the specimen is received by the laboratory (Fig. 1). There are several fixatives available, and these include formalin, sodium acetate-acetic acid-formalin (SAF), Schaudinn's fluid, and polyvinyl alcohol (PVA) (Table 3). Regardless of the fixative selected, adequate mixing of the specimen and preservative is mandatory (Fig. 1 to 8) (23, 26, 28, 35, 53, 57).

When selecting one or more stool fixatives, remember that a permanent-stained smear is mandatory for a complete examination for parasites. You may also want to perform methods such as fluorescence, EIA, or a method using the new cartridge immunoassay devices, and you will need to confirm that the fixative you are using is compatible with the diagnostic procedures you have selected (1–7, 10, 11, 14, 16, 18, 22, 24, 25, 29–32, 40, 45, 47, 49, 50–52, 58). It is also important to remember that disposal regulations for compounds containing mercury are becoming more restrictive; each laboratory will have to check applicable state and federal regulations to help determine fixative options.

TABLE 1 Body sites and specimen collection[a]

Site(s)	Specimen option	Collection method recommendations
Blood	Smears of whole blood	Thick and thin films: fresh (first choice) or with anticoagulant (second choice)
	Anticoagulated blood	EDTA (first choice)
		Heparin (second choice)
Bone marrow	Aspirate	Sterile
Central nervous system	Spinal fluid	Sterile
	Brain biopsy specimen	Sterile
Cutaneous ulcer	Aspirates from below surface	Sterile plus air-dried smears
	Biopsy specimen	Sterile, nonsterile to histopathology (formalin acceptable)
Eye	Biopsy specimen	Sterile (in saline), nonsterile to histopathology
	Scrapings	Sterile (in saline)
	Contact lens	Sterile (in saline)
	Lens solution	Sterile; unopened commercial solutions not acceptable
Intestinal tract	Fresh stool	Half-pint waxed or plastic container
	Preserved stool	5 or 10% formalin, MIF, SAF, Schaudinn's fluid, PVA, modified PVA, single-vial systems
	Sigmoidoscopy material	Fresh, PVA, or Schaudinn's smears
	Duodenal contents	Entero-Test or aspirates
	Anal impression smear	Cellulose tape (pinworm exam) or other collection device
	Adult worm or worm segments	Saline, 70% alcohol
Liver and/or spleen	Aspirates	Sterile, collected in four separate aliquots (liver)
	Biopsy specimen	Sterile, nonsterile to histopathology
Lung	Sputum	True sputum (not saliva)
	Induced Sputum	No preservative (10% formalin if time delay)
	Bronchoalveolar lavage fluid	Sterile, air-dried smears
	Transbronchial aspirate	Sterile, air-dried smears
	Tracheobronchial aspirate	Sterile, air-dried smears
	Brush biopsy	Sterile, air-dried smears
	Open lung biopsy	Sterile
	Aspirate	Sterile, air-dried smears
Muscle	Biopsy specimen	Fresh, squash preparation, nonsterile to histopathology (formalin acceptable)
Skin	Scrapings	Aseptic, smear or vial
	Skin snip	No preservative
	Biopsy specimen	Sterile (in saline), nonsterile to histopathology
Urogenital system	Vaginal discharge	Saline swab, transport swab (no charcoal), culture medium, plastic envelope culture; air-dried smear for fluorescent antibody
	Urethral discharge	Saline swab, transport swab (no charcoal), culture medium, plastic envelope culture; air-dried smear for fluorescent antibody
	Prostatic secretions	Saline swab, transport swab (no charcoal), culture medium, plastic envelope culture; air-dried smear for fluorescent antibody
	Urine	Single unpreserved specimen, 24-h unpreserved specimen, early morning

[a]Most products used for specimen collection are available from any major medical supply house. Adapted from reference 20.

Formalin

Formalin is used as an all-purpose fixative that is appropriate for helminth eggs and larvae and for protozoan cysts. Two concentrations are commonly used: 5%, which is recommended for preservation of protozoan cysts, and 10%, which is recommended for helminth eggs and larvae. Although 5% is often recommended for all-purpose use, most commercial manufacturers provide 10%, which is more likely to kill all helminth eggs. To help maintain organism morphology, the formalin can be buffered with sodium phosphate buffers, i.e., neutral formalin. Selection of specific formalin formulations is at the user's discretion. Aqueous formalin will permit the examination of the specimen as a wet mount only, a technique much less accurate

TABLE 2 Stool collection: fresh versus preserved[a]

Stool type	Pros	Cons
Fresh	1. No requirements for stool fixatives. 2. Ability to see motile trophozoites. 3. Lower cost. 4. Can perform direct wet exam, concentration, and permanent stained smear. 5. Relevant only if time from stool passage to laboratory is acceptable (liquid or watery stool, 30 min; semiformed stool, 1 h; formed stool, 24 h); in symptomatic patients, the trophozoite form of the intestinal protozoa is present and will not encyst when outside the body.	1. May have excessive lag time between stool passage and fixation or processing; trophozoites may disintegrate, thus giving a false-negative result. 2. O&P examination (direct wet exam, concentration, and permanent-stained smear) may be negative due to lack of organism preservation and morphology integrity.
Preserved	1. Organism morphology is preserved when lag time between stool passage and fixation is short; this can be accomplished by allowing the patients to collect and fix stool specimens at home. Once the specimen is mixed with the preservative, the delivery time to the laboratory is not critical. 2. Can perform concentration and permanent stain.	1. Cost of collection vials may represent a cost increase; however, this may be less expensive overall because of a much more accurate prediction of result (patient outcome). 2. Disposal of vials may be a problem if the laboratory is using preservatives containing mercuric chloride.

[a]Most products used for specimen collection are available from any major medical supply house. Adapted from reference 20.

TABLE 3 Preservatives and procedures commonly used in diagnostic parasitology (stool specimens)[a]

Preservative	Suitable for:	
	Concentration	Permanent-stained smear
5 or 10% formalin	Yes	No
5 or 10% buffered formalin	Yes	No
MIF[b]	Yes	Yes, with polychrome IV
SAF[c]	Yes	Yes, with iron hematoxylin
PVA[d]	Yes	Yes, with trichrome or iron hematoxylin
Modified PVA[e]	Yes	Yes, with trichrome or iron hematoxylin
Modified PVA[f]	Yes	Yes, with trichrome or iron hematoxylin
Single-vial systems[g]	Yes	Yes, with trichrome or iron hematoxylin
Schaudinn's fluid (without PVA)[d]	No	Yes, with trichrome, iron hematoxylin, or a "matched" stain developed by the same manufacturer

[a]Suppliers of stool fixatives: Alpha-Tec, Evergreen Scientific, Hardy Diagnostics, Medical Chemical Corporation, Meridian Diagnostics, MML Diagnostic Packaging, PML Microbiologicals, Remel, and Volu-Sol, Inc. **Note**: Not all companies supply every fixative contained in this table; however, those listed above offer a wide selection of fixatives and collection vials.

[b]MIF is a stain preservative for most parasites; however, most laboratories using MIF examine the material only as a wet preparation (direct smear and/or concentration sediment); another fixative such as PVA could be used, and the permanent-stained smear could be prepared from this.

[c]SAF is a liquid fixative from which both a concentration and permanent-stained slide can be prepared; however, the concentrate sediment must be placed on albumin-coated slides prior to staining (e.g., iron hematoxylin) so the stool material adheres to the slide. Trichrome is an alternate stain.

[d]This fixative uses the mercuric chloride base in the Schaudinn's fluid; this formulation is still considered to be the gold standard against which all other fixatives are evaluated (organism morphology after permanent staining). Additional fixatives prepared with non-mercuric chloride-based compounds continue to be developed, tested, and marketed.

[e]This modification uses a copper sulfate base rather than mercuric chloride.

[f]This modification uses a zinc base rather than mercuric chloride and works well with trichrome stain; iron hematoxylin can also be used.

[g]These modifications use a combination of ingredients (including zinc), but are prepared from proprietary formulas; in some cases the manufacturer has available a specific stain that has been designed to go with their particular fixative.

than a stained smear for the identification of intestinal protozoa. Most companies offer both 5 and 10% formalin (most are buffered); most laboratories use 10% formalin (Fig. 5 and 6) (19, 20, 39).

Protozoan cysts (not trophozoites), coccidian oocysts, helminth eggs, and larvae are well preserved for long periods of time in 10% aqueous formalin. Hot (60°C) formalin can be used for specimens containing helminth eggs, since in cold formalin some thick-shelled eggs may continue to develop, become infective, and remain viable for long periods. Several grams of fecal material should be thoroughly mixed in 5 or 10% formalin.

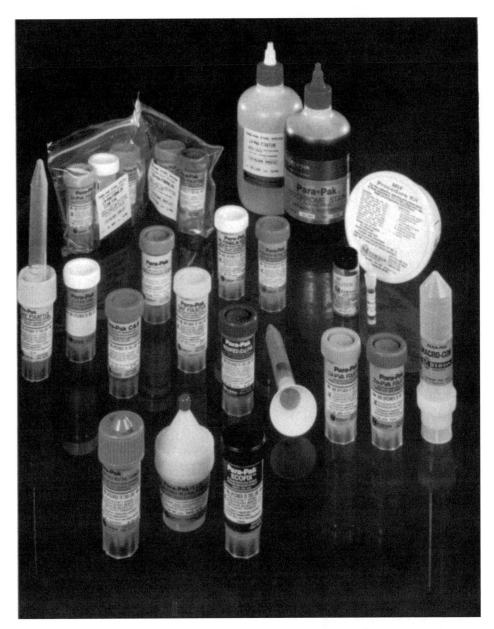

FIGURE 1   Sample of Para-Pak parasitology products (courtesy of Meridian Diagnostics).

MIF

Merthiolate-iodine-formalin (MIF) is a good stain preservative for most kinds and stages of parasites found in feces; it is especially useful for field surveys. It is used with all common types of stools and aspirates; protozoa, eggs, and larvae can be diagnosed without further staining in temporary wet mounts, either made immediately after fixation or prepared several weeks later. Although some laboratories maintain that a permanent-stained smear can be prepared from specimens preserved in MIF, most laboratories using such a fixative examine the material only as a wet preparation (direct smear and/or concentration sediment). The MIF preservative is prepared in two stock solutions, stored separately and mixed immediately before use (19).

SAF

Both a concentration and a permanent-stained smear can be performed from specimens preserved in the liquid fixative SAF; this fixative does not contain mercuric chloride, as is found in Schaudinn's fluid and PVA (Fig. 2). The sediment is used to prepare the permanent smear, and the stool material must be placed on an albumin-coated slide to improve adherence to the glass (19, 24).

SAF is considered to be a "softer" fixative than mercuric chloride. The organism morphology will not be quite as sharp after staining as will that of organisms originally fixed in solutions containing mercuric chloride. The pairing of SAF-fixed material with iron hematoxylin staining provides better organism morphology than does staining SAF-fixed

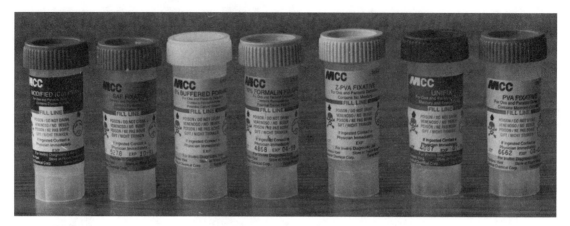

FIGURE 2   Representative sample of stool collection vials. From left to right they are modified PVA (prepared with a copper base), SAF fixative; 5% buffered formalin, 10% buffered formalin, ZPVA (prepared with a zinc base), Unifix (single-vial collection system), and PVA (prepared with a mercury base) (courtesy of Medical Chemical Corporation).

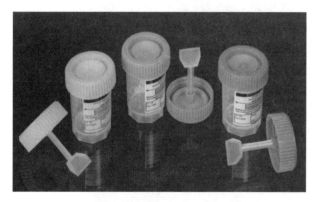

FIGURE 3   Fecal collection and transport vials (courtesy of Evergreen Scientific).

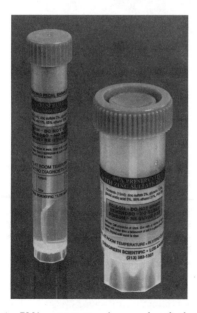

FIGURE 4   PVA preservative (prepared with the zinc base) (courtesy of Evergreen Scientific).

material with trichrome (personal observation). Although SAF has a long shelf life and is easy to prepare, the smear preparation technique may be a bit more difficult for less experienced personnel who are not familiar with fecal specimen techniques. Laboratories that have considered using only a single preservative have selected this option. Helminth eggs and larvae, protozoan trophozoites and cysts, and coccidian oocysts and microsporidian spores are preserved using this method; a number of distributors provide prepared SAF, ready to use.

Schaudinn's Fluid

Schaudinn's fluid is used with fresh stool specimens or samples from the intestinal mucosal surface and is prepared using mercuric chloride. Many institutions that do not have problems meeting recommended specimen delivery times select this approach. Permanent-stained smears are then prepared from fixed material. When using this approach, a second vial is required for the concentration and/or various screening methods (19, 20, 39, 41).

PVA

PVA is a plastic resin that is incorporated into Schaudinn's fixative (Fig. 2). The PVA powder serves as an adhesive for the stool material; i.e., when the stool-PVA mixture is spread onto the glass slide, it adheres because of the PVA component. Fixation is still accomplished by the Schaudinn's fluid itself. Perhaps the greatest advantage in the use of PVA is the fact that a permanent-stained smear can be prepared. PVA fixative solution is highly recommended as a means of preserving cysts and trophozoites for examination at a later time. The use of PVA also permits specimens to be shipped (by regular mail service) from any location in the world to a laboratory for subsequent examination. PVA is particularly useful for liquid specimens and should be used in the ratio of 3 parts PVA to 1 part fecal specimen (9, 12, 19, 20, 39, 41).

Modified PVA

Although preservatives have been developed that do not contain mercury compounds, substitute compounds have not provided the quality of preservation necessary for good protozoan morphology on the permanent-stained smear.

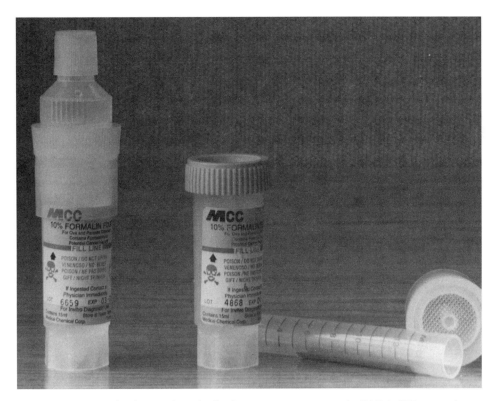

FIGURE 5 Example of a 10% formalin fecal concentration system, the PARA-SED system (courtesy of Medical Chemical Corporation).

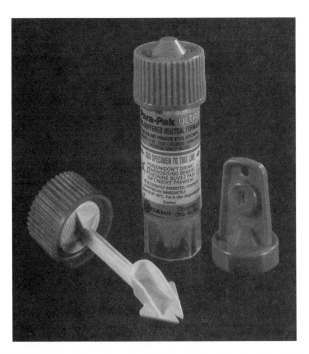

FIGURE 6 Example of a 10% formalin vial, the Para-Pak Ultra (courtesy of Meridian Diagnostics).

FIGURE 7 Example of a single-vial fecal collection system, Para-Pak ECOFIX (courtesy of Meridian Diagnostics).

Copper sulfate has been tried but does not provide results equal to those seen with mercuric chloride (Fig. 2). Zinc sulfate has recently proven to be an acceptable mercury substitute and is used with trichrome stain (Fig. 2). Although zinc substitutes have become widely available, each manufacturer has a proprietary formula for the fixative. Although there are different opinions, some feel that the zinc sulfate substitute is somewhat better than PVA prepared with copper sulfate. Most manufacturers who supply stool collection vials offer both of these options (9, 28, 35).

FIGURE 8 Example of a single-vial fecal collection system, ProtoFix (courtesy of Alpha-Tec).

Single-Vial Collection Systems

Several manufacturers now have available single-vial stool collection systems, similar to SAF or modified PVA methods. From the single vial, both the concentration and permanent-stained smear can be prepared. It is also possible to perform immunoassay procedures from some of these vials. Make sure to ask the manufacturer about all three capabilities (concentrate, permanent-stained smear, and immunoassay procedures) and for specific information indicating there are no formula components that would interfere with any of the three methods. Like the zinc substitutes, these formulas are proprietary. Several examples are Unifix (Fig. 2), ECOFIX (Fig. 7), and ProtoFix (Fig. 8) (19, 20, 23).

Stool Preservatives: User Notes and Suggestions

A. Most of the commercially available kits have a "fill-to" line on the vial label to indicate how much fecal material to add to ensure adequate preservation of the fecal material. However, patients often overfill the vials; remember to open the vials with the vials turned away from your face.

B. Although the two-vial system (one vial of 5 or 10% buffered formalin [concentration] and one vial of PVA [permanent-stained smear]) has always been the "gold standard," laboratories are beginning to use other options. Changes in the selection of fixatives are based on the following considerations.
  1. Problems with disposal of mercury-based fixatives and lack of multilaboratory contracts for disposal of such products
  2. The cost of a two-vial system compared with the cost of a single collection vial
  3. Selection of specific stains (e.g., trichrome or iron hematoxylin) to use with specific fixatives

  4. Whether the newer immunoassay kits can be used with stool specimens preserved with that particular fixative

Stool Preservatives: Procedure Limitations

A. Adequate fixation still depends on the following parameters.
  1. Meeting recommended time limits for lag time between passage of the specimen and fixation
  2. Use of the correct ratio of fixative to specimen (3:1)
  3. Thorough mixing of the fixative and specimen. Once the specimen is received in the laboratory, any additional mixing at that time will not counteract the lack of fixative-specimen mixing and contact prior to that time.

B. Unless the appropriate stain is used with each fixative, the final permanent-stained smear may be difficult to examine (organisms hard to see and/or identify). Examples of appropriate combinations are
  1. Schaudinn's or PVA fixative with trichrome or iron hematoxylin stains
  2. SAF fixative with iron hematoxylin stain (trichrome is second choice)
  3. Single-vial mercuric chloride substitute systems with trichrome, iron hematoxylin, or company-developed proprietary stains matched to their specific fixatives (an example would be ECOFIX coupled with ECO-STAIN [Fig. 1 and 7] [23]).

C. If you want to investigate one of the products for potential use in your laboratory, ask to see the package inserts and ask for actual laboratory user contacts with whom you can discuss product performance. Make sure to check the references to see if clinical studies have been published in peer-reviewed journals. If not, ask why not.

Intestinal Tract Specimens (Stool)

Immunoassay Methods

The use of immunoassay methods for fecal specimens has become incorporated into many laboratories, particularly during the last couple of years (Table 4). The selection and use of screening procedures depend on a number of variables, including geographic area, population served, positivity rate, client test preference, number of test requests, laboratory staffing, personnel training and experience, equipment availability, turnaround time requirements, epidemiology considerations, clinical relevance of test results, and cost. Like any type of diagnostic test option, there are pros and cons. Certainly the specificity of these methods is very high; however, the tests are limited to specific organisms. A negative test does not rule out many other possible etiologic agents.

To date, test formats include fluorescent-antibody assay, EIA and the newer immunochromatographic immunoassays. The reagents have been used for direct fluorescent-antibody (DFA) microscopy, microplate immunoassays, and the newer cartridge test systems. The sensitivity and specificity of all of these reagents have been comparable; selection of a particular format will vary from laboratory to laboratory. Specific format examples can be seen in Fig. 9 to 17 (see also Table 8).

Organisms for which these commercial products have been developed include *G. lamblia*, *Cryptosporidium parvum*, *Entamoeba histolytica*/*Entamoeba dispar* group, *E. histolytica*, and *Trichomonas vaginalis*. These procedures are more sensi-

TABLE 4 Rapid diagnostic procedures (nontraditional approaches)

Fecal specimen procedure	Traditional procedure(s)	Alternative options	Comments
Concentration	Specimen sampling (pipettes), use of slides and coverslips	Specimen sampling (DiaSys semiautomated system), DiaSys glass viewing chambers	The traditional sampling approach using pipettes and the preparation of wet smears using glass slides and coverslips can be replaced with a semiautomated sampling and viewing system from DiaSys Corporation. The specimen is drawn through tubing from the mixed concentrated stool sediment into two viewing chambers that fit onto the microscope stage. The quality of the glass is excellent and organism morphology can be easily seen within the viewing chambers.
Permanent-stained smear[a]	Microscopic examination of stained fecal smear using oil immersion objectives	The Genzyme Combo Rapid Test is a solid-phase qualitative immunochromatographic procedure that can be used for the detection of C. parvum and G. lamblia. The test can be performed on fresh, frozen, unfixed, or formalin-fixed (5%, 10%, SAF) fecal specimens. This assay is distributed by Becton Dickinson as ColorPAC. The ImmunoCard STAT!, a similar cartridge, is available from Meridian Diagnostics.	Although these tests are limited to C. parvum and G. lamblia, they can be very beneficial in the absence of trained microscopists. However, in patients who remain symptomatic after a negative result, the O&P examination should always remain as an option.
		The BIOSITE Diagnostics Triage Parasite Panel is a qualitative EIA procedure for the rapid detection of G. lamblia, the E. histolytica/E. dispar group, and C. parvum. The test can be performed on fresh or fresh frozen, unfixed human fecal specimens.	Although this test is limited to the use of fresh or fresh frozen fecal specimens, it can be helpful in the absence of trained microscopists. It is important to remember that the test does not differentiate between E. histolytica (pathogen) and E. dispar (nonpathogen). In patients who remain symptomatic after a negative result, the O&P examination should always remain as an option.

[a]The three alternative options listed are in a cartridge format that is simple and rapid to use. Batch testing is possible; the methods use very few steps, and the visual assessment of positive versus negative is quite easy. Remember that the total O&P examination is recommended when the immunoassay tests (limited to specific organisms) are negative and the patient remains asymptomatic.

tive than the routine O&P examination. However, as mentioned above, the test results are limited in scope. Either the organism is present or it is not, and no other possible etiologic agents can be ruled in or out.

It is very important to know what stool fixatives can or cannot be used for stool specimens that will be tested using one of the immunoassay methods. Some procedures require fresh or frozen specimens only, while others can be used with formalinized specimens. With the newer single-vial stool collection systems, it is mandatory that you confirm with the manufacturer the applicability of their collection vial fixatives to these new immunoassay procedures.

O&P Examination

The specimen most commonly submitted to the diagnostic parasitology laboratory is the stool specimen, and the most commonly performed procedure in parasitology is the O&P examination, which comprises three separate protocols: the direct wet mount, the concentration, and the permanent-stained smear. The direct wet mount requires fresh stool, is designed to allow detection of motile protozoan trophozoites, and is examined microscopically at low and high dry magnifications ($\times 100$, entire 22-by-22-mm coverslip; $\times 400$, one-third to one-half of the 22-by-22-mm coverslip). Iodine can also be used to examine the wet mount (Fig. 18). However, due to potential problems with lag time between the time of specimen passage and receipt in the laboratory, the direct wet examination has been eliminated from the routine O&P examination in favor of receipt of specimens collected in stool preservatives; if specimens are received in the laboratory in stool collection preservatives, the direct wet preparation is not performed.

The second part of the O&P examination is the concentration that is designed to facilitate recovery of protozoan cysts, coccidian oocysts, microsporidial spores, and helminth eggs and larvae. Both flotation and sedimentation methods are available, the most common procedure being the formalin-ethyl acetate sedimentation method (formerly used was the formalin-ether method). The concentrated specimen is examined as a wet preparation, with or without iodine, using low and high dry magnifications

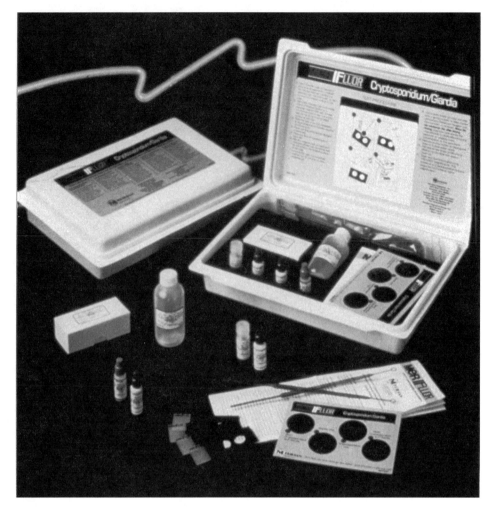

FIGURE 9 MERI/FLUOR DFA kit for the detection of *Cryptosporidium* and *Giardia* (courtesy of Meridian Diagnostics).

(×100 and ×400) as indicated for the direct wet-smear examination. Various commercial concentration devices are available and provide results that are comparable with the gold standard gauze-and-funnel method (Table 5; Fig. 19 to 27). A semiautomated system for sampling and examination of concentration sediments is also available (Diasys Corporation) (Fig. 28 and 29) (Table 5).

The third part of the O&P examination is the permanent-stained smear that is designed to facilitate identification of intestinal protozoa. Several staining methods are available, the two most common being the Wheatley modification of the Gomori tissue trichrome and the iron hematoxylin stains (Fig. 30 to 33). This part of the O&P examination is critical for the confirmation of suspicious objects seen in the wet examination and identification of protozoa that might not have been seen in the wet preparation. The permanent-stained smear is the most important procedure performed for the identification of intestinal protozoan infections; the permanent-stained smears are examined using oil immersion objectives (60× for screening and 100× for final review of a minimum of 300 oil immersion fields). Specific modified trichrome stains for the microsporidia can be seen in Fig. 30 to 34. Those work-

ing with SAF-preserved specimens will require the use of albumin for the preparation of permanent-stained slides (Fig. 35). Commercially available quality control slides are also available (Tables 4 and 6).

Other Diagnostic Methods
Several other diagnostic techniques are available for the recovery and identification of parasitic organisms. Most laboratories do not routinely offer all of these techniques, but many are relatively simple and inexpensive to perform. The clinician should be aware of the possibilities and the clinical relevance of information obtained from using such techniques. Occasionally, it is necessary to examine stool specimens for the presence of scolices and proglottids of cestodes and adult nematodes and trematodes to confirm the diagnosis and/or for identification to the species level.

BLOOD COLLECTION
There are a number of parasites that may be recovered in a blood specimen, either whole blood, buffy coat prepara-

tions, or various types of concentrations (9, 17, 31, 44, 46, 54). These parasites include *Plasmodium*, *Babesia*, and *Trypanosoma* species; *Leishmania donovani*; and microfilariae. Although some organisms may be motile in fresh, whole blood, usually organism identification is accomplished from the examination of permanent-stained blood films, both thick and thin films. Blood films can be prepared from fresh, whole blood collected with no anticoagulants, anticoagulated blood, or sediment from the various concentration procedures (19, 20, 36, 37, 42).

Many laboratories receive blood for these procedures in a tube rather than relying on finger stick blood. EDTA anticoagulant is preferred, and the tube should be filled with blood to provide the proper blood/anticoagulant ratio. For detection of stippling, the smears should be prepared within 1 h after the specimen is drawn. After that time, stippling may not be visible on stained films; however, the overall organism morphology will still be excellent. Most laboratories routinely use commercially available blood collection tubes that are readily available through most supply houses.

The time the specimen was drawn should be clearly indicated on the tube of blood and also on the result report. The physician will then be able to correlate the results with any fever pattern or other symptoms that the patient may have. There should also be some comments on the test result report that is sent back to the physician that one negative specimen does not rule out the possibility of a parasitic infection.

Although microscopy of Giemsa-stained thick and thin blood films has been used to diagnose malaria for decades, alternative approaches have been developed that are sensitive, specific, and rapid and might not require laboratory equipment. Various methods include fluorescent microscopy (acridine orange), fluorescent microscopy after centrifugation (QBC), flow cytometry, automated blood cell analyzers, biochemical methods (OptiMAL), antigen detection assays (Parasight-F and ICT Malaria P.f.), and PCR. Additional information on these products can be seen in Tables 7 and 8 (31). Dipstick antigen detection is probably the most practical complement to microscopy, although it is currently recommended as a backup. However, increased sensitivity, detection of all four malaria species, and quantification might eliminate the need for microscopy in the future. Not all products are Food and Drug Administration (FDA) approved, and the FDA status is pending on some of the available products. Each laboratory considering using these methods should check with the specific companies listed in Table 7. An excellent review and bibliography can be found in reference 31.

COLLECTION OF SPECIMENS FROM OTHER BODY SITES

Although clinical specimens for examination can be obtained from many other body sites, these specimens and appropriate diagnostic methods are not as commonly performed as those used for the routine stool specimen (Fig. 36). The majority of specimens from other body sites (Table 1) would be submitted as fresh specimens for further testing. Another example would be examination of paddles or devices designed to sample perianal skin for the presence of pinworm eggs (Fig. 37). Another simplified culture option has been developed for the isolation and identification of *T. vaginalis*. This approach has proven to be much more sensitive than the examination of wet preparations alone and has been incorporated into use in many institutions with dramatic increases in the number of positive specimens identified (Fig. 38 and 39). CHEMICON International, Inc., also has a *T. vaginalis* DFA kit available (Table 8) (11, 13, 16, 30, 34, 48, 55).

TABLE 5 Stool processing[a]

Procedure	Commercial options	Comments
Concentration	1. Alexon-Trend 2. Evergreen Scientific 3. Medical Chemical Corporation 4. Meridian Diagnostics	Some of these options use the entire contents of the collection vial, minimizing sample error and allowing detection of parasites even when present in very low numbers. Like most available commercial products, the kit contains the concentrator units, centrifuge vials, caps, surfactant (if relevant) and instruction sheet.
Preserved stool	1. Alpha-Tec 2. Evergreen Scientific 3. Hardy Diagnostics 4. Medical Chemical Corporation 5. Meridian Diagnostics 6. MML Diagnostic Packaging 7. PML Microbiologicals 8. Remel 9. Volu-Sol, Inc.	1. Cost of collection vials represents an overall cost increase; however, this may be less expensive overall because of a much more accurate prediction of result (patient outcome). 2. Disposal of vials may be a problem if the laboratory is using preservatives containing mercuric chloride.

[a]These products are available from those manufacturers indicated; this list is not all-inclusive but represents examples of the types of products that are used in the diagnostic microbiology or parasitology laboratory.

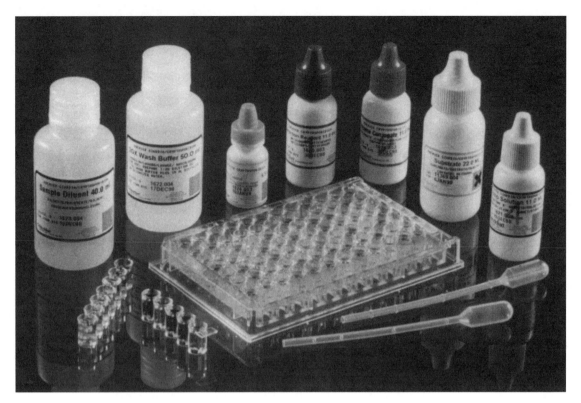

FIGURE 10 Immunoassay for *Giardia* (courtesy of Meridian Diagnostics).

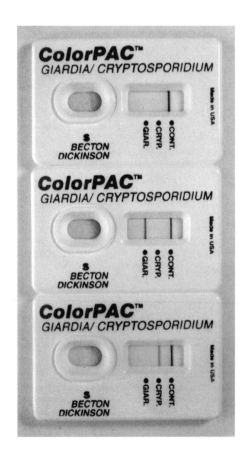

FIGURE 11 ColorPAC cartridge for the detection of *Cryptosporidium* and *Giardia*. This system can be used with fresh, frozen, or formalinized stool specimens. The top cartridge shows a positive flow control, the middle cartridge shows a specimen positive for G. *lamblia*, and the bottom cartridge shows a specimen positive for C. *parvum*. (Courtesy of Genzyme Diagnostics, distributed by Becton Dickinson.)

1. Add 2 drops of Sample Treatment Buffer to tube.

4. Add 2 drops of Conjugate B to tube. Mix.

2. Add patient specimen to tube using transfer pipet. Mix.

5. Pour contents of tube into sample well of test device.

3. Add 2 drops of Conjugate A to tube.

6. Visually read results after 10 min.

FIGURE 12 *Giardia/Cryptosporidium* Combo (ColorPac) assay procedure (courtesy of Genzyme Corporation, distributed by Becton Dickinson).

REF 750830
30 Test Kit
Coffret de Test
Tests
Pruebas

For *in vitro* Diagnostic Use **ImmunoCard®**
STAT!
CRYPTO/GIARDIA

For Detection of *Cryptosporidium/Giardia*
Pour la détection de *Cryptosporidium/Giardia*
Zum Nachweis von *Cryptosporidium/Giardia*
Per la ricerca di *Cryptosporidium/Giardia*
Para la detección de *Cryptosporidium/Giardia*

Manufactured for / Fabriqué pour /
Hergestellt für / Prodotto per / Fabricado para:

Meridian
Diagnostics, Inc.

AR Authorized Representative:
Meridian Diagnostics Europe
Via dell' Industria, 7
20020 Villa Cortese (MI)
ITALY

FIGURE 13 ImmunoCard STAT! *Giardia/Cryptosporidium* assay procedure (courtesy of Meridian Diagnostics).

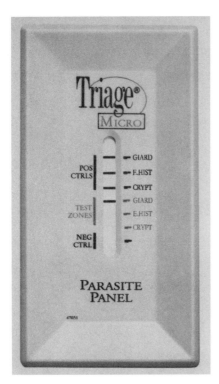

FIGURE 14 Triage parasite panel for the detection of *Giardia, Cryptosporidium*, and the *E. histolytica/E. dispar* group. It is important to remember that this system requires fresh or frozen stool specimens; formalinized specimens cannot be used in this procedure. The specimen shown is positive for *G. lamblia*. (Courtesy of BIOSITE Diagnostics.)

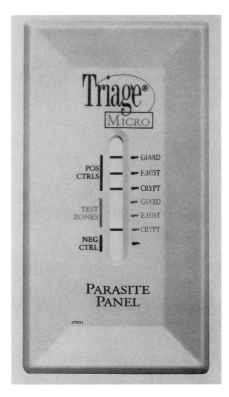

FIGURE 16 Triage parasite panel. The specimen shown is positive for *C. parvum*. (Courtesy of BIOSITE Diagnostics.)

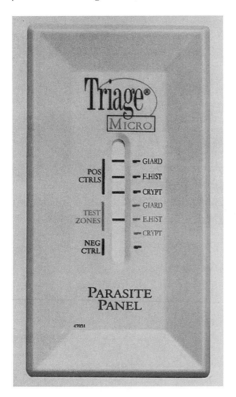

FIGURE 15 Triage parasite panel. The specimen shown is positive for the *E. histolytica/E. dispar* group. (Courtesy of BIOSITE Diagnostics.)

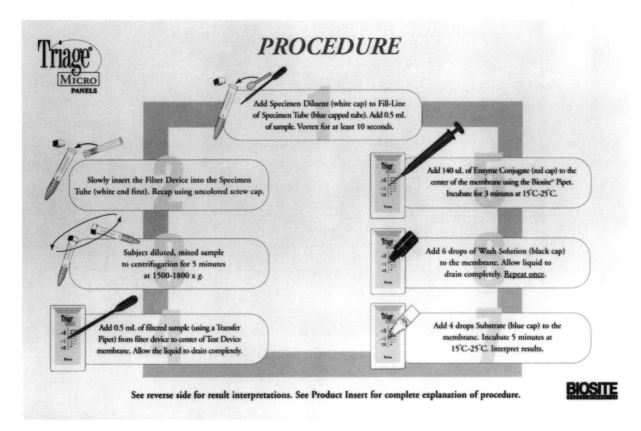

FIGURE 17 Triage assay procedure (courtesy of BIOSITE Diagnostics).

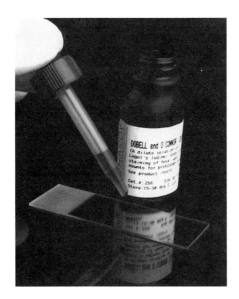

FIGURE 18 D'Antoni's iodine (used during the examination of fecal wet preparations) (courtesy of Hardy Diagnostics).

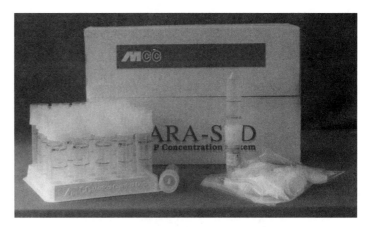

FIGURE 19 PARA-SED fecal concentration system (courtesy of Medical Chemical Corporation).

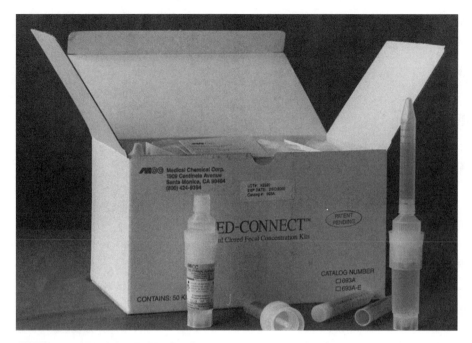

FIGURE 20 SED-CONNECT closed concentration system for the recovery of helminth eggs, larvae, protozoan cysts, coccidian oocysts, and microsporidial spores (courtesy of Medical Chemical Corporation).

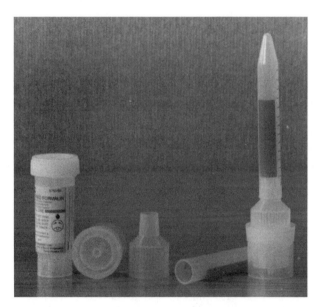

FIGURE 21 Formalin stool collection vial (SED-CON-NECT and PARA-SED sedimentation concentration systems) (courtesy of Medical Chemical Corporation).

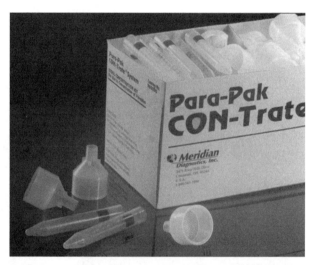

FIGURE 22 Para-Pak CON-Trate fecal concentration system (courtesy of Meridian Diagnostics).

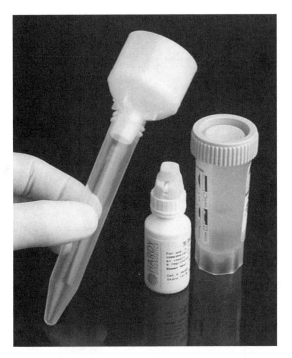

FIGURE 23 Para Kit, fecal concentrator kit (courtesy of Hardy Diagnostics).

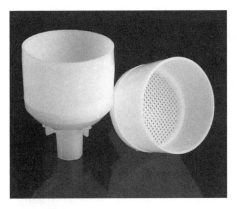

FIGURE 24 Para Kit, fecal concentrator funnel (courtesy of Hardy Diagnostics).

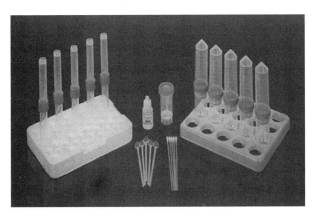

FIGURE 25 Fecal parasite concentrators (FPC [left] and FPC JUMBO [right]). These are designed for use with a 15- or 50-ml centrifuge tube. (Courtesy of Evergreen Scientific).

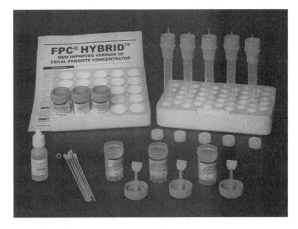

FIGURE 26 FPC HYBRID improved version of the fecal parasite concentrator (courtesy of Evergreen Scientific).

FIGURE 27 CONSED solution used for concentration procedures (courtesy of Alpha-Tec).

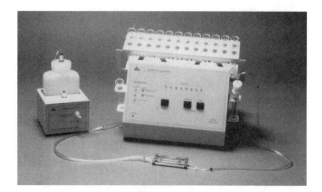

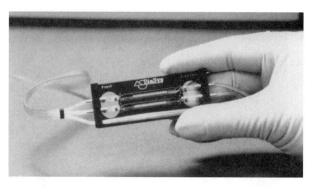

FIGURE 28 Semiautomated sampling and examination system for fecal concentration sediments (courtesy of DiaSys Corporation).

FIGURE 29 Detailed view (from Fig. 27) of the glass viewing chamber that holds a portion of the concentration sediment and rests on the microscope stage. The fecal material is viewed through the microscope focused on the contents in the chamber. (Courtesy of DiaSys Corporation).

FIGURE 30 Various stains used in diagnostic parasitology (courtesy of Medical Chemical Corporation).

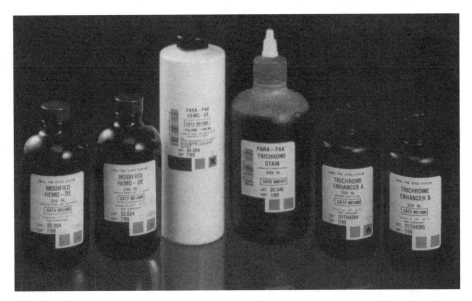

FIGURE 31 Various reagents and stains used in the ECO-STAIN system (courtesy of Meridian Diagnostics).

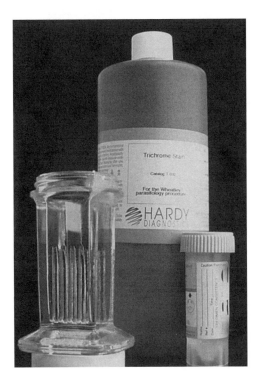

FIGURE 32 Trichrome stain (courtesy of Hardy Diagnostics).

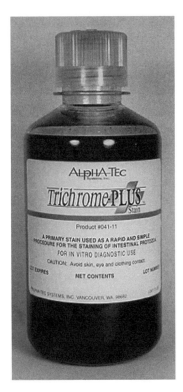

FIGURE 33 Trichrome-Plus stain (courtesy of Alpha-Tec).

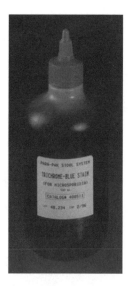

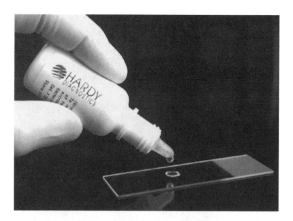

FIGURE 35 Mayer's albumin used as an adhesive for preparation of permanent stains from SAF fixative (courtesy of Hardy Diagnostics).

FIGURE 34 Trichrome-Blue stain used for the detection of microsporidial spores (courtesy of Meridian Diagnostics). See also Fig. 29.

TABLE 6 Sources of commercial reagents and supplies[a,b]

Reagent or supply	Alexon-Trend	AJP Scientific	Alpha-Tec	J.T. Baker	Bio-Spec	BIOSITE	Chemicon International, Inc.	City Chemical Corp.
Specimen collection kits								
Formalin-PVA			X		X			
PVA, modified (copper)			X					
PVA, modified (zinc)			X					
MIF			X					
SAF			X					
Single-vial system			X					
Pinworm paddles								
Parasite screening systems (immunoassays)[c]	X					X	X	
Preservatives (bulk)								
Schaudinn's fixative			X + −		X + −			
PVA fixative		X	X		X			
PVA powder[d]				X	X			
MIF		X	X					
SAF			X					
Concentration systems								
Sedimentation			X					
Flotation			X	X	X			
Stains								
Trichrome solution	X (W)	X (W)	X (W)		X (W)			
Trichrome, modified[e]	X							
Trichrome, dye powders								
Chromotrope 2R				X				
Fast green FCF				X				
Light green SF yellowish	X			X				
Hematoxylin solution					X			
Hematoxylin powder	X			X				
Chlorazol-black E powder								X
Giemsa solution	X	X		X	X			
Giemsa powder				X				
Carbol fuchsin-Kinyoun's	X	X	X		X			
Modified acid-fast, DMSO	X		X					
Auramine-rhodamine	X							
Acridine orange	X							
Toluidine blue O	X							
Miscellaneous								
Lugol's iodine, dilute (1:5)	X	X	X		X			
Dobell & O'Connor's iodine			X					
Triton X-100			X	X				
Eosin-saline solution (1%)	X				X			
Mayer's albumin			X					
Control slides or suspensions	X		X					

(Continued on next page)

TABLE 6 Sources of commercial reagents and supplies[a,b] *(Continued)*

Reagent or supply	Evergreen Scientific	Fisher Scientific	Genzyme Becton Dickinson	Hardy Diagnostics	Harleco	Medical Chemical Corp.	Meridian Diagnostics	MML Diagnostic Packaging
Specimen collection kits								
Formalin-PVA	X			X		X	X	X
PVA, modified (copper)				X		X	X	X
PVA, modified (zinc)	X			X		X	X	
MIF				X		X	X	
SAF	X			X		X	X	X
Single-vial system	X			X		X	X	
Pinworm paddles	X			X				
Parasite screening systems (immunoassays)[c]			X	X		X		
Preservatives (bulk)								
Schaudinn's fixative				X +		X +	X +	
PVA fixative				X		X	X	
PVA powder[d]		X						
MIF						X	X	
SAF								
Concentration systems								
Sedimentation	X			X		X	X	
Flotation						X		
Stains								
Trichrome solution				X (W)		X (W)	X (W)	
Trichrome, modified[e]				X		X	X	
Trichrome, dye powders								
Chromotrope 2R					X			
Fast green FCF		X			X			
Light green SF yellowish		X			X			
Hematoxylin solution				X		X		
Hematoxylin powder		X			X			
Chlorazol-black E powder								
Giemsa solution		X		X	X	X		
Giemsa powder		X			X			
Carbol fuchsin-Kinyoun's		X		X	X	X		
Modified acid-fast, DMSO								
Auramine-rhodamine		X		X		X		
Acridine orange				X				
Toluidine blue O				X		X		
Miscellaneous								
Lugol's iodine, dilute (1:5)				X	X	X	X	
Dobell & O'Connor's iodine		X		X				
Triton X-100				X		X	X	
Eosin-saline solution (1%)		X				X		
Mayer's albumin				X				
Control slides or suspensions				X		X	X	

(Continued on next page)

TABLE 6 *(Continued)*

Reagent or supply	PML Microbiologicals	Poly Science	Remel	Rohm & Haas	Rowley Biochem	Scientific Device Lab	TechLab	Volu-Sol
Specimen collection kits								
Formalin-PVA	X		X					
PVA, modified (copper)	X		X					X
PVA, modified (zinc)	X		X					
MIF	X		X					
SAF	X		X					
Single-vial system								
Pinworm paddles						X		
Parasite screening systems (immunoassays)[c]							X	
Preservatives (bulk)								
Schaudinn's fixative			X −		X −			X −
PVA fixative	X		X			X		X
PVA powder[d]								
MIF								
SAF	X							
Concentration systems								
Sedimentation								
Flotation						X		
Stains								
Trichrome solution	X		X (W)		X (W, G)	X (W)		X (W, G)
Trichrome, modified[e]			X			X		X
Trichrome, dye powders								
Chromotrope 2R					X			
Fast green FCF					X			
Light green SF yellowish					X			
Hematoxylin solution	X				X			X
Hematoxylin powder					X			
Chlorazol-black E powder		X			X			
Giemsa solution					X			X
Giemsa powder					X			
Carbol fuchsin-Kinyoun's	X		X		X			X
Modified acid-fast, DMSO	X							X
Auramine-rhodamine	X		X					X
Acridine orange	X		X					X
Toluidine blue O								
Miscellaneous								
Lugol's iodine, dilute (1:5)	X		X		X			X
Dobell & O'Connor's iodine			X			X		
Triton X-100	X		X	X		X		
Eosin-saline solution (1%)								
Mayer's albumin			X					
Control slides or suspensions	X		X			X		

[a]Adapted from references 19 and 36.

[b]Abbreviations: X, reagent or supply available; W, Wheatley; G, Gomori; +, with acetic acid; −, without acetic acid.

[c]Immunoassay procedures include enzyme, nonenzymatic, fluorescent, and cartridge assays for organism antigen or organisms.

[d]Use grade with high hydrolysis and low viscosity for parasite studies.

[e]Used for the identification of microsporidium spores in stool or other specimens.

TABLE 7 Alternative approaches to malaria diagnosis[a]

Principle	Method	Manufacturer or comment(s)[b]
Giemsa-stained light microscopy	Thick or thin films	Traditional method; sensitive; expertise required; individual diagnosis; parasite count
	After Cyto-centrifugation	Histopathology sources
Fluorescent stains of DNA RNA	Thick or thin films	
	After centrifugation, QBC (sensitivity, 88–98%; specificity, 58–90%; PPV, 40–68%; NPV, 94–99%)	Becton Dickinson Europe
	Flow cytometry	Instrument manufacturers
Molecular methods	DNA-RNA hybridization	Poor sensitivity
	PCR	Often used as gold standard when comparing other methods
Malaria pigment detection	Dark-field microscopy	Poor sensitivity
	Automated blood cell analyzers	Various companies
Antigen detection	Histidine-rich protein 2	Becton Dickinson Europe
	Parasight-F (sensitivity, 90–92%; specificity, 99%; PPV, 96%; NPV, 97–98%)	Rapid tests for *P. falciparum*; no parasite count
	ICT Malaria P.f. (sensitivity, 93%; specificity, 98%; PPV, 94%; NPV, 98%)	AMRAD ICT
	Parasite lactate dehydrogenase	Flow Incorporated
	OptiMAL (*P. falciparum*: sensitivity, 89–92%; specificity, 92–99%; PPV, 87–98%; NPV, 95–97%); (*P. vivax*: sensitivity, 62–94%; specificity, 100%; PPV, 100%; NPV, 98%)	Rapid test for *P. vivax* and *P. falciparum*; no parasite count

[a]Adapted from reference 31 with permission.
[b]ICT Malaria P.f., AMRAD ICT, P.O. Box 228, Brookvale, New South Wales 2100, Australia; OptiMAL, Flow, Inc., 6127 S.W. Corbett, Portland, OR 97201; Parasight-F and QBC, Becton Dickinson Europe, 5 Chemin des Sources, BP 37, 38241 Meylan Cedex, France.

TABLE 8 Summary of commercially available kits for immunodetection of parasitic organisms or antigens[a]

Organism and kit name[b]	Manufacturer and/or distributor	Cost ($)/no. of tests	Type of test	Sensitivity (%)[c]	Specificity (%)[c]	PPV[c]	NPV[c]	Comment(s)	Reference(s)
C. parvum									
ProSpecT Microplate Assay	Alexon-Trend	637.50/96 199/24	EIA	94–100	98–100			Different EIA formats (contact company for additional information)	1, 6, 24, 45, 52, 58
Rapid Assay Color Vue Combination with *G. lamblia*		316.50/20 535/96 811/96	EIA EIA EIA	93–95 76–93 99	98–100 93–100 99+				1
MeriFluor (combination with *G. lamblia*)	Meridian Diagnostics	280/50	DFA	100	99	100	100	Both DFA and EIA formats available; DFA is combination reagent (*Giardia/Cryptosporidium*)	2, 4, 18, 24, 29, 56, 58
Premier EIA	Meridian Diagnostics	528/96	EIA	100	99				7, 24
Crypto-CELISA	Cellabs		EIA	89–99	93–98				
Crypto'Giardia-Cel	Cellabs		DFA	100	100				
Cryptosporidium	Novocastra		DFA					Contact manufacturer	
RIM Cryptosporidium	Remel		EIA	97–100	98–100			Contact manufacturer	
Triage Parasite Panel (combination with *G. lamblia* and *E. histolytica/E. dispar* group)	BIOSITE Diagnostics, Inc.	510/20 (8.50/ reportable test)	Cartridge device, EIA	98, 100, 92	98, 99, >99[e]	NA, NA, NA[e]	99, 100, 99[e]	Requires fresh or frozen stool; combination test with *Giardia* and *E. histolytica/E. dispar* group	25
ColorPAC *Giardia/ Cryptosporidium* Rapid Assay	Genzyme/ Becton Dickinson	554/30	Cartridge device, immunoassay	97–100, 100[f]	100, >99[f]			Can be used with fresh, frozen, or formalin-preserved stool; combination test with *Giardia*	14, 22
ImmunoCard STAT! *Giardia/ Cryptosporidium*	Meridian Diagnostics		Cartridge device, immunoassay					Can be used with fresh, frozen, or formalin-preserved stool; combination test with *Giardia*	
TechLab *Crypto/ Giardia* IF Test	TechLab	245/50 patient tests	DFA	100	100			Can be used with fresh, frozen, or formalin-preserved stool;	24

(Continued on next page)

TABLE 8 Summary of commercially available kits for immunodetection of parasitic organisms or antigens[a] (*Continued*)

Organism and kit name[b]	Manufacturer and/or distributor	Cost ($)/no. of tests	Type of test	Sensitivity (%)[c]	Specificity (%)[c]	PPV[c]	NPV[c]	Comment(s)	Reference(s)
TechLab *Cryptosporidium* Test	TechLab	575/96	EIA	89–99	93–98	100, 93–97	100, 86–96	combination test with *Giardia*	
E. histolytica TechLab *Entamoeba* Test	TechLab	575/96	EIA	93	97	96	94	Requires fresh or frozen stool; reagents *do* differentiate between *E. histolytica* and *E. dispar*	32, 40
E. histolytica/E. dispar group									
TechLab *E. histolytica/E. dispar* Test	TechLab	350/48	EIA	93	97			Does not differentiate *E. histolytica* from *E. dispar*; requires fresh or frozen stool	32, 40
E. histolytica/E. dispar	Wampole, Alexon-Trend-Seradyn		EIA	93	97				
ProSpecT Microplate Assay	Remel	637.50/96, 403/48	EIA	87	99	99	92		
Entamoeba-CELISA	Cellabs		EIA	93	97				
G. lamblia ProSpecT	Alexon-Trend	637.50/96	EIA					Different EIA formats (contact company for additional information)	2, 3, 10, 24, 49, 51, 56, 58
EZ Microplate Assay		637.50/96,	EIA	96–98	98				
MicroplateAssay		199/24	EIA	98–100	98–100	69–90	100		
Rapid assay		316.50/20,	EIA	93	98				
Combination with *Cryptosporidium* Color Vue		811/96	EIA	99.1	99.6				
GiardEIA	Antibodies, Inc.	535/96	EIA						
		176.50/24	EIA						
MeriFluor (combination with *Cryptosporidium* spp.)	Meridian Diagnostics	280/50	DFA	100	99			Both DFA and EIA formats available; DFA is combination	2, 4, 18, 24, 29, 56

TABLE 8 (Continued)

Organism and kit name[b]	Manufacturer and/or distributor	Cost ($)/no. of tests	Type of test	Sensitivity (%)[c]	Specificity (%)[c]	PPV[c]	NPV[c]	Comment(s)	Reference(s)
Premier EIA	Meridian Diagnostics	528/96	EIA	98	98			reagent (Giardia/Cryptosporidium)	58
Giardia-CELISA	Cellabs		DFA	100	99	79–100	64–100		5
Giardia-Cel	Cellabs		EIA	100	98				5
Giardia	Novocastra		EIA	98–100	98–100			Contact manufacturer	
RIM Giardia	Remel		EIA	98				Contact manufacturer	
Wampole Giardia	Wampole	EIA	98					See TechLab	
Triage Parasite Panel (combination with C. parvum and E. histolytica/E. dispar group)	BIOSITE Diagnostics, Inc.	510/20 (8.50/reportable test)	Cartridge device, EIA	100, 92[g]	99, >99[g]			Requires fresh or frozen stool; combination test with Cryptosporidium and E. histolytica/E. dispar group	25, 49
ColorPAC Giardia/Cryptosporidium Rapid Assay	Genzyme/Becton Dickinson		Cartridge device, immunoassay	97–100, 100[b]	98–100, >99[h]	100	100	Can be used with fresh, frozen, or formalin-preserved stool; combination test with Cryptosporidium	14, 22
ImmunoCard STAT! Giardia/Cryptosporidium	Meridian Diagnostics		Cartridge device, immunoassay					Can be used with fresh, frozen, or formalin-preserved stool; combination test with Giardia	
TechLab Crypto/Giardia IF Test	TechLab, Wampole	245/50 patient tests	DFA	100	100			Can be used with fresh, frozen, or formalin-preserved stool; combination test with Cryptosporidium	24
TechLab GIARDIA TEST	TechLab, Wampole	575/96	EIA	91–100	98–100	80–100	96–100		10, 18, 24
T. vaginalis Affirm VP₃	MicroProbe/Becton Dickinson	504/24 (three analytes)	DNA probe	90	100			T. vaginalis, G. vaginalis, Candida spp. (seven species)	11, 16
Quik-Trich	Integrated Diagnostics Inc.	90/12; 225/30	Latex agglutination	95	100			Trichomonas only	13, 34, 55
Quik-Tri/Can		165/12; 395/30	Dual latex agglutination					Combination with Candida	30

(Continued on next page)

TABLE 8 Summary of commercially available kits for immunodetection of parasitic organisms or antigens[a] *(Continued)*

Organism and kit name[b]	Manufacturer and/or distributor	Cost ($)/no. of tests	Type of test	Sensitivity (%)[c]	Specificity (%)[c]	PPV[c]	NPV[b]	Comment(s)	Reference(s)
T. VAG DFA kit	Light Diagnostics-Chemicon International	195/50	DFA	97	99			96.1% agreement with wet mount, but more positives found with DFA (8/231)	38, 55
Plasmodium spp.									
ICT Malaria P.f.	Chemicon International		Rapid	80–100				A complete review is found in reference 31; additional references for specific products are listed and discussed.	31
ICT Malaria P.f./P.v.	Chemicon International		Rapid	80–100					
Parasight-F	Becton Dickinson		Rapid	87–97	81–99				
Malaria-Ag	Cellabs		EIA						
OptiMAL	Flow		Rapid	93					
Helminths									
Wuchereria bancrofti ICT Filariasis	Chemicon		Rapid	100					9, 44, 46
Filariasis Ag-CELISA	Cellabs		EIA	96	100				
TropBio	JCU Tropical		Rapid	100					54
			EIA	100					54
Taenia specific Ag	Genzyme Virotech	130/kit	ELISA	2.5× microscopy, >95%					GC[d]

[a]Reference 20 contains an excellent review of diagnostic products for the diagnosis of malaria (other than routine thick or thin blood films). Reference 35 contains information on immunodiagnostic products related to all relevant human parasitic infections. Both of these references are highly recommended as review documents for an excellent overview of possible test options. Not every reference has been included; however, those cited will provide the reader with a comprehensive review of these commercially available kits. Unfortunately, as new kits come to market and various companies merge, it may be somewhat difficult to identify all the kits regarding distributors. If in doubt, contact the manufacturer for additional information. Manufacturer addresses are as follows: Alexon-Trend-Seradyn, 14000 Unity St., NW, Ramsey, MN 55303-9115; Antibodies Inc., P.O. Box 1560, Davis, CA 95617; BIOSITE Diagnostics, Inc., 11030 Roselle St., San Diego, CA 92121; Chemicon International, Inc., 28835 Single Oak Dr., Temecula, CA 92590; Flow, Inc., 6127 S.W. Corbett, Portland, OR 97201; Genzyme Diagnostics, 1531 Industrial Rd., San Carlos, CA 94070; Genzyme Virotech, GmbH, Lowenplatz 5, 66428 Russelheim, Germany; Integrated Diagnostics, 1756 Sulphur Springs Rd., Baltimore, MD 21227; Meridian Diagnostics, Inc., 3471 River Hills Dr., Cincinnati, OH 45244; MicroProbe Corp., 1725 220th St., NE, Bothell, WA 98021; Novocastra, 30 Ingold Rd., Burlingame, CA 94010; Remel, 12076 Santa Fe Dr., Lenexa, KS 66215; TechLab Inc., 1861 Pratt Dr., Suite 1030, Blacksburg, VA 24060-6364; TropBio Pty Ltd., James Cook University, Townsville, Queensland 4811, Australia; Wampole Laboratories, Half Acre Rd., P.O. Box 1001, Cranbury, NJ 08512.

[b]A number of the kits are manufactured by a single manufacturer but labeled under different company names; consequently, some of the data for sensitivity and specificity may be identical to data for kits with another name or distributed by another company.

[c]Percentages have been rounded off to the nearest number; specific references provide exact percentages. Abbreviations: PPV, positive predictive value; NPV, negative predictive value; NA, not available.

[d]GC, general comments, no specific references identified.

[e]Data are for *Giardia*, *Entamoeba* spp., and *Cryptosporidium*, respectively.

[f]Data are for C. *parvum* and G. *lamblia*, respectively.

[g]Data are for *Entamoeba* spp. and *Cryptosporidium*, respectively.

[h]Data are for C. *parvum* and G. *lamblia*, respectively.

FIGURE 36 Sputocol brand sputum collection kit (courtesy of Evergreen Scientific).

FIGURE 38 In Pouch TV culture pouch for the growth and identification of *T. vaginalis* (courtesy of BIOMED Diagnostics).

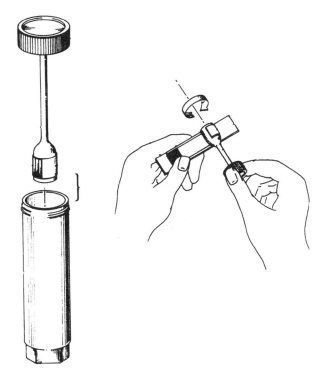

FIGURE 37 PCA pinworm egg collection-transport apparatus. The PCA combines the utility of the cellulose tape with the ready-to-use convenience of the adhesive paddle (courtesy of Evergreen Scientific).

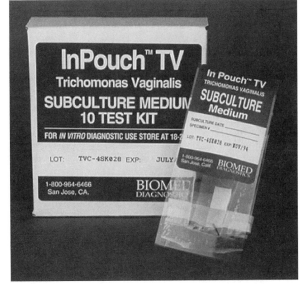

FIGURE 39 In Pouch TV subculture medium for the growth and identification of *T. vaginalis* (courtesy of BIOMED Diagnostics).

REFERENCES

1. **Aarnaes, S. L., J. Blanding, S. Speier, D. Forthal, L. M. de la Maza, and E. M. Peterson.** 1994. Comparison of the ProSpecT and Color Vue enzyme-linked immunoassays for the detection of *Cryptosporidium* in stool specimens. *Diagn. Microbiol. Infect. Dis.* **19:**221–225.

2. **Aldeen, W. E., K. Carroll, A. Robison, M. Morrison, and D. Hale.** 1998. Comparison of nine commercially available enzyme-linked immunosorbent assays for detection of *Giardia lamblia* in fecal specimens. *J. Clin. Microbiol.* **36:**1338–1340.

3. **Aldeen, W. E., D. Hale, A. J. Robison, and K. Carroll.** 1995. Evaluation of a commercially available ELISA assay for detection of *Giardia lamblia* in fecal specimens. *Diagn. Microbiol. Infect. Dis.* **21:**77–79.

4. **Alles, A. J., M. A. Waldron, L. S. Sierra, and A. R. Mattia.** 1995. Prospective comparison of direct immunofluorescence and conventional staining methods for detection of *Giardia* and *Cryptosporidium* spp. in human fecal specimens. *J. Clin. Microbiol.* **33:**1632–1634.

5. **Aretio, R., M. J. Perez, and E. Martin.** 1994. Evaluation of 2 immunologic techniques (ELISA, DIF) for the diagnosis of *Giardia lamblia* in feces. *Enferm. Infecc. Microbiol. Clin.* **12:**337–340.

6. **Arrowood, M. J.** 1997. Diagnosis, p. 43–64. *In* R. Fayer (ed.), *Cryptosporidium and Cryptosporidiosis.* CRC Press, Boca Raton, Fla.

7. **Baveja, U. K.** 1998. Acid fast staining versus ELISA for detection of *Cryptosporidium* in stool. *J. Commun. Dis.* **30:**141–144.

8. **Beaver, P. C., R. C. Jung, and E. W. Cupp.** 1984. *Clinical Parasitology,* 9th ed. Lea & Febiger, Philadelphia, Pa.

9. **Bhumiratana, A., S. Koyadun, S. Suvannadabba, K. Karnjanopas, J. Rojanapremsuk, P. Buddhirakkul, and W. Tantiwattanasup.** 1999. Field trial of the ICT filariasis for diagnosis of *Wuchereria bancrofti* infections in an endemic population of Thailand. *Southeast Asian J. Trop. Med. Public Health* **30:**562–568.

10. **Boone, J. H., T. D. Wilkins, T. E. Nash, J. E. Brandon, E. A. Macias, R. C. Jerris, and D. M. Lyerly.** 1999. TechLab and Alexon *Giardia* enzyme-linked immunosorbent assay kits detect cyst wall protein 1. *J. Clin. Microbiol.* **37:**611–614.

11. **Briselden, A. M., and S. L. Hillier.** 1994. Evaluation of Affirm VP Microbial Identification Test for *Gardnerella vaginalis* and *Trichomonas vaginalis. J. Clin. Microbiol.* **32:**148–152.

12. **Brooke, M. M., and M. Goldman.** 1949. Polyvinyl alcohol-fixative as a preservative and adhesive for protozoa in dysenteric stools and other liquid material. *J. Lab. Clin. Med.* **34:**1554–1560.

13. **Carney, J. A., P. Unadkat, A. Yule, R. Rajakumar, C. J. Lacey, and J. P. Ackers.** 1988. New rapid latex agglutination test for diagnosing *Trichomonas vaginalis* infection. *J. Clin. Pathol.* **41:**806–808.

14. **Chan, R., J. Chen, M. K. York, N. Setijono, R. L. Kaplan, F. Graham, and H. B. Tanowitz.** 2000. Evaluation of a combination rapid immunoassay for detection of *Giardia* and *Cryptosporidium* antigens. *J. Clin. Microbiol.* **38:**393–394.

15. **DeMeo, L. R., D. L. Draper, J. A. McGregor, D. F. Moore, C. R. Peter, P. S. Kapernick, and W. M. McCormack.** 1996. Evaluation of a deoxyribonucleic acid probe for the detection of *Trichomonas vaginalis* in vaginal secretions. *Am. J. Obstet. Gynecol.* **174:**1339–1342.

16. **Doing, K. M., J. L. Hamm, J. A. Jellison, J. A. Marquis, and C. Kingsbury.** 1999. False-positive results obtained with the Alexon ProSpecT *Cryptosporidium* enzyme immunoassay. *J. Clin. Microbiol.* **37:**1582–1583.

17. **Federal Register.** 1991. Occupational exposure to bloodborne pathogens. *Fed. Regist.* 29CFR1910.1030.

18. **Fedorko, D. P., E. C. Williams, N. A. Nelson, L. B. Calhoun, and S. S. Yan.** 2000. Performance of three enzyme immunoassays and two direct fluorescence assays for detection of *Giardia lamblia* in stool specimens preserved in ECOFIX. *J. Clin. Microbiol.* **38:**2781–2783.

19. **Garcia, L. S.** 2001. *Diagnostic Medical Parasitology,* 4th ed. ASM Press, Washington, D.C.

20. **Garcia, L. S.** 1999. *Practical Guide to Diagnostic Parasitology.* ASM Press, Washington, D.C.

21. **Garcia, L. S.** 1999. Parasitology: stool collection and processing. *Rev. Med. Microbiol.* **10:**51–63.

22. **Garcia, L. S., and R. Y. Shimizu.** 2000. Detection of *Giardia lamblia* and *Cryptosporidium parvum* antigens in human fecal specimens using the ColorPAC combination rapid solid-phase qualitative immunochromatographic assay. *J. Clin. Microbiol.* **38:**1267–1268.

23. **Garcia, L. S., and R. Y. Shimizu.** 1998. Evaluation of intestinal protozoan morphology in human fecal specimens preserved in EcoFix: comparison of Wheatley's trichrome stain and EcoStain. *J. Clin. Microbiol.* **36:**1974–1976.

24. **Garcia, L. S., and R. Y. Shimizu.** 1997. Evaluation of nine immunoassay kits (enzyme immunoassay and direct fluorescence) for detection of *Giardia lamblia* and *Cryptosporidium parvum* in human fecal specimens. *J. Clin. Microbiol.* **35:**1526–1529.

25. **Garcia, L. S., R. Y. Shimizu, and C. N. Bernard.** 2000. Detection of *Giardia lamblia, Entamoeba histolytica/Entamoeba dispar,* and *Cryptosporidium parvum* antigens in human fecal specimens using the Triage parasite panel enzyme immunoassay. *J. Clin. Microbiol.* **38:**3337–3340.

26. **Garcia, L. S., R. Y. Shimizu, T. C. Brewer, and D. A. Bruckner.** 1983. Evaluation of intestinal parasite morphology in polyvinyl alcohol preservative: comparison of copper sulfate and mercuric chloride base for use in Schaudinn's fixative. *J. Clin. Microbiol.* **17:**1092–1095.

27. **Garcia, L. S., R. Y. Shimizu, and D. A. Bruckner.** 1986. Blood parasites: problems in diagnosis using automated differential instrumentation. *Diagn. Microbiol. Infect. Dis.* **4:**173–176.

28. **Garcia, L. S., R. Y. Shimizu, A. Shum, and D. A. Bruckner.** 1993. Evaluation of intestinal protozoan morphology in polyvinyl alcohol preservative: comparison of zinc sulfate- and mercuric chloride-based compounds for use in Schaudinn's fixative. *J. Clin. Microbiol.* **31:**307–310.

29. **Garcia, L. S., A. C. Shum, and D. A. Bruckner.** 1992. Evaluation of a new monoclonal antibody combination reagent for direct fluorescence detection of *Giardia* cysts and *Cryptosporidium* oocysts in human fecal specimens. *J. Clin. Microbiol.* **30:**3255–3257.

30. **Gombosova, A., M. Valent, and M. Klobusicky.** 1992. Super duo latex fixation test—an alternative diagnostic method for trichominiasis. *Bratisl. Lek. Listy* **93:**538–540.

31. **Hanscheid, T.** 1999. Diagnosis of malaria: a review of alternatives to conventional microscopy. *Clin. Lab. Haematol.* **21:**235–245.

32. **Haque, R., I. K. M. Ali, S. Akther, and W. A. Petri, Jr.** 1998. Comparison of PCR, isoenzyme analysis, and

antigen detection for diagnosis of *Entamoeba histolytica* infection. *J. Clin. Microbiol.* **36:**449–452.

33. **Hiatt, R. A., E. K. Markell, and E. Ng.** 1995. How many stool examinations are necessary to detect pathogenic intestinal protozoa? *Am. J. Trop. Med. Hyg.* **53:**36–39.

34. **Hopwood, V., E. G. Evans, and J. A. Carney.** 1985. Rapid diagnosis of vaginal candidosis by latex particle agglutination. *J. Clin. Pathol.* **38:**455–458.

35. **Horen, W. P.** 1981. Modification of Schaudinn fixative. *J. Clin. Microbiol.* **13:**204–205.

36. **Isenberg, H. D. (ed.).** 1992. *Clinical Microbiology Procedures Handbook.* American Society for Microbiology, Washington, D.C.

37. **Isenberg, H. D. (ed.).** 1995. *Essential Procedures for Clinical Microbiology.* ASM Press, Washington, D.C.

38. **Krieger, J. N., M. R. Tam, C. E. Stevens, I. O. Nielsen, J. Hale, N. B. Kiviat, and K. K. Holmes.** 1988. Diagnosis of trichomoniasis. Comparison of conventional wet-mount examination with cytologic studies, cultures, and monoclonal antibody staining of direct specimens. *JAMA* **259:**1223–1227.

39. **Melvin, D. M., and M. M. Brooke.** 1982. *Laboratory Procedures for the Diagnosis of Intestinal Parasites,* 3rd ed. U.S. Department of Health, Education, and Welfare publication no. (CDC) 82-8282. Government Printing Office, Washington, D.C.

40. **Mirelman, D., Y. Nuchamowitz, and T. Stolarsky.** 1997. Comparison of use of enzyme-linked immunosorbent assay-based kits and PCR amplification of rRNA genes for simultaneous detection of *Entamoeba histolytica* and *E. dispar. J. Clin. Microbiol.* **35:**2405–2407.

41. **National Committee for Clinical Laboratory Standards.** 1997. Procedures for the recovery and identification of parasites from the intestinal tract. Proposed guideline M28-A. National Committee for Clinical Laboratory Standards, Wayne, Pa.

42. **National Committee for Clinical Laboratory Standards.** 1999. Use of blood film examination for parasites. Approved Guideline M15-A. National Committee for Clinical Laboratory Standards, Wayne, Pa.

43. **Palmer, C. J., J. F. Lindo, W. I. Klaskala, J. Quesada, R. Kaminsky, and A. L. Ager.** 1998. Evaluation of the OptiMAL test for rapid diagnosis of *Plasmodium vivax* and *Plasmodium falciparum* malaria. *J. Clin. Microbiol.* **36:**203–206.

44. **Pani, S. P., S. L. Hoti, A. Elango, J. Yuvaraj, R. Lall, and K. D. Ramaiah.** 2000. Evaluation of the ICT whole blood antigen card test to detect infection due to nocturnally periodic *Wuchereria bancrofti* in South India. *Trop. Med. Int. Health* **5:**359–363.

45. **Parisi, M. T., and P. M. Tierno, Jr.** 1995. Evaluation of new rapid commercial enzyme immunoassay for detection of *Cryptosporidium* oocysts in untreated stool specimens. *J. Clin. Microbiol.* **33:**1963–1965.

46. **Phantana, S., S. Sensathein, J. Songtrus, S. Klagrathoke, and K. Phongnin.** 1999. ICT filariasis test: a new screening test for Bancroftian filariasis. *Southeast Asian J. Trop. Med. Public Health* **30:**47–51.

47. **Pillai, D. R., and K. C. Kain.** 1999. Immunochromatographic strip-based detection of *Entamoeba histolytica—E. dispar* and *Giardia lamblia* coproantigen. *J. Clin. Microbiol.* **37:**3017–3019.

48. **Rajakumar, R., C. J. Lacey, E. G. Evans, and J. A. Carney.** 1987. Use of slide latex agglutination test for rapid diagnosis of vaginal candidosis. *Genitourin. Med.* **63:**192–195.

49. **Rocha, M. O., R. T. Mello, T. M. Guimaraes, V. D. Toledo, M. D. Moreira, and C. A. Costa.** 1999. Detection of a *Giardia lamblia* coproantigen by using a commercially available immunoenzymatic assay, in Belo Horizonte, Brazil. *Rev. Inst. Med. Trop. Sao Paulo* **41:**151–154.

50. **Rosenblatt, J. E., and L. M. Sloan.** 1993. Evaluation of an enzyme-linked immunosorbent assay for detection of *Cryptosporidium* spp. in stool specimens. *J. Clin. Microbiol.* **31:**1468–1471.

51. **Rosenblatt, J. E., L. M. Sloan, and S. K. Schneider.** 1993. Evaluation of an enzyme-linked immunosorbent assay for the detection of *Giardia lamblia* in stool specimens. *Diagn. Microbiol. Infect. Dis.* **16:**337–341.

52. **Scheffler, E. H., and L. L. Van Etta.** 1994. Evaluation of rapid commercial enzyme immunoassay for detection of *Giardia lamblia* in formalin-preserved stool specimens. *J. Clin. Microbiol.* **32:**1807–1808.

53. **Scholten, T. H., and J. Yang.** 1974. Evaluation of unpreserved and preserved stools for the detection and identification of intestinal parasites. *Am. J. Clin. Pathol.* **62:**563–567.

54. **Simonsen, P. E., and S. K. Dunyo.** 1999. Comparative evaluation of three new tools for diagnosis of bancroftian filariasis based on detection of specific circulating antigens. *Trans. R. Soc. Trop. Med. Hyg.* **93:**278–282.

55. **Smith, R. F.** 1986. Detection of *Trichomonas vaginalis* in vaginal specimens by direct immunofluorescence assay. *J. Clin. Microbiol.* **24:**1107–1108.

56. **Wilson, M., and P. M. Schantz.** 2000. Parasitic immunodiagnosis, p. 1117–1123. *In* G. T. Strickland (ed.), *Hunter's Tropical Medicine and Emerging Infectious Diseases,* 8th ed. W.B. Saunders Company, Philadelphia, Pa.

57. **Yang, J., and T. Scholten.** 1977. A fixative for intestinal parasites permitting the use of concentration and permanent staining procedures. *Am. J. Clin. Pathol.* **67:**300–304.

58. **Zimmerman, S. K., and C. A. Needham.** 1995. Comparison of conventional stool concentration and preserved-smear methods with Merifluor *Cryptosporidium/Giardia* direct immunofluorescence assay and ProSpecT *Giardia* EZ microplate assay for detection of *Giardia lamblia. J. Clin. Microbiol.* **33:**1942–1943.

Molecular Methods for Diagnosis of Infectious Diseases

DONALD JUNGKIND AND HARALD H. KESSLER

12

This chapter discusses both home brew assays and commercially available kits and related instrumentation for the detection of infectious agents by molecular methods. Table 1 shows the companies that have worked on developing commercial nucleic acid amplification diagnostic kits, with varying degrees of success. The goal of the chapter is to discuss commercially available reagents and kits primarily from a point of view other than sensitivity, specificity, and clinical utility. Discussions of the organism-specific and clinical-utility aspects of these products may be found in other sections of this book and in the scientific literature. Much of the material about general characteristics contained in this chapter is not published in peer-reviewed journals and reflects information made available by the companies in product literature, as well as personal communications and the opinions of the authors. While we have attempted to include a good representation of companies that produce molecular diagnostic kits and of their products, we make no claim that we have been able to review and include every company and every product of each company. We apologize in advance to the companies and our readers for any omissions, errors, or interpretations that are not consistent with their current understanding of the products. The vast amount of material and the fast pace of development will soon date a single chapter. The reader is advised to consider the material and opinions summarized here and then to continue to search for additional product information and to form opinions about the characteristics and serviceability of products in this rapidly expanding field.

This chapter is subdivided into two sections. The first section deals with molecular amplification tests that are performed under home brew assay guidelines using reagents that are commercially available. It is recognized that the use of molecular methods for diagnosis of infectious diseases is a rapidly evolving field, and much of the initial diagnostic work is done with home brew assays developed and validated within each laboratory. There can be a considerable lag time before U.S. Food and Drug Administration (FDA)-approved test kits are available. Also, for some infections, there may not yet be commercial interest in developing an FDA-approved assay. Specialized products have emerged to meet the immediate clinical needs when commercial kit assays are not available. The second section of this chapter reviews the commercially available products that make it easier to assemble home brew assays using partial kits and

analyte-specific reagents. The assays derived from commercially available kit components and analyte-specific reagents may have the potential for a higher degree of standardization and consistency than the original home brew assays. The home brew kit assays must pass a careful evaluation of accuracy, linearity (if quantitative), within-run precision, day-to-day reproducibility, comparison to reference methods, and clinical correlation before clinical use.

The second section also deals with the commercially available kits and instrumentation that have been approved by the FDA and complete kits and instrumentation that have received equivalent approval or clinical acceptance for use in Europe or Asia. We review both the direct-detection methods and the amplified kits that are available. Where possible, peer-reviewed publications were used as references. However, due to the rapid developments in this area, much of the data for this chapter is not yet published and was taken from product literature made available by the respective companies.

Another category of kits and instrumentation comprises those approved in the United States for research use only. Some of these are discussed briefly. There may be important test kits in use elsewhere in the world whose FDA submission and approval process is not complete. Such kits are made available for research use only and may not be used in the United States as the sole means for making the diagnosis of the disease. They can, when used in conjunction with other tests and clinical information as part of an approved clinical study, provide valuable information to the researcher and clinician if they have passed at least the first four aspects of test validation listed above.

HOME BREW MOLECULAR ASSAYS IN THE UNITED STATES AND EUROPE

Over the last few years, several commercially available molecular assays have been introduced. Home brew assays, however, are still the majority. In contrast to the situation in the United States (see above), there are presently no limitations on the use of home brew molecular assays (except for assays for detection of human immunodeficiency virus [HIV] infection) in most parts of the European Union. The recently announced European Parliament and Council Directive 98/79/EC, however, is concerned with in vitro diagnostic medical devices. It will lead to a change in the

TABLE 1 Commercial target amplification technologies and tests[a]

Technology	Company	Yr started	No. of tests
PCR	Kodak, Ortho	1986	0
Qβ	Gen-Trak	1986	0
bDNA	Chiron, Bayer	1987	4
LCR	Abbott	1988	3
PCR	Roche	1989	16
	Abbott	1994	1
NASBA	bioMérieux (Organon)	1991	2
SDA	Becton Dickinson	1992	5
TMA	Gen-Probe	1993	8

[a]Not all companies are still developing assays, not all assays are still available, and not all assays are FDA approved.

present situation: an obligatory performance evaluation of in vitro diagnostic medical devices will be established, including obligatory registration, license number, declaration of the manufacturer, relevant documentation, production according to quality management guidelines, and application of harmonized standards.

For detection of infectious agents in the routine molecular diagnostic laboratory, molecular assays must include three basic steps: (i) extraction of nucleic acid from clinical samples, (ii) amplification of DNA or cDNA, and (iii) hybridization and detection of amplification products.

Historically, home brew assays have involved complicated and time-consuming extraction and hybridization procedures which required toxic and radioactive reagents. Another serious challenge for a routine molecular diagnostic laboratory is to prevent contamination of negative specimens. To reduce these risks, commercially available components should be used whenever possible.

Nucleic Acid Extraction

The extraction of nucleic acid includes lysis of the nucleic-acid-containing specimen and removal of substances which might inhibit subsequent steps of reverse transcription (RT) and/or amplification while protecting target DNA or RNA from degradation. Furthermore, the risk of contamination and potential hazards caused by microorganisms and toxic reagents should be kept to a minimum during specimen preparation.

Classic extraction protocols are usually based on phenol-chloroform extraction, either with or without proteinase K digestion, followed by alcohol precipitation (13, 26, 70, 115). RNA extraction protocols are even more laborious than those for DNA (44, 103). The protocols have usually been time-consuming, labor-intensive, and susceptible to contamination. It has been demonstrated that the probability of false-positive results because of contamination increases in relation to the number of manipulations involved in sample processing (28, 74).

Today, more rapid nucleic acid extraction protocols including fewer manipulation steps have largely replaced the classic protocols. Specially designed components for extraction of DNA and/or RNA have been introduced and found useful for nucleic acid isolation from a variety of materials (4, 11, 25, 38, 47, 62, 66, 90, 105, 109, 128, 131, 145, 152) (Table 2). Several companies have recently brought prepackaged extraction kits to the market. They utilize, for instance, silica gel membranes, glass fiber surfaces, or magnetic beads. Successful nucleic acid extraction from a variety of materials has been demonstrated (6, 14, 31, 42, 43, 64, 86, 91, 96, 100, 107, 118, 120, 150, 151) (Table 3).

Sensitivity

Detection limits have been described for home brew molecular assays employing commercially available components. There are a large variety of techniques but a lack of sufficient validation. Studies describe modifications of commercially available components and extraction kits and combinations with extraction reagents from commercially available molecular assays, which are exclusively optimized as the specific assay may require (47, 89, 130). Any combination, however, would require optimization. An overall meta-analysis of those studies is not possible.

In comparison with classic nucleic acid extraction protocols, use of the cation exchanger Chelex 100 has been shown to increase the sensitivity of molecular assays for detection of HIV type 1 (HIV-1) proviral DNA in blood (40, 143), cytomegalovirus (CMV) in cultures and clinical samples (159), and *Legionella pneumophila* in bronchoalveolar lavage fluids (54).

An evaluation of eight different extraction methods for isolation of HIV-1 RNA from plasma showed slightly superior sensitivity and reproducibility of the classic extraction

TABLE 2 Examples of commercially available nucleic acid isolation components

Manufacturer	Component	Basic reagent or principle of procedure	Selected references (material/pathogen)
Bio-Rad	Chelex 100	Cationic chelating resin	62 (bronchoalveolar lavage fluid; induced sputum/*M. pneumoniae*); 66 (bronchoalveolar lavage fluid/*Legionella* spp.), 90 (urine/*B. burgdorferi*), 145 (cerebrospinal fluid/CMV)
BioVentures	Gene Releaser	Anionic binding resin	4 (sputum/*M. tuberculosis*), 109 (bronchoalveolar lavage fluid/*P. carinii*), 131 (whole blood/HIV-1)
Iowa Biotechnology	Catrimox-14	Cationic surfactant	105 (cell wall disruption), 128 (whole blood, plasma/HCV)
Life Technologies	DNAzol, TRIzol	Guanidine thiocyanate	25 (whole blood/*Ehrlichia* spp.), 152 (cell culture/rabies and rabies-related viruses)
Millipore (Amicon)	Microcon	Ultrafiltration	47 (*C. trachomatis* transport medium/*C. trachomatis*)
Roche	TriPure	Guanidine thiocyanate	38 (*C. pneumoniae*)

TABLE 3 Examples of commercially available prepackaged extraction kits

Manufacturer	Extraction kit	Principle of procedure	Selected references (material/pathogen)
Bio 101	Geneclean	Silica matrix	86 (urine/*B. burgdorferi*)
Biotecx	Ultraspec	Chloroform plus precipitation	118 (nasopharyngeal aspirate/rhinoviruses)
Dynal	Dynabeads DNA Direct	Magnetic separation	6 (cerebrospinal fluid, blood/*Neisseria meningitidis*, *H. influenzae*, *S. pneumoniae*, *S. agalactiae*, *L. monocytogenes*), 31 (whole blood, bone marrow, cell culture/genomic DNA)
Epicentre	MasterPure	Salt precipitation	150 (whole blood/HCV)
Gentra	Puregene	Cell lysis plus protein precipitation	47 (*C. trachomatis* transport medium/ *C. trachomatis*)
Macherey-Nagel	Nucleospin	Silica column	42 (blood culture, pleural fluid, bile, abdominal fluid, bronchoalveolar lavage fluid, skin biopsy specimen/*C. albicans*)
Orca	IsoQuick	Extraction matrix	91 (cerebrospinal fluid/herpes simplex virus)
Organon Teknika	NucliSens RNA Extraction Kit	Silica beads	96 (plasma/HIV-1)
Promega	Wizard	Spin column	100 (sputum/*P. carinii*)
Qiagen	QIAamp	Silica column	14 (whole blood/HIV-1), 107 (stool, ocular swab, throat swab/human adenoviruses), 120 (stool/ hepatitis A virus), 151 (amniotic fluid, fetal blood, ascitic fluid, fetal biopsy specimen, placenta tissue/human parvovirus B19)
Roche	High Pure	Glass fiber	43 (plasma/HIV-1), 64 (whole blood/HCV)

protocol compared with commercially available reagents and kits (142). The authors, however, state that excessive purification procedures mean a high workload and higher risk of contamination or sample mix-up. Therefore, they recommend for the routine extraction of HIV-1 RNA from plasma the single-step method involving TRIzol. In another study, QIAamp hepatitis C virus (HCV) kit spin columns, silica beads, and phenol-chloroform were compared, and no statistically significant difference in the minimal plasma HIV-1 RNA concentration was found (123). A comparison of four commercially available extraction kits to noncommercial kits for extraction of HIV-1 RNA from plasma showed that the QIAamp Viral RNA Kit gave the highest sensitivity (43). The High Pure Viral RNA Kit and the noncommercial extraction kits were also very sensitive. The RNA Insta-Pure LS kit and the PANext RNA extraction kit were not suitable for extraction of HIV-1 RNA in plasma. Extractions with commercially available kits proved to be fast and easy to perform. The cost per sample was less than that of the classic extraction protocols. In a recent study, HIV-1 RNA was recovered from low-titer plasma samples, and the NucliSens RNA Extraction Kit was found to be superior to the QIAamp Viral RNA Kit (96).

For extraction of hepatitis B virus (HBV) DNA in serum, a classic phenol-chloroform extraction protocol including proteinase K digestion, the Gene Releaser method, and the QIAamp blood kit were compared (72). The classic and Gene Releaser methods gave the same detection limit. Use of the commercially available components accelerated and simplified the extraction procedure.

When nine different methods for extraction of adenovirus DNA in urine were compared, the protocol employing Chelex 100 and the QIAamp blood kit yielded the highest sensitivity and proved superior for removing inhibitors (34). Furthermore, they were technically easy to perform.

Comparison of the conventional proteinase K-phenol-chloroform extraction method and the Ultraspec protocol

for extraction of rhinovirus RNA in nasopharyngeal aspirates showed equal sensitivities (118). The classic method proved more time-consuming and laborious. Therefore, the commercially available kit was preferred.

For detection of legionella DNA in urine samples, a commercially available extraction kit (Geneclean) proved superior to classic methods (85). Similar results were found for extraction of *Borrelia burgdorferi* DNA in urine (86).

For detection of *Pneumocystis carinii* DNA in blood specimens, extraction with proteinase K-phenol-chloroform, the QIAamp Tissue Kit, and the ReadyAmp Genomic DNA Purification System showed equal sensitivities (108). Because of simpler handling, the QIAamp kit was preferred.

When six different methods for extraction of the DNA of fungal pathogens were compared, use of commercial kits shortened the duration of sample preparation significantly (83). In that study, the detection limits of commercial kits (QIAamp Tissue Kit and Gene Releaser) proved similar to the classic protocol. The cost per sample, however, was considerably higher when commercial kits were used. Similar results were found in another study of rapid detection of *Candida albicans* in clinical specimens (42).

A comparison of six different methods for DNA extraction from archival Giemsa-stained bone marrow slides showed that rough DNA extraction methods had decreased efficiencies compared to complete DNA extraction protocols (144).

Contamination Control

A serious challenge for a molecular diagnostic facility is to avoid false positive results due to carryover contamination (68, 73). Risk of contamination is less of a concern when prepackaged kits are used. However, it is essential always to be aware of the potential for contamination and to prevent its occurrence rather than to deal with these problems retrospectively.

Contamination control can be classified in two basic approaches: physical and enzymatic. Physical approaches to contamination control have been described in detail elsewhere (5). The introduction of uracil-N-glycosylase provides an enzymatic means to prevent carryover contamination. Uracil-N-glycosylase inactivates previously amplified DNA that may be present in the new sample so that it can no longer serve as a template for PCR (102, 134, 138). The high denaturation temperature used in PCR conveniently destroys any residual uracil-N-glycosylase, thus keeping it from neutralizing the next generation of newly formed amplified products. Uracil-N-glycosylase, also known as AmpErase, has been successfully used in home brew PCR assays (67, 71, 82).

Laboratory and Personnel Requirements

For classic extraction protocols, special laboratory equipment and protection of personnel are required. Implementation of commercially available DNA isolation components or use of prepackaged extraction kits minimizes hazards to personnel and helps to simplify the procedure. For most prepackaged extraction kits, only basic laboratory equipment, including a sterile working bench, heating block, vortex, and microcentrifuge, is required. Because extraction with the kits is less time-consuming and easier to perform, savings in personnel are possible. All in all, lower cost is usually expected when commercially available DNA extraction components or prepackaged extraction kits are used.

Nucleic Acid Amplification

Home brew molecular assays usually employ PCR technology for nucleic acid amplification. The whole procedure is carried out in a programmable thermal cycler. Usually, 30 to 40 cycles are done, which results in a theoretically exponential increase in the total number of DNA copies synthesized (35, 153).

For amplification of RNA targets, RNA must be converted to cDNA by RT prior to PCR. Conventional RT-PCR is a two-step process requiring separate enzymes and buffer conditions. The reopening of the reaction tubes is considered to be the major contamination source. Therefore, the DNA polymerase of the thermophilic bacterium *Thermus thermophilus*, which possesses enhanced reverse transcriptase activity in the presence of manganese ions, was introduced (94). RT is carried out at 60°C, which increases specificity as well as efficiency by destabilizing secondary structures in the RNA template. Furthermore, there is no need to change or add reagents between the RT and PCR steps. The reaction tubes remain closed during the entire procedure, which minimizes the contamination risk and reduces the hands-on work required.

To enhance sensitivity, nested PCR is frequently used. This widely used technique, however, is technically difficult, a permanent source for carryover contamination, and poorly reproducible between laboratories (158). Enzymatic decontamination is not possible when nested PCR is used.

Therefore, nested PCR is not recommended for routine diagnostic assays.

Hybridization and Detection of Amplification Products

In the routine diagnostic laboratory, hybridization of amplification products is an obligatory step. Compared to gel electrophoresis alone, nucleic acid hybridization leads to increased sensitivity and excludes unspecific amplification products (110, 112, 122). A classic method for analyzing amplification products is agarose gel electrophoresis followed by Southern blot probe hybridization, which is a time-consuming and labor-intensive process and usually involves radioactive reagents (18, 117, 132, 135). To avoid problems related to radioactivity, nonradioactive hybridization techniques are being used more and more. Among them, the biotin-avidin (or biotin-streptavidin) and the digoxigenin-antidigoxigenin systems have been widely applied as DNA enzyme immunoassays for the detection of PCR amplification products (8, 60, 61, 87). The digoxigenin-antidigoxigenin technique requires labeled amplification products, which are generated by specifically labeled primers or by incorporation of digoxigenin-dUTP instead of dTTP. Several manufacturers have brought nonradioactive hybridization and colorimetric detection techniques in microtiter plate format to the market, and additional efforts have been made to better standardize the digoxigenin-antidigoxigenin technique by automation (10, 16, 32, 65, 77, 78, 79, 92, 111, 121, 129, 131, 133, 139, 160, 162) (Table 4).

Sensitivity

The detection limits of commercially available hybridization and detection systems have been compared to those of classic Southern blot hybridization and detection employing radioactive or nonradioactive reagents as a reference assay. Only a limited number of studies which compare commercially available hybridization and detection systems with classic methods have been published.

A comparative evaluation of colorimetric microtiter plate systems based on the biotin-streptavidin or the digoxigenin-antidigoxigenin system for detection of herpes simplex virus in cerebrospinal fluid showed that three of the four microtiter plate systems gave sensitivities equal to that of Southern blot hybridization (133). Comparison of microtiter plate assays based on either the biotin-streptavidin or the digoxigenin-antidigoxigenin system with Southern blot hybridization for detection of CMV DNA or legionella DNA showed complete correlation (39, 55). When a microtiter plate hybridization based on the biotin-avidin technique, a fully automated system based on the digoxigenin-antidigoxigenin technique, and Southern blot hybridization were compared, concordant results were obtained with cerebrospinal fluid samples from patients with and without herpes simplex encephalitis (116).

TABLE 4 Commercially available hybridization and detection kits

Manufacturer	Detection kit	Principle of procedure	Selected references
Abbott (formerly Digene)	SHARP Signal System	Biotin-streptavidin	16, 79, 129, 139, 160, 162
DiaSorin	Gen-Eti-K	Biotin-avidin	10, 82, 131
Roche (formerly Boehringer Mannheim)	PCR ELISA, Enzymmun	Digoxigenin-antidigoxigenin	32, 65, 77, 78, 111, 121

For detection of rabies and rabies-related viruses, a microtiter plate hybridization based on the digoxigenin-antidigoxigenin technique proved 100 times more sensitive than a technology based on Southern blot hybridization (152). When a fully automated system based on the digoxigenin-antidigoxigenin technique and the classical Southern blot hybridization were compared for detection of hepatitis A virus RNA in stool specimens, the automated hybridization and detection system showed two-times-better sensitivity (120).

In all studies, microtiter plate techniques and the fully automated system proved superior to classical Southern blotting, allowing simultaneous analysis of multiple samples in a significantly shorter time. Furthermore, these assays do not require toxic chemical agents and may be adapted for automation.

Contamination Control

Hybridization and detection steps are done in the postamplification area, which must be strictly separated from the preamplification area. To avoid carryover contamination, materials used in the postamplification area must never be used in the preamplification area.

Compared to Southern blot hybridization, microwell titer plate-based methods require fewer manual steps and thus may be less prone to contamination problems. When the fully automated system is used, contamination during the hybridization and detection steps has never been reported.

Laboratory and Personnel Requirements

Unlike Southern blot hybridization, which requires special laboratory equipment and, in the case of radioactivity, protection of personnel, microtiter plate-based methods help to simplify the hybridization and detection steps. The equipment of the routine diagnostic laboratory (incubator, microtiter plate washer, and microtiter plate reader) is sufficient. Microtiter plate-based methods are less hazardous, less time-consuming, and easier to perform. Lower cost is expected when commercially available microtiter plate-based methods are used in the routine diagnostic laboratory.

Quality Control

Because of the major variation in the performance of home brew molecular assays, quality control is difficult to address. When commercially available components for home brew molecular assays are used, the variability between assays may be reduced. Attention must be paid to internal controls, external controls, and interlaboratory quality assurance.

Internal controls are positive controls to detect a technical failure of the assay, which could be produced, for example, by the presence of an inhibitor. A competitive internal standard is therefore introduced into the assay; it has primer-binding regions identical to those of the target sequence, a randomized internal sequence with a length and a base composition similar to those of the target sequence, and a unique probe-binding region that differentiates the internal-control amplification product from the target amplification product (114).

External controls include positive and negative controls. They must be included in the entire test process. External controls are often provided by the test kit manufacturer to perform with one specific test. External controls, which are independent of the specific test kit, may also be used. They are provided by unbiased control manufacturers and are designed to perform with a large number of test kits and procedures.

Interlaboratory quality assurance assesses the results of a certain laboratory in comparison with those of other laboratories. International proficiency studies have showed that the quality of performance in laboratories using home brew molecular assays was worse than that in laboratories using commercially available assays (29, 76, 158).

Conclusions

The integration of commercially available components into home brew molecular assays for detection of pathogens helps to avoid human error and enables standardization of testing procedures, thus increasing the precision and reproducibility of the results. To guarantee the analytical quality of home brew molecular assays in the routine diagnostic laboratory, the use of commercially available components and automation must be considered seriously. The next section deals with nucleic acid tests (NAT) that have been taken to the next stage of development: complete commercial kit assays for the most frequently used tests.

COMMERCIALLY AVAILABLE NUCLEIC ACID DETECTION ASSAYS

The commercially available nucleic acid detection assays have several important advantages over home brew assays (80). These assays fall into two general categories. The first to be discussed will be the direct nonamplified nucleic acid detection methods, and the second will be the amplified tests.

Nonamplified Methods

Gen-Probe Nonamplified Direct Nucleic Acid Detection Methods

DNA probes to detect specific rRNA targets for various organisms are available from Gen-Probe, Inc. [10210 Genetic Center Dr., San Diego, CA 92121; phone, (858) 410-8000]. rRNA has an advantage as a target because there are many thousands of copies in each cell, adding to test sensitivity. This is in contrast to DNA targets, which typically have only one or a few copies per cell. Moreover, the ribosome has unique regions associated with many species, allowing greater specificity. Gen-Probe has three main lines of nucleic acid diagnostic products: culture identification tests, direct DNA probe detection products, and nucleic acid amplification products. The molecular amplification products will be discussed in a separate section. The AccuPROBE culture identification tests are used to identify organisms grown in culture. The direct DNA probe assays are used to identify organisms directly in specimens. The two direct-probe products are the PACE 2 test for the dual detection of *Chlamydia trachomatis* and *Neisseria gonorrhoeae* from cervical and male urethral swabs, and the Group A Streptococcus Direct test for the detection of *Streptococcus pyogenes* from throat swabs (49, 53).

The AccuPROBE culture identification tests are used to identify mycobacteria, fungi, and other bacteria grown in culture. In the mycobacterium line, there are probes to identify *Mycobacterium tuberculosis* complex, *Mycobacterium avium* complex, *M. avium*, *Mycobacterium intracellulare*, *Mycobacterium gordonae*, and *Mycobacterium kansasii* that have been grown on common agar or broth media (3, 125, 136). In the fungal line, there are AccuPROBE tests for

Blastomyces dermatitidis, *Coccidioides immitis*, and *Histoplasma capsulatum* (48, 101).

The FDA-approved bacterial AccuPROBE tests are those for identification of *Campylobacter* species, including *Campylobacter jejuni*, *Campylobacter coli*, and *Campylobacter lari*, isolated from culture. The AccuPROBE kit for *Enterococcus* species will identify *Enterococcus avium*, *Enterococcus casseliflavus*, *Enterococcus durans*, *Enterococcus faecalis*, *Enterococcus faecium*, *Enterococcus gallinarum*, *Enterococcus hirae*, *Enterococcus mundtii*, *Enterococcus pseudoavium*, *Enterococcus malodorous*, and *Enterococcus raffinosus*. There are also AccuPROBE tests for culture isolates of *Haemophilus influenzae*, *Listeria monocytogenes*, *N. gonorrhoeae*, *Staphylococcus aureus*, *Streptococcus agalactiae*, *S. pyogenes*, and *Streptococcus pneumoniae* (17, 154).

The principle of both assays is based on hybridization of rRNA from the organism with an acridinium ester-labeled DNA hybridization probe complementary to a definitive rRNA region for each organism. A chemical process (hybrid protection assay) is used to distinguish between hybridized and unhybridized probe instead of a more difficult physical separation method. The final step of the assay is to put the tubes in a luminometer, which automatically injects the proper reagents into the tubes to produce and detect the chemiluminescent signal. The assay detection signal is a qualitative chemiluminescent reaction using an acridinium ester label that is present in the hybrid in proportion to the amount of target. This is a homogeneous assay format, which requires no wash steps that could spread contamination. The PACE 2 assay uses a similar procedure but also includes magnetic particles, which specifically capture hybridized molecules and decrease the background levels of the test. The assays use a standard format and reagents. For the smaller laboratory, there is the manual pipetting procedure. For larger batch sizes, the MultiPROBE automated liquid-handling system, a fully automated sample and reagent processor, can be used with the Gen-Probe assays. It interfaces with the LEADER 450i luminometer. This instrument stores up to 30 protocols, allows the operator to program up to 11 different assays in one run, and offers automated processing of up to 250 samples.

The advantage of these assays is their simplicity. They are nonisotopic and, depending on the particular assay, can be completed in 30 to 120 min. They do not have the potential contamination problems of the target amplification assays. When the quantity of target is sufficiently large, the direct-probe tests can allow for accurate detection of the target organism. With regard to the culture identification tests, rapid identification of mycobacteria can have a significant clinical impact in the areas of epidemiology and therapy. For the systemic fungi, the turnaround time for identification can be significantly improved. Rapid and definitive identification of problem bacterial isolates can be useful in certain situations. Direct-probe tests may have less sensitivity than a well-done culture or an amplified nucleic acid assay (75).

Becton Dickinson and Company Direct Nonamplified DNA Detection (Affirm)

Becton Dickinson Biosciences [7 Loveton Circle, Sparks, MD 21152-0999; phone, (410) 316-4000] has a DNA probe-based test, the Affirm VPIII, which uses complementary sequences of DNA that hybridize with the targeted organisms. Their first test is for vaginitis; it can differentiate *Candida*, *Gardnerella*, and *Trichomonas*, the three major causes of vaginitis, from a single sample with a total time to results of under 1 h (19, 33). There is a vaginal-fluid sample preparation step requiring 10 min. The probe hybridization and color development occurs in 33 min and requires a small walkaway instrument. A positive reaction results in a blue-colored bead, detectable visually. The total hands-on time is approximately 5 min. The ability to multiplex three analytes into a single easy-to-perform test is the advantage of this system.

Amplified Methods

This section deals with the tests for specific nucleic acid analytes that utilize some method to amplify either the target, the probe, or the signal in order to increase the sensitivity of the assay.

Signal Amplification Assays
Bayer bDNA Assay

The branched-DNA (bDNA) assay [Bayer Diagnostics: World, 511 Benedict Ave., Tarrytown, NY 10591; phone, (914) 524-3837] achieves higher analytical sensitivity by enhancing the signal of a hybridization-based assay (98). bDNA assays are based on sandwich hybridization in which probes combine with specific complementary targets consisting of viral nucleic acid. The bDNA assay has been commercialized as the Quantiplex test for quantification of HIV-1. In the first-generation assay, signal amplification was achieved by having 15 branches on the probe. Each branch contained binding sites for three separate label probes. Theoretically, each amplifier molecule bound to a target can bind up to 45 separate label probes. The second-generation HIV-1 assay was made to achieve attachment of up to 10,080 alkaline phosphatase-labeled probes to a single viral RNA target through a more complex branching scheme.

In the third-generation Quantiplex bDNA HIV-1 assay, more signal sensitivity was achieved by reducing background by incorporation of nonnatural nucleotides (iso-C and iso-G), which hybridize to the target with higher affinity than their natural counterparts, and through the redesign of target probes. Increasing the number of binding sites for the preamplifier probes gives rise to increased signal amplification. By these techniques, the sensitivity of the bDNA assay has been improved from 10,000 virus copies per ml in the first-generation assay to 500 copies per ml in the second-generation assay and to 50 copies per ml in the third-generation assay. The upper limit of detection is 1 million copies per ml in the HIV-1 version 2.0 bDNA assay. In the version 3.0 assay, the upper range is 500,000 copies per ml (36, 93).

In the Quantiplex assay, a 1-ml aliquot of plasma is centrifuged to sediment HIV-1 particles. The pelleted virus is lysed to release HIV-1 RNA from virions by incubation with proteinase K and detergent. Viral RNA is captured on the surface of a 96-microwell plate by hybridization with 10 oligonucleotide target probes that hybridize to the viral RNA *pol* region and to capture probes chemically linked to the microwell plate. After the plates are washed, bDNA amplifier molecules are hybridized to the bound nucleic acid by a second set of capture probes, and alkaline phosphatase-labeled probes are bound to the bDNA. These probes mediate binding of amplifier to the target in the next step of the assay. Finally, a chemiluminescent dioxethane substrate is added. The quantitation is determined by a four-point standard curve assayed in each microplate. Similar assays have been developed for HCV and HBV. A new automated version (Quantiplex 340 Automated bDNA System) greatly reduced the workload, with 168 samples per run.

The Bayer HCV RNA 2.0 Quantitative Assay (bDNA) is a signal amplification nucleic acid probe assay for the quantitation of HCV RNA in human serum or plasma. It is also automated. The assay can be run on an instrument with up to 84 results per run. Recently, Bayer introduced an HCV RNA 3.0 Quantitative Assay (bDNA) with a dynamic range reported to be more than 4 log units and with up to a 100-fold-increased sensitivity compared to the second-generation assay.

The Bayer HBV DNA Quantitative Assay (bDNA) is a signal amplification nucleic acid probe assay for the quantitation of HBV DNA in human serum. The assay has a wide dynamic range of HBV DNA that can be achieved without repeat testing of diluted specimens. It uses a 96-microwell format run on a system with up to 84 results per run.

The bDNA instrument system can process from 12 to 168 samples, with simultaneous processing of two microtiter plates. There is data management software that receives data from the System 340, graphs the standard curve, and generates the results. These assays are available for research use only and not FDA approved for diagnostic procedures in the United States.

The advantages of the quantitative bDNA assay are good tolerance of target sequence variability, simple sample preparation, less sample-to-sample variation, and a wider dynamic range than can be achieved with traditional first-generation PCR-based quantitative assays (98). For HIV, the bDNA assay can quantify subtypes A to F with equal accuracy. The relative costs of labor, disposables, and biohazardous wastes were not determined in most cases. In a study by Elbeik et al. using matched clinical plasma samples to compare the quantification and costs of the bDNA 3.0 HIV-1 assay and the AMPLICOR version 1.5 HIV-1 assay [Roche Molecular Systems, 4300 Hacienda Dr., Pleasanton, CA 94588; phone, (925) 730-8000], a cost analysis based on labor, disposables, and biohazardous wastes showed savings with the bDNA 3.0 assay compared to the costs of the nonautomated AMPLICOR 1.5 assay (36).

A disadvantage of the assay is that it requires an overnight incubation step. Also, the bDNA reports higher copy numbers per milliliter than some RT-PCR methods. This may not be an issue, as the assays are adjusted to report in standardized international units. Another disadvantage at the time of this writing is that the tests are available in the United States as kits for research use only. The Quantiplex HIV 3.0 assay has received Agence de Médicament approval in France.

Digene Hybrid Capture System

The Hybrid Capture II system [Digene Corporation, 1201 Clopper Rd., Gaithersburg, MD 20878; phone, (301) 944-7000] uses signal amplification to detect and quantitate viral nucleic acid. Digene has FDA-approved assays for human papillomavirus and for CMV (27, 50). The human papillomavirus assays are ViraPap and ViraType Plus. ViraType Plus is a quantitative assay. The test for CMV is qualitative at this time but has been successfully configured as a quantitative assay. Assays are available for *C. trachomatis* and *N. gonorrhoeae*, and a research-use-only quantitative assay is available for HBV (24).

The basic steps of the assay start with a clinical specimen being combined with a base solution which disrupts the virus or bacteria and releases target DNA. The target DNA is denatured and hybridized with an RNA probe, creating RNA-DNA hybrids, which are captured with an antibody specific for them. Alkaline phosphatase conjugated to antibodies is combined with the hybrids, and the bound alkaline phosphatase is detected with a chemiluminescent dioxetane substrate. Upon cleavage by alkaline phosphatase, the substrate produces light that is measured in a luminometer in relative light units.

Digene's CMV DNA test is the first molecular diagnostic test to be FDA cleared for the qualitative detection of human CMV DNA in peripheral white blood cells isolated from whole blood. The RNA probe covers 17% of the total CMV genome. There is reported to be no cross-reactivity to herpesviruses or human DNA. The test can provide results within 6 h, and the lower limit of detection is 700 copies/ml of blood. It is reported to detect significantly more active CMV infections than culture techniques (50). A blood sample (EDTA) can be stored at room temperature for up to 24 h and for an additional 24 h at 4°C, for a total storage time of 48 h. The advantage of the Digene hybrid capture system is that target amplification is not required. It is more sensitive than the first-generation direct-detection systems. It has good linearity and precision. The hybrid capture system can work in small or large specimen batches. It has a relatively simple workflow, taking approximately 4 h to complete.

Its disadvantage is that although it can be quantitative, the theoretical lower limit of sensitivity is not as great as with the nucleic acid amplification assays (104). However, not all types of assays require the ultimate in sensitivity to differentiate active disease from subclinical quiescent infection.

Nucleic Acid Target Amplification Assays
Roche AMPLICOR PCR

PCR was the first NAT and has been the most successfully implemented technology. It has had widespread use in the research and clinical areas. As a result, there are published assays for the majority of the infectious agents. Research and commercial assays have been developed for DNA and RNA targets, both qualitative and quantitative.

The principle of PCR is well known. A DNA target is required. Two primers specific for the target region and a thermostable DNA polymerase are key components in the amplification process. PCR requires alternating temperatures to anneal and amplify, creating double-stranded DNA. A higher temperature is required to separate the strands and allow the amplification process to continue in a geometric manner until millions of copies are made from a single target. If the original source target is RNA, a reverse transcriptase is used to create a DNA copy that can be amplified and detected. Quantitation can be achieved by at least two methods. The first-generation method was to make dilutions of the amplicon, measure each of them, and compare them to similarly amplified standards. The TaqMan tests are being developed as second-generation real-time quantitative tests. Commercial kits for commonly required assays plus related instrumentation are in development and trials. The homogeneous TaqMan assay has the advantage of providing answers in real time rather then having to wait for additional detection steps. It requires no postamplification detection steps, uses a closed system, and has a wide dynamic range for quantitative detection of either RNA or DNA targets. The mechanics of the instrumentation are also simpler. This second-generation Roche TaqMan system is not yet FDA approved in a commercial clinical kit format, but it is available worldwide for home brew PCR assay development using the Roche Light Cycler and basic reagents available commercially. The Roche COBAS TaqMan system, coupled with an automated specimen preparation system, is under development, with evaluations scheduled for 2002. The initial launch menu for this system will be quantitative tests for HCV, HIV-1, and HBV.

The first commercial PCR kit assay was the Roche AMPLICOR microwell assay for *C. trachomatis* (7). This assay was the prototype for a series of assays for *M. tuberculosis*, *N. gonorrhoeae*, HIV-1, HBV, HCV, CMV, and an internal control (9). The assay used a Perkin-Elmer 9600 thermocycler and an enzyme-linked immunosorbent assay-like postamplification colorimetric enzymatic reaction involving oligonucleotide capture probes attached to a detection microwell plate. Positive results were read on a spectrophotometer. An especially useful group of assays has been the quantitative tests for chronic virus infections. The first FDA-approved HIV quantitative assays were developed by Roche Molecular Systems.

The COBAS AMPLICOR was the first system to automate both the amplification and detection steps of the PCR testing process by combining five instruments (thermal cycler, automatic pipetter, incubator, washer, and reader) into one (59). The COBAS AMPLICOR system is a benchtop system that makes PCR testing a routine diagnostic procedure suitable for most clinical laboratories. The COBAS AMPLICOR automation maximizes hands-off time for technologists (58, 59, 69). The software allows Multiplex (multiple detections from an amplified sample) and ReFlex (automated preselected next-test detection) testing. The COBAS AMPLICOR system amplifies and detects up to 24 samples simultaneously, with parallel amplification and detection of up to 48 samples. Up to six detections can be performed on each amplified sample, and the system delivers 50 detection results per h. For the MONITOR assays, quantitative test dilutions are performed automatically onboard. The system is controlled by AmpliLink software, which connects to a laboratory information system, allows bar coding, and automatically monitors all aspects of the instrument and testing process (63). The software also controls the new COBAS AmpliPrep instrument.

Specimen preparation can be performed using conveniently formatted manual methods. Another option exists for the most labor-intensive specimen preparation processes of purification of nucleic acid from plasma or serum for virus load assays. The COBAS AmpliPrep is a fully automated instrument that processes specimens and saves up to 76% of the hands-on time required for manual preparation (57). On a typical day shift schedule, 144 specimens can be easily processed, and 192 specimens or more per day can be processed if the instrument is allowed to run overnight. The prepared samples are compatible with both the semiautomated microwell plate and the automated platforms, COBAS AMPLICOR and the new COBAS TaqMan instrument, which is nearing commercial release. The initial test menu for the COBAS AmpliPrep is HCV MONITOR, HCV Qualitative, and HIV-1 MONITOR, UltraSensitive.

The COBAS AmpliPrep system is self-contained, housing both mechanics and software, so specimen contamination is prevented. The system has an onboard capacity of four assays and 72 specimens, with the possibility of continuous loading. All reactions take place in disposable high-walled specimen-processing units to ensure the integrity of results. Specimens are never touched directly by any part of

the instrument. After a bar-coded specimen is loaded in the instrument, the system opens the specimen tube, and a single-use pipette transfers the specimen to its specimen-processing unit. Lysing solution and biotinylated probes are automatically added and incubated. After hybridization, magnetic beads coated with streptavidin bind to the probe and form a complex with the captured target. At that point, the entire specimen is moved to a magnetic wash wheel. Now ready for the amplification and detection part of the PCR process, the prepared sample is sealed and returned to the sample rack in the instrument. In one study, there was no statistically significant variance in accuracy between automated and manually prepared test specimens made by a very experienced technologist (57).

The test menu is the most extensive of any commercial system. For the microwell assay series, it includes qualitative tests for HIV-1 components (proviral DNA), HCV, human T-cell leukemia virus types 1 and 2, CMV, *C. trachomatis*, *N. gonorrhoeae*, *M. tuberculosis*, *M. avium*, *M. intracellulare*, and an internal control, plus quantitative microwell assays for HIV, HCV, and HBV. For the COBAS AMPLICOR system, qualitative tests are included for HCV, *C. trachomatis*, *N. gonorrhoeae*, *M. tuberculosis*, *M. intracellulare*, and *M. avium*, plus the quantitative tests HCV MONITOR, HIV MONITOR, CMV MONITOR, and HBV MONITOR and an internal control (2, 97, 124, 140, 157).

Workload testing studies done according to College of American Pathologists workload measurement guidelines measured all aspects of the test process. This included initial handling of the specimen, specimen testing, recording and reporting of results, daily and periodic activities, and maintenance and repair, plus direct technical supervision. Using this system, it was found that a single-analyte test, such as that for *Chlamydia*, took 2.9 min of hands-on time when performed by the microwell assay (7). A multiplex test, such as the *C. trachomatis*-*N. gonorrhoeae* test plus an internal control run on the COBAS AMPLICOR, took 3.3 min for three results (59). The manual HCV assay took 7.8 min, with the majority of that time being specimen preparation. Automated specimen preparation could reduce that portion of the hands-on labor to less than 1 min (57). For the home brew assays, Roche has the MagNAPure specimen preparation instrument, which works well for the Light Cycler instrument.

The advantages of the Roche AMPLICOR system include the choice of microwell plates for high-volume single-analyte assays, such as that for *Chlamydia*, or the use of the automated COBAS AMPLICOR system. A single COBAS AMPLICOR instrument can produce up to 70 *C. trachomatis*-*N. gonorrhoeae* results, 22 *M. tuberculosis* results, and 10 HCV results with 8 h of hands-on time (69). If the workload is higher, more COBAS AMPLICOR instruments can be used. The Roche AmpliLink software can coordinate the activities of multiple COBAS AMPLICOR instruments and link to the laboratory information system (63).

PCR is a mature and versatile process that works well with both DNA and RNA targets. Quantitative tests are available for several of the most important viruses.

An area of possible concern for high-volume users is the moderate throughput for a single automated COBAS AMPLICOR instrument. Multiple instruments can increase throughput.

Abbott Laboratories PCR

Abbott Laboratories, Diagnostics Division [100 Abbott Park Rd., Abbott Park, IL 60064-3500; phone, (847) 937-6100] is developing a PCR-based qualitative assay that detects both HIV-1 and HIV-2, with a sensitivity of 20 to 50 copies/ml. The assay has a specificity of 99.6% and an inhibition rate of 1.7% (1). A 1-ml sample is processed with a manifold system and Qiagen columns. The extracted sample is used for PCR amplification. An internal control sequence, processed and amplified with each sample, monitors for amplification inhibition. Samples are reverse transcribed and are then amplified by RT-PCR. HIV-1- and HIV-2-specific probes are hybridized to the amplified products. Following hybridization, the amplified target is detected in the LCx instrument by microparticle enzyme immunoassay techniques. The detection system has an automated inactivation step that helps guard against PCR contamination. The HIV-1 and -2 qualitative RNA assay detects HIV-1 group M subtypes A, B, C, D, E, F, and G and group O. The assay is not FDA approved and has not been released for testing.

Gen-Probe TMA

The transcription-mediated amplification (TMA) assay system is an amplification system with an amplicon detection process very similar in format to other Gen-Probe assays. TMA is an isothermal amplification process and can be performed in a heat block or water bath. A thermocycler is not required. The kinetics of TMA are very rapid, and up to 10 billion RNA amplicons can be produced from a single target molecule in 30 min. TMA can be used with any type of nucleic acid target, including rRNA, mRNA, and DNA. TMA uses two primers and two enzymes. One is an RNA polymerase, and the other is a reverse transcriptase. The entire process is performed at a single temperature. RNA amplicons may be more labile in this environment than wild-type DNA amplicons, and that may help to lessen the chance of carryover contamination. The detection of amplicons produced by the TMA reaction is performed by the same hybrid protection assay process used in the direct Gen-Probe assays. The process is started by the addition of acridinium ester-labeled DNA probes, which specifically bind to the target amplicon. The hybridized probe is measured by the same type of chemiluminescent signal process and instrumentation as in the AccuPROBE system.

The advantages of the TMA process are good sensitivity, speed, and instrumentation simplicity when working with smaller workloads. The TMA process is easy to implement in laboratories already doing AccuPROBE or PACE 2 assays. There is a single-tube format with no wash steps. Reagents are added to the amplification tube but never removed or transferred. This helps lower the chance of cross contamination. Experience and good technique are still required to minimize cross contamination when using the TMA system without automation. Recommended contamination control for TMA involves the use of a single tube for the entire process after sample preparation, no pipetting of amplicons, oil overlay to prevent the amplicon from escaping, and no wash steps to decrease aerosol production; at the hybrid protection assay step, the detection reagents destroy the amplicon, and bleach is used to clean up and degrade the amplicon.

One possible disadvantage is that internal controls are not used in the first-generation Gen-Probe assays. If an amplification control is desired, a sample can be split, one aliquot can be spiked with target rRNA, and the aliquots can be run together to determine if the sample is inhibitory. Gen-Probe is working on specimen preparation as a way to decrease the need for internal controls. For future amplification tests, they plan to remove inhibitors from specimens rather than monitoring them with internal controls. They are developing new amplification tests that will use a target capture specimen-processing step that will purify the sample before amplification to completely remove inhibitors from the samples.

The current test menu of FDA-approved tests includes direct tests for M. tuberculosis complex and C. trachomatis (20, 45, 119). A TMA test for hepatitis B virus is being sold in Japan. A dual kinetic assay has been developed for TMA enabling the simultaneous detection of two analytes. The first test of this type to be approved is the APTIMA Combo 2 assay. It is a TMA assay for C. trachomatis and N. gonorrhoeae using the new target capture specimen-processing technology designed to remove inhibitors of TMA. It allows the simultaneous detection of both organisms without any need for reflex testing. This test will detect both organisms from cervical and male urethral swabs and from urine samples from both males and females.

Chiron and Gen-Probe have formed an alliance to produce tests to detect HIV-1, HIV-2, HCV, and HBV in blood products. These tests will be fully automated by an instrument developed by Gen-Probe (TIGRIS). TIGRIS will be a high-volume instrument able to perform multiplex assays with an internal amplification control. The Gen-Probe TIGRIS System will replace the manual and semiautomated tasks required for first-generation amplified systems because all steps of the assay are automated, from sample processing to detection. Unlike the first-generation TMA system, the TIGRIS system will not require the manual pipetting of samples from transport containers, spinning specimens to form pellets, vortexing swab samples, transferring specimens to and from thermal cyclers, or transferring samples and reaction containers to different workstations.

Becton Dickinson tSDA

The second-generation thermophilic strand displacement amplification (tSDA) test [Becton Dickinson Microbiology Systems, 54 Loveton Circle, Sparks, MD 21152; phone, (410) 316-4000] is well suited to isothermal amplification of DNA targets. Like other NAT tests, the tSDA test has gone through a number of refinements. However, it can still be thought of as occurring in two segments: first, a target generation phase, and second, an exponential amplification. tSDA is based on the primer-directed nicking activity of a thermophilic restriction enzyme and the strand displacement activity of a thermophilic exonuclease-deficient polymerase. In target generation, double-stranded DNA is heat denatured, creating two single-stranded copies. A series of primers combine with DNA polymerase, and amplification occurs. Since this is an isothermal amplification, bumper primers are used to displace the newly created strands to form altered targets capable of further exponential amplification. The use of a thermophilic restriction endonuclease allows amplification and strand displacement to proceed at 52 to 60°C, resulting in reduced background amplification with an increase in sensitivity and specificity (52, 126, 127, 146–149).

The detection of the amplified product is done with fluorogenic reporter probes which can be extended and displaced by the DNA polymerase. This system permits real-time, sequence-specific detection of targets amplified during tSDA (81). The new probes possess target binding sequences plus a fluorophore at a BsoBI recognition site and quencher maintained in sufficiently close proximity that fluorescence is quenched in the intact single-stranded probe. If the target is present during tSDA, the probe is converted into a fully double-stranded form and is cleaved by the restriction enzyme BsoBI, which also serves as the nicking agent for tSDA. The fluorophore and quencher diffuse apart upon probe cleavage, causing increased fluorescence. Target replication may be followed in real time during the tSDA reaction. Probe performance may be enhanced by embedding the fluorogenic BsoBI site within the loop of a folded hairpin structure. The new probe designs permit detection of as few as 10 target copies within 30 min in a closed-tube, real-time format, eliminating the likelihood of carryover contamination (95). The probes may be used to detect RNA targets in tSDA mixtures containing reverse transcriptase. Furthermore, a two-color competitive tSDA format permits accurate quantification of target levels from the real-time fluorescence data.

The BDProbe Tec ET real-time fluorescence detection method using closed tubes in a sealed microwell plate gave an amplification and detection time of 1 h. The 1-h assay time allows up to six runs per shift (180 patient results for C. trachomatis and N. gonorrhoeae or 276 patient results for C. trachomatis). A run can be from 1 to 94 specimens per h, generating up to 564 results in 8 h. The hands-on time to process each specimen, not including specimen preparation, is approximately 1 min.

The BDProbe Tec ET system has been used for detection of C. trachomatis and N. gonorrhoeae in swab specimens or in urine. An amplification control is included with each specimen. A direct test for M. tuberculosis is available outside the United States. Assays for M. tuberculosis complex, M. avium complex, and M. kansasii are available as culture confirmation (81, 106).

Another molecular platform which will complement the BDProbeTec ET system is under development by Becton Dickinson, the BDAutoPrep. This system will automate sample preparation and all reagent pipetting. It accepts a primary specimen tube, extracts nucleic acids, and pipettes specimens into microwell trays.

The new second-generation tSDA system has several advantages. The first is the rapidity with which the simultaneous amplification and detection can take place. Positive results can be detected in an hour. The second is the automated detection and high throughput. The microwell format allows for large batches, and the breakaway wells can be used for smaller runs. There is no minimum batch size. Runs require only one positive and one negative control per 94 results. Finally, the sealed system should help limit contamination if all procedures are correctly followed. The BDProbe Tec ET instrument has only one moving part. This should result in a long mean time between failures. Other conveniences are predispensed dry reagents stable for 1 year, a color-coded assay design, amplification control to monitor each specimen, bar code support, and laboratory information system connectivity.

An original disadvantage of the tSDA system was that RNA targets were more difficult for this DNA-oriented assay. However, an RT-SDA has been developed for amplifying RNA targets, with the HIV gag region as a model system. This broadens the potential applications of SDA so that it is no longer restricted to DNA targets. The higher-temperature amplification system has almost completely removed specificity issues related to earlier enzyme versions of tSDA. Quantitative assays are not currently FDA approved but are under development (99). The system has manual specimen preparation, but the BDAutoPrep instrument should answer

that need. At present, the first step of amplification, primer binding, is a manual step. The sample is transferred manually from the priming plate to the amplification plate, and the plate is sealed with a manual device. After that, the assay is placed into the BDProbe Tec ET instrument, where all steps are completed automatically. Tests for Chlamydia and N. gonorrhoeae have been approved by the FDA.

bioMérieux NASBA

Nucleic acid sequence-based amplification (NASBA) [bioMérieux Corporation, 100 Akzo Ave., Durham, NC 27712; phone, (919) 620-2000] is well suited for amplification of RNA targets and mRNA expression targets. NASBA, which was formerly marketed by Organon Teknika, relies on the simultaneous activity of three enzymes: avian myeloblastosis virus reverse transcriptase, RNase H, and T7 RNA polymerase. Two oligonucleotide primers specific for each analyte are required. The coordinated interaction of the enzymes and primers allows amplification of a single target nucleic acid sequence to $>10^9$ copies in 90 min or less. The fact that these reactions all occur at a constant temperature of 41°C eliminates the need for thermocycling equipment because the three enzymes create a self-sustaining target sequence replication process. In the NASBA reaction, the enzymatic process involves binding a T7 promoter sequence and reverse transcriptase to the RNA target. A copy of cDNA complementary to the target molecule is produced. This cDNA is released from the original RNA target strand by the action of RNase H. A key feature of this cDNA is that it has the T7 promoter DNA sequence linked to the cDNA copy of the target RNA. This transcription-competent free cDNA strand and T7 RNA polymerase are used to create multiple antisense RNA copy transcripts of the original target. These antisense RNA copies are used to make more cDNA, which is freed from the RNA-cDNA strand by RNase H, and the cycles repeat exponentially, shortly producing enough cDNA to be measured in a detection process (113, 141).

bioMérieux has implemented the NASBA technology in the NucliSens system. An important initial instrument of the NucliSens system is the NucliSens Extractor, which prepares the specimen. It uses a very versatile nucleic acid extraction system referred to as the Boom method (11, 41, 56, 142, 155). This is a solid-phase extraction method with silica particles that capture all nucleic acids after lysis of the target organism in a lysis buffer. The freed nucleic acids are stabilized after attachment to the capture particles. At the appropriate time, an extraction procedure is used to release purified nucleic acid from the silica particles. The extraction procedure involves an initial wash followed by addition of guanidine isothiocyanate, ethanol, and acetone to produce a highly purified product, which can be eluted for use in a variety of nucleic acid procedures or amplification assays. The compatible specimens are reported to include serum, plasma, whole blood, cerebrospinal fluid, genital secretions, brain tissue, other organ tissues, white cells, sputum, urine, feces, throat swabs, dermal lesion swabs, and pus samples. Plasma with citrate, heparin, and EDTA are reported to be equally well extracted and amplified in the NASBA reaction. The system can perform 100 to 120 extractions per working day. It can process up to 10 extractions per run, taking approximately 35 or 45 min per run (for low- and high-volume sample extraction, respectively). Although it can handle a variety of specimen types, some types of samples may require pretreatment.

Another key instrument of the NucliSens system is the NucliSens Reader. The postamplification detection process involves an electrochemiluminescent (ECL) reaction (141). The NucliSens system can perform qualitative and quantitative assays, and both types of assay can be combined in the same run. There is a one-step hybridization reaction between a capture probe and the amplified nucleic acid. The reader separates free probes from bound probes and reads the luminescence of the bound probes. After amplification of the target RNA by NASBA, the NucliSens Reader automatically detects the amplified material and calculates the results.

For rapid and sensitive detection of amplified target RNA, a sandwich hybridization method is used. The ECL detection technology involves two components: the ECL label Tris (2,2'-bipyridine) ruthenium (Ru), which is coupled to a detection oligonucleotide, and tripropylamine present in the reaction buffer. When a voltage is applied to an electrode, both components are activated by oxidation. The oxidized tripropylamine is transformed into a highly reducing agent, which reacts with activated Ru to create an excited state. This form returns to its ground state with emission of a photon at a 620-nm wavelength. In practice, two specific probes are employed, a capture probe bound to paramagnetic beads and a detection probe labeled with Ru. Hybridized sample is drawn into the ECL flow cell, where a magnetic arm beneath the cell automatically traps all magnetic particles on the electrode. Unbound nucleic acid and the ECL label are washed away. The ECL reaction is induced by applying a voltage trigger to the electrode. The resulting emitted light is detected by the photomultiplier tube, and signals are interpreted by the computer connected to the NucliSens Reader. Finally, the magnetic arm is moved away from the cell, which is flushed with cleaning solution. Measurement of 50 tubes takes approximately 50 min. The amount of light produced is directly proportional to the amount of Ru present in the reaction.

Each patient sample has calibrators that follow it throughout the entire process of sample preparation, amplification, and detection, so any variation in the process or suboptimal amplification affecting the patient's RNA target will also be reflected in the calibrator readings. Data analysis software will correct for these changes, reporting quantitative values with improved accuracy. For quantitative assays, use is made of three artificial RNAs, or calibrators, which have the same length and base composition as the target RNA and thus highly similar characteristics in all processing steps (isolation, amplification, and detection). For qualitative assays, one artificial RNA is used. All controls differ only in a 20-base randomized nucleotide stretch, used for separate detection in the NucliSens Reader. Control RNAs are added to each sample before extraction, isolated and amplified in the same tube, and separately detected. When amplification for the target RNA is negative, a valid result is reported only when the internal control is positive.

The advantages of this complete system are flexibility for sample type, inactivation of virus upon addition to the lysis buffer, stabilization of nucleic acid during extraction, contamination control by a contained extraction environment, single-tube amplification, and fully automated detection. The incorporation of three calibrators gives a higher degree of process control for the fast and somewhat more complex three-enzyme NASBA reaction. There can be up to a 6-log-unit dynamic range in the assay without the need to make dilutions of amplified samples.

There are some possible disadvantages. The system is currently designed for RNA and mRNA targets, although the specimen preparation system will work with DNA and RNA. There is no direct chemical sterilization of the amplified product, so contamination control depends on adhering to the steps of the process and ensuring that the system and disposables function as expected. There is some evidence that sensitivity, while good, may not be quite as high as in some other NAT tests. In one early version of the CMV pp67 assay, the sensitivity was reported to be somewhat lower than that of other tests (15). Part of this issue may be due to differences that might be expected when an mRNA target is used rather than DNA. NASBA sensitivity was reported to be equivalent to those of other good NAT assays such as HCV (30). Clinical utility should be a key consideration in deciding how much specificity and sensitivity are necessary.

The NASBA tests that are available are qualitative assays for the detection of CMV late pp67 mRNA (15, 46, 161). The CMV pp67 mRNA test is FDA approved. Quantitative assays for HIV-1 RNA have been reported (21, 51, 93, 156). For home brew assays, a basic kit is available, which can be modified to work with numerous analytes. This could help standardize the development of home brew assays.

Nucleic Acid Probe Amplification Assays
Abbott Laboratories LCR (LCx)

Abbott has implemented the ligase chain reaction (LCR) amplification assay with its LCx assay system. The LCR utilizes two pairs of complementary oligonucleotide probes to bind to a DNA target sequence. The probe pairs are designed to bind adjacent to each other at complementary sites on the target molecule. If perfect base pairing of both the 3' and 5' ends of the oligonucleotide probes occurs, a thermophilic DNA ligase will covalently link the two probes into one molecule that can then mimic one strand of the original target sequence. After strand separation, the newly ligated oligonucleotide probe serves as a target template for annealing of new complementary probes, followed by ligation, separation, and repetition of the probe amplification process until millions of copies are made. Using labeled probes during the entire amplification process allows the ligation product to be detected by the microparticle enzyme immunoassay in an LCx probe system analyzer.

The LCR test has very good sensitivity, with a theoretical detection limit of one target molecule. The amplified material from each assay could be a source of contamination of subsequent assays. However, experience has shown that the contamination barriers designed into the system keep actual contamination to acceptably low levels if proper laboratory procedures are followed. The disadvantages include the absence of an internal amplification control, a limited test menu, and a system with a first-generation degree of automation.

The test menu is limited to *C. trachomatis* and *N. gonorrhoeae* in the United States (22). An *M. tuberculosis* test is available in Europe (23, 84, 137).

CONCLUSIONS

Where possible, the use of commercial FDA (or equivalent)-approved NAT kits and instrumentation is advantageous because of the standardization of results, convenience of use for greater numbers of clinical laboratories, and benefits related to technologist training, support, and savings of hands-on time related to a higher degree of automation. The importance of having choices and the competitive pricing that results from having competition from different NAT systems is recognized. However, for the individual laboratory, it may be useful to consider the financial benefits of adopting a basic home brew system and a basic commercial system. Having the smallest possible number of platforms will result in long-term savings in maintenance, amortization of costs of instrumentation, and costs related to personnel training. Because the test menus from all companies are limited at this time, larger laboratories may need several NAT platforms to provide complete service.

REFERENCES

1. **Abravaya, K., C. Esping, R. Hoenle, J. Gorzowski, R. Perry, P. Kroeger, J. Robinson, and R. Flanders.** 2000. Performance of a multiplex qualitative PCR LCx assay for detection of human immunodeficiency virus type 1 (HIV-1) group M subtypes, group O, and HIV-2. *J. Clin. Microbiol.* **38:**716–723.
2. **Afonso, A. M., J. Didier, E. Plouvier, B. Falissard, M. P. Ferey, M. Bogard, and E. Dussaix.** 2000. Performance of an automated system for quantification of hepatitis C virus RNA. *J. Virol. Methods* **86:**55–60.
3. **Alcaide, F., M. A. Benitez, J. M. Escriba, and R. Martin.** 2000. Evaluation of the BACTEC MGIT 960 and the MB/BacT systems for recovery of mycobacteria from clinical specimens and for species identification by DNA AccuProbe. *J. Clin. Microbiol.* **38:**398–401.
4. **Amicosante, M., L. Richeldi, G. Trenti, G. Paone, M. Campa, A. Bisetti, and C. Saltini.** 1995. Inactivation of polymerase inhibitors for *Mycobacterium tuberculosis* DNA amplification in sputum by using capture resin. *J. Clin. Microbiol.* **33:**629–630.
5. **Anonymous.** 1999. Association for Molecular Pathology statement. Recommendations for in-house development and operation of molecular diagnostic tests. *Am. J. Clin. Pathol.* **111:**449–463.
6. **Backman, A., P. Lantz, P. Radstrom, and P. Olcen.** 1999. Evaluation of an extended diagnostic PCR assay for detection and verification of the common causes of bacterial meningitis in CSF and other biological samples. *Mol. Cell. Probes* **13:**49–60.
7. **Bass, C. A., D. L. Jungkind, N. S. Silverman, and J. M. Bondi.** 1993. Clinical evaluation of a new polymerase chain reaction assay for detection of *Chlamydia trachomatis* in endocervical specimens. *J. Clin. Microbiol.* **31:**2648–2653.
8. **Bayer, E. A., and M. Wilchek.** 1990. Introduction to avidin-biotin technology. *Methods Enzymol.* **184:**5–13.
9. **Beavis, K. G., M. B. Lichty, D. L. Jungkind, and O. Giger.** 1995. Evaluation of Amplicor PCR for direct detection of *Mycobacterium tuberculosis* from sputum specimens. *J. Clin. Microbiol.* **33:**2582–2586.
10. **Becker, K., R. Roth, and G. Peters.** 1998. Rapid and specific detection of toxigenic *Staphylococcus aureus*: use of two multiplex PCR enzyme immunoassays for amplification and hybridization of staphylococcal enterotoxin genes, exfoliative toxin genes, and toxic shock syndrome toxin 1 gene. *J. Clin. Microbiol.* **36:**2548–2553.
11. **Beld, M., M. R. Habibuw, S. P. Rebers, R. Boom, and H. W. Reesink.** 2000. Evaluation of automated RNA-extraction technology and a qualitative HCV assay

for sensitivity and detection of HCV RNA in pool-screening systems. *Transfusion* **40:**575–579.
12. **Bergmann, J. S., and G. L. Woods.** 1998. Clinical evaluation of the BDProbeTec strand displacement amplification assay for rapid diagnosis of tuberculosis. *J. Clin. Microbiol.* **36:**2766–2768.
13. **Bernet, C., M. Garret, B. Barbeyrac, C. Bebear, and J. Bonnet.** 1989. Detection of *Mycoplasma pneumoniae* by using the polymerase chain reaction. *J. Clin. Microbiol.* **27:**2492–2496.
14. **Blaak, H., A. B. van't Wout, M. Brouwer, M. Cornelissen, N. A. Kootstra, N. Albrecht-van Lent, R. P. Keet, J. Goudsmit, R. A. Coutinho, and H. Schuitemaker.** 1998. Infectious cellular load in human immunodeficiency virus type 1 (HIV-1)-infected individuals and susceptibility of peripheral blood mononuclear cells from their exposed partners to non-syncytium-inducing HIV-1 as major determinants for HIV-1 transmission in homosexual couples. *J. Virol.* **72:**218–224.
15. **Blank, B. S., P. L. Meenhorst, J. W. Mulder, G. J. Weverling, H. Putter, W. Pauw, W. C. van Dijk, P. Smits, S. Lie-A-Ling, P. Reiss, and J. M. Lange.** 2000. Value of different assays for detection of human cytomegalovirus (HCMV) in predicting the development of HCMV disease in human immunodeficiency virus-infected patients. *J. Clin. Microbiol.* **38:**563–569.
16. **Boivin, G., J. Handfield, G. Murray, E. Toma, R. Lalonde, J. G. Lazar, and M. G. Bergeron.** 1997. Quantitation of cytomegalovirus (CMV) DNA in leukocytes of human immunodeficiency virus-infected subjects with and without CMV disease by using PCR and the SHARP signal detection system. *J. Clin. Microbiol.* **35:**525–526.
17. **Bourbeau, P. P., B. J. Heiter, and M. Figdore.** 1997. Use of Gen-Probe AccuProbe Group B streptococcus test to detect group B streptococci in broth cultures of vaginal-anorectal specimens from pregnant women: comparison with traditional culture method. *J. Clin. Microbiol.* **35:**144–147.
18. **Brice, S. L., D. Krzemien, W. L. Weston, and J. C. Huff.** 1989. Detection of herpes simplex virus DNA in cutaneous lesions of erythema multiforme. *J. Investig. Dermatol.* **93:**183–187.
19. **Briselden, A. M., and S. L. Hillier.** 1994. Evaluation of affirm VP Microbial Identification Test for *Gardnerella vaginalis* and *Trichomonas vaginalis*. *J. Clin. Microbiol.* **32:**148–152.
20. **Brown, T. J., E. G. Power, and G. L. French.** 1999. Evaluation of three commercial detection systems for *Mycobacterium tuberculosis* where clinical diagnosis is difficult. *J. Clin. Pathol.* **52:**193–197.
21. **Burgisser, P., P. Vernazza, M. Flepp, J. Boni, Z. Tomasik, U. Hummel, G. Pantaleo, and J. Schupbach.** 2000. Performance of five different assays for the quantification of viral load in persons infected with various subtypes of HIV-1. Swiss HIV Cohort Study. *J. Acquir. Immune Defic. Syndr.* **23:**138–144.
22. **Carroll, K. C., W. E. Aldeen, M. Morrison, R. Anderson, D. Lee, and S. Mottice.** 1998. Evaluation of the Abbott LCx ligase chain reaction assay for detection of *Chlamydia trachomatis* and *Neisseria gonorrhoeae* in urine and genital swab specimens from a sexually transmitted disease clinic population. *J. Clin. Microbiol.* **36:**1630–1633.

23. **Cavirani, S., F. Fanti, S. Conti, A. Calderaro, E. Foni, G. Dettori, C. Chezzi, and F. Scatozza.** 1999. Detection of *Mycobacterium bovis* in bovine tissue samples by the Abbott LCx *Mycobacterium tuberculosis* assay and comparison with culture methods. *New Microbiol.* **22:**343–349.

24. **Chan, H. L., N. W. Leung, T. C. Lau, M. L. Wong, and J. J. Sung.** 2000. Comparison of three different sensitive assays for hepatitis B virus DNA in monitoring of responses to antiviral therapy. *J. Clin. Microbiol.* **38:**3205–3208.

25. **Chu, F. K.** 1998. Rapid and sensitive PCR-based detection and differentiation of aetiologic agents of human granulocytotropic and monocytotropic ehrlichiosis. *Mol. Cell. Probes* **12:**93–99.

26. **Class, H. C. J., J. H. T. Wagenvoort, H. G. M. Niesters, T. T. Tio, J. H. van Rijsoort-Vos, and W. G. V. Quint.** 1991. Diagnostic value of the polymerase chain reaction for chlamydia detection as determined in a follow-up study. *J. Clin. Microbiol.* **29:**42–45.

27. **Clavel, C., M. Masure, M. Levert, I. Putaud, C. Mangeonjean, M. Lorenzato, P. Nazeyrollas, R. Gabriel, C. Quereux, and P. Birembaut.** 2000. Human papillomavirus detection by the hybrid capture II assay: a reliable test to select women with normal cervical smears at risk for developing cervical lesions. *Diagn. Mol. Pathol.* **9:**145–150.

28. **Clewley, J. P.** 1989. The polymerase chain reaction, a review of the practical limitations for human immunodeficiency virus diagnosis. *J. Virol. Methods* **25:**179–188.

29. **Damen, M., H. T. M. Cuypers, H. L. Zaaijer, H. W. Reesink, W. P. Schaasberg, W. H. Gerlich, H. G. Niesters, and P. N. Lelie.** 1996. International collaborative study on the second EUROHEP HCV-RNA reference panel. *J. Virol. Methods* **58:**175–185.

30. **Damen, M., P. Sillekens, H. T. Cuypers, I. Frantzen, and R. Melsert.** 1999. Characterization of the quantitative HCV NASBA assay. *J. Virol. Methods* **82:**45–54.

31. **Deggerdal, A., and F. Larsen.** 1997. Rapid isolation of PCR-ready DNA from blood, bone marrow and cultured cells, based on paramagnetic beads. *BioTechniques* **22:**554–557.

32. **DeMedina, M., M. Ashby, V. Schluter, M. Hill, B. Leclerq, J. P. Pennell, L. J. Jeffers, K. R. Reddy, E. R. Schiff, G. Hess, and G. O. Perez.** 1998. Prevalence of hepatitis C and G virus infection in chronic hemodialysis patients. *Am. J. Kidney Dis.* **31:**224–226.

33. **DeMeo, L. R., D. L. Draper, J. A. McGregor, D. F. Moore, C. R. Peter, P. S. Kapernick, and W. M. McCormack.** 1996. Evaluation of a deoxyribonucleic acid probe for the detection of *Trichomonas vaginalis* in vaginal secretions. *Am. J. Obstet. Gynecol.* **174:**1339–1342.

34. **Echavarria, M., M. Forman, J. Ticehurst, J. S. Dumler, and P. Charache.** 1998. PCR method for detection of adenovirus in urine of healthy and human immunodeficiency virus-infected individuals. *J. Clin. Microbiol.* **36:**3323–3326.

35. **Eisenstein, B. I.** 1990. The polymerase chain reaction: a new method of using molecular genetics for medical diagnosis. *N. Engl. J. Med.* **322:**178–183.

36. **Elbeik, T., E. Charlebois, P. Nassos, J. Kahn, F. M. Hecht, D. Yajko, V. Ng, and K. Hadley.** 2000. Quantitative and cost comparison of ultrasensitive human immunodeficiency virus type 1 RNA viral load assays: Bayer bDNA Quantiplex versions 3.0 and 2.0 and

Roche PCR Amplicor Monitor version 1.5. *J. Clin. Microbiol.* **38:**1113–1120.

37. **Erice, A., D. Brambilla, J. Bremer, J. B. Jackson, R. Kokka, B. Yen-Lieberman, and R. W. Coombs.** 2000. Performance characteristics of the QUANTIPLEX HIV-1 RNA 3.0 assay for detection and quantitation of human immunodeficiency virus type 1 RNA in plasma. *J. Clin. Microbiol.* **38:**2837–2845.

38. **Esposito, G., F. Blasi, L. Allegra, R. Chiesa, G. Melissano, R. Cosentini, P. Tarsia, L. Dordoni, C. Cantoni, C. Arosio, and L. Fagetti.** 1999. Demonstration of viable Chlamydia pneumoniae in atherosclerotic plaques of carotid arteries by reverse transcriptase polymerase chain reaction. *Ann. Vasc. Surg.* **13:**421–425.

39. **Espy, M. J., and T. F. Smith.** 1995. Comparison of SHARP signal system and Southern blot hybridization analysis for detection of cytomegalovirus in clinical specimens by PCR. *J. Clin. Microbiol.* **33:**3028–3030.

40. **Essary, L. R., S. J. Kinard, A. Butcher, H. Wang, K. A. Laycock, E. Donegan, B. McCreedy, S. Connell, J. Batchelor, J. Harris, J. Spadoro, and J. S. Pepose.** 1996. Screening potential corneal donors for HIV-1 by polymerase chain reaction and a colorimetric microwell hybridization assay. *Am. J. Ophthalmol.* **122:** 526–534.

41. **Fahle, G. A., and S. H. Fischer.** 2000. Comparison of six commercial DNA extraction kits for recovery of cytomegalovirus DNA from spiked human specimens. *J. Clin. Microbiol.* **38:**3860–3863.

42. **Flahaut, M., D. Sanglard, M. Monod, J. Bille, and M. Rossier.** 1998. Rapid detection of *Candida albicans* in clinical samples by DNA amplification of common regions from *C. albicans*-secreted aspartic proteinase genes. *J. Clin. Microbiol.* **36:**395–401.

43. **Fransen, K., D. Mortier, L. Heyndrickx, C. Verhofstede, W. Janssens, and G. van der Groen.** 1998. Isolation of HIV-1 RNA from plasma: evaluation of seven different methods for extraction. *J. Virol. Methods* **76:**153–157.

44. **Garson, J. A., R. S. Tedder, M. Briggs, P. Tuke, J. A. Glazebrook, A. Trute, D. Parker, J. A. J. Barbara, M. Contreras, and S. Aloysius.** 1990. Detection of viral hepatitis C sequences in blood donations by 'nested' polymerase chain reaction and prediction of infectivity. *Lancet* **335:**1419–1422.

45. **Goessens, W. H., J. W. Mouton, W. I. van der Meijden, S. Deelen, T. H. van Rijsoort-Vos, N. Lemmens-den Toom, H. A. Verbrugh, and R. P. Verkooyen.** 1997. Comparison of three commercially available amplification assays, AMP CT, LCx, and COBAS AMPLICOR, for detection of *Chlamydia trachomatis* in first-void urine. *J. Clin. Microbiol.* **35:**2628–2633.

46. **Goossens, V. J., M. J. Blok, M. H. Christiaans, J. P. van Hooff, P. Sillekens, K. Hockerstedt, I. Lautenschlager, J. M. Middeldorp, and C. A. Bruggeman.** 1999. Diagnostic value of nucleic-acid-sequence-based amplification for the detection of cytomegalovirus infection in renal and liver transplant recipients. *Intervirology* **42:**373–381.

47. **Gossack, J. P., and J. L. Beebe.** 1998. Use of DNA purification kits for polymerase chain reaction testing of Gen-Probe *Chlamydia trachomatis* PACE 2 specimens. *Sex. Transm. Dis.* **25:**265–271.

48. **Gromadzki, S. G., and V. Chaturvedi.** 2000. Limitation of the AccuProbe *Coccidioides immitis* culture iden-

tification test: false-negative results with formaldehyde-killed cultures. *J. Clin. Microbiol.* **38:**2427–2428.

49. **Hale, Y. M., M. E. Melton, J. S. Lewis, and D. E. Willis.** 1993. Evaluation of the PACE 2 *Neisseria gonorrhoeae* assay by three public health laboratories. *J. Clin. Microbiol.* **31:**451–453.

50. **Ho, S. K., F. K. Li, K. N. Lai, and T. M. Chan.** 2000. Comparison of the CMV Brite Turbo assay and the Digene Hybrid Capture CMV DNA (version 2.0) assay for quantitation of cytomegalovirus in renal transplant recipients. *J. Clin. Microbiol.* **38:**3743–3745.

51. **Hodara, V., A. Monticelli, S. Pampuro, H. Salomon, H. Jauregui Rueda, and O. Libonatti.** 1998. HIV-1 viral load: comparative evaluation of three commercially available assays in Argentina. *Acta Physiol. Pharmacol. Ther. Latinoam.* **48:**107–113.

52. **Ichiyama, S., Y. Ito, F. Sugiura, Y. Iinuma, S. Yamori, M. Shimojima, Y. Hasegawa, K. Shimokata, and N. Nakashima.** 1997. Diagnostic value of the strand displacement amplification method compared to those of Roche Amplicor PCR and culture for detecting mycobacteria in sputum samples. *J. Clin. Microbiol.* **35:**3082–3085.

53. **Iwen, P. C., R. A. Walker, K. L. Warren, D. M. Kelly, S. H. Hinrichs, and J. Linder.** 1995. Evaluation of a nucleic acid-based test (PACE 2C) for simultaneous detection of *Chlamydia trachomatis* and *Neisseria gonorrhoeae* in endocervical specimens. *J. Clin. Microbiol.* **33:**2587–2591.

54. **Jaulhac, B., M. Reyrolle, Y. K. Sodahlon, S. Jarraud, M. Kubina, H. Monteil, Y. Piemont, and J. Etienne.** 1998. Comparison of sample preparation methods for detection of *Legionella pneumophila* in culture-positive bronchoalveolar lavage fluids by PCR. *J. Clin. Microbiol.* **36:**2120–2122.

55. **Jonas, D., A. Rosenbaum, S. Weyrich, and S. Bhakdi.** 1995. Enzyme-linked immunoassay for detection of PCR-amplified DNA of legionellae in bronchoalveolar fluid. *J. Clin. Microbiol.* **33:**1247–1252.

56. **Jongerius, J. M., M. Bovenhorst, C. L. van der Poel, J. A. van Hilten, A. C. Kroes, J. A. van der Does, E. F. van Leeuwen, and R. Schuurman.** 2000. Evaluation of automated nucleic acid extraction devices for application in HCV NAT. *Transfusion* **40:**871–874.

57. **Jungkind, D. Automation of clinical microbiology: past, present, and future.** *Eur. J. Clin. Virol.*, in press.

58. **Jungkind, D. L.** 1996. Evaluation of an automated COBAS AMPLICOR PCR system for detection of *Chlamydia trachomatis/Neisseria gonorrhoeae* and the impact on laboratory management, p. 272–273. *In* A. Stary (ed.), *Proceedings of the Third Meeting of the European Society for Chlamydia Research.* Societa Editrice Esculapio, Bologna, Italy.

59. **Jungkind, D. L., S. DiRenzo, K. G. Beavis, and N. S. Silverman.** 1996. Evaluation of an automated COBAS AMPLICOR PCR system for detection of several infectious agents and the impact on laboratory management. *J. Clin. Microbiol.* **34:**2778–2783.

60. **Kessler, C.** 1991. The digoxigenin:anti-digoxigenin (DIG) technology—a survey on the concept and realization of a novel bioanalytical indicator system. *Mol. Cell. Probes* **5:**161–205.

61. **Kessler, C.** 1992. The digoxigenin:anti-digoxigenin (DIG) system, p. 35–69. *In* C. Kessler (ed.), *Nonradioactive Labeling and Detection of Biomolecules.* Springer, New York, N.Y.

62. **Kessler, H. H., D. E. Dodge, K. Pierer, K. K. Y. Young, Y. Liao, B. I. Santner, E. Eber, M. G. Roeger, D. Stuenzner, B. Sixl-Voigt, and E. Marth.** 1997. Rapid detection of *Mycoplasma pneumoniae* by an assay based on PCR and probe hybridization in a nonradioactive microwell plate format. *J. Clin. Microbiol.* **35:**1592–1594.

63. **Kessler, H. H., D. Jungkind, E. Stelzl, S. Direnzo, S. K. Vellimedu, K. Pierer, B. Santner, and E. Marth.** 1999. Evaluation of AMPLILINK software for the COBAS AMPLICOR system. *J. Clin. Microbiol.* **37:**436–437.

64. **Kessler, H. H., K. Pierer, B. I. Santner, S. K. Vellimedu, E. Stelzl, E. Marth, P. Fickert, and R. E. Stauber.** 1998. Evaluation of molecular parameters for routine assessment of viremia in patients with chronic hepatitis C who are undergoing antiviral therapy. *J. Hum. Virol.* **1:**314–319.

65. **Kessler, H. H., K. Pierer, B. Weber, A. Sakrauski, B. Santner, D. Stuenzner, E. Gergely, and E. Marth.** 1994. Detection of herpes simplex virus DNA from cerebrospinal fluid by PCR and a rapid, nonradioactive hybridization technique. *J. Clin. Microbiol.* **32:**1881–1886.

66. **Kessler, H. H., F. F. Reinthaler, A. Pschaid, K. Pierer, B. Kleinhappl, E. Eber, and E. Marth.** 1993. Rapid detection of *Legionella* species in bronchoalveolar lavage fluids with the EnviroAmp Legionella PCR amplification and detection kit. *J. Clin. Microbiol.* **31:**3325–3328.

67. **King, J. A., and J. K. Ball.** 1993. Detection of HIV-1 by digoxigenin-labelled PCR and microtitre plate solution hybridization assay and prevention of PCR carry-over by uracil-N-glycosylase. *J. Virol. Methods* **44:**67–76.

68. **Kitchin, P. A., Z. Szotyori, C. Fromholc, and N. Almond.** 1990. Avoidance of false positives. *Nature* **344:**201.

69. **Klapper, P., D. Jungkind, T. Fenner, R. Antinozzi, J. Schirm, and C. Blanckmeister.** 1998. Multicenter international work flow study of an automated polymerase chain reaction instrument. *Clin. Chem.* **44:**1737–1739.

70. **Klapper, P. E., G. M. Cleator, C. Dennett, and A. G. Lewis.** 1990. Diagnosis of herpes encephalitis via Southern blotting of cerebrospinal fluid DNA amplified by polymerase chain reaction. *J. Med. Virol.* **32:**261–264.

71. **Kox, L. F., D. Rhienthong, A. M. Miranda, N. Udomsantisuk, K. Ellis, J. van Leeuwen, S. van Heusden, S. Kuijper, and A. H. Kolk.** 1994. A more reliable PCR for detection of *Mycobacterium tuberculosis* in clinical samples. *J. Clin. Microbiol.* **32:**672–678.

72. **Kramvis, A., S. Bukofzer, and M. C. Kew.** 1996. Comparison of hepatitis B virus DNA extractions from serum by the QIAamp blood kit, GeneReleaser, and the phenol-chloroform method. *J. Clin. Microbiol.* **34:**2731–2733.

73. **Kwok, S.** 1990. Procedures to minimize PCR-product-carryover, p. 142–145. *In* M. A. Innis, D. H. Gelfand, and J. J. Sninsky (ed.), *PCR Protocols. A Guide to Methods and Applications.* Academic Press, San Diego, Calif.

74. **Kwok, S., and R. Higuchi.** 1989. Avoiding false positives with PCR. *Nature* (London) **339:**237–238.

75. **Lauderdale, T. L., L. Landers, I. Thorneycroft, and K. Chapin.** 1999. Comparison of the PACE 2 assay, two amplification assays, and Clearview EIA for detection

of *Chlamydia trachomatis* in female endocervical and urine specimens. *J. Clin. Microbiol.* **37:**2223–2229.

76. **Lelie, P. N., H. T. M. Cuypers, A. A. J. van Drimmelen, and W. G. V. Quint.** 1998. Quality assessment of hepatitis C virus nucleic acid amplification methods. *Infusionsther. Transfusionsmed.* **25:**102–110.

77. **Lichtinghagen, R., and R. Glaubitz.** 1995. A competitive polymerase chain reaction assay for reliable identification of *Bordetella pertussis* in nasopharyngeal swabs. *Eur. J. Clin. Chem. Clin. Biochem.* **33:**87–93.

78. **Lichtinghagen, R., and R. Glaubitz.** 1996. A principle of quality assessment using a competitive polymerase chain reaction assay for the detection of *Chlamydia trachomatis* in cervical specimens. *Eur. J. Clin. Chem. Clin. Biochem.* **34:**765–770.

79. **Lin, H. J., T. Tanwandee, and F. B. Hollinger.** 1997. Improved method for quantification of human immunodeficiency virus type 1 RNA and hepatitis C virus RNA in blood using spin column technology and chemiluminescent assays of PCR products. *J. Med. Virol.* **51:**56–63.

80. **Lisby, G.** 1999. Application of nucleic acid amplification in clinical microbiology. *Mol. Biotechnol.* **12:**75–99.

81. **Little, M. C., J. Andrews, R. Moore, S. Bustos, L. Jones, C. Embres, G. Durmowicz, J. Harris, D. Berger, K. Yanson, C. Rostkowski, D. Yursis, J. Price, T. Fort, A. Walters, M. Collis, O. Llorin, J. Wood, F. Failing, C. O'Keefe, B. Scrivens, B. Pope, T. Hansen, K. Marino, K. Williams, et al.** 1999. Strand displacement amplification and homogeneous real-time detection incorporated in a second-generation DNA probe system, BDProbeTecET. *Clin. Chem.* **45:**777–784.

82. **Loewy, Z. G., J. Mecca, and R. Diaco.** 1994. Enhancement of *Borrelia burgdorferi* PCR by uracil-*N*-glycosylase. *J. Clin. Microbiol.* **32:**135–138.

83. **Löffler, J., H. Hebart, U. Schumacher, H. Reitze, and H. Einsele.** 1997. Comparison of different methods for extraction of DNA of fungal pathogens from cultures and blood. *J. Clin. Microbiol.* **35:**3311–3312.

84. **Lumb, R., K. Davies, D. Dawson, R. Gibb, T. Gottlieb, C. Kershaw, K. Kociuba, G. Nimmo, N. Sangster, M. Worthington, and I. Bastian.** 1999. Multicenter evaluation of the Abbott LCx *Mycobacterium tuberculosis* ligase chain reaction assay. *J. Clin. Microbiol.* **37:**3102–3107.

85. **Maiwald, M., M. Schill, C. Stockinger, J. H. Helbig, P. C. Luck, W. Witzleb, and H. G. Sonntag.** 1995. Detection of Legionella DNA in human and guinea pig urine samples by polymerase chain reaction. *Eur. J. Clin. Microbiol. Infect. Dis.* **14:**25–33.

86. **Maiwald, M., C. Stockinger, D. Hassler, M. von Knebel Doeberitz, and H. G. Sonntag.** 1995. Evaluation of the detection of *Borrelia burgdorferi* DNA in urine samples by polymerase chain reaction. *Infection* **23:**173–179.

87. **Mantero, G., Z. Antonella, and A. Albertini.** 1991. DNA enzyme immunoassay: general method for detecting products of polymerase chain reaction. *Clin. Chem.* **37:**422–429.

88. **Martell, M., J. Gomez, J. I. Esteban, S. Sauleda, J. Quer, B. Cabot, R. Esteban, and J. Guardia.** 1999. High-throughput real-time reverse transcription-PCR quantitation of hepatitis C virus RNA. *J. Clin. Microbiol.* **37:**327–332.

89. **Mathis, A., R. Weber, H. Kuster, and R. Speich.** 1997. Simplified sample processing combined with a sensitive one-tube nested PCR assay for detection of *Pneumocystis carinii* in respiratory specimens. *J. Clin. Microbiol.* **35:**1691–1695.

90. **Mercier, G., A. Burckel, and G. Lucotte.** 1997. Detection of *Borrelia burgdorferi* DNA by polymerase chain reaction in urine specimens of patients with erythema migrans lesions. *Mol. Cell. Probes* **11:**89–94.

91. **Mitchell, P.. S., M. J. Espy, T. F. Smith, D. R. Toal, P. N. Rys, E. F. Berbari, D. R. Osmon, and D. H. Persing.** 1997. Laboratory diagnosis of central nervous system infections with herpes simplex virus by PCR performed with cerebrospinal fluid specimens. *J. Clin. Microbiol.* **35:**2973–2877.

92. **Muller, N., V. Zimmermann, B. Hentrich, and B. Gottstein.** 1996. Diagnosis of *Neospora caninum* and *Toxoplasma gondii* infection by PCR and DNA hybridization immunoassay. *J. Clin. Microbiol.* **34:**2850–2852.

93. **Murphy, D. G., L. Cote, M. Fauvel, P. Rene, and J. Vincelette.** 2000. Multicenter comparison of Roche COBAS AMPLICOR MONITOR version 1.5, Organon Teknika NucliSens QT with Extractor, and Bayer Quantiplex version 3.0 for quantification of human immunodeficiency virus type 1 RNA in plasma. *J. Clin. Microbiol.* **38:**4034–4041.

94. **Myers, T. W., and D. H. Gelfand.** 1991. Reverse transcription and DNA amplification by a *Thermus thermophilus* DNA polymerase. *Biochemistry* **30:**7661–7666.

95. **Nadeau, J. G., J. B. Pitner, C. P. Linn, J. L. Schram, C. H. Dean, and C. M. Nycz.** 1999. Real-time, sequence-specific detection of nucleic acids during strand displacement amplification. *Anal. Biochem.* **276:**177–187.

96. **Niubo, J., W. Li, K. Henry, and A. Erice.** 2000. Recovery and analysis of human immunodeficiency virus type 1 (HIV) RNA sequences from plasma samples with low HIV RNA levels. *J. Clin. Microbiol.* **38:**309–312.

97. **Noborg, U., A. Gusdal, E. K. Pisa, A. Hedrum, and M. Lindh.** 1999. Automated quantitative analysis of hepatitis B virus DNA by using the Cobas Amplicor HBV monitor test. *J. Clin. Microbiol.* **37:**2793–2797.

98. **Nolte, F. S.** Branched DNA signal amplification for direct quantitation of nucleic acid sequences in clinical specimens. *Adv. Clin. Chem.* **33:**201–235.

99. **Nycz, C. M., C. H. Dean, P. D. Haaland, C. A. Spargo, and G. T. Walker.** 1998. Quantitative reverse transcription strand displacement amplification: quantitation of nucleic acids using an isothermal amplification technique. *Anal. Biochem.* **259:**226–234.

100. **Olsson, M., K. Elvin, C. Lidman, S. Lofdahl, and E. Linder.** 1996. A rapid and simple nested PCR assay for the detection of *Pneumocystis carinii* in sputum samples. *Scand. J. Infect. Dis.* **28:**597–600.

101. **Padhye, A. A., G. Smith, D. McLaughlin, P. G. Standard, and L. Kaufman.** 1992. Comparative evaluation of a chemiluminescent DNA probe and an exoantigen test for rapid identification of *Histoplasma capsulatum*. *J. Clin. Microbiol.* **30:**3108–3111.

102. **Pang, J., J. Modlin, and R. Yolken.** 1992. Use of modified nucleotides and uracil-DNA glycosylase (UNG) for the control of contamination in the PCR-based amplification of RNA. *Mol. Cell. Probes* **6:**251–256.

103. **Parvaz, P., E. Guichard, P. Chevallier, J. Ritter, C. Trepo, and M. Sepetjan.** 1994. Hepatitis C: description of a highly sensitive method for clinical detection of viral RNA. *J. Virol. Methods* **47:**83–94.

104. **Pawlotsky, J. M., A. Bastie, C. Hezode, I. Lonjon, F. Darthuy, J. Remire, and D. Dhumeaux.** 2000. Routine detection and quantification of hepatitis B virus DNA in clinical laboratories: performance of three commercial assays. *J. Virol. Methods* **85:**11–21.

105. **Payton, M., and K. Pinter.** 1999. A rapid novel method for the extraction of RNA from wild-type and genetically modified kanamycin resistant mycobacteria. *FEMS Microbiol. Lett.* **180:**141–146.

106. **Pfyffer, G. E., P. Funke-Kissling, E. Rundler, and R. Weber.** 1999. Performance characteristics of the BDProbeTec system for direct detection of *Mycobacterium tuberculosis* complex in respiratory specimens. *J. Clin. Microbiol.* **37:**137–140.

107. **Pring-Akerblom, P., F. E. Trijssenaar, T. Adrian, and H. Hoyer.** 1999. Multiplex polymerase chain reaction for subgenus-specific detection of human adenoviruses in clinical samples. *J. Med. Virol.* **58:**87–92.

108. **Rabodonirina, M., L. Cotte, A. Boibieux, K. Kaiser, M. Mayencon, D. Raffenot, C. Trepo, D. Peyramond, and S. Picot.** 1999. Detection of *Pneumocytis carinii* DNA in blood specimens from human immunodeficiency virus-infected patients by nested PCR. *J. Clin. Microbiol.* **37:**127–131.

109. **Rabodonirina, M., D. Raffenot, L. Cotte, A. Boibieux, M. Mayencon, G. Bayle, F. Persat, F. Rabatel, C. Trepo, D. Peyramond, and M. A. Piens.** 1997. Rapid detection of *Pneumocystis carinii* in bronchoalveolar lavage specimens from human immunodeficiency virus-infected patients: use of a simple DNA extraction procedure and nested PCR. *J. Clin. Microbiol.* **35:**2748–2751.

110. **Rand, K. H., and H. Houck.** 1990. Taq polymerase contains bacterial DNA of unknown origin. *Mol. Cell. Probes* **4:**445–450.

111. **Reischl, U., R. Ruger, and C. Kessler.** 1994. Nonradioactive labeling and high-sensitive detection of PCR products. *Mol. Biotechnol.* **1:**229–240.

112. **Rogers, B. B., S. L. Josephson, S. K. Mak, and P. J. Sweeney.** 1992. Polymerase chain reaction amplification of herpes simplex virus DNA from clinical samples. *Obstet. Gynecol.* **79:**464–469.

113. **Romano, J. W., K. G. Williams, R. N. Shurtliff, C. Ginocchio, and M. Kaplan.** 1997. NASBA technology: isothermal RNA amplification in qualitative and quantitative diagnostics. *Immunol. Investig.* **26:**15–28.

114. **Rosenstraus, M., Z. Wang, S. Y. Chang, D. DeBonville, and J. P. Spadoro.** 1998. An internal control for routine diagnostic PCR: design, properties, and effect on clinical performance. *J. Clin. Microbiol.* **36:**191–197.

115. **Rozenberg, F., and P. Lebon.** 1991. Amplification and characterization of herpesvirus DNA in cerebrospinal fluid from patients with acute encephalitis. *J. Clin. Microbiol.* **29:**2412–2417.

116. **Sakrauski, A., B. Weber, H. H. Kessler, K. Pierer, and H. W. Doerr.** 1994. Comparison of two hybridization assays for the rapid detection of PCR amplified HSV genome sequences from cerebrospinal fluid. *J. Virol. Methods* **50:**175–184.

117. **Saldanha, J., and P. Minor.** 1994. A sensitive PCR method for detecting HCV RNA in plasma pools, blood products, and single donations. *J. Med. Virol.* **43:**72–76.

118. **Santti, J., T. Hyypiä, and P. Halonen.** 1997. Comparison of PCR primer pairs in the detection of human rhinoviruses in nasopharyngeal aspirates. *J. Virol. Methods* **66:**139–147.

119. **Scarparo, C., P. Piccoli, A. Rigon, G. Ruggiero, M. Scagnelli, and C. Piersimoni.** 2000. Comparison of enhanced *Mycobacterium tuberculosis* amplified direct test with COBAS AMPLICOR Mycobacterium tuberculosis Assay for direct detection of *Mycobacterium tuberculosis* complex in respiratory and extrapulmonary specimens. *J. Clin. Microbiol.* **38:**1559–1562.

120. **Schalasta, G., U. Engels, and L. Lindemann.** 1995. A rapid and simple PCR assay for detection of hepatitis A virus RNA in stool specimens. *Clin. Lab.* **41:**233–238.

121. **Schlueter, V., S. Schmolke, K. Stark, G. Hess, B. Ofenloch-Haehnle, and A. M. Engel.** 1996. Reverse transcription-PCR detection of hepatitis G virus. *J. Clin. Microbiol.* **34:**2660–2664.

122. **Schmidt, T. M., B. Pace, and N. R. Pace.** 1991. Detection of DNA contamination in Taq polymerase. *BioTechniques* **11:**176–177.

123. **Shafer, R. W., D. J. Levee, M. A. Winters, K. L. Richmond, D. Huang, and T. C. Merigan.** 1997. Comparison of QIAamp HCV kit spin columns, silica beads, and phenol-chloroform for recovering human immunodeficiency virus type 1 RNA from plasma. *J. Clin. Microbiol.* **35:**520–522.

124. **Sia, I. G., J. A. Wilson, M. J. Espy, C. V. Paya, and T. F. Smith.** 2000. Evaluation of the COBAS AMPLICOR CMV MONITOR test for detection of viral DNA in specimens taken from patients after liver transplantation. *J. Clin. Microbiol.* **38:**600–606.

125. **Somoskovi, A., J. E. Hotaling, M. Fitzgerald, V. Jonas, D. Stasik, L. M. Parsons, and M. Salfinger.** 2000. False-positive results for *Mycobacterium celatum* with the AccuProbe *Mycobacterium tuberculosis* complex assay. *J. Clin. Microbiol.* **38:**2743–2745.

126. **Spargo, C. A., M. S. Fraiser, M. Van Cleve, D. J. Wright, C. M. Nycz, P. A. Spears, and G. T. Walker.** 1996. Detection of M. tuberculosis DNA using thermophilic strand displacement amplification. *Mol. Cell. Probes* **10:**247–256.

127. **Spears, P. A., C. P. Linn, D. L. Woodard, and G. T. Walker.** 1997. Simultaneous strand displacement amplification and fluorescence polarization detection of *Chlamydia trachomatis* DNA. *Anal. Biochem.* **247:**130–137.

128. **Stapleton, J. T., D. Klinzman, W. N. Schmidt, M. A. Pfaller, P. Wu, D. R. Labrecque, J. Q. Han, M. J. Perino Phillips, R. Woolson, and B. Alden.** 1999. Prospective comparison of whole-blood- and plasma-based hepatitis C virus RNA detection systems: improved detection using whole blood as the source of viral RNA. *J. Clin. Microbiol.* **37:**484–489.

129. **Strand, A., S. Andersson, I. Zehbe, and E. Wilander.** 1999. HPV prevalence in anal warts tested with the MY09/MY11 SHARP signal system. *Acta Derm. Venereol.* **79:**226–229.

130. Sweet, D., M. Lorente, A. Valenzuela, J. A. Lorente, and J. C. Alvarez. 1996. Increasing DNA extraction yield from saliva stains with a modified Chelex method. *Forensic Sci. Int.* **83:**167–177.

131. Tagliaferro, L., M. Corbelli, G. Maietta, V. Pellegrino, and P. Pignatelli. 1995. Use of a rapid and simple method to extract proviral DNA in the identification of HIV-1 by PCR. *New Microbiol.* **18:**303–306.

132. Talley, A. R., F. Garcia-Ferrer, K. A. Laycock, M. Loeffelholz, and J. S. Pepose. 1992. The use of polymerase chain reaction for the detection of chlamydial keratoconjunctivitis. *Am. J. Ophthalmol.* **114:**685–692.

133. Tang, Y. W., P. N. Rys, B. J. Rutledge, P. S. Mitchell, T. F. Smith, and D. H. Persing. 1998. Comparative evaluation of colorimetric microtiter plate systems for detection of herpes simplex virus in cerebrospinal fluid. *J. Clin. Microbiol.* **36:**2714–2717.

134. Thornton, C. G., J. L. Hartley, and A. Rashtchian. 1992. Utilizing uracil DNA glycosylase to control carryover contamination in PCR: characterization of residual UDG activity following thermal cycling. *BioTechniques* **13:**180–184.

135. Tjhie, J. H. T., F. J. M. van Kuppeveld, R. Roosendaal, W. J. G. Melchers, R. Gordijn, D. M. MacLaren, J. M. M. Walboomers, C. J. L. M. Meijer, and A. J. C. van den Brule. 1994. Direct PCR enables detection of *Mycoplasma pneumoniae* in patients with respiratory tract infections. *J. Clin. Microbiol.* **32:**11–16.

136. Tortoli, E., M. T. Simonetti, and F. Lavinia. 1996. Evaluation of reformulated chemiluminescent DNA probe (AccuProbe) for culture identification of *Mycobacterium kansasii*. *J. Clin. Microbiol.* **34:**2838–2840.

137. Tortoli, E., M. Tronci, C. P. Tosi, C. Galli, F. Lavinia, S. Natili, and A. Goglio. 1999. Multicenter evaluation of two commercial amplification kits (Amplicor, Roche and LCx, Abbott) for direct detection of Mycobacterium tuberculosis in pulmonary and extrapulmonary specimens. *Diagn. Microbiol. Infect. Dis.* **33:**173–179.

138. Udaykumar, J., S. Epstein, and I. K. Hewlett. 1993. A novel method employing UNG to avoid carry-over contamination in RNA-PCR. *Nucleic Acids Res.* **21:**3917–3918.

139. Valentine-Thon, E. 1995. Evaluation of SHARP signal system for enzymatic detection of amplified hepatitis B virus DNA. *J. Clin. Microbiol.* **33:**477–480.

140. Van Der Pol, B., T. C. Quinn, C. A. Gaydos, K. Crotchfelt, J. Schachter, J. Moncada, D. Jungkind, D. H. Martin, B. Turner, C. Peyton, and R. B. Jones. 2000. Multicenter evaluation of the AMPLICOR and automated COBAS AMPLICOR CT/NG tests for detection of *Chlamydia trachomatis*. *J. Clin. Microbiol.* **38:**1105–1112.

141. Van Gemen, B., R. van Beuningen, A. Nabbe, D. van Strijp, S. Jurriaans, P. Lens, and T. Kievits. 1994. A one-tube quantitative HIV-1 RNA NASBA nucleic acid amplification assay using electrochemiluminescent (ECL) labelled probes. *J. Virol. Methods* **49:**157–167.

142. Verhofstede, C., K. Fransen, D. Marissens, R. Verhelst, G. van der Groen, S. Lauwers, G. Zissis, and J. Plum. 1996. Isolation of HIV-1 RNA from plasma: evaluation of eight different extraction methods. *J. Virol. Methods* **60:**155–159.

143. Vignoli, C., X. de Lamballerie, C. Zandotti, C. Tamalet, and P. de Micco. 1995. Advantage of a rapid extraction method of HIV1 DNA suitable for polymerase chain reaction. *Res. Virol.* **146:**159–162.

144. Vince, A., M. Poljak, and K. Seme. 1998. DNA extraction from archival Giemsa-stained bone-marrow slides: comparison of six rapid methods. *Br. J. Haematol.* **101:**349–351.

145. Vogel, J. U., J. Cinatl, A. Lux, B. Weber, A. J. Driesel, and H. W. Doerr. 1996. New PCR assay for rapid and quantitative detection of human cytomegalovirus in cerebrospinal fluid. *J. Clin. Microbiol.* **34:**482–483.

146. Walker, G. T., and C. P. Linn. 1996. Detection of *Mycobacterium tuberculosis* DNA with thermophilic strand displacement amplification and fluorescence polarization. *Clin. Chem.* **42:**1604–1608.

147. Walker, G. T., C. P. Linn, and J. G. Nadeau. 1996. DNA detection by strand displacement amplification and fluorescence polarization with signal enhancement using a DNA binding protein. *Nucleic Acids Res.* **24:**348–353.

148. Walker, G. T., J. G. Nadeau, and C. P. Linn. 1995. A DNA probe assay using strand displacement amplification (SDA) and filtration to separate reacted and unreacted detector probes. *Mol. Cell. Probes* **9:**399–403.

149. Walker, G. T., J. G. Nadeau, C. P. Linn, R. F. Devlin, and W. B. Dandliker. 1996. Strand displacement amplification (SDA) and transient-state fluorescence polarization detection of *Mycobacterium tuberculosis* DNA. *Clin. Chem.* **42:**9–13.

150. Watson, J., J. Schanke, H. Grunenwald, R. Meis, L. Hoffman, M. Lewandowska-Skarbek, and E. Moan. 1998. A new method for DNA and RNA purification. *J. Clin. Ligand Assay* **21:**394–403.

151. Wattre, P., A. Dewilde, L. Subtil, L. Andreoletti, and V. Thirion. 1998. A clinical and epidemiological study of human parvovirus B19 infection in fetal hydrops using PCR, Southern blot hybridization, and chemiluminescence detection. *J. Med. Virol.* **54:**140–144.

152. Whitby, J. E., P. R. Heaton, H. E. Whitby, E. O'Sullivan, and P. Johnstone. 1997. Rapid detection of rabies and rabies-related viruses by RT-PCR and enzyme-linked immunosorbent assay. *J. Virol. Methods* **69:**63–72.

153. White, T. J., R. Madej, and D. H. Persing. 1992. The polymerase chain reaction: clinical applications. *Adv. Clin. Chem.* **29:**161–196.

154. Williams-Bouyer, N., B. S. Reisner, and G. L. Woods. 2000. Comparison of Gen-Probe AccuProbe group B streptococcus culture identification test with conventional culture for the detection of group B streptococci in broth cultures of vaginal-anorectal specimens from pregnant women. *Diagn. Microbiol. Infect. Dis.* **36:**159–162.

155. Witt, D. J., and M. Kemper. 1999. Techniques for the evaluation of nucleic acid amplification technology performance with specimens containing interfering substances: efficacy of boom methodology for extraction of HIV-1 RNA. *J. Virol. Methods* **79:**97–111.

156. Witt, D. J., M. Kemper, A. Stead, C. C. Ginocchio, and A. M. Caliendo. 2000. Relationship of incremental specimen volumes and enhanced detection of human immunodeficiency virus type 1 RNA with

nucleic acid amplification technology. *J. Clin. Microbiol.* **38:**85–89.

157. **Yang, Y., M. H. Wisbeski, M. Mendoza, S. Dorf, D. Xu, M. Nguyen, S. Yeh, and R. Sun.** 1999. Performance characteristics of the AmpliScreen(TM) HIV-1 test, an assay designed for screening plasma minipools. *Biologicals* **27:**315–323.

158. **Zaaijer, H. L., H. T. M. Cuypers, H. W. Reesink, I. N. Winkel, G. Gerken, and P. N. Lelie.** 1993. Reliability of polymerase chain reaction for detection of hepatitis C virus. *Lancet* **341:**722–724.

159. **Zandotti, C., X. de Lamballerie, C. Guignole-Vignoli, C. Bollet, and P. de Micco.** 1993. A rapid DNA extraction method from culture and clinical samples suitable for the detection of human cytomegalovirus by the polymerase chain reaction. *Acta Virol.* **37:**106–108.

160. **Zehbe, I., and E. Wilander.** 1997. Nonisotopic ELISA-based detection of human papillomavirus-amplified DNA. *Mod. Pathol.* **10:**188–191.

161. **Zhang, F., S. Tetali, X. P. Wang, M. H. Kaplan, F. V. Cromme, and C. C. Ginocchio.** 2000. Detection of human cytomegalovirus pp67 late gene transcripts in cerebrospinal fluid of human immunodeficiency virus type 1-infected patients by nucleic acid sequence-based amplification. *J. Clin. Microbiol.* **38:**1920–1925.

162. **Zhong, K. J., and K. C. Kain.** 1999. Evaluation of a colorimetric PCR-based assay to diagnose *Plasmodium falciparum* malaria in travelers. *J. Clin. Microbiol.* **37:**339–341.

Automated Immunoassay Analyzers

RICHARD L. HODINKA

13

Enzyme immunoassays (EIA) have great potential for automation in diagnostic microbiology laboratories, and a number of semiautomated and fully automated immunoassay analyzers are now commercially available for the performance of a variety of viral, bacterial, fungal, and parasitic antibody and/or antigen detection tests (Table 1). These systems have found particular utility in the area of clinical virology, where extensive test menus now exist for the automated performance of serologic assays for human immunodeficiency virus (HIV), viral agents of hepatitis, the so-called TORCH agents implicated in congenital and peripartum infections, and other viruses of clinical importance. The technologies used in the instruments involve the capture of a particular analyte using coated polystyrene beads, magnetic particles, microplate wells, or other solid-phase receptacles, or fiber matrix filters. The captured analytes are then detected by colorimetric, chemiluminescence, or fluorescence measurements. The systems are either compact bench units that take up a limited amount of space in the laboratory or freestanding units with larger footprints requiring more space. Many of the systems are formatted for batch testing only, but some instruments provide continuous random-access and STAT-processing capabilities that allow the addition of tests and reagents without disruption of the test being performed. The majority of the automated immunoassay analyzers provide walk-away simplicity to perform assays from sample processing through interpretation of results. The instruments can automatically generate work lists of specimens to be tested, pipette specimens from primary tubes and dilute the samples, dispense all reagents, time the incubations at a desired temperature, shake the assay vessels if needed, perform washes, and read and store the final results. A few analyzers have no sample-handling capabilities and perform only the post-sample analytical steps. Most manufacturers of automated instruments also provide software for the analysis and management of patient data and for monitoring the quality of the testing performed. Many of the instruments can also interface with computer-based hospital laboratory information systems for seamless reporting of results. The quantity and choice of automated instruments used depends mainly on the volume of specimens for testing and the number of individual tests to be performed. Some analyzers will run only assay protocols developed and preset by the manufacturer, while others permit the incorporation of user-defined procedures or are completely open systems that can be programmed to perform all analytical steps of a wide variety of manufacturer's EIA kits.

The use of automated immunoassay analyzers can be advantageous to the laboratory that has a shortage of trained medical technologists or that needs to reduce costs or improve the throughput and turnaround time for test results. These instruments allow the users to consolidate multiple workstations and improve productivity by offering speed, efficiency, standardization, less hands-on time, and ease of use. For those laboratories that are consolidating departments, most of the analyzers also have large onboard menus to perform assays for cardiac biomarkers, thyroid function, reproductive endocrinology, anemia, drugs of abuse, diabetes, allergy, tumor markers, and autoimmune screening.

The information presented in Table 1 about each automated immunoassay analyzer was supplied by the individual manufacturers through their computer websites, written materials, and/or direct conversations with company representatives. I have attempted to provide information that will allow the reader to quickly compare and contrast the various instruments with respect to their basic features and the available menus of tests that can be performed on each analyzer for clinical microbiology. The reader should contact the manufacturer listed in the appendix for a more comprehensive description and the current list price of a particular immunoassay analyzer.

TABLE 1 Automated immunoassay systems for viral, bacterial, fungal, and parasitic antibody and antigen detection

Manufacturer[a]	Product name	Selected features	Test menu[b] Available in United States	In development	Only available outside United States	Selected reference(s)[c]
Abbott Laboratories	IMx	MEIA, FPIA, and ion capture technologies; benchtop unit; batch testing; 20 tests/h; no onboard capacity to run different assays simultaneously; onboard dilution of samples; onboard data management; closed system for Abbott products only	Rubella virus IgG and IgM, CMV IgG and IgM, *Toxoplasma gondii* IgG and IgM, anti-HAV IgM, anti-HAV total, anti-HBc IgM	Hepatitis and retrovirus assays	HBsAg, HBsAg v2, HBsAg confirmatory, anti-HBsAg, anti-HBc total, anti-HBeAg, HBeAg, anti-HCV 2.0, anti-HCV 3.0, anti-HIV-1/2 3rd Generation Plus, *Chlamydia trachomatis* Ag, Lyme IgG and IgM	1, 5, 8, 16, 18, 19, 29, 36, 38, 42–44, 47, 50
	AxSYM	MEIA, FPIA and ion capture technologies; freestanding unit; 120 tests/h; onboard capacity to run 20 different assays simultaneously; continuous random access; STAT processing; primary tube sampling; onboard dilution of samples; automatic sample rerun capability; onboard data management; closed system for Abbott products only	Rubella virus IgG and IgM, *T. gondii* IgG and IgM, CMV IgG	Hepatitis and retrovirus assays, CMV IgM	Anti-HIV-1 and -2; anti-HIV-1 and -2 gO, CMV IgM, anti-HAV total, anti-HAV total 2.0, anti-HAV total Quant, anti-HAV IgM, anti-HAV IgM 2.0, HBsAg, HBsAg confirmatory, anti-HBsAg, anti-HBc total, anti-HBc IgM, anti-HBeAg 2.0, HBeAg, HBeAg 2.0, HBeAg Quant, anti-HCV 2.0, anti-HCV 3.0	9, 17, 27, 29, 32, 33, 35, 49, 59
	Architect i2000	Chemiluminesce with flexible protocols; freestanding unit; 200 tests/h per unit; modularity with seamless integration of multiple instruments to single workstations; supports batch and continuous random-access testing; onboard capacity to run 25 different assays simultaneously per module; primary tube sampling; onboard dilution of samples; automatic sample rerun capability; onboard data management; closed system for Abbott products only	None	Hepatitis assays, HIV-1 and -2 gO, Rubella virus IgG and IgM, *T. gondii* IgG and IgM, CMV IgG and IgM, HSV-2 IgG	HBsAg, HBsAg confirmatory, anti-HBsAg, anti-HBc total, HCV	28, 37
Bayer Diagnostics	Immuno 1	EIA using colorimetric measurement; freestanding unit; random access; 120 tests/h; onboard capacity to run 22 different assays simultaneously; primary tube sampling; onboard data management; closed system for Bayer products only	Rubella virus IgG and IgM, *T. gondii* IgG and IgM	CMV IgG	None	40

(Continued on next page)

TABLE 1 Automated immunoassay systems for viral, bacterial, fungal, and parasitic antibody and antigen detection (*Continued*)

Manufacturer[a]	Product name	Selected features	Test menu[b] Available in United States	In development	Only available outside United States	Selected reference(s)[c]
	Advia Centaur	Magnetic particle separation with chemiluminescence detection; freestanding unit; continuous random access; up to 240 tests/h; onboard capacity to run 30 different assays simultaneously; primary tube sampling; onboard dilution of samples; automatic sample rerun capability; onboard data management; closed system for Bayer products only	None	Rubella virus IgG and IgM, *T. gondii* IgG and IgM, CMV IgG and IgM	None	28
Beckman Coulter	Access	Magnetic particle separation with chemiluminescence detection; benchtop unit; up to 100 tests/h; onboard capacity to run 24 different assays simultaneously; random and continuous access; primary tube sampling; onboard dilution of samples; onboard data management; closed system for Beckman Coulter products only	Rubella virus IgG, *T. gondii* IgG, *C. trachomatis* Ag, *C. trachomatis* Ag confirmatory, urine *C. trachomatis* Ag	CMV IgG and IgM, anti-HBeAg, HBeAg	Rubella virus IgM, *T. gondii* IgM, anti-HIV-1 and -2, HBsAg, HBsAg confirmatory, anti-HBsAg, anti-HBc total, anti-HBc IgM, anti-HCV (the HIV-1 and -2 and anti-HCV assays are available through partnership with Sanofi Bio-Rad Pasteur)	34
BioChem Immuno-Systems	Labotech	EIA colorimetric measurement; benchtop unit; fully automated microplate processor; can handle up to three 96-well microtiter plates; onboard capacity to run eight different assays simultaneously; primary tube sampling; onboard dilution of samples; onboard data management; open system reagent configuration; best suited for mid to high specimen volumes	Can be programmed to perform all analytical steps of a wide variety of manufacturers' EIA kits. This includes sample handling, dispensing reagents, incubating and washing microwells, reading results, data reduction, and printing customized reports. Examples of assays that have been performed on the Labotech include rubella virus IgG and IgM, measles virus IgG and IgM, *T. gondii* IgG and IgM, VZV IgG and IgM, mumps virus IgG, CMV IgG and IgM, HSV-1 and -2 IgG and IgM, EBV VCA IgG and IgM, EBV EA IgG, EBV EBNA IgG, parvovirus B19 IgG and IgM, anti-HAV total, anti-HAV IgM, anti-HBc total, anti-HBc IgM, HBsAg, anti-HBsAg, HBeAg, anti-HBeAg, syphilis IgG and IgM, Lyme IgG and IgM, *Helicobacter pylori* IgG, *C. trachomatis* Ag and Ab, *Legionella pneumophila* Ag and IgG and IgM Ab, *Mycoplasma pneumoniae* IgG and IgM, *Clostridium difficile* toxin A, *Candida albicans* IgG, IgM, and IgA, *Giardia lamblia* Ag, and *Entamoeba histolytica* Ag and Ab.			51, 52

TABLE 1 *(Continued)*

Manufacturer[a]	Product name	Selected features	Test menu[b] Available in United States	In development	Only available outside United States	Selected reference(s)[c]
	PersonalLAB	EIA colorimetric measurement; benchtop unit; fully automated microplate processor; can handle up to two 96-well microtiter plates; onboard capacity to run six different assays simultaneously; primary tube sampling; onboard dilution of samples; onboard data management; open system reagent configuration; best suited for low to midsize volumes of specimens	Can be programmed to perform all analytical steps of a wide variety of manufacturers' EIA kits. This includes sample handling, dispensing reagents, incubating and washing microwells, reading results, data reduction, and printing customized reports. Examples of assays that have been performed on the PersonalLAB include rubella IgG and IgM, measles IgG and IgM, T. gondii IgG and IgM, VZV IgG and IgM, mumps IgG, CMV IgG and IgM, HSV-1 and -2 IgG and IgM, EBV VCA IgG and IgM, EBV EA IgG, EBV EBNA IgG, parvovirus B19 IgG and IgM, anti-HAV total, anti-HAV IgM, anti-HBc total, anti-HBc IgM, HBsAg, anti-HBsAg, HBeAg, anti-HBeAg, syphilis IgG and IgM, Lyme IgG and IgM, H. pylori IgG, C. trachomatis Ag and Ab, L. pneumophila Ag and IgG and IgM Ab, M. pneumoniae IgG and IgM, C. difficile toxin A, C. albicans IgG, IgM, and IgA, G. lamblia Ag, and E. histolytica Ag and Ab.			None found
BioMerieux-Vitek	VIDAS; Mini-VIDAS	Enzyme-linked fluorescence assay; automated benchtop unit; random-access and batch testing; calibration stored for 14 days; minimum maintenance; onboard data management; closed system for BioMerieux products only; VIDAS contains five sections accepting 6 tests each and can handle 60 tests/h; Mini-VIDAS has two sections accepting 6 tests each and can handle 24 tests/h; number of different assays tat can be run simultaneously is 30 for VIDAS and 12 for Mini-VIDAS	(Same for both instruments.) H. pylori IgG, CMV IgG and IgM, measles virus IgG, mumps virus IgG, rubella virus IgG, VZV IgG, T. gondii IgG and IgM, T. gondii IgG and IgM competition, Lyme IgG and IgM, C. trachomatis Ag, C. trachomatis blocking assay, rotavirus Ag, RSV Ag, C. difficile toxin A, industrial screening assays for detection of Escherichia coli O157, Listeria monocytogenes, Salmonella spp., or enterotoxigenic coagulase-negative staphylococci in foods	H. pylori IgG, quantitative T. gondii IgG, VIDAS PROBE system to perform TMA to detect amplified C. trachomatis DNA, amplified Neisseria gonorrhoeae DNA, combination-amplified C. trachomatis and N. gonorrhoeae, amplified Mycobacterium tuberculosis DNA, quantitative HIV-1 RNA	T. gondii IgG avidity, anti-HIV-1 and -2, HIV p24 Ag, HIV DUO (combined detection of p24 Ag and anti-HIV-1 and anti-HIV-2 Ab), anti-HAV total, anti-HAV IgM, HBsAg, anti-HBsAg, anti-HBc total, anti-HBc IgM, HBeAg, anti-HBeAg	2, 3, 7, 11, 14, 15, 21, 26, 30, 39, 41, 45, 46, 48, 54, 55, 58
Bio-Rad Laboratories	CODA	EIA colorimetric measurement; benchtop unit; fully automated microplate processor; batch testing; can handle as many as 270 or as few as 1 or 2 samples per run; onboard capacity to run nine different assays simultaneously; onboard dilution of samples; onboard data management; open system reagent configuration	Can be programmed to perform all analytical steps of a wide variety of manufacturers' EIA kits. This includes sample dispensing, reagent addition, plate shaking, incubating and washing microwells, reading results, data reduction, and printing customized reports. The CODA system also runs the current product line of Bio-Rad EIA kits, including T. gondii IgG and IgM, rubella virus IgG and IgM, CMV IgG and IgM, HSV-1 and -2 IgG, HSV IgM, syphilis IgG and IgM, and H. pylori IgG, IgM, and IgA. Anti-HIV-1 and -2 and HBsAg are available outside the United States.			52

(Continued on next page)

TABLE 1 Automated immunoassay systems for viral, bacterial, fungal, and parasitic antibody and antigen detection (*Continued*)

Manufacturer[a]	Product name	Selected features	Test menu[b]			Selected reference(s)[c]
			Available in United States	In development	Only available outside United States	
Bio-Tek Instruments	Omni	EIA colorimetric measurement; benchtop unit; fully automated microplate processor; batch testing; positive plate and reagent tracking; can handle up to 17 96-well microtiter plates with a throughput of 5 plates/h; onboard capacity to run 10 different assays simultaneously; onboard data management; open system reagent configuration	Can be programmed to perform all postsampling analytical steps of a wide variety of manufacturers' EIA kits. This includes dispensing reagents, incubating and washing microwells, reading results, data reduction, and printing customized reports. A list of assays that have been performed on the Omni includes rubella virus IgG and IgM, measles virus IgG and IgM, *T. gondii* IgG and IgM, VZV IgG, mumps virus IgG, CMV IgG and IgM, HSV-1 and -2 IgG and IgM, EBV VCA IgG and IgM, EBV EA IgG, EBV EBNA IgG, HBsAg, anti-HBc total, HBeAg, anti-HBeAg, anti-HCV, anti-HTLV-2, anti-HIV-1 and -2, HIV-1 p24 Ag, syphilis IgG, Lyme IgG and IgM, *H. pylori* IgG, and Chagas disease IgG.			None found
Diagnostic Products Corp.	Immulite	Enzyme-amplified chemiluminescence bead chemistry; benchtop unit; 60 tests/h; onboard capacity to run 12 different assays simultaneously; continuous random access; STAT processing; onboard data management; closed system for Diagnostic Products assays only	CMV IgG, rubella virus IgG (quantitative), *T. gondii* IgG (quantitative)	CMV IgM, EBV serology, anti-HAV, HBeAg, anti-HBeAg, Chagas disease virus IgG, Lyme IgG and IgM screen, *C. difficile* toxin A	Rubella virus IgM, *T. gondii* IgM, HBsAg, HBsAg confirmatory, anti-HBsAg, anti-HBc total, anti-HBc IgM, Lyme IgG, *H. pylori* IgG	10, 12, 53
	Immulite 2000	Enzyme-amplified chemiluminescence bead chemistry; freestanding unit; continuous random access; STAT processing; 200 tests/h; onboard capacity to run 24 different assays simultaneously; primary tube sampling; onboard dilution of samples; automatic sample rerun capability; onboard data management; designed for medium- and high-volume laboratories; closed system for Diagnostic Products assays only	CMV IgG, rubella virus IgG (quantitative), *T. gondii* IgG (quantitative), *H. pylori* IgG	CMV IgM, HBeAg, anti-HBeAg	Rubella virus IgM, *T. gondii* IgM, HBsAg, HBsAg confirmatory, anti-HBsAg, anti-HBc total, anti-HBc IgM	28

TABLE 1 (Continued)

Manufacturer[a]	Product name	Selected features	Test menu[b] Available in United States	In development	Only available outside United States	Selected reference(s)[c]
Diamedix	MAGO Plus	EIA colorimetric measurement; bench unit; fully automated microplate processor; batch or random-access testing; can process up to four 96-well microtiter plates; onboard capacity to run nine different assays simultaneously; primary tube sampling; onboard dilution of samples; onboard data management; system supports Diamedix EIA kits and has 20 open user-programmable channels for additional custom tests	Rubella virus IgG, *T. gondii* IgG, CMV IgG, HSV-1 and -2 IgG, measles virus IgG, VZV IgG, EBV VCA IgG and IgM, EBV EA-D IgG and IgM, EBNA-1 IgG and IgM, mumps virus IgG, *H. pylori* IgG, anti-*Borrelia burgdorferi* IgG and IgM, anti-*B. burgdorferi* IgM	Rubella virus capture IgM, *T. gondii* capture IgM, CMV capture IgM, HSV-1 and -2 IgM, VZV IgM, mumps virus IgM, measles virus IgM, *M. pneumoniae* IgG and IgM	Rubella virus capture IgM, *T. gondii* IgA and capture IgM, CMV capture IgM, HSV-1 and -2 IgM, VZV capture IgM, mumps virus capture IgM, measles virus capture IgM, syphilis IgG and IgM, *H. pylori* IgA	None found
DiaSorin, Inc.[d]	ETI-LAB	ETI-LAB is a Labotech instrument. See features described under BioChem ImmunoSystems.	ETI-LAB has been programmed to perform all analytical steps of DiaSorin's EIA kits. This includes sample handling, dispensing reagents, incubating and washing microwells, reading results, data reduction, and printing customized reports. A list of supported assays includes anti-HIV-1 and -2, anti-HAV total, anti-HAV IgM, anti-HBc total, anti-HBc IgM, HBsAg, HBsAg confirmatory, anti-HBsAg, HBeAg, anti-HCV, anti-HDV, HDVAg, rubella virus IgG and IgM, measles virus IgG and IgM, *T. gondii* IgG and IgM, VZV IgG and IgM, mumps virus IgG and IgM, CMV IgG and IgM, HSV-1 and -2 IgG and IgM, EBV VCA IgG and IgM, EBV EBNA-1 IgG, influenza virus A IgG and IgM, influenza virus B IgG and IgM, adenovirus IgG and IgM, syphilis IgG and IgM, anti-*B. burgdorferi* IgG and IgM, *H. pylori* IgG and IgA, *Chlamydia* IgG, IgM, and IgA, *M. pneumoniae* IgG and IgM, *Brucella* IgG and IgM, *Bordetella pertussis* IgG, IgM, and IgA, diphtheria toxin IgG and IgM, tetanus toxoid IgG, C. *albicans* IgG, IgM, and IgA, and *Aspergillus* IgG, IgM, and IgA.			None found
Dynex Technologies[e]	DSX	EIA colorimetric measurement; benchtop unit; modular design; fully automated microplate processor; batch testing; can process up to four 96-well microtiter plates; onboard capacity to run 12 different assays simultaneously; primary tube sampling; onboard dilution of samples; onboard data management; open system reagent configuration	Can be programmed to perform all analytical steps of different manufacturers' EIA kits. This includes sample dispensing, reagent addition, plate shaking, incubating and washing microwells, reading results, data reduction, and printing customized reports. A list of assays that have been performed on the DSX system includes rubella virus IgG and IgM, *T. gondii* IgG and IgM, CMV IgG and IgM, HSV-1 and -2 IgG and IgM, measles virus IgG, mumps virus IgG, VZV IgG, EBV EA IgG, EBV EBNA IgG, *H. pylori* IgG, and *L. pneumophila* IgG, IgM, and IgA.			None found

(Continued on next page)

TABLE 1 Automated immunoassay systems for viral, bacterial, fungal, and parasitic antibody and antigen detection (*Continued*)

Manufacturer[a]	Product name	Selected features	Test menu[b]			Selected reference(s)[c]
			Available in United States	In development	Only available outside United States	
	DIAS	EIA colorimetric measurement; benchtop unit; modular design; microplate processor; batch testing; can process up to 24 96-well microtiter plates; onboard capacity to run 12 different assays simultaneously; onboard data management; open system reagent configuration	Can be programmed to perform all postsampling analytical steps of different manufacturers' EIA kits. This includes dispensing reagents, incubating and washing microwells, reading results, data reduction, and printing customized reports. A list of assays that have been performed on the DIAS system includes rubella virus IgG and IgM, *T. gondii* IgG and IgM, CMV IgG and IgM, HSV-1 and -2 IgG and IgM, measles virus IgG and IgM, mumps virus IgG, VZV IgG, viral hepatitis markers, HIV, and EBV.			None found
GRIFOLS-Quest	Triturus	EIA colorimetric measurement; benchtop unit; fully automated microplate processor; batch or random-access testing; can process up to four 96-well microtiter plates; onboard capacity to run eight different assays simultaneously; primary tube sampling; onboard dilution of samples; onboard data management; open system reagent configuration that supports SeraQuest EIA kits and/or a user-defined test menu	Can be programmed to perform all analytical steps of available SeraQuest assays and a wide variety of other manufacturers' EIA kits. This includes sample dispensing, reagent addition, incubating and washing microwells, reading results, data reduction, and printing customized reports. A list of assays that have been performed on the Triturus system includes rubella virus IgG and IgM, *T. gondii* IgG and IgM, CMV IgG and IgM, HSV IgG and IgM, VZV IgG, measles virus IgG, mumps virus IgG, EBV VCA IgG and IgM, EBV EA IgG, EBV EBNA IgG, and Chagas IgG.			None found
Meridian Diagnostics	DUET	EIA colorimetric measurement; benchtop unit; fully automated microplate processor; batch testing; can process up to two 96-well microtiter plates; onboard capacity to run eight different Meridian assays or one non-Meridian assay simultaneously; primary tube sampling; onboard dilution of samples; onboard data management; system formatted primarily for Meridian EIA kits but can be programmed to process other manufacturers' assays.	Runs Meridian infectious-disease EIA products, including CMV IgG and IgM, rubella virus IgG and IgM, mumps virus IgG, EBV VCA IgG and IgM, EBV EA IgG, EBV EBNA IgG, HSV-1 and -2 IgG and IgM, type-specific HSV-1 IgG, type-specific HSV-2 IgG, *T. gondii* IgG and IgM, Lyme IgG and IgM, syphilis IgG and IgM, Chagas IgG, G. *lamblia* IgG, *Cryptosporidium* IgG, *H. pylori* IgG and IgA, and *C. difficile* toxin A. Can be programmed to process other manufacturers' EIA kits.			None found
Ortho-Clinical Diagnostics	VITROS ECi	Enhanced chemiluminescence measurement; freestanding unit; continuous random access; STAT processing; up to 90 tests/h; onboard capacity to run 20 different assays simultaneously; primary tube sampling; onboard dilution of samples; automatic sample rerun capability; onboard data management; closed system for Ortho-Clinical Diagnostics products only	None	None	Anti-HIV-1 and -2, anti-HAV IgM, HBsAg, HBsAg confirmatory, anti-HBsAg, anti-HBc total, anti-HBc IgM, HBeAg, anti-HBeAg, anti-HCV	28

TABLE 1 (*Continued*)

Manufacturer[a]	Product name	Selected features	Test menu[b]			Selected reference(s)[c]
			Available in United States	In development	Only available outside United States	
Roche Diagnostics	ELECSYS 2010	Magnetic particle electro-chemiluminescence chemistry; benchtop unit; fully automated; continuous random access; STAT processing; up to 88 tests/h; onboard capacity to run 15 different assays simultaneously; primary tube sampling; onboard dilution of samples; onboard data management; closed system for Roche Diagnostics products only	None	Anti-HAV total, anti-HAV IgM	HBsAg, anti-HBsAg, anti-HBc total, anti-HBc IgM, anti-HBeAg, HBeAg	20, 28, 56, 57
	ELECSYS 1010	Magnetic particle electro-chemiluminescence chemistry; benchtop unit; fully automated; continuous random access; STAT processing; up to 50 tests/h; onboard capacity to run six different assays simultaneously; primary tube sampling; onboard dilution of samples; onboard data management; closed system for Roche Diagnostics products only	None	None	HBsAg, anti-HBsAg, anti-HBc total	None found
	Cobas Core II	EIA polystyrene bead chemistry; benchtop unit; continuous random access; batch and STAT processing; 150 tests/h; onboard capacity to run 18 different assays simultaneously; primary tube sampling; onboard dilution of samples; automatic sample rerun capability; onboard data management; closed system for Roche Diagnostics products only	None	HSV-2 IgG	Rubella virus IgG and IgM, T. gondii IgG and IgM, CMV IgG and IgM, anti-HSV-1 and -2, anti-HIV-1 and -2, anti-HTLV-1 and -2, anti-HAV total, anti-HAV IgM, HBsAg, HBsAg confirmatory, anti-HBsAg, anti-HBsAg Quant, anti-HBc total, anti-HBc IgM, HBeAg, anti-HBeAg, anti-HCV, anti-H. pylori Quant	4, 6, 13, 22–25, 31, 50
Rosys Anthos	Plato 1300; Plato 3300; Plato 7	EIA colorimetric or luminescence measurement; benchtop units; fully automated microplate processors; throughput of 3 plates/h (Plato 1300), 10 plates/h (Plato 3300), 20 plates/h (Plato 7); primary tube sampling; several different tests can be run simultaneously; onboard dilution of samples; onboard data management; open system reagent configuration	All instruments can be programmed to perform all analytical steps of different manufacturers' EIA kits. This includes sample diluting and dispensing, reagent addition, plate shaking, incubating and washing microwells, reading results, data reduction, and printing customized reports. A list of assays performed on these instruments was not available from the manufacturer.			None found

(*Continued on next page*)

TABLE 1 Automated immunoassay systems for viral, bacterial, fungal, and parasitic antibody and antigen detection (*Continued*)

Manufacturer[a]	Product name	Selected features	Test menu[b]			Selected reference(s)[c]
			Available in United States	In development	Only available outside United States	
Sigma Diagnostics	APTUS 4E	EIA colorimetric measurement; benchtop unit; fully automated microplate processor; supports batch or random-access testing; can process up to four 96-well microtiter plates; onboard capacity to run nine different assays simultaneously; primary tube sampling; onboard dilution of samples; onboard data management; system formatted primarily for Sigma EIA kits but has open channels to process other manufacturers' assays	Runs all Sigma infectious-disease serologies, including rubella virus IgG and IgM, CMV IgG and IgM, *T. gondii* IgG and IgM, HSV-1 IgG, HSV-2 IgG, HSV-1 and/or -2 IgM, VZV IgG, mumps virus IgG, measles virus IgG, EBV VCA IgG and IgM, EBV EA IgG, EBV EBNA IgG, M. *pneumoniae* IgG and IgM, *L. pneumophila* IgG, IgM, and IgA, *H. pylori* IgG, and *B. burgdorferi* IgG and IgM. Can be programmed to process other manufacturers' EIA kits.			None found
Tosoh Medics, Inc.	AIA Nex•IA	Immunofluorescence measurement; freestanding unit; 120 tests/h; onboard capacity to run 20 different assays simultaneously; continuous random access; STAT processing; primary tube sampling; onboard dilution of samples; onboard data management; closed system for Tosoh products only	None	None	Rubella virus IgG, *T. gondii* IgG and IgM, HBsAg, anti-HBsAg, HBeAg	None found
	AIA-600 II	Immunofluorescence measurement; benchtop unit; 60 tests/h; no onboard capacity to run different assays simultaneously; continuous random access; STAT processing; primary tube sampling; onboard dilution of samples; onboard data management; closed system for Tosoh products only	None	None	Rubella virus IgG, *T. gondii* IgG and IgM, HBsAg, anti-HBsAg, HBeAg	None found

[a]The Intracel Alpha 4 automated immunoassay system from Intracel (formerly Bartels) has been discontinued.
[b]Abbreviations: Ab, antibody; Ag, antigen; CMV, cytomegalovirus; EBV, Epstein-Barr virus; EBNA, Epstein-Barr nuclear antigen; EA, early antigen; FPIA, fluorescence polarization immunoassay; gO, group O: HIV-1, HIV type 1; HAV, hepatitis A virus; HBc, hepatitis B core; HBeAg, hepatitis B e antigen; HBsAg, hepatitis B surface antigen; HCV, hepatitis C virus; HDV, hepatitis delta virus; HSV, herpes simplex virus; HTLV-1, human T-cell leukemia virus type 1; Ig, immunoglobulin; MEIA, microparticle enzyme immunoassay; TMA, transcription-mediated amplification; v2, version 2; VZV, varicella-zoster virus; VCA, viral capsid antigen.
[c]An electronic search of citations from MEDLINE and additional life sciences journals from the National Library of Medicine was conducted to provide pertinent references on the use of automated immunoassay systems in the setting of a clinical microbiology laboratory. In some cases, no references could be found for the described use of certain instruments.
[d]DiaSorin no longer supports the ETI-STAR (PersonalLAB) analyzer and has discontinued production of the Copalis 1 automated immunoassay instrument.
[e]Dynex Technologies has recently changed its name to Thermo Labsystems. The manufacturer also discontinued the production of the DIAS automated immunoassay instrument in June 2001.

REFERENCES

1. **Abbott, G. G., J. W. Safford, R. G. MacDonald, M. C. Craine, and R. R. Applegren.** 1990. Development of automated immunoassays for immune status screening and serodiagnosis of rubella virus infection. *J. Virol. Methods* **27:**227–239.

2. **Azevedo-Pereira, J. M., M. H. Lourenco, F. Barin, R. Cisterna, F. Denis, P. Moncharmont, R. Grillo, and M. O. Santos-Ferreira.** 1994. Multicenter evaluation of a fully automated screening test, VIDAS HIV 1+2, for antibodies to human immunodeficiency virus type 1 and 2. *J. Clin. Microbiol.* **32:**2559–2563.

3. **Blackburn, C. W., L. M. Curtis, L. Humpheson, and S. B. Petitt.** 1994. Evaluation of the Vitek Immunodiagnostic Assay System (VIDAS) for the detection of *Salmonella* in foods. *Lett. Appl. Microbiol.* **19:**32–36.

4. **Bonanni, P., G. C. Icardi, A. M. Raffo, A. Ferrari Bravo, A. Roccatagliata, and P. Crovari.** 1996. Analytical and laboratory evaluation of a new fully-automated third generation enzyme immunoassay for the detection of antibodies to the hepatitis C virus. *J. Virol. Methods* **2:**113–122.

5. **Brokenshire, M. K., P. J. Say, A. H. van Vonno, and C. Wong.** 1997. Evaluation of the microparticle enzyme immunoassay Abbott IMx Select Chlamydia and the importance of urethral site sampling to detect *Chlamydia trachomatis* in women. *Genitourin. Med.* **73:**498–502.

6. **Burgisser, P., F. Simon, M. Wernli, T. Wust, M. F. Beya, and P. C. Frei.** 1996. Multicenter evaluation of new double-antigen sandwich enzyme immunoassay for measurement of anti-human immunodeficiency virus type 1 and type 2 antibodies. *J. Clin. Microbiol.* **34:**634–637.

7. **Candolfi, E., R. Ramirez, M. P. Hadju, C. Shubert, and J. S. Remington.** 1994. The Vitek immunodiagnostic assay for detection of immunoglobulin M toxoplasma antibodies. *Clin. Diagn. Lab. Immunol.* **1:**401–405.

8. **Chernesky, M. A., D. Jang, J. Sellors, P. Coleman, J. Bodner, I. Hrusovsky, S. Chong, and J. B. Mahoney.** 1995. Detection of *Chlamydia trachomatis* antigens in male urethral swabs and urines with a microparticle enzyme immunoassay. *Sex. Transm. Dis.* **22:**55–59.

9. **Costongs, G. M., R. J. van Oers, B. Leerkens, W. Hermans, and P. C. Janson.** 1995. Evaluation of the Abbott automated random, immediate and continuous access immunoassay analyzer, the AxSYM. *Eur. J. Clin. Chem. Clin. Biochem.* **33:**105–111.

10. **Costongs, G. M., R. J. van Oers, B. Leerkes, and P. C. Janson.** 1995. Evaluation of the DPC IMMULITE random access immunoassay analyzer. *Eur. J. Clin. Chem. Clin. Biochem.* **33:**887–892.

11. **Dennehy, P. H., T. E. Schutzbank, and G. M. Thorne.** 1994. Evaluation of an automated immunodiagnostic assay, VIDAS Rotavirus, for detection of rotavirus in fecal specimens. *J. Clin. Microbiol.* **32:**825–827.

12. **Diepersloot, R. J., Y. van Zantvliet-Van Osstrom, and C. A. Gleaves.** 2000. Comparison of a chemiluminescent immunoassay with two microparticle enzyme immunoassays for detection of hepatitis B virus surface antigen. *Clin. Diagn. Lab. Immunol.* **7:**865–866.

13. **Doche, C., M. Thome, I. Dimet, and J. Bienvenu.** 1996. Evaluation of the fully automated Cobas Core enzyme immunoassay for the quantification of antibodies against hepatitis B virus surface antigen. *Eur. J. Clin. Chem. Clin. Biochem.* **34:**365–368.

14. **Doern, G. V., L. Robbie, and L. Marrama.** 1994. Comparison of two enzyme immunoassays and two latex agglutination assays for detection of cytomegalovirus antibody. *Diagn. Microbiol. Infect. Dis.* **20:**109–112.

15. **Doern, G. V., L. Robbie, and R. St. Amand.** 1997. Comparison of the Vidas and Bio-Whittaker enzyme immunoassays for detecting IgG reactive with varicella-zoster virus and mumps virus. *Diagn. Microbiol. Infect. Dis.* **28:**31–34.

16. **Eble, K., J. Clemens, C. Krenc, M. Rynning, J. Stojak, J. Stuckmann, P. Hutten, L. Nelson, L. DuCharme, S. Hojvat, and L. Mimms.** 1991. Differential diagnosis of acute viral hepatitis using rapid, fully automated immunoassays. *J. Med. Virol.* **33:**139–150.

17. **Evans, R., and D. O. Ho-Yen.** 2000. Evidence-based diagnosis of toxoplasma infection. *Eur. J. Clin. Microbiol. Infect. Dis.* **19:**829–833.

18. **Fayol, V., and G. Ville.** 1991. Evaluation of automated enzyme immunoassays for several markers for hepatitis A and B using the Abbott IMx analyzer. *Eur. J. Clin. Chem. Clin. Biochem.* **29:**67–70.

19. **Fiore, M., J. Mitchell, T. Doan, R. Nelson, G. Winter, C. Grandone, K. Zeng, R. Haraden, J. Smith, K. Harris, J. Leszczynski, D. Berry, S. Safford, G. Barnes, A. Scholnick, and K. Ludington.** 1988. The Abbott IMx automated benchtop immunochemistry analyzer system. *Clin. Chem.* **34:**1726–1732.

20. **Forest, J. C., J. Masse, and A. Lane.** 1998. Evaluation of the analytical performance of the Boehringer Mannheim Elecsys 2010 immunoanalyzer. *Clin. Biochem.* **31:**81–88.

21. **Gangar, V., M. S. Curiale, A. D'Onorio, A. Schultz, R. L. Johnson, and V. Atrache.** 2000. VIDAS enzyme-linked immunofluorescent assay for detection of *Listeria* in foods: collaborative study. *J. AOAC Int.* **83:**903–918.

22. **Goossens, H., Y. Glupczynski, A. Burette, C. Van den Borre, and J. P. Butzler.** 1992. Evaluation of a commercially available second-generation immunoglobulin G enzyme immunoassay for detection of *Helicobacter pylori* infection. *J. Clin. Microbiol.* **30:**176–180.

23. **Grangeot-Keros, L., and G. Enders.** 1997. Evaluation of a new enzyme immunoassay based on recombinant rubella virus-like particles for detection of immunoglobulin M antibodies to rubella virus. *J. Clin. Microbiol.* **35:**398–401.

24. **Grangeot-Keros, L., B. Pustowoit, and T. Hobman.** 1995. Evaluation of Cobas Core Rubella IgG EIA recomb, a new enzyme immunoassay based on recombinant rubella-like particles. *J. Clin. Microbiol.* **33:**2392–2394.

25. **Groen, J., B. Hersmus, H. G. Niesters, W. Roest, G. van Dijk, W. van der Meijden, and A. D. Osterhaus.** 1999. Evaluation of a fully automated glycoprotein G-2 based assay for the detection of HSV-2 specific IgG antibodies in serum and plasma. *J. Clin. Virol.* **12:**193–200.

26. **Hadziyannis, E., W. Sholtis, S. Schindler, and B. Yen-Lieberman.** 1999. Comparison of VIDAS with direct immunofluorescence for the detection of respiratory syncytial virus in clinical specimens. *J. Clin. Virol.* **14:**133–136.

27. **Heijtink, R. A., J. Kruining, P. Honkoop, M. C. Kuhns, W. C. Hop, A. D. Osterhaus, and S. W. Schalm.** 1997. Serum HBeAg quantitation during antiviral therapy for chronic hepatitis B. *J. Med. Virol.* **53:**282–287.

28. **Hendriks, H. A., W. Kortlandt, and W. M. Verweij.** 2000. Standardized comparison of processing capacity and efficiency of five new-generation immunoassay analyzers. *Clin. Chem.* **46:**105–111.

29. **Hennig, H., P. Schlenke, H. Kirchner, I. Bauer, B. Schulte-Kellinghaus, and H. Bludau.** 2000. Evaluation of newly developed microparticle enzyme immunoassays for the detection of HCV antibodies. *J. Virol. Methods* **84:**181–190.

30. **Kerdahi, K. F., and P. F. Istafanos.** 2000. Rapid determination of *Listeria monocytogenes* by automated enzyme-linked immunoassay and nonradioactive DNA probe. *J. AOAC Int.* **83:**86–88.

31. **Lavanchy, D., J. Steinmann, A. Moritz, and P. C. Frei.** 1996. Evaluation of a new automated third-generation anti-HCV enzyme immunoassay. *J. Clin. Lab. Anal.* **10:**269–276.

32. **Lazzarotto, T., C. Galli, R. Pulvirenti, R. Rescaldani, R. Vezzo, A. La Gioia, C. Martinelli, S. La Rocca, G. Agresti, L. Grillner, M. Nordin, M. van Ranst, B. Combs, G. T. Maine, and M. P. Landini.** 2001. Evaluation of the Abbott AxSYM cytomegalovirus (CMV) immunoglobulin M (IgM) assay in conjunction with other CMV IgM tests and a CMV IgG avidity assay. *Clin. Diagn. Lab. Immunol.* **8:**196–198.

33. **Lin, D. B., T. P. Tsai, C. C. Yang, H. M. Wang, S. C. Yuan, M. H. Cheng, S. L. You, and C. J. Chen.** 2000. Current seroprevalence of hepatitis A virus infection among kindergarten children and teachers in Taiwan. *Southeast Asian J. Trop. Med. Public Health* **31:**25–28.

34. **Liu, X., B. P. Turner, C. E. Peyton, B. S. Reisner, A. O. Okorodudu, A. A. Mohammad, G. D. Hankins, A. S. Weissfeld, and J. R. Petersen.** 2000. Prospective study of IgM to *Toxoplasma gondii* on Beckman Coulter's Access ™ immunoassay system and comparison with Zeus ELISA and Gull IFA assays. *Diagn. Microbiol. Infect. Dis.* **36:**237–239.

35. **Maine, G. T., R. Stricker, M. Schuler, J. Spesard, S. Brojanac, B. Iriarte, K. Herwig, T. Gramins, B. Combs, J. Wise, H. Simmons, T. Gram, J. Lonze, D. Ruzicki, B. Byrne, J. D. Clifton, L. E. Chovan, D. Wachta, C. Holas, D. Wang, T. Wilson, S. Tomazic-Allen, M. A. Clements, G. L. Wright, T. Lazzarotto, A. Ripalti, and M. P. Landini.** 2000. Development and clinical evaluation of a recombinant-antigen-based cytomegalovirus immunoglobulin M automated immunoassay using the Abbott AxSYM analyzer. *J. Clin. Microbiol.* **38:**1476–1481.

36. **Matter, L., and D. Germann.** 1995. Detection of human immunodeficiency virus (HIV) type 1 antibodies by new automated microparticle enzyme immunoassay for HIV types 1 and 2. *J. Clin. Microbiol.* **33:**2338–2341.

37. **Ognibene, A., C. J. Drake, K. Y. Jeng, T. E. Pascucci, S. Hsu, F. Luceri, and G. Messeri.** 2000. A new modular chemiluminescence immunoassay analyzer evaluated. *Clin. Chem. Lab. Med.* **38:**251–260.

38. **Ostrow, D. H., B. Edwards, D. Kimes, J. Macioszek, H. Irace, L. Nelson, K. Bartko, J. Neva, C. Krenc, and L. Mimms.** 1991. Quantitation of hepatitis B surface antibody by an automated microparticle enzyme immunoassay. *J. Virol. Methods* **32:**265–276.

39. **Pelloux, H., E. Brun, G. Vernet, S. Marcillat, M. Jolivet, D. Guergour, H. Fricker-Hidalgo, A. Goullier-Fleuret, and P. Ambroise-Thomas.** 1998. Determination of anti-Toxoplasma gondii immunoglob-

ulin G avidity: adaptation to the Vidas system (bioMerieux). *Diagn. Microbiol. Infect. Dis.* **32:**69–73.

40. **Rao, L. V., O. A. James, L. M. Mann, A. A. Mohammad, A. O. Okorodudu, M. G. Bissell, and J. R. Petersen.** 1997. Evaluation of Immuno-1 toxoplasma IgG assay in the prenatal screening of toxoplasmosis. *Diagn. Microbiol. Infect. Dis.* **27:**13–15.

41. **Ratnam, S., V. Gadag, R. West, J. Burris, E. Oates, F. Stead, and N. Bouilianne.** 1995. Comparison of commercial enzyme immunoassay kits with plaque reduction neutralization test for detection of measles virus antibody. *J. Clin. Microbiol.* **33:**811–815.

42. **Robbins, D., T. Wright, C. Coleman, L. Umhoefer, B. Moore, A. Spronk, C. Douville, I. K. Kuramoto, M. Rynning, D. Gracey, V. Salbilla, F. Nehmadi, and L. T. Mimms.** 1992. Serological detection of HBeAg and anti-HBe using automated microparticle enzyme immunoassays. *J. Virol. Methods* **38:**267–281.

43. **Robbins, D. J., J. Krater, W. Kiang, X. Alcalde, S. Helgesen, J. Carlos, and L. Mimms.** 1991. Detection of total antibody against hepatitis A virus by an automated microparticle enzyme immunoassay. *J. Virol. Methods* **32:**255–263.

44. **Safford, J. W., G. G. Abbott, M. C. Craine, and R. G. MacDonald.** 1991. Automated microparticle enzyme immunoassays for IgG and IgM antibodies to *Toxoplasma gondii*. *J. Clin. Pathol.* **44:**238–242.

45. **Sandin, R. L., C. C. Knapp, G. S. Hall, J. A. Washington, and I. Rutherford.** 1991. Comparison of the Vitek Immunodiagnostic Assay System with an indirect immunoassay (Toxostat Test Kit) for detection of immunoglobulin G antibodies to *Toxoplasma gondii* in clinical specimens. *J. Clin. Microbiol.* **29:**2763–2767.

46. **Schachter, J., R. B. Jones, R. C. Butler, B. Rice, D. Brooks, B. Van der Pol, M. Gray, and J. Moncada.** 1997. Evaluation of the Vidas Chlamydia test to detect and verify *Chlamydia trachomatis* in urogenital specimens. *J. Clin. Microbiol.* **35:**2102–2106.

47. **Schaefer, L. E., J. W. Dyke, F. D. Meglio, P. R. Murray, W. Crafts, and A. C. Niles.** 1989. Evaluation of microparticle enzyme immunoassays for immunoglobulins G and M to rubella virus and *Toxoplasma gondii* on the Abbott IMx automated analyzer. *J. Clin. Microbiol.* **27:**2410–2413.

48. **Shanholtzer, C. J., K. E. Willard, J. J. Holter, M. M. Olson, D. N. Gerding, and L. R. Peterson.** 1992. Comparison of the VIDAS *Clostridium difficile* toxin A immunoassay with *C. difficile* culture and cytotoxin and latex tests. *J. Clin. Microbiol.* **30:**1837–1840.

49. **Smith, J., and G. Osikowicz.** 1993. Abbott AxSYM random and continuous access immunoassay system for improved workflow in the clinical laboratory. *Clin. Chem.* **39:**2063–2069.

50. **Steinmann, J., and T. Weigel.** 1994. Comparison of two Cobas Core enzyme immunoassays with other test systems for the detection of cytomegalovirus-specific IgG and IgM antibodies. *J. Clin. Lab. Anal.* **8:**191–199.

51. **Tang, Y. W., J. A. Helgason, A. D. Wold, and T. F. Smith.** 1998. Detection of Epstein-Barr virus-specific antibodies by an automated enzyme immunoassay. Performance evaluation and cost analysis. *Diagn. Microbiol. Infect. Dis.* **31:**549–554.

52. **Tholcken, C. A. and G. L. Woods.** 2000. Evaluation of the Bio-Rad syphilis IgG test performed on the CODA system for serologic diagnosis of syphilis. *Diagn. Microbiol. Infect. Dis.* **37:**157–160.

53. **Van Der Ende, A., R. W. van Der Hulst, P. Roorda, G. N. Tytgat, and J. Dankert.** 1999. Evaluation of three commerical serological tests with different methodologies to assess *Helicobacter pylori* infection. *J. Clin. Microbiol.* **37:**4150–4152.

54. **Vernozy-Rozand, C., C. Mazuy, G. Prevost, C. Lapeyre, M. Bes, Y. Brun, and J. Fleurette.** 1996. Enterotoxin production by coagulase-negative staphylococci isolated from goats' milk and cheese. *Int. J. Food Microbiol.* **30:**271–280.

55. **Vernozy-Rozand, C., C. Mazuy, S. Ray-Gueniot, S. Boutrand-Loei, A. Meyrand, and Y. Richard.** 1998. Evaluation of the VIDAS methodology for detection of *Escherichia coli* O157 in food samples. *J. Food Prot.* **61:**917–920.

56. **Weber, B., A. Bayer, P. Kirch, V. Schluter, D. Schlieper, and W. Melchior.** 1999. Improved detection of hepatitis B virus surface antigen by a new rapid automated assay. *J. Clin. Microbiol.* **37:**2639–2647.

57. **Weber, B., A. Muhlbacher, U. Michl, G. Paggi, V. Bossi, C. Sargento, R. Camacho, E. H. Fall, A. Berger, U. Schmitt, and W. Melchior.** 1999. Multicenter evaluation of a new rapid automated human immunodeficiency virus antigen detection assay. *J. Virol. Methods* **78:**61–70.

58. **Weber, B., E. H. Fall, A. Berger, and H. W. Doerr.** 1998. Reduction of diagnostic window by new fourth-generation human immunodeficiency virus screening assay. *J. Clin. Microbiol.* **36:**2235–2239.

59. **Weber, B., N. Behrens, and H. W. Doerr.** 1997. Detection of human immunodeficiency virus type 1 and type 2 antibodies by a new automated microparticle immunoassay AxSYM HIV-1/HIV-2. *J. Virol. Methods* **63:**137–143.

Commercial Methods in Clinical Veterinary Microbiology

THOMAS J. INZANA, DAVID S. LINDSAY, XIANG-JIN MENG, AND KAREN W. POST

14

Clinical veterinary microbiology is currently a rapidly progressing discipline. Procedures and regulations are becoming more standardized, as evidenced by the recent publication of documents by the National Committee for Clinical Laboratory Standards: *Development of In Vitro Susceptibility Testing Criteria and Quality Control Parameters for Veterinary Antimicrobial Agents* (document M37-A) (150) and *Performance Standards for Antimicrobial Disk and Dilution Susceptibility Tests for Animals* (document M31-A) (151). Furthermore, zoonotic agents are more commonly being isolated from patients whose immune systems are severely compromised due to AIDS or other infections, or due to chemotherapy for neoplasia (81). Many of the agents currently listed as responsible for emerging infectious diseases are also of zoonotic origin or are derived from animals acting as intermediate hosts (121–123). Therefore, the need to identify and determine the antibiotic susceptibility of microbial pathogens of animal origin has never been greater.

Despite the need to adequately identify microorganisms of animal origin, commercial tests developed specifically for the animal health care market are far less common than those for detecting microorganisms of human origin. The limitation in commercial tests available is due largely to lower demand for a specific product and the cost that is acceptable to a patient's owner. While the cost of diagnostic tests in human medicine is covered largely by insurance, most owners must pay the full cost of such tests for their animals. Costs are particularly limited in the case of farm animals, in which veterinary care is almost always an economic decision based on the value of the animal(s) versus the cost of care. In many cases veterinary diagnosticians utilize products made for human clinical microbiology. The fungi that are responsible for animal infections are essentially identical to and cause the same diseases as those that cause human infections. The bacterial pathogens causing infections in animals are also either identical, as in the case of enteric bacteria, or differ only at the species level in the case of some bacteria that inhabit mucosal sites (e.g., *Bordetella*, *Haemophilus*, and *Chlamydia* species, etc.). The viruses and parasites that commonly cause infections in animals, however, are more distinct from those isolated from human clinical specimens.

In the absence of commercial diagnostic products, many diagnostic laboratories offer tests for specific pathogens, although these tests are largely unregulated. These tests are primarily based on PCR or other molecular identification systems or on serology. The Veterinary Virology section below also reviews the spectrum of tests and services that are provided by veterinary diagnostic laboratories, as well as recommendations on collection and transport of clinical specimens. These procedures are widely applicable to other areas of veterinary microbiology as well as virology. The overall purpose of this chapter is to describe, and in some cases report evaluations of, those commercial identification systems that are used by veterinary diagnostic laboratories. Tests that are primarily used in human clinical medicine may be referred to but will not be described further.

VETERINARY VIROLOGY

From the use of cowpox as a vaccine to prevent human smallpox by Edward Jenner over 200 years ago to the recent isolation of the deadly zoonotic Nipah virus infecting swine and humans (16), veterinary virology represents a significant part of the field of virology. However, due to the consideration for economic return on investment and the wide range of diagnostic tests required for each animal species, the number and types of commercially available diagnostic tests for veterinary viral diseases are very limited compared to those commercially available for human viral diseases. It is estimated that for the eight major domestic animal species (cat, chicken, cattle, dog, goat, horse, sheep, and swine), there exist more than 200 viruses, belonging to more than 25 different virus families (95). The number of known veterinary viruses will become incredibly large if other animal species such as turkey, duck, and goose and laboratory, aquatic, and wildlife animals are included. Therefore, it represents a tremendous challenge for veterinary diagnosticians to detect and identify all potential pathogenic viruses and their variants. It is impossible to discuss all available tests for each individual virus of veterinary importance in a single chapter. Therefore, the scope of this section is to discuss the general principle and application of currently available tests for the diagnosis of viral diseases of veterinary importance. The reader is referred to other references (95) for the biological and genetic characteristics, pathogenesis, disease association, prevention, and control of viruses of veterinary importance.

Collection and Transportation of Clinical Samples

The successful detection of viruses in the clinical laboratory depends critically on the proper collection and transportation of clinical samples by the submitting veterinarian. In general, care should be taken to protect the virus in specimens from environmental damage and maintain virus infectivity by using the proper transport system (55). The materials intended for viral diagnosis should be collected as soon as possible after the onset of clinical disease. The samples collected during the acute phase of viral infection usually contain adequate amounts of viral particles, which are readily detectable with available assays. Samples collected at later times, such as the convalescent phase of infection, will consume more laboratory time and often yield poor or negative results even though a virus was the cause of the disease. Certain viral infections (such as porcine reproductive and respiratory syndrome virus) could predispose animals to secondary viral or bacterial infections (137). Therefore, samples collected late in the disease process may lead to a misdiagnosis when infection with a secondary pathogen is involved. The types of samples to be collected are also important, but will largely depend upon the diagnostic tests selected and the suspected viruses or clinical diseases manifested. Selection of the appropriate clinical specimens also requires knowledge of the pathogenesis of a particular viral disease by the submitting veterinarian. In general, the appropriate samples for virus detection are nasal swabs; body fluids; fecal and blood samples from infected, live animals; and relevant tissues and organs from necropsied animals. More specifically, it is recommended to collect nasal or nasopharyngeal swab materials from animals with respiratory diseases; fecal samples from animals with enteric diseases; cerebrospinal fluid, fecal samples, and nasal swab materials from animals with central nervous system signs; genital swab materials from animals with genital diseases; conjunctival swab materials from animals with eye diseases; and vesicle swab materials and biopsies from animals with skin lesions (70, 95). As a general practice, blood samples should always be collected from animals with a suspected viral disease. When collecting blood into syringes, the needle should be removed from the syringe and the blood should be slowly expelled into the tube to reduce hemolysis. The whole blood should not be frozen, as this will cause complete hemolysis and render the samples unsuitable for serology. The whole blood should be allowed to clot at room temperature, and the serum and buffy coat cells should be separated by centrifugation. In most cases, paired serum samples collected several weeks apart are needed for a definitive diagnosis. A fourfold or greater increase in antiviral antibody titer in serum is indicative of current infection associated with the suspected virus. The amount of clinical material collected is also important for a definitive diagnosis. Whenever possible, adequate amounts of sample—especially feces, sera, and other body fluids—should be taken. Insufficient amounts of the sample could lead to an inconclusive diagnosis or a false-negative result.

Many viruses are labile and could quickly become inactivated if stored or transported improperly. When collecting swab materials, nonbleached swabs should be used to avoid toxicity to either the virus or cell cultures (55). Samples collected for virus detection should be kept cold and moist. Swabs should be kept moist with a virus transport medium that contains a buffered salt solution, proteins to prevent virus inactivation, and antibiotics to prevent bacterial and fungal growth. A variety of transport media are suitable for virus transport (55). Many companies (such as VWR and Fisher) manufacture commercial swabs made of Dacron or calcium alginate or viral transport systems (such as BD Biosciences) that can be used to collect, store, and transport clinical materials containing viruses. Samples intended for bacterial identification should be collected separately, and this transport medium should not contain antibiotics. Clinical materials intended for virus identification should be refrigerated at 4°C after collection and transported to a diagnostic laboratory as soon as possible. During transportation, samples should be carefully packed in an appropriate shipping container with an adequate amount of commercial refrigerant packs. Care should be taken to prevent any potential leakage and safety hazards. Most viruses will remain stable at 4°C for up to 2 to 3 days (55). However, if there are anticipated delays in submitting samples, the clinical material should be immediately frozen in a −70°C freezer. Storage of samples intended for virus detection at −10 to −20°C, such as the freezer sections of standard refrigerators, is detrimental and should be avoided, as most viruses rapidly lose infectivity when stored at this temperature due to water crystal formation (70, 88). Frozen samples should be transported to a diagnostic laboratory on dry ice, which is commercially available. Special permits are required for international transportation of infectious materials, and appropriate authorizations from respective countries should be obtained prior to transportation. Interstate transportation of certain veterinary viruses within the United States also requires a permit from the U.S. Department of Agriculture's (USDA's) Animal and Plant Health Inspection Service.

Proper labeling of samples and recording relevant clinical information prior to submission are as important as sample collection and transportation. Each veterinary diagnostic laboratory has its own standard sample submission form available upon request or via downloading through the Internet. Submitting veterinarians are usually asked to provide information such as animal species or source, age, body weight, sex, herd or flock size, number of animals in the herd or flock affected or dead, location of animals, major clinical signs, date of disease onset, date of samples collected, presumptive clinical diagnosis, vaccination history, types and numbers of samples submitted, and type(s) of diagnostic tests requested. The clinical information, especially the presumptive diagnosis, is important for laboratory diagnosticians not only to select the most appropriate and sensitive tests for the submitted specimens but also to evaluate and interpret the test results.

Selection of Diagnostic Laboratories and Tests

Most states in the United States have their own veterinary diagnostic laboratory system, which is often associated with their respective veterinary school. Each of these laboratories usually specializes in one or more areas, such as poultry diseases, food animal diseases, companion animal diseases, or laboratory animal diseases. A few laboratories such as the USDA's National Veterinary Service Laboratories (NVSL) have the capacity to provide more comprehensive diagnostic services and reagents for viruses of veterinary importance. NVSL-Ames (Ames, Iowa) provides reagents and comprehensive diagnostic services. The NVSL-Foreign Animal Disease Diagnostic Laboratory has the capacity of diagnosing animal diseases that are foreign to the United States. The selection of a particular diagnostic laboratory depends largely on the submitting veterinarians' preference, test availability of a particular reference laboratory, animal species involved, and type of disease suspected. Some veterinarians tend to submit clinical samples to local or regional diagnostic laboratories for convenience, but this preference is often limited by the availability of certain tests at local or regional diagnostic laboratories. Overall, the selection of a particular reference laboratory and test is at the discretion of the submitting veterinarian. For lists of available viral diagnostic tests from veterinary diagnostic laboratories in the United States, readers are referred to nearby veterinary schools for locations of reference laboratories and their available viral diagnostic services. Each laboratory usually has copies of published lists of services that are available upon request or are already available on public domains such as the Internet.

Selection of a viral diagnostic test depends on many factors such as the clinical specimens available for submission, the type of viral disease suspected, test availability, and cost consideration. There are often several diagnostic tests available for a given disease. The principles and selection of these tests are discussed below. In many cases, a confirmatory test is also performed on the same sample or different samples from the same diseased animal. Many traditional tests and procedures are still important in the diagnosis of veterinary viral diseases, such as pathological and histopathological examination, electron microscopy, and virus isolation. Most modern viral diagnostic tests are rapid assays (44, 52, 71, 95) aimed at detection of antiviral antibodies (enzyme-linked immunosorbent assay [ELISA], Western immunoblot analysis, immunofluorescence, serum virus neutralization [SVN], hemagglutination inhibition [HI], agar gel immunodiffusion [AGID], and complement fixation [CF]), detection of viral antigens (immunohistochemistry [IHC], antigen-capture ELISA, and frozen tissue immunofluorescence), or detection (or amplification) of viral nucleic acids (PCR and its related techniques, in situ hybridization [ISH], Southern blot hybridization, and dot blot hybridization). In addition, modern molecular tests and procedures (such as PCR-restriction fragment length polymorphism [PCR-RFLP] analysis, sequencing, and sequence analysis) are being used more commonly for further genetic characterization of viruses and for differential diagnosis. The basic principles and general application of the commonly used laboratory procedures and diagnostic assays in veterinary diagnostic virology are discussed below.

Pathology and Histopathology

Pathology and histopathology form integral parts of laboratory procedures in the veterinary diagnostic laboratories. Veterinary pathologists often play an important role in the diagnosis of viral diseases, especially in the identification of previously unrecognized viruses. Specimens from live diseased animals with acute clinical signs and from dead animals with no postmortem autolysis can be submitted for pathological evaluation. Specimens submitted for histopathological examination should be fixed in 10% buffered formalin. Veterinary pathologists can evaluate fresh tissues for the presence of gross lesions and formalin-fixed tissues for the presence of characteristic microscopic lesions. Proper selection and preservation of samples by the submitting veterinarian are critical for histopathological examination. Generally, samples submitted for histopathology should be fresh slices of affected organs fixed in 10% neutral buffered formalin. The organ slices should include the lesion and transitional areas, as

well as the adjacent normal tissues. If the submitting veterinarian is not certain what organ samples are appropriate, samples from multiple organs should be collected and submitted. The veterinary pathologist can select the most appropriate samples for examination. Based on characteristic lesions or inclusion bodies due to certain viral diseases, the veterinary pathologist can make a tentative diagnosis or establish a differential list of potential pathogens. However, the pathologist will usually have to rely on laboratory diagnostic tests for a definitive diagnosis.

Assays for Detection of Viruses

EM

Despite its limitation due to low sensitivity (usually $>10^5$ to 10^6 virus particles per ml of sample required), electron microscopy (EM) remains an important diagnostic tool for virus detection in veterinary medicine since it affords an opportunity for direct visualization of viruses in clinical materials (98). EM is particularly appealing for identification of viruses that cannot be propagated in vitro, for identification of viruses from clinical samples that are toxic to cell cultures (such as urine, feces, and semen), or for characterization and classification of viruses. Based on the morphological characteristics of viruses, such as virion structure, dimension, and presence or absence of an envelope, an electron microscopist or diagnostic virologist can readily identify the virus family it belongs to. Based upon EM results, the veterinary diagnostician can make a tentative diagnosis or come up with a differential list of potential pathogens. However, a definitive diagnosis usually cannot be made, since EM is generally considered a nonspecific test. Samples appropriate for EM examination include, but are not limited to, secretions, excretions, thin tissue sections, vesicular fluids, and cell culture materials.

The most commonly used EM technique for viral detection is negative staining. Viruses present in clinical materials have little contrast with their surroundings. The negative-staining technique overcomes this problem by staining the surrounding background with a heavy metal stain such as phosphotungstate or uranyl acetate to increase the contrast between viruses and the background. After negative staining, the structural outlines and surface projections of viruses can be visualized by an electron microscope. EM following negative staining is commonly used to detect viruses associated with enteric diseases of domestic animals (6, 12, 17, 23, 26, 83). As mentioned above, EM generally can detect but cannot definitively identify viruses. Additional tests such as immuno-EM and immunogold EM are usually required for a definitive diagnosis. Negative-staining immuno-EM is achieved by adding an antiviral antibody to the clinical specimen. If the antibody is specific to the virus in the specimen, the antibody will aggregate the virus and form virion-antibody complexes that can be visualized by negative-staining EM. For immunogold EM, gold (5-nm-diameter particles)-labeled protein G (or protein A) is usually added to the virion-antibody mixture. The gold-labeled protein G (or protein A) binds to the Fc fragment of the antibody. The gold-labeled virus particles are sedimented by centrifugation, washed to remove unbound gold particles, sprayed on EM grids, and visualized by an electron microscope. The immunogold EM technique has been used for the identification of many viruses of veterinary importance, such as porcine rotavirus (79), porcine reproductive and respiratory syndrome virus (80, 91), and bovine coronavirus (41).

VI and Virus Identification

Virus isolation (VI) followed by virus identification as a diagnostic tool is usually slow but is still considered the "gold standard" in veterinary viral diagnosis. VI is still used in veterinary diagnostic laboratories for routine virus isolation for many diseases (such as swine influenza and porcine reproductive and respiratory syndrome). VI is particularly appealing for isolation of viruses from previously unrecognized diseases, for detection of viruses for which there are no other available assays, and for production of an adequate amount of virus required for further analyses such as RFLP analysis, sequencing, genotyping, and serotyping. Most clinical materials are appropriate for VI, including samples of biopsy and necropsy tissues, blood, secretions, and excretions from diseased animals. However, some clinical samples (such as urine, feces, and semen) are potentially toxic to cell cultures. When collecting tissue samples for VI, it is recommended to collect the cleanest tissues (such as lymph nodes and liver tissues) first and the most contaminated tissues (such as enteric tissues) last to prevent bacterial contamination of the clean samples. The amount of material required for VI is usually more than that needed for rapid diagnostic assays. Therefore, submitting veterinarians are advised to collect and submit an adequate amount of clinical material intended for VI.

Not all known viruses can be propagated in cell culture, and no single cell culture can support the growth of all viruses. Therefore, the choice of a cell culture for VI will largely depend upon the particular virus suspected (95). At least three major types of cell cultures are routinely used for VI: primary cultures, diploid cell lines, and established cell lines (88). Primary cell cultures are prepared from organs of freshly killed animals. Diploid cell lines usually have a finite life span of 50 to 60 in vitro passages. A variety of established or continuous cell lines from various animal species are known to support the propagation of many veterinary viruses, and are available from commercial vendors such as the American Type Culture Collection (ATCC), Manassas, Va., and BioWhittaker, Inc. The capability of certain viruses to grow in cell cultures can be greatly enhanced by biochemical or physical treatments (47). For example, the growth of rotaviruses and reoviruses can be enhanced in the presence of trypsin-containing cell culture maintenance medium and by treatment of clinical samples with trypsin. Isolation of an unrecognized virus in cell cultures often represents a tremendous challenge to veterinary virologists. When attempting to isolate a virus from a previously unrecognized disease, primary cell cultures derived from tissues of the same species as the diseased animal are usually the first choice.

Many viruses, when replicating in cell cultures, produce characteristic cytopathic effects that can be visualized under a microscope. However, recognition of viral growth in cell culture is sometime difficult, as many viruses are noncytopathic. The observation of a cytopathic effect in cell culture is not sufficient for a definitive viral diagnosis, and additional tests such as HI, hemadsorption, SVN, immunodetection techniques, or other antigenic and genetic characterizations must be performed for identification of viruses growing in cell cultures. Sometimes, chicken embryos and live animals are also used for VI. For example, chicken embryos are often used to isolate avian viruses and influenza virus, and suckling mice are useful for isolation of viruses such as rabies, enteroviruses, flaviviruses, and alphaviruses. Evidence for viral growth in embryonated chicken eggs can be visualized on the chorioallantoic membrane for certain viruses such as poxviruses and avian herpesviruses, but for most viruses additional tests such as hemagglutination, immunofluorescence, and PCR are usually required to confirm viral growth. In mice inoculated with virus, clinical signs and characteristic pathological lesions are often indicative of virus infection, but additional laboratory tests such as serology, immunofluorescence, and IHC must also be performed for a definitive diagnosis. Natural host animals may also be used to isolate viruses when other methods fail to produce results. When a new virus is isolated from a previously unrecognized disease, Koch's postulates must be fulfilled: the new virus must be inoculated into its natural animal host to see if it can reproduce the natural disease and if the same virus can be reisolated from experimentally infected animals.

Assays for Detection of Antiviral Antibodies

ELISA

ELISA is an assay commonly used for detection of antiviral antibodies in veterinary diagnostic laboratories. The basic principle of ELISA is that an antigen-antibody reaction can be measured by a colorimetric reaction (optical density) between enzyme-labeled antibody and substrate. This relatively inexpensive test is sensitive and rapid, and depending upon the viral disease suspected, a diagnosis could be made within hours. A large number of samples can be processed in a single run in a fully automated fashion using an automated plate washer and reader. With a known viral antigen coated on a 96-well ELISA plate, viral antibodies in serum can be measured both qualitatively (positive or negative) and quantitatively (antibody titers). Competitive ELISA (cELISA) or blocking ELISA utilizes a known specific viral antibody to compete against specific antibodies in the test serum sample. A viral antigen is coated onto a solid-phase surface, the test serum and an enzyme-labeled antibody specific for the coated viral antigen are added together, and they compete for binding to the immobilized viral antigen. The color developed (by adding a chromogenic enzyme substrate) is inversely proportional to the amount of antiviral antibody present in the test serum. A serum sample is considered positive in the cELISA or blocking ELISA if the optical density is reduced by a defined percentage compared with that of an unblocked sample. The availability of standard reagents and commercial kits renders quality control relatively simple. ELISA kits for detection of antiviral antibodies are commercially available for a variety of veterinary viruses from numerous vendors (Table 1). These commercial ELISA kits are approved by regulatory agencies, and the results are usually reproducible from laboratory to laboratory.

SVN

SVN is an assay to detect virus-neutralizing antibodies in sera. SVN not only can detect but also can quantitate antiviral neutralizing antibodies. SVN is usually performed by mixing a standard amount of virus with serial dilutions of a serum sample. Alternatively, a standard dilution of a serum sample is mixed with serial dilutions of the test virus. The serum-virus mixtures are then inoculated onto monolayers of susceptible cell cultures, and the inoculated cells are monitored for evidence of virus infection. The highest

TABLE 1 Selected commercial diagnostic kits for antigen or antibody detection of veterinary viruses of major domestic animal species.

Species	Virus	Assay	Vendor(s)[a]
Avian	Avian rotavirus	Antigen-capture ELISA[b]	Meridian
	Newcastle disease virus	ELISA[c]	Svanova, IDEXX, Synbiotics
	Infectious bursal disease virus	ELISA	IDEXX, Synbiotics
	Infectious bronchitis virus	ELISA	Svanova, IDEXX, Synbiotics
	Avian reovirus	ELISA	IDEXX, Synbiotics
	Avian encephalomyelitis virus	ELISA	Synbiotics, IDEXX
	Chicken anemia virus	ELISA	Synbiotics, IDEXX
	Reticuloendotheliosis virus	ELISA	IDEXX
	Avian leukosis virus	ELISA, antigen-capture ELISA	Synbiotics, IDEXX
	Avian rhinotracheitis virus	ELISA	IDEXX
	Avian pneumovirus	ELISA	Svanova, IDEXX
	Avian influenza virus	ELISA	Synbiotics, IDEXX
	Hemorrhagic enteritis virus	ELISA	Synbiotics
Bovine	Bovine rotavirus	Antigen-capture ELISA	Meridian
	Bovine viral diarrhea virus	Antigen-capture ELISA	SBI, IDEXX
		ELISA	IDEXX, Svanova
	Bluetongue virus	AGID[b,c]	VMRD, VDT
		cELISA[c]	VMRD, VDT
		CF[c]	VDT
	Bovine leukemia virus	ELISA	Svanova, IDEXX
		AGID	Synbiotics
	Infectious bovine rhinotracheitis virus (bovine herpesvirus 1)	ELISA	Svanova, IDEXX
	Pestivirus (cattle)	Antigen-capture ELISA	Svanova
	Bovine coronavirus	ELISA	Svanova
	Bovine respiratory syncytial virus	ELISA	Svanova
	Parainfluenza 3 virus	ELISA	Svanova
	Epizootic hemorrhagic disease virus	AGID	VDT
Canine	Canine parvovirus	Antigen-capture ELISA	IDEXX, Synbiotics
	Canine rotavirus	Antigen-capture ELISA	Meridian
Caprine or ovine	Pestivirus (sheep)	Antigen-capture ELISA	Svanova
	Border disease virus	ELISA	Svanova
	Caprine arthritis-encephalitis virus	AGID	VDT
	Ovine progressive pneumonia virus	AGID	VDT
Equine	Equine influenza virus	Antigen-capture ELISA	Centaur
	Equine rotavirus	Antigen-capture ELISA	Meridian
	Equine infectious anemia virus	AGID	VMRD, IDEXX, Synbiotics
		ELISA	VAI, Centaur, IDEXX, Synbiotics
		cELISA	IDEXX
	Equine herpesvirus	ELISA (type 1 and 4 differential test)	Svanova

(Continued on next page)

dilution of serum that protects susceptible cells from virus infection is defined as the SVN titer. One of the drawbacks of SVN is the requirement of a susceptible cell culture for the test virus. For viruses that do not replicate in cell cultures, live animals or embryonated chicken eggs could be used to determine antiviral neutralizing antibodies by inoculating the in vitro serum-virus mixtures into live animals or embryonated eggs for evidence of virus infection. The in vivo neutralization approach is expensive and time-consuming, and is of little practical value. Commercial SVN kits are not available and veterinary diagnostic laboratories usually have in-house SVN procedures for each test virus. Neutralizing antiviral antibodies or reference viruses used for SVN controls are commercially available from

TABLE 1 *(Continued)*

Species	Virus	Assay	Vendor(s)^a
Feline	Feline leukemia virus	Antigen-capture ELISA	IDEXX, Synbiotics
	Feline immunodeficiency virus	ELISA	Avecon, IDEXX
	Feline infectious peritonitis virus	ELISA	Avecon
Porcine	Porcine rotavirus	Antigen-capture ELISA	Meridian
	Porcine reproductive and respiratory syndrome virus	ELISA	IDEXX
	Pseudorabies virus	Latex agglutination^c	VAI
		ELISA	Svanova, IDEXX
	Classical swine fever virus	ELISA, antigen-capture ELISA	IDEXX
	Swine influenza virus	Antigen-capture ELISA	Centaur
	Porcine parvovirus	ELISA	Svanova
	Transmissible gastroenteritis virus, porcine respiratory coronavirus	ELISA (differential test)	Svanova

^{a} Vendor abbreviations and locations: Avecon: Avecon Diagnostics, Inc., 405 South Main St., Coopersburg, PA 18036; Centaur: Centaur, Inc., P.O. Box 25667, Overland Park, KS 66225-5667; IDEXX: IDEXX Laboratories, Inc., One IDEXX Dr., Westbrook, ME 04092; Meridian: Meridian Diagnostics, Inc., 3471 River Hills Dr., Cincinnati, OH 45244; Svanova: Svanova Biotech, Uppsala Science Park, Glunten, S-751 83 Uppsala, Sweden (Svanova Biotech's U.S. distributor: Diagnostic Chemical Limited, Marketing and Product Manager, Veterinary Diagnostics, 160 Christian St., Oxford, CT 06478); Synbiotics: Synbiotics Corp., 11011 Via Frontera, San Diego, CA 92127; SBI: Syracuse Bioanalytical, Inc., Langmuir Laboratory Box 1013, 95 Brown Rd., Suite 144, Ithaca, NY 14850; VDT: Veterinary Diagnostic Technology, Inc., 4890 Van Gordon St., Suite 101, Wheat Ridge, CO 80033; VAI: Viral Antigens, Inc., 5171 Wilfong Rd., Memphis, TN 38134; VMRD: VMRD, Inc., 4641 Pullman-Albion Rd., Pullman, WA 99163.
^{b} For detection of viral antigen.
^{c} For detection of antiviral antibody (serology).

many veterinary diagnostic laboratories (such as NVSL) or commercial vendors such as VMRD, Inc. (Pullman, Wash.), and ATCC.

AGID

AGID is a simple assay to detect antiviral antibodies in serum samples. With a known reference antiviral antibody, AGID can also be used to detect viral antigen. The basic principle of AGID is the reaction between diffusing antiviral antibody and viral antigen in an agar plate. Viral antigen preparations are placed in a central agar well, and positive control and test sera are placed in peripheral wells. Precipitin lines form when the diffusing antiviral antibodies meet and react at optimal conditions with the diffusing viral antigen in the agar. Commercial AGID kits are available for detection of antibodies to many veterinary viruses, such as equine infectious anemia virus and bluetongue virus (Table 1). AGID has also been used in many diagnostic laboratories for detection of antiviral antibodies to many other viruses for which commercial AGID kits are not yet available, such as caprine arthritis-encephalitis virus (112), swine encephalomyocarditis virus (60), avian influenza virus (159), and bovine leukemia virus (128). AGID is usually considered to be a relatively specific test, but the sensitivity may be low, resulting in false-negative results. A confirmatory test may have to be used to validate AGID results.

HI Assay

A simple and inexpensive assay, the HI assay has been widely used to detect antiviral antibodies in serum for many veterinary viruses that hemagglutinate red blood cells (RBCs), such as influenza virus and paramyxovirus. The test serum is diluted serially in 96-well plates and incubated with a stan-

dard amount of virus. After addition of RBCs from certain animal species, hemagglutination is inhibited in the presence of sufficient amounts of serum antiviral antibodies that react with the virus. The HI titer in serum is determined as the highest dilution of test serum that inhibits the aggregation of RBCs. HI reagents (such as reference viruses or antibodies) are commercially available from many veterinary diagnostic laboratories (such as NVSL) and commercial vendors such as the ATCC and VMRD, Inc. With a known reference antiviral antibody, the HI assay can also be used to detect viral antigen.

FA Assay

The fluorescent antibody (FA) assay is a rapid and specific test for detection of antiviral antibody. A known viral antigen is immobilized on a slide or plate (such as virus-infected cells fixed in wells of a culture slide or plate). The test serum containing antiviral antibody is then added to the wells, and the antigen-antibody complex is visualized by adding a fluorochrome (such as fluorescein isothiocyanate [FITC] or Texas red)-conjugated secondary antibody that reacts with the specific antiviral antibody. A few commercial FA kits for human and zoonotic viruses (such as western and eastern equine encephalitis viruses, parainfluenza virus, respiratory syncytial virus, herpes simplex virus, and adenovirus) are available from commercial vendors such as Kirkegaard & Perry Laboratories, Inc. (KPL). Commercial FA kits for veterinary viruses are rare, but FA reagents such as specific antiviral antibodies, viral antigens, and conjugated secondary antibodies are readily available from commercial vendors such as VMRD, Inc., ATCC, and KPL and from veterinary diagnostic laboratories such as NVSL. Each veterinary diagnostic laboratory usually has its own in-house FA procedures for each virus. With a known specific antiviral antibody, immunofluorescence assay can also be used to detect viral antigens (see below).

CF Test

The basic principle of the CF test is the ability of complement to lyse RBCs in the presence of anti-RBC antibody (hemolysin) (43). The test serum is mixed with defined amounts of a known viral antigen, guinea pig complement, sheep RBCs, and hemolysin. In the absence of specific antiviral antibody, sheep RBCs and hemolysin activate complement, resulting in lysis of RBCs. If the test serum contains antiviral antibody, the antigen-specific antiviral antibody forms immune complexes with the viral antigen. The antibody-antigen complexes bind to and activate complement, thus preventing RBCs from lysis by hemolysin. The amount of CF antiviral antibody in the test serum is inversely proportional to the amount of RBC lysis. The CF test has been used to detect antibodies of many veterinary viruses, such as bovine leukemia virus (73), bovine respiratory syncytial virus (31), bovine herpesvirus (34), and malignant catarrhal fever virus (126). The CF test is relatively insensitive and labor-intensive and is less commonly used in modern veterinary diagnostic laboratories (43). Nevertheless, CF test kits for certain veterinary viruses are commercially available and still in use in viral diagnostics (Table 1). With a known antiviral antibody, the CF test can also be used to detect viral antigen such as that of foot-and-mouth disease virus (97).

Western Immunoblot Analysis

Western immunoblot analysis affords the opportunity to detect serum antiviral antibodies to a specific viral protein. A purified known viral protein or the entire purified virion is denatured with the anionic detergent sodium dodecyl sulfate (SDS). The denatured viral proteins are separated by SDS-polyacrylamide gel electrophoresis according to their molecular weights, transferred and immobilized to a nitrocellulose or nylon membrane, and hybridized with the test serum containing antiviral antibodies. The presence of an antibody-antigen complex is demonstrated by adding an enzyme- or radioisotope-labeled secondary antibody. Western immunoblot analysis is often used as a confirmatory test to clarify doubtful results obtained with serological assays. This assay has been used for detection of antiviral antibodies of many veterinary viruses such as equine lentivirus (14), bovine immunodeficiency virus (157), African horse sickness virus (10), and bovine leukemia virus (56). With a known specific antiviral antibody, Western blot analysis can also be used to detect viral antigens in clinical samples. Reagents for Western immunoblot analysis are available from commercial companies such as Bio-Rad Laboratories. Reference viral antigens and antiviral antibodies for Western blot analysis can be purchased from commercial vendors such as VMRD, Inc., and veterinary diagnostic laboratories such as NVSL.

Assays for Detection of Viral Antigen

Antigen-Capture ELISA

Antigen-capture ELISA is a modified ELISA for detection of viral antigen (both infectious and inactivated viruses) from a variety of clinical materials such as tissues, secretions, and excretions. Antigen-specific antibody is coated onto a solid-phase surface (usually a 96-well plate), and the clinical specimen containing viral antigen is then added. After addition of an enzyme-labeled antibody, a chro-

mogenic enzyme substrate is added for color development. The color developed is proportional to the amount of viral antigen present in the specimen. The interest in antigen-capture ELISA for rapid diagnosis of veterinary viral infections remains high, as this assay is simple and easy to perform. The availability of commercial reagents and kits is advantageous for quality control. Antigen-capture ELISA kits are commercially available for many veterinary viruses (Table 1). The assay has also been used in veterinary diagnostic laboratories for detection of viral antigens of many viruses for which commercial antigen-capture ELISA kits are not yet available, such as bluetongue virus (39), bovine coronavirus (124), bovine viral diarrhea virus (119), bovine respiratory syncytial virus (153), and infectious bursal disease virus (38).

Immunofluorescence Assay for Antigen Detection

The direct immunofluorescence assay utilizes a specific antiviral antibody labeled with a fluorochrome such as FITC to detect viral antigen using a fluorescence microscope. For detection of virus-antigen by indirect immunofluorescence assay, the specific antiviral antibody is not labeled and the antibody-antigen complex is visualized by adding a fluorochrome (such as FITC)-labeled secondary antibody that reacts with the specific antiviral antibody. Many clinical materials such as frozen tissue sections, cell smears, or monolayers are appropriate for this assay, but fresh clinical samples are the key for success. The limitation of the immunofluorescence assay for viral antigen detection is the requirement for obtaining cells or tissues containing viral antigens; this assay is not useful for viral antigen detection in specimens such as serum or feces. Immunofluorescence assay is routinely used in combination with VI for identification of viruses growing in cell cultures. The reagents for detection of viral antigen by immunofluorescence assay are commercially available from many vendors, including VMRD, Inc., KPL, and ATCC, or from veterinary diagnostic laboratories (e.g., NVSL). Each veterinary diagnostic laboratory usually has its own in-house immunofluorescence assay procedures for each virus.

IHC

IHC is an attractive technique for veterinary diagnosticians because it conveniently utilizes formalin-fixed tissues for detection of viral antigen. IHC affords the opportunity to visually localize specific viral antigens within the lesions. The specific viral antigen within tissue sections can be visualized by adding a virus-specific antibody, followed by a biotinylated secondary antibody, an avidin-biotin-peroxidase complex, and a substrate such as diaminobenzidine tetrahydrochloride for color development (Fig. 1). Over the years, IHC has gained increasing popularity among veterinary pathologists for confirmation of histopathological diagnoses, since the formalin-fixed tissue sections can be used for both histopathological examination and IHC. Most IHC tests in veterinary use employ the avidin-biotin-peroxidase complex method, although other systems such as a peroxidase-antiperoxidase complex, or a streptavidin-biotinylated immunoperoxidase complex, are also used. A major drawback for IHC is its nonspecific reaction due to the presence of endogenous peroxidase activity in tissues. This problem can be minimized by blocking the endogenous peroxidase activity with hydrogen peroxide and by including adequate controls. IHC has been successfully used to detect antigens of many veterinary viruses, including but not limited to feline leukemia virus (51),

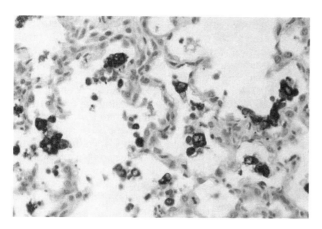

FIGURE 1 IHC of porcine reproductive and respiratory syndrome virus (PRRSV) antigen in the alveolar macrophages of a lung from a pig infected with PRRSV. Alveolar macrophages show dark brown staining of PRRSV antigen. Shown is the result of avidin-biotin-peroxidase complex staining and hematoxylin counterstain. Courtesy of Patrick G. Halbur, Iowa State University.

porcine reproductive and respiratory syndrome virus (36), eastern equine encephalomyelitis virus (102), swine vesicular disease virus (94), swine influenza virus (139), and bovine coronavirus (23). Many commercial companies, such as DAKO, Vector Laboratories, Inc., and KPL, manufacture universal IHC kits that usually contain all necessary IHC reagents except for the specific viral antibody. Specific antiviral antibody can be purchased separately from veterinary diagnostic laboratories such as NVSL or commercial vendors such as VMRD, Inc.

LAT

The Latex agglutination test (LAT) is a very simple test in which viral antigen present in clinical specimens is detected by agglutination with small latex beads coated with a specific antiviral antibody. The antibody-coated latex beads are stable for a relatively long period, and the results can be read by eye or can be measured photometrically within 30 min. This assay is traditionally used to detect the presence of viral antigen in feces of many veterinary viruses, such as porcine rotavirus (120), bovine rotavirus (37, 96), and canine parvovirus (8). With a known viral antigen, LAT can also be used to detect antiviral antibody, such as that for swine pseudorabies virus. LAT kits are commercially available for a few veterinary viruses (Table 1). The major concern with the use of LAT is its low sensitivity compared to those of other immunoassays.

Assays for Detection of Viral Nucleic Acids

Southern Blot Hybridization

The landmark paper by Edwin Southern in 1975 (132) signaled the dawn of a new era of molecular biology. The Southern blot detects specific sequences among hundreds and thousands of restriction fragments of genomic DNA. The technique utilizes restriction enzymes that recognize specific cleavage sites of DNA and cleave the DNA into fragments of various size and number. After separation of the cleaved DNA fragments by gel electrophoresis, the DNA fragments are transferred to a membrane such as nitrocellulose, denatured, and hybridized with a labeled,

specific probe. In veterinary medicine, Southern blot analysis is particularly useful for detecting nucleic acids of DNA viruses such as porcine parvovirus (63), bovine leukemia virus (28), ovine papillomavirus (82), and bovine herpesvirus (74). Traditionally, ^{32}P has been the isotope of choice for labeling probes by methods such as nick translation and end labeling. As a strong β-particle emitter, ^{32}P provides a strong detection signal and high sensitivity for labeled probes. However, due to safety concerns, many diagnostic laboratories now tend to use nonradioisotope labeling methods such as biotin (64) and digoxigenin (58) systems. The digoxigenin system is particularly attractive because nonspecific background can be minimized. The modified digoxigenin only occurs in nature in the *Digitalis* plant; thus, nonspecific reactions are greatly reduced. Several companies such as DAKO, Roche Molecular Biochemicals, and Amersham-Pharmacia Biotech, Inc. (Piscataway, N. J.), manufacture commercial reagents and/or kits for the biotin, digoxigenin, and radioisotope labeling systems, respectively. However, the specific viral genomic fragments to be labeled as probes are not commercially available and have to be generated in the research or diagnostic laboratory for each virus of interest.

Dot Blot Hybridization

Dot blot hybridization is a relatively simple molecular technique for detecting viral DNA or RNA immobilized onto a filter such as a nitrocellulose membrane. Total DNA or RNA extracted from clinical samples is immobilized, denatured, and hybridized with a radioisotope- or nonradioisotope-labeled single-strand viral nucleic acid probe. After the unbound probe is washed away, the membrane is exposed to a film for autoradiography or, in the case of a nonradioisotope probe, the signal of bound probes in the membrane is measured by color development with a substrate. The bound probes can be stripped from the membrane by denaturation methods such as boiling the membrane at 100°C for 5 min, and the same membrane can then be reused with other virus-specific probes. Dot blot hybridization has been used to detect nucleic acids of many veterinary viruses, such as porcine and bovine enteric coronavirus (127), infectious bursal disease virus (42), bovine rotavirus (100), swine encephalomyocarditis virus (90), feline infectious peritonitis virus (84), and avian reoviruses (75). Like Southern blotting, specific probes for dot blot hybridization have to be generated in the research or diagnostic laboratories, but reagents for probe labeling and hybridization are commercially available such as from Amersham-Pharmacia.

ISH

ISH is one of the molecular techniques that have gained popularity among veterinary diagnosticians, especially veterinary pathologists, as it detects viral nucleic acids in affected tissues at the genomic level. As with other hybridization techniques, the principle of ISH is base pairing (hybridization) between a labeled viral DNA or RNA probe and the complementary base sequences of the target viral genomic DNA or RNA that are present in tissues or cells. Over the years, nonradioactive labels, such as biotin and digoxigenin, have gradually replaced traditional radioactive labels, such as ^{35}S and ^{32}P, in labeling DNA or RNA probes used in ISH. For example, biotin-labeled viral DNA or RNA probes are now commonly used to detect viral nucleic acids in affected tissues or cells by adding

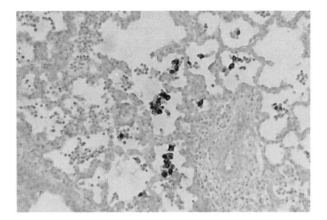

FIGURE 2 ISH detection of porcine reproductive and respiratory syndrome virus (PRRSV) nucleic acid in the alveolar macrophages of a lung from a pig infected with PRRSV. PRRSV nucleic acid is detected in the cytoplasm of macrophages. Shown is the result of ISH with a digoxigenin-labeled PRRSV riboprobe detected by antidigoxigenin-alkaline phosphatase complex, and nuclear fast red counterstain. Courtesy of Patrick G. Halbur, Iowa State University.

enzyme-labeled avidin and a substrate (Fig. 2). Under a light microscope, a colorimetric reaction within a particular virus-infected cell can be visualized in the context of other surrounding microscopic structures. Many commercial companies such as Roche Molecular Biochemicals and DAKO manufacture universal ISH kits that usually contain all the necessary reagents for probe labeling and ISH procedures. However, the commercial ISH kits do not include specific viral DNA or RNA fragments to be used as probes. A specific viral genomic fragment or a synthetic oligonucleotide for each virus of interest has to be generated in a research or diagnostic laboratory for labeling. ISH has now been widely used to detect the nucleic acids of many veterinary viruses, including, but not limited to, porcine reproductive and respiratory syndrome virus (40, 67), porcine circovirus (89), porcine epidemic diarrhea virus (61), avian reoviruses (75), chicken anemia virus (116), avian adenovirus (69), bovine coronavirus (158), and rabbit hemorrhagic disease virus (29).

PCR

The Nobel prize-winning method of PCR in 1983 by Kary Mullis has revolutionized the field of molecular biology. The unmatched sensitivity and quick turnaround time of PCR are extremely attractive for rapid detection of low levels of viral nucleic acids from virtually any clinical samples. Compared to its use in the diagnosis of human viral diseases, PCR has not yet been widely used in veterinary viral diagnostics largely due to the relatively high cost of PCR compared to other tests. Few commercial PCR diagnostic kits for detection of viral nucleic acids are available in veterinary medicine, and the current PCR tests used in veterinary diagnostic laboratories are primarily based on in-house reagents and procedures. Nevertheless, PCR-based viral diagnostics promise to become more widely used in diagnostic veterinary virology.

The basic principle of PCR is the enzymatic synthesis of a specific DNA sequence by using two short synthetic

oligonucleotide primers, usually 18 to 24 bp in length. The primers bind to the complementary sequences of the target DNA, and the viral genomic region flanking the two primers is amplified repeatedly by a thermostable *Taq* polymerase in an automated cycling fashion in a Thermocycler. A typical PCR cycle consists of denaturation of target DNA (usually at 94°C for 30 to 60 s), annealing of primers to the target DNA at an appropriate temperature (usually between 42 and 60°C) for 30 to 60 s, and extension of the products at 72°C for a certain length of time (depending upon the size of the amplified fragments). After 25 to 35 cycles, the target DNA is usually amplified to a level that can be detected as a visible band by gel electrophoresis. Determining the identity of the PCR product(s) of the expected size usually requires further confirmation, such as restriction enzyme digestion, nucleotide sequencing, or Southern blot analysis. The design of PCR primers is a key to the successful amplification of the target viral nucleic acid. The target virus nucleotide sequence must be known in order to design specific primers for PCR detection. Many computer programs, such as MacVector (Accelrys) and Oligo (Molecular Biology Insights, Inc.), can aid in designing optimal PCR primers. Since the introduction of PCR in 1983, numerous modifications of the original method have been applied for the diagnosis of viral diseases, such as reverse transcriptase PCR (RT-PCR), nested PCR, multiplex PCR, quantitative PCR, in situ PCR, and real-time PCR. Reagents for these PCR-related assays are readily available from many commercial vendors such as Perkin-Elmer (Norwalk, Conn.), GIBCO-BRL (Life Technologies), Promega, and Invitrogen.

RT-PCR

RT-PCR is designed to detect viral RNA from clinical samples. The first step of the assay involves the extraction of viral genomic RNA or total RNA from clinical samples. Many commercial kits or reagents are available for RNA extraction from a variety of clinical specimens, such as Qiagen RNA extraction kits (Qiagen Inc.) and TriZol reagents (Life Technologies). The RNA template is reverse-transcribed into cDNA with an RT, and the resulting cDNA is then amplified by PCR with specific viral primers. A limitation of PCR has been that the enzymatic activity of the traditional retrovirus RT does not function in temperatures over 42°C. Therefore, the presence of strong secondary structures in the single-strand RNA template during cDNA synthesis at 42°C often limits specific cDNA synthesis. The discovery of a DNA polymerase with RT activity from a thermophilic bacterium has overcome this problem, and cDNA synthesis can now be performed at temperatures up to 65 to 70°C. Commercial RT-PCR kits in which cDNA synthesis is carried out at temperatures up to 70°C are now available, such as the rTth RT-PCR kit (Perkin-Elmer). RT-PCR has been widely used to detect nucleic acids of many RNA viruses, such as porcine epidemic diarrhea virus (49), infectious bronchitis virus (142), avian pneumovirus (2), porcine reproductive and respiratory syndrome virus (140), and bovine viral diarrhea virus (110).

Nested PCR and Nested RT-PCR

Nested PCR and nested RT-PCR are extremely sensitive assays that are frequently used for detection of nucleic acids of both DNA viruses such as canine herpesvirus (125) and bovine herpesvirus (113) and RNA viruses such as feline infectious peritonitis virus (27) and canine coronavirus

(107). The assays utilize two sets of specific PCR primers. In the first round of PCR, the external sets of primers amplify a product that serves as the template for the second round of PCR amplification. The second round of PCR (nested PCR) utilizes an internal set of PCR primers located within the region of the amplified product from the first round. Therefore, the second round of PCR amplification often serves as a confirmation for the specificity of the PCR products amplified in the first round. Because of the extreme sensitivity, care must be taken to avoid carryover or cross-contamination, which can result in false-positive results.

Multiplex PCR

Multiplex PCR is designed to detect two or more target viral sequences in a single PCR; therefore, it is more cost-effective compared to other PCR methods. Usually two or more sets of primers specific for different target viral sequences are included in the same amplification reaction, and each specific set of primers produce different-size PCR products that can be distinguished by gel electrophoresis. Currently, multiplex PCR is being used in veterinary medicine for the detection of multiple virus infections in the same specimen and for simultaneous detection and differentiation of different genotypes or serotypes of related viruses (30, 68, 111, 143).

In Situ PCR and In Situ RT-PCR

In situ PCR and in situ RT-PCR are newly developed PCR methods that can directly detect viral nucleic acids in paraffin-embedded, formalin-fixed tissue sections. These methods directly incorporate a labeled deoxynucleoside triphosphate, such as digoxigenin-11-dUTP, into the amplicon during the in situ PCR, resulting in the visualization of viral DNA or RNA in tissues by a colorimetric reaction. In situ PCR or in situ RT-PCR is still at an experimental stage in veterinary diagnostics, although it has been successfully used to detect nucleic acids of many viruses, including, but not limited to, equine herpesvirus (9), avian reoviruses (75), and infectious bursal disease virus (76).

Negative-Strand RT-PCR

Negative-strand RT-PCR is designed to detect nucleic acids of replicating RNA viruses from clinical samples. During replication, positive-strand RNA viruses produce an intermediate negative-strand RNA. A specific primer complementary to the negative-strand RNA of the target virus is used for reverse transcription and cDNA synthesis. The resulting cDNA is subsequently amplified by PCR. This method allows distinction between virus replicating in the tissues and circulating virus. A positive result of a negative-strand RT-PCR assay indicates the presence of nucleic acid from a virus replicating in the particular tissue, not due to contamination of tissues by virus circulating through the bloodstream. Negative-strand RT-PCR is primarily used for viral pathogenesis studies, for identification of virus replication sites, and for confirmation of virus replication in cell cultures in combination with VI.

Quantitative PCR and Semiquantitative PCR

Quantitative PCR and semiquantitative PCR are used to measure the amount of target viral DNA or RNA in clinical samples. A known quantity of a competitive DNA or RNA template is mixed with the sample containing the target virus nucleic acid. The same set of primers is used to amplify both the target viral template and the competitor template, which differs slightly in size for differentiation. The amount of target viral template is quantitated by interpolation of the equivalence point between target and competitor templates. For semiquantitative PCR, serial dilutions of clinical samples containing the target virus nucleic acid are tested, and the result is often expressed as the number of copies of genome equivalents per milliliter of a sample. Despite the recent development of advanced real-time PCR technology, quantitative PCR is still a useful method for virus quantitation because it requires less sophisticated instrumentation than does real-time PCR.

Real-Time PCR

Real-time PCR is a new technology that combines PCR amplification and detection into a single step, thereby allowing rapid, high-throughput quantitative analysis of target nucleic acids of interest. The capability of real-time quantitative PCR for large-scale rapid detection and quantification of pathogens in clinical samples makes it particularly appealing to clinical diagnostic laboratories. The accumulated PCR product is monitored continuously during amplification, and detection in the early exponential phase of the reaction permits rapid and reproducible quantitation of the template. The basic principle of real-time quantitative PCR is the detection of target nucleic acid sequences using a fluorogenic 5′ nuclease assay. A fluorogenic probe (often called the TaqMan probe) complementary to the target sequence is added to the PCR mixture. If the target nucleic acids of interest are present, the probe anneals specifically between the forward and reverse primer sites during PCR. This annealing and detection process occurs in every cycle (thus the name "real-time" quantitative PCR). The fluorescence intensity of the annealed probe provides a direct measurement of the number of target molecules in the reaction mixture, and the results are instantly displayed on a computer screen. Other advantages of real-time quantitative PCR for clinical diagnosis include reproducibility from sample to sample and from laboratory to laboratory, increased capacity for screening large numbers of clinical samples, and significantly decreased time for reporting results. Real-time quantitative PCR can detect the nucleic acid of a pathogen from clinical samples in a matter of minutes (5). Several commercial companies now manufacture real-time quantitative PCR instruments, such as the ABI Prism 7700 Sequence Detection System (Perkin-Elmer Applied Biosystems, Foster City, Calif.). One limitation of this assay is the relatively high cost of the instrumentation. Nevertheless, real-time quantitative PCR has been used to detect nucleic acids of several veterinary viruses, including, but not limited to, classical swine fever virus (87), feline coronaviruses (35), feline immunodeficiency virus (72), and infectious bursal disease virus of chickens (92).

Assays for Genetic Characterization and Differentiation

PCR-RFLP Assay

The PCR-RFLP assay combines PCR amplification and restriction enzyme digestion techniques, enabling two related viruses to be distinguished at the molecular level without having to determine their nucleotide sequences. By carefully selecting unique restriction enzymes that recognize specific sites in the genomes of target viruses, PCR-RFLP analysis can differentiate minor genomic differences between two viral isolates or genotypes that would otherwise not be distinguished by other assays except sequencing

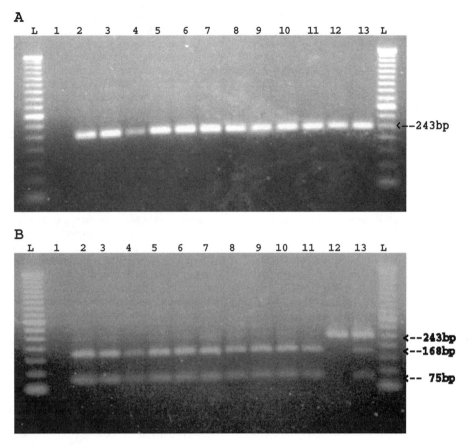

FIGURE 3 Detection and differentiation of porcine circovirus type 1 and 2 infections by a PCR-RFLP assay. (A) PCR amplification of a 243-bp fragment. Lane 1, negative control sample; lanes 2 to 13, samples containing porcine circovirus (both type 1 and 2). (B) RFLP analysis of the PCR products. Lanes L, 50-bp DNA ladder; lane 1, a sample of liver tissue from a control specific-pathogen-free pig; lanes 2 to 11, clinical samples from 10 cases of porcine circovirus type 2; lane 12, PK15 cells containing porcine circovirus type 1; lane 13, a sample containing porcine circoviruses (both type 1 and 2). The expected PCR fragment (A) and three RFLP fragments, 243, 168, and 75 bp, respectively (B) are indicated with arrows. Modified from Fenaux et al. (25).

(Fig. 3). PCR-RFLP analysis requires post-PCR manipulation in which the amplified PCR products are digested with a chosen restriction enzyme(s). After separation of the digested PCR products by electrophoresis, a definitive or differential diagnosis can often be made on the basis of the restriction fragment patterns. Reagents for PCR and RFLP restriction enzyme digestions are commercially available from vendors such as Life Technologies and Perkin-Elmer. PCR-RFLP analysis has been used to detect and differentiate the genomic sequences of many veterinary viruses, such as feline panleukopenia virus (46), infectious bursal disease virus (136), porcine reproductive and respiratory syndrome virus (152), feline immunodeficiency virus (45), and porcine circovirus (25). The assay provides rapid reports of results and is less expensive than other post-PCR manipulations such as sequencing and sequence analysis.

Sequencing and Sequence Analysis

PCR amplification coupled with sequencing and sequence analysis is generally used to confirm the specificity of the amplified PCR products, differential diagnosis, genotyping, and genetic characterization of the target virus. This method is often too expensive and time-consuming and requires sophisticated sequencing instrumentation and sequence analysis software that prohibits its use for routine diagnostics.

Quality Control of Veterinary Diagnostic Assays
Sensitivity and Specificity

Accurate diagnosis of a veterinary viral disease is critically important for effective and timely treatment of infected animals, for preventing spread of the agent within and between herds, and, in cases where the virus may have zoonotic potential, for preventing spread of disease to humans. The evaluation and interpretation of test results is largely dependent upon two critical components of a particular diagnostic assay: sensitivity and specificity (43). For any given viral diagnostic test, it must be sensitive enough to detect at an acceptable detection threshold (analytical sensitivity) and minimize false-negative results (diagnostic sensitivity). The test must also be specific enough to minimize nonspecific detection resulting in false-positive results. There are many factors that may affect the specificity and sensitivity of a test, such as the purity of the reagents, the cutoff values used for immunoassays, the quality of clinical samples, instrument, and technical skill of the technicians. Sensitivity and specificity also vary from test to test. Therefore, the selection of a particular diagnostic assay for a given virus is important. A confirmatory test, if available, may also have to be performed.

The commercial viral diagnostic kits in veterinary use have set sensitivities and specificities that are standardized from lot to lot, thereby rendering quality control relatively easy. Compared to human viral diagnostics, however, there are not many commercial kits available for veterinary use. Consequently, veterinary diagnostic laboratories are dependent upon in-house assays, in which the sensitivity and specificity may not be known. Each laboratory usually follows its own standard for quality control, evaluation, and interpretation of test results. Therefore, reproducibility of test results from laboratory to laboratory is often not known for the in-house assays. The specificity and sensitivity of a test can usually be improved by including adequate control specimens and by careful handling and processing of samples. The importance of including positive and negative controls for each run of specimens cannot be overemphasized. Even for commercial kits that come with standard controls, inclusion of additional in-house positive and negative controls is sometimes still necessary for quality assurance. In addition to the routine quality control procedures for a clinical virology laboratory, extra care must be taken in a molecular diagnostic laboratory setting. Many factors that may affect the quality of a molecular diagnostic test must be controlled, such as routine calibration of pipettors, the temperatures and ramping times of thermal cyclers, and accurate temperatures of water baths, heating blocks, and incubators (104). To minimize contamination and to ensure the accuracy of test results, supplies and reagents intended for PCR use should be delivered directly to a designated area or laboratory and stored in a separate refrigerator or freezer. Nonspecific amplification due to carryover or cross-contamination in PCR-related methods can be significantly reduced by performing procedures (such as reagent preparation, RNA extraction, cDNA synthesis, and PCR assembly) in a laboratory separate from the one in which PCR products are analyzed.

VETERINARY BACTERIOLOGY

Bacteriology Identification Systems

Bacterial identification systems have not been developed specifically for veterinary pathogens. This is because many bacteria isolated from animals are identical to or are of the same genus as those that cause disease in humans and because it is not cost-effective to develop systems that would only identify the relatively few host-specific bacteria that are isolated in clinical veterinary laboratories. Therefore, systems developed for isolation of bacteria from human clinical specimens are often used for the identification of bacteria from animal specimens. For a detailed discussion of rapid systems and instruments used in bacteriology, please see chapter 3. This section, rather, will focus on reviews of studies that have evaluated the applicability of bacterial identification systems for identification of bacteria isolated from animal specimens.

To facilitate the identification of bacterial isolates from animals it is essential for one to know the animal species and the body site from which the specimen was collected. For identification, most veterinary diagnostic laboratories still use conventional macrotube biochemical tests (15, 108). There are two reasons for this. First, the cost of rapid identification systems compared to the cost of macrotube biochemicals may prohibit their use in many veterinary laboratories. Additionally, while these systems generally have comprehensive databases, they usually have not included a sufficient number of veterinary strains to generate accurate biocodes within their databases (1, 21, 114). Many studies have evaluated the miniaturized biochemical test systems and semiautomated systems for their ability to identify isolates of veterinary origin. A review of these has been written (149). To summarize, no one system was capable of accurately identifying all the diverse organisms encountered in the veterinary laboratory. With minor modifications, however, the clinical microbiologist may be able to effectively utilize many of these systems.

Enterobacteriaceae

Studies have been performed to evaluate the feasibility of using commercial miniaturized biochemical test systems to identify gram-negative bacilli of veterinary origin. Identification of members of the family *Enterobacteriaceae* have been evaluated in two studies. Both sets of authors utilized the API 20E kit (BioMerieux-Vitek). Swanson and Collins, using veterinary enteric pathogens, found this system to accurately identify 96% of the isolates (133), while Peele et al. found that only 66% of veterinary enteric bacteria were correctly identified to genus and species level (103). The latter study evaluated primarily ATCC-derived strains and only 38 members of the family *Enterobacteriaceae*, whereas the former study utilized 503 clinical isolates. The study of Peele et al. also included an evaluation of the Crystal E/NF (BD Biosciences), which correctly identified 68% of the enteric, bacterial isolates from animals.

Enterotoxigenic *Escherichia coli* strains are common pathogens of newborn calves. Expression of pili is commonly associated with disease-causing enterotoxigenic strains. A simple and rapid latex agglutination test is available from VMRD that detects the presence of K99 pili on *E. coli* strains isolated from calves with scours. Although not an identification test, exposure of poultry to *Salmonella* spp. can be determined by an ELISA for antibodies to *Salmonella* flagella (manufactured by IDEXX). Evaluations of these kits have not been reported.

Weakly Fermentative and Nonfermentative Gram-Negative Bacteria

Studies have also been undertaken to evaluate the capability of miniaturized biochemical test systems to correctly identify gram-negative weakly fermentative and nonfermentative bacilli. This diverse group comprises a large portion of veterinary significant pathogens and includes the genera *Actinobacillus*, *Pasteurella*, *Haemophilus*, *Mannheimia*, and *Bordetella*. Collins et al. compared the abilities of the API 20E, Minitek (BD Biosciences), and Oxi/Ferm tube (Roche Diagnostics) systems to correctly identify *Pasteurella multocida* and *Mannheimia* (*Pasteurella*) *haemolytica* (21). They concluded that neither of these systems was satisfactory for the identification of these two important veterinary pathogens. In another study conducted by Collins and Swanson, the capability of the API 20E system to correctly identify 272 nonfermenting and weakly fermenting gram-negative bacilli was evaluated (20). The accuracy of this system was determined to be only 62% with this group of organisms. Two studies performed by Salmon et al. were conducted to evaluate systems designed to identify weakly fermentative or nonfermentative veterinary isolates. The first study concluded that although the RapID NH system (Remel) did not include the veterinary pathogens *Haemophilus somnus*, *P. multocida*, *M. haemolytica*, and *Actinobacillus pleuropneumoniae* in its database, with minor modifications such as animal source and growth requirements, the system would permit correct identification of these isolates (115). The second study evaluated the RapID NF Plus system (Remel) using 273 bacterial isolates in the genera *Actinobacillus*, *Bordetella*, *Haemophilus*, *Mannheimia*, and *Pasteurella* (114). It was concluded that with the addition of the unique biocodes that were generated by the study to the system database, 85.5% of the isolates could be correctly identified. For the isolates that were already in the system database, only 20% were correctly identified. Another study was undertaken using the RapID NF Plus system to evaluate its feasibility in identifying the avian respiratory pathogen *Ornithobacterium rhinotracheale* (105). The identification system did not include this organism in its database; however, five unique biocodes were obtained, and the authors concluded that the system did appear suitable for use in a diagnostic setting for identification.

A commercial ELISA for the detection of antibodies to *P. multocida* is available from IDEXX but has not been independently evaluated.

Gram-Positive Bacteria

Several studies have evaluated miniaturized biochemical test systems for identification of gram-positive organisms of veterinary significance. For the staphylococci the following commercially available systems have been evaluated: The API Staph (formerly Staph-Ident [BioMerieux-Vitek]), API Staph-Trac (BioMerieux-Vitek) and the Minitek gram-positive set (BD Biosciences). The results of these studies were variable, but staphylococci isolated from bovine mammary glands were identified incorrectly more often when using the Staph-Ident system. Studies by a variety of investigators (53, 65, 109, 146, 148) indicated that the Staph-Ident system correctly identified 54, 80.5, 40.6, 89.2, and 45.2% of isolates, respectively. A major reason attributing to the inaccuracy of this system was its inability to adequately identify *Staphylococcus hyicus*, a common mastitis and swine isolate (53). The API Staph-Ident system performed better on isolates from nonbovine udder sources. Kloos and Wolfshohl (62) used strains of *S. hyicus* and *Staphylococcus intermedius* in a study that compared a conventional identification scheme to the Staph-Ident. Greater than 90% agreement was found between these two methodologies. In two other studies involving characterization of animal staphylococci, this system was used to mainly discriminate *S. aureus* from *S. intermedius* with apparent success (7, 22). The Staph-Trac system (a 24-h ID system) has been evaluated primarily for identification of *Staphylococcus* strains of bovine origin. Overall this system performed better than the Staph-Ident system, correctly identifying 91.2% (66), 80.8% (86), and 66.1% (146) of isolates. Another study reported that this system could also be used to identify swine strains of *S. hyicus* (78). Lastly, another commercial system, the Minitek gram-positive set (BD Biosciences) has been evaluated using staphylococci of bovine origin (147). An overall accuracy of 79.2% was found, and the authors concluded that the high misidentification rate was due to database deficiencies of the animal-associated isolates of *S. intermedius* and *S. hyicus*.

Studies to evaluate the efficacy of the commercial miniaturized biochemical systems for the identification of animal isolates of streptococci have also been done. The capability of the API 20 Strep system (BioMerieux-Vitek) to identify streptococci of bovine origin has been evaluated by several authors (54, 106, 145). This system successfully identified 71.4 to 96.5% of isolates to the species level. There were difficulties with the identification of several *Streptococcus* species, including *Streptococcus uberis*. The Rapid Strep system (BioMerieux-Vitek) was evaluated for its capability to correctly identify 85 group C streptococcal equine isolates (4). The system was able to accurately identify all *Streptococcus equi* subsp. *equi* and *S. equi* subsp. *zooepidemicus* isolates within 24 h without the use of any supplemental tests. For identification of "*Streptococcus equisimilis*," specific grouping sera was necessary to confirm the identification.

Several authors have utilized the API Rapid Strep system to identify swine isolates of *Streptococcus suis*. Gottschalk et al. (33) found that only 54% of the strains were correctly identified. In another study by Gottschalk et al. (32) only 65% of *S. suis* isolates were correctly identified using this system. Since there are currently over 30 capsular serotypes, identification of this organism by biochemical means alone and without capsular typing is not recommended (32).

Another commercial system for the identification of streptococci is the Minitek gram-positive kit, which has been evaluated in one study for its ability to identify organisms isolated from bovine mammary glands (147). Only 34% of isolates were correctly identified. The API 20 Strep system has also been utilized for the identification of *Arcanobacterium* (*Actinomyces*) *pyogenes* (93). Although this organism is not currently in the database of the system, 62 strains from bovine milk samples were tested, and four unique biocodes for this organism were generated. The authors suggested that the API 20 Strep system could be used along with characteristic microscopic and colonial morphology for the routine identification of this organism.

The rapid latex agglutination kits that are marketed for identification of streptococci from humans may also be applicable to the identification of streptococci from animals. Reagents for group C and G streptococci, commonly isolated from horses and dogs, respectively, are usually included in these kits. These tests are very accurate for the typing of isolated streptococci. However, the control group C reagent of the PathoDx Strep A latex agglutination test (Diagnostic Products Corp.), which is designed to identify group A streptococci from human throat swabs, could identify group C streptococci other than *Streptococcus dysgalactiae* from horses with a sensitivity of 95.3% and a specificity of 100%. However, only 25% of *S. dysgalactiae* isolates could be identified (48). Similar agglutination tests are available for identification of *Staphylococcus aureus*, but have not been tested on isolates specifically from veterinary specimens. An ELISA for detection of antibodies to *S. aureus* in milk is commercially available from VMRD.

Anaerobic Bacteria

Several studies have evaluated commercial microbiochemical systems for the identification of veterinary anaerobic bacteria. Adney and Jones (1) evaluated the RapID ANA system (Remel) using 183 clinical isolates. With the additional tests recommended by the manufacturer, the system correctly identified 81% of isolates to the genus level. The majority of the errors that were incurred were a result of biocodes not appearing within the codebook of the system.

A study of four commercial systems—the RapID ANA (Remel), An-Ident and API 20A (BioMerieux-Vitek), and Minitek (BD Biosciences) systems—for their usefulness in identifying *Actinobaculum* (*Eubacterium*) *suis*, a cause of cystitis and pyelonephritis in sows, was conducted by Walker et al. (141). It was concluded that both the RapID ANA and the An-Ident systems were useful adjuncts for identification of *A. suis*, but that colonial morphology and Gram stain reaction must be taken into account.

Fluorescent antibody reagents for identification of *Clostridium chauvoei*, *Clostridium novyi*, *Clostridium septicum*, and *Clostridium sordellii* are available from VMRD.

Mycoplasmas

The An-Ident and API-ZYM systems (BioMerieux-Vitek), which are enzyme activity assays, were evaluated for their ability to generate enzyme profiles of seven species of mycoplasma isolated from swine—*Mycoplasma hyosynoviae*, *Mycoplasma arginini*, *Mycoplasma salivarium*, *Mycoplasma hyorhinis*, *Mycoplasma flocculare*, *Mycoplasma hyopneumoniae*, and *Acholeplasma laidlawii* (59). Two common isolates from pigs, *M. hyorhinis* and *M. hyosynoviae*, as well as *A. laidlawii*, could easily be differentiated by these enzyme assay systems. However, *M. flocculare* and *M. hyopneumoniae* could not be differentiated. Overall, the systems showed promise for identification of mycoplasmas, but confirmation may require an FA assay.

A DNA probe combined with PCR amplification for detection of *Mycoplasma gallisepticum* and *Mycoplasma synoviae* genomic DNA from chicken and turkey tracheal swab samples is available from IDEXX but has not been independently evaluated.

Automated Systems

Four automated systems have been evaluated for use in the identification of veterinary isolates: the Vitek gram-positive system, the Radiometer/Sensititre AP 80 system (now Trek Diagnostics), the Sceptor system (BD Biosciences), and the Quantum II system (Abbott Laboratories). The Vitek gram-positive identification system (BioMerieux-Vitek) was evaluated using staphylococci of bovine origin and streptococci and had overall accuracy rates of 44.6 and 94.4%, respectively (54, 86). The authors felt that the poor performance of the system in the identification of the staphylococci could be attributed to the low number of veterinary strains in the database. Patten et al. conducted a trial using the Sensititre system to evaluate gram-negative clinical veterinary isolates with emphasis placed on the genera *Actinobacillus*, *Bordetella*, and *Pasteurella* and the family *Enterobacteriaceae* (101). After 18 h of incubation in test plates, 85% of the bacteria were correctly identified to the species level. The Sceptor system (99) was evaluated for the capability to identify 605 isolates, which included the *Enterobacteriaceae* (315 isolates), gram-negative nonenteric bacteria (191 isolates), and gram-positive bacteria (99 isolates). Correct identifications were obtained for 92.7% of the *Enterobacteriaceae*, 86.4% of gram-negative nonenteric bacteria, and 77.8% of gram-positive bacteria. The unacceptable specificity of the system for gram-negative nonenteric and gram-positive bacteria was felt to be due to an inadequate database for veterinary-specific isolates. The Quantum II system, which is no longer commercially available, was evaluated in two studies. Jones et al. (57) found that the system correctly identified 89.9% of isolates in the family *Enterobacteriaceae* but was less accurate (81.3%) in the identification of nonfermentative bacte-

ria. More importantly, however, was the failure of the system to identify *P. multocida*, a common veterinary pathogen. The ability of the Quantum II system to correctly identify common gram-negative pathogens of fish was evaluated by Teska et al. (135). Most of these organisms were not in the database. However, unique biocodes were generated for *Edwardsiella ictaluri*, *Yersinia ruckeri*, *Serratia liquefaciens*, *Aeromonas hydrophila*, and *Aeromonas salmonicida*, enabling the system to be utilized for these bacteria.

In conclusion, there have been no published reports of any commercial system, either microbiochemical, semiautomated, or automated that is capable of identifying the diverse array of organisms encountered in a clinical veterinary setting. Many of these systems can be partially utilized successfully by the laboratorian in conjunction with supplemental tests in the identification process. The veterinary diagnostic bacteriologist should be aware of these pitfalls and be prepared to perform ancillary tests when questionable identifications occur.

Mycobacterium avium subsp. *paratuberculosis*

Identification of animals infected with *M. avium* subsp. *paratuberculosis* can be difficult, because a substantial number of infected animals in a herd may be in the early stages of infection, and therefore clinically healthy, having not yet mounted an immune response or started shedding bacteria. Methods currently available for the diagnosis of *M. avium* subsp. *paratuberculosis* has been reviewed by Collins (18). Standard culture methods for *M. avium* subsp. *paratuberculosis* have been the mainstay for paratuberculosis diagnosis. Culture methods for *M. avium* subsp. *paratuberculosis* have been reviewed by Whipple et al. (154), and will not be discussed here. Standard culture methods for *M. avium* subsp. *paratuberculosis* require 12 to 16 weeks or more due to the low growth rate of this organism. Animals with *M. avium* subsp. *paratuberculosis* infection do not recover, and this bacterium lacks the siderophore mycobactin, making *M. avium* subsp. *paratuberculosis* an obligate pathogen of animal cells. Therefore, culture of the agent is 100% specific, but is only about 50% sensitive (131). An alternative technique is the commercial BACTEC 460 system (BD Biosciences), which has been used to improve the detection time for *M. avium* subsp. *paratuberculosis* to 4 to 7 weeks. In this system a liquid medium (BACTEC 12B) containing $^{14}CO_2$-labeled palmitate is used. Metabolism of palmitate releases $^{14}CO_2$, which is then detected by the BACTEC 460 instrument. When combined with filter concentration of fecal samples, identification of *M. avium* subsp. *paratuberculosis* by the BACTEC system was more sensitive and rapid than conventional culture (19). The BACTEC system has also been applied to culture of *M. avium* subsp. *paratuberculosis* from sheep. Whittington et al. (155) showed that culture of *M. avium* subsp. *paratuberculosis* was positive from the intestinal tissues of all sheep with multibacillary disease and with paucibacillary disease. Culture of *M. avium* subsp. *paratuberculosis* was positive from the feces of 98% of animals with multibacillary disease and 48% of animals with paucibacillary disease. However, most veterinary laboratories do not have the BACTEC system. Therefore, nonculture methods have been established to more rapidly identify this bacterium.

A genetic probe specific to an insertion element (IS900) unique to *M. avium* subsp. *paratuberculosis* in combination with PCR to amplify the DNA is available through IDEXX Laboratories. In comparison to culture the DNA probe only requires 3 days and is also 100% specific, provided false-positive results do not occur due to contamination of samples by gene products in the laboratory. However, the sensitivity is lower than culture (10^4 M. *avium* subsp. *paratuberculosis* bacteria/g of specimen is the lower level of detection), it is relatively expensive, and skilled technicians and specialized equipment are required. It is possible that a combination of culture and the DNA probe would both increase sensitivity and shorten the time of diagnosis (18).

Several assays have been used to detect serum antibodies to M. *avium* subsp. *paratuberculosis* in cattle, the most common being ELISA, AGID, and CF. Of these ELISA is considered the most sensitive and quantitative and CF is considered the least sensitive. However, all of the USDA-licensed tests have >99% specificity. The sensitivity and specificity of the ELISA are optimized when cross-reacting antibodies are first removed by adsorption with an extract of *Mycobacterium phlei* (18). The ELISA is commercially available through IDEXX, CSL Laboratories, Pourquier, and ImmuCell. A commercial AGID kit is also available through ImmuCell. However, the sensitivity of any antibody-mediated test depends on the time course of the infection. In the early stage of infection, the sensitivity of ELISA in low-level fecal shedders was only 15%, whereas in animals with clinical signs of disease it was 87%, with an overall sensitivity of 45% (134).

A cell-mediated immune response is common following mycobacterial infections and can be detected by skin testing with an extract of the bacteria. However, skin testing is not recommended for the diagnosis of paratuberculosis, due to low specificity. Recently, however, a commercial assay that detects the cytokine gamma interferon became available from IDEXX Laboratories as an indicator of a cell-mediated immune response to M. *avium* subsp. *paratuberculosis*. Preliminary tests appear encouraging as this test may be able to detect fecal shedders earlier than assays for serum antibodies or fecal culture. Furthermore, it can be used to test for infections in cattle, sheep, and goats. Overall, however, the most effective way to diagnose clinical paratuberculosis from the standpoint of speed, accuracy, and cost is by a USDA-licensed ELISA, followed by fecal culture (18).

Chlamydia psittaci

Culture of C. *psittaci* in McCoy cells is the accepted procedure for diagnosis of this agent. The use of FA reagents can improve the speed and specificity of culture. However, in areas where local laboratories with cell culture facilities are not available, specimens may need to be transported through the mail. This may result in loss of viability and/or contamination of specimens, reducing sensitivity and specificity. Other methods of diagnosis include serology, histochemistry of tissues using various pathology stains (hematoxylin and eosin, Giemsa, or Gimenez methods), immunohistochemistry using a colorimetric substrate, or direct FA testing of tissue. IHC using a monoclonal antibody to the lipopolysaccharide antigen in combination with avidin-biotin alkaline phosphatase has been shown to be more sensitive for diagnosis of C. *psittaci* in psittacine birds than other nonculture methods (24).

Diagnosis of C. *psittaci* based on antigen detection is an attractive alternative. Commercial kits for identification of C. *psittaci* are not available, but ELISAs and FA assays for detection of the genus-specific lipopolysaccharide have been developed for detection of C. *trachomatis* from urogenital and conjunctival swabs. These kits have not been approved for diagnosis of C. *psittaci*. However, several studies have evaluated the capability of some of these kits to detect C. *psittaci* in various veterinary specimens (117, 118,

138, 156). Unfortunately, most of the kits that are currently available have not been examined. A complete list of these kits can be found in chapter 7. The semiautomated Chlamydiazyme system (Abbott Laboratories) was reported to have an overall sensitivity of 85.7% and a specificity of 85.7% for detection of C. *psittaci* from vaginal secretions, placentas, and fetal tissues from aborting ewes compared to cell culture (117). Sensitivity and specificity varied depending on the specimen, but the authors felt that vaginal swabs collected within 3 days of abortion were the best samples for detecting *Chlamydia* infection. In a later study the same authors reported a sensitivity of 88.1% and a specificity of 96% when using the Kodak Surecell test kit (118). Again, the best specimens were vaginal swabs or placenta. However, the Surecell kit is no longer available.

An FA kit for detection of antibodies to C. *psittaci* in cats is available from VMRD.

Miscellaneous Agents

A serological ELISA kit and FA assays for detection of *Brucella abortus* antibodies in milk and serum are available from IDEXX. An FA kit for detection of antibodies to *Brucella ovis* in dogs is available from VMRD. FA substrate slides for detection of antibody to *Ehrlichia canis*, *Borrelia burgdorferi*, *Coxiella burnetii*, and *Rickettsia rickettsii* by the indirect FA technique are also available from VMRD. These tests may also be used to differentiate the antibody class (immunoglobulin G [IgG] or IgM) with a suitable fluoresceinated second antibody conjugate. Synbiotics Corp. sells a micro-ELISA for detection of antibody to B. *burgdorferi* in dogs.

VETERINARY MYCOLOGY

The fungi that cause diseases in animals are essentially identical to those that cause diseases in humans. An exception is *Malassezia pachydermatis*, which is an important cause of otitis media and interdigital dermatitis in dogs, and an uncommon cause of catheter-associated sepsis in neonatal intensive care units. Therefore, the information in chapter 9 is as applicable to most fungi isolated from animal specimens as that isolated from human specimens. For instance, all the tests described for cryptococcal capsular polysaccharide antigen testing, identification testing, and susceptibility testing would be applicable to isolates from animals.

The identification of dermatophytes from animal specimens will be emphasized in this chapter because dermatophytes are causes of zoonotic infections as well as common infections in animals. Some dermatophyte species predominate in a specific host, such as *Microsporum canis* in cats, *Microsporum nanum* in swine, *Trichophyton equinum* in horses, and *Trychophyton verrucosum* in cattle. Commercial methods are primarily limited to culture systems, with identification still based primarily on recognition of specific spores and hyphae or nutritional requirements. Culture is most effectively done by inoculating skin scrapings from the edge of the lesion and some plucked hairs onto dermatophyte test medium (DTM). DTM is available from a variety of sources, such as Remel and BD Biosciences. The InTray culture system from Tridelta Development Ltd. allows dermatophytes to be cultured and identified in a completely closed culture system. Once the InTray is inoculated with hair, skin, or nails it is resealed and needs not be reopened. A dermatophyte will turn the medium from yellow to red, as it does in DTM, and the tray can be placed directly on a microscope stage for identification of the agent through a clear plastic window. Most dermatophytes of the genus

Microsporum can be identified by distinctive macroconidia. For *Trichophyton* spp. the species-specific nutritional requirements of some species can be determined using *Trichophyton* media (Remel and BD Biosciences). These media consist of seven different agar preparations which are nutritionally distinct. Three zoophilic agents that can be identified using these media are *T. verrucosum*, which requires inositol and/or thiamine HCl for growth; *T. equinum*, which requires nicotinic acid for growth; and *Trychophyton gallinae*, which requires ammonium nitrate for growth (Difco Manual, 10th ed., BD Biosciences, Sparks, Md.).

Serological tests are also sometimes used to confirm a questionable systemic fungal diagnosis. Commercial kits are not available that will test for animal antibodies (most often canine or feline) to fungal pathogens, but some laboratories offer such testing. Antech Diagnostics (Farmingdale, N.Y.) will test for serum antibodies for histoplasmosis, blastomycosis, coccidioidomycosis, and aspergillosis.

VETERINARY PARASITOLOGY

Veterinary diagnostic parasitology is much different from human diagnostic parasitology. Veterinarians and veterinary diagnosticians rely on examination of fecal flotations and microscopic identification of parasite ova or cysts in fresh, unstained preparations (50, 129). Fecal samples are usually examined fresh. If fecal samples are preserved they are usually preserved in 10% neutral buffered formalin solution. Fecal flotation is a common concentration technique seldom used by human diagnostic parasitologists. Sedimentation techniques common in human diagnostic parasitology are limited to examination of fecal samples for fluke eggs.

Examination of blood for microfilariae or antigens of *Dirofilaria immitis* is the second most common parasitological diagnostic test conducted in veterinary clinics and diagnostic laboratories. Detecting *D. immitis* antigens in blood has gained major importance with the advent of the macrolide heartworm preventatives. Use of these preventatives sterilizes adult female *D. immitis*, making detection of microfilariae in blood impossible.

With the exception of canine and feline heartworm diagnostic products, there are few test kits marketed specifically for use in animals. Many of the products approved for use in human diagnostic parasitology laboratories (see chapter 11) can be used in the veterinary diagnostic parasitology laboratory. These include tests for *Cryptosporidium parvum*, *Giardia lamblia*, and the amoebas.

Ectoparasitology is much more important in veterinary medicine than in human medicine. Examination of skin scrapings or earwax for mites and the identification of fleas, lice, and ticks on the host animal are also common parasitological examinations conducted in veterinary practices (11, 129). In areas of tick-transmitted diseases, the proper identification of ticks is essential in many cases to help rule in or rule out tick-borne pathogens (i.e., the Lyme disease spirochete).

Equipment

A good microscope with a calibrated ocular micrometer is essential for examining fecal flotations for parasite ova and cysts. Differential diagnosis of many coccidial oocysts and helminth ova depends on measuring their size in fecal flotations. Objectives of 10× and 40× are used most often, while the oil immersion objective is only needed in the examination of blood smears for protozoal parasites. A stereomicroscope is often needed to aid in the identification

of helminths and arthropods. A swinging-bucket centrifuge is necessary to conduct fecal flotations. Appropriate test tubes, sieves, disposable gloves, applicator sticks, tongue blades, solution containers and bottles, scalpel blades, microscope slides, coverslips, chemicals, weigh boats, scales, and a refrigerator are also needed in the laboratory. An epifluorescent microscope is required to examine feces for some of the zoonotic protozoa when using the commercial kits approved for use with fecal samples from humans. A funnel, hose clamp, and ring stand are needed to conduct the Baermenn technique for lungworm larvae.

Fecal Specimen Collection and Examination

Fecal collection systems and fecal flotation solutions used for human parasitology (see chapter 11) can also be used in veterinary diagnostic parasitology laboratories. These have not gained wide usage in veterinary medicine because of the cost of these products.

Fecal samples should be obtained from the rectum when possible, using a fecal loop for small animals or a gloved hand for large animals. It is best if 1 to 5 g of fecal material can be obtained. This is often impossible for small animals, but is usually readily obtainable from large animals. If less than 1 g of feces is available the diagnostician's choices of procedures are limited. Samples should never be collected from feces that have fallen to the ground because the samples may become contaminated with free-living nematodes or their products. Samples should not be collected and then submitted in examination gloves (i.e., palpation gloves) or other similar inappropriate containers. Samples should be placed in plastic containers with screw top lids and identified as to animal name or number, species, age, date, owner's name, and veterinarian's name. It is important to know the animal's age and species because the prevalence of parasites often differs in different age groups of animals. Samples should be kept on ice until processing or until they can be refrigerated. Fecal samples that are to be mailed to a diagnostic laboratory should be fixed in 10% neutral buffered formalin solution. Diarrheic feces to be examined for living trophozoites should be examined as soon as possible after being collected.

Fecal flotation using zinc sulfate solution (specific gravity, 1.18) is the method most often used to detect cysts of *G. lamblia* in the feces of animals. Some people also add a drop of iodine to the coverslip to stain the *Giardia* cysts a dark blue color. A single negative fecal exam is not sufficient to consider a patient negative, because *Giardia* cysts are shed intermittently. It is best to examine three fecal samples collected over a 7- to 10-day period. If they are all negative then the patient can be considered negative. A fecal flotation is done by mixing a teaspoon or so of feces with 15 to 20 ml of zinc sulfate solution and straining it through two or three layers of cheesecloth. The mixture is then placed in a 15-ml test tube, and additional zinc sulfate solution is added until a positive meniscus is produced. The test tube is placed in a centrifuge (swinging-bucket type) and a 22-mm^2 coverslip is placed on top of the meniscus. It is centrifuged at about 500 × *g* for 5 min. The coverslip, which will have parasite ova and cysts attached, is removed and placed on a glass microscope slide for examination. This procedure will work with all types of flotation solutions.

Sugar, zinc sulfate, and sodium nitrate solutions are common flotation mediums. Sugar solution (specific gravity, 1.33) is often used for coccidial oocysts and has the advantage that it does not form crystals in the flotation preparations. Sheather's sugar solution (227 g of table sugar, 178 ml of water, and 3 ml of 37% formalin) is the most common sugar solution used. Commercially available zinc sulfate preparations include Feca-Dry II (Life Science), Ova Float ZN 1.18 (Butler), OvaSol (Vedco Inc.), and Ovum floatation dry (PHoenix Diagnostics). Sodium nitrate solutions available include FECA-MED, FECA-MIX, and Fecasol (Vedco Inc.).

Several gravity-based flotation apparatuses are commercially available to aid in the conduct of fecal flotations (Fecalyzer [Evsco Pharmaceuticals], Ovassay plus [Synbiotics], and Fecal flotation device [Vedco Inc.]). These kits usually have a built-in specimen container, a mixing blade, a flotation tube, and a sieving device to remove large fecal particles. Because these kits are not centrifuged, they must stand for a minimum of 10 min before the coverslip can be examined.

The Baermann technique is used to detect lungworm larvae in the feces. A large portion of feces (5 to 15 g) is placed in cheesecloth and tied at the top with a rubber band. This fecal pack is placed in a funnel that has a collecting tube and a clamp attached. Warm water is added to the funnel, and the nematode larvae migrate out of the pack and down to the bottom of the collecting tube. After a set time (1 h to overnight) the liquid in the collecting tube is obtained, centrifuged, and examined for larvae.

A sieving-based apparatus (Fluke Finder [Visual Difference]) can be used instead of a standard sedimentation test to demonstrate the ova of the liver fluke *Fasciola hepatica* in cattle. Feces are forced through the apparatus, and the *F. hepatica* ova are trapped on a screen.

ELISA-based detection kits are available to detect antigens of *C. parvum* and *G. lamblia* in animal feces (SafePath Laboratories LLC). The immunofluorescent-antibody (IFA)-based tests approved for use in humans for detecting *C. parvum*, *G. lamblia*, and amebas can also be used to detect these organisms in animal feces.

Canine Heartworm (*D. immitis*)

Examination of Blood for Microfilariae

Microfilariae can be found in the blood of dogs with patent heartworm infections that are not on macrolide heartworm preventatives. In North America, the microfilariae of *D. immitis* must be differentiated from microfilariae of *Dipetalonema reconditum*, a filarial parasite of connective tissue.

Direct Examination

A drop of blood is placed on a microscope slide, covered with a coverslip, and examined for microfilariae. Microfilariae will be motile in fresh samples. They will also be motile in samples that have been collected in anticoagulant, refrigerated, and warmed up to room temperature. This is the least sensitive method of detecting microfilariae of *D. immitis*. This technique can be used as a rapid screening test, but a negative test should be followed up by a concentration technique. If more that 5 to 10 microfilariae are present in a drop, they usually are *D. immitis* (11).

Concentration Techniques

Concentration techniques such as Knott's test and membrane filter tests (Difil-Test [Evsco]) are far superior to direct examination (11). This is especially true in cases where few microfilariae are present in the sample. Concentration tests rely on the lysis of RBCs so a larger volume of blood can be examined for microfilariae. Knott's test uses 2% formalin (2 ml of 37% formalin plus 98 ml of distilled

water) to cause lysis of host RBCs. Knott's test requires adding 10 ml of 2% formalin to 1 ml of EDTA- or heparin-treated blood and allowing 3 min for the RBCs to lyse. The sample is then concentrated by centrifugation, and a drop of 0.1% methylene blue is added to the sediment. The sediment is then examined for microfilariae. Membrane filter tests (Difil-Test [Evsco]) are similar, except the blood is mixed with lysing solution (PHoenix Diagnostics or Vedco Inc.) and then concentrated on a filter connected to a syringe. The blood solution is filtered, and the filter is removed from the housing and examined for microfilariae.

Examination of Serum, Plasma, or Whole Blood for *D. immitis* Antigens

All antigen detection tests currently marketed detect antigens associated with the adult female *D. immitis* reproductive tract (Fig. 4; Table 2). This causes inherent problems with these tests. The problems are (i) male-only infections are not detected, (ii) the tests do not perform well if few adult females are present (C. H. Courtney and Q. Y. Zeng, *Proc. 45th Annu. Meet. Am. Assoc. Vet. Parasitol.*, abstr. 51, p. 53, 2000; J. W. McCall, N. Supakorndej, A. R. Donoghue, and R. K. Turnbull, *Proc. 45th Annu. Meet. Am. Assoc. Vet. Parasitol.*, abstr. 52, p. 53, 2000), and (iii) tests will not detect infections until female worms have matured and begin releasing antigen into the circulation. Another consideration is that adult antigens may circulate for up to 4 months after the heartworms have died. Additionally, there are rare cases where dogs have circulating microfilariae but their antigen tests are negative. Antigen tests are usually of little value in diagnosing heartworm infections in cats because cats usually have few heartworms.

Examination of Serum, Plasma, or Whole Blood for Antibodies to *D. immitis*

Currently there are two commercially available antibody test kits to detect *D. immitis* antibodies in cats. One is based on flow technology (HESKA Solo Step FH [Heska Corp.]) and the other is an ELISA (Assure/FH [Synbiotics Corp.]). A negative antibody test indicates that the cat is not infected with *D. immitis* or has been infected for less than 60 days. A positive test indicates that (i) adult heartworms are present in the heart or pulmonary arteries; (ii) the cat is

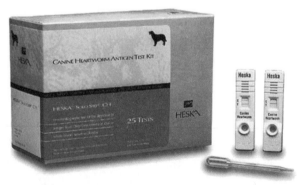

FIGURE 4 HESKA Solo Step canine heartworm antigen diagnostic test. The test is a flow technology ELISA-based test. A positive test results in two bands on the membrane (control band and test band), whereas a negative test will result in only one band (control band).

TABLE 2 Heartworm antigen tests approved for use in dogs

Test name	Manufacturer	Test type
ASSURE	Synbiotics Corporation	ELISA
DiroCHEK	Synbiotics Corporation	ELISA
HESKA Solo Step CH	Heska Corporation	Flow technology
ITC Gold HW	Synbiotics Corporation	Flow technology
PetCheck HTWM PF	IDEXX Laboratories	ELISA
Snap Canine Heartworm PF	IDEXX Laboratories	Flow technology
Uni-Tec CHW	Synbiotics Corporation	ELISA
VetRED	Synbiotics Corporation	Hemagglutination
Witness *Dirofilaria*	Synbiotics Corporation	Flow technology

infected with late-L4-stage larvae or adults; (iii) the cat was infected, but worms have been cleared; or (iv) an ectopic infection may be present. It has been suggested that antibody tests be used in conjunction with antigen tests and thoracic radiography to increase the efficacy of diagnosing heartworm infection in cats (130).

Detection of Other Nematode Parasites

Antibodies to the muscle nematode *Trichinella spiralis* can be detected in the serum and meat juice of swine by ELISA (*Trichinella* Antibody test kit [SafePath Laboratories]). Because *T. spiralis* does not cause clinical disease in pigs, this test is primarily used to ensure the safety of pork.

Serological Tests for *Neospora caninum,* *Toxoplasma gondii,* and *Leishmania* and *Babesia* Species

There are three commercially available serological tests for *N. caninum*: an ELISA (HerdChek *Neospora caninum* antibody test kit [IDEXX Laboratories]), a direct agglutination test (*Neospora caninum* serological test [Vétoquinol]), and an IFA test (*Neospora caninum* fluorescent antibody substrate slide [VMRD, Inc.]). The ELISA and agglutination tests detect IgG antibodies, and the IFA test can be modified to detect IgM. The tests are intended for use in cattle (all three tests) and dogs (IFA test). All three tests use the tachyzoite stage of the parasite as antigen. Several serological tests are available for detecting antibodies to *T. gondii* in humans. Any of the agglutination-based tests will also work with animal sera. Slides labeled with tachyzoites of *T. gondii* are available for use with animal sera in conducting IFA tests (*Toxoplasma gondii* fluorescent antibody substrate slide [VMRD, Inc.]). An ELISA kit is available to test dog serum for antibodies to *Leishmania donovani* (Helica Biosystems Inc.). Antigen slides for IFA testing are also available commercially for *Babesia bigemina* (*Babesia bigemina* positive blood slide [VMRD, Inc.]) and *Babesia bovis* (*Babesia bovis* positive blood slide [VMRD, Inc.]).

Tritrichomonas foetus

The advent of a self-contained pouch system (In Pouch TF, BIOMED Diagnostics, Inc., San Jose, Calif.) for the transportation of samples and in vitro culture of samples for *T. foetus* has greatly improved the diagnosis of this agent. This is a culture system for detecting *T. foetus* in smegma from bulls or cervical mucus from cows. The pouch system is as effective as culture in Diamond's medium or liver infusion broth (77),

and the medium is equivalent to Diamond's medium for the transportation of samples from the field to the laboratory (13). The sample is placed in the top chamber of a pouch containing culture medium. The top chamber is rolled down, and the sample is placed in the bottle chamber and incubated at 37°C. The pouch is placed on a holder and viewed daily for up to 5 days by light microscopy for the presence of motile trichomonads. Growth in positive samples is usually present by 2 days. The shelf life of the pouch system is 1 year.

IHC Staining

Primary antisera (goat origin) for IHC to detect *N. caninum* and *T. gondii* in animal tissues are available (VMRD). The sera can be used with any of the commercial immunodetection kits on the market.

Ectoparasitology

Identification of mites, lice, ticks, fleas, and fly larvae infecting animals is a common occurrence for veterinary parasitologists. Skin scrapings are used to demonstrate mites. The lesion is scraped with a scalpel blade, and the tissue is placed on a slide in a drop of glycerol or mineral oil. The larger ectoparasites are manually removed from the host and immobilized by fixation in 70% alcohol or 10% formalin solution. Ectoparasites are identified using dichotomous keys (3).

We thank Sabrina Swenson of the NVSL-Ames; Peter Woolcock of the California Animal Health and Food Safety Laboratory System, University of California—Davis, Fresno; K.-J. Yoon of Iowa State University Veterinary Diagnostic Laboratory, Ames; Michael T. Collins of the University of Wisconsin; and Anne Zajac of Virginia Polytechnic Institute and State University for providing valuable information and advice.

REFERENCES

1. **Adney, W. S., and R. L. Jones.** 1985. Evaluation of the RapID-ANA sytem for the identification of anaerobic bacteria of veterinary origin. *J. Clin. Microbiol.* **22:**980–983.
2. **Ali, A., and D. L. Reynolds.** 1999. A reverse transcription-polymerase chain reaction assay for the detection of avian pneumovirus (Colorado strain). *Avian Dis.* **43:**600–603.
3. **Anonymous.** 1984. *Pictorial Keys to Arthropods of Reptiles, Birds and Mammals of Public Health Significance.* Public Health Service, Bureau of Disease Prevention and Environmental Control, National Communicable Disease Center, Atlanta, Ga.

4. **Bannister, M. F., C. E. Benson, and C. R. Sweeney.** 1985. Rapid species identification of group C streptococci from horses. *J. Clin. Microbiol.* **21:**524–526.

5. **Belgrader, P., W. Benett, D. Hadley, J. Richards, P. Stratton, R. Mariella, and F. Milanovich.** 1999. PCR detection of bacteria in seven minutes. *Science* **284:**449–450.

6. **Benfield, D. A., I. J. Stotz, E. A. Nelson, and K. S. Groon.** 1984. Comparison of a commercial enzyme-linked immunosorbent assay with electron microscopy, fluorescent antibody, and virus isolation for the detection of bovine and porcine rotavirus. *Am. J. Vet. Res.* **45:**1998–2002.

7. **Biberstein, E. L., S. S. Jang, and D. W. Hirsch.** 1984. Species distribution of coagulase-positive staphylococci in animals. *J. Clin. Microbiol.* **19:**610–615.

8. **Bodeus, M., C. Cambiaso, M. Surleraux, and G. Burtonboy.** 1988. A latex agglutination test for the detection of canine parvovirus and corresponding antibodies. *J. Virol. Methods* **19:**1–12.

9. **Borchers, K., U. Wolfinger, B. Lawrenz, A. Schellenbach, and H. Ludwig.** 1997. Equine herpesvirus 4 DNA in trigerminal ganglia of naturally infected horses detected by direct in situ PCR. *J. Gen. Virol.* **78:**1109–1114.

10. **Bougrine, S. I., O. F. Fihri, and M. M. Fehri.** 1998. Western immunoblotting as a method for the detection of African horse sickness virus protein-specific antibodies: differentiation between infected and vaccinated horses. *Arch. Virol. Suppl.* **14:**329–336.

11. **Bowman, D. D.** 1999. *Georgis' Parasitology for Veterinarians.* W. B. Saunders Company, Philadelphia, Pa.

12. **Bridger, J. C.** 1980. Detection by electron microscopy of caliciviruses, astroviruses and rotavirus-like particles in the feces of piglets with diarrhea. *Vet. Rec.* **107:**532–533.

13. **Bryan, L. A., J. R. Campbell, A. A. Gajadhar.** 1999. Effects of temperature on the survival of *Tritrichomonas foetus* in transport, Diamond's and InPouch TF media. *Vet. Rec.* **144:**227–232

14. **Burki, F., W. Rossmanith, and E. Rossmanith.** 1992. Equine lentivirus, comparative studies on four serological tests for the diagnosis of equine infectious anemia. *Vet. Microbiol.* **33:**353–360.

15. **Carter, G. R., and J. R. Cole.** 1990. *Diagnostic Procedures in Veterinary Bacteriology and Mycology,* 5th ed. Academic Press, San Diego, Calif.

16. **Chua, K. B., W. J. Bellini, P. A. Rota, B. H. Harcourt, A. Tamin, S. K. Lam, T. G. Ksiazek, P. E. Rollin, S. R. Zaki, W. Shieh, C. S. Goldsmith, D. J. Gubler, J. T. Roehrig, B. Eaton, A. R. Gould, J. Olson, H. Field, P. Daniels, A. E. Ling, C. J. Peters, L. J. Anderson, and B. W. Mahy.** 2000. Nipah virus: a recently emergent deadly paramyxovirus. *Science* **288:**1432–1435.

17. **Collins, J. K., C. A. Riegel, J. D. Olson, and A. Fountain.** 1987. Shedding of enteric coronavirus in adult cattle. *Am. J. Vet. Res.* **48:**361–365.

18. **Collins, M. T.** 1996. Diagnosis of paratubersulosis. *Vet. Clin. N. Am. Food Anim. Pract.* **12:**357–371.

19. **Collins, M. T., K. B. Kenefick, D. C. Sockett, R. S. Lambrecht, J. McDonald, and J. B. Jorgensen.** 1990. Enhanced radiometric detection of *Mycobacterium paratuberculosis* using filter concentrated fecal specimens. *J. Clin. Microbiol.* **28:**2514–2519.

20. **Collins, M. T., and E. C. Swanson.** 1981. Use of the API 20E system to identify non-enterobacteriaceae from veterinary medical sources. *Am. J. Vet. Res.* **42:**1269–1273.

21. **Collins, M. T., N. Weaver, and R. P. Ellis.** 1981. Identification of *Pasteurella multocida* and *Pasteurella haemolytica* by API 20E, Minitek and Oxi-Ferm systems. *J. Clin. Microbiol.* **23:**433–437.

22. **Cox, H. U., J. D. Hoskins, S. S. Newman, G. H. Turnwald, C. S. Foil, A. F. Roy, and M. T. Kearney.** 1985. Distribution of staphylococcal species on clinically healthy cats. *Am. J. Vet. Res.* **46:**1824–1828.

23. **Dar, A. M., S. Kapil, and S. M. Goyal.** 1998. Comparison of immunohistochemistry, electron microscopy, and direct fluorescent antibody test for the detection of bovine coronavirus. *J. Vet. Diagn. Investig.* **10:**152–157.

24. **Elder, J., and C. Brown.** 1999. Review of techniques for the diagnosis of *Chlamydia psittaci* infection in psittacine birds. *J. Vet. Diagn. Investig.* **11:**539–541.

25. **Fenaux, M., P. G. Halbur, M. Gill, T. E. Toth, and X.-J. Meng.** 2000. Genetic characterization of type 2 porcine circovirus (PCV-2) from pigs with postweaning multisystemic wasting syndrome in different geographic regions of North America and development of a differential PCR-restriction fragment length polymorphism assay to detect and differentiate between infections with PCV-1 and PCV-2. *J. Clin. Microbiol.* **38:**2494–2503.

26. **Finlaison, D. S.** 1995. Fecal viruses of dogs—an electron microscope study. *Vet. Microbiol.* **46:**295–305.

27. **Gamble, D. A., A. Lobbiani, M. Gramegna, L. E. Moore, and G. Colucci.** 1997. Development of a nested PCR assay for detection of feline infectious peritonitis virus in clinical specimens. *J. Clin. Microbiol.* **35:**673–675.

28. **Gaudi, S., W. Ponti, A. Agresti, R. Meneveri, M. Malcovati, L. Bonizzi, G. Poli, A. Amato, and E. Ginelli.** 1990. Detection of bovine leukaemia virus (BLV) infection by DNA probe technology. *Mol. Cell. Probes* **4:**163–174.

29. **Gelmetti, D., V. Grieco, C. Rossi, L. Capucci, and A. Lavazza.** 1998. Detection of rabbit haemorrhagic disease virus (RHDV) by in situ hybridization with a digoxigenin labeled RNA probe. *J. Virol. Methods* **72:**219–226.

30. **Gilbert, S. A., K. M. Burton, S. E. Prins, and D. Deregt.** 1999. Typing of bovine viral diarrhea viruses directly from blood of persistently infected cattle by multiplex PCR. *J. Clin. Microbiol.* **37:**2020–2023.

31. **Gillette, K. G.** 1983. Enzyme-linked immunosorbent assay for serum antibody to bovine respiratory syncytial virus: comparison with complement-fixation and neutralization tests. *Am. J. Vet. Res.* **44:**2251–2255.

32. **Gottschalk, M., R. Higgins, M. Jacques, M. Beaudoin, and J. Henrichsen.** 1991. Characterization of six new capsular types (23-28) of *Streptococcus suis. J. Vet. Diagn. Investig.* **29:**2590–2594.

33. **Gottschalk, M., R. Higgins, M. Jacques, M. Beaudoin, and J. Henrichsen.** 1991. Isolation and characterization of *Streptococcus suis* capsular types 9–22. *J. Vet. Diagn. Investig.* **3:**60–65.

34. **Guo, W. Z., D. T. Shen, J. F. Evermann, and J. R. Gorham.** 1988. Comparison of an enzyme-linked immunosorbent assay and a complement-fixation test for the detection of IgG to bovine herpesvirus type 4 (bovine cytomegalovirus). *Am. J. Vet. Res.* **49:**667–670.

35. **Gut, M., C. M. Leutenegger, J. B. Huder, N. C. Pedersen, and H. Lutz.** 1999. One-tube fluorogenic reverse transcription-polymerase chain reaction for the

quantitation of feline coronaviruses. *J. Virol. Methods* **77:**37–46.

36. **Halbur, P. G., J. J. Andrews, E. L. Huffman, P. S. Paul, X.-J. Meng, and Y. Niyo.** 1994. Development of a streptavidin-biotin immunoperoxidase procedure for the detection of porcine reproductive and respiratory syndrome virus antigen in porcine lung. *J. Vet. Diagn. Investig.* **6:**254–257.

37. **Hammami, S., A. E. Castro, and B. I. Osburn.** 1990. Comparison of polyacrylamide gel electrophoresis, an enzyme-linked-immunosorbent assay, and an agglutination test for the direct identification of bovine rotavirus from feces and coelectrophoresis of viral RNAs. *J. Vet. Diagn. Investig.* **2:**184–190.

38. **Hassan, M. K., Y. M. Saif, and S. Shawky.** 1996. Comparison between antigen-capture ELISA and conventional methods used for titration of infectious bursal disease virus. *Avian Dis.* **40:**562–566.

39. **Hawkes, R. A., P. D. Kirkland, D. A. Sanders, F. Zhang, Z. Li, R. J. Davis, and N. Zhang.** 2000. Laboratory and field studies of an antigen capture ELISA for bluetongue virus. *J. Virol. Methods* **85:**137–149.

40. **Haynes, J. S., P. G. Halbur, T. Sirinarumitr, P. S. Paul, X.-J. Meng, and E. L. Huffman.** 1997. Temporal and morphologic characterization of the distribution of porcine reproductive and respiratory syndrome virus (PRRSV) by in situ hybridization in pigs infected with isolates of PRRSV that differ in virulence. *Vet. Pathol.* **34:**39–43.

41. **Heckert, R. A., L. J. Saif, and G. W. Myers.** 1989. Development of protein A-gold immunoelectron microscopy for detection of bovine coronavirus in calves: comparison with ELISA and direct immunofluorescence of nasal epithelial cells. *Vet. Microbiol.* **19:**217–231.

42. **Henderson, K. S., and D. J. Jackwood.** 1990. Comparison of the dot blot hybridization assay with antigen detection assays for the diagnosis of infectious bursal disease virus infections. *Avian Dis.* **34:**744–748.

43. **Herrmann, J. E.** 1995. Immunoassays for the diagnosis of infectious diseases, p. 110–122. *In* P. R. Murray, E. J. Baron, M. A. Pfaller, F. C. Tenover, and R. H. Yolken (ed.), *Manual of Clinical Microbiology*, 6th ed. American Society for Microbiology, Washington, D.C.

44. **Hierholzer, J. C.** 1991. Rapid diagnosis of viral infection, p. 556–590. *In* A. Vaheri and A. Balows (ed.), *Rapid Methods and Automation in Microbiology and Immunology.* Springer-Verlag, New York, N.Y.

45. **Hohdatsu, T., K. Motokawa, M. Usami, M. Amioka, S. Okada, and H. Koyama.** 1998. Genetic subtyping and epidemiological study of feline immunodeficiency virus by nested polymerase chain reaction-restriction fragment length polymorphism analysis of the gag gene. *J. Virol. Methods* **70:**107–111.

46. **Horiuchi, M., K. Yuri, T. Soma, H. Katae, H. Nagasawa, and M. Shinagawa.** 1996. Differentiation of vaccine virus from field isolates of feline panleukopenia virus by polymerase chain reaction and restriction fragment length polymorphism analysis. *Vet. Microbiol.* **53:**283–293.

47. **Hughes, J. H.** 1993. Physical and chemical methods for enhancing rapid detection of viruses and other agents. *Clin. Microbiol. Rev.* **6:**150–175.

48. **Inzana, T. J., and B. Iritani.** 1989. Rapid detection of group C streptococci from animals by latex agglutination. *J. Clin. Microbiol.* **27:**309–312.

49. **Ishikawa, K., H. Sekiguchi, T. Ogino, and S. Suzuki.** 1997. Direct and rapid detection of porcine epidemic diarrhea virus by RT-PCR. *J. Virol. Methods* **69:**191–195.

50. **Ivens, V. R., D. L. Mark, and N. D. Levine.** 1978. *Principal Parasites of Domestic Animals in the United States. Biological and Diagnostic Information.* University of Illinois Press, Urbana-Champaign.

51. **Jackson, M. L., D. M. Haines, S. M. Meric, and V. Misra.** 1993. Feline leukemia virus detection by immunohistochemistry and polymerase chain reaction in formalin-fixed, paraffin-embedded tumor tissue from cats with lymphosarcoma. *Can. J. Vet. Res.* **57:**269–276.

52. **James, K.** 1990. Immunoserology of infectious diseases. *Clin. Microbiol. Rev.* **3:**132–152.

53. **Jasper, D. E., F. Infante, and J. D. Dellinger.** 1985. Efficacy of the API Staph-Ident system for identification of *Staphylococcus* species from milk. *Am. J. Vet. Res.* **46:**1263–1267.

54. **Jayarao, B., S. P. Oliver, K. R. Matthews, and S. H. King.** 1991. Comparative evaluation of Vitek gram-positive identification system and API Rapid Strep system for identification of *Streptococcus* species of bovine origin. *Vet. Microbiol.* **26:**301–308.

55. **Johnson, F. B.** 1990. Transport of viral specimens. *Clin. Microbiol. Rev.* **3:**120–131.

56. **Johnson, M., F. Rommel, and J. Mone.** 1998. Development of a syncytia inhibition assay for the detection of antibodies to bovine leukemia virus in naturally infected cattle; comparison with Western blot and agar gel immunodiffusion. *J. Virol. Methods* **70:**177–182.

57. **Jones, R. L., W. S. Adney, M. A. Davis, H. Vonbyren, and G. Thompson.** 1987. Evaluation of Quantum II microbiology system for identification of gram-negative bacteria of veterinary origin. *J. Clin. Microbiol.* **25:**2071–2074.

58. **Kessler, C., H. J. Holtke, R. Seibl, J. Burg, and K. Muhlegger.** 1990. Non-radioactive labeling and detection of nucleic acids. I. A novel DNA labeling and detection system based on digoxigenin: anti-digoxigenin ELISA principle (digoxigenin system). *Biol. Chem. Hoppe-Seyler* **371:**917–927.

59. **Kies, L. M., M. C. DeBey, and R. F. Ross.** 1991. Identification of porcine mycoplasmas using commercial enzyme assay systems. *J. Vet. Diagn. Investig.* **3:**348–350.

60. **Kim, H. S., H. S. Joo, W. T. Christianson, and R. B. Morrison.** 1991. Evaluation of serologic methods for the detection of antibodies to encephalomyocarditis virus in swine fetal thoracic fluids. *J. Vet. Diagn. Investig.* **3:**283–286.

61. **Kim, O., and C. Chae.** 2000. In situ hybridization for the detection and localization of porcine epidemic diarrhea virus in the intestinal tissues from naturally infected piglets. *Vet. Pathol.* **37:**62–67.

62. **Kloos, W. E., and J. F. Wolfshohl.** 1982. Identification of *Staphylococcus* species with the API STAPH-IDENT system. *J. Clin. Microbiol.* **16:**509–516.

63. **Krell, P. J., T. Salas, and R. P. Johnson.** 1988. Mapping of porcine parvovirus DNA and development of a diagnostic DNA probe. *Vet. Microbiol.* **17:**29–43.

64. **Langer, P. R., A. A. Waldrop, and D. C. Ward.** 1981. Enzymatic synthesis of biotin-labeled polynucleotides: novel nucleic acid affinity probes. *Proc. Natl. Acad. Sci. USA* **78:**6633–6637.

65. **Langlois, B. E., R. J. Harmon, and K. Akers.** 1983. Identification of *Staphylococcus* species of bovine origin

with the API Staph-Ident system. *J. Clin. Microbiol.* **18:**1212–1219.

66. **Langlois, B. E., R. J. Harmon, and K. Akers.** 1984. Identification of *Staphylococcus* species of bovine origin with the DMS Staph-Trac system. *J. Clin. Microbiol.* **20:**227–230.

67. **Larochelle, R., H. Mardassi, S. Dea, and R. Magar.** 1996. Detection of porcine reproductive and respiratory syndrome virus in cell cultures and formalin-fixed tissues by in situ hybridization using a digoxigenin-labeled probe. *J. Vet. Diagn. Investig.* **8:**3–10.

68. **Larochelle, R., M. Antaya, M. Morin, and R. Magar.** 1999. Typing of porcine circovirus in clinical specimens by multiplex PCR. *J. Virol. Methods* **80:**69–75.

69. **Latimer, K. S., F. D. Niagro, O. C. Williams, A. Ramis, M. A. Goodwin, B. W. Ritchie, and R. P. Campagnoli.** 1997. Diagnosis of avian adenovirus infections using DNA in situ hybridization. *Avian Dis.* **41:**773–782.

70. **Lennette, D. A.** 1995. Collection and preparation of specimens for virological examination, p. 868–875. *In* P. R. Murray, E. J. Baron, M. A. Pfaller, F. C. Tenover, and R. H. Yolken (ed.), *Manual of Clinical Microbiology,* 6th ed. American Society for Microbiology, Washington, D.C.

71. **Lennette, E. H., P. Halonen, and F. A. Murphy.** 1988. *Laboratory Diagnosis of Infectious Diseases. Principles and Practices,* vol. 2. Springer-Verlag, New York, N.Y.

72. **Leutenegger, C. M., D. Klein, R. Hofmann-Lehmann, C. Mislin, U. Hummel, J. Boni, F. Boretti, W. H. Guenzburg, and H. Lutz.** 1999. Rapid feline immunodeficiency virus provirus quantitation by polymerase chain reaction using the TaqMan fluorogenic real-time detection system. *J. Virol. Methods* **78:**105–116.

73. **Levy, D., L. Deshayes, B. Guillemain, and A. L. Parodi.** 1977. Bovine leukemia virus specific antibodies among French cattle. I. Comparison of complement fixation and hematological tests. *Int. J. Cancer* **19:**822–827.

74. **Lin, T. M., G. Y. Shi, S. J. Jiang, C. F. Tsai, B. J. Hwang, C. T. Hsieh, and H. L. Wu.** 1999. Persistent infection of bovine herpesvirus type 4 in bovine endothelial cell cultures. *Vet. Microbiol.* **70:**41–53.

75. **Liu, H. J., M. H. Liao, C. D. Chang, J. H. Chen, M. Y. Lin, and M. C. Tung.** 1999. Comparison of two molecular techniques for the detection of avian reoviruses in formalin-fixed, paraffin-embedded chicken tissues. *J. Virol. Methods* **80:**197–201.

76. **Liu, X., J. J. Giambrone, and E. J. Hoerr.** 2000. In situ hybridization, immunohistochemistry, and in situ reverse transcription-polymerase chain reaction for detection of infectious bursal disease virus. *Avian Dis.* **44:**161–169.

77. **Lun, Z., S. Parker, and A. A. Gajadhar.** 2000. Comparison of growth rates of *Tritrichomonas foetus* isolates from various geographic regions using three different culture media. *Vet. Parasitol.* **89:**199–208.

78. **Maddox, R. L., and G. Koehne.** 1982. Identification of *Staphylococcus hyicus* with the API Staph strip. *J. Clin. Microbiol.* **15:**984–986.

79. **Magar, R., and R. Larochelle.** 1992. Immunohistochemical detection of porcine rotavirus using immunogold silver staining (IGSS). *J. Vet. Diagn. Investig.* **4:**3–7.

80. **Magar, R., R. Larochelle, Y. Robinson, and C. Dubuc.** 1993. Immunohistochemical detection of porcine reproductive and respiratory syndrome virus using colloidal gold. *Can. J. Vet. Res.* **57:**300–304.

81. **Mandell, G. L., J. E. Bennett, and R. Dolin.** 1995. *Mandell, Douglas, and Bennett's Principles and Practice of Infectious Diseases,* 4th ed., vol 1. Churchill Livingstone, New York, N.Y.

82. **Manni, V., F. Roperto, G. Di Guardo, D. Galati, R. U. Condoleo, and A. Venuti.** 1998. Presence of papillomavirus-like DNA sequences in cutaneous fibropapillomas of the goat udder. *Vet. Microbiol.* **61:**1–6.

83. **Marshall, J. A., M. L. Kennett, S. M. Rodger, M. J. Studdert, W. L. Thompson, and I. D. Gust.** 1987. Virus and virus-like particles in the feces of cats with and without diarrhea. *Aust. Vet. J.* **64:**100–105.

84. **Martinez, M. L., and R. C. Weiss.** 1993. Detection of feline infectious peritonitis virus infection in cell cultures and peripheral blood mononuclear leukocytes of experimentally infected cats using a biotinylated cDNA probe. *Vet. Microbiol.* **34:**259–271.

85. **Martini, M., G. Capelli, G. Poglayen, F. Bertotti, and C. Turilli.** 1996. The validity of some haematological and ELISA methods for the diagnosis of canine heartworm disease. *Vet. Res. Commun.* **20:**331–339.

86. **Matthews, K. R., S. P. Oliver, and S. H. King.** 1990. Comparison of Vitek gram-positive identification system with API Staph-Trac for species identification of staphylococci of bovine origin. *J. Clin. Microbiol.* **28:**1649–1651.

87. **McGoldrick, A., J. P. Lowings, G. Ibata, J. J. Sands, S. Belak, and D. J. Paton.** 1998. A novel approach to the detection of classical swine fever virus by RT-PCR with a fluorogenic probe (TaqMan). *J. Virol. Methods* **72:**125–135.

88. **McIntosh, K.** 1996. Diagnostic virology, p. 401–430. *In* B. N. Fields, D. M. Knipe, P. M. Howley, R. M. Chanock, J. L. Melnick, T. P. Monath, B. Roizman, and A. E. Straus (ed.), *Fields Virology,* 3rd ed. Lippincott Williams & Wilkins, Philadelphia, Pa.

89. **McNeilly, F., S. Kennedy, D. Moffett, B. M. Meehan, J. C. Foster, E. G. Clarke, J. A. Ellis, D. M. Haines, B. M. Adair, and G. M. Allan.** 1999. A comparison of in situ hybridization and immunohistochemistry for the detection of a new porcine circovirus in formalin-fixed tissues from pigs with post-weaning multisystemic wasting syndrome (PMWS). *J. Virol. Methods* **80:**123–128.

90. **Meng, X.-J., P. S. Paul, E. M. Vaughn, and J. J. Zimmerman.** 1993. Development of a radiolabeled nucleic acid probe for the detection of encephalomyocarditis virus of swine. *J. Vet. Diagn. Investig.* **5:**254–258.

91. **Meng, X.-J., P. S. Paul, P. G. Halbur, and M. A. Lum.** 1996. Characterization of a high-virulence US isolate of porcine reproductive and respiratory syndrome virus in a continuous cell line, ATCC CRL11171. *J. Vet. Diagn. Investig.* **8:**374–381.

92. **Moody, A., S. Sellers, and N. Bumstead.** 2000. Measuring infectious bursal disease virus RNA in blood by multiplex real-time quantitative RT-PCR. *J. Virol. Methods* **85:**55–64.

93. **Morrison, J. R. A., and G. S. Tillotson.** 1988. Identification of *Actinomyces* (*Corynebacterium*) *pyogenes* with the API 20 Strep system. *J. Clin. Microbiol.* **26:**1865–1866.

94. **Mulder, W. A., F. van Poelwijk, R. J. Moormann, B. Reus, G. L. Kok, J. M. Pol, and A. Dekker.** 1997. Detection of early infection of swine vesicular disease virus in porcine cells and skin sections. A comparison of immunohistochemistry and in-situ hybridization. *J. Virol. Methods* **68:**169–175.

95. Murphy, F. A., E. P. J. Gibbs, M. C. Horzinek, and M. J. Studdert. 1999. *Veterinary Virology*, 3rd ed. Academic Press, New York, N.Y.

96. Nussbaum, D. J., J. R. Salord, and D. D. Rimmele. 1999. Evaluation of quantitative latex agglutination for detection of Cryptosporidium parvum, E. coli K99, and rotavirus in calf feces. *J. Vet. Diagn. Investig.* 11:314–318.

97. Oliver, R. E., A. I. Donaldson, C. F. Gibson, P. L. Roeder, P. M. Le Blanc Smith, and C. Hamblin. 1988. Detection of foot-and-mouth disease antigen in bovine epithelial samples: comparison of sites of sample collection by an enzyme linked immunosorbent assay (ELISA) and complement fixation test. *Res. Vet. Sci.* 44:315–319.

98. Palmer, E. L., and M. L. Martin. 1988. *Electron Microscopy in Viral Diagnosis*. CRC Press, Boca Raton, Fla.

99. Papp, J. R., and C. A. Muckle. 1991. Evaluation of the Sceptor system for identification of bacteria of veterinary origin. *J. Clin. Microbiol.* 29:10–15.

100. Parwani, A. V., B. I. Rosen, J. Flores, M. A. McCrae, M. Gorziglia, and L. J. Saif. 1992. Detection and differentiation of bovine group A rotavirus serotypes using polymerase chain reaction-generated probes to the VP7 gene. *J. Vet. Diagn. Investig.* 4:148–158.

101. Patten, V. H., S. J. Shin, J. Cole, C. W. Watson, and W. H. Fales. 1995. Evaluation of a commercial automated system and software for the identification of veterinary bacterial isolates. *J. Vet. Diagn. Investig.* 7:506–508.

102. Patterson, J. S., R. K. Maes, T. P. Mullaney, and C. L. Benson. 1996. Immunohistochemical diagnosis of eastern equine encephalomyelitis. *J. Vet. Diagn. Investig.* 8:156–160.

103. Peele, D., J. Bradford, W. Pryor, and S. Vore. 1997. Comparison of identifications of human and animal source gram-negative bacteria by API 20E and Crystal E/NF systems. *J. Clin. Microbiol.* 35:213–216.

104. Podzorski, R. P., and D. H. Persing. 1995. Molecular detection and identification of microorganisms, p. 130–157. *In* P. R. Murray, E. J. Baron, M. A. Pfaller, F. C. Tenover, and R. H. Yolken (ed.), *Manual of Clinical Microbiology*, 6th ed. American Society for Microbiology, Washington, D.C.

105. Post, K. W., S. C. Murphy, J. B. Boyette, and P. M. Resseguie. 1999. Evaluation of a commercial system for the identification of Ornithobacterium rhinotracheale. *J. Vet. Diagn. Investig.* 11:97–99.

106. Poutrel, B., and H. Z. Ryniewicz. 1984. Evaluation of the API 20 Strep system for species identification of streptococci isolated from bovine mastitis. *J. Clin. Microbiol.* 19:213–214.

107. Pratelli, A., M. Tempesta, G. Greco, V. Martella, and C. Buonavoglia. 1999. Development of a nested PCR assay for the detection of canine coronavirus. *J. Virol. Methods* 80:11–15.

108. Quinn, P. J., J. C. Donnelly, M. E. Carter, B. K. J. Markey, P. R. Torgerson, and R. M. S. Breathnach. 1997. *Microbial and Parasitic Diseases of the Dog and Cat*. W. B. Saunders, London, United Kingdom.

109. Rather, P. N., A. P. Davis, and B. J. Wilkinson. 1986. Slime production by bovine milk Staphylococcus aureus and identification of coagulase-negative staphylococcal isolates. *J. Clin. Microbiol.* 23:858–862.

110. Renshaw, R. W., R. Ray, and E. J. Dubovi. 2000. Comparison of virus isolation and reverse transcription polymerase chain reaction assay for detection of bovine viral diarrhea virus in bulk milk tank samples. *J. Vet. Diagn. Investig.* 12:184–186.

111. Reubel, G. H., B. S. Crabb, and M. J. Studdert. 1995. Diagnosis of equine gammaherpesvirus 2 and 5 infections by polymerase chain reaction. *Arch. Virol.* 140:1049–1060.

112. Rimstad, E., N. East, E. DeRock, J. Higgins, and N. C. Pedersen. 1994. Detection of antibodies to caprine arthritis-encephalitis virus using recombinant gag proteins. *Arch. Virol.* 134:345–356.

113. Rocha, M. A., E. F. Barbosa, S. E. Guimaraes, E. D. Neto, and A. M. Gouveia. 1998. A high sensitivity-nested PCR assay for BHV-1 detection in semen of naturally infected bulls. *Vet. Microbiol.* 63:1–11.

114. Salmon, S. A., J. L. Watts, R. D. Walker, and R. J. Yancey. 1995. Evaluation of a commercial system for the identification of gram-negative, nonfermenting bacteria of veterinary importance. *J. Vet. Diagn. Investig.* 7:161–164.

115. Salmon, S. A., J. L. Watts, and R. J. Yancey. 1993. Evaluation of the RapID NH system for identification of *Haemophilus somnus, Pasteurella multocida, Pasteurella haemolytica* and *Actinobacillus pleuropneumoniae* isolated from cattle and pigs with respiratory disease. *J. Clin. Microbiol.* 31:1362–1363.

116. Sander, J., R. Williams, R. Novak, and W. Ragland. 1997. In situ hybridization on blood smears for diagnosis of chicken anemia virus in broiler breeder flocks. *Avian Dis.* 41:988–992.

117. Sanderson, T. P., and A. A. Andersen. 1992. Evaluation of a commercial solid-phase enzyme immunoassay for the detection of ovine *Chlamydia psittaci. J. Vet. Diagn.* 4:192–193.

118. Sanderson, T. P., and A. A. Andersen. 1989. Evaluation of an enzyme immunoassay for detection of *Chlamydia psittaci* in vaginal secretions, placentas, and fetal tissues from aborting ewes. *J. Vet. Diagn.* 1:309–315.

119. Sandvik, T., and J. Krogsrud. 1995. Evaluation of an antigen-capture ELISA for detection of bovine viral diarrhea virus in cattle blood samples. *J. Vet. Diagn. Investig.* 7:65–71.

120. Sanekata, T., E. Kishimoto, K. Sato, H. Honma, K. Otsuki, and M. Tsubokura. 1991. Detection of porcine rotavirus in stools by a latex agglutination test. *Vet. Microbiol.* 27:245–251.

121. Scheld, W. M., D. Armstrong, and J. M. Hughes (ed.). 1997. *Emerging Infections 1*. ASM Press, Washington, D.C.

122. Scheld, W. M., W. A. Craig, and J. M. Hughes (ed.). 1998. *Emerging Infections 2*. ASM Press, Washington, D.C.

123. Scheld, W. M., W. A. Craig, and J. M. Hughes (ed.). 1999. *Emerging Infections 3*. ASM Press, Washington, D.C.

124. Schoenthaler, S. L., and S. Kapil. 1999. Development and applications of a bovine coronavirus antigen detection enzyme-linked immunosorbent assay. *Clin. Diagn. Lab. Immunol.* 6:130–132.

125. Schulze, C., and W. Baumgartner. 1998. Nested polymerase chain reaction and in situ hybridization for diagnosis of canine herpesvirus infection in puppies. *Vet. Pathol.* 35:209–217.

126. **Sentsui, H., T. Nishimori, I. Nagai, and N. Nishioka.** 1996. Detection of sheep-associated malignant catarrhal fever virus antibodies by complement fixation tests. *J. Vet. Med. Sci.* **58:**1–5.

127. **Shockley, L. J., P. A. Kapke, W. Lapps, D. A. Brian, L. N. Potgieter, and R. Woods.** 1987. Diagnosis of porcine and bovine enteric coronavirus infections using cloned cDNA probes. *J. Clin. Microbiol.* **25:**1591–1596.

128. **Simard, C., S. Richardson, P. Dixon, C. Belanger, and P. Maxwell.** 2000. Enzyme-linked immunosorbent assay for the diagnosis of bovine leukosis: comparison with the agar gel immunodiffusion test approved by the Canadian Food Inspection Agency. *Can. J. Vet. Res.* **64:**101–106.

129. **Sloss, M. W., R. L. Kemp, and A. M. Zajac.** 1994. *Veterinary Clinical Parasitology.* Iowa State University Press, Ames.

130. **Snyder, P. S., J. K. Levy, M. E. Salute, S. P. Gorman, P. S. Kubilis, P. W. Smail, and L. L. George.** 2000. Performance of serologic tests used to detect heartworm infection in cats. *J. Am. Vet. Med. Assoc.* **216:**693–700.

131. **Sockett, D. C., D. J. Carr, and M. T. Collins.** 1992. Evaluation of conventional and radiometric fecal culture and a commercial DNA probe for diagnosis of *Mycobacterium paratuberculosis* infections in cattle. *Can. J. Vet. Res.* **56:**148–153.

132. **Southern, E. M.** 1975. Detection of specific sequences among DNA fragments separated by gel electrophoresis. *J. Mol. Biol.* **98:**503–517.

133. **Swanson, E. C., and M. T. Collins.** 1980. Use of the API 20E system to identify veterinary *Enterobacteriaceae. J. Clin. Microbiol.* **12:**10–14.

134. **Sweeney, R. W., R. H. Whitlock, C. L. Buckley, and P. A. Spencer.** 1995. Evaluation of a commercial enzyme-linked immunosorbent assay for the diagnosis of paratubersulosis in dairy cattle. *J. Vet. Diagn. Investig.* **7:**488–493.

135. **Teska, J. D., E. B. Shotts, and T. Hsu.** 1989. Automated biochemical identification of bacterial fish pathogens using the Abbott Quantum II. *J. Wildl. Dis.* **25:**103–107.

136. **Ture, O., Y. M. Saif, and D. J. Jackwood.** 1998. Restriction fragment length polymorphism analysis of highly virulent strains of infectious bursal disease viruses from Holland, Turkey, and Taiwan. *Avian Dis.* **42:**470–479.

137. **Van Reeth, K.** 1997. Pathogenesis and clinical aspects of a respiratory porcine reproductive and respiratory syndrome virus infection. *Vet. Microbiol.* **55:**223–230.

138. **Vanrompay, D., A. van Nerom, R. Ducatelle, and F. Haesebrouck.** 1994. Evaluation of five immunoassays for detection of *Chlamydia psittaci* in cloacal and conjunctival specimens from Turkeys. *J. Clin. Microbiol.* **32:**1470–1474.

139. **Vincent, L. L., B. H. Janke, P. S. Paul, and P. G. Halbur.** 1997. A monoclonal-antibody-based immunohistochemical method for the detection of swine influenza virus in formalin-fixed, paraffin-embedded tissues. *J. Vet. Diagn. Investig.* **9:**191–195.

140. **Wagstrom, E. A., K. J. Yoon, C. Cook, and J. J. Zimmerman.** 2000. Diagnostic performance of a reverse transcription-polymerase chain reaction test for porcine reproductive and respiratory syndrome virus. *J. Vet. Diagn. Investig.* **12:**75–78.

141. **Walker, R. L., R. J. Greene, and T. M. Gerig.** 1990. Evaluation of four commercial anaerobic systems for identification of *Eubacterium suis. J. Vet. Diagn. Investig.* **2:**318–322.

142. **Wang, C., B. Miguel, F. W. Austin, and R. W. Keirs.** 1999. Comparison of the immunofluorescent assay and reverse transcription-polymerase chain reaction to detect and type infectious bronchitis virus. *Avian Dis.* **43:**590–596.

143. **Wang, X., and M. I. Khan.** 1999. A multiplex PCR for Massachusetts and Arkansas serotypes of infectious bronchitis virus. *Mol. Cell. Probes* **13:**1–7.

144. **Watts, J. L.** 1989. Evaluation of the Minitek grampositive set for identification of streptococci isolated from bovine mammary glands. *J. Clin. Microbiol.* **27:**1008–1010.

145. **Watts, J. L.** 1989. Evaluation of the Rapid Strep system for identification of gram-positive, catalase negative cocci isolated from bovine intramammary infections. *J. Dairy Sci.* **72:**2728–2732.

146. **Watts, J. L., and S. C. Nickerson.** 1986. A comparison of the Staph-Ident and the Staph-Trac systems to conventional methods in the identification of staphylococci isolated from bovine udders. *Vet. Microbiol.* **12:**179–187.

147. **Watts, J. L., W. E. Owens, and S. C. Nickerson.** 1986. Evaluation of the Minitek gram-positive set for identification of staphylococci isolated from the bovine mammary gland. *J. Clin. Microbiol.* **23:**873–875.

148. **Watts, J. L., J. W. Pankey, and S. C. Nickerson.** 1984. Evaluation of the Staph-Ident and STAPH-ase systems for identification of staphylococci from bovine mammary infections. *J. Clin. Microbiol.* **20:**448–452.

149. **Watts, J. L., and R. J. Yancey.** 1994. Identification of veterinary pathogens by use of commercial identification systems and new trends in antimicrobial susceptibility testing of veterinary pathogens. *Clin. Microbiol. Rev.* **7:**346–356.

150. **Watts, J. L., M. M. Chengappa, J. R. Cole, J. M. Cooper, T. J. Inzana, M. R. Plaunt, T. R. Shrycok, C. Thornsberry, R. D. Walker, and C. C. Wu.** 1999. *Development of In Vitro Susceptibility Testing Criteria and Quality Control Parameters for Veterinary Antimicrobial Agents.* Approved Guideline. Document M37-A. National Committee for Clinical Laboratory Standards, Wayne, Pa.

151. **Watts, J. L., M. M. Chengappa, J. R. Cole, J. M. Cooper, T. J. Inzana, M. R. Plaunt, T. R. Shrycok, C. Thornsberry, R. D. Walker, and C. C. Wu.** 1999. *Performance Standards for Antimicrobial Disk and Dilution Susceptibility Tests for Bacteria Isolated from Animals.* Approved Standard. Document M31-A. National Committee for Clinical Laboratory Standards, Wayne, Pa.

152. **Wesley, R. D., W. L. Mengeling, K. M. Lager, D. F. Clouser, J. G. Landgraf, and M. L. Frey.** 1998. Differentiation of a porcine reproductive and respiratory syndrome virus vaccine strain from North American field strains by restriction fragment length polymorphism analysis of ORF 5. *J. Vet. Diagn. Investig.* **10:**140–144.

153. **West, K., J. Bogdan, A. Hamel, G. Nayar, P. S. Morley, D. M. Haines, and J. A. Ellis.** 1998. A comparison of diagnostic methods for the detection of bovine respiratory syncytial virus in experimental clinical specimens. *Can. J. Vet. Res.* **62:**245–250.

154. **Whipple, D. L., D. R. Callahan, and J. L. Jarnigan.** 1991. Cultivation of *Mycobacterium paratuberculosis* from bovine fecal specimens and a suggested standardized procedure. *J. Vet. Diagn. Investig.* **3:**368–373.

155. **Whittington, R. J., I. Marsh, S. McAllister, M. J. Turner, D. J. Marshall, and C. A. Fraser.** 1999. Evaluation of modified BACTEC 12B radiometric medium and solid media for culture of *Mycobacterium avium* subsp. *paratuberculosis* from sheep. *J. Clin. Microbiol.* **37:**1077–1083.

156. **Wood, M. M., and P. Timms.** 1992. Comparison of nine antigen detection kits for diagnosis of urogenital infections due to *Chlamydia psittaci* in koalas. *J. Clin. Microbiol.* **30:**3200–3205.

157. **Zhang, S., W. Xue, C. Wood, Q. Chen, S. Kapil, and H. C. Minocha.** 1997. Detection of bovine immunodeficiency virus antibodies in cattle by western blot assay with recombinant gag protein. *J. Vet. Diagn. Investig.* **9:**347–351.

158. **Zhang, Z., G. A. Andrews, C. Chard-Bergstrom, H. C. Minocha, and S. Kapil.** 1997. Application of immunohistochemistry and in situ hybridization for detection of bovine coronavirus in paraffin-embedded, formalin-fixed intestines. *J. Clin. Microbiol.* **35:**2964–2965.

159. **Zhou, E. M., M. Chan, R. A. Heckert, J. Riva, and M. F. Cantin.** 1998. Evaluation of a competitive ELISA for detection of antibodies against avian influenza virus nucleoprotein. *Avian Dis.* **42:**517–522.

A Comparison of Current Laboratory Information Systems

CHARLES T. LADOULIS, ALLAN L. TRUANT, AND RAY D. ALLER

15

It is becoming increasingly important in today's clinical laboratory medicine environment to maximize the quality, accuracy, and efficiency of one's laboratory (8). Further, many laboratories are experiencing increased testing due in part to the burgeoning area of molecular probe testing, molecular amplification testing, increasing managed care environments, and laboratory mergers and acquisitions. The increase in volume, therefore, adds increasing pressure on laboratory resources, placing additional emphasis on laboratory information systems (LISs) to be optimally designed and utilized.

The clinical laboratory computer, or as it is now commonly called, the LIS, has rapidly become one of the most important features of a modern laboratory. Order entry, patient and physician demographics, preliminary reporting, instrument interfacing, final reporting, laboratory statistics, workload recording, billing, external fax reporting, and public health reporting are only some of the important functions for which many laboratories now commonly use LISs (5). In the clinical microbiology laboratory, computers have been used in the organization of the laboratory's functions and in the reporting of results for more than 30 years, but in the past decade, computers have become an essential component of the modern clinical microbiology laboratory. The relatively recent introduction of bar coding for specimen accession and reporting has been a further improvement and significant time-saving feature of many LISs (7).

One of the most important criteria for choosing an LIS is selecting a company that will be a reliable long-term business partner that supports and tailors its products to meet a laboratory's changing needs. Occasionally, laboratory professionals may be entranced by flashy features, leading-edge technology, or charming salespeople. Therefore, there is no substitute for speaking with several of an LIS vendor's short- and long-term clients to assess how willingly and enthusiastically the company supports the evolving business model of each client laboratory. In some instances a laboratory's capabilities are limited by its information system because it selected a system based on features, technology, or popularity, not on the company's track record of support.

Another common theme that lingers through the years is the tendency of laboratory professionals to use only a small portion of the features of their systems. We are all very busy, and increasingly more so in recent years, and often we are unable to commit to educational activities related to our own LIS. We encourage laboratories to take advantage of vendor consultants, occasionally at no-cost options, to derive increased value from their products.

Despite being interested in increasing the value of their systems and improving their service, many hospital administrators still do not comprehend the benefit of sending laboratory staff to LIS user group meetings. Responsible system managers will bring to the workplace and implement technology improvements worth 10 to 100 times the cost of their attendance at such meetings. In addition, implementation of many functions will significantly improve patient care and infection control activities. Offered at many user group meetings are details of dozens of "wraparound" techniques for extending the capabilities of LISs beyond the vendor's design. These include creative use of full-keyboard and function key bar coding, scripting, macros, and print stream parsing and extraction tools.

The systems profiled in the present chapter (2, 6) vary wildly in scope and price (Table 1 to 13). They range from basic single-user products for less than $1,000 to complex multisite systems that cost millions of dollars. Some laboratories may get more benefit from a simple lower-cost system than a complex one that takes years to configure and install and for staff to learn to use. The major cost factor in switching from one large LIS to another is not the purchase or installation fees the vendor charges but the immense amount of work the laboratory must perform to train staff and remap its workload and laboratory description from one information model to a different one.

With regard to recent industry developments, the information provided herein indicates that a growing number of vendors are providing LOINC support (see Abbreviations below; also see references 3 and 4). Some vendors are developing, and a few are implementing, capabilities to automatically transmit reports of notifiable diseases to public health agencies.

Because the vast majority of LISs incorporate many, if not most, of the sections of laboratory medicine and other areas of surgical and anatomic pathology, we attempt in this chapter to detail the LIS characteristics of each system for all laboratories from which such data are available. Features are included for chemistry, hematology, microbiology, blood bank donor center, blood bank transfusion service, surgical pathology, cytology, and outreach or commercial laboratories.

In very select circumstances, a stand-alone microbiology LIS may be considered and judged optimal by certain laboratories (1). Public health laboratories or specialty microbiology laboratories in academic medical centers or in industry may be examples of laboratories that find such stand-alone LISs optional. It is usually advisable, in our opinion, to consider consolidating efforts with other laboratories and having a single LIS. If, however, a general-purpose LIS has been chosen and the microbiology module has been found to be inefficient or inadequate to support the needs of the microbiology laboratory, a stand-alone microbiology system is occasionally considered.

It is important to note that all data listed in this chapter have been provided by the vendors. We encourage the reader not only to verify the information provided herein but also to learn of all new or added functions and capabilities before making purchasing decisions.

ABBREVIATIONS

The following abbreviations are used in Table 1 to 13: A/D/T, admissions, discharge, transfer data; ANSI, American National Standards Institute; ASTM, American Society for Testing and Materials; CDC, Centers for Disease Control and Prevention; HIS, hospital information system; HL7, an information system interface communication standard, the Health Level Seven; HTML, hypertext markup language; LOINC, the Laboratory Observation Identifier Names and Codes database and its test-naming convention (a standard set of names and codes for laboratory test results for automated data pooling, including multi-institutional research and cross-facility patient care); MS, Microsoft; NCCLS, National Committee for Clinical Laboratory Standards; RDBMS, relational database management system; SNOMED, the Systematized Nomenclature of Medical Disease; SQL, structured query language.

TABLE 1 LISs from Accent Information Systems, Antek, and Cerner

Parameter	Data for LIS from:			
	Accent Information Systems	Antek	Cerner Citation	Cerner Corp.
Contact information	Kurt Balhorn 2707 N. 108th St., Ste 100 Omaha, NE 68164 (402) 733-2700 http://www.nelsondata.com	Richard Jefferson 220 Business Center Dr. Reisterstown, MD 21136 (800) 359-0911 http://www.LabDAQ.com	Julie Hull 2800 Rockcreek Pkwy. Kansas City, MO 64117-2551 (816) 201-1024 http://www.cerner.com	Julie Hull 2800 Rockcreek Pkwy. Kansas City, MO 64117-2551 (816) 201-1024 http://www.cerner.com
Name of system	AccentLab	LabDAQ	LIS, CLAB, CLAB Plus	PathNet HNA Classic
No. of contracts for sites operating system	21	944	278	387
U.S. hospital contracts	21	122		363
U.S. independent laboratory contracts	0	822		18
Clinic or group practice contracts	0	0		6
Other contracted sites in the United States	0	0		0
Contracts for foreign sites	0	0		0
Contracts signed as of 1 August 2000 but not yet installed (hospitals/ independent laboratories/ other sites)	4 (4/0/0)	16 (3/13/0)		8 (8/0/0)
No. of sites operating system	21	944	278	666
First system installation yr	1996	1990	1979	1984
No. of staff to develop/ install/support/other in entire firm	6/5/7/9	—	90 total	961/1,211/186/321
No. of staff to develop/ install/support/other in LIS division	3/3/4/2	9/19/19/17		187/80/118/30
No. of terminals and/or workstations in live sites (minimum–maximum)	1–6	1–60	3–250	7–620

(Continued on next page)

TABLE 1 LISs from Accent Information Systems, Antek, and Cerner *(Continued)*

Parameter	Data for LIS from:			
	Accent Information Systems	Antek	Cerner Citation	Cerner Corp.
Hardware				
Central	HP E60 NetServer	AMD/Intel Pentium III	Compaq, HP	Compaq, IBM
Terminals and/or workstations	HP Vectra	AMD/Intel Pentium III	Compaq, HP	VTs, PCs
Central hardware redundant or fault tolerant	Yes	Yes	Yes	Yes
Software				
Programming language(s)	Delphi	Delphi	C++, Visual Basic	COBOL, C++
Operating system(s)	Windows 95, 98, 2000, NT 4.0	Windows NT	Windows NT	Open VMS, AIX
Database system(s)	MS SQL	Ctrieve, Oracle	Pervasive Btrieve, MS SQL	Proprietary
System includes full transaction logging	Yes	Yes	Yes	Yes
Features (% of live installations, available but no installations, or not available)				
Chemistry and/or hematology	100	100	—	98
Bar-coded collection labels	65	15	—	80
Microbiology	9	100	—	86
Blood bank donor center	Not available	—	—	17
Blood bank transfusion service	Not available	—	—	73
Surgical pathology	Not available	—	—	58
Cytology	Not available	—	—	58
HIS interface				
A/D/T	62	15	—	98
Order entry	9	5	—	90
Result reporting	100	15	—	98
Ad hoc reporting	9	—	—	65
Rules-based systems	19	—	—	31
Utilization management	19	100	—	10
Outreach or commercial laboratory	38	—	—	22
Compliance checking	38	—	—	15
Accounts receivable	Not available	—	—	Not available
Materials management or inventory	Not available	—	—	—
Test partition	Not available	—	—	80
Remote faxing or printing	57	30	—	37
Physician office outreach	19	15	—	5

TABLE 1 *(Continued)*

Parameter	Data for LIS from:			
	Accent Information Systems	Antek	Cerner Citation	Cerner Corp.
LIS provides surveillance data to public health agencies using CDC/ HL7/LOINC/SNOMED standard:				
Microbiology data	Available but not yet installed	Not available	Available but not yet installed	Not available
Tumor diagnosis or registry data	Not available	Not available	Available but not yet installed	Not available
ASTM-HL7 interface	Yes	Yes	—	Yes
Interfaces to hospital or integrated health care systems	Nelson Data Resources	SMS, LabCorp, Quest, SKB, Meditech	Major vendors	Major vendors
Interfaces to physician office management systems	Versys, Wismer Martin, and HL7-compliant systems	Medic, Medical Manager, MedicaLogic, Practice Partners, PDS, PCN, Neusoft IDX, Dairyland, others	HL7-compliant vendors	Major vendors
Interface to automated laboratory transportation system(s)	No	No	Planned	Lab InterLink/ Labotix, Beckman, Coulter IDS, Roche/BMC/ Hitachi
Software provides indexed field in each test definition for LOINC code	Yes	Yes	Yes	Yes
Provide LOINC dictionary for each new installation	No	—	No	No
LIS permits use of SNOMED II/SNOMED International	Yes/no	No/no	Yes/yes	Yes/yes
Market modules for other hospital departments	Yes	Yes	Yes	Yes
% of LIS installations stand-alone	38	90	95	40
No. of different laboratory instruments interfaced with LIS	≥250	≥200	>300	≥400
Source code	Escrow	Escrow	Escrow	Escrow
User programming in separate partition	Yes	No	No	Yes
User group	Yes	No	Yes	Yes
Cost (hardware/software/ monthly maintenance)				
Smallest	$4,000/$21,000/$300	$2,500/$8,000/$56	—	—
Largest	$20,000/$40,000/$60	$30,000/$50,000/$200	—	—

TABLE 2 LISs from Cerner, Clinical Information Systems Inc., Clinical Software Solutions, and ClinLab Inc.

Parameter	Data for LIS from:			
	Cerner Corp.	Clinical Information Systems Inc.	Clinical Software Solutions	ClinLab Inc.
Contact information	Julie Hull 2800 Rockcreek Pkwy. Kansas City, MO 64117-2551 (816) 201-1074 http://www.cerner.com	V. L. Flaxbeard 18805 Willamette Dr. West Linn, OR 97068 (800) 869-0680 http://www.cislab.com	Bryan Jones 219 South Fir St. Chandler, AZ 85226 (800) 570-0474 http://www.clinsoft.com	K. Anckarstrom-Bohm 2411 W. Graves Ave., Ste. 1 Orange City, FL 32763 (904) 774-0030 http://www.clinlabinc.com
Name of system	**PathNet HNA Millennium**	**CISlab**	**CSSWIN**	**ClinLab LIS**
No. of contracts for sites operating system	**36**	**95**	**≥200**	**25**
U.S. hospital contracts	26	8	56	4
U.S. independent laboratory contracts	1	80	≥160	3
Clinic or group practice contracts	2	4	2	10
Other contracted sites in the United States	2	0	2	8
Contracts for foreign sites	5	3	0	0
Contracts signed as of 1 August 2000 but not yet installed (hospitals/ independent laboratories/ other sites)	11 (10/0/1)	3 (0/3/0)	4 (2/1/1)	2 (0/2/0)
No. of sites operating system	**52**	**101**	**≥200+**	**25**
First system installation yr	1997	1988	1987	1987
No. of staff to develop/ install/support/other in entire firm	961/1,211/186/321	6/7/7/1	—	4/3/4/2
No. of staff to develop/ install/support/other in LIS division	187/80/118/30	—	—	—
No. of terminals and/or workstations in live sites (minimum–maximum)	7–620	1–43	1–20	4–70
Hardware				
Central	IBM, Compaq	Compaq, HP, Dell, generic PCs	IBM compatible	IBM, Dell, Compaq server
Terminals and/or workstations	PCs	PCs, Wyse, Link	IBM compatible	IBM, Dell, Compaq
Central hardware redundant or fault-tolerant	Yes	Yes	Yes	Yes
Software				
Programming language(s)	Visual C++, Visual Basic, Java	Delphi, C++, COBOL	4GL	Clipper, Visual FoxPro, Delphi
Operating system(s)	Windows, Windows NT	Windows NT, SCO UNIX	Windows	Novell, Windows NT, Windows 2000
Database system	Oracle RDBMS	Interbase RDBMS, C-ISAM, MS SQL 7.0	SQL	dBase, FoxPro, MS, Apollo
System includes full transaction logging	Yes	No	Yes	Yes

TABLE 2 *(Continued)*

Parameter	Data for LIS from:			
	Cerner Corp.	Clinical Information Systems Inc.	Clinical Software Solutions	ClinLab Inc.
Features (% of live installations, available but no installations, or not available)				
Chemistry or hematology	100	100	95	100
Bar-coded collection labels	100	85	30	80
Microbiology	40	100	15	100
Blood bank donor center	Not available	1	—	Not available
Blood bank transfusion service	12	—	—	Not available
Surgical pathology	42	21	5	Not available
Cytology	30	25	5	Not available
HIS interface				
A/D/T	80	5	20	90
Order entry	80	5	20	70
Result reporting	80	5	20	70
Ad hoc reporting	100	1	20	100
Rules-based systems	80	—	90	50
Utilization management	5	Available but not yet installed	10	100
Outreach or commercial laboratory	10	7	10	40
Compliance checking	10	85	60	70
Accounts receivable	Not available	98	5	Available but not yet installed
Materials management or inventory	Not available	Available but not yet installed	5	Available but not yet installed
Test partition	100	100	100	100
Remote faxing or printing	50	100	25	40
Physician office outreach	8	60	10	40
LIS provides surveillance data to public health agencies using CDC/ HL7/LOINC/SNOMED standard:				
Microbiology data	Not available	Four sites	Available but not installed	Four sites
Tumor diagnosis or registry data	Not available	Three sites	Available but not installed	—
ASTM-HL7 interface	Yes	Yes	Yes	Yes
Interfaces to hospital or integrated health care systems	Major vendors	HBOC, PCS, Tower Systems	SMS, Dairyland, Pearl, LabCorp, ASTM-HL7 compliant systems	MEDITECH
Interfaces to physician office management systems	Major vendors	Medical Manager	Medical Manager, other ASTM-HL7-compliant systems, others	Medical Manager, IDX, Medic, SoftAid, Neusoft, Medstar, MedicaLogic
Interface to automated laboratory transportation system(s)	Available but not yet operational	Planned	No	Planned

(Continued on next page)

TABLE 2 LISs from Cerner, Clinical Information Systems Inc., Clinical Software Solutions, and ClinLab Inc. *(Continued)*

Parameter	Data for LIS from:			
	Cerner Corp.	Clinical Information Systems Inc.	Clinical Software Solutions	ClinLab Inc.
Software provides indexed field in each test definition for LOINC code	Yes	Yes	Yes	Yes
Provide LOINC dictionary for each new installation	No	No	No	No
LIS permits use of SNOMED II/SNOMED International	Yes/yes	Yes/yes	Yes/yes	No/no
Market modules for other hospital departments	Yes	No	Yes	No
% of LIS installations stand-alone	40	—	25	—
No. of different laboratory instruments interfaced with LIS	200	≥150	≥300	105
Source code	Escrow	Escrow	No	Escrow
User programming in separate partition	Yes	No	Yes	No
User group	Yes	No	No	No
Cost (hardware/software/ monthly maintenance)				
Smallest	—	$3,000/$5,000/$200	—	—/$15,000/$190
Largest	—	$100,000/$150,000/ $1,500	—	—/$170,000/$2,125

TABLE 3 LISs from Comp Pro Med Inc., Computer Service and Support Inc., CPSI, and Creative Computer Applications, Inc.

Parameter	Data for LIS from:			
	Comp Pro Med Inc.	Computer Service & Support Inc.	CPSI	Creative Computer Applications Inc.
Contact information	Avarie Fisher 5550 Skylane Blvd., Ste. N Santa Rosa, CA 95403 (707) 578-0239 http://www.comppromed.com	2106 New Rd., Bldg. E-6 Linwood, NJ 08221 (800) 336-4277 http://www.csslis.com	James Bouchard 6600 Wall St. Mobile, AL 36695 (800) 711-CPSI http://www.cpsinet.com	Chris Coleman 26115-A Mureau Rd. Calabasas, CA 91302 (800) 437-9000 http://www.ccainc.com
Name of system	**Polytech**	**CLS-2000**	**CPSI System**	**CyberLAB II**
No. of contracts for sites operating system	**114**	**81**	**166**	**220**
U.S. hospital contracts	50	0	165	105
U.S. independent laboratory contracts	33	75	1	61
Clinic or group practice contracts	20	6	0	40
Other contracted sites in the United States	0	0	0	10
Contracts for foreign sites	11	0	0	4
No. of contracts signed as of 1 August 2000 but not yet installed (hospitals/independent laboratories/other sites)	2 (0/2/0)	10 (0/10/0)	10 (10/0/0)	6 (4/2/0)

TABLE 3 *(Continued)*

Parameter	Data for LIS from:			
	Comp Pro Med Inc.	Computer Service & Support Inc.	CPSI	Creative Computer Applications Inc.
No. of sites operating system	**114**	—	**166**	**≥400**
First system installation yr	1990	1980	1986	1982
No. of staff to develop/install/support/other in entire firm	5/5/9/4	6/3/4/4	55/94/281/45	12/4/19/38
No. of staff to develop/install/support/other in LIS division	—	—	24/29/43/9	6/4/13/38
No. of terminals and/or workstations in live sites (minimum–maximum)	1–65	3–≥50	14–161	3–≥250
Hardware				
Central	Dell, IBM, HP	IBM RS/6000	UNIX server, HP LH6000, HP E60	Compaq, HP, IBM RS
Terminals and/or workstations	Dell, IBM, HP	Windows-based PCs, workstations	Wyse50 compatible, IBM compatible	PCs, ASCII terminals
Central hardware redundant or fault tolerant	—	Yes	Yes	Yes
Software				
Programming language(s)	C, C++, Assembler	C++	COBOL, C++	Microfocus COBOL, C, C++, HTML
Operating system(s)	Windows 95, 98, NT; others	AIX	Unixware7, Windows NT, Windows 2000 server	UNIX
Database system	SQL, Btrieve	LABase	Proprietary	Multi-indexed sequential access
System includes full transaction logging	Yes	Yes	No	Yes
Features (% of live installations, available but no installations, or not available)				
Chemistry or hematology	100	100	100	100
Bar-coded collection labels	90	95	100	100
Microbiology	10	60	100	85
Blood bank donor center	Not available	—	Not available	Not available
Blood bank transfusion service	Not available	20	Not available	Not available
Surgical pathology	Not available	70	Not available	5
Cytology	Not available	—	Not available	100
HIS interface				
A/D/T	25	—	100	65
Order entry	25	15	100	50
Result reporting	25	5	100	55

(Continued on next page)

TABLE 3 LISs from Comp Pro Med Inc., Computer Service and Support Inc., CPSI, and Creative Computer Applications, Inc. (*Continued*)

Parameter	Data for LIS from:			
	Comp Pro Med Inc.	Computer Service & Support Inc.	CPSI	Creative Computer Applications Inc.
Ad hoc reporting	100	80	100	100
Rules-based systems	100	100	100	100
Utilization management	100	50	100	100
Outreach or commercial laboratory	15	100	100	100
Compliance checking	100	90	100	100
Accounts receivable	60	90	100	20
Materials management or inventory	Not available	60	100	Not available
Test partition	100	20	45	100
Remote faxing or printing	85	95	100	100
Physician office outreach	Not available	50	67	35
LIS provides surveillance data to public health agencies using CDC/ HL7/LOINC/ SNOMED standard:				
Microbiology data	Two sites	Available but not yet installed	Not available	Available but not yet installed
Tumor diagnosis or registry data	Not available	Available but not yet installed	Not available	Available but not yet installed
ASTM-HL7 interface	Yes	—	No	Yes
Interfaces to hospital or integrated health care systems	Sunquest, SMS, CHC, Intermed, Cerner, First Data, HDS, Antrim, Infostat, others	PCN, IDX, Medic, HBOC	Not available	Quadramed, HBOC, SMS, Dairyland, others
Interfaces to physician office management systems	MedicaLogic, others	—	Not available	Atlas, Medical Manager, Medic, others
Interface to automated lab transportation system(s)	Planned	Available but not yet operational	No	Available, installed, operational
Software provides indexed field in each test definition for LOINC code	Yes	Yes	No	Yes
Provide LOINC dictionary for each new installation	Yes	No	—	No
LIS permits use of SNOMED II/SNOMED International	No/no	No/no	No/no	Yes/yes
Market modules for other hospital departments	No	—	Yes	Yes
% of LIS installations stand-alone	—	—	<1	76
No. of different laboratory instruments interfaced with LIS	≥200	450	194	—
Source code	Escrow	Yes	Escrow	Escrow

TABLE 3 *(Continued)*

Parameter	Data for LIS from:			
	Comp Pro Med Inc.	Computer Service & Support Inc.	CPSI	Creative Computer Applications Inc.
User programming in separate partition	No	No	—	No
User group	No	No	Yes	Yes
Cost (hardware/software/ monthly maintenance)				
Smallest	$3,000/—/—	$10,000/$10,000/ $250	$61,000/$138,000/ $2,000	$30,000/$50,000/ 1.5% per month
Largest	$125,000/—/—	$30,000/$150,000/ $2,500	$186,000/$458,000/ $8,500	$300,000/$700,000/ 1.5% per month

TABLE 4 LISs from Diamond Computing, Dynamic Healthcare Technologies, and DynaMedix Corp.

Parameter	Data for LIS from:			
	Diamond Computing	Dynamic Healthcare Technologies	DynaMedix Corp.	DynaMedix Corp.
Contact information	James T. Campbell 2345 Fourth St. Tucker, GA 30084 (770) 496-0286	Todd Charest 615 Crescent Executive Ct., Ste. 500 Lake Mary, FL 32746 (800) 832-3020 http://www.dht.com	222 W. Las Colinas Blvd., 7th Fl. Irving, TX 75039 (972) 725-9100 http://www.dynamedix.com	222 W. Las Colinas Blvd., 7th Fl. Irving, TX 75039 (972) 725-9100 http://www.dynamedix.com
Name of system	**LabGEM**	**Premier Series**	**Genesys Plus**	**Genesys PRO**
No. of contracts for sites operating system	**48**	**76**	**165**	**897**
U.S. hospital contracts	14	61	112	540
U.S. independent laboratory contracts	34	3	6	30
Clinic or group practice contracts	0	12	40	313
Other contracted sites in the United States	0	0	0	0
Contracts for foreign sites	6	0	7	14
Contracts signed as of 1 August 2000 but not yet installed (hospitals/ independent laboratories/ other sites)	2 (1/1/0)	12 (11/1/0)	5 (4/0/1)	—
No. of sites operating system	**48**	**≥100**	**165**	**897**
First system installation yr	1984	1990	1982	1982
No. of staff to develop/ install/support/other in entire firm	—	39/61/48/66	7/4/8/11	7/4/8/11
No. of staff to develop/ install/support/other in LIS division	—	4/11/9/8	2/4/4/—	2/4/4/—

(Continued on next page)

TABLE 4 LISs from Diamond Computing, Dynamic Healthcare Technologies, and DynaMedix Corp. *(Continued)*

Parameter	Data for LIS from:			
	Diamond Computing	Dynamic Healthcare Technologies	DynaMedix Corp.	DynaMedix Corp.
No. of terminals and/or workstations in live sites (minimum–maximum)	6–500	6–500	1–12	1–12
Hardware				
Central	IBM RS/6000, DEC Alpha, Intel	IBM AS/400	ATX, Micro	ATX, Micro
Terminals and/or workstations	Intel PC, Wyse	PC able to run IBM Client Access or standard Web browser	ATX, Micro	ATX, Micro
Central hardware redundant or fault tolerant	Yes	Yes	Yes	Yes
Software				
Programming language(s)	M Caché	RPG CL 4GL, Java	C	C
Operating system(s)	AIX, UNIX, MSM, Windows NT	IBM OS/400	DynaGen OS	DynaGen OS
Database system	—	DB2/400 integrated with OS	Relational database	Relational database
System includes full transaction logging	Yes	Yes	No	No
Features (% of live installations, available but no installations, or not available)				
Chemistry or hematology	100	100	100	100
Bar-coded collection labels	100	95	25	25
Microbiology	80	100	0	0
Blood bank donor center	—	—	0	0
Blood bank transfusion service	—	—	0	0
Surgical pathology	30	20	0	0
Cytology	50	20	0	0
HIS interface				
A/D/T	100	99	10	0
Order entry	100	85	—	—
Result reporting	100	85	—	—
Ad hoc reporting	90	100	10	10
Rules-based systems	—	100	—	—
Utilization management	80	100	—	—
Outreach or commercial laboratory	100	85	—	—
Compliance checking	100	60	100	100
Accounts receivable	50	Not available	15	0
Materials management or inventory	—	Not available	0	0
Test partition	—	100	0	0
Remote faxing or printing	100	90	15	15
Physician office outreach	100	Available but not yet installed	0	0

TABLE 4 *(Continued)*

Parameter	Data for LIS from:			
	Diamond Computing	Dynamic Healthcare Technologies	DynaMedix Corp.	DynaMedix Corp.
LIS provides surveillance data to public health agencies using CDC/ HL7/LOINC/SNOMED standard:				
Microbiology data	—	Not available	—	—
Tumor diagnosis or registry data	—	Not available	—	—
ASTM-HL7 interface	Yes	Yes	Yes	Yes
Interfaces to hospital or integrated health care systems	HBOC, Cerner, Sunquest, IBAX, MEDITECH, others	HBOC, SMS, FirstCoast, Dairyland, Compucare, others	via gateway + ASTM, MEDITECH, SMS, Dairyland, others	ASTM
Interfaces to physician office management systems	Doctor Chart, Clinscan, Fiscal, Medical Manager, others	Available	Medical Manager, Logician, QRS, Paradigm, Dairyland, Practice Partners	—
Interface to automated laboratory transportation system(s)	Available but not operational	No	Planned	Planned
Software provides indexed field in each test definition for LOINC code	Yes	No	No	No
Provide LOINC dictionary for each new install	Yes	No	—	—
LIS permits use of SNOMED II/SNOMED International	Yes/yes	Yes/yes	No/no	No/no
Market modules for other hospital departments	—	Yes	No	Yes
% of LIS installations stand-alone	—	15	—	—
No. of different laboratory instruments interfaced with LIS	150	≥180	—	—
Source code	Yes	Escrow	Yes	Yes
User programming in separate partition	Yes	Yes	No	No
User group	—	Yes	No	No
Cost (hardware/software/ monthly maintenance)				
Smallest	—	$7,500/$125,000/ $19,800	$18,000[a]/$110	$11,000[a]/$110
Largest	—	$75,000/$650,000/ $40,000	≥$60,000[a]/$375	$16,000[a]/$187

[a]Cost for hardware plus software.

TABLE 5 LISs from Fletcher-Flora Computer Products, HEX Laboratory Systems, Informatica Tesi de Italia, SA de CV, and Information Data Management, Inc.

Parameter	Data for LIS from:			
	Fletcher-Flora Computer Products	HEX Laboratory Systems	Informatica Tesi de Italia, SA de CV	Information Data Management, Inc.
Contact information	Robert M. Gibson 1580 Orangethorpe Way Anaheim, CA 92801 (800) 777-1471, ext. 232 http://www.labpak.com	Susan Bollinger 1042B El Camino Real, Ste. 308 Encinitas, CA 92024 (800) 729-2085 http://www.hexlab.com	Fernando Estrada Bosques de Ciruelos 168 Posi 8, Col. Bosques de las Lomas Mexico, DF 11700, Mexico 52-55-96-66-16	Susan L. McBride 9701 W. Higgins Rd., Ste. 500 Rosemont, IL 60018 (800) 249-4276 http://www.idm.com
Name of system	LabPak Series	**LAB/HEX**	**WINLAB**	Surround
No. of contracts for sites operating system	>465	113	540	2
U.S. hospital contracts	—	9	0	0
U.S. independent laboratory contracts	—	45	0	0
Clinic or group practice contracts	—	40	0	0
Other contracted sites in the United States	—	9	0	2
Contracts for foreign sites	—	10	540	0
Contracts signed as of 1 August 2000 but not yet installed (hospitals/independent laboratories/other sites)	—	3 (0/0/3)	10 (7/3/0)	2 (0/0/2)
No. of sites operating system	**465**	**123**	**≥540**	**2**
First system installation yr	1980	1981	1981	2000
No. of staff to develop/install/support/other in entire firm	7/5/4/9	5/5/5/3	12/13/4/12	21/3/13/9
No. of staff to develop/install/support/other in LIS division	—	—	8/13/4/12	—
No. of terminals and/or workstations in live sites (minimum–maximum)	1–20	2–≥40	1–35	3–20
Hardware				
Central	IBM, Intel PC	Intel Pentium III, Dell, HP 9000, IBM, DEC, others	Intel Pentium II server	Windows NT server, 4.0 or later
Terminals and/or workstations	—	PCs, Wyse terminals	Pentium III PCs	Windows NT workstations
Central hardware redundant or fault tolerant	—	Yes	Yes	Yes

TABLE 5 *(Continued)*

Parameter	Data for LIS from:			
	Fletcher-Flora Computer Products	HEX Laboratory Systems	Informatica Tesi de Italia, SA de CV	Information Data Management, Inc.
Software				
Programming language(s)	C, C++, Java, Visual Basic	TBred Basic, Inquire IV	Visual Basic	Java
Operating system(s)	Windows 98, NT	SCO, UNIX, Linux	Windows NT, 2000, 95, 98	Windows NT
Database system	Btrieve, SQL	IDOL IV, 4GL, SQL	MS SQL server	Oracle 7, Oracle 8
System includes full transaction logging	No	Yes	Yes	Yes
Features (% of live installations, available but no installations, or not available)				
Chemistry or hematology	100	100	80	100
Bar-coded collection labels	75	100	30	100
Microbiology	45	100	8	Not available
Blood bank donor center	Not available	—	20	Not available
Blood bank transfusion service	Not available	Partner with another company	20	Not available
Surgical pathology	Not available	75	Not available	Not available
Cytology	Not available	100	Not available	Not available
HIS interface				
A/D/T	40	15	10	Not available
Order entry	40	10	5	Not available
Result reporting	40	10	10	Not available
Ad hoc reporting	65	100	100	Not available
Rules-based systems	100	100	5	Not available
Utilization management	100	100	Not available	Not available
Outreach or commercial laboratory	55	40	Not available	Not available
Compliance checking	90	75	10	50
Accounts receivable	Not available	65	5	Not available
Materials management or inventory	Available December 2001	Not available	3	Not available
Test partition	70	100	2	100
Remote faxing or printing	75	100	2	100
Physician office outreach	35	75	2	Not available
LIS provides surveillance data to public health agencies using CDC/HL7/LOINC/SNOMED standard:				
Microbiology data	Available but not yet installed	Not available	Not available	Not available
Tumor diagnosis or registry data	Not available	Not available	Not available	Not available
ASTM-HL7 interface	Yes	Yes	No	No

(Continued on next page)

TABLE 5 LISs from Fletcher-Flora Computer Products, HEX Laboratory Systems, Informatica Tesi de Italia, SA de CV, and Information Data Management, Inc. *(Continued)*

Parameter	Data for LIS from:			
	Fletcher-Flora Computer Products	HEX Laboratory Systems	Informatica Tesi de Italia, SA de CV	Information Data Management, Inc.
Interfaces to hospital or integrated health care systems	HBOC, IDX, HSI, GPMS, Dairyland, QSI, MEDITECH, others	SMS, Sunquest, PSI, HBOC, Antrim, Cerner, Experior, Logician	—	None
Interfaces to physician office management systems	Medical Manager, PCN, Infocue, Medic, CCA, MedicaLogic, others	MedicaLogic, Medical Manager, Medic, IDX	—	None
Interface to automated laboratory transportation system(s)	—	Planned	No	No
Software provides indexed field in each test definition for LOINC code	Yes	Yes	No	No
Provide LOINC dictionary for each new installation	No	No	No	No
LIS permits use of SNOMED II/SNOMED International	No/no	Yes/yes	No/no	No/no
Market modules for other hospital departments	No	No	No	No
% of LIS installations stand-alone	—	—	—	—
No. of different laboratory instruments interfaced with LIS	296	≥250	≥200	3
Source code	Escrow	Escrow	No	Escrow
User programming in separate partition	No	No	Yes	No
User group	No	No	No	No
Cost (hardware/software/ monthly maintenance)				
Smallest	$1,000/$1,790/$59	$5,000/$15,000/$200	—	—/$25,000/$500
Largest	—	$100,000/$180,000/ $2,800	—	—/$120,000/$1,000

TABLE 6 LISs from Integrated Management Solutions, Intellidata Inc., ISYS/Biovation, and Keane, Inc.

Parameter	Data for LIS from:			
	Integrated Management Solutions	Intellidata Inc.	ISYS/Biovation	Keane, Inc.
Contact information	Donna Whetstone 4900 Bradford Dr. Huntsville, AL 35805 (888) 946-5227 http://www.imswinlab.com	Gary Rex 1483 Old Bridge Rd. Woodbridge, VA 22192 (800) 711-5525 http://www.intellidata.com	Jim LeClair 6925 Lake Ellenor Dr., Ste. 135 Orlando, FL 32809 (800) 527-4797 http://www.isysbiov.com	Leslie Johnson 290 Broadhollow Rd. Melville, NY 11747 (800) 699-6773 http://www.keane.com/hsd
Name of system	**WinLAB**	**IntelliLab**	**Messenger**	**LabFUSION**
No. of contracts for sites operating system	—	**71**	**12**	**39**
U.S. hospital contracts	—	18	2	36
U.S. independent laboratory contracts	—	8	0	1
Clinic or group practice contracts	—	42	5	1
Other contracted sites in the United States	—	3	2	0
Contracts for foreign sites	—	0	3	1
Contracts signed as of 1 August 2000 but not yet installed (hospitals/ independent laboratories/ other sites)	—	10 (2/2/6)	12 (5/7/0)	—
No. of sites operating system	**≥50**	**210**	**12**	**26**
First system installation yr	1991	1988	1999	1989
No. of staff to develop/ install/support/other in entire firm	3/1/2/4	6/4/6/8	3/4/4/4	145/100/140/65
No. of staff to develop/ install/support/other in LIS division	—	—	—	6/1/2/0
No. of terminals and/or workstations in live sites (minimum–maximum)	1	1–≥1,300	2–75	6–50
Hardware				
Central	Dell	Intel-based servers, IBM RS/6000, DEC, HP, DG, others	Platform-independent PCs	IBM RS/6000, AS/400, DG, HP, Compaq
Terminals and/or workstations	Dell	Windows 95, 98, NT PCs	Platform-independent PCs	PCs, Wyse terminals
Central hardware redundant or fault tolerant	Yes	Yes	Yes	Yes
Software				
Programming language(s)	Soft Velocity Clarion version 5.0, MS Visual C++	Visual Basic, Basic, C, C++	Delphi, C++	Progress
Operating system(s)	Windows 2000, NT 4.0, 98	Windows NT, Linux, UNIX, AIX	Windows NT, 98, OS independent	UNIX

(Continued on next page)

TABLE 6 LISs from Integrated Management Solutions, Intellidata Inc., ISYS/Biovation, and Keane, Inc. *(Continued)*

Parameter	Data for LIS from:			
	Integrated Management Solutions	Intellidata Inc.	ISYS/Biovation	Keane, Inc.
Database system	Soft Velocity TPS, Pervasive Btrieve 2000	mvBase, multidimensional database	SQL server preferred, DB independent	Progress
System includes full transaction logging	Yes	Yes	Yes	Yes
Features (% of live installations, available but no installations, or not available)				
Chemistry or hematology	100	100	100	100
Bar-coded collection labels	100	85	100	100
Microbiology	Available 2001	40	Available but not yet installed	80
Blood bank donor center	—	Available but not yet installed	Not available	0
Blood bank transfusion service	—	Available but not yet installed	Not available	0
Surgical pathology	Available 2001	40	Available 2001	5
Cytology	—	50	Available 2001	5
HIS interface				
A/D/T	—	40	50	90
Order entry	—	25	50	80
Result reporting	—	25	50	80
Ad hoc reporting	—	100	100	100
Rules-based systems	—	100	100	100
Utilization management	—	100	Available but not yet installed	100
Outreach or commercial laboratory	—	40	Available 2001	10
Compliance checking	100	—	Available but not yet installed	20
Accounts receivable	—	10	Not available	10
Materials management or inventory	—	Not available	Not available	Not available
Test partition	100	100	100	100
Remote faxing or printing	100	100	100	70
Physician office outreach	100	—	Not available	10
LIS provides surveillance data to public health agencies using CDC/ HL7/LOINC/SNOMED standard:				
Microbiology data	—	Available but not yet installed	Available but not yet installed	Not available
Tumor diagnosis or registry data	—	Available but not yet installed	Available but not yet installed	Not available
ASTM-HL7 interface	Yes	Yes	Yes	Yes
Interfaces to hospital or integrated health care systems	Versys, IDX	SMS, CPL, Antrim, MPMS, Vanguard, SmithKline, LabCorp, others	HBOC, HDN	HBOC, SMS, IMN, MEDITECH

TABLE 6 *(Continued)*

Parameter	Data for LIS from:			
	Integrated Management Solutions	Intellidata Inc.	ISYS/Biovation	Keane, Inc.
Interfaces to physician office management systems	Vital, Medical Manager	Medic, MedicaLogic, Medical Mgr, IDX, Data Medic, Autochart, Disc, others	Epic, Medical Manager, others	—
Interface to automated laboratory transportation system(s)	—	Planned	Available but not yet operational	No
Software provides indexed field in each test definition for LOINC code	No	Yes	Yes	No
Provide LOINC dictionary for each new installation	No	No	No	—
LIS permits use of SNOMED II/SNOMED International	No/no	Yes/yes	Yes/yes	Yes/no
Market modules for other hospital departments	—	Yes	No	Yes
% of LIS installations stand-alone	—	80	—	25
No. of different laboratory instruments interfaced with LIS	≥50	≥300	≥450	100
Source code	No	Escrow	Escrow	No
User programming in separate partition	No	Yes	No	No
User group	No	No	No	Yes
Cost (hardware/software/ monthly maintenance)				
Smallest	$4,000/$8,000/$150	$5,000/$25,000/ 1.5% after 1st yr	$30,000/$50,000/ $1,200	—
Largest	$10,000/$23,000/$410	$75,000/$175,000/ 1.5% after 1st yr	$200,000/$500,000/ $8,500	—

TABLE 7 LISs from LabSoft Inc.[a]

Parameter	Data for LIS from LabSoft Inc.			
Name of system	**Beethoven II LIS**	**BloodHound II LIS**	**EZLink LIS**	**LABNET LIS**
No. of contracts for sites operating system	**147**	**539**	**57**	**105**
U.S. hospital contracts	32	110	16	35
U.S. independent laboratory contracts	9	0	7	18
Clinic or group practice contracts	102	409	34	52
Other contracted sites in U.S.	0	0	0	0
Contracts for foreign sites	4	20	0	0
Contracts signed as of 1 August 2000 but not yet installed (hospitals/ independent laboratories/ other sites)	9 (4/5/0)	—	6 (1/5/0)	5 (3/1/1)

(Continued on next page)

TABLE 7 LISs from LabSoft Inc.[a] (*Continued*)

Parameter	Data for LIS from LabSoft, Inc.			
No. of sites operating system	**147**	**539**	**57**	**125**
First system installation yr	1995	1992	1995	1995
No. of staff to develop/install/ support/other in entire firm	2/3/3/4	2/3/3/4	2/3/3/4	2/3/3/4
No. of staff to develop/install/ support/other in LIS division	—	—	—	—
No. of terminals and/or workstations in live sites (minimum–maximum)	—	1	2–3	3–61
Hardware				
Central	Compaq, Acer, Dell	Compaq, Acer (all Pentium)	Compaq, Acer, HP (all Pentium)	Compaq, Acer, Dell
Terminals and/or workstations	Compaq, Acer, Dell	—	Compaq, Acer, HP (all Pentium)	Compaq, Acer, Dell
Central hardware redundant or fault tolerant	Yes	Yes	Yes	Yes
Software				
Programming language(s)	Delphi	C++	Delphi	Delphi, Visual C++
Operating system(s)	Windows 2000, NT, 98, 95	MS-DOS	Windows 2000, NT, 98, 95	Windows 2000, NT, 98, 95
Database system	MS SQL server	—	SQL server	SQL server
System includes full transaction logging	Yes	Yes	Yes	Yes
Features (% of live installations, available but no installations, or not available)				
Chemistry or hematology	100	100	100	100
Bar-coded collection labels	90	3	100	100
Microbiology	30	Not available	20	15
Blood bank donor center	Available but not yet installed	Not available	Available but not yet installed	Available but not yet installed
Blood bank transfusion service	Available but not yet installed	Not available	Available but not yet installed	Available but not yet installed
Surgical pathology	Available but not yet installed	Not available	Available but not yet installed	Not available
Cytology	Available but not yet installed	Not available	Available but not yet installed	Not available
HIS interface				
A/D/T	50	Available but not yet installed	40	50
Order entry	50	Available but not yet installed	40	40
Result reporting	50	Available but not yet installed	40	40
Ad hoc reporting	50	Not available	100	100
Rules-based systems	50	Available but not yet installed	100	50
Utilization management	100	—	100	100
Outreach or commercial laboratory	30	—	30	30
Compliance checking	50	—	50	100

TABLE 7 *(Continued)*

Parameter	Data for LIS from LabSoft, Inc.			
Accounts receivable	Available but not yet installed	—	Available but not yet installed	Not available
Materials management or inventory	Available but not yet installed	—	Available but not yet installed	Not available
Test partition	Available but not yet installed	10	Available but not yet installed	Available but not yet installed
Remote faxing or printing	—	—	80	100
Physician office outreach	Not available	—	20	50
LIS provides surveillance data to public health agencies using CDC/HL7/LOINC/ SNOMED standard:				
Microbiology data	Available but not yet installed	Not available	Available but not yet installed	Available but not yet installed
Tumor diagnosis or registry data	Not available	Not available	Available but not yet installed	Not available
ASTM-HL7 interface	Yes	Yes	Yes	Yes
Interfaces to hospital or integrated health care systems	HBOC, SMS, Dairyland, Epicare	—	HBOC, IDX, Dairyland, others	HBOC, SMS, Dairyland, Epicare
Interfaces to physician office management systems	Medical Manager, HealthCareData, STC, Physix, Medident, custom	—	Medical Manager, Epic, MegaWest	Medical Manager, HealthCareData, STC, Physix, Medident, custom
Interface to automated laboratory transportation system(s)	Planned	No	Planned	Planned
Software provides indexed field in each test definition for LOINC code	Yes	No	Yes	Yes
Provide LOINC dictionary for each new installation	No	—	No	No
LIS permits use of SNOMED II/ SNOMED International	Yes/yes	No/no	Yes/yes	Yes/yes
Market modules for other hospital departments	No	No	No	No
% of LIS installations stand-alone	—	—	—	—
No. of different laboratory instruments interfaced with LIS	152	152	152	152
Source code	Escrow	Escrow	Escrow	Escrow
User programming in separate partition	Yes	No	Yes	Yes
User group	No	No	—	No
Cost (hardware/software/ monthly maintenance)				
Smallest	$2,500/$6,000/$70	$1,500/$4,500/$380	$5,000/$11,000/$110	$0/$31,000/$500
Largest	$6,000/$15,000/$132	$2,700/$15,000/$50	$12,000/$21,000/$150	$130,000/$270,000/ $3,500

[a]Contact information: Gary Yancich, 8402 Laurel Fair Cir., Ste. 207, Tampa, FL 33610. Phone: (800) 767-3279. URL: http://www.labsoftlis.com.

TABLE 8 LISs from LabConnect.com Inc., M/MGMT Systems Inc., McKesson HBOC, and MEDCOM Information Systems Inc.

Parameter	Data for LIS from:			
	LabConnect.com Inc.	M/MGMT Systems Inc.	McKesson HBOC	MEDCOM Information Systems Inc.
Contact information	Will Campbell (Intellidata Inc.) 1483 Old Bridge Rd. Woodbridge, VA 22192 (866) 239-6048 http://www.lab-connect.com	Robert Mann 9261 Folsom Blvd., Ste. 503 Sacramento, CA 95826 (916) 856-1961 or (800) 664-8797 http://www.mmgmt.com	Mary E. Beach 5995 Windward Pkwy. Alpharetta, GA 30005 (404) 338-2081 http://www.mckhboc.com	David Baird 2117 Stonington Ave. Hoffman Estates, IL 60195 (847) 885-1553 http://idt.net/~medcom19
Name of system	**Lab-Connect.com (ASP)**	**M/LAB**	**Pathways Laboratory**	**MEDCOM Laboratory Information Manager**
No. of contracts for sites operating system	1	24	20	418
U.S. hospital contracts	—	0	17	16
U.S. independent laboratory contracts	—	17	0	59
Clinic or group practice contracts	—	6	2	343
Other contracted sites in the United States	—	1	0	0
Contracts for foreign sites	—	0	1	0
Contracts signed as of 1 August 2000 but not yet installed (hospitals/ independent laboratories/other sites)	1 (0/0/1)	2 (0/2/0)	18 (17/1/0)	5 (1/3/2)
No. of sites operating system	1	30	31	430
First system installation yr	1999	1987	1998	1992
No. of staff to develop/ install/support/other in entire firm	6/4/6/8	4/4/6/2	—	4/6/4/7
No. of staff to develop/ install/support/other in LIS division	—	3/3/5/2	—	3/6/4/5
No. of terminals and/or workstations in live sites (minimum–maximum)	—	8–64	10–350	1–11
Hardware				
Central	Intel-based servers, IBM RS/6000, DEC, HP, DG, others	Pentium	DG, HP	MEDCOM IBM PC compatible
Terminals and/or workstations	Windows 95, 98, NT PCs	Pentium	PCs	MEDCOM IBM PC compatible
Central hardware redundant or fault tolerant	Yes	Yes	Yes	Yes
Software				
Programming language(s)	Visual Basic, C++	M (MUMPS)	Delphi, ANSI standard C	C++
Operating system(s)	Windows NT, Linux, UNIX, AIX	DOS, Windows NT, VAX VMS	UNIX	DOS; Windows 95, 98, NT

TABLE 8 *(Continued)*

Parameter	Data for LIS from:			
	LabConnect.com Inc.	M/MGMT Systems Inc.	McKesson HBOC	MEDCOM Information Systems Inc.
Database system	mvBase, multidimensional database	M (MUMPS), Caché	Oracle	dBase compatible
System includes full transaction logging	Yes	Yes	Yes	Yes
Features (% of live installations, available but no installations, or not available)				
Chemistry or hematology	100	20	100	100
Bar-coded collection labels	100	10	100	70
Microbiology	0	0	100	—
Blood bank donor center	0	Not available	Through Mediware	—
Blood bank transfusion service	0	Not available	Through Mediware	—
Surgical pathology	0	Not available	Through Tamtron	—
Cytology	0	Available but not yet installed	Through Tamtron	—
HIS interface				
A/D/T	100	Not available	100	3
Order entry	100	20	100	2
Result reporting	100	30	100	4
Ad hoc reporting	100	100	100	100
Rules-based systems	100	50	100	—
Utilization management	100	100	100	—
Outreach or commercial laboratory	0	Not available	100	—
Compliance checking	0	Not available	100	Available but not yet installed
Accounts receivable	0	85	15	—
Materials management or inventory	0	Available but not yet installed	Available in 2002	—
Test partition	100	100	100	100
Remote faxing or printing	100	35	100	5
Physician office outreach	0	Not available	100	—
LIS provides surveillance data to public health agencies using CDC/ HL7/LOINC/ SNOMED standard:				
Microbiology data	Available but not yet installed	Three sites	Not available	Not available
Tumor diagnosis or registry data	Available but not yet installed	Not available	—	Not available
ASTM-HL7 interface	Yes	Yes	Yes	Yes
Interfaces to hospital or integrated health care systems	SMS, CPL, Antrim, MPMS, Vanguard, SmithKline, LabCorp, others	Not available	McKesson HBOC, SMS, IDX, homegrown, MEDITECH, Last Word	—

(Continued on next page)

TABLE 8 LISs from LabConnect.com Inc., M/MGMT Systems Inc., McKesson HBOC, and MEDCOM Information Systems Inc. (*Continued*)

Parameter	Data for LIS from:			
	LabConnect.com Inc.	M/MGMT Systems Inc.	McKesson HBOC	MEDCOM Information Systems Inc.
Interfaces to physician office management systems	Medic, MedicaLogic, Medical Manager, IDX, Datamedic, Disc, Penchart, Centrex, others	SMS, others with HL7 interface	iMcKesson, Clinscan	Medical Manager, MedicaLogic, AMS, Edimis, Wismer Martin, others
Interface to automated laboratory transportation system(s)	Planned	No	Available but not yet operational	—
Software provides indexed field in each test definition for LOINC code	Yes	No	Yes	No
Provide LOINC dictionary for each new installation	No	No	Yes	No
LIS permits use of SNOMED II/SNOMED International	Yes/yes	No/no	Not available	No/no
Market modules for other hospital departments	—	No	Yes	No
% of LIS installations stand-alone	—	—	25	—
No. of different laboratory instruments interfaced with LIS	4 (≥300 available)	10	112	≥200
Source code	Escrow	Yes	Escrow	No
User programming in separate partition	Yes	Yes	Yes	No
User group	No	Yes	Yes	No
Cost (hardware/software/ monthly maintenance)				
Smallest	—	—/$64,000/$500	—	$2,300/$7,000/$95
Largest	—	—/$240,000/$1,200	—	$35,000/$65,000/$725

TABLE 9 LISs from MediSolution, Ltd., MEDITECH, and Modulus Data Systems Inc.

Parameter	Data for LIS from:			
	MediSolution, Ltd.	MEDITECH	MEDITECH	Modulus Data Systems Inc.
Contact information	Nathalie Huot Montreal, Quebec H2P 1B9 Canada (514) 850-5000 http://www.medisolution.com	Paul Berthiaume MEDITECH Cir. Westwood, MA 02090 (781) 821-3000 http://www.meditech.com	Paul Berthiaume MEDITECH Cir. Westwood, MA 02090 (781) 821-3000 http://www.meditech.com	George Liviakis 2348 Walsh Ave. Santa Clara, CA 95051 (408) 567-9333 http://www.modulusdata systems.com
Name of system	**MEDILAB**	**Client/server LIS**	**MAGIC LIS**	**MOD-U-LIS**
No. of contracts for sites operating system	>510	426	751	12
U.S. hospital contracts	0	—	—	9
U.S. independent laboratory contracts	0	—	—	3
Clinic or group practice contracts	0	—	—	0

TABLE 9 *(Continued)*

Parameter	Data for LIS from:			
	MediSolution, Ltd.	MEDITECH	MEDITECH	Modulus Data Systems Inc.
Other contracted sites in the United States	0	—	—	0
Contracts for foreign sites	>510	—	—	0
Contracts signed as of 1 August 2000 but not yet installed (hospitals/ independent laboratories/other sites)	40 (35/5/0)	—	—	1 (0/1/0)
No. of sites operating system	**>510**	**426**	**751**	**—**
First system installation yr	1972	1969	1969	1972
No. of staff to develop/ install/support/other in entire firm	500 total	1,739 total	1,739 total	9/2/4/1
No. of staff to develop/ install/support/other in LIS division	70/13/10/2	—	—	—
No. of terminals and/or workstations in live sites (minimum– maximum)	4–300	—	—	15–200
Hardware				
Central	Sun, DG UNIX, IBM, HP	DG, Compaq	DG, Compaq	HP 9000, IBM RS/ 6000, NCR 526
Terminals and/or workstations	IBM-compatible PCs	—	—	ADDS, Wyse, PCs
Central hardware redundant/fault tolerant	Yes	Yes	Yes	Yes
Software				
Programming language(s)	C++, Java	MAGIC CS	MAGIC	C
Operating system(s)	Sun OS, Windows NT, Windows 2000, Windows 98, UNIX, AIX, others	MAGIC	MAGIC	UNIX
Database system	SQL Server, Oracle	SQL	MAGIC	Unity 4GL, Accell
System includes full transaction logging	Yes	Yes	Yes	Yes
Features (% of live installations, available but no installations, or not available)				
Chemistry or hematology	95	100	100	100
Bar-coded collection labels	100	100	100	100
Microbiology	100	100	100	100
Blood bank donor center	10	—	—	10

(Continued on next page)

TABLE 9 LISs from MediSolution, Ltd., MEDITECH, and Modulus Data Systems Inc. *(Continued)*

Parameter	Data for LIS from:			
	MediSolution, Ltd.	MEDITECH	MEDITECH	Modulus Data Systems Inc.
Blood bank transfusion service	Not available	100	100	60
Surgical pathology	15	100	100	20
Cytology	15	100	100	20
HIS interface				
A/D/T	100	25	25	80
Order entry	10	25	—	80
Result reporting	100	25	—	80
Ad hoc reporting	50	100	100	100
Rules-based systems	100	100	100	100
Utilization management	30	100	100	100
Outreach or commercial laboratory	50	25	25	80
Compliance checking	Available December 2001	—	—	60
Accounts receivable	25	100	100	10
Materials management/ inventory	Not available	100	100	Not available
Test partition	100	—	—	100
Remote faxing or printing	100	100	100	100
Physician office outreach	25	—	—	80
LIS provides surveillance data to public health agencies using CDC/HL7/ LOINC/SNOMED standard:				
Microbiology data	Available but not yet installed	—	—	Zero sites
Tumor diagnosis or registry data	Two sites	—	—	Zero sites
ASTM-HL7 interface	Yes	Yes	Yes	Yes
Interfaces to hospital or integrated health care systems	—	Sunquest, HBOC, SMS, others	Sunquest, HBOC, SMS, others	Gerber Alley, HBOC, SMS, Hemocare, AMI, others
Interfaces to physician office management systems	European systems	—	—	—
Interface to automated laboratory transportation system(s)	Planned	Sysmex, Roche/ BMC/Hitachi	Sysmex, Roche/ BMC/Hitachi	Sysmex, Roche/ BMC/Hitachi
Software provides indexed field in each test definition for LOINC code	No	No	No	Yes
Provide LOINC dictionary for each new installation	No	No	No	Yes
LIS permits use of SNOMED II/ SNOMED International	Yes/yes	Yes/yes	Yes/yes	Yes/—

TABLE 9 (Continued)

Parameter	Data for LIS from:			
	MediSolution, Ltd.	MEDITECH	MEDITECH	Modulus Data Systems Inc.
Market modules for other hospital departments	Yes	Yes	Yes	No
% of LIS installations stand-alone	50	25	25	—
No. of different laboratory instruments interfaced with LIS	290	Hundreds	Hundreds	≥200
Source code	Escrow	Yes	Yes	Yes
User programming in separate partition	Yes	No	No	Yes
User group	Yes	Yes	Yes	No
Cost (hardware/ software/monthly maintenance)				
Smallest	$15,000/$45,000/$750	—	—	$80,000/$100,000/ $2,000
Largest	$600,000/$3,000,000/ $50,000	—	—	$600,000/$1,000,000/ $9,000

TABLE 10 LISs from Multidata Computer Systems Inc., Opus Healthcare Solutions, Orchard Software Corp., and Psyché Systems Corp.

Parameter	Data for LIS from:			
	Multidata Computer Systems Inc.	Opus Healthcare Solutions Inc. (formerly Health Science Systems)	Orchard Software Corp.	Psyché Systems Corp.
Contact information	Michael Slater 330 Seventh Ave. New York, NY 10001 (212) 967-6700 http://www.mul.com	Angie Evans 10713 N. Ranch Rd. 620, Ste. 400 Austin, TX 78726 (800) 676-3371 http://www.opushealth care.com	Sales Department 701 Congressional Blvd., Ste. 360 Carmel, IN 46032 (800) 856-1948 http://www.orchard soft.com	Patricia Salem 10 Laurel Ave. Wellesley, MA 02481 (800) 345-1514 http://www.psyche systems.com
Name of system	**Multi Tech LIS**	**OpusLab**	**Harvest (Costello)**	**Lab Web**
No. of contracts for sites operating system	**38**	**39**	**247**	**25**
U.S. hospital contracts	7	32	65	21
U.S. independent laboratory contracts	30	2	21	4
Clinic or group practice contracts	1	2	159	2
Other contracted sites in the United States	0	3	2	0
Contracts for foreign sites	0	0	0	4
Contracts signed as of 1 August 2000 but not yet installed (hospitals/ independent laboratories/ other sites)	1 (0/1/0)	3 (1/1/1)	25 (4/5/16)	1 (0/1/0)

(Continued on next page)

TABLE 10 LISs from Multidata Computer Systems Inc., Opus Healthcare Solutions, Orchard Software Corp., and Psyché Systems Corp. (*Continued*)

Parameter	Data for LIS from:			
	Multidata Computer Systems Inc.	Opus Healthcare Solutions Inc. (formerly Health Science Systems)	Orchard Software Corp.	Psyché Systems Corp.
No. of sites operating system	38	39	257	29
First system installation yr	1983	1987	1993	1976
No. of staff to develop/ install/support/other in entire firm	4/2/3/2	15/11/5/6	11/20 (installation + support)/16	8/8/4/12
No. of staff to develop/ install/support/other in LIS division	—	11/11/5/6	—	4/5/3/6
No. of terminals and/or workstations in live sites (minimum–maximum)	4–100	6–100	1–135	5–≥100
Hardware				
Central	Pentium compatible, DEC Alpha, DEC VAX, UNIX RISC	HP 9000 (L series)	HP Pentium compatible	Compaq
Terminals and/or workstations	DEC VT series or compatible or PC with WRQ Reflection software	PC clients	HP Pentium compatible	PC
Central hardware redundant or fault tolerant	Yes (optional)	Yes	Yes	Yes
Software				
Programming language(s)	M, Visual Basic, HTML	C++, C, Java, Dynamic HTML, PERL, HTML, Fortran	ACI Fourth Dimension, C++	Fortran, Visual Basic
Operating system(s)	Windows 2000, Windows NT, DEC VMS, UNIX, Linux	UNIX	Windows 2000, NT, 98; MacOS 8.0	Open VMS; Windows 98, 2000, NT
Database system	M	Opus, DBMS, Access, SQL	ACI Fourth Dimension, 4D client/server	Oracle RDB, proprietary
System includes full transaction logging	Yes (optional)	Yes	Yes	Yes
Features (% of live installations, available but no installations, or not available)				
Chemistry or hematology	90	100	100	100
Bar-coded collection labels	80	100	85	100
Microbiology	80	80	35	33
Blood bank donor center	—	Not available	Not available	Not available
Blood bank transfusion service	—	Not available	Not available	33
Surgical pathology	10	30	<10	33
Cytology	40	20	<10	33
HIS interface				

TABLE 10 *(Continued)*

Parameter	Data for LIS from:			
	Multidata Computer Systems Inc.	Opus Healthcare Solutions Inc. (formerly Health Science Systems)	Orchard Software Corp.	Psyché Systems Corp.
A/D/T	20	80	Not available	90
Order entry	20	75	50	50
Result reporting	10	75	50	50
Ad hoc reporting	40	100	100	100
Rules-based systems	90	100	20	Not available
Utilization management	20	100	90	100
Outreach or commercial laboratory	90	75	30	100
Compliance checking	70	25	50	Available but not yet installed
Accounts receivable	90	Not available	Not available	Not available
Materials management or inventory	20	Not available	Not available	Not available
Test partition	100	75	25	100
Remote faxing or printing	80	100	80	100
Physician office outreach	80	10	Available 2001	100
LIS provides surveillance data to public health agencies using CDC/ HL7/LOINC/SNOMED standard:				
Microbiology data	Not available	Available but not yet installed	Three sites	Available but not yet installed
Tumor diagnosis or registry data	Not available	Available but not yet installed	Zero sites	Available but not yet installed
ASTM-HL7 interface	Yes	Yes	Yes	Yes
Interfaces to hospital or integrated health care systems	SMS, CSM, custom	SMS, HBOC, Cycare, First Coast, Western Star, Hemocare	HBOC, IDX, Cerner, SMS, M2 Dairyland, Antrim, Sunquest, others	Keane, HBOC, SMS, MEDITECH, homegrown, others
Interfaces to physician office management systems	Medical Manager	—	Medical Manager, MedicaLogic, Clinitec, Medic, Millbrook, Abaton, etc.	Medical Manager and other HL7-compliant systems
Interface to automated laboratory transportation system(s)	Planned	No	Planned	No
Software provides indexed field in each test definition for LOINC code	Yes	Yes	Yes	Yes
Provide LOINC dictionary for each new installation	Yes (on request)	No	No	No
LIS permits use of SNOMED II/SNOMED International	No/no	Yes/yes	No/no	Yes/yes
Market modules for other hospital departments	No	Yes	No	No

(Continued on next page)

TABLE 10 LISs from Multidata Computer Systems Inc., Opus Healthcare Solutions, Orchard Software Corp., and Psyché Systems Corp. (*Continued*)

Parameter	Data for LIS from:			
	Multidata Computer Systems Inc.	Opus Healthcare Solutions Inc. (formerly Health Science Systems)	Orchard Software Corp.	Psyché Systems Corp.
% of LIS installations stand-alone	—	80	—	—
No. of different laboratory instruments interfaced with LIS	≥100	≥200	≥200	≥100
Source code	Escrow	Escrow	No	Yes
User programming in separate partition	Yes	Yes	No	No
User group	No	Yes	No	Yes
Cost (hardware/software/ monthly maintenance)				
Smallest	$15,000/$30,000/ $600	$30,000/$100,000/ $900	$4,000/$15,000/$200	$5,900/$7,500/$200
Largest	$250,000/$300,000/ $5,000	$300,000/$2,500,000/ $10,000	≥$100,000/≥$800,000/ $5,000	$200,000/$30,000/ $6,500

TABLE 11 LISs from SSC Soft Computer, Schuyler House, SIA, and SMS

Parameter	Data for LIS from:			
	SCC Soft Computer	Schuyler House	SIA (a Sysmex company)	SMS
Contact information	Ellie Vahman 34350 U.S. Highway 19 North Palm Harbor, FL 34684 (727) 789-0100 http://www.softcomputer.com	Steve Allen 14635 Titus St. Panorama City, CA 91402 (800) 706-0266 http://www.schuylab.com	Bill Blair 5210 E. Williams Cir. Tucson, AZ 85711 (520) 790-4624	Maryellen Kaytrosh 51 Valley Stream Pkwy. Malvern, PA 19355 (610) 219-6300 http://www.smed.com
Name of system	SoftLab	SchuyLab	**MOLIS**	**NOVIUS Lab**
No. of contracts for sites operating system	220	348	89	≥190
U.S. hospital contracts	161	47	0	≥190
U.S. independent laboratory contracts	32	138	1	5
Clinic or group practice contracts	3	145	0	—
Other contracted sites in the United States	0	14	0	—
Contracts for foreign sites	24	4	88	—
Contracts signed as of 1 August 2000 but not yet installed (hospitals/independent laboratories/other sites)	16 (15/1/0)	6 (2/0/4)	6 (3/3/0)	≥25
No. of sites operating system	310	348	89	≥170
First system installation yr	1985	1993	1989	1984

TABLE 11 *(Continued)*

Parameter	Data for LIS from:			
	SCC Soft Computer	Schuyler House	SIA (a Sysmex company)	SMS
No. of staff to develop /install/support/other in entire firm	185/47/67/141	4/5/6/5	300/200/400/300	1,300/1,400/900/ ≥2,000
No. of staff to develop/ install/support/other in LIS division	85/20/21/34	—	35/31/25/25	40/30/40/10
No. of terminals and/or workstations in live sites (minimum–maximum)	3–210	1–41	5–200	15–≥250
Hardware				
Central	IBM RS/6000, HP 9000	Dell Pentium PCs	Compaq (Alpha, Intel), IBM (Intel, RS/6000), IBM 3090, HP 9000	Alpha, RS/6000
Terminals and/or workstations	ASCII terminals, PCs	Dell Pentium PCs	Windows PCs, VT 320 or equivalent	Intel-based PCs
Central hardware redundant or fault tolerant	Yes	No	Yes	Yes
Software				
Programming language(s)	Visual C, C++, Java	C, MASM	Uniface	C, C++
Operating system(s)	UNIX, Windows NT	Novell; Windows 95, 98, 2000, NT	Windows NT, Windows 2000, AIX, HP-UX, TRU 64, Alpha, Open VMS, others	OS/2, AIX, UNIX, Windows NT
Database system	db-VISTA, RDM++, Oracle	Btrieve	Oracle, Sybase, Informix, C-ISAM	Sybase, SQL server, others
System includes full transaction logging	Yes	No	Yes	—
Features (% of live installations, available but no installations, or not available)				
Chemistry or hematology	100	100	100	100
Bar-coded collection labels	100	20	100	100
Microbiology	98	80	95	100
Blood bank donor center	Pending FDA clearance	0	Through Wyndgate	—
Blood bank transfusion service	45	0	Through Wyndgate	55
Surgical pathology	34	0	16	25
Cytology	34	10	16	21
HIS interface				
A/D/T	90	5	55	100
Order entry	100	0	55	100
Result reporting	80	1	55	100

(Continued on next page)

TABLE 11 LISs from SSC Soft Computer, Schuyler House, SIA, and SMS (*Continued*)

Parameter	Data for LIS from:			
	SCC Soft Computer	Schuyler House	SIA (a Sysmex company)	SMS
Ad hoc reporting	100	0	100	100
Rules-based systems	100	0	100	100
Utilization management	100	100	100	100
Outreach or commercial laboratory	65	0	90	25
Compliance checking	100	20	Available 2001	—
Accounts receivable	20	20	90	12
Materials management or inventory	Available in 2001	0	0	—
Test partition	100	0	100	100
Remote faxing or printing	100	20	100	90
Physician office outreach	87	5	30	—
LIS provides surveillance data to public health agencies using CDC/ HL7/LOINC/ SNOMED standard:				
Microbiology data	≥30 sites	—	Available but not yet installed	Not available
Tumor diagnosis or registry data	≥30 sites	—	Available but not yet installed	Not available
ASTM-HL7 interface	Yes	Yes	Yes	Yes
Interfaces to hospital or integrated health care systems	SMS, HBOC, Quadramed, IDX, Keane, Perot, CPSI, custom, others	HBOC, Dairyland, Wismer-Martin, Nelson Data Resources, Tricord, SMS	SMS, SAP, CEPAGE, LaLisa, Sandra, BHIS	—
Interfaces to physician office management systems	Medical Manager, MedicaLogic, Healthworks Alliance, Trizetto, others	Apex, QSI, PCN, Medical Manager	MCS, Medistar, Healthone	—
Interface to automated laboratory transport system(s)	Lab InterLink/ Labotix, Beckman, Coulter IDS, Roche/BMC/ Hitachi	Planned	Sysmex, Roche/ BMC/Hitachi, PVT, Streamline	—
Software provides indexed field in each test definition for LOINC code	Yes	Yes	Yes	Yes
Provide LOINC dictionary for new installation	Yes	No	No	—
LIS permits use of SNOMED II/SNOMED International	Yes/yes	No/no	Yes/yes	No/no
Market modules for other hospital departments	Yes	No	No	Yes
% of LIS installations stand-alone	90	—	—	15

TABLE 11 (*Continued*)

Parameter	Data for LIS from:			
	SCC Soft Computer	Schuyler House	SIA (a Sysmex company)	SMS
No. of different laboratory instruments interfaced with LIS	≥300	≥150	243	≥300
Source code	No	Escrow at user request	Escrow	Yes
User programming in separate partition	Yes	No	Yes	Yes
User group	Yes	No	No	Yes
Cost (hardware/software/ monthly maintenance)				
Smallest	$30,000/$50,000/ 1.25% per month	$1,800/$5,000/ $68	$25,000/$150,000/ $950	—
Largest	$1,000,000/$3,000,000/ 1.25% per month	$32,000/$64,000/ $1,000	$300,000/$1,200,000/ $15,000	—

TABLE 12 LISs from STARLAB Custom Software Systems Inc., Sunquest, and Sunquest/e-Suite

Parameter	Data for LIS from:			
	STARLAB Custom Software Systems Inc.	Sunquest Information Systems	Sunquest Information Systems	Sunquest/e-Suite, Inc.
Contact information	George Widuch or DeWitt Rhaly 7012 Westbelt Dr. Nashville, TN 37209 (800) 344-8053 http://www.css-corporate.com	Carrie Hunnicutt 4801 E. Broadway Blvd. Tucson, AZ 85711 (520) 570-2000 http://www.sunquest.com	Carrie Hunnicutt 4801 E. Broadway Blvd. Tucson, AZ 85711 (520) 570-2000 http://www.sunquest.com	Monica Hawkins 101 East Park Blvd, 12th Fl. Plano, TX 75074 (800) 726-8746 http://www.getesuite.com
Name of system	**STARLAB**	FlexiLab	Sunquest Commercial Lab	**e-Hospital Laboratory**
No. of contracts for sites operating system	19	609	46	4
U.S. hospital contracts	14	578	8	3
U.S. independent laboratory contracts	3	10	37	1
Clinic or group practice contracts	2	0	1	0
Other contracted sites in U.S.	0	0	0	0
Contracts for foreign sites	0	21	0	0
Contracts signed as of 1 August 2000 but not yet installed (hospitals/independent laboratories/other sites)	0	8 (8/0/0)	3 (1/2/0)	4 (0/4/0)
No. of sites operating system	20	950	89	32
First system installation yr	1984	1980	1982	1996

(*Continued on next page*)

TABLE 12 LISs from STARLAB Custom Software Systems Inc., Sunquest, and Sunquest/e-Suite (*Continued*)

Parameter	Data for LIS from:			
	STARLAB Custom Software Systems Inc.	Sunquest Information Systems	Sunquest Information Systems	Sunquest/e-Suite, Inc.
No. of staff to develop/install/support/other in entire firm	8/4/5/12	195/150/242/193	28/12/14/30	195/150/242/193
No. of staff to develop/install/support/other in LIS division	4/2/4/3	165/125/235/187	14/5/5/10	5/7/12/4
No. of terminals and/or workstations in live sites (minimum–maximum)	4–70	32–500	10–≥1,000	4–≥900
Hardware				
Central	HP 9000 Series 800, IBM RS/6000	IBM RS/6000, DEC Alpha	IBM RS/6000, DEC Alpha, Compaq	DEC Alpha, Compaq
Terminals and/or workstations	HP 700/70, Pentium workstation	SQ 420, Pentium PCs	Compaq, DEC VT480 or compatible	Pentium PCs
Central hardware redundant/fault tolerant	—	Yes	Yes	Yes
Software				
Programming language(s)	COBOL	M, C, C++	Open M	Open M
Operating system(s)	HP-UX, AIX, Linux	AIX, OSF, Open VMS	VMS, UNIX	VMS
Database system	T-ISAM	InterSystems M	Open M	Open M
System includes full transaction logging	Yes	Yes	Yes	Yes
Features (% of live installations, available but no installations, or not available)				
Chemistry or hematology	100	100	100	100
Bar-coded collection labels	80	100	100	100
Microbiology	10	98	100	100
Blood bank donor center	—	5	Not available	Not available
Blood bank transfusion service	—	75	Not available	Not available
Surgical pathology	Available but not yet installed	70	100	25
Cytology	Available but not yet installed	70	95	25
HIS interface				
A/D/T	80	100	100	75
Order entry	50	100	100	75
Result reporting	50	100	100	75
Ad hoc reporting	45	70	100	100
Rules-based systems	Not available	100	100	100
Utilization management	10	100	100	100
Outreach or commercial laboratory	50	75	100	100
Compliance checking	100	25	100	Available but not yet installed
Accounts receivable	10	5	90	50

TABLE 12 *(Continued)*

Parameter	Data for LIS from:			
	STARLAB Custom Software Systems Inc.	Sunquest Information Systems	Sunquest Information Systems	Sunquest/e-Suite, Inc.
Materials management/ inventory	Not available	20	13	Not available
Test partition	Available but not yet installed	100	100	100
Remote faxing or printing	75	100	100	100
Physician office outreach	100	—	Not available	25
LIS provides surveillance data to public health agencies using CDC/ HL7/LOINC/ SNOMED standard:				
Microbiology data	—	—	Available but not yet installed	Four sites
Tumor diagnosis or registry data	—	—	Available but not yet installed	One site
ASTM-HL7 interface	Yes	Yes	Yes	Yes
Interfaces to hospital or integrated health care systems	Dairyland, HMS	SMS, HBOC, MEDITECH, others	Cerner, SMS, IBAX, ARUP, Compucare, Specialty Labs, MEDITECH, IDX, others	FAMS, IDX
Interfaces to physician office management systems	MegaWest, IDX	Medical Manager	—	Clinscan
Interface to automated laboratory transportation system(s)	Planned	Planned	Planned	No
Software provides indexed field in each test definition for LOINC code	Yes	No	No	No
Provide LOINC dictionary for each new installation	No	No	No	No
LIS permits use of SNOMED II/SNOMED International	No/no	Yes/yes	Yes/no	Yes/no
Market modules for other hospital departments	Yes	Yes	No	Yes
% of LIS installations stand-alone	42	95	—	50
No. of different laboratory instruments interfaced with LIS	≥60	≥250	≥300	40
Source code	Escrow	Escrow	Yes	No
User programming in separate partition	No	Yes	Yes	No
User group	Yes	Yes	Yes	Yes
Cost (hardware/software/ monthly maintenance)				
Smallest	$15,000/$25,000/ $600	$100,000/$250,000/ 1.5% per month	$35,000/$75,000/ $1,600	$3,500 (ASP subscription fees only)
Largest	$200,000/$250,000/ $4,000	$500,000/≥$1,000,000/ 1.5% per month	$1,800,000/$2,250,000/ $60,750	$12,500 (ASP subscription fees only)

TABLE 13 LISs from Triple G Corp.

Parameter	Data for LIS from Triple G Corp.	
Contact information	Tena Bennett 3100 Steeles Ave., Ste. 600 Markham, Ontario L3R 8T3 Canada (905) 305-0041 http://www.tripleg.com	Angelica Lau 3100 Steeles Ave., Ste. 600 Markham, Ontario L3R 8T3 Canada (905) 305-0041 http://www.tripleg.com
Name of system	**TriWin**	**ULTRA**
No. of contracts for sites operating system	**83**	**40**
U.S. hospital contracts	1	5
U.S. independent laboratory contracts	1	6
Clinic or group practice contracts	0	0
Other contracted sites in U.S.	0	0
Contracts for foreign sites	81	29
Contracts signed as of 1 August 2000 but not yet installed (hospitals/independent laboratories/other sites)	6 (6/0/0)	7 (5/2/0)
No. of sites operating system	**136**	**230**
First system installation yr	1991	1990
No. of staff to develop/install/support/other in entire firm	61/34/18/13	61/34/18/13
No. of staff to develop/install/support/other in LIS division	4/3/1/5	57/31/17/8
No. of terminals and/or workstations in live sites (minimum–maximum)	3–125	21–≥500
Hardware		
Central	Hardware independent	UNIX server
Terminals and/or workstations	Hardware independent	Windows NT
Central hardware redundant or fault tolerant	Yes	Yes
Software		
Programming language(s)	Visual Basic	C, Unify Vision, Unify 4GL, Unify Accell
Operating system(s)	Windows 98, NT	UNIX
Database system	Advantage	Unify Dataserver
System includes full transaction logging	Yes	Yes
Features (% of live installations, available but no installations, or not available)		
Chemistry or hematology	85	100
Bar-coded collection labels	70	100
Microbiology	20	80
Blood bank donor center	—	Not available
Blood bank transfusion service	10	10
Surgical pathology	5	40
Cytology	5	40
HIS interface		
A/D/T	80	50
Order entry	5	50
Result reporting	5	50
Ad hoc reporting	15	75
Rules-based systems	Available 2001	75
Utilization management	Available but not yet installed	75
Outreach or commercial laboratory	20	75
Compliance checking	Available 2001	10
Accounts receivable	20	80
Materials management or inventory	Available but not yet installed	Not available

TABLE 13 *(Continued)*

Parameter		Data for LIS from Triple G Corp.
Test partition	Not available	100
Remote faxing or printing	70	100
Physician office outreach	Available but not yet installed	—
LIS provides surveillance data to public health agencies using CDC/HL7/LOINC/SNOMED standard:		
Microbiology data	Not available	Available but not yet installed
Tumor diagnosis or registry data	Not available	Available but not yet installed
ASTM-HL7 interface	Yes	Yes
Interfaces to hospital or integrated health care systems	MEDITECH, Cerner, HBOC	MEDITECH, EDS, EPIC, Hemocare, Ulticare, HealthVision, others
Interfaces to physician office management systems	—	—
Interface to automated laboratory transportation system(s)	Lab InterLink, MDS AutoLab	MDS AutoLab, Lab InterLink/Labotix
Software provides indexed field in each test definition for LOINC code	Yes	Yes
Provide LOINC dictionary for each new installation	No	No
LIS permits use of SNOMED II/SNOMED International	No/no	Yes/yes
Market modules for other hospital departments	No	No
% of LIS installations stand-alone	—	—
No. of different laboratory instruments interfaced with LIS	≥250	≥200
Source code	Escrow	Yes
User programming in separate partition	No	Yes
User group	Yes	Yes
Cost (hardware/software/monthly maintenance)		
Smallest	$10,000/$20,000/1.5% of software total	$100,000/$150,000/$2,000
Largest	$150,000/$250,000/1.5% of software total	—/≥$1,500,000/≥$28,000

Some of the information and the tables were adapted from CAP Today (1, 2), copyrighted by the College of American Pathologists, CAP Press, and are reprinted with permission. Updated information should be obtained by prospective users form each vendor.

REFERENCES

1. **Aller, R., and M. Weilert.** 1996. Microbiology information systems. When it's appropriate to 'stand-alone.' *CAP Today* **10:**30–34.
2. **Aller, R., and K. Carey.** 1999. Are we the way we were? *CAP Today* **13:**64–88.
3. **Aller, R. D.** 1996. Software standards and the laboratory information system. *Am. J. Clin. Pathol.* **105**(Suppl.)**:**S48.
4. **Huff, S. M.** 1998. Clinical data exchange standards and vocabularies for messages, p. 62–67. In *Proceedings of the AMIA Symposium.*
5. **Truant, A. L., J. Moghaddas, and E. D. Williamson.** 2000. Using College of American Pathologists workload units for intralaboratory comparison. *Lab. Med.* **31:**266–270.
6. **Weiner, H., and R. D. Aller.** 2000. Death of a system: how to cope with sunsetting. *CAP Today* **14:**50–78.
7. **Willard, K. E., and C. H. Schanholtzer.** 1995. User interface reengineering: innovative applications of bar coding in a clinical microbiology laboratory. *Arch. Pathol. Lab. Med.* **119:**706–712.
8. **Wilson, M. L.** 1997. Clinically relevant, cost-effective clinical microbiology: strategies to decrease unnecessary testing. *Am. J. Clin. Pathol.* **107:**154–167.

Emerging Infectious Diseases

BYUNGSE SUH, JAY R. KOSTMAN, AND RICHARD MANIGLIA

16

Despite our constant efforts to prevent and control infectious diseases by the introduction of various vaccines and antibiotics, microbes continue to surprise us by either their unexpected appearance as the cause of brand new maladies or their appearance as modified diseases. During the past century, we accomplished a great deal in that a global eradication of smallpox has been achieved and eradication of polio and dracunculiasis is expected in the near future. In the meantime, however, many more new infectious diseases have emerged. Whereas human immunodeficiency virus (HIV) represents the most dramatic and obvious example, legionellosis, Lyme disease, *Heliobacter pylori* infection, *Escherichia coli* O157:H7, hantavirus pulmonary syndrome, and human ehrlichiosis represent well-recognized emerging infections. Although important factors that are accountable for emerging infectious diseases are not always obvious, advanced medical technology, changes in human behavior, altered ecologic balance, increases in travel and trade, and changes in pathogenic potential of microbes can be listed as major contributing factors.

This chapter is devoted to select emerging and reemerging infectious disease entities that are not yet well described in standard texts. Since they are so new, definitive diagnostic laboratory tests are often not readily available. Therefore, wherever applicable, typical clinical manifestations and laboratory findings will be described.

STREPTOCOCCUS INIAE

S. iniae is a fish pathogen that was first reported in 1976 to cause subcutaneous abscesses in Amazon freshwater dolphins (*Inia geoffrensis*) (145). It has been recognized as a new human pathogen to cause invasive infections ranging from cellulitis to bacteremia, endocarditis, meningitis, and arthritis (108, 199). *S. iniae* infection represents a new zoonotic infection that is clearly related to aquaculture, which is one of the fastest-growing commercial food production industries to supplement wild fish and plants. Although *S. iniae* may colonize the surfaces of fish without causing disease, it has been recognized as an important etiologic agent of epizootic meningoencephalitis involving tilapia (*Oreochromis* species), yellow tail (*Serida quinqueradiata*), rainbow trout, and coho salmon (108).

In a recent report of 11 patients diagnosed with culture-confirmed invasive *S. iniae* infection, consisting of a cluster of 9 patients over a 1-year period (December 1995 to December 1996) and 2 additional patients from a retrospective analysis, tilapia was proven to be the source of the etiologic organism in 6 patients (199). Tilapias, also known as St. Peter's fish or Hawaiian sunfish, belong to the cichlid group of freshwater fish (Fig. 1) and are commonly used in Asian cooking. Although they are indigenous to Africa, Madagascar, southern India, Sri Lanka, and South and Central America, tilapia are now farmed in many different countries, including China, the Philippines, Thailand, Indonesia, Egypt, and the United States.

The summary of the clinical features of a cluster of nine patients was as follows: all of the patients handled live or freshly killed fish; their median age was 69 years; the female/male ratio was 2:1; they all were of Asian descent; and eight of the patients recalled percutaneous injuries with the dorsal fin, fish bone, or a knife used in preparing fish. In addition, all the patients developed bacteremia, and the majority (eight of nine) developed hand cellulitis with fever and lymphangitis, whereas one developed endocarditis with clinical and laboratory findings of meningitis and arthritis (108, 199). All recovered without sequelae with appropriate antibiotic treatment.

S. iniae strains produce a β-hemolytic reaction on Trypticase soy agar with 5% sheep blood and are nongroupable with Lancefield A through V antisera. They are pyrrolidonyl aminopeptidase positive and Voges-Proskauer negative (108, 159). The molecular typing pattern obtained by pulsed-field gel electrophoresis on clinical isolates of *S. iniae* from patients with invasive disease as well as on standard strains obtained from the American Type Culture Collection strongly suggests that an invasive clone of the organism is responsible for causing clinically significant infections in humans and fish (108, 159, 199). When all the clinical isolates were tested, the MICs of penicillin, cefazolin, ceftriaxone, erythromycin, clindamycin, and trimethoprim-sulfamethoxazole were ≤ 0.25 µg/ml, while those of ciprofloxacin and gentamicin were 0.5 and 16 µg/ml, respectively (199).

CYCLOSPORA

Spherical organisms in diarrheal stool previously referred to as blue-green algae, cyanobacterium-like bodies, large *Cryptosporidium* or coccidian-like bodies have recently been

FIGURE 1 Tilapias (also known as St. Peter's fish or Hawaiian sunfish). Courtesy of Donald E. Low.

identified as *Cyclospora* (133). The human-associated *Cyclospora* (*Cyclospora cayetanensis*) shows variable acid-fast staining properties, is 8 to 10 μm in diameter (twice the size of *Cryptosporidium* [4 to 6 μm]), and is closely related to the *Eimeria* genus but not to *Cryptosporidium* (152). *C. cayetanensis* displays limited host species range, and humans appear to be the only known host for the organism. Upon excretion by the host *Cyclospora* oocysts are not immediately infective and need to be exposed to the external environment for 7 to 10 days before they become infective (152). Although a molecular detection method for *Cyclospora* based on PCR has been described (154), the diagnosis of cyclosporiasis is best made by direct acid-fast staining (107) or using a modified safranin stain with microwave heating (191) of fecal specimens.

Human cyclosporiasis occurs worldwide. However, infection is much more prevalent in the area of endemicity, including Southeast Asia, Central and South America, the Caribbean, and certain regions of Africa. Much of the information pertaining to the epidemiology of cyclosporiasis comes from studies performed in Peru and Nepal. Positive detection rates of *Cyclospora* reported from various areas are as follows: 1.1% of children (<18 years of age) in Lima, Peru (111); 12% of symptomatic children (18 to 60 months of age) in Nepal (74); and 3 to 4% of the general population in Guatemala (C. Bern, D. B. Hernandez, M. J. Arrowood, M. B. Lopez, A. M. De Merida, B. L. Herwaldt, and R. Klein, *Prog. Abstr. Int. Conf. Emerg. Infect. Dis.*, p. 47, 1998). Infection affects not only immunocompromised hosts but also normal subjects, with higher prevalence rate and more symptomatic cases seen in children than in adults. The presenting symptoms are predominantly prolonged watery diarrhea, sometimes relapsing, with weight loss, and other commonly associated clinical features include anorexia, fatigue, abdominal pain, and vomiting (152). The duration of diarrheal illness in patients infected with human immunodeficiency virus (HIV) is usually longer (136).

The natural reservoir for *Cyclospora* has not yet been well defined, although contaminated water has been incriminated with no definitive proof. Recently, three more *Cyclospora* species have been isolated from stool samples of Ethiopian monkeys: *Cyclospora papionis* from baboons (*Papioanubis* sp.), *Cyclospora colobi* from colobus monkeys (*Colobus guereza*), and *Cyclospora cercopitheci* from green monkeys (*Cercopithecus aethiops*) (54). These four species, one from

humans and three from monkeys, form a single branch that falls between avian and mammalian *Eimeria* spp.

It appears that the organism can survive fully chlorinated tap water. The first outbreak of acute watery diarrheal illness caused by *Cyclospora* was reported in 1995 involving 11 house staff members from a physician's dormitory where the tap water may have been contaminated by stagnant water (77). In May and June of 1996 and April through June of 1997, two separate large outbreaks of cyclosporiasis were reported. The first, involving 1,465 cases (978 laboratory confirmed) from the United States (20 states and the District of Columbia) and Canada (Quebec and Ontario), and the second, involving 370 patients (140 laboratory confirmed) from eight states and Ontario, both were associated with consumption of raspberries imported from Guatemala (27–30, 71).

Unlike cryptosporidiosis, for which no effective therapeutic agent is currently available, effective treatment for cyclosporiasis is achieved both in adults and children with the use of trimethoprim-sulfamethoxazole (75, 111). Better understanding of the natural reservoir, precise means of transmission, and tolerance of the organism to various physical and chemical agents as well as the development of an in vitro cultivation method and an animal model is needed for better management of the infection.

EHRLICHIA

Human ehrlichioses represent newly recognized zoonotic infections usually transmitted to humans following a tick bite. The etiologic agents, ehrlichiae, are obligate intracellular, gram-negative bacteria that infect and propagate within cytoplasmic vacuoles of mononuclear and granulocytic phagocytes. The genus *Ehrlichia* belongs to the family *Rickettsiaceae*. There are currently 14 different species within this genus, which are divided into three distinct clades based on their serologic cross-reactivity, characteristics of major immunogenic surface proteins, and cellular tropism (51, 55, 81). The *Ehrlichia phagocytophila* genogroup, *Ehrlichia canis* genogroup (i.e., *Ehrlichia chaffeensis*), and *Ehrlichia sennetsu* genogroup (i.e., *E. sennetsu*) represent the three clades to which the three currently known human ehrlichia pathogens, the human granulocyte ehrlichia (HGE) agent, the human monocytic ehrlichia (HME) agent *E. chaffeensis*, and the sennetsu ehrlichia agent *E. sennetsu*, belong, respectively. More recently another *Ehrlichia* species, *Ehrlichia ewingii*, has also been shown to infect humans (73).

The known reservoirs for *E. chaffeensis* include white-tailed deer and domestic dogs, and those for HGE appear to be white-footed mice, chipmunks, and voles. The reservoir for *E. sennetsu* is unclear, but it is closely associated with consumption of raw fish, presumably fish infested with ehrlichia-infected flukes (51, 193). The tick vectors involved in human ehrlichioses include *Ixodes scapularis*, *Ixodes pacificus*, and *Ixodes ricinus* for HGE (110, 155, 187), which are the same as Lyme disease-transmitting ticks; *Amblyomma americanum* (lone star tick) for *E. chaffeensis* (5); and *Rhipicephalus sanguineus* (brown dog tick) for *Ehrlichia canis* (173).

From 1986 through 1997, there have been 1,223 cases of human ehrlichioses (742 HME, 449 HGE, and 32 unspecified); HME is most commonly reported from southeastern and south central states, including Arkansas, North Carolina, Missouri, and Oklahoma, and HGE is most commonly reported from northeastern and upper midwestern states,

TABLE 1. Clinical and laboratory abnormalities in human monocytotropic and granulocytotropic ehrlichioses[a]

Sign, symptom, or laboratory finding	% Patients with abnormal findings	
	HME	HGE
Fever	97	94–100
Headache	81	61–85
Chills or rigors	67	39–98
Myalgia	68	78–98
Malaise	84	98
Nausea	48	39
Anorexia	66	37
Vomiting	37	34
Diarrhea	25	22
Abdominal pain	22	
Rash	36	2–11
Cough	26	29
Dyspnea	23	
Lymphadenopathy	25	
Confusion	20	17
Leukopenia	60	50–59
Thrombocytopenia	68	59–92
Elevated AST[b]	86	69–91
Elevated ALT[c]	80	61
Elevated urea nitrogen	38	
Elevated creatinine	29	70

[a]Reprinted from reference 195 with permission.
[b]AST, aspartate aminotransferase.
[c]ALT, alanine aminotransferase.

including Connecticut, New York, Wisconsin, and Minnesota (8). These epidemiologic data reflect the distribution of corresponding tick vectors for each type of ehrlichiosis. Both HME and HGE are highly seasonal, and nearly 70% have an onset in May or June, which coincides with the peak seasonal activity of the nymphal stage of the vector ticks.

Human ehrlichioses HME and HGE are multisystem diseases which are clinically indistinguishable from each other. A typical case scenario would be a previously healthy male patient 40 to 55 years of age who presents with an acute influenza-like illness manifested by fever, myalgia, rigors, headache, anorexia, nausea, vomiting, cough, pharyngitis, diarrhea, and abdominal pain. Lymphadenopathy and skin rash may also be present. Clinical and laboratory abnormalities in HME and HGE are summarized in Table 1 (196). The majority of patients (>80%) give a history of recent tick exposure, and common laboratory abnormalities observed in more than two-thirds of the patients include leukopenia, thrombocytopenia, and elevated hepatic transaminases (8, 56, 193, 196).

The diagnostic approach should begin with a thorough analysis of the case history, including the epidemiology and tick exposure, physical findings, and pertinent laboratory test results that suggest the diagnosis. Attempts to make a specific diagnosis should not delay institution of appropriate antibiotic therapy, because the diagnostic test results usually do not become available in a timely fashion and the disease process may progress rapidly. A peripheral smear of buffy coat leukocytes should be subjected to Romanowsky staining (Giemsa or Wright stain) to identify the ehrlichia morulae. Although this is a low-yield procedure in patients with HME (1%), the positive yield in patients with HGE has been reported to be between 20 and 80% (194, 195). The isolation of the etiologic agent is, of course, the "gold standard" and most specific, but it is labor-intensive, time-consuming, and insensitive.

The majority of cases of HME and HGE are diagnosed by retrospective serologic testing, and the most commonly used serologic test is the indirect fluorescent-antibody test, which is available through commercial laboratories and public health laboratories. During the acute phase, the serum specimen should be obtained and initially tested for the presence of antibodies against *E. chaffeensis* or *E. phagocytophila* group ehrlichiae, and an aliquot should be saved to be tested with the convalescent serum in 3 to 6 weeks. Diagnosis is made by finding either a rise or fall of fourfold or more in antibody titer during the time period, and a single elevated titer of ≥1:80 is considered significant (8, 56, 195, 196).

PCR assays that utilize ehrlichial 16S rRNA genes as target sequences for amplification represent a sensitive and rapid method by which patients with HME or HGE can be identified. Those assays are currently performed in commercial laboratories and at the Centers for Disease Control and Prevention (CDC) (6, 51, 55). A negative PCR assay, however, should be interpreted with caution, because factors that would clear ehrlichial antigenemia may lead to a false-negative result. Effective antimicrobial therapy and spontaneous recovery may lower the antigen level below that which can be detected by the PCR method.

In vitro susceptibility testing using *E. chaffeensis* and the HGE agent in cell culture systems confirms that doxycycline and rifamycins are rapidly ehrlichicidal, whereas chloramphenicol is ineffective (19, 91). Trovafloxacin is another antibiotic that has been shown to be very active in vitro against the HGE agent, while ciprofloxacin and ofloxacin are only moderately active (91). Gentamicin, macrolides, clindamycin, trimethoprim-sulfamethoxazole, and β-lactam antibiotics are not active. Thus far, clinical efficacy has been well demonstrated only with tetracycline and doxycycline in human ehrlichiosis. In clinical situations where tetracyclines are contraindicated, trovafloxacin may be an option, but clinical efficacy has not yet been demonstrated.

E. COLI O157:H7

E. coli serotype O157:H7 was first identified as a human pathogen in 1983 following an investigation of two separate outbreaks, one in Michigan and the other in Oregon, of an unusual gastrointestinal illness, characterized by severe crampy abdominal pain, watery diarrhea followed by bloody diarrhea, and little or no fever, involving 47 patients during the first half of 1982 (156). These outbreaks were linked to the consumption of poorly cooked hamburger patties at restaurants of a single fast-food chain. Since then, the organism has been recognized as a major cause of hemorrhagic colitis and the hemolytic-uremic syndrome (HUS) and represents the most commonly identified etiologic agent to cause bloody diarrhea in North America and Europe (14). In a 1-year (1985–1986) prospective, population-based study in the Puget Sound area of Washington State, it was estimated that the incidence of *E. coli* O157:H7 infection would be 8/100,000 person-years or approximately 20,000 cases with 250 deaths each year in the United States (109).

E. coli O157:H7 is an emerging food-borne human pathogen which is often referred to as either the enterohemorrhagic *E. coli* for its propensity to cause bloody diarrhea or hemorrhagic colitis or the Shiga toxin-producing *E. coli* because strains all produce one or more bacteriophage-coded Shiga-like toxins (130, 135, 180; A. D. O'Brien, T. A. Lively, T. W. Chang, and S. L. Gorbach, Letter, *Lancet* **ii**:573, 1983). Although *E. coli* O157:H7 represents only one of nearly 200 serotypes of *E. coli* isolated from patients with human diarrheal illnesses, this specific serotype of *E. coli* is recognized as the most important cause of hemorrhagic colitis and HUS (67, 181).

It has been demonstrated that the gastrointestinal tract of healthy animals—including cattle (120), sheep (96), goats (166), deer (87), horses (190), and dogs (190)—can be colonized with *E. coli* O157:H7. Many outbreaks of hemorrhagic colitis and/or HUS have been traced back to the consumption of food or beverage or contact with water contaminated with animal feces. Ground beef (10, 105), unpasteurized milk (M. L. Martin, L. D. Shipman, J. G. Wells, M. E. Potter, K. Hedberg, and I. K. Wachsmuth, Letter, *Lancet* **ii**:1043, 1986) or apple juice (39), apple cider (12), deer meat jerky (87), unchlorinated water (182), and lake water (86) are some examples that have been incriminated as sources of *E. coli* O157:H7. It is currently believed that the origin of *E. coli* O157:H7 is the gastrointestinal tract of animals.

Infection with *E. coli* O157:H7, even when caused by a single strain of the organism, may lead to a remarkably wide spectrum of clinical manifestations ranging from being asymptomatic to fatal cases (67), with patients at the extremes of age at greater risk for more-symptomatic illness, such as HUS and thrombotic thrombocytopenic purpura. It is estimated that *E. coli* O157:H7 may be responsible for 0.6 to 2.4% of all cases of diarrhea and 15 to 36% of cases of bloody diarrhea or hemorrhagic colitis. In addition 38 to 61% of patients infected with *E. coli* O157:H7 may develop hemorrhagic colitis. The incidence of *E. coli* O157:H7 in HUS may be as high as 46 to 58%, and 2 to 7% of *E. coli* O157:H7 infection may progress to HUS (181).

Infection with *E. coli* O157:H7 carries an incubation period of 1 to 9 days (mean, 4 to 6 days) (10, 24, 182), which is followed by abdominal cramps, diarrhea, and bloody diarrhea or hemorrhagic colitis in the majority of patients. A case-control nationwide, population prevalence study involving 10 U.S. hospitals over a 2-year period revealed that 118 (0.39%) of the 30,463 fecal specimens tested positive for *E. coli* O157:H7 (171). The highest age-specific positive yields for *E. coli* O157:H7 from fecal specimens were in patients 5 to 9 years (0.90%) and 50 to 59 years (0.89%) of age. Clinical symptoms observed in patients infected with *E. coli* O157:H7 were as follows: diarrhea (98.0%), bloody diarrhea (91.3%), abdominal cramps (90.5%), seven or more bowel movements per day (65.5%), reported fever (35.0%), vomiting (35.6%), and hospitalization (47.1%). On the other hand, clinical signs and laboratory findings were objective fever (41.4%), abdominal tenderness (72.0%), visible blood in stool specimens (63%), stool specimen positive for occult blood (82.8%), and fecal leukocytes in stool specimens (70.5%), with a lower incidence of specimens being ≥10 leukocytes per high power field (23.9%). Approximately 2 to 7% of patients infected with *E. coli* O157:H7 may progress to develop HUS, which is a distinct clinical entity characterized by microangiopathic hemolytic anemia, thrombocytopenia, and varying degrees of renal failure. This entity is seen more commonly in infancy and younger children, in whom *E. coli* O157:H7 is considered to be the commonest etiologic agent (181). Although the overall mortality rate of patients infected with *E. coli* O157:H7 is rather low at 1.2% (66), once complications such as HUS develop, the case fatality rate in children is 5 to 10% (82), whereas the rate may be as high as 88% among elderly patients (24).

The diagnostic approach to the *E. coli* O157:H7 infection should begin with a high index of suspicion in any patient with bloody diarrhea. In addition it is imperative that a special request be made to indicate that a specific serotype of *E. coli* is being sought, because routine screening for this pathogen may not be performed in all clinical laboratories. A presumptive identification of *E. coli* O157:H7 is based on the fact that *E. coli* O157:H7 does not ferment sorbitol rapidly, unlike most other strains of *E. coli*. When plated on sorbitol-MacConkey agar, *E. coli* O157:H7 appears sorbitol negative at 24 h, and this procedure is 100% sensitive and 85% specific for detecting the organism (113). The probability of positive culture is highest during the first week of illness (75 to 100%); otherwise, the yield falls precipitously afterward (0 to 33%) (186, 201). The sorbitol-negative colonies can then be subjected to further characterization to confirm the presence of serotype-specific O157 and/or H7 antigen with a commercially available latex slide agglutination test (37, 114), followed by a serologic confirmation by direct immunofluorescence assay (37, 114, 188). More recently, assay methods that allow a rapid detection of *E. coli* O157:H7 directly from stool specimens have also been developed by employing an immunofluorescence stain (indirect fluorescent antibody) (137) and enzyme-linked immunosorbent assay (ELISA)

(138). Another sensitive method of diagnosing *E. coli* O157:H7 infection is to test for the presence of the organism-specific Shiga-like toxin(s) also known as Vero toxin, which can be detected in culture broth filtrate or stool extracts (83; A. D. O'Brien, T. A. Lively, M. E. Chen, S. W. Rothman, and S. B. Formal, Letter, *Lancet* **i:**702, 1983). This approach, although more cumbersome and laborious to perform, is specific for toxin-producing *E. coli*, including serotype O157:H7, and also is useful in the diagnosis of *E. coli* O157:H7 infection during the late phase of the infection when the culture may be negative. Additional information relating to *E. coli* and toxin detection can be found in the bacteriology chapter of this manual.

Although most strains are susceptible to antimicrobial agents, antibiotic therapy for *E. coli* O157:H7 infection has not been shown to modify the clinical course (134, 156), and in fact antimicrobial therapy may be deleterious (134, 205). It would be prudent to hold off on antimicrobial therapy in patients with *E. coli* O157:H7 infection until more definitive data become available with a large multicenter, randomized therapeutic trial study.

BARTONELLA INFECTIONS

The genus *Bartonella* as is now described consists of members of the former genus *Rochalimaea*, as well as the previous member of the *Bartonella* genus, *Bartonella bacilliformis*. The merger of the two genera was based on similarities in DNA sequence, particularly those of the 16S rRNA gene (17). There are now four species of *Bartonella* that are well documented to be human pathogens, *B. bacilliformis*, *Bartonella quintana*, *Bartonella henselae*, and *Bartonella elizabethae*. In addition, several species of *Bartonella* have been described as pathogens in a variety of animal species, including dogs, cats, and voles (7, 15, 38, 99). All *Bartonella* species are gram-negative, oxidase-negative, fastidious, aerobic rods (13, 16).

The understanding of the variety of clinical presentations due to *Bartonella* species has expanded as a result of the AIDS epidemic, the ability to detect DNA by PCR amplification, and the different manifestations of disease in immunocompetent and immunocompromised hosts.

Immunocompromised Hosts

The initial report of *Bartonella* infection in a patient with AIDS was in 1983 with multiple subcutaneous vascular nodules that when visualized with electron microscopy demonstrated multiple rod-like organisms (178). The vascular proliferative lesions that were similar in appearance to Kaposi's sarcoma were subsequently called bacillary angiomatosis (94). Bacillary angiomatosis usually occurs in late-stage HIV infection, when the CD4 lymphocyte count is less than 100 cells. The skin lesions are usually erythematous and may be pedunculated. There may be different appearances in multiple lesions in the same patient. In addition to the skin, lesions of bacillary angiomatosis have been found in the respiratory and gastrointestinal tracts, reticuloendothelial system, vascular system, bone, brain, and lymph nodes (61, 88, 153, 163, 169, 177, 185). Although most common in patients with AIDS, bacillary angiomatosis has also been described in organ transplant recipients and other immunosuppressed patients (126, 170, 189).

The most common extracutaneous manifestation of bacillary angiomatosis are peliosis hepatis and splenitis (139). The venous lakes within the liver parenchyma that characterize peliosis hepatis are best visualized by computed tomography (176). Both *B. henselae* and *B. quintana* have been implicated as causes of bacillary angiomatosis and peliosis hepatis in immunocompromised hosts.

Immunocompetent Hosts

Cat scratch disease is the most common manifestation of *Bartonella* infection. In children and young adults, cat scratch disease typically presents as regional adenopathy after an inoculation (9, 23, 116). Five to twenty percent of patients may present with complications other than regional lymphadenopathy (115). These complications include Parinaud's oculoglandular syndrome, consisting of unilateral conjuntivitis and preauricular adenopathy (22); hepatic or splenic involvement (43, 104, 112); and central nervous system involvement with encephalopathy (117, 129). The majority of cat scratch disease is caused by *B. henselae*; the different clinical presentations associated with *B. henselae* infection appear to predominantly be due to the host immune system although there may be differences in virulence between strains as well (4).

Bartonella species have recently been recognized as causes of bacteremia and endocarditis as well (49, 76, 79, 147, 175). Endocarditis has been described with *B. henselae* and *B. elizabethae* as well as *B. quintana*. These complications have been described in both immunocompetent and immunocompromised hosts. Profound weight loss is a common feature, and the bacteremia may be relapsing.

The histologic appearance of the lesions of bacillary angiomatosis are characteristic and consist of lobular proliferation of blood vessels which should distinguish the lesion from Kaposi's sarcoma (101). In cat scratch disease, histology is also characteristic with so-called stellate granulomas and microabscesses (80). In both diseases, Warthin-Starry staining will identify the characteristic bacillary forms of *Bartonella*, usually in clumps (102, 198).

Bartonella species may be isolated from blood, lymph nodes, skin biopsy, or organ biopsy specimens from infected individuals. All methods of growth require extended incubation times. The optimal recovery of *Bartonella* species from blood appears to be through the use of lysis-centrifugation methods (18). Samples are subsequently plated onto fresh blood or chocolate agar plates and incubated for prolonged periods in 5% carbon dioxide (93). Several reports describe the use of acridine orange staining of BACTEC bottles after at least 8 days of incubation with subsequent subculture on chocolate agar (97). Other culture media that have been used for *Bartonella* species include cell culture systems and the recent development of a defined liquid medium (200, 206). Endothelial cell lines have been used for both tissue and blood from infected patients, with growth occurring after at least several weeks of incubation. The overall sensitivity and specificity of culture techniques for *Bartonella* species have not been well described.

Identification of isolates to species level is usually of limited clinical value. When identification to species level is performed, it is usually accomplished through genetic methods, either by DNA-DNA hybridization or by PCR amplification of specific genes with restricton endonuclease analysis. The citrate synthase gene has been used for this purpose where a given clinical isolate can undergo DNA amplification of the citrate synthase gene with subsequent nuclease digestion and comparison of the obtained pattern with those obtained from standard strains for species identification (150).

Nucleic acid detection techniques for *Bartonella* species were first described by Relman based on amplification of the 16S rRNA gene (151). This work was the first to identify DNA sequences closely related to *B. quintana* from patients with bacillary angiomatosis and led to further work identifying these sequences from other patients with other manifestations of *Bartonella* infection (3, 11, 119, 149). Although the initial work in this area was done with broad-range 16S rRNA primers for PCR, there are now numerous reports of *Bartonella*-specific PCR primers that can identify organisms from blood and tissue of infected patients (2, 65, 100).

An indirect imunofluorescence assay has been developed to detect antibodies to *B. henselae* in patients with bacillary angiomatosis and cat scratch disease. Several studies have now demonstrated that in patients who meet strict clinical definitions for cat scratch disease, >90% of patients will have antibody titers of >1:64 against *B. henselae* (42, 98, 148). There is occasional cross-reactivity to the causative agents of brucellosis and Lyme disease (98). Several commercial laboratories offer antibody testing against *B. henselae*; although most patients with cat scratch disease or bacillary angiomatosis will have a positive antibody titer, the role of routine serologic testing in these entities is unclear.

BOVINE SPONGIFORM ENCEPHALOPATHY AND VARIANT CREUTZFELDT-JAKOB DISEASE

Bovine spongiform encephalopathy was described as part of an epizootic in the United Kingdom in 1986 (203). The disease is speculated to have arisen from the practice of feeding cattle the carcasses of scrapie-infected sheep and then propagated by feeding young calves with bovine meat and meal (128; J. Gibbs and C. Debate, Presentation, International Symposium on Spongiform Encephalopathies, 1996). Cases peaked in the early 1990s and have subsequently declined as a result of control measures (128).

In the mid-1990s, 10 cases of a variant form of Creutzfeldt-Jakob disease were reported in the United Kingdom, and fear spread that this variant was related to the epizootic of bovine spongiform encephalopathy (Gibbs and Debate, presentation). These affected patients were from geographically disparate areas in the United Kingdom and did not have epidemiologic evidence of increased risk of contact with affected cattle. No bovine spongiform encephalopathy has been reported in the United States, and active surveillance for cases of Creutzfeldt-Jakob disease in the United States has not revealed any cases with a clinical course compatible with the variant form of Creutzfeldt-Jakob disease described from the United Kingdom (see below) (164).

The patients affected by variant Creutzfeldt-Jakob disease in the United Kingdom have been unusually young, with a mean age at onset of 28 years, and have lived for a mean of 14 months after diagnosis, compared with a mean survival of less than 6 months for patients with Creutzfeldt-Jakob disease in the United States (United Kingdom Spongiform Encephalopathy Advisory Committee statement to House of Commons, 20 March 1996). The atypical clinical manifestations of the new variant disease include prominent behavioral features at the time of diagnosis, with the subsequent development of ataxia followed by dementia and myoclonus (Gibbs and Debate, presentation).

The diagnosis of the new variant of Creutzfeldt-Jakob disease rests on the demonstration of the typical neuropathologic pattern in brain tissue (Gibbs and Debate, presentation). The brains of affected individuals reveal accumulations of amyloid-like plaques in the cerebrum and cerebellum; these plaques are surrounded by material that by immunocytochemical analysis contains large amounts of prion protein, the transmissable agent which consists of an abnormal form of a host glycoprotein (204). Molecular studies have revealed that the prion protein associated with new-variant Creutzfeldt-Jakob disease is distinct from those that cause typical Creutzfeldt-Jakob disease (40). This analysis is based on Western blot analysis of the prion protein from affected brains and requires a research laboratory and brain tissue.

Further identification of prospective cases of new-variant Creutzfeldt-Jakob disease will require ongoing surveillance and the development of further strain typing studies to assess the relationship between prions from patients with new-variant disease and those from animals with spongiform encephalopathy.

HHV-8 AND KAPOSI'S SARCOMA

Human herpesvirus 8 (HHV-8), also known as Kaposi's sarcoma-associated herpes virus, was originally identifed at a molecular level using a technique called representational difference analysis (36). This method is able to clone the molecular difference between tissues that contain small amounts of DNA sequences in diseased tissues and was applied to patients with Kaposi's sarcoma. HHV-8, like Epstein-Barr virus, is a gamma-herpesvirus, able to establish latent infection in lymphocytes and associated with cell proliferation. Many gene products from HHV-8 have been described, including a viral thymidylate kinase, dihydrofolate reductase, cytokine genes, cyclin-like proteins, and a G-protein-coupled receptor (126). Several of the gene products mimic cytokine pathways and may induce malignant cellular transformation.

There are three subtypes of HHV-8, with all three strains occurring in the United States (210). Serological prevalence studies suggest that the most likely mode of transmission of HHV-8 is through sexual transmission (85). Different immunofluorescent assays have been developed against either latent or lytic antigens of HHV-8 (60, 84). Antibody to lytic antigens appears to be more common, with prevalence rates ranging from 100% in African Kaposi's sarcoma patients to 90 to 100% of male homosexual HIV-infected patients to 20 to 25% in HIV-infected injection drug users and HIV-infected women (103).

Using molecular amplification techniques, HHV-8 DNA has been detected in genital secretions from patients with AIDS-related Kaposi's sarcoma, and to a lesser degree from HIV-infected and uninfected men without Kaposi's sarcoma (124). Other studies have found HHV-8 sequences in oral secretions or in tissues in the oral cavity, suggesting that saliva contact may be another mode of transmission of HHV-8 infection (45).

The strongest association between HHV-8 and human disease is with Kaposi's sarcoma (those associated with AIDS as well as other types). The initial report of the high prevalence of HHV-8 DNA in lesions from patients with Kaposi's sarcoma (36) has now been followed by the detection of HHV-8 DNA from Kaposi's sarcoma lesions in a variety of organ sites (25, 41, 57). In addition, the association appears to be specific, with only a few other associations with other conditions (see below). From a pathogenic standpoint, HHV-8 DNA has been found in peripheral blood mononuclear cells prior to the onset of Kaposi's sarcoma, and patients with Kaposi's sarcoma have high levels of antibody to HHV-8, whereas HIV-infected individuals without Kaposi's sarcoma do not (172, 202). Finally, HHV-8 has been propagated from a human epithelial cell line by coculture with Kaposi's sarcoma cells (58).

In addition to Kaposi's sarcoma, HHV-8 has been associated with other cancers, usually in patients with abnormal immune systems. The strongest non-Kaposi's sarcoma association is with body cavity-based lymphomas (35). In these lymphomas, there is a male predominance with primary localization to a body cavity and rapid clinical progression. Finally, HHV-8 DNA sequences have been identified in peripheral blood mononuclear cells and lesions from patients with multicentric Castleman's disease. This entity is characterized by lesions (lymph node or reticuloendothelial system) with angiofollicular hyperplasia and clinical signs of fever, adenopathy, and splenomegaly (53).

The diagnosis of the clinical entities associated with HHV-8 relies on the histologic characterization of the involved tissue. However, there have been recent advances in the detection of infection with HHV-8 by serologic or molecular methods that could lead to more directed attempts at preventing the oncologic complication of viral infection in the future.

The initial serologic assays for HHV-8 were indirect immunofluorescence assays. The distribution of HHV-8 antibodies using this assay varied from 100% in Italian patients with classic (non-HIV-associated) Kaposi's sarcoma (106) to 83% in U.S. AIDS patients with Kaposi's sarcoma, 35% in homosexual men with HIV but not Kaposi's sarcoma, and 8% in patients with syphilis but not HIV infection (123). In general, the seroprevalence of HHV-8 infection in HIV-infected cohorts parallels the relative risk of Kaposi's sarcoma. A more sensitive immunofluorescence assay has recently been developed that detects antibodies in 100% of African patients with endemic Kaposi's sarcoma and 95% of American patients with AIDS-related Kaposi's sarcoma. Using this assay, 90% of American HIV-infected homosexual men, 23% of HIV-infected injecting drug users in the United States, and 21% of HIV-infected American women have antibodies to HHV-8. In the general population in the United States, the seroprevalence rate is 25% using this assay (174).

Immunoblot assays have also been developed for the detection of antibodies to HHV-8. These assays are, in general, less sensitive than the immunofluorescence assays. It appears that the antibodies detected with these assays might be present only after the virus has been reactivated as may occur with immunosuppression (174). Thus, the immunoblot assay may not detect past exposure to the virus but may detect only markers of reactivation.

WNV

West Nile encephalitis virus (WNV) is a member of the flavivirus family, antigenically related to Murray Valley encephalitis virus, Japanese encephalitis virus, and St. Louis encephalitis virus. WNV was first isolated in the West Nile province of Uganda in 1937 (121). The first recorded epidemics occurred in Israel in the 1950s. The largest recorded epidemic occurred in South Africa in 1974 (31). In the summer of 1999, an outbreak of infection was reported for the first time in North America, centered in New York City and the surrounding suburban areas (32).

WNV is maintained in nature by the *Culex* mosquito and transmitted to wild birds who develop high-grade persistent viremia. Humans are incidental hosts and not involved in the transmission cycle. In areas of endemicity, seroprevalence rates range between 10 and 95%. In areas where the virus has been recently introduced with nonimmune hosts (such as the New York City outbreak), the outbreak may be explosive. WNV can infect a wide range of vertebrates, but in humans it usually produces either asymptomatic infection or mild febrile disease. Within its normal geographic distribution of Africa, the Middle East, western Asia, and Europe, WNV has not been documented to cause epizootics in birds; crows with antibodies to WNV are common, suggesting that asymptomatic or mild infection usually occurs among crows in those regions. Therefore, an epizootic producing high mortality in crows and other bird species is unusual for WNV and may represent introduction to a native bird population or a new virulent strain. Migratory birds may play an important role in the natural transmission cycles.

The incubation period after exposure is short, between 2 and 6 days. The initial manifestations of disease include low-grade fever, myalgias, and weakness and other nonspecific findings, including abdominal pain, nausea, and diarrhea. Rash is a common finding and may occur in up to 50% of infected individuals. The rash, if present, is truncal with involvement of the face or arms in a livedo-reticularis pattern. Other physical findings include adenopathy (particularly submental), pharyngitis, and conjunctivitis. Hepatosplenomegaly may be more common than in other etiologies of viral encephalitis. The central nervous system involvement is usually without seizures; tremors, flaccid paralysis, and aseptic meningitis may also occur. Of the 56 confirmed and probable cases reported from the New York City outbreak, there were seven deaths (33).

In the New York outbreak, the genomic sequences identified to date from a human brain, virus isolates from zoo birds, and viruses isolated from a dead crow and two mosquito pools from Connecticut appear identical. Based on preliminary serologic testing, this outbreak was originally believed to be caused by the St. Louis encephalitis virus, a virus antigenically related to WNV. Results of PCR-based sequencing that identified WNV prompted more specific testing. The immunoglobulin M (IgM) capture ELISA used by the CDC in testing serum and cerebrospinal fluid samples in the outbreak is rapid, sensitive, and quantitative and also suggested WNV as the culprit. The limitations of some serologic assays emphasize the importance of isolating the flavivirus from entomologic, clinical, or veterinary material. The availability of virus isolates and genomic sequences from birds and human brain tissue confirmed the identity of this virus as WNV. Isolates are determined to be WNV based on an antigenic mapping procedure using a panel of monoclonal antibodies that is capable of distinguishing West Nile from the related Kunjin and St. Louis encephalitis viruses (33).

Although it is not known when and how a WNV was introduced into North America, international travel of infected persons to New York or transport by imported infected birds may have played a role. WNV is transmitted principally by *Culex* species mosquitoes, but also can be transmitted by *Aedes*, *Anopheles*, and other species. The predominance of urban *Culex* mosquitoes trapped during this outbreak suggests an important role for this species. Enhanced monitoring through surveillance for early detection of this virus outside of the affected area will be crucial to guide extension of control measures. Ongoing monitoring of mosquito and bird populations in New York City and its environs over the winter of 1999–2000 has indicated that the virus was able to survive the winter and has the potential to cause an ongoing epidemic in the coming year (34).

EBOLA AND MARBURG VIRUSES

Ebola and Marburg viruses are in the family *Filoviridae*. While they can cause devastating human disease with significant mortality, relatively little is known about their natural history. While primates are the only known hosts susceptible to infection with these filoviruses, they are not believed to be reservoirs for the virus (141). During epidemiologic investigations following known epidemics, however, the reservoir has not been determined. Spiders, soft ticks, bats, and monkeys have all been suggested as possible reservoirs (141). This has contributed to the mystery and fear associated with these agents.

Filoviruses consist of a single strand of negative-sense RNA. There are four different strains of Ebola virus (Zaire, Sudan, Côte d'Ivoire, and Reston), and today there are known to be two strains of Marburg virus (161, 162). There is significant variation in genetic structure and surrounding structural proteins between the various subtypes and between Ebola and Marburg viruses (144). Serologic cross-reactivity has not been shown between the two viruses.

There have been several distinct outbreaks of Marburg and Ebola viruses. In 1967, Marburg viruses was first seen in Europe. This outbreak was linked to importation of African green monkeys for use in medical research (118). Additional cases of Marburg virus were documented in Kenya in 1980 and again in 1987.

Ebola virus was first seen in 1976. It caused two epidemics in Zaire and the Sudan, which appeared to be unrelated in terms of origin (207, 208). The rates of mortality in these two epidemics were appallingly high, with 88% of 318 documented patients in Zaire eventually dying; in the Sudan 53% of 284 total patients succumbed to their disease (207, 208). Limited outbreaks occurred in Zaire in 1977 and 1979. However, another major epidemic occurred in 1995-1996 in Gabon. This outbreak had a case-fatality rate of close to 80% (59, 63, 89).

Finally in 1989 and 1990 an outbreak occurred in Reston, Va., related again to the importation of primates for medical research. The origin of these primates, however, was the Philippines (142). While this strain of Ebola virus caused significant mortality in the monkeys there were no human deaths directly attributed to the virus. In addition, four humans were found to have seroconverted without developing any signs of clinical disease (142, 158).

The modes of transmission of these viruses are unclear to date. Clearly contact with monkeys or monkey blood was a risk factor in the 1967 Marburg virus outbreak. The 1976 Zaire epidemic was in part related to improperly sterilized needles used in medical procedures. In the Reston outbreak aerosol and/or droplet spread was noted. In general, when secondary cases occur, whether it be in medical personnel or in close family contacts, blood exposure to skin or mucous membranes appears to be an important factor.

The clinical signs of evolving filovirus infection can be multisystemic and catastrophic. There is an incubation period of 5 to 10 days which is followed by the patient developing fever, myalgias, and headache (21, 48, 62, 89, 118, 207, 208). Additional systemic signs that can develop include gastrointestinal complaints, chest discomfort, photophobia, lymphadenopathy, jaundice, pancreatitis, delirium, and even coma (141). Cutaneous signs of bleeding disorders may include the development of petechiae, hemorrhages, and ecchymoses. By the second week of infection, patients will either improve or succumb to multiorgan system failure. Liver failure and disseminated intravascular coagulation are common terminal events (141). For those who survive, recovery may take a prolonged period of time. In addition, there may be further systemic complications during this convalescent period, including orchitis, recurrent hepatitis, uveitis, and transverse myelitis (118, 143). Mortality rates for the various subtypes have been listed above. In general the Reston strain is the only one known to cause subclinical disease. There are several diagnostic options for Ebola and Marburg viruses. A complete social and travel history is very important in identifying individuals at risk for filovirus infection. The viruses can be cultured during the acute phases of the disease. In addition, antigen detection and PCR techniques are available to aid in the diagnosis of filoviruses. An ELISA exists for the detection of IgM that can be found in the early convalescence phase of the disease. The IgG serologic test is not reliable however (141).

HHV-6

The search for viruses in patients with lymphoproliferative disorders led to the discovery of HHV-6 in 1986 (160). The virus was isolated from lymphocytes of patients and was initially designated human B-cell lymphotropic virus. Further study has determined, however, that it actually is a T-cell lymphotropic virus (D. V. Ablashi, S. Z. Salahuddin, S. F. Josephs, F. Iman, P. Lusso, R. C. Gallo, F. V. Di Marzo, and P. D. Markham, Letter, *Nature* **329:**207, 1987).

HHV-6 is a beta-herpesvirus and is worldwide in its distribution (132). HHV-6 is a DNA virus that is approximately 160 to 200 nm in diameter. There are believed to be two distinct viral forms (HHV-6A and HHV-6B) which are similar in genetic structure but differ in biologic and epidemiologic features (44). HHV-6A has been found in indiviuals who are immunosuppressed but not in patients with roseola. HHV-6B has only been found in individuals with roseola (47).

The virus has an affinity for CD4 lymphocytes (160) but may also affect CD8 lymphocytes, natural killer cells, macrophages, megakaryocytes, epithelial cells, and neural cells (1, 168). Perhaps more importantly, HHV-6 may change the natural history of other viral infections. This may be through a direct mechanism whereby HHV-6 acts as a coinfectious agent with another virus or through indirect means whereby another virus is potentiated by the HHV-6-induced immunosuppression in the host (183, 184). Epstein-Barr virus, cytomegalovirus and HIV have all been associated with HHV-6 infection.

Exposure to HHV-6 is very common and occurs worldwide. Ninety percent of children less than 2 years old have been found to be seropositive for HHV-6 in many studies (157). Maternal antibody provides protection early in life, with natural infection occurring almost uniformly by age 2 years. Transmission mechanisms are unclear, but it is believed to be related to contact with sick household members (47). The virus is known to be shed in urine and saliva.

Individuals who are immunosuppressed are at high risk for HHV-6 complications. It is not known whether this is from reactivation of latent virus or superinfection (47). In the transplant population, different organs are associated with differing rates of HHV-6 infection as follows: liver transplants, 24 to 31%; renal transplants, 38 to 66%; heart transplants, 14%; bone marrow transplants, 38 to 60% (46, 47, 70, 131, 168). The transplanted organ, blood product transfusions, and posttransplant immunosuppression are all important factors in promoting infection after transplantation (78, 168, 197).

Infection with HHV-6 is usually asymptomatic. Symptomatic infection in infants and the development of roseola infantum (sixth disease) has been well established (209). There is a wide spectrum of disease in infants who are symptomatic from HHV-6 infection. In addition to the characteristic rash, high fever, otitis, lower respiratory tract disease and gastrointestinal disease have all been described (68, 146). HHV-6 infection may play a role in a significant number of fever-induced seizures in young children.

Infection in immunocompromised individuals is being recognized as more of a problem, particularly in transplant and HIV-infected patients. This occurs either through direct mechanisms of the virus on the host or via indirect mechanisms. Symptomatic disease is felt to be more likely from primary infection rather than from reactivation in this population. Primary infection is noted to cause fever, leukopenia, rash, encephalitis, and interstitial pneumonitis. There has also been some association with HHV-6 infection and white matter demyelination seen in AIDS dementia complex (92). Reactivation disease is less likely to be symptomatic; however, it may potentiate the effects of other viral pathogens like cytomegalovirus (46, 70).

Serologic tests exist to detect antibody to HHV-6. In normal hosts primary infection can be determined by seroconversion from negative to positive IgG or by detecting IgM. However, these assays do not distinguish between variants A or B. In an adult, where the prevalence of seropositivity is quite high, serologic tests may be difficult to interpret. IgM can be produced in acute or reactivation disease and may be of questionable clinical significance in an adult.

Qualitative and quantitative PCR studies exist for detection of HHV-6 in cellular and acellular specimens. These studies are still investigational, however (165).

The definitive diagnostic study for HHV-6 is felt to be culturing the virus from mononuclear blood cells. When the virus is recovered from a specimen from an individual with an acute illness, such as roseola, this is felt to be diagnostic (179). In immunosuppressed patients viral replication rates are high regardless of clinical findings, and a positive culture may be difficult to interpret.

HANTAVIRUS PULMONARY SYNDROME

The virus responsible for causing hantavirus pulmonary syndrome is a member of the genus *Hantavirus* within the family *Bunyaviridae*. Species identified in the Old World include Hantaan, Seoul, Puumala, and Dobrava viruses. These viruses are well known for causing hemorrhagic fevers and renal failure. In North America several distinct hantaviruses have been identified, but the most important is Sin Nombre virus (SNV). This is the virus that was identified in 1993 during the outbreak of hantavirus pulmonary syndrome in the Four Corners area of the southwestern United States. Since 1993, hantavirus has been found throughout the United States (167). Since 1996, only 4 or 5 of the 155 known cases of hantavirus pulmonary syndrome have been caused by strains other than SNV (122).

Hantaviruses are negative-sense RNA viruses. Like most of the other bunyaviruses, hantaviruses are all associated with a specific rodent reservoir. What distinguishes hantaviruses from the other bunyavirus groups is that transmission occurs when the virus is aerosolized from the excrement of infected animals and not by an arthropod vector (167). The major rodent responsible for transmission of hantavirus pulmonary syndrome is *Peromyscus maniculatus* (the deer mouse). This rodent is found throughout the continental United States but not in the Southeast or Atlantic seaboard (125).

While most cases of hantavirus pulmonary syndrome have occurred during the spring and summer there is not the distinct seasonality seen with the other hantaviruses (122). The vast majority of patients can easily recall exposure to rodents or rodent excrement before the onset of symptoms (90). This usually occurs in a household setting. The density of the rodent population in a household is directly correllated with the risk of developing hantavirus pulmonary syndrome. Occupational exposures in agricultural and livestock workers have also been noted to be risk factors for acquiring hantavirus pulmonary syndrome. To date there have been no documented cases of transmission of SNV from patients to health care providers (192). Traveling to areas of high endemicity does not appear to place individuals at high risk if potential rodent exposure is minimized.

The hantavirus pulmonary syndrome begins with a prodromal phase consisting of sudden onset of fever, myalgias, headache, and backache. Other associated symptoms that may occur include gastrointestinal upset such as nausea, vomiting, and diarrhea. This prodrome phase typically lasts for 3 to 4 days (20, 26, 50, 72, 95, 125). The prodromal phase is then followed by a symptom complex consisting of sudden onset of cough and shortness of breath which rapidly progresses to noncardiogenic pulmonary edema and hemodynamic collapse. A large percentage of patients will require intubation and mechanical ventilation. Hemodynamic monitoring will frequently reveal a normal wedge pressure, low cardiac index, and an elevated supraventricular rhythm (69).

After the cardiopulmonary phase, which typically lasts 2 to 4 days, a diuresis phase begins in those patients who are to survive, and they rapidly improve. There is a convalescent phase that can last for months, but there do not appear to be any long-term sequelae related to hantavirus pulmonary syndrome in those who survive (122). Overall mortality is approximately 50% however.

During the prodromal phase, the chest radiograph will show only mild pulmonary edema, perhaps with Kerley B lines. Extensive bilateral pulmonary infiltrates may develop and progress rapidly as the patient's clinical status worsens. Laboratory abnormalities will include thrombocytopenia and leukocytosis, with an increased number of immature granulocytes on the differential. Individuals with severe disease will also have hemoconcentration, hypoalbuminemia, and lactic acidosis (122). Liver function abnormalities, primarily a transaminitis, are also noted in many cases.

Capillary leak is the mechanism by which most of the pathologic actions of hantavirus pulmonary syndrome occur. This can happen in many organs. While hantavirus antigens can be detected on endothelial cells, the exact mechanism of the capillary leak process is unknown.

Serologic assays exist today for the diagnosis of hantavirus pulmonary syndrome (Centers for Disease Control and Prevention [http://www.cdc.gov.ncidod/diseases/hanta/hps/noframes/phys/diag.htm]). These assays look for antibodies to specific viral antigens of SNV. The CDC and various state laboratories use an ELISA to detect IgM to SNV. An IgG assay also is used in some settings. The diagnosis of hantavirus disease is made when there is a documented fourfold rise in IgG titer between acute and convalescent titers. Finding IgM to SNV is also felt to be diagnostic of hantavirus infection (140). Western blot assays and a rapid immunoblot strip assay are also available.

Immunohistochemistry staining of tissue samples can also be used to look for hantavirus antigens. This study is usually done as a confirmatory test in appropriate patients. PCR testing on tissue or blood cells can also be used to detect hantavirus RNA. At the present time, however, this is an experimental study due to its low sensitivity and potential for cross contamination. Viral culture (isolation) is very difficult and is not part of the diagnostic workup for hantavirus pulmonary syndrome.

REFERENCES

1. **Ablashi, D. V., P. Lusso, C. L. Hung, S. Z. Salahuddin, S. F. Josephs, T. Liana, B. Kramarsky, and P. Biberfield.** 1988. Utilization of human hematopoietic cell lines for the propagation and characterization of HBLV (human herpesvirus 6). *Int. J. Cancer* **42:**787–791.

2. **Anderson, B., K. Sims, R. Regnery, L. Robinson, M. J. Schmidt, S. Goral, C. Huger, and K. Edwards.** 1994. Detection of *Rochalimaea henselae* DNA in specimens from cat scratch disease patients by PCR. *J. Clin. Microbiol.* **32:**942–948.

3. **Anderson, B., C. Kelly, R. Threlkel, and K. Edwards.** 1993. Detection of *Rochalimaea henselae* in cat-scratch disease skin test antigens. *J. Infect. Dis.* **168:**1034–1036.

4. **Anderson, B. E., and M. A. Neuman.** 1997. *Bartonella* spp. as emerging human pathogens. *Clin. Microbiol. Rev.* **10:**203–219.

5. **Anderson, B. E., K. G. Sims, J. G. Olson, J. E. Childs, J. F. Piesman, C. M. Happ, G. O. Maupin, and B. J. B. Johnson.** 1993. *Amblyomma americanum*: a potential vector of human ehrlichiosis. *Am. J. Trop. Med. Hyg.* **49:**239–244.

6. **Anderson, B. E., J. W. Summer, J. E. Dawson, T. Tzianabos, C. R. Greene, J. G. Olson, D. B. Fishbein, M. Olsen-Rasmussen, B. P. Hollowau, E. H. George, and A. F. Azad.** 1992. Detection of the etiologic agent of human ehrlichiosis by polymerase chain reaction. *J. Clin. Microbiol.* **30:**775–780.

7. **Baker, J. A.** 1946. A rickettsial infection in Canadian voles. *J. Exp. Med.* **84:**37–55.

8. **Bakken, J. S., J. Krueth, C. Wilson-Nordskog, R. L. Tilden, K. Asanovich, and J. S. Dumler.** 1996. Clinical and laboratory characteristics of human granulocytic ehrlichiosis. *JAMA* **275:**199–205.

9. **Bass, J. W., J. M. Vincent, and D. A. Person.** 1997. The expanding spectrum of Bartonella infections: II. Cat-scratch disease. *Pediatr. Infect. Dis. J.* **16:**163–179.

10. **Bell, B. P., M. Goldoft, P. M. Griffin, M. A. Davis, D. C. Gordon, P. I. Tarr, C. A. Bartleson, J. H. Lewis, T. J. Barrett, J. G. Wells, R. Barson, and J. Kobayashi.** 1994. A multistate outbreak of *Escherichia coli* O157:H7-associated bloody diarrhea and hemolytic uremic syndrome from hamburgers: the Washington experience. *JAMA* **272:**1349–1353.

11. **Bergmans, A. M. C., J.-W. Groothedde, J. F. P. Schellekens, J. D. A. Vanemden, J. M. Ossewaarde, and L. M. Schouls.** 1995. Etiology of cat-scratch disease: comparison of polymerase chain reaction detection of *Bartonella* (formerly *Rochalimaea*) and *Afipia felis* DNA with serology and skin tests. *J. Infect. Dis.* **171:**916–923.

12. **Besser, R. E., S. M. Lett, J. T. Weber, M. P. Doyle, T. J. Barrett, J. G. Wells, and P. M. Griffin.** 1993. An outbreak of diarrhea and hemolytic uremic syndrome from *Escherichia coli* O157:H7 in fresh-pressed apple cider. *JAMA* **269:**2217–2220.

13. **Birtles, R. J., T. G. Harrison, N. A. Saunders, and D. Molyneux.** 1995. Proposals to unify the genera *Grahamella* and *Bartonella*, with descriptions of *Bartonella talpae* comb. nov., *Bartonella peromysci* comb. nov., and three new species, *Bartonella grahamii* sp. nov., *Bartonella taylorii* sp. nov., and *Bartonella doshiae* sp. nov. *Int. J. Syst. Bacteriol.* **45:**1–8.

14. **Bopp, C. A., F. W. Brenner, J. G. Wells, and N. A. Strockbine.** 1999. *Escherichia, Shigella,* and *Salmonella,* p. 459–474. *In* P. R. Murray, E. J. Baron, M. A. Pfaller, F. C. Tenover, and R. H. Yolken (ed.), *Manual of Clinical Microbiology,* 7th ed. American Society for Microbiology, Washington, D.C.

15. **Breitschwerdt, E. B., D. L. Kordick, D. E. Malarkey, B. Keene, T. L. Hadfield, and K. Wilson.** 1995. Endocarditis in a dog due to infection with a novel *Bartonella* subspecies. *J. Clin. Microbiol.* **33:**154–160.

16. **Brenner, D. J., S. P. O'Connor, D. G. Hollis, R. E. Weaver, and A. G. Steigerwalt.** 1991. Molecular characterization and proposal of neotype strain for *Bartonella bacilliformis. J. Clin. Microbiol.* **29:**1299–1302.

17. **Brenner, D. J., S. P. O'Connor, H. H. Winkler, and A. G. Steigerwalt.** 1993. Proposals to unify the genera *Bartonella* and *Rochalimaea*, with descriptions of *Bartonella quintana* comb. nov., *Bartonella vinsonii* comb. nov., *Bartonella henselae* comb. nov., and *Bartonella elizabethae* comb. nov., and to remove the family *Bartonellaceae* from the order *Rickettsiales. Int. J. Syst. Bacteriol.* **43:**777–786.

18. **Brenner, S. A., J. A. Rooney, P. Manzewitsch, and R. L. Regnery.** 1997. Isolation of *Bartonella* (*Rochalimaea*) *henselae*: effects of methods of blood collecton and handling. *J. Clin. Microbiol.* **35:**544–547.

19. **Brouqui, P., and D. Raoult.** 1992. In vitro antibiotic susceptibility of the newly recognized agent of ehrlichiosis in humans. *Ehrlichia chaffeensis. Antimicrob. Agents Chemother.* **36:**2799–2803.

20. **Butler, J. C., and C. J. Peters.** 1994. Hantavirus and hantavirus pulmonary syndrome. *Clin. Infect. Dis.* **19:**387–394.

21. **Bwaka, M. A., M. Bonnet, P. Calain, R. Colebunders, A. DeRoo, Y. Guimard, K. R. Katwiki, K. Kibadi, M. A. Kipasa, K. J. Kuvula, B. B. Mapanda, M. Massamba, K. D. Mupapa, J. J. Muyembe-Tamfun, E. Ndaberey, C. J. Peters, P. E. Rollin, and E. VandenEnden.** 1999. Ebola hemorrhagic fever in Kikwit, Democratic Republic of Congo (former Zaire): clinical observations. *J. Infect. Dis.* **179**(Suppl. 1):S1–S7.

22. **Carithers, H. A.** 1978. Oculoglandular disease of Parinaud: a manifestation of cat-scratch disease. *Am. J. Dis. Child.* **132:**1195–1200.

23. **Carithers, H. A.** 1985. Cat-scratch disease: an overview based on a study of 1,200 patients. *Am. J. Dis. Child.* **139:**1124–1133.

24. **Carter, A. O., A. A. Borczyk, J. A. K. Carlson, B. Harvey, J. C. Hockin, M. A. Karmali, C. Krishnan, D. A. Korn, and H. Lior.** 1987. A severe outbreak of *Escherichia coli* O157:H7-associated hemorrhagic colitis in a nursing home. *N. Engl. J. Med.* **317:**496–500.

25. **Cathomas, G., M. Tamm, C. Mcgandy, A. P. Perruchoud, M. J. Mihatsch, and P. Dalaven.** 1996. Detection of herpesvirus-like DNA in the bronchoalveolar

lavage fluid of patients with pulmonary Kaposi's sarcoma. *Eur. Respir. J.* **9:**1743–1746.

26. **Centers for Disease Control.** 1993. Outbreak of an acute illness—Southwestern United States, 1993. *Morb. Mortal. Wkly. Rep.* **42:**441–443.

27. **Centers for Disease Control and Prevention.** 1996. Outbreaks of *Cyclospora cayetanensis* infection—United States. *Morb. Mortal. Wkly. Rep.* **45:**549–551.

28. **Centers for Disease Control and Prevention.** 1996. Update: outbreaks of *Cyclospora cayetanensis* infection—United States. *Morb. Mortal. Wkly. Rep.* **45:**611–612.

29. **Centers for Disease Control and Prevention.** 1997. Update: outbreaks of cyclosporiasis—United States and Canada, 1997. *Morb. Mortal. Wkly. Rep.* **46:**461–462.

30. **Centers for Disease Control and Prevention.** 1997. Update: outbreaks of cyclosporiasis—United States. *Morb. Mortal. Wkly. Rep.* **46:**521–523.

31. **Centers for Disease Control and Prevention.** 1999. Update: West Nile-like viral encephalitis—New York. *Morb. Mortal. Wkly. Rep.* **48:**890–892.

32. **Centers for Disease Control and Prevention.** 1999. Update: West Nile-like viral encephalitis—New York. *Morb. Mortal. Wkly. Rep.* **48:**944–946.

33. **Centers for Disease Control and Prevention.** 2000. Update: West Nile-like viral encephalitis—New York. *Morb. Mortal. Wkly. Rep.* **49:**100–102.

34. **Centers for Disease Control and Prevention.** 2000. Update: West Nile Virus isolated from mosquitoes—New York, 2000. *Morb. Mortal. Wkly. Rep.* **49:**211.

35. **Cesarman, E., Y. Chang, P. Moore, J. W. Said, and D. M. Knowles.** 1995. Kaposi's sarcoma-associated herpesvirus-like DNA sequences in AIDS-related body-cavity-based-lymphomas. *N. Engl. J. Med.* **332:**1186–1191.

36. **Chang, Y., E. Cesarman, M. Pessin, F. Lee, J. Culpeper, D. M. Knowles, and P. S. Moore.** 1994. Identification of herpesvirus-like DNA sequences in AIDS-associated Kaposi's sarcoma. *Science* **266:**1865–1869.

37. **Chapman, P. A.** 1989. Evaluation of commercial latex slide test for identifying *Escherichia coli* O157:H7. *J. Clin. Pathol.* **42:**1109–1110.

38. **Clarridge, J. E., T. J. Raich, D. Pirwani, B. Simon, L. Tsai, M. C. Rodriguez-Barradas, R. Regnery, A. Zollo, D. C. Jones, and C. Rambo.** 1995. Strategy to detect and identify *Bartonella* species in routine clinical laboratory yields *Bartonella henselae* from human immunodeficiency virus-positive patient and unique *Bartonella* strain from his cat. *J. Clin. Microbiol.* **33:**2107–2113.

39. **Cody, S. H., M. K. Glynn, J. A. Farrar, K. L. Cairns, P. M. Griffin, J. Kobayashi, M. Fyfe, R. Hoffman, A. S. King, J. H. Lewis, B. Swaminathan, R. G. Bryant, and D. J. Vugia.** 1999. An outbreak of *Escherichia coli* O157:H7 infection from unpasteurized commercial apple juice. *Ann. Intern. Med.* **130:**202–209.

40. **Collinge, J., K. C. L. Sidle, J. Meads, J. Ironside, and A. F. Hill.** 1996. Molecular analysis of prion strain variation and the aetiology of "new variant" CJD. *Nature* **383:**685–690.

41. **Corbellino, M., C. Parravicini, J. Aubin, and E. Berti.** 1996. Kaposi's sarcoma and herpesvirus-like DNA sequences in sensory ganglia. *N. Engl. J. Med.* **334:**1341–1342.

42. **Dalton, M. J., L. E. Robinson, J. Cooper, R. L. Regnery, J. G. Olsen, and J. E. Childs.** 1995. Use of *Bartonella* antigens for serologic diagnosis of cat-scratch disease at a national referral center. *Arch. Intern. Med.* **155:**1670–1676.

43. **Delahoussaye, P. M., and B. M. Osborne.** 1990. Cat-scratch disease presenting as abdominal visceral granulomas. *J. Infect. Dis.* **161:**71–78.

44. **Dewhurst, S.** 1994. Newly discovered human herpesvirus (HHV-6 and HHV-7). *Curr. Opin. Infect. Dis.* **7:**238–245.

45. **Di Alberti, L., S. Ngui, S. Porter, P. M. Speight, C. M. Scully, J. M. Zakrewska, I. G. Williams, L. Artese, A. Piattelli, and C. G. Teo.** 1997. Presence of human herpesvirus 8 variants in the oral tissues of human immunodeficiency virus-infected patients. *J. Infect. Dis.* **175:**703–707.

46. **Dockrell, D. H., J. Prada, M. F. Jones, R. Patel, A. D. Badley, W. S. Hamsen, D. M. Ilstrup, R. H. Wiesner, R. A. Kron, T. F. Smith, and C. V. Paya.** 1997. Seroconversion to human herpesvirus 6 following liver transplantation is a marker of cytomegalovirus disease. *J. Infect. Dis.* **76:**1135–1140.

47. **Dockrell, D. H., T. F. Smith, and C. V. Paya.** 1999. Human herpesvirus 6. *Mayo Clin. Proc.* **74:**163–170.

48. **Dowell, S. F., R. Mukuna, T. G. Ksiazek, A. S. Khan, P. E. Rollin, and C. J. Peters.** 1999. Transmission of Ebola hemorrhagic fever: a study in risk factors in family members, Kikwit Zaire 1995. *J. Infect. Dis.* **179**(Suppl. 1):S87–S91.

49. **Drancourt, M., J. L. Mainardi, P. Bronqui, F. Vandenesch, A. Carta, F. Lehnert, J. Etienne, F. Goldstein, J. Acar, and D. Raoult.** 1995. *Bartonella (Rochalimaea) quintana* endocarditis in three homeless men. *N. Engl. J. Med.* **332:**419–423.

50. **Duchin, J. S., F. T. Koster, C. J. Peters, G. L. Simpson, B. Tempest, S. R. Zaki, T. G. Ksiazek, P. E. Rollin, S. Nichol, E. T. Umland, R. L. Moolenaar, S. E. Reef, K. B. Nolte, M. M. Gallaher, J. C. Butler, and R. F. Breiman.** 1994. Hantavirus pulmonary syndrome: a clinical description of 17 patients with a newly recognized disease. *N. Engl. J. Med.* **330:**949–955.

51. **Dumler, J. S.** 1999. Ehrlichia, p. 821–822. *In* P. R. Murray, E. J. Baron, M. A. Pfaller, F. C. Tenover, and R. H. Yolken (ed.), *Manual of Clinical Microbiology*, 7th ed. American Society for Microbiology, Washington, D.C.

52. **Dumler, J. S., K. M. Asanovich, J. S. Bakken, P. Richter, R. Kimsey, and J. E. Madigan.** 1995. Serologic cross-reaction among *Ehrlichia equi, Ehrlichia phagocytophila,* and human granulocytic ehrlichia. *J. Clin. Microbiol.* **33:**1098–1103.

53. **Dupin, N., I. Gorin, J. Deleuze, H. Agut, J. M. Huraux, and J. P. Escande.** 1995. Herpes-like DNA sequences, AIDS-related tumors, and Castleman's disease. *N . Engl. J. Med.* **333:**798.

54. **Eberhard, M. D., A. J. da Silva, B. G. Lilley, and N. J. Pieniazek.** 1999. Morphologic and molecular characterization of new *Cyclospora* species from Ethiopian monkeys: *C. cercopitheci* sp. n., *C. colobi* sp. n., and *C. papionis* sp. n. *Emerg. Infect. Dis.* **5:**651–658.

55. **Everett, E. D., K. A. Evans, R. B. Henry, and G. McDonald.** 1994. Human ehrlichiosis in adults after tick exposure: diagnosis using polymerase chain reaction. *Ann. Intern. Med.* **120:**730–735.

56. **Fishbein, D. B., J. E. Dawson, and L. E. Robinson.** 1994. Human ehrlichiosis in the United States, 1985–1990. *Ann. Intern. Med.* **20:**736–743.

57. **Flaitz, C., Y. Jin, M. Hicks, C. M. Nichols, Y. W. Yang, and I. J. Su.** 1997. Kaposi's sarcoma-associated herpesvirus-like DNA sequences (KSHV-HHV-8) in oral

AIDS-Kaposi's sarcoma: a PCR and clinicopathologic study. *Oral Surg. Oral Med. Oral Pathol.* **83:**259–264.

58. **Foreman, K., J. Friborg, W. Kong, C. Woffendin, P. Polverini, B. J. Nickoloff, and G. J. Nabel.** 1997. Propagation of a human herpesvirus from AIDS-associated Kaposi's sarcoma. *N. Engl. J. Med.* **336:**163–171.

59. **Formenty P., C. Hatz, B. LeGuenno, A. Stoll, P. Rogenmoser, and A. Widmer.** 1999. Human infection due to Ebola Cote d'Ivoire: clinical and biological presentations. *J. Infect. Dis.* **179**(Suppl. 1)**:**S48–S53.

60. **Gao, S. J., L. Kingsley, M. Li, W. Zheng, C. Parravicini, J. Ziegler, R. Newton, C. R. Rinaldo, A. Saah, J. Phair, R. Detels, Y. Chang, and P. S. Moorel.** 1996. KSHV antibodies among Americans, Italians and Ugandans with and without Kaposi's sarcoma. *Nat. Med.* **2:**925–928.

61. **Garcia-Tsao, G., L. Panzini, M. Toselevitz, and A. B. West.** 1992. Bacillary peliosis hepatitis as a cause of acute anemia in a patient with the acquired immunodeficiency syndrome. *Gastroenterology* **102:**1065–1070.

62. **Gear, J. S. S., G. A. Cassel, A. J. Gear, B. Trappler, L. Clausen, and A. M. Meyers, M. C. Kew, T. H. Bothwell, R. Sher, G. B. Miller, J. Schneider, H. J. Koornhof, E. D. Gomperts, M. Isaacson, and J. H. Gear.** 1975. Outbreak of Marburg virus disease in Johannesburg. *BMJ* **4:**489–493.

63. **Georges, A., E. B. Leroy, A. A. Renout, C. T. Benissan, R. J. Nabias, M. T. Ngoc, P. I. Obiang, J. P. M. LePage, E. J. Bertherat, D. D. Benoni, E. J. Wickings, J. P. Amblard, J. M. Lansound-Soukate, J. M. Milleliri, S. Bize, and M. C. Georges-Courbot.** 1999. Recent Ebola outbreaks in Gabon from 1994 to 1997: epidemiologic and health control issues. *J. Infect. Dis.* **179**(Suppl. 1)**:**S65–S75.

64. Reference deleted.

65. **Goral, S., B. Anderson, C. Hager, and K. M. Edwards.** 1994. Detection of *Rochalimaea henselae* DNA by polymerase chain reaction from suppurative nodes of children with cat scratch disease. *Pediatr. Infect. Dis. J.* **13:**994–997.

66. **Griffin, P. M.** 1995. *Escherichia coli* O157:H7 and other enterohemorrhagic *Escherichia coli*, p. 739–761. *In* M. J. Blaser, P. D. Smith, J. I. Ravdin, H. B. Greenberg, and R. L. Guerrant (ed.), *Infections of the Gastrointestinal Tract.* Raven Press, Ltd., New York, N.Y.

67. **Griffin, P. M., S. M. Ostroff, R. V. Tauxe, K. D. Greene, J. G. Wells, J. H. Lewis, and P. A. Blake.** 1988. Illnesses associated with *Escherichia coli* O157:H7 infections: a broad clinical spectrum. *Ann. Intern. Med.* **109:**705–712.

68. **Hall, C. B., C. E. Long, K. C. Schnabel, M. T. Caserta, K. M. McIntyre, M. A. Costanzo, A. Knott, S. Dewhurst, R. A. Insel, and L. G. Epstein.** 1994. Human herpesvirus-6 infection in children: a prospective study of complications and reactivation. *N. Engl. J. Med.* **331:**432–438.

69. **Hallin, G. W., S. Q. Simpson, R. E. Crowell, D. S. James, F. T. Koster, G. J. Merz, and H. Levy.** 1996. Cardiopulmonary manifestations of the hantavirus pulmonary syndrome. *Crit. Care Med.* **24:**252–258.

70. **Herbein, G., J. Strasswimmer, M. Altieri, M. L. Woehl-Jaegle, P. Wolf, and G. Obert.** 1996. Longitudinal study of human herpesvirus 6 infection in organ transplant recipients. *Clin. Infect. Dis.* **22:**171–173.

71. **Herwaldt, B. L., M. L. Ackers, and the Cyclospora Working Group.** 1997. An outbreak in 1996 of cyclosporins associated with imported raspberries. *N. Engl. J. Med.* **336:**1548–1556.

72. **Hjelle, B., S. Jenison, G. Mertz, F. Koster, and K. Fovear.** 1994. Emergence of hantavirus disease in the southwestern United States. *West J. Med.* **161:**467–473.

73. **Hmiel, S. P., R. Buller, M. Arens, M. Gaudreault-Keener, and G. A. Storch.** 1998. Human infection with *Ehrlichia ewingii*, the agent of Ozark canine granulocytic ehrlichiosis. *Proceedings of the First International Conference on Emerging Infectious Diseases.* Addendum 4.

74. **Hoge, C. W., P. Echeverria, R. Rajah, J. Jacobs, S. Malthouse, E. Chapman, L. M. Jimenez, and D. R. Shlim.** 1995. Prevalence of *Cyclospora* species and other enteric pathogens among children less than 5 years of age in Nepal. *J. Clin. Microbiol.* **33:**3058–3060.

75. **Hoge, C. W., D. R. Shlim, M. Ghimire, J. G. Rabold, P. Pandey, A. Walch, R. Rajah, P. Gaudio, and P. Echeverria.** 1995. Placebo-controlled trial of co-trimoxazole for cyclospora infections among travellers and foreign residents in Nepal. *Lancet* **345:**691–693.

76. **Holmes, A. H., T. C. Greenough, G. J. Balady, R. Regnery, B. Anderson, J. C. O'Keane, J. Fonger, and E. McCrone.** 1995. *Bartonella henselae* endocarditis in an immunocompetent adult. *Clin. Infect. Dis.* **21:**1004–1007.

77. **Huang, P., J. T. Weber, D. M. Sosin, P. M. Griffin, E. G. Long, J. J. Murphy, F. Kocka, C. Peters, and C. Kallick.** 1995. The first reported outbreak of diarrheal illness associated with Cyclospora in the United States. *Ann. Intern. Med.* **123:**409–414.

78. **Jacobs, U., J. Ferber, and H. U. Klehr.** 1994. Severe allograft dysfunction after OKT3-induced human herpesvirus-6 reactivation. *Transplant. Proc.* **26:**3121.

79. **Jalava, J., P. Kotilainen, S. Nikkari, M. Skurnik, E. Vanttinen, O.-P. Lehtonen, E. Eerola, and P. Toivanen.** 1995. Use of the polymerase chain reaction and DNA sequencing for detection of Bartonella quintana in the aortic valve of a patient with culture-negative infective endocarditis. *Clin. Infect. Dis.* **21:**891–896.

80. **Johnson, W. T., and E. B. Helwig.** 1969. Cat-scratch disease (histopathologic changes in the skin). *Arch. Dermatol.* **100:**148–154.

81. **Jongejan, F., L. A. Wassink, M. J. C. Thielemans, N. M. Perie, and G. Uilenberg.** 1989. Serotypes in *Cowdria ruminantium* and their relationship with *Ehrlichia phagocytophila* determined by immunofluorescence. *Vet. Microbiol.* **21:**31–40.

82. **Karmali, M. A., M. Petric, C. Lim, P. C. Fleming, G. S. Arbus, and H. Lior.** 1985. The association between idiopathic hemolytic-uremic syndrome and infection by verotoxin-producing *Escherichia coli*. *J. Infect. Dis.* **151:**77–82.

83. **Karmali, M. A., B. T. Steele, M. Petric, and C. Lim.** 1983. Sporadic cases of haemolytic-uraemic syndrome associated with faecal cytotoxin and cytotoxin-producing *Escherichia coli* in stools. *Lancet.* **i:**619–620.

84. **Kedes, D., D. Ganem, N. Ameli, P. Bacchetti, and R. Greenblatt.** 1997. The prevalence of serum antibody to human herpesvirus 8 (Kaposi's sarcoma-associated herpesvirus) among HIV-seropositive and high-risk HIV-seronegative women. *JAMA* **277:**478–481.

85. **Kedes, D., E. Operskalski, M. Busch, R. Kohn, J. Flood, and D. Ganem.** 1996. The seroepidemiology of human herpesvirus 8 (Kaposi's sarcoma-associated her-

pesvirus): distribution of infection in KS risk groups and evidence for sexual transmission. *Nat. Med.* **2:**918–924.

86. **Keene, W. E., J. M. McAnulty, F. C. Hoesly, L. P. Williams, K. Hedberg, G. L. Oxman, T. J. Barrett, M. A. Pfaller, and D. W. Fleming.** 1994. A swimming-associated outbreak of hemorrhagic colitis caused by *Escherichia coli* O157:H7 and *Shigella sonnei. N. Engl. J. Med.* **331:**579–584.

87. **Keene, W. E., E. Sazie, J. Kok, D. H. Rice, D. D. Hancock, V. K. Balan, T. Zhao, and M. P. Doyle.** 1997. An outbreak of *Escherichia coli* O157:H7 infections traced to jerky made from deer meat. *JAMA* **277:**1229–1231.

88. **Kemper, C. A., C. M. Lombard, S. C. Deresinsksi, and L. S. Tompkins.** 1990. Visceral bacillary epitheliod angiomatosis: possible manifestations of disseminated cat-scratch disease in the immunocompromised host. A report of two cases. *Am. J. Med.* **89:**216–222.

89. **Khan, A., T. F. Keweteminga, D. H. Heymann, B. LeGuenno, P. Nabeth, B. Kerstiens, Y. Fleerackers, P. H. Kilmarx, G. R. Rodier, O. Nkuku, P. E. Rollin, A. Sanchez, S. R. Zaki, R. Swanepoel, O. Tomori, S. T. Nichol, C. J. Peters, J. J. Muyembe-Tamfun, and T. G. Ksiazek.** 1999. The re-emergence of Ebola hemorrhagic fever (EHF) Zaire, 1995. *J. Infect. Dis.* **179**(Suppl. 1)**:**S76–S86.

90. **Khan, A., R. F. Khabbaz, L. R. Armstrong, R. C. Holman, S. P. Bauer, J. Graber, T. Strine, G. Miller, S. Reef, J. Tappero, P. E. Rolin, S. T. Nichol, S. R. Zaki, R. T. Bryan, L. E. Chapman, C. J. Peters, and T. G. Ksiazek.** 1996. Hantavirus pulmonary syndrome: the first 100 U.S. cases. *J. Infect. Dis.* **173:**1297–1303.

91. **Klein, M. D., C. M. Nelson, and J. L. Goodman.** 1997. Antibiotic susceptibility of the newly cultivated agent of human granulocytic ehrlichiosis: promising activity of quinolones and rifamycins. *Antimicrob. Agents Chemother.* **41:**76–79.

92. **Knox, K. K., and D. R. Carrigan.** 1995. Active human herpesvirus (HHV-6) infection of the central nervous system in patients with AIDS. *J. Acquir. Immune Defic. Syndr. Hum. Retrovirol.* **9:**69–73.

93. **Koehler, J. E., F. D. Quinn, T. G. Berger, P. E. LeBoit, and J. W. Tappero.** 1992. Isolation of *Rochalimaea* species from cutaneous and osseous lesions of bacillary angiomatosis. *N. Engl. J. Med.* **327:**1625–1631.

94. **Koehler, J. E., and J. W. Tappero.** 1993. Bacillary angiomatosis and bacillary peliosis in patients infected with the human immunodeficiency virus. *Clin. Infect. Dis.* **17:**612–624.

95. **Koster, F. T., and S. Jenison.** 1997. The hantavirus. *In* S. L. Gorbach, J. G. Bartlett, and N. R. Blacklow (ed.), *Infectious Diseases*, 2nd ed. W. B. Saunders, Philadelphia, Pa.

96. **Kudva, I. T., P. G. Hatfield, and C. J. Hovde.** 1996. *Escherichia coli* O157:H7 in microbial flora of sheep. *J. Clin. Microbiol.* **34:**431–433.

97. **Larson, A. M., M. J. Dougherty, D. J. Nowowiejski, D. F. Welch, G. M. Matar, B. Swaminathan, and M. B. Coyle.** 1994. Detection of *Bartonella* (*Rochalimaea*) *quintana* by routine acridine orange staining of broth blood cultures. *J. Clin. Microbiol.* **32:**1492–1496.

98. **La Scola, B., and D. Raoult.** 1996. Serologic cross-reactions between *Bartonella quintana*, *Bartonella henselae*, and *Coxiella burnetii. J. Clin. Microbiol.* **34:**2270–2274.

99. **Lawson, P. A., and M. D. Collins.** 1996. Description of *Bartonella clarridgeiae* sp. nov. isolated from the cat

of a patient with *Bartonella henselae* septicemia. *Med. Microbiol. Lett.* **5:**64–73.

100. **Le, H. H., D. A. Palay, B. Anderson, and J. P. Steinberg.** 1994. Conjunctival swab to diagnose ocular cat scratch disease. *Am. J. Ophthalmol.* **118:**249–250.

101. **LeBoit, P. E., T. G. Berger, B. M. Egbert, J. H. Beckstead, T. S. Yen, and M. H. Stoler.** 1989. Bacillary angiomatosis: the histopathology and differential diagnosis of a pseudoneoplastic infection in patients with human immunodeficiency virus disease. *Am. J. Surg. Pathol.* **13:**909–920.

102. **LeBoit, P. E., T. G. Berger, B. M. Egbert, S. B. Yen, M. H. Stoler, T. A. Bonfiglio, J. A. Strauchen, C. K. English, and D. J. Wear.** 1988. Epitheliod hemangioma-like vascular proliferation in AIDS: manifestation of cat-scratch disease bacillus infection. *Lancet* **i:**960–963.

103. **Lennette, E., D. Blackbourn, and J. Levy.** 1996. Antibodies to human herpesvirus type 8 in the general population and in Kaposi's sarcoma patients. *Lancet* **348:**858–861.

104. **Lenoir, A. A., G. A. Storch, K. Deschryver-Kecskemeti, G. S. Shackleford, R. J. Rothbaum, D. J. Wear, and J. L. Rosenblum.** 1988. Granulomatous hepatitis associated with cat-scratch disease. *Lancet* **i:**1132–1136.

105. **Le Saux, N., J. S. Spika, B. Friesen, I. Johnson, D. Melnychuck, C. Anderson, R. Dion, M. Rahman, and W. Tostowaryk.** 1993. Ground beef consumption in noncommercial settings is a risk factor for sporadic *Escherichia coli* O157:H7 infection in Canada. *J. Infect. Dis.* **167:**500–501.

106. **Lin, J.-C., S.-C. Lin, E.-C. Mar, P. E. Pellett, F. R. Stamey, J. A. Stewart, and T. J. Spira.** 1995. Is Kaposi's sarcoma-associated herpesvirus detectable in semen of HIV-infected homosexual men? *Lancet* **346:**1601–1602.

107. **Long, E. G., E. H. White, W. W. Carmichael, P. M. Quinlisk, R. Raja, B. L. Swisher, H. Daugharty, and M. T. Cohen.** 1991. Morphologic and staining characteristics of a cyanobacterium-like organism associated with diarrhea. *J. Infect. Dis.* **164:**199–202.

108. **Low, D. E., E. Liu, J. Fuller, and A. McGeer.** 1999. *Streptococcus iniae:* an emerging pathogen in the aquaculture industry, p. 53–65. *In* W. M. Scheld, W. A. Craig, and J. M. Hughes (ed.), *Emerging Infections 3.* American Society for Microbiology, Washington, D.C.

109. **MacDonald, K. L., M. J. O. Leary, M. L. Cohen, P. Norris, J. G. Wells, E. Noll, J. M. Kobayashi, and P. A. Blake.** 1988. *Escherichia coli* O157:H7, an emerging gastrointestinal pathogen: results of a one-year, prospective, population-based study. *JAMA* **259:**3567–3570.

110. **MacLeod, J. R., and W. S. Gordon.** 1993. Studies in tick-borne fever of sheep. Transmission by the tick, *Ixodes ricinus*, with a description of the disease produced. *Parasitology* **25:**273–285.

111. **Madico, G., J. McDonald, R. H. Gilman, L. Cabrera, and C. R. Sterling.** 1997. Epidemiology and treatment of *Cyclospora cayetanensis* infection in Peruvian children. *Clin. Infect. Dis.* **24:**977–981.

112. **Malatack, J. J., and R. Jaffe.** 1993. Granulomatous hepatitis in three children due to cat-scratch disease without peripheral adenopathy. *Am. J. Dis. Child.* **147:**949–953.

113. **March, S. B., and S. Ratnam.** 1986. Sorbitol-MacConkey medium for detection of *Escherichia coli*

O157:H7 associated with hemorrhagic colitis. *J. Clin. Microbiol.* **23:**869–872.

114. **March, S. B., and S. Ratnam.** 1989. Latex agglutination test for detection of *Escherichia coli* serotype O157. *J. Clin. Microbiol.* **27:**1675–1677.

115. **Margileth, A. M.** 1993. Cat-scratch disease. *Adv. Pediatr. Infect. Dis.* **8:**1–21.

116. **Margileth, A. M., D. J. Wear, T. L. Hadfield, C. J. Schlagel, G. T. Spigel, and J. E. Muhlbauer.** 1984. Cat-scratch disease: bacteria in skin at the primary inoculation site. *JAMA* **252:**928–931.

117. **Marra, C. M.** 1995. Neurologic complications of *Bartonella henselae* infection. *Curr. Opin. Neurol.* **8:**164–169.

118. **Martini, G. A., and R. Siegert (ed.).** 1971. *Marburg Virus Disease*, p. 1–230. Springer-Verlag, Berlin, Germany.

119. **Maurin, M., V. Roux, S. Stein, F. Ferrier, R. Viraben, and D. Raoult.** 1994. Isolation and characterization by imunofluorescence, sodium dodecyl sulfate-polyacrylamide gel electrophoresis, Western Blot, restriction fragment length polymorphism PCR, 16S rRNA gene sequencing, and pulsed-field gel electrophoresis of *Rochalimaea quintana* from a patient with bacillary angiomatosis. *J. Clin. Microbiol.* **32:**1166–1171.

120. **McDonough, P. L., C. A. Rossiter, R. B. Rebhun, S. M. Stehman, D. H. Lein, and S. J. Shin.** 2000. Prevalence of *Escherichia coli* O157:H7 from cull dairy cows in New York state and comparison of culture methods used during preharvest food safety investigations. *J. Clin. Microbiol.* **38:**318–322.

121. **McIntosh, B. M., P. G. Jupp, I. DosSantos, and G. M. Meenehan.** 1976. Epidemics of West Nile and Sindbis viruses in South Africa with Culex (Culex) univittatus Theibald as a vector. *S. Afr. J. Sci.* **72:**295–300.

122. **Mertz, G. J., B. Hjelle, and R. T. Bryon.** 1998. Hantavirus infection. *Disease-a-Month* **44(3):**85–138.

123. **Miller, G., M. O. Rigsby, L. Heston, E. Grogan, R. Sun, C. Metroka, J. A. Levy, S. J. Gao, Y. Chang, and P. Moore.** 1996. Antibodies to butyrate-inducible antigens of Kaposi's sarcoma-associated herpesvirus in patients with HIV-1 infection. *N. Engl. J. Med.* **334:**1292–1297.

124. **Monini, P., L. de Lellis, M. Fabris, F. Rigolin, and E. Cassai.** 1996. Kaposi's sarcoma associated herpesvirus DNA sequences in prostate tissue and human semen. *N. Engl. J. Med.* **334:**1168–1172.

125. **Moolenaar, R. L., C. Dalton, H. B. Lipman, E. T. Umland, M. Gallaher, J. S. Duehn, L. Chapman, S. R. Zaki, T. G. Ksiazek, and P. E. Rollin.** 1995. Clinical features that differentiate hantavirus pulmonary syndrome from three other acute respiratory illness. *Clin. Infect. Dis.* **21:**643–649.

126. **Moore, P., C. Boshoff, R. Weiss, and Y. Chang.** 1996. Molecular mimicry of human cytokine and cytokine response pathway genes by KSHV. *Science* **274:**1739–1744.

127. **Myers, S. A., N. S. Prose, J. A. Garcia, K. H. Wilson, K. P. Dunsmore, and H. Kamino.** 1992. Bacillary angiomatosis in a child undergoing chemotherapy. *J. Pediatr.* **121:**574–578.

128. **Nathanson, N., J. Wilesmith, and C. Griot.** 1997. Bovine spongiform encephalopathy (BSE): causes and consequences of a common source epidemic. *Am. J. Epidemiol.* **145:**959–969.

129. **Noah, D. L., J. S. Bresee, M. J. Gorenseck, J. A. Rooney, J. L. Cresanta, R. L. Regnery, J. Wong, J. DelToro, J. G. Olson, and J. F. Childs.** 1995. Cluster of five children with acute encephalopathy associated with cat-scratch disease in south Florida. *Pediatr. Infect. Dis. J.* **14:**866–869.

130. **Noda, M., T. Yutsudo, N. Nakabayashi, T. Hirayama, and Y. Takeda.** 1987. Purification and some properties of Shiga-like toxin from *Escherichia coli* O157:H7 that is immunologically identical to Shiga toxin. *Microb. Pathog.* **2:**339–349.

131. **Okuno, T., K. Higashi, K. Shiraki, K. Yamanishi, M. Takahashi, Y. Kokado, M. Ishibashi, S. Takahara, T. Sonoda, and K. Tanaka.** Human herpesvirus 6 infection in renal transplantation. *Transplantation* **49:**519–522.

132. **Oren, I., and J. D. Sobel.** 1992. Human herpesvirus 6: review. *Clin. Infect. Dis.* **14:**741–746.

133. **Ortega, Y. R., C. R. Sterling, R. H. Gilman, V. A. Cama, and F. Diaz.** 1993. Cyclospora species—a new protozoan pathogen of humans. *N. Engl. J. Med.* **328:**1308–1312.

134. **Ostroff, S. M., J. M. Kobayashi, and J. H. Lewis.** 1989. Infections with *Escherichia coli* O157:H7 in Washington State. The first year of statewide disease surveillance. *JAMA* **262:**355–359.

135. **Padhye, V. V., F. B. Kittell, and M. P. Doyle.** 1986. Purification and physicochemical properties of a unique Vero cell cytotoxin from *Escherichia coli* O157:H7. *Biochem. Biophys. Res. Commun.* **139:**424–430.

136. **Pape, J. W., R. I. Verdier, M. Boncy, J. Boncy, and W. D. Johnson, Jr.** 1994. Cyclospora infection in adults infected with HIV. Clinical manifestations, treatment, and prophylaxis. *Ann. Intern. Med.* **121:**654–657.

137. **Park, C. H., D. L. Hixon, W. L. Morrison, and C. B. Cook.** 1994. Rapid diagnosis of enterohemorrhagic *Escherichia coli* O157:H7 directly from fecal specimens using immunofluorescence stain. *Am. J. Clin. Pathol.* **101:**91–94.

138. **Park, C. H., N. M. Vandel, and D. L. Hixon.** 1996. Rapid immunoassay for detection of *Escherichia coli* O157:H7 directly from stool specimens. *J. Clin. Microbiol.* **34:**988–990.

139. **Perkocha, L. A., S. M. Geaghan, T. S. B. Yen, S. L. Nishimura, S. P. Chan, R. Garcia-Kennedy, G. Honda, A. C. Stoloff, H. Z. Klein, R. L. Goldman, S. VanMeter, L. D. Ferrel, and P. E. LeBoit.** 1990. Clinical and pathological features of bacillary peliosis hepatitis in association with human immunodeficiency virus infection. *N. Engl. J. Med.* **323:**1581–1586.

140. **Peters, C. J.** 2000. California encephalitis, hantavirus pulmonary syndrome, and bunyaviridae hemorrhagic fevers, p. 1849–1855. *In* G. L. Mandell, J. E. Bennett, and R. Dolin (ed.) *Principles and Practice of Infectious Diseases*, 5th ed. Churchill Livingstone, Philadelphia, Pa.

141. **Peters, C. J.** 2000. Marburg and ebola virus hemorrhagic fevers, p. 1821–1823. *In* G. L. Mandell, J. E. Bennett, and R. Dolin (ed.), *Principles and Practice of Infectious Diseases*, 5th ed., Churchill Livingstone, Philadelphia, Pa.

142. **Peters, C. J., E. D. Johnson, and P. B. Jahrling.** 1991. Filoviruses, p. 159-175. *In* S. Morse (ed.), *Emerging Viruses*. Oxford University Press, New York, N.Y.

143. **Peters, C. J., and J. W. LeDuc (ed.).** 1999. Ebola: the virus and the disease. *J. Infect. Dis.* **179**(Suppl. 1): S1–S288.

144. **Peters, D., J. A. Sanchez, and P. E. Rollin.** 1999. Filoviridae: Marburg and Ebola viruses, p. 1161–1176. *In* B. N. Fields and D. N. Knipe (ed.), *Virology*, 3rd ed. Raven, New York, N.Y.

145. **Pier, G. B., and S. H. Madin.** 1976. *Streptococcus iniae* sp. nov., a beta-hemolytic streptococcus isolated from an Amazon freshwater dolphin, *Inia geoffrensis. Int. J. Syst. Bacteriol.* **26**:545–553.

146. **Pruksanamonda, P., C. B. Hall, R. A. Insel, K. M. McIntyre, P. E. Pellet, C. E. Long, K. Schnabel, P. H. Pincus, F. R. Stamey, and T. R. Dambaugh.** 1990. Primary human herpesvirus-6 infection in young children. *N. Engl. J. Med.* **326**:1445–1450.

147. **Raoult, D., P. E. Fournier, M. Drancourt, T. J. Marrie, J. Etienne, J. Cosserat, P. Cacoub, Y. Poinsignon, P. Leclerg, and A. M. Sefton.** 1996. Diagnosis of 22 new cases of *Bartonella* endocarditis. *Ann. Intern. Med.* **125**:646–652.

148. **Regnery, R., J. G. Olson, B. A. Perkins, and W. Bibb.** 1992. Serological response to "Rochalimaea henselae" antigen in suspected cat-scratch disease. *Lancet* **339**:1443–1445.

149. **Regnery, R., and J. Tappero.** 1995. Unraveling mysteries associated with cat-scratch disease, bacillary angiomatosis, and related syndromes. *Emerg. Infect. Dis.* **1**:16–21.

150. **Regnery, R. L., B. E. Anderson, J. E. Clarridge III, M. Rodriguez-Barradas, D. C. Jones, and J. H. Carr.** 1992. Characterization of a novel *Rochalimaea* species, *R. henselae* sp. nov., isolated from the blood of a febrile, human immunodeficiency virus-positive patient. *J. Clin. Microbiol.* **30**:265–274.

151. **Relman, D. A.** 1993. The identification of uncultured microbial pathogens. *J. Infect. Dis.* **168**:1–8.

152. **Relman, D. A.** 1998. *Cyclospora*: whence and where to?, p. 185–194. *In* W. M. Scheld, W. A. Craig, and J. M. Hughes (ed.). *Emerging Infections 2.* American Society for Microbiology, Washington, D.C.

153. **Relman, D. A., J. S. Loutit, T. M. Schmidt, S. Falkow, and L. S. Tompkins.** 1990. The agent of bacillary angiomatosis: an approach to the identification of uncultured pathogens. *N. Engl. J. Med.* **23**:1573–1580.

154. **Relman, D. A., T. M. Schmidt, A. Gajadhar, M. Sogin, J. Cross, K. Yoder, O. Sethabutr, and P. Echeverria.** 1996. Molecular phylogenetic anlaysis of *Cyclospora*, the human intestinal pathogens, suggests that it is closely related to *Eimeria* species. *J. Infect. Dis.* **173**:440–445.

155. **Rickter, P. J., R. B. Kimsey, J. E. Madigan, J. E. Barlough, J. S. Dumler, and D. L. Brooks.** 1996. *Ixodes pacificus* as a vector of *Ehrlichia equi. J. Med. Entomol.* **33**:1–5.

156. **Riley, L. W., R. S. Remis, S. D. Helgerson, H. B. McGee, J. G. Wells, B. R. Davis, R. J. Hebert, E. S. Olcott, L. M. Johnson, N. T. Hargrett, P. A. Blake, and M. L. Cohen.** 1983. Hemorrhagic colitis associated with a rare *Escherichia coli* serotype. *N. Engl. J. Med.* **308**:681–685.

157. **Robinson, S. W.** 1994. Human herpesvirus 6. *Curr. Clin. Top. Infect. Dis.* **14**:159–169.

158. **Rollin, P. E., J. Williams, D. Bressler, S. Pearson, M. Cottingham, G. Pucak, A. Sanchez, S. G. Trappier, R. L. Peters, P. W. Greer, S. Zaki, T. Demarcus, K. Kelley, D. Simpson, T. W. Geisbert, P. B. Jahrlina, C. J. Peters, and T. G. Ksiazek.** 1999. Ebola (subtype Reston) virus among quarantined non-human primates recently imported from the Philippines to the United States. *J. Infect. Dis.* **179**(Suppl. 1):S108–S114.

159. **Ruoff, K. L., R. A. Whiley, and D. Beighton.** 1999. Streptococcus, p. 285–286. *In* P. R. Murray, E. J. Baron, M. A. Pfaller, F. C. Tenover, and R. H. Yolken (ed.), *Manual of Clinical Microbiology*, 7th ed. American Society for Microbiology, Washington, D.C.

160. **Salahuddin, S. Z., D. V. Ablashi, P. D. Markham, S. F. Josephs, M. Sturzengger, M. Kaplan, G. Halligan, P. Biberfield, F. Wong-Staal, and B. Kramarsky.** 1986. Isolation of a new virus, HBLV, in patients with lymphoproliferative disorders. *Science* **234**:596–601.

161. **Sanchez, A., S. G. Trappier, B. W. J. Mahy, C. J. Peters, and S. T. Nichol.** 1996. The virion glycoproteins of Ebola viruses are encoded in two reading frames and are expressed through transcriptional editing. *Proc. Natl. Acad. Sci. USA* **93**:3602–3607.

162. **Sanchez, A., S. G. Trappier, U. Stroker , S. T. Nichol, M. D. Bowen, and H. Feldmann.** 1998. Variation in the glycoprotein and VP35 genes of Marburg virus strains. *Virology* **240**:138–146.

163. **Schlossberg, D., Y. Morad, T. B. Krouse, D. J. Wear, C. K. English, and M. Littman.** 1989. Culture proved disseminated cat scratch disease in acquired immunodeficiency syndrome. *Arch. Intern. Med.* **149**:1437–1439.

164. **Schonberger, L. B.** 1998. New variant Creutzfeldt-Jacob disease and bovine spongiform encephalopathy. *Infect. Dis. Clin. N. Am.* **12**:111–121.

165. **Secchiero, P., D. Zella, R. W. Crowley, R. C. Gallo, and P. Lusso.** 1995. Quantitative PCR for human herpesvirus 6 and 7. *J. Clin. Microbiol.* **33**:2124–2130.

166. **Shukla, R., R. Slack, R. George, T. Cheasty, B. Rowe, and J. Scutter.** 1995. *Escherichia coli* O157:H7 infection associated with a farm visitor centre. *Commun. Dis. Rep. CDR Rev.* **5**:R86–90.

167. **Simpson, S. Q.** 1998. Hantavirus pulmonary syndrome. *Heart Lung* **27**:51–57.

168. **Singh, N., and D. R. Carrigan.** 1996. Human herpesvirus 6 in transplantation: an emerging pathogen. *Ann. Intern. Med.* **124**:1065–1071.

169. **Slater, L. N., D. F. Welch, D. Hensel, and D. W. Coody.** 1990. A newly recognized fastidious gram-negative pathogen as a cause of fever and bacteremia. *N. Engl. J. Med.* **23**:1587–1593.

170. **Slater, L. N., D. F. Welch, and K.-W. Min.** 1992. Rochalimaea henselae causes bacillary angiomatosis and peliosis hepatitis. *Arch. Intern. Med.* **152**:602–606.

171. **Slutsker, L., A. A. Ries, K. D. Greene, J. G. Wells, L. Hutwagner, and P. M. Griffin for The *Escherichia coli* O157:H7 Study Group.** 1997. *Escherichia coli* O157:H7 diarrhea in the United States: clinical and epidemiologic features. *Ann. Intern. Med.* **126**:505–513.

172. **Smith, M., C. Bloomer, R. Horvat, E. Goldstein, J. M. Casparian, and B. Chandran.** 1997. Detection of human herpesvirus 8 DNA in Kaposi's sarcoma lesions and peripheral blood of human immunodeficiency virus-positive patients and correlation with serologic measurements. *J. Infect. Dis.* **176**:84–93.

173. **Smith, R. D., D. M. Sells, E. H. Stephenson, M. Ristic, and D. L. Huxsoll.** 1997. Development of *Ehrlichia canis*, causative agent of canine ehrlichiosis, in the tick *Rhipicephalus sanguineus* and its differentiation from a symbiotic rickettsia. *Am. J. Vet. Res.* **37:**119–126.

174. **Smithburn, K. C., T. P. Hughes, A. W. Burke, and J. H. Paul.** 1940. A neurotropic virus isolated from the blood of a native of Uganda. *Am. J. Trop. Med. Hyg.* **20:**471–492.

175. **Spach, D. H., A. S. Kanter, N. A. Daniels, D. J. Nowowiejski, A. M. Larson, R. A. Schmidt, B. Swaminathan, and D. J. Brenner.** 1995. *Bartonella (Rochalimaea)* species as a cause of apparent "culture-negative" endocarditis. *Clin. Infect. Dis.* **20:**1044–1047.

176. **Spach, D. H., and J. E. Koehler.** 1998. *Bartonella*-associated infections. *Infect. Dis. Clin. N. Am.* **12:**137–155.

177. **Steeper, T. A., H. Rosentine, J. Weiser, S. Inampudi, and D. C. Snover.** 1992. Bacillary epitheloid angiomatosis involving the liver, spleen and skin in AIDS patients with concurrent Kaposi's sarcoma. *Am. J. Clin. Pathol.* **97:**713–718.

178. **Stoler, M. H., T. A. Bonfiglio, R. T. Steigbigel, and M. Pereira.** 1983. An atypical subcutaneous infection associated with the acquired immunodeficiency syndrome. *Am. J. Clin. Pathol.* **80:**714–718.

179. **Straus, S. E.** 2000. Human herpesvirus types 6 and 7, p. 1613–1616. *In* G. L. Mandell, J. E. Bennett, and R. Dolin (ed.), *Principles and Practice of Infectious Diseases.* Churchill Livingstone, Philadelphia, Pa.

180. **Strockbine, N. A., L. R. Marques, J. W. Newland, H. W. Smith, R. K. Holmes, and A. D. O'Brien.** 1986. Two toxin-converting phages from *Escherichia coli* O157:H7 strain 933 encode antigenically distinct toxins with similar biologic activities. *Infect. Immun.* **53:**135–140.

181. **Su, C., and L. J. Brandt.** 1995. *Escherichia coli* O157:H7 infection in humans. *Ann. Intern. Med.* **123:**698–714.

182. **Swerdlow, D. L., B. A. Woodruff, R. C. Brady, P. M. Griffin, S. Tippen, H. D. Donnell, E. Geldreich, B. J. Payne, A. Meyer, J. G. Wells, K. D. Greene, M. Bright, N. H. Bean, and P. A. Blake.** 1992. A waterborne outbreak in Missouri of *Escherichia coli* O157:H7 associated with bloody diarrhea and death. *Ann. Intern. Med.* **117:**812–819.

183. **Takasaki, T., N. Ohkana, K. Sano, S. Morimatsu, T. Nakano, M. Nakai, J. Yamaguchi, and I. Kurane.** 1997. Electron microscopic study of human herpesvirus 6 infected human T-cell lines superinfected with human immunodeficiency virus type 1. *J. Acta Virol.* **41:**221–229.

184. **Tang, Y. W., M. J. Espy, D. H. Persing, and T. F. Smith.** 1997. Molecular evidence and clinical significance of herpesvirus 6 infection in the central nervous system. *J. Clin. Microbiol.* **35:**2869–2872.

185. **Tappero, J. W., J. E. Koehler, T. G. Berger, C. J. Cockerell, T.-H. Lee, M. P. Busch, D. P. Sites, J. Mohle-Boetan, A. L. Reingold, and P. E. LeBoit.** 1993. Bacillary angiomatosis and bacillary splenitis in immunocompetent adults. *Ann. Intern. Med.* **118:**363–365.

186. **Tarr, P. I., M. A. Neill, C. R. Clausen, S. L. Watkins, D. L. Christie, and R. O. Hickman.** 1990. *Escherichia coli* O157:H7 and the hemolytic-uremic syndrome: importance of early cultures in establishing the etiology. *J. Infect. Dis.* **162:**553–556.

187. **Telford, S. R., III, J. E. Dawson, P. Katavolos, C. K. Warner, C. P. Kolbert, and D. H. Persing.** 1996. Perpetuation of the agent of human granulocytic ehrlichiosis in a deer tick-rodent cycle. *Proc. Natl. Acad. Sci.USA* **93:**6209–6214.

188. **Tison, D. L.** 1990. Culture confirmation of *Escherichia coli* serotype O157:H7 by direct immunofluorescence. *J. Clin. Microbiol.* **28:**612–613.

189. **Torok, L., S. Z. Viragh, I. Borka, and M. Tapai.** 1994. Bacillary angiomatosis in a patient with lymphocytic leukemia. *Br. J. Dermatol.* **130:**665–668.

190. **Trevena, W. B., R. S. Hooper, C. Wray, G. A. Willshaw, T. Cheasty, and G. Domingue.** 1996. Verocytotoxin-producing *Escherichia coli* O157:H7 associated with companion animals. *Vet. Rec.* **138:** 400.

191. **Visvesvara, G. S., H. Moura, E. Kovacs-Nace, S. Wallace, and M. L. Eberhard.** 1997. Uniform staining of *Cyclospora* oocysts in fecal smears by a modified safranin technique with microwave heating. *J. Clin. Microbiol.* **35:**730–733.

192. **Vitek, C. R., R. F. Breiman, T. G. Ksiazek, P. E. Rollin, J. C. McLaughlin, E. T. Umland, K. B. Nolte, A. Loera, C. M. Sewell, and C. J. Peters.** 1996. Evidence against person-to-person transmission of hantavirus to health care workers. *Clin. Infect. Dis.* **22:**824–826.

193. **Walker, D. H.** 1998. Emerging human ehrlichioses: recently recognized, widely distributed, life-threatening tick-borne diseases, p. 81–91. *In* W. M. Scheld, D. Armstrong, and J. M. Hughes (ed.), *Emerging Infections 1.* American Society for Microbiology, Washington, D.C.

194. **Walker, D. H., and J. S. Dumler.** 1996. Emergence of ehrlichioses as human health problems. *Emerg. Infect. Dis.* **2:**18–29.

195. **Walker, D. H., and J. S. Dumler.** 1997. Human monocytic and granulocytic ehrlichioses. Discovery and diagnosis of emerging tick-borne infections and the critical role of the pathologist. *Arch. Pathol. Lab. Med.* **121:**785–791.

196. **Walker, D. H., and J. S. Dumler.** 2000. *Ehrlichia chaffeensis* (human mococytotropic ehrlichiosis), *Ehrlichia phagocytophila* (human granulocytotropic ehrlichiosis), and other ehrlichiae, p. 2057–2064. *In* G. L. Mandell, J. E. Bennett, and R. Dolin (ed.), *Mandell, Douglas, and Bennett's Principles and Practice of Infectious Diseases,* 5th ed. Churchill Livingstone, Philadelphia, Pa.

197. **Ward, K. N., J. J. Gray, and S. Efstathiou.** 1989. Brief report: primary human herpesvirus 6 infection in a patient following liver transplantation from a seropositive donor. *J. Med. Virol.* **28:**69–72.

198. **Wear, D. J., A. M. Margileth, T. L. Hadfield, and G. W. Fischer, C. J. Schlagel, and F. M. King.** 1983. Cat-scratch disease:a bacterial infection. *Science* **221:**1403–1404.

199. **Weinstein, M. R., M. Litt, D. A. Kertesz, P. Wyper, D. Rose, M. Coulter, A. McGeer, R. Facklam, C. Ostach, B. M. Willey, A. Borczyk, D. E. Low, and The S. iniae Study Group.** 1997. Invasive infections due to a fish pathogen, *Streptococcus iniae.* *N. Engl. J. Med.* **337:**589–594.

200. **Weiss, E., and G. A. Dasch.** 1982. Differential characteristics of strains of *Rochalimaea: Rochalimaea vinsonni* sp. nov., the Canadian vole agent. *Int. J. Syst. Bacteriol.* **32:**305–314.

201. **Wells, J. G., B. R. Davis, I. K. Wachsmuth, L. W. Riley, R. S. Remis, R. Sokolow, and G. K. Morris.** 1983. Laboratory investigation of hemorrhagic colitis outbreaks associated with a rare *Escherichia coli* serotype. *J. Clin. Microbiol.* **18:**512–520.

202. **Whitby, D., M. Howard, M. Tenant-Flowers, N. S. Brink, A. Copas, C. Boshoff, T. Hatzioannov, F. E. Suggett, D. M. Aldam, A. S. Denton, R. F. Miler, I. V. D. Weller, R. A. Weiss, R. S. Teder, and T. F. Schulz.** 1995. Detection of Kaposi's sarcoma associated herpesvirus in peripheral blood of HIV-infected individuals and progression to Kaposi's sarcoma. *Lancet* **346:**799–802.

203. **Wilesmith, J. W., G. A. H. Wells, M. P. Cranwell, and J. B. Ryan.** 1988. Bovine spongiform encephalopathy: epidemiological studies. *Vet. Rec.* **123:**638–644.

204. **Will, R. G., J. W. Ironside, M. Zeidler, S. N. Cousens, K. Estibiero, A. Alperovitch, S. Poser, M. Pocchiari, A. Hofman, and P. G. Smith.** 1996. A new variant of Creutzfeldt-Jacob disease in the UK. *Lancet* **347:**921–925.

205. **Wong, C. S., S. Jelacic, R. L. Babeeb, S. L. Watkins, and P. I. Tarr.** 2000. The risk of the hemolytic-uremic syndrome after antibiotic treatment of *Escherichia coli* O157:H7 infections. *N. Engl. J. Med.* **342:**1930–1936.

206. **Wong, M. T., D. C. Thorton, R. C. Kennedy, and M. J. Dolan.** 1995. A chemically defined liquid medium that supports primary isolation of *Rochalimaea* (*Bartonella*) *henselae* from blood and tissue specimens. *J. Clin. Microbiol.* **33:**742.

207. **World Health Organization.** 1976. Ebola hemorrhagic fever in Zaire, 1976. Report of an international commission. *Bull. W. H. O.* **56:**271–293.

208. **World Health Organization.** 1978. Ebola hemorrhagic fever in Sudan, 1976. Report of a WHO/international study. *Bull. W. H. O.* **56:**274–277.

209. **Yamanishi, K., T. Okuno, K. Shiraki, M. Takahashi, T. Kondo, Y. Asano, and T. Kurata.** 1988. Identification of human herpesvirus-6 as a casual agent for exanthem subitum. *Lancet* **i:**1065–1067.

210. **Zhong, J.-C., C. Metroka, M. Reitz, J. Nicholas, and G. S. Hayward.** 1997. Strain variability among Kaposi sarcoma-associated herpesvirus (human herpesvirus 8) genomes: evidence that a large cohort of United States AIDS patients may have been infected by a single common isolate. *J. Virol.* **71:**2505–2511.

Rapid Systems and Instruments for Antimicrobial Susceptibility Testing of Bacteria

ALAN T. EVANGELISTA AND ALLAN L. TRUANT

17

INTRODUCTION

Antimicrobial susceptibility testing (AST) can be performed by four basic methods: disk diffusion, broth dilution, agar dilution, and gradient diffusion. Clinical microbiology laboratories have used primarily combinations of the disk, broth, and gradient methods, depending on testing preference, the type of organism, and commercial availability of the antimicrobial agent. The more complex agar dilution method has served as a research tool and is often used as a reference method for selected organisms and in comparative antimicrobial studies. Detailed procedural descriptions of these four methods can be found in the *Manual of Clinical Microbiology*, 7th ed. (14, 26), in NCCLS documents for antimicrobial susceptibility testing (31, 32), and in a published review by Jorgensen and Ferraro (23).

A gradual shift in the type of AST method used by clinical microbiology laboratories in the United States has occurred over the past three decades (1970 to 2000). In the 1970s, the manual disk diffusion method, using commercially prepared disks, was the predominant method of AST. Results were reported as interpretive categories: susceptible (S), intermediate (I), and resistant (R). In the 1980s, the number of laboratories utilizing a commercial broth dilution AST method gradually increased, with results reported as the MIC of the antimicrobial agent with the interpretive S-I-R designation. In the 1990s, commercial broth-based systems providing MIC results accounted for two-thirds of all testing (23).

Commercial systems for AST can be divided into automated, semiautomated, and manual systems. An automated testing system consists of computer-assisted incubation, reading, interpretation, and reporting functions that do not require manual intervention. A semiautomated system requires off-line incubation. The panels of a semiautomated system are automatically read with computer-assisted interpretation and reporting, but manual loading of each panel into the system is required. Manual AST panels are read visually by laboratory personnel, and results are either recorded by hand or manually entered into a computer for interpretation and reporting. Each automated, semiautomated, or manual commercial system provides a selection of AST panels with a fixed configuration of antimicrobial agents. Upon request, several companies will prepare custom AST panels for clinical or investigational use.

We recognize that susceptibility testing methods and panels are subject to periodic modifications and that new systems and panels are introduced into various global markets as an ongoing process. The purpose of this chapter is to review recently published evaluations of currently used commercial systems, and any omission of a specific method or published evaluation is unintentional.

Evaluation of Commercial AST Performance

Guidelines for evaluating AST performance have been established by the U.S. Food and Drug Administration (15). Error rates for commercial AST panels are determined by comparing AST results with the reference NCCLS broth dilution method using at least 100 unselected clinical isolates of a single genus or species. The detection rate for very major errors (false susceptibility) should be <1.5% of the test isolates, and major errors (false resistance) should be <3% of the test isolates. Essential agreement with the reference method (± 1 dilution) should be $\geq 90\%$.

AUTOMATED BROTH DILUTION AST SYSTEMS

Commercial automated broth dilution AST systems currently available in the United States include the Vitek System and the Vitek 2 (bioMerieux Inc., St. Louis, Mo.), the MicroScan WalkAway (Dade Behring Inc., West Sacramento, Calif.), and the Sensititre ARIS (Trek Diagnostic Systems, Inc., Westlake, Ohio). An additional automated system to be released in the United States in 2003 is the Phoenix System (BD Diagnostic Systems, Sparks, Md.). The most common automated systems from 1980 to 2000 have been the Vitek System and the MicroScan WalkAway. Both systems have provided periodic software and panel upgrades during that time, with most test panels showing good overall essential agreement with the NCCLS reference broth dilution method. A comparison of the five automated systems is presented in Table 1.

The automated and semiautomated systems discussed in this chapter have the capability of producing standardized or customized patient test reports generated by computer software packages that are referred to as data management systems (DMS). The DMS package usually contains an epidemiology component which can archive results, thereby providing specialized reports such as summary reports, infection control reports, organism trending reports, and antibiograms. To optimize the availability and transcription accuracy of rapid patient reports, the integrated data from automated and semiautomated systems may be transferred through a computer interface to the laboratory information system (LIS). Specialized reports may then be generated by the LIS in addition to reports generated by the automated system microcomputer.

An additional level of enhancement in automated systems is referred to as "expert software," which examines and validates the antimicrobial susceptibility profile or phenotype of an individual isolate. These expert systems use specific rules or algorithms (preprogrammed or user defined) to flag unlikely resistance patterns and recommend changes. Expert software may also predict cross-resistance to other antimicrobial agents and can facilitate the addition of footnotes or comments to a patient's report regarding the resistance pattern. An additional advantage of some of the automated systems is a software package that enables interfacing with a pharmacy system so that the microbiology result can be matched with the patient's record of antimicrobial therapy. Thus, the early detection of in vitro resistance could potentially affect patient outcome if the pharmacy system was alerted to the need to modify therapy. The names of the expert systems and pharmacy software packages available for the automated systems are listed in Table 1.

The relationship between an automated rapid susceptibility report and the clinical and economic impact on patient care has been the subject of several investigations (11, 13, 24, 34, 43). While empiric therapy is commonly initiated in a significant number of hospitalized patients, a physician's modification to a more appropriate therapy in response to rapid susceptibility reports has been recorded for about 10 to 20% of cases (11, 13). Modification of therapy may result in a direct cost savings to the hospital (11, 13) and lower mortality rates (11). Thus, automated systems have the potential to offer improved patient care in addition to some degree of labor savings. Along with the advantages of automated systems, the purchaser should also be mindful of the limitations of each system. In this chapter, each of the five current automated systems is discussed individually and evaluations (advantages and limitations) are presented as an overview of the current literature.

The Vitek System

The Vitek System, also referred to as the Vitek 1, is a broth dilution-based AST system which uses a 45-well plastic card containing 16 to 18 antimicrobial agents (Table 1). During 1998 and 1999, all Vitek Systems were transitioned to newer software compatible with the 45-well card format. The first-generation 30-well cards were discontinued as of March 2000. The Vitek instrument measures changes in turbidity over time (growth curve), comparing a growth control well with wells with various drug concentrations. Results are reported in 4 to 18 h as MIC and S-I-R category, depending on the growth rate and susceptibility parameters of the organism. Five Vitek instruments with different card capacities are available: Vitek 32 (32 cards) and Vitek 60, 120, 240, and 480. Table 1 lists a variety of susceptibility cards, which are available for gram-positive and gram-negative organisms. The Vitek System 120 is shown in Fig. 1.

In an evaluation by Rittenhouse et al., 500 gram-negative clinical isolates were tested by the Vitek System in comparison with the reference broth microdilution method of the NCCLS (35). The 500 isolates consisted of 100 isolates each of *Escherichia coli*, *Klebsiella pneumoniae*, *Proteus mirabilis*, *Enterobacter* species, and *Pseudomonas aeruginosa*,

with a total of nine antimicrobial agents evaluated by using the Vitek F5 card, a first-generation 30-well card (35). Very major errors were detected with the Vitek System at a rate of 0.8% (4 of 500), and major errors were found at a rate of 3.4% (17 of 500).

The first-generation 30-well gram-positive susceptibility (GPS) card was evaluated by Tenover et al. to determine the ability to detect vancomycin resistance in 50 selected isolates of enterococci, including 20 isolates with the *vanB* genotype (41). This evaluation was part of a larger study, which evaluated 10 commercially available AST methods in comparison to the reference broth microdilution method of the NCCLS. The Vitek GPS card yielded a very major error rate of 10.3% and a major error rate of 0% in this study. Upon repeat testing of discrepant isolates, the very major error rate dropped to 3.4%. This study demonstrated the difficulty with the 30-well first-generation cards in detecting low-level vancomycin resistance in *vanB*-containing enterococci.

The newer, 45-well cards contain an additional four to eight antimicrobial agents per card and no longer have the limitations for *P. aeruginosa*-piperacillin and *Klebsiella* spp.-ceftriaxone. In addition, the 45-well cards, GNS-120 and -121, are the only panels for automated systems that contain confirmatory tests to detect an extended-spectrum beta-lactamase (ESBL) (Table 1). The GPS 45-well cards contain additional concentrations of oxacillin to comply with the new lower NCCLS MIC breakpoints for coagulase-negative *Staphylococcus* spp. which were revised in 1999 (30). An evaluation of the GPS 45-well card also reported 100% sensitivity for the detection of both *vanA* and *vanB* genotypes of vancomycin-resistant enterococci (12). While the fixed format of a 45-well card imposes some limitations in the selection of antimicrobial agents for testing, several card configurations are currently available, with future plans for additional cards to accommodate newer antimicrobial agents.

Vitek 2

A more automated version of the Vitek System, the Vitek 2, was introduced in the United States in 1999 (Fig. 2). The Vitek 2 uses a 64-well plastic card containing 19 to 20 antimicrobial agents (Fig. 3). The system contains computer software to deduce the results for an additional 4 to 10 antimicrobial agents, resulting in the reporting of 23 to 30 antimicrobial agents per card (Table 1). The Vitek 2 system incorporates test setup and sample verification with a Smart Carrier component. All AST inoculum dilutions are performed by the instrument, as well as card sealing and incubator loading functions. Optical reading of cards is performed every 15 min in the Vitek 2, with a multichannel fluorometer and photometer to record fluorescence, turbidity, and colorimetric signals. Susceptibility results are reported in 4 to 18 h, depending on the organism and susceptibility parameters.

An evaluation of the Vitek 2 tested 327 clinical isolates of *Enterobacteriaceae*, *P. aeruginosa*, and *Acinetobacter* spp. with six cephalosporins, and the results were compared with those of an agar dilution reference method (M. M. Tracewski, A. L. Barry, S. D. Brown, J. A. Hindler, D. A. Bruckner, and D. F. Sahm, *Abstr. 98th Gen. Meet. Am. Soc. Microbiol.*, abstr. C-37, p. 137, 1998). For each of six cephalosporins, the MIC essential agreement was 94 to 97% of the initial tests, with 0.6% very major errors and 0.9% major errors. A Vitek 2 AST-GP card was evaluated with 150 clinical isolates of

TABLE 1. Comparison of automated AST systems[a]

Item	Vitek System (bioMerieux)	Vitek 2 (bioMerieux)	MicroScan WalkAway (Dade Behring)	Sensititre ARIS (Trek Diagnostic Systems)	Phoenix[b] (BD Diagnostic Systems)
No. of wells/panel	45	64	96	96	136 (51 for ID, 85 for AST)
No. of drugs/panel	16–18	19–20 tested / 4–10 deduced / 23–30 reported	22–24 on Combo panel / 32 on Bkpt panel / 30 on MIC panel	18–22 on MIC panel / 15–17 on Bkpt panel	18–21 on MIC panel / 25 on Combo panel
Gram-negative panels	GNS-108–115, GNS-118, GNS-122–127	AST-GN04, AST-GN07–09	Neg Combo[c] 12, 13, 15, 16, 20–27 / Neg Bkpt Combo[c] 5, 7, 11–16 / Neg MIC[c] 3 / Rapid Neg Combo 6–9 / Rapid Neg Bkpt 3 / rapID/S Plus—TBD	MIC: MH, MI, MJ / Bkpt: BC, BD	Combo panels / TBD
Gram-negative urine panels	GNS-203, -204, -206	AST-GN05	Neg Urine Combo[c] 4, 5–7 / Neg Urine MIC[c] 8, 10, 11 / Rapid Urine Combo 4, 5 / Rapid Bkpt; Rapid MIC 2	BE, BG	TBD
ESBL screen	GNS-120, -121, -127	AST-GN04, -05; AST-GN07–10	Neg Combo w/cefpoxodime	None designated	TBD
ESBL confirmation	GNS-120, -121	None designated	None designated	None designated	TBD
Nonfermenter panels	None designated	AST-GN10	None designated	None designated	None designated
Gram-positive panels	GPS-105–107	AST-GP55	Pos Combo[c] 10–14 / Pos Bkpt[c] 9–13 / Pos MIC[c] 8, 10–13 / Rapid Pos Combo 1 / Rapid Bkpt 1; Rapid MIC 2 / rapID/S Plus—TBD	MIC: MG / Bkpt: BH	TBD
S. pneumoniae panels	None	AST-GP56	MICroSTREP / MICroSTREP Plus[b]	HPB	None designated
DMS	bioLiason and DataTrac	bioLiason and DataTrac	DMS	SAMS Data Management	DMS
Resistance evaluation expert software	Expert System	AES	Expert software rules for DMS	Expert software rules for SAMS	BDXpert and Advanced BDXpert
Pharmacy software	TheraTrac 2	TheraTrac 2	PharmLINK	None designated	None designated

[a] Abbreviations: Bkpt, breakpoint; Neg, negative; Pos, positive; TBD, panel identification to be designated.
[b] Currently available only outside the United States.
[c] MicroScan conventional overnight panels.

FIGURE 1 Vitek System 120 (courtesy of bioMerieux).

enterococci, which included *vanA*, *vanB*, and *vanC* strains and six species of enterococci (16). The essential agreement results for ampicillin, vancomycin, teicoplanin, and high-level gentamicin resistance were 93, 95, 97, and 97%, respectively.

The detection of antimicrobial resistance is also facilitated in the Vitek 2 by the Advanced EXPERT System (AES). This software system validates MIC results by a set of in vitro testing rules, provides result interpretations and corrections, and adds footnotes (NCCLS or laboratory defined) for communication to the physician. The AES was evaluated for its ability to identify beta-lactam phenotypes in a test panel of 196 isolates of *Enterobacteriaceae* and *P.*

aeruginosa (37). Overall, the AES was able to ascertain the beta-lactam phenotype for 93.4% of the isolates tested.

MicroScan WalkAway

The automated AST system MicroScan WalkAway is a broth microdilution method utilizing a standard 96-microwell panel (Fig. 4 and 5). The microwells contain serial dilutions of dehydrated antimicrobial agents. Three types of panel configurations for gram-positive bacteria and three for gram-negative bacteria are available: (i) Combo panels containing both antimicrobial dilutions and identification (ID) substrates, (ii) Breakpoint Combo panels containing antimicrobials in breakpoint dilutions and

FIGURE 2 Vitek 2 (courtesy of bioMerieux).

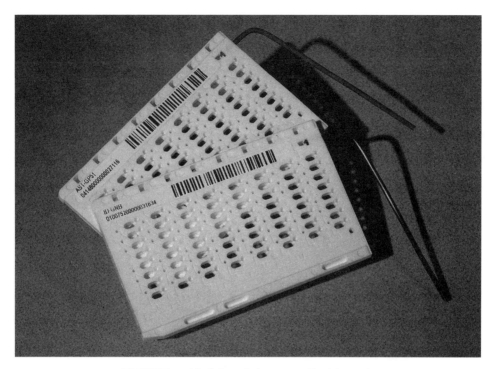

FIGURE 3 Vitek 2 cards (courtesy of bioMerieux).

ID substrates, and (iii) MIC panels containing only antimicrobial dilutions. Breakpoint panels contain only two or three concentrations of each antimicrobial agent, and the resulting MIC corresponds to the interpretive category S, I, or R. In contrast, standard MIC panels contain a wide range of doubling dilution antimicrobial concentrations for the determination of the MIC. The MicroScan panel types are listed in Table 1. The panels are manually inoculated and rehydrated by using a RENOK inoculator and then placed into the incubator-reader component of the WalkAway. Results are obtained after 15 to 18 h by turbidimetric readings of overnight conventional panels

and after 3.5 to 7 h by fluorometric readings of rapid panels. Three instruments with different panel capacities are currently available: the WalkAway 40, the WalkAway 96, and the WalkAway SI. All three instruments are capable of reading either conventional or rapid panels. However, no new antimicrobials have been added to the rapid panels since 1993. New combination panels called rapID/S Plus panels are planned for 2002 for the WalkAway SI system and for retrofitted WalkAway 40 and 96 systems. The new 96-well panels will be configured with 36 wells containing fluorogenic substrates for rapid organism identification and 60 wells with antimicrobials in Mueller-Hinton broth for the turbidimetric determination of MICs utilizing a proprietary technique (processing station) to optimize growth and allow readings from 4.5 to 18 h. The WalkAway SI system is shown in Fig. 4.

An evaluation of the MicroScan WalkAway system was conducted by Rittenhouse et al., using the 500 gram-negative clinical isolates mentioned above in the Vitek section (35). The conventional MicroScan MIC Type 5 panel was compared to the reference broth microdilution method, and a total of five organisms and nine antimicrobials were evaluated in the study. Very major errors (false susceptibility) were detected at a rate of 3.0% (15 of 500), and no major errors (false resistance) were detected. MicroScan Pos MIC conventional (overnight) panels were evaluated by Tenover et al. for the ability to detect vancomycin-resistant enterococci (41). The MicroScan conventional panels were read manually and produced no very major or major errors. In the same study by Tenover et al., MicroScan Rapid Pos MIC panels were evaluated with the MicroScan WalkAway system (41). The MicroScan rapid panels produced 20.7% very major errors and 13.3% major errors for vancomycin and enterococci. Upon repeat testing of discrepant isolates, the very major error rate increased to 27.6%, with no change in the major error rate. This study demonstrated the

FIGURE 4 MicroScan WalkAway SI (courtesy of Dade Behring).

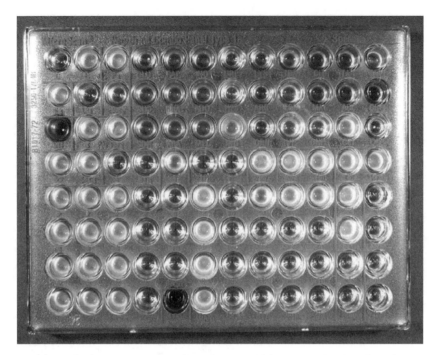

FIGURE 5 MicroScan Combo ID/MIC panel (courtesy of Dade Behring).

difficulty with the MicroScan rapid panels in detecting low-level vancomycin resistance in enterococci. Until new algorithms could be established for the MicroScan rapid panels, Tenover and colleagues recommended the use of alternate methods for the testing of vancomycin resistance in enterococci.

Sensititre ARIS

The automated Sensititre ARIS system is a broth microdilution method utilizing a standard 96-microwell panel containing serial dilutions of dehydrated antimicrobial agents. Panel configurations are currently available for gram-positive and gram-negative bacteria in either a MIC-only format or a breakpoint format and are listed in Table 1. The Sensititre ARIS and Sensititre AutoReader instruments use fluorescence technology to read the panels. Panels either are inoculated by an automated device, the Sensititre AutoInoculator, or are inoculated manually. A fluorogenic substrate is added to the organism inoculum broth and dispensed into the microwells at the same time as the test organism. Growth is determined by generating a fluorescent product from a nonfluorescent substrate. Up to 64 panels may be placed in the Sensititre ARIS, which automatically incubates, reads, and reports results in 18 to 24 h. In addition to the clinical panels listed in Table 1, Trek Diagnostic Systems provides custom MIC panels for investigators performing surveillance studies and evaluations of new antimicrobials. The Sensititre ARIS instrument is shown in Fig. 6.

Evaluations of Sensititre panels using only the semiautomated Sensititre AutoReader and not the Sensititre ARIS system have been published. Staneck and colleagues evaluated Sensititre MIC panels by testing 17 antimicrobial agents against 828 isolates of gram-negative rods (38). The AutoReader 18-h MIC results produced agreement within ±1 twofold dilution of the reference broth method in 95.3% of the cases. In the same study, 11 antimicrobial agents were tested against 148 gram-positive cocci, with 93.5% agreement with the reference method. Doern and colleagues evaluated Sensititre breakpoint panels, testing 17 antimicrobial agents against numerous species of the family *Enterobacteriaceae* (10). A total of 6,086 organism-antimicrobial agent combinations were evaluated with the Sensititre AutoReader, and concordance was noted in 97.2% of the comparisons after 18 h of incubation. In the same study, a total of 1,377 *P. aeruginosa*-antimicrobial agent combinations were evaluated, and agreement was achieved in 92.2% of the cases after 18 h of incubation. These published evaluations did not include very major or major error percentages.

Daly and colleagues evaluated imipenem stability in predried Sensititre MIC and breakpoint panels using 11 clinical strains and 1 control strain of *P. aeruginosa* (6). Imipenem concentrations measured by high-pressure liquid chromatography were stable for a 15-month period, but the breakpoint panels declined by approximately 50% by 18 months. Entry of moisture into the predried panels coincided with loss of imipenem activity, as indicated by pink desiccants. The authors concluded that the labeled expiration date of 18 months for dried Sensititre MIC panels was acceptable providing that panels containing pink desiccant were not used.

The Phoenix System

The Phoenix Automated Microbiology System utilizes combination panels for bacterial ID and AST. The combination panels contain a total of 136 wells, of which 51 are utilized for ID and up to 85 are utilized for AST (Table 1). A bacterial inoculum for ID and a second inoculum for AST are poured into receptacles of the respective ID and AST sections of the single combination panel. The panels are tilted and each dehydrated well is then self-inoculated. Susceptibility results are determined in the Phoenix system by utilizing a Resazurin-based redox dye as well as kinetic

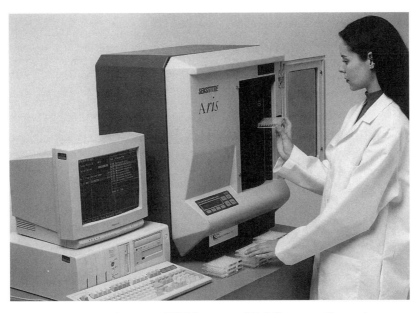

FIGURE 6 Sensititre ARIS (courtesy of Trek Diagnostic Systems).

measurements of turbidity to detect bacterial growth in the presence of an antimicrobial agent. Up to 100 panels can be placed into the Phoenix system for ID, AST, or a combination of ID and AST. The panels are automatically incubated for up to 18 h, and the results are read and reported. The average times to results in the Phoenix have been reported as 6.5 h for *Enterobacteriaceae*, 12 h for pseudomonads, 5.5 h for *Enterococcus* spp., and 7 h for *Staphylococcus* spp. (T. Wiles, D. Turner, W. B. Brasso, J. Hong, and J. Reuben, *Abstr. 99th Gen. Meet. Am. Soc. Microbiol.*, abstr. C-94, p.

123, 1999). The Phoenix system is shown in Fig. 7, and the Phoenix Combo ID/AST panels are shown in Fig. 8.

An evaluation of the Phoenix system was performed by Meyer and colleagues, using 174 gram-negative clinical isolates, including 32 nonfermenters, and 31 antimicrobial agents (W. A. Meyer, D. Lockwood, W. Brasso, and J. Reuben, *Abstr. 100th Gen. Meet. Am. Soc. Microbiol.*, abstr. C-309, p. 203, 2000). The results were compared with those for a MicroScan standard broth microdilution panel (Combo Type 11) run on a MicroScan WalkAway 96

FIGURE 7 Phoenix system (courtesy of BD Diagnostic Systems).

FIGURE 8 Phoenix ID/AST panel (courtesy of BD Diagnostic Systems).

instrument. Overall categorical agreement between Phoenix and MicroScan for the 4,871 organism-antimicrobial agent pairs was reported as 96.0%. The percent categorical agreement for beta-lactams, cephalosporins, aminoglycosides, fluoroquinolones, and miscellaneous classes was reported as 94.9, 95.4, 97.5, 97.5, and 95.7, respectively. This analysis did not report percentages of very major and major errors.

Susceptibility testing of 246 gram-positive cocci with 10 antimicrobial agents was conducted by Wiles et al., comparing the Phoenix system with a reference NCCLS broth microdilution procedure (T. Wiles, W. Brasso, D. Turner, D. Holliday, and K. Fischbein, *Abstr. 9th Eur. Congr. Clin. Microbiol. Infect. Dis.*, abstr. P1156, p. 396, 1999). With a challenge set of 125 isolates of *Staphylococcus* spp., the essential agreement ranged from 85 to 95%. Lower essential agreements of 85 and 86% were noted when the staphylococci were tested with penicillin and ampicillin, respectively. Significant very major errors (false susceptibility) and major errors (false resistance) were reported for oxacillin as 4 and 2.4%, respectively, and for ampicillin as 2.4 and 3.2%, respectively. A challenge set of 121 isolates of *Enterococcus* spp. yielded an essential agreement range from 85 to 98%, with the lower essential agreement of 85% noted for vancomycin. A very major error rate of 1.7% was noted for *Enterococcus* spp. when tested with both penicillin and vancomycin. The study reported a major error rate of 0.8% for penicillin, with no major errors for vancomycin. The authors concluded that further enhancements in the development process were needed for the improvement of Phoenix susceptibility results for staphylococci and enterococci.

SEMIAUTOMATED AND MANUAL BROTH DILUTION AST SYSTEMS

Manufacturers of microwell broth dilution MIC panels usually offer both semiautomated instruments and manual devices to facilitate the reading of the panels. The choice of a test system may be influenced by the laboratory budget. However, the upgrading of a manual system to a semiautomated or automated system may also depend on the need to handle a larger volume of tests and/or the need to provide rapid computer-assisted reports. The basic manual system relies on the observer to read, record, and interpret MIC results. A computer-assisted manual system provides the observer with a touch screen, keypad, or light pen to enter and record results. The next level of instrumentation is a semiautomated system, which automatically reads, records, and interprets the MIC results from an AST panel that has been manually loaded into the reader. Both manual and semiautomated systems can provide a variety of patient and laboratory reports, depending on the type of microcomputer and software that is chosen, and these computers often have the ability to be interfaced with an LIS. The systems to be reviewed are listed in Table 2.

MicroScan

In addition to the automated WalkAway system, MicroScan has two instruments that are semiautomated, the AutoScan 3 and the AutoScan 4, which are listed in Table 2. Use of the AutoScan instruments requires off-line incubation of the conventional 96-well broth microdilution MicroScan AST panels. The panels are then manually loaded one at a time into the AutoScan instrument and read automatically. The MicroScan DMS also is coupled with the AutoScan 3 or 4 for the computer-assisted application of AST interpretive criteria and the generation of patient and epidemiologic reports. For smaller institutions or laboratories that choose manual testing, the TouchScan SR, a manual reader, is available from MicroScan. The manual TouchScan SR can also be coupled with the MicroScan DMS for the generation of various reports. Another component of the MicroScan systems is a device called the RENOK, which can be used for the hydration and inoculation of all types of MicroScan panels used by any of the MicroScan instruments (Table 2). Evaluations of the MicroScan conventional AST panels that are read by the AutoScan 3, AutoScan 4, and TouchScan SR can be found above in the MicroScan WalkAway section of this chapter. Additional panels available from MicroScan for susceptibility testing of *Streptococcus pneumoniae* include a frozen panel available in the United States called the MICroSTREP and a dried panel not available in the United States called the MICroSTREP Plus (Table 1).

Sensititre

The Sensititre susceptibility system uses dehydrated 96-microwell panels which can be read by automated, semiautomated, and manual Sensititre instruments. In addition to the automated Sensititre ARIS, which was discussed above, panels may be read either by the semiautomated Sensititre AutoReader or by the manual SensiTouch Reader or Sensititre manual viewer (Table 2). When the AutoReader is utilized, a fluorogenic substrate is added to the bacterial inoculum. Panels are incubated off-line for 18 h and then loaded one at a time into the AutoReader, where bacterial growth is detected by the presence of fluorescence. The Sensititre automated microbiology system (SAMS) software then interprets and records the MIC results, and reports can be generated.

With the SensiTouch Reader, panels are read visually by the aid of a specially designed light box and MIC endpoints are typed on a keypad, which expedites data entry into the SAMS software described above. Fluorogenic substrate is not added to panels that are read manually. Manual reading without computer-assisted software also is offered by the use of the Sensititre Manual Viewer, whereby the observer reads and records the MIC results. Panels may be inoculated man-

TABLE 2. Semiautomated and manual AST systems

System	Manufacturer	Semiautomated reader	Computer-assisted manual reader	Other specialized component(s)	Result[a]			
					GP	GN	FAS	ANA
Broth microdilution AST systems								
MicroScan	Dade Behring	AutoScan 3, AutoScan 4	TouchScan SR	RENOK, DMS	+	+	+	−
Sensititre	Trek Diagnostic Systems	AutoReader	SensiTouch Reader	Manual viewer, AutoInoculator, DMS	+	+	+	+
Pasco	BD Diagnostic Systems	None	Pasco Reader	DMS	+	+	+	+
MicroMedia	Medical Specialties	None	None	Manual viewer	+	+	+	+
PML	PML Microbiologicals	None	None	Manual viewer	−	−	+	+
MicroTech	MicroTech Medical Systems	None	None	None	−	−	+	+
IDS	IDS/Remel	None	None	None	−	−	+	+
ATB[b]	bioMerieux	AutoReader	None	Manual viewer, DMS	+	+	+	+
Micronaut[b]	Merlin Diagnostika	AutoReader	None	Manual viewer, AutoInoculator, DMS	+	+	+	−
Non-broth AST methods								
Etest	AB Biodisk	None	None	Etest strip for agar gradient diffusion	+	+	+	+
BIOMIC	Giles Scientific	Video Reader for disk diffusion	None	DMS	+	+	+	−
SIRSCAN	i2a	Image Analyzer for disk diffusion	None	DMS	+	+	+	−

[a]GP, gram positive; GN, gram negative; FAS, fastidious (*H. influenzae* and *S. pneumoniae*); ANA, anaerobic.
[b]Currently available only outside the United States.

ually with a multitip pipettor or automatically with the Sensititre AutoInoculator. The AutoInoculator provides a consistent standardized inoculum for panels read by any of the Sensititre instruments. Evaluations of Sensititre gram-negative AST panels, read by the AutoReader, were presented above in the Sensititre ARIS section of this chapter. An evaluation of Sensititre gram-positive AST panels was performed by Tenover and colleagues, using 50 well-characterized strains of enterococci (41). In a comparison of 10 commercial methods, no very major or major errors were detected with the Sensititre panels in evaluating vancomycin susceptibility in enterococci. Additional Sensititre AST panels available from Trek include HP panels for testing *Haemophilus* spp. and *S. pneumoniae* and Anaerobic MIC panels, which are for "research use only" in the United States. The HP panel for the testing of *S. pneumoniae* is reviewed below (see "AST of *S. pneumoniae*"). JustOne strips for testing a single antimicrobial agent are also produced by Trek.

Pasco

The Pasco System (BD Diagnostic Systems) utilizes frozen panels that contain 104 microwells. Panels are thawed, inoculated, and read manually after an 18-h incubation period. Pasco panels are available for the testing of gram-negative and gram-positive organisms and have the following configurations: MIC only, MIC/ID Combination, and Breakpoint/ID Combination. All panels are read manually with the Pasco Reader, which registers the MIC when the observer touches the respective microwell with a light pen. The Pasco Reader is also coupled to a microcomputer containing the Pasco DMS, which applies interpretive criteria to each result and is capable of generating a variety of reports. Evaluations of Pasco panels for gram-negative susceptibility testing have not been recently published. In an evaluation of vancomycin susceptibility results for enterococci, Tenover and colleagues reported no very major or major errors when Pasco gram-positive AST panels were used (41). An additional panel for the testing of *S. pneumoniae* is called the Strep Plus panel, which is reviewed below (see "AST of *S. pneumoniae*").

MicroMedia

The MicroMedia System (Medical Specialties, Inc., Cleveland, Ohio) utilizes frozen panels with 96 microwells that are smaller than the standard commercial panel and are called Fox panels. The panels are thawed, inoculated, and read manually after an 18-h incubation period (Table 2).

MicroMedia panels are available for the susceptibility testing of gram-negative, gram-positive, fastidious, and anaerobic organisms. The anaerobic panels are designated for research use only. Evaluations of the MicroMedia Fox panels for gram-negative susceptibility testing have not been recently published. In an evaluation of vancomycin susceptibility results for enterococci, Tenover and colleagues reported no very major or major errors when MicroMedia Fox gram-positive AST panels were used (41). Error rates and percent agreement for the MicroMedia Fox fastidious panel for *S. pneumoniae* compared to the NCCLS reference method are listed in Table 3.

PML

Commercial PML AST panels (PML Microbiologicals, Wilsonville, Oreg.) are available only for the testing of fastidious organisms such as *S. pneumoniae* and *H. influenzae* and for the testing of anaerobes. PML utilizes frozen panels with a 96-well configuration. After inoculation, the panels

TABLE 3. Comparison of MIC methods for testing *S. pneumoniae*

Antimicrobial agent	Method	n	No. of interpretive category discrepancies vs reference method (%)			Essential agreement (% ± 1 dilution of reference method)	Reference
			Very major	Major	Minor		
Penicillin	Pasco	55	0	0	1 (1.8)	97.4	32
		157	0	0	16 (10.2)	95.5	34
		75	0	0	1 (1.3)	99.3	35
	Sensititre	55	0	0	7 (12.7)	96.8	32
		157	0	0	18 (11.5)	89.2	34
	MicroTech	55	0	0	7 (12.7)	100	32
	MicroMedia	55	0	0	9 (16.4)	97.4	32
	Vitek 2	407	0	0	38 (9.3)	89.8	Jorgensen et al.[a]
		100	0	0	0	99.0	Fremaux et al.[b]
	MicroScan	205	0	0	1 (0.5)	100	33
	MICroSTREP	157	0	0	18 (11.5)	96.2	34
	Etest	55	0	0	2 (3.6)	94.2	32
Erythromycin	Pasco	55	0	0	1 (1.8)	100	32
		157	1 (1.5)	1 (1.1)	1 (0.6)	96.8	34
		75	1 (1.3)	1 (1.3)	0	99.3	35
	Sensititre	55	0	0	1 (1.8)	100	32
		157	1 (1.5)	1 (1.1)	3 (1.9)	93.6	34
	MicroTech	55	0	0	0	100	32
	MicroMedia	55	1 (3.9)	0	3 (5.5)	91.7	32
	Vitek 2	407	1/91 (1/1)	0	4 (1.0)	Not done	Jorgensen et al.
		100	0	0	0	100	Fremaux et al.
	MicroScan	205	0	0	1 (0.5)	100	33
	MICroSTREP	157	0	0	1 (0.6)	98.7	34
	Etest	55	0	0	2 (3.6)	22.5	32
Ceftriaxone	Pasco	55	0	0	8 (14.5)	100	32
		157	0	0	25 (15.9)	97.5	34
		75	0	0	8 (10.7)	100	35
	Sensititre	55	0	0	7 (12.7)	100	32
		157	0	1 (0.9)	39 (24.8)	94.9	34
	MicroTech	55	1	0	8 (14.5)	96.0	32
	MicroMedia	55	1	0	12 (21.8)	92.9	32
	Vitek 2	407	0	0	32 (7.9)	96.2	Jorgensen et al.
		100	0	0	0	100	Fremaux et al.
	MicroScan	205	0	0	20 (9.8)	100	33
	MICroSTREP	157	0	0	19 (12.1)	96.8	34
	Etest	55	0	0	7 (12.7)	98.1	32

[a]Jorgensen et al., *Abstr. 98th Gen. Meet. Am. Soc. Microbiol.*
[b]Fremaux et al., *Abstr. 37th Intersci. Conf. Antimicrob. Agents Chemother.*

are incubated overnight and read with a manual viewer (Table 2).

MicroTech

Frozen 96-well MicroTech MIC panels (MicroTech Medical Systems, Aurora, Colo.) are available for fastidious organisms and for anaerobes only. One half of the fastidious panel contains 12 antibiotics in *Haemophilus* Test Medium for the MIC testing of *Haemophilus* spp., and the other half contains 11 antibiotics in Mueller-Hinton broth supplemented with 3% lysed horse blood for the MIC testing of *S. pneumoniae*. MicroTech supplies a plastic manual inoculator but does not supply a manual viewer or computer-assisted programs (Table 2). MicroTech also manufactures these panels for IDS/Remel (Lenexa, Kans.). Error rates and percent agreement for the MicroTech fastidious panel for *S. pneumoniae* compared with the NCCLS reference method are listed in Table 3.

IDS

Manual IDS MIC panels (IDS/Remel) are available as frozen 96-well panels with the same configuration as MicroTech panels. The IDS FAS MIC panel is an overnight incubation panel used for fastidious organisms, and the IDS ANA MIC panel is for anaerobic organisms. A manual inoculator is available from IDS, but components such as a manual viewer or computer-assisted programs are not supplied by IDS (Table 2).

The ATB System

The ATB System (bioMerieux, Marcy l'Etoile, France) is available in Europe and in other locations outside the United States. The ATB System uses dried panels with 32 cupules, which can be read after 18 to 24 h by semiautomated and manual methods in the ATB System (Table 2). ATB panels are available for the susceptibility testing of gram-negative, gram-positive, fastidious, and anaerobic organisms. The system is supported by a DMS.

The Micronaut System

The Micronaut System (Merlin Diagnostika, Bornheim-Hersel, Germany) is available in Europe and in other locations outside the United States. The Micronaut System uses dehydrated panels with either a 96- or a 386-well configuration. Panels for AST are available for performing MIC or breakpoint testing of gram-negative, gram-positive, and fastidious organisms and are read by semiautomated or manual methods. The system is supported by a DMS with "expert" rule analysis and has the ability to be interfaced with a LIS.

NON-BROTH DILUTION AST METHODS

Etest

The agar gradient diffusion method of susceptibility testing is performed by the epsilometer test, or Etest (AB Biodisk, Solna, Sweden). The Etest is a plastic strip with a predefined gradient of approximately 15 concentrations of antimicrobial agent on the bottom surface and a MIC interpretive scale on the upper surface. The Etest procedure requires the inoculation of the entire surface of a large, 150-mm-diameter agar plate with a standard suspension of the test organism similar to that used in the disk diffusion procedure. Up to six Etest strips may be placed in a radial fashion on the inoculated plate. The plates are incubated overnight, and an elliptical zone of growth inhibition due to the antibiotic gradient formed in the agar is visualized.

FIGURE 9 Etest (courtesy of AB Biodisk and Remel).

The MIC is read as the point of intersection of the elliptical zone with the edge of the strip (Fig. 9). Studies comparing the Etest to broth and agar dilution MIC methods using selected sets of gram-negative and gram-positive bacteria have been published (2, 21). In these studies, the MIC reported by Etest compared favorably to the reference methods within 1 twofold dilution interval. In an evaluation of vancomycin susceptibility results for enterococci, Tenover and colleagues reported no very major or major errors with the Etest method (41). Use of the Etest for the testing of *S. pneumoniae* is reviewed below (see "AST of *S. pneumoniae*").

The BIOMIC System

The BIOMIC System (Giles Scientific, New York, N.Y.) uses a video-assisted plate reader to read and interpret disk diffusion susceptibility agar plates (Table 2). After a disk diffusion plate is placed in the video reader and an image appears on the screen, which is reviewed for possible adjustment by the operator, quantitative zone diameters are calculated by digital image analysis and both qualitative (S-I-R) and quantitative (MIC) results are provided. Korgenski and Daly performed an evaluation of disk diffusion susceptibility plates for rapidly growing gram-negative and gram-positive organisms, comparing the BIOMIC System with visually measured zones for 3,339 organism-antimicrobial agent combinations (27). The BIOMIC results demonstrated 0.1% very major errors (false-susceptible reads) and 0.2% major errors (false-resistant reads). Berke and Tierno compared the BIOMIC System with the Vitek System and reported overall MIC agreements between the two systems of 91.8% for gram-positive bacteria and 91.7% for gram-negative bacteria (3).

The SIRSCAN System

The SIRSCAN System (i2a, Intelligence Artificielle Applications, Montpellier, France) is a semiautomated image analyzer that measures zone diameters of disk diffusion susceptibility plates (Table 2). The SIRSCAN reports both interpretive category and MIC results and is coupled with a user-programmed expert system. Medeiros and Crellin compared readings of disk diffusion plates by the SIRSCAN with visual readings by experienced observers using a hand-held caliper (28). A total of 368 clinical isolates, including 241 gram-negative rods and 127 *Staphylococcus* spp., were tested. For all organisms and antibiotics tested, the SIRSCAN results yielded 0.3% very major errors. Individual organism-antibiotic very major errors occurred most often with staphylococci and amoxicillin-clavulanic acid (3.9%), *Pseudomonas* spp. and piperacillin (3.6%), and coagulase-negative staphylococci (CoNS) and oxacillin (2.7%).

AST OF *S. PNEUMONIAE*

Susceptibility testing of *S. pneumoniae* is reviewed separately in this chapter since this organism has fastidious growth requirements and presents a special challenge to commercial AST methods. *S. pneumoniae* is a well-recognized, leading pathogen of community-acquired pneumonia, acute otitis media, and sinusitis as well as a major cause of bacteremia and meningitis. During the past decade (1990 to 2000), surveillance studies of *S. pneumoniae* have documented increasing resistance to penicillin, cephalosporins, and macrolides (1, 4, 9, 17, 36). An inexpensive manual method to screen for penicillin-resistant pneumococci is the oxacillin disk screening test. According to NCCLS guidelines (33), if the oxacillin zone size is ≤19 mm, a follow-up MIC method should be used to confirm resistance to penicillin and other beta-lactam agents. A confirmatory MIC test is required since many penicillin-susceptible pneumococci produce oxacillin zone diameters of ≤19 mm and would be judged falsely penicillin resistant if the oxacillin disk test were used alone (8). This section describes published evaluations of several commercial MIC methods for *S. pneumoniae*. The comparative reference method in most evaluations was the NCCLS reference broth microdilution method, which utilized cation-adjusted Mueller-Hinton broth supplemented with 2 to 5% lysed horse blood (33).

MicroScan MICroSTREP

The MicroScan MICroSTREP panel (Dade Behring) is a frozen broth microdilution panel that contains 10 antimicrobial agents in Mueller-Hinton broth supplemented with 3% lysed horse blood. Two published evaluations of the MICroSTREP panel are presented in Table 3, including values for the essential agreement (percentage of test results ±1 dilution of the reference broth microdilution method). Jorgensen and colleagues compared the MICroSTREP panel with the NCCLS reference method, using 205 selected pneumococcal isolates, and reported no very major or major errors for penicillin, erythromycin, ceftriaxone, or vancomycin (25). The essential agreement for the MICroSTREP panel was 100% for penicillin, erythromycin, ceftriaxone, and vancomycin. The authors concluded that the MICroSTREP panel had accuracy comparable to that of the NCCLS broth microdilution reference method. A study by Guthrie et al. evaluated the MICroSTREP panel against 157 pneumococcal isolates and found no very major or major errors for penicillin and ceftriaxone and one very major error and one major error for erythromycin (18). This study reported essential agreements of 96.2% for penicillin, 98.7% for erythromycin, and 96.8% for ceftriaxone. The antibiotic configuration of the frozen MICroSTREP panel includes chloramphenicol, which is infrequently used in clinical practice for *S. pneumoniae*, and does not include fluoroquinolones, which are more frequently used. However, the second-generation panel, the dried MICroSTREP Plus, does include fluoroquinolones but is not yet available in the United States.

The MICroSTREP Plus panel, a dried panel available in Europe, Canada, and other areas outside the United States, was evaluated by Tjhio and colleagues, using 50 selected pneumococcal isolates (J. T. Tjhio, P. C. Schreckenberger, and W. M. Janda, *Abstr. 100th Gen. Meet. Am. Soc. Microbiol.*, abstr. C-313, p. 204, 2000). Essential agreement with the NCCLS reference method was reported as ≥98% for penicillin, ceftriaxone, levofloxacin, and vancomycin and 96% for erythromycin.

Vitek 2 (AST-GP56)

The Vitek 2 *S. pneumoniae* susceptibility card (AST-GP56; bioMerieux) contains 11 antibiotics in a freeze-dried Wilkens-Chalgren broth. A prototype of this card was evaluated by Jorgensen and colleagues, using a collection of 407 clinical isolates of pneumococci with known resistance properties (J. H. Jorgensen, A. L. Barry, M. M. Traczewski, D. F. Sahm, M. L. McElmeel, and S. A. Crawford, *Abstr. 98th Gen. Meet. Am. Soc. Microbiol.*, abstr. C-422, p. 201, 1998). No very major or major errors were detected for penicillin and ceftriaxone, and the essential agreement was 89.8 and 96.2%, respectively (Table 3). For erythromycin there was one very major error and no major errors, but there were too few MICs within the range of the panel for the calculation of an essential agreement with the reference method. Fremaux and colleagues also evaluated the Vitek 2 *S. pneumoniae* susceptibility card, using 100 pneumococcal isolates, and compared the Vitek 2 card results with those of a standard agar dilution method (A. Fremaux, G. Sissia, P. Geslin, and J. Zindel, *Abstr. 37th Intersci. Conf. Antimicrob. Agents Chemother.*, abstr. D-49, p. 92, 1997). No very major or major errors were detected for penicillin, erythromycin, ceftriaxone, ofloxacin, and vancomycin (Table 3). A 99% essential agreement was reported for penicillin, and a 100% essential agreement was found for erythromycin, ceftriaxone, ofloxacin, and vancomycin. The inclusion of newer fluoroquinolones is planned for the next phase of Vitek 2 pneumococcal AST cards.

Pasco

Frozen Pasco MIC panels for testing pneumococci were evaluated by Tenover et al., using a challenge set of 55 pneumococcal isolates (40). No very major or major errors were reported for penicillin, erythromycin, and ceftriaxone, and the percent essential agreement was 97.4, 100, and 100, respectively (Table 3). A study by Guthrie and colleagues tested Pasco panels against 157 selected pneumococcal isolates and found no very major or major errors for penicillin and ceftriaxone and one very major error and one major error for erythromycin (18). Essential agreement was reported as 95.5% for penicillin, 96.8% for erythromycin, and 97.5% for ceftriaxone (Table 3).

An evaluation of the more recent frozen Pasco Strep Plus panel, which contains 26 antimicrobials, was performed by Mohammed and Tenover with 75 selected pneumococcal isolates (29). No very major or major errors were found with penicillin, ceftriaxone, levofloxacin, or vancomycin, and only one very major error and one major error were found with erythromycin (Table 3). The study reported essential agreements of 99.3% for penicillin, erythromycin, and vancomycin and 100% for ceftriaxone and levofloxacin. The authors concluded that the Pasco Strep Plus panel had accuracy comparable to that of the NCCLS broth microdilution reference method.

Sensititre HP

The dried Sensititre HP MIC panels (Trek Diagnostic Systems) for fastidious organisms contain 10 reportable antimicrobial agents for *S. pneumoniae* and 21 reportable agents for *Haemophilus* spp. When pneumococci are tested, Mueller-Hinton broth supplemented with 2% lysed horse blood is used as the organism suspension diluent. Tenover and colleagues evaluated the dried Sensititre HP MIC panels using a challenge set of 55 pneumococcal isolates (40). No very major or major errors were reported for penicillin, erythromycin, and ceftriaxone, and the percent essential agreements were 96.8, 100, and 100, respectively (Table 3). Guthrie et al. also evaluated Sensititre dried panels using 157 selected pneumococcal isolates and found no very major errors for penicillin and ceftriaxone, no major errors for penicillin, and one major error for ceftriaxone (18). For erythromycin, the study reported one very major error and one major error. Essential agreement was reported as 89.2% for penicillin, 93.6% for erythromycin, and 94.9% for ceftriaxone (Table 3). Guthrie and colleagues also had evaluated MicroScan MICroSTREP and Pasco panels along with Sensititre and concluded that all three panels provided interpretive results comparable to one another and to the NCCLS reference method.

Etest

Tenover and colleagues evaluated the Etest using a challenge set of 55 pneumococcal isolates (40). The Etest was performed with Mueller-Hinton agar containing 5% sheep blood and incubated overnight at 35^0C in 5% CO_2 as recommended by the manufacturer. In comparison with the NCCLS reference broth microdilution method, no very major or major errors were reported for penicillin, erythromycin, and ceftriaxone, and the essential agreement was 94.2% for penicillin and 98.1% for ceftriaxone but only 22.5% for erythromycin (Table 3). Tjhio and colleagues performed another evaluation of the Etest using 50 selected pneumococcal isolates (Tjhio et al., *Abstr. 100th Gen. Meet. Am. Soc. Microbiol.*). Essential agreement with the NCCLS reference method was reported as ≥98% for penicillin, ceftriaxone, levofloxacin, and vancomycin but only 42% for erythromycin. The authors reported that the Etest showed higher MICs of erythromycin due to the CO_2 incubation and that the MICs for 29 of 50 isolates differed by more than 1 twofold dilution compared with the reference NCCLS broth microdilution method.

According to the manufacturer of Etest, CO_2 incubation lowers agar pH, thereby yielding lower MICs for tetracyclines and some beta-lactams while increasing the MICs of macrolides, clindamycin, aminoglycosides, glycopeptides, and quinolones (*Susceptibility Testing of Pneumococci*, Etest technical guide 5C; AB Biodisk, Solna, Sweden). The Etest CO_2 effect of increasing MICs for fluoroquinolones has been noted by some investigators. Thornsberry and colleagues tested *S. pneumoniae* isolates in a U.S. surveillance study by the NCCLS reference broth microdilution method and reported a levofloxacin MIC mode of 0.5 µg/ml and a MIC at which 90% of the isolates tested are inhibited (MIC_{90}) of 1.0 µg/ml (42). When these isolates were tested by Etest, the MIC mode increased to 1.0 µg/ml and the MIC_{90} increased to 2.0 µg/ml. In another surveillance study of *S. pneumoniae*, Jones et al. reported that the Etest tended to elevate the MICs of gatifloxacin compared to the reference broth microdilution method (22). The results showed that 32% of the isolates had elevated Etest MICs, with 4.3% of the isolates showing an Etest MIC increase of four to eight times the MIC determined by the reference broth method.

DETECTION OF MRSA

The prevalence of methicillin-resistant *S. aureus* (MRSA) has been increasing in the last decade and has been presenting patient treatment and infection control problems in hospitals throughout the world (46). Methicillin or oxacillin resistance in *S. aureus* is due to the acquisition of a chromosomal *mecA* gene, which codes for the production of an additional penicillin-binding protein, PBP2a, that has a low affinity for beta-lactam antibiotics (19, 44). The supplemental PBP2a continues to function when PBP1, -2, and -3 have been inactivated by beta-lactam agents, thus producing intrinsic resistance to all beta-lactams, including cephalosporins. Detection of MRSA presents special diagnostic challenges since phenotypic expression of heterogeneous resistance occurs in many strains (7, 20). Methods for the detection of MRSA include growth-dependent phenotypic methods, latex agglutination assays to detect PBP2a, and molecular methods such as PCR and DNA hybridization to detect the *mecA* gene. Molecular methods have emerged as the "gold standard" for MRSA detection; however, many clinical laboratories do not have the resources to routinely perform PCR or DNA hybridization techniques. This section reviews several commercial methods for the rapid detection of MRSA.

Rapid Phenotypic Assays To Detect MRSA

The 45-well Vitek GPS-106 card (bioMerieux) was evaluated by Yamazumi et al. for its ability to detect MRSA among 200 bloodstream isolates of *S. aureus* and was found to have a 98.0% sensitivity and 100% specificity compared with detection of the *mecA* gene by PCR (48). Shubert et al. evaluated the Vitek 2 AST card (bioMerieux) and reported an overall agreement with *mecA* PCR of 100% for the detection of MRSA and 98.7% for the detection of methicillin-resistant CoNS (C. Shubert, R. Griffith, W. McLaughlin, M. Ullery, and M. Peyret, *Abstr. 98th Gen. Meet. Am. Soc. Microbiol.*, abstr. C-478, p. 211, 1998). Struelens et al. evaluated the Rapid ATB Staph system (bioMerieux) compared to *mecA* PCR and reported a sensitivity of 97% for the detection of MRSA (39).

MicroScan gram-positive rapid panels (Dade Behring) were evaluated by Woods and colleagues for their ability to detect oxacillin resistance in 92 MRSA isolates and 103 CoNS using the MicroScan WalkAway 96 system (47). The MIC results obtained with the dried MicroScan Pos MIC type 1 rapid panels were compared with those for the conventional overnight MicroScan Pos MIC type 6 panels. The rapid panels detected 96.7% of MRSA and 72% of methicillin (oxacillin)-resistant CoNS. In this study, the overnight MicroScan panels were not evaluated but served as the reference method.

The Phoenix system was evaluated by Brasso et al. for detection of oxacillin resistance in staphylococci (W. Brasso, D. Holliday, B. Turng, J. Sinha, and C. Yu, *Abstr. 9th Eur. Congr. Clin. Microbiol. Infect. Dis.*, abstr. P37, p. 94, 1999). The Phoenix Pos ID/AST panels were compared to the NCCLS standard broth microdilution method, and the essential agreement for oxacillin testing was 96.5% for *S. aureus* and 90.5% for *S. epidermidis*. The Phoenix panels detected 98.5% of MRSA and 100% of methicillin-resistant CoNS.

The manual BBL Crystal MRSA ID test (BD Diagnostic Systems) detects MRSA based on growth in the presence of 4 µg of oxacillin per ml. Growth is detected by using a fluorescence indicator that detects oxygen consumption due to bacterial metabolism. Zambardi and colleagues evaluated the BBL Crystal MRSA ID test and reported a 92.9% sensitivity and a 99.4% specificity for the detection of MRSA compared with *mecA* PCR (49).

Latex Agglutination Tests To Detect PBP2a

Two commercial latex agglutination methods have been developed for the detection of MRSA, and both tests use a latex-bound monoclonal antibody directed toward the

PBP2a antigen. Neither test is available in the United States. The Oxoid Penicillin-Binding Protein Latex Agglutination Test (Oxoid, Nepean, Ontario, Canada) was evaluated by Verma et al. against 15 isolates of *S. aureus* and 61 CoNS (P. Verma, G. Dewalt, and M. K. Hayden, *Abstr. 100th Gen. Meet. Am. Soc. Microbiol.*, abstr. C-326, p. 207, 2000). The latex test showed 100% agreement for *S. aureus* compared to conventional methods, but *mecA* PCR was not used as the reference standard. For CoNS, the sensitivity was 77% and the specificity was 100%.

The MRSA Screen Test (Denka Seiken Co., Ltd., Tokyo, Japan) is a 20-min slide latex agglutination test directed towards PBP2a extracted from test colonies of *S. aureus*. Cavassini et al. evaluated the MRSA Screen Test in comparison with *mecA* PCR and reported a sensitivity of 100% for 80 isolates of MRSA and a specificity of 99.2% for 120 isolates of methicillin-susceptible *S. aureus* (MSSA) (5). van Griethuysen and colleagues evaluated the MRSA Screen Test against 267 *mecA*-positive and 296 *mecA*-negative isolates of *S. aureus* and reported a 98.5% sensitivity and 100% specificity (45).

Detection of the *mecA* Gene by DNA Probe Technology

The Velogene Rapid MRSA Identification Assay (Alexon-Trend/Remel) is a DNA probe test based on cycling probe technology that detects the *mecA* gene in MRSA. Cycling steps are performed in heat blocks and therefore thermal cyclers are not necessary. The test that is available in the United States uses an enzyme immunoassay format in a microwell plate and is read either visually or by the use of a spectrophotometer. Arbique et al. compared the Velogene Rapid MRSA Identification Assay with *mecA* PCR, using 179 MRSA and 68 MSSA isolates, and reported a 99% sensitivity and 99% specificity (J. C. Arbique, K. R. Forward, D. J. Haldane, T. F. Hatchette, and R. J. Davidson, *Abstr. 100th Gen. Meet. Am. Soc. Microbiol.*, abstr. C-323, p. 206, 2000).

A second Velogene assay that uses a lateral flow strip instead of a microwell plate for the detection step is in development and is called the Velogene Strip MRSA Assay. Hall and colleagues evaluated the Velogene Strip MRSA Assay in comparison with *mecA* PCR using 49 MRSA and 25 MSSA isolates and reported a 98.6% sensitivity and 97.5% specificity (G. S. Hall, M. Tuohy, D. Wilson, R. Lankford, and G. W. Procop, *Abstr. 100th Gen. Meet. Am. Soc. Microbiol.*, abstr. C-324, p. 207, 2000).

REFERENCES

1. **Applebaum, P. C.** 1996. Epidemiology and *in vitro* susceptibility of drug-resistant *Streptococcus pneumoniae*. *Pediatr. Infect. Dis. J.* **15**:932–939.
2. **Baker, C. N., S. A. Stocker, D. M. Culver, and C. Thornsberry.** 1991. Comparison of the E-test to agar dilution, broth microdilution, and agar diffusion susceptibility testing techniques by using a special challenge set of bacteria. *J. Clin. Microbiol.* **29**:533–538.
3. **Berke, I., and P. M. Tierno, Jr.** 1996. Comparison of efficacy and cost effectiveness of BIOMIC VIDEO and Vitek antimicrobial susceptibility test systems for use in the clinical microbiology laboratory. *J. Clin. Microbiol.* **34**:1980–1984.
4. **Butler, J. C., J. Hoffman, M. S. Cetron, J. A. Elliot, R. R. Facklam, R. F. Breiman, and the Pneumococcal Sentinel Surveillance Working Group.** 1996. The continued emergence of drug-resistant *Streptococcus pneumoniae* in the United States: an update from the Centers for Disease Control and Prevention pneumococcal sentinel surveillance system. *J. Infect. Dis.* **174**:986–993.
5. **Cavassini, M., A. Wenger, K. Jaton, D. S. Blanc, and J. Bille.** 1999. Evaluation of MRSA-Screen, a simple anti-PBP 2a slide latex agglutination kit, for rapid detection of methicillin resistance in *Staphylococcus aureus*. *J. Clin. Microbiol.* **37**:1591–1594.
6. **Daly, J. S., B. A. DeLuca, S. R. Hebert, R. A. Dodge, and D. T. Soja.** 1994. Imipenem stability in a predried susceptibility panel. *J. Clin. Microbiol.* **32**:2584–2587.
7. **de Lancastre, H., A. M. Sa Figueiredo, C. Urban, J. Rahal, and A. Tomasz.** 1991. Multiple mechanisms of methicillin resistance and improved methods for detection in clinical isolates of *Staphylococcus aureus*. *Antimicrob. Agents Chemother.* **35**:632–639.
8. **Doern, G. V., A. B. Brueggemann, and G. Pierce.** 1997. Assessment of the oxacillin disk screening test for determining penicillin resistance in *Streptococcus pneumoniae*. *Eur. J. Clin. Microbiol. Infect. Dis.* **16**:311–313.
9. **Doern, G. V., M. A. Pfaller, K. Kugler, J. Freeman, and R. N. Jones.** 1998. Prevalence of antimicrobial resistance among respiratory tract isolates of *Streptococcus pneumoniae* in North America: 1997 results from the SENTRY antimicrobial surveillance program. *Clin. Infect. Dis.* **27**:764–770.
10. **Doern, G. V., J. L. Staneck, C. Needham, and T. Tubert.** 1987. Sensititre AutoReader for same-day breakpoint broth microdilution susceptibility testing of members of the family *Enterobacteriaceae*. *J. Clin. Microbiol.* **25**:1481–1485.
11. **Doern, G. V., R. Vantour, M. Gaudet, and B. Levy.** 1994. Clinical impact of rapid in vitro susceptibility testing and bacterial identification. *J. Clin. Microbiol.* **32**:1757–1762.
12. **Endtz, H. P., N. Van Den Braak, A. Van Belkum, W. H. Goessens, D. Kreft, A. B. Stroebel, and H. A. Verbrugh.** 1998. Comparison of eight methods to detect vancomycin resistance in enterococci. *J. Clin. Microbiol.* **36**:592–594.
13. **Evangelista, A. T.** 1989. The clinical impact of automated susceptibility reporting using a computer interface, p. 131–142. *In* B. Kleger, D. Jungkind, E. Hinks, and L. A. Miller (ed.), *Rapid Methods in Clinical Microbiology: Present Status and Future Trends*. Plenum Press, New York, N.Y.
14. **Ferraro, M. J., and J. H. Jorgensen.** 1999. Susceptibility testing instrumentation and computerized expert systems for data analysis and interpretation, p. 1593–1600. *In* P. R. Murray, E. J. Baron, M. A. Pfaller, F. C. Tenover, and R. H. Yolken (ed.), *Manual of Clinical Microbiology*, 7th ed. American Society for Microbiology, Washington, D.C.
15. **Food and Drug Administration.** 1991. Review criteria for assessment of antimicrobial susceptibility devices. Revised draft document. Based on CFR 21 part 807 subpart E for 510(k) and CFR 21 part 814 for PMA. U.S. Food and Drug Administration, Rockville, Md.
16. **Garcia-Garrotte, F., E. Cercenado, and E. Bouza.** 2000. Evaluation of a new system, Vitek 2, for identification and antimicrobial susceptibility testing of enterococci. *J. Clin. Microbiol.* **38**:2108–2111.
17. **Gay, K., W. Baughman, Y. Miller, D. Jackson, C. G. Whitney, A. Schuchat, M. M. Farley, and F. Tenover.** 2000. The emergence of *Streptococcus pneu-*

moniae resistant to macrolide antimicrobial agents: a 6-year population-based assessment. *J. Infect. Dis.* **182:**1417–1427.

18. **Guthrie, L. L., S. Banks, W. Setiawan, and K. B. Waites.** 1999. Comparison of MicroScan MICroSTREP, PASCO, and Sensititre MIC panels for determining antimicrobial susceptibilities of *Streptococcus pneumoniae. Diagn. Microbiol. Infect. Dis.* **33:**267–273.

19. **Hartman, B. J., and A. Tomasz.** 1984. Low-affinity penicillin-binding protein associated with β-lactam resistance in *Staphylococcus aureus. J. Bacteriol.* **158:**513–516.

20. **Hartman, B. J., and A. Tomasz.** 1986. Expression of methicillin resistance in heterogeneous strains of *Staphylococcus aureus. Antimicrob. Agents Chemother.* **29:**85–92.

21. **Huang, M. B., C. N. Baker, S. Banerjee, and F. C. Tenover.** 1992. Accuracy of the Etest for determining antimicrobial susceptibilities of staphylococci, enterococci, *Campylobacter jejuni,* and gram-negative bacteria resistant to antimicrobial agents. *J. Clin. Microbiol.* **30:**3243–3248.

22. **Jones, R. N., D. M. Johnson, M. E. Erwin, M. L. Beach, D. J. Biedenbach, M. A. Pfaller, and The Quality Control Study Group.** 1999. Comparative antimicrobial activity of gatifloxacin tested against *Streptococcus* spp. including quality control guidelines and Etest method validation. *Diagn. Microbiol. Infect. Dis.* **34:**91–98.

23. **Jorgensen, J. H., and M. J. Ferraro.** 1998. Antimicrobial susceptibility testing: general principles and contemporary practices. *Clin. Infect. Dis.* **26:**973–980.

24. **Jorgensen, J. H., and J. M. Matsen.** 1987. Physician acceptance and application of rapid microbiology instrument test results, p. 209–212. *In* J. H. Jorgensen (ed.), *Automation in Clinical Microbiology.* CRC Press, Inc., Boca Raton, Fla.

25. **Jorgensen, J. H., M. L. McElmeel, and S. A. Crawford.** 1998. Evaluation of the Dade MicroScan MICro STREP antimicrobial susceptibility testing panel with selected *Streptococcus pneumoniae* challenge strains and recent clinical isolates. *J. Clin. Microbiol.* **36:**788–791.

26. **Jorgensen, J. H., J. D. Turnidge, and J. A. Washington.** 1999. Antibacterial susceptibility tests: dilution and disk diffusion methods, p. 1526–1543. *In* P. R. Murray, E. J. Baron, M. A. Pfaller, F. C. Tenover, and R. H. Yolken (ed.), *Manual of Clinical Microbiology.* American Society for Microbiology, Washington, D.C.

27. **Korgenski, E. K., and J. A. Daly.** 1998. Evaluation of the BIOMIC Video Reader System for determining interpretive categories of isolates on the basis of disk diffusion susceptibility results. *J. Clin. Microbiol.* **36:**302–304.

28. **Medeiros, A. A., and J. Crellin.** 2000. Evaluation of the SIRSCAN automated zone reader in a clinical microbiology laboratory. *J. Clin. Microbiol.* **38:**1688–1693.

29. **Mohammed, M. J., and F. C. Tenover.** 2000. Evaluation of the PASCO Strep Plus broth microdilution antimicrobial susceptibility panels for testing *Streptococcus pneumoniae* and other streptococcal species. *J. Clin. Microbiol.* **38:**1713–1716.

30. **National Committee for Clinical Laboratory Standards.** 1999. *Performance Standards for Antimicrobial Susceptibility Testing.* Ninth informational supplement, M100-S9. National Committee for Clinical Laboratory Standards, Wayne, Pa.

31. **National Committee for Clinical Laboratory Standards.** 2000. *Methods for Dilution Antimicrobial Susceptibility Tests for Bacteria That Grow Aerobically,* 5th ed. Approved standard M7-A5. National Committee for Clinical Laboratory Standards, Wayne, Pa.

32. **National Committee for Clinical Laboratory Standards.** 2000. *Performance Standards for Antimicrobial Disk Susceptibility Tests,* 7th ed. Approved standard M2-A7. National Committee for Clinical Laboratory Standards, Wayne, Pa.

33. **National Committee for Clinical Laboratory Standards.** 2001. *Performance Standards for Antimicrobial Susceptibility testing.* Eleventh informational supplement, M100-S11. National Committee for Clinical Laboratory Standards, Wayne, Pa.

34. **Pestotnik, S. L., R. S. Evans, J. P. Burke, P. M. Gardner, and D. C. Classen.** 1990. Therapeutic antibiotic monitoring: surveillance using computerized expert system. *Am. J. Med.* **88:**43–48.

35. **Rittenhouse, S. F., L. A. Miller, L. J. Utrup, and J. A. Poupard.** 1996. Evaluation of 500 Gram negative isolates to determine the number of major susceptibility interpretation discrepancies between the Vitek and MicroScan WalkAway for 9 antimicrobial agents. *Diagn. Microbiol. Infect. Dis.* **26:**1–6.

36. **Sahm, D. F., J. A. Karlowsky, L. J. Kelly, I. A. Critchley, M. E. Jones, C. Thornsberry, Y. Mauriz, and J. Kahn.** 2001. Need for annual surveillance of antimicrobial resistance in *Streptococcus pneumoniae* in the United States: 2-year longitudinal analysis. *Antimicrob. Agents Chemother.* **45:**1037–1042.

37. **Sanders, C. C., M. Peyret, E. S. Moland, C. Shubert, K. S. Thomson, J.-M. Boeuforas, and W. E. Sanders, Jr.** 2000. Ability of the Vitek 2 advanced expert system to identify β-lactam phenotypes in isolates of *Enterobacteriaceae* and *Pseudomonas aeruginosa. J. Clin. Microbiol.* **38:**570–574.

38. **Staneck, J. L., S. D. Allen, E. E. Harris, and R. C. Tilton.** 1985. Automated reading of MIC microdilution trays containing fluorogenic enzyme substrates with the Sensititre AutoReader. *J. Clin. Microbiol.* **22:**187–191.

39. **Struelens, M. J., C. Nonhoff, P. Van Der Auwera, R. Mertens, E. Serruys, and Groupement Pour Le Dépistage, L'Etude et la Prévention des Infections Hospitaliéres-Groep ter Opsporing, Studie en Preventie van Infecties in de Ziekenhuizen.** 1995. Evaluation of Rapid ATB Staph for 5-hour antimicrobial susceptibility testing of *Staphylococcus aureus. J. Clin. Microbiol.* **33:**2395–2399.

40. **Tenover, F. C., C. N. Baker, and J. M. Swenson.** 1996. Evaluation of commercial methods for determining antimicrobial susceptibility of *Streptococcus pneumoniae. J. Clin. Microbiol.* **34:**10–14.

41. **Tenover, F. C., J. M. Swenson, C. M. O'Hara, and S. A. Stocker.** 1995. Ability of commercial and reference antimicrobial susceptibility testing methods to detect vancomycin resistance in enterococci. *J. Clin. Microbiol.* **33:**1524–1527.

42. **Thornsberry, C., P. Ogilvie, J. Kahn, Y. Mauriz, and the Laboratory Investigator Group.** 1997. Surveillance of antimicrobial resistance in *Streptococcus pneumoniae, Haemophilus influenzae,* and *Moraxella catarrhalis* in the United States in 1996–1997 respiratory season. *Diagn. Microbiol. Infect. Dis.* **29:**249–257.

43. **Trenholme, G. M., R. L. Kaplan, P. H. Karakusis, T. Stine, J. Fuhrer, W. Landau, and S. Levin.** 1989. Clinical impact of rapid identification and susceptibility testing of bacterial blood culture isolates. *J. Clin. Microbiol.* **27:**1342–1345.

44. **Utsui, Y., and T. Yokota.** 1985. Role of an altered penicillin-binding protein in methicillin- and cephem-resistant *Staphylococcus aureus*. *Antimicrob. Agents Chemother.* **38:**345–347.

45. **van Griethuysen, A., M. Pouw, N. van Leeuwen, M. Heck, P. Willemse, A. Butting, and J. Kluytmans.** 1999. Rapid slide latex agglutination test for detection of methicillin resistance in *Staphylococcus aureus*. *J. Clin. Microbiol.* **37:**2789–2792.

46. **Waldvogel, F. A.** 2000. *Staphylococcus aureus* (including toxic shock syndrome), p. 2069–2092. *In* G. L. Mandell, J. E. Bennett, and R. Dolin (ed.), *Mandell, Douglas and Bennett's Principles and Practice of Infectious Diseases*, 5th ed. Churchill Livingstone, New York, N.Y.

47. **Woods, G. L., D. LaTemple, and C. Cruz.** 1994. Evaluation of MicroScan rapid gram-positive panels for detection of oxacillin-resistant staphylococci. *J. Clin. Microbiol.* **32:**1058–1059.

48. **Yamazumi, T., S. A. Marshall, W. W. Wilke, D. J. Diekema, M. A. Pfaller, and R. N. Jones.** 2001. Comparison of the Vitek Gram-Positive Susceptibility 106 Card and the MRSA-Screen latex agglutination test for determining oxacillin resistance in clinical bloodstream isolates of *Staphylococcus aureus*. *J. Clin. Microbiol.* **39:**53–56.

49. **Zambardi, G., J. Fleurette, G. C. Schito, R. Auckenthaler, E. Bergogne-Berezin, R. Hone, A. King, W. Lenz, C. Lohner, A. Makristhatis, F. Marco, C. Muller-Serieys, C. Nonhoff, I. Phillips, P. Rohner, M. Rotter, K. P. Schaal, M. Struelens, and A. Viebahn.** 1996. European multicentre evaluation of a commercial system for identification of methicillin-resistant *Staphylococcus aureus*. *Eur. J. Clin. Microbiol. Infect. Dis.* **15:**747–749.

Clinical Microbiology: Looking Ahead

DONNA M. WOLK AND DAVID H. PERSING

18

What will the future hold in terms of the laboratory diagnosis of infectious disease? During the coming decades, the expectations and responsibilities of clinical and public health microbiology laboratories will continue to increase. Clinical laboratories will be challenged to respond to the need for detection of infectious disease agents introduced by bioterrorism or medical advances, such as xenotransplantation. Furthermore, if microbial-diversity studies are predictive, laboratories will be called upon to respond to the diagnostic challenges presented by many organisms whose disease associations are presently unknown. As clinical microbiologists, we will find ourselves not only detecting and isolating pathogens but also screening for mutations indicative of drug resistance and gene expression related to pathogenicity. As the ability to identify molecular evidence of host immunologic response or pathogen-induced disruption of host cell signaling patterns becomes a reality, diagnostic assays to identify and monitor those changes will become possible. Inevitably, completion of the Human Genome Project will facilitate the study of human genetic variation as it relates to susceptibility and resistance to infectious diseases. The possibilities and expectations for the future of laboratory diagnosis of disease are immeasurable.

Current technology does not allow for all of these expectations to be met in most health care workplaces, which are currently limited by hospital mergers, downsizing, and budget reductions. However, as scientists and biotechnology manufacturers respond to the needs of the health care workplace, new high-throughput, cost-effective technologies are beginning to surface. Many of these technologies incorporate molecular methods that hold promise to help bridge the gap between what is theoretically possible and those technologies which are practical for patient care.

At present, a few molecular tests are commercially available for clinical microbiology laboratories. These kit-based tests typically make use of nonisotopic detection methods that rely on colorimetric, fluorescent, or chemiluminescent detection of nucleic acids. Nucleic acid testing has already made possible the quantification and therapeutic monitoring of viral agents and the detection of fastidious organisms, such as *Neisseria gonorrhoeae*, *Chlamydia trachomatis*, and hepatitis C virus (26, 27, 110, 123). Methods like these have become commonplace in most large diagnostic microbiology laboratories, and their use is reviewed in previous chapters.

One of the most powerful methods for nucleic acid testing, PCR, promises a substantial improvement over traditional culture-based disease diagnosis for some infectious agents. PCR is useful when rapid identification is indicated for slow-growing pathogens or when disease is caused by uncultivable or unknown agents (4, 26, 92). Unfortunately, PCR procedures are mostly "home brew" designs and can require tedious nucleic acid extraction procedures and laborious detection methods, such as gel electrophoresis with hybridization blots. PCR is still a diagnostic tool that is performed mostly in large reference, academic, or research-oriented clinical laboratories. Laboratories like these can maintain the infrastructure and resources to support self-validation of their home brew assays as required or recommended by the College of American Pathologists and the NCCLS. Unfortunately, many clinical laboratories do not have the resources or the engineered space to perform PCR in a timely, reproducible, contamination-free manner. Moreover, the proprietary interests of most manufacturers have made the cost of PCR technology somewhat prohibitive.

Regardless of the challenges faced by clinical laboratories, PCR, and adaptations of it, have already facilitated advances in the science of infectious-disease diagnostics. Since its introduction, improvements have been made to the enzymes, reaction components, and instrumentation, allowing traditional PCR to evolve and better integrate with the routine diagnostics laboratory. Developments such as PCR–enzyme-linked immunosorbent assay for colorimetric amplicon detection, multiplex PCR for amplification and detection of multiple targets, and in situ PCR for the amplification of DNA targets in fixed tissue specimens are proving to be beneficial (110). Changes to laboratory practice, workflow design, and standards of patient care will become apparent as scores of new technologies are marketed to the clinical diagnostic industry. Many of these technologies are already available and are beginning to affect the way we think, plan, and educate our laboratory workforce.

PCR and other molecular testing methods have already enabled new paradigms and strategies for diagnosing human infections. Advances in PCR and molecular sequencing technology have facilitated genetic analysis of a variety of novel organisms, such as the etiologic agents of Whipple's disease (94), cat scratch disease (5, 6), bacillary angiomatosis (50, 93), and various others (7, 19, 46, 78, 120). The use of consensus primers, such as those designed to amplify the 16S ribosomal gene, has enabled us to identify the disease etiologies of uncultivable organisms and previously unidentified agents of disease. Additionally, use of PCR has allowed the identification of genetic polymorphisms that encode drug resistance phenotypes, such as those found in the thymidine kinase gene of herpes simplex virus (58), the *rpoB* gene of *Mycobacterium tuberculosis* (18, 37, 70, 89, 112), the *mecA* gene of *Staphylococcus aureus*, and the *vanA* and *vanB* genes of the enterococci (15, 86, 119). In addition, PCR has made possible the identification of atypical presentations of an illness and the development of testing algorithms that will provide rapid confirmation of disease, ultimately improving infectious-disease diagnosis and reducing the total resource commitments to diagnostic and therapeutic interventions (14, 90).

Bacterial genomic sequencing efforts are also likely to have an impact on clinical practice. To date, complete genome sequences for at least 20 species have been published in their entirety. Most can be found at http://www.tigr.org, the web site for The Institute for Genomic Research, a nonprofit research institute in Rockville, Md. These sequences, and those to follow, will prove to be a useful foundation for the design of novel diagnostic and therapeutic approaches. The combination of molecular-sequence data with traditional biochemical, morphological, and phenotypic characteristics of organisms has created the "polyphasic" approach to the classification of newly discovered and existing pathogens (82, 116, 117). In certain circumstances, sequence data may also be used as the source of stand-alone molecular detection protocols. The number of molecularly characterized organisms continues to increase and adds to our understanding of the phylogenetic diversity and pathogenic potential of human pathogens (71, 111).

This chapter is intended to provide an overview of new and emerging molecular diagnostic technologies. While not all techniques can be described here, examples of relevant technologies for specimen preparation, PCR, post-PCR detection, and analysis are presented. Several non-PCR methods will also be discussed.

SPECIMEN PREPARATION

Since even the most advanced and accurate technologies are only as good as the sample that is used for testing, the development of DNA and RNA extraction techniques may be the most important predictor of the overall effectiveness of molecular technologies. Regardless of the downstream detection method, the isolation and cleanup of DNA can be formidable tasks. Many of today's newest technologies rely on DNA hybridization or detection of PCR-amplified DNA or RNA. Both are often difficult because sample substrates, such as blood, tissue, stool, and the like, may contain inhibitors or low-abundance targets. Since extraction of DNA has been a persistent problem for molecular diagnosis, novel fixatives and nucleic acid purification and enrichment schemes need to be developed in order to integrate these specimen sources into the routine molecular diagnostics laboratory.

Novel nucleic acid capture technologies, such as magnetic-bead capture of nucleic acids or membrane-based nucleic acid capture, will enable more timely and cost-effective sample preparation and remove most common inhibitors. Successful systems will need to integrate automated high-throughput extraction platforms, designed to ensure contamination-free extraction of DNA and RNA, with flexible, user-friendly software that can be integrated with laboratory information management systems.

Existing systems, like the QIAamp 96 DNA blood kit, used in conjunction with the BioRobot 9604 (Qiagen, Valencia, Calif.), offer the promise of automated DNA isolation from blood, plasma, serum, bone marrow, body fluids, lymphocytes, and cultured cells. The method uses silica gel membrane technology for isolation of nucleic acids without the use of phenol, chloroform, or ethanol precipitation. The buffer system allows for selective binding of nucleic acids to the membrane, washing, and subsequent elution of the nucleic acids. The BioRobot (Fig. 1) is reported to process up to 96 samples in microwell plate format within 2 h and uses an automated tip change system to avoid cross contamination and a bar code reader for sample identification (Fig. 2). A capacitance-based liquid level sensing system is used to confirm accurate pipetting volume and monitors for the adequacy of reagent levels. The same automation can be combined with the RNeasy 96 BioRobot kit to purify total human and animal RNA for use in reverse transcription-PCR and other procedures. The system provides integration with flexible, programmable, user-friendly software.

Another high-throughput extraction system combines the Biomek 2000 Laboratory Automation Workstation from Beckman Coulter (Fullerton, Calif.) with the Dynal (Oslo, Norway) DNA DIRECT Auto 96 method and is based on magnetic-bead capture of DNA from whole blood. The Beckman Coulter system is also computer integrated and allows the simultaneous processing of up to 96 samples in 10 min. The Dynabeads magnetic-bead technology eliminates the centrifugation steps required by some other systems. In addition, the Dynal-MPC–Auto 96 Magnet can be used with other liquid-handling robots and can be directed from a variety of different software platforms.

Several other manufacturers are marketing moderate-throughput walk-away DNA extraction systems. Organon Teknika (Durham, N.C.) uses solid-phase technology in the form of silicon dioxide particles to bind nucleic acids. The automated Nuclisens Extractor is a closed system designed for contamination-free nucleic acid extraction from plasma and can be coupled with the nucleic acid sequence-based amplification system described in previous chapters. In addition, Autogen (Framingham, Mass.) produces various models of automated DNA and RNA extraction systems, which are based on traditional nucleic acid extraction chemistry.

Many challenges to nucleic acid extraction remain. Automated extraction instruments will need to meet the quality control requirements of other existing laboratory robotics before routine implementation of this technology can be achieved. The ability to interface with high-throughput PCR systems will be an important factor for any new automated extraction system. Both Roche Molecular Biochemicals (Indianapolis, Ind.) and Applied Biosystems, Inc. (ABI; Foster City, Calif.) have announced plans to launch automated nucleic acid extraction systems that interface with their corresponding systems for PCR and PCR product detection. The ABI PRISM 6700 instrument (Fig. 3) promises integration with the ABI PRISM 5700, and the Roche MagNApure (Fig. 4) will integrate with their LightCycler PCR instrument. Systems such as these can be programmed to automate both extraction and PCR setup, providing a benchmark for high-throughput extraction technology.

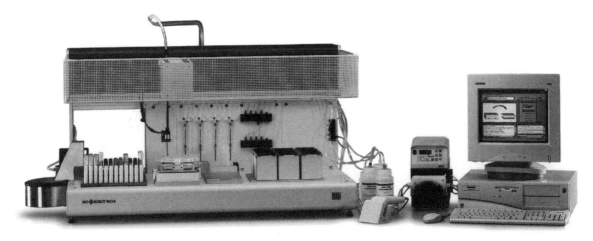

FIGURE 1   Qiagen Biorobot 9604 automated DNA extraction system. (Courtesy of Qiagen.)

PCR AND PCR PRODUCT ANALYSIS

Detection Formats: Fluorescent Monitoring of Real-Time PCR

Advances in fluorescent chemistry have facilitated fluorescent detection of nucleic acids or hybridized nucleic acid probes and have enabled technology to advance from traditional end product analysis to real-time monitoring of PCR amplicons within a closed system (35, 36, 39, 56, 59, 64). This strategy may very well be the most exciting advance for clinical microbiology since the advent of PCR itself. Fluorescent product detection has already begun to further the use of PCR in the clinical laboratory (23).

There are several approaches to real-time PCR that are likely to exert an influence on the future of diagnostic microbiology. First is the real-time monitoring of the incorporation of fluorescent dye (35, 36), such as SYBR-Green I (Molecular Probes, Eugene, Oreg.). SYBR-Green I dye binds to double-stranded DNA, generated during PCR, and emits fluorescence, which can be measured by various detection systems. The use of dye incorporation produces an intense signal that can be measured in real time or by reaction end point analysis; however, since the dyes bind to all products produced in the PCR, dye incorporation will not ensure the measurement of a specific PCR product. To ensure specific hybridization and product identification, protocols that are designed to detect a specific probe, bound to a known target, will merit more attention in a diagnostic laboratory.

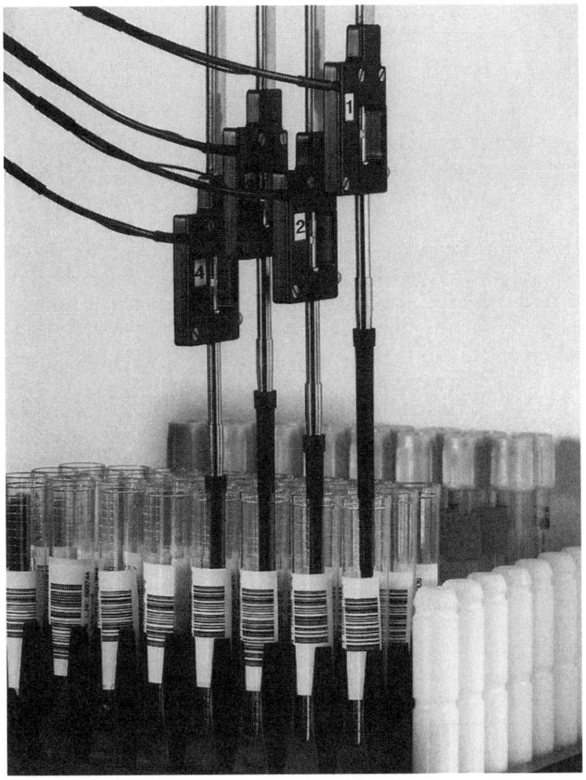

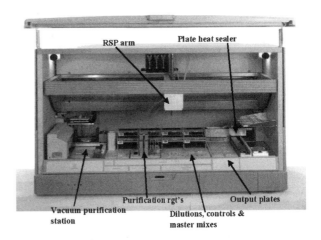

FIGURE 3   The ABI PRISM 6700 instrument combines extraction technology with preparation of PCR master mix. Specimens and reagents (rgt's) are mixed in plates, which are sealed and PCR ready. RSP, robotic sample preparation. (Courtesy of ABI.)

FIGURE 2   Qiagen Biorobot 9604 Tip-Change System and bar-coded samples. (Courtesy of Qiagen.)

FIGURE 4 The Roche MagNA Pure LC uses magnetic-bead technology for nucleic acid extraction and is programmable to load extracted samples into glass LightCycler capillary tubes in their rotor housing. (Courtesy of Roche Molecular Biochemicals.)

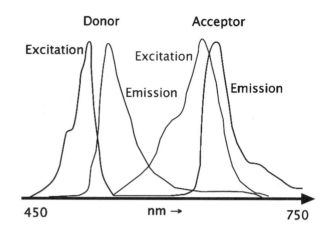

FIGURE 5 Fluorescent excitation and emission spectra of a sample donor and acceptor dye pair. (Courtesy of Roche Molecular Biochemicals.)

FRET

An alternative to fluorescent-dye incorporation, fluorescence resonance energy transfer (FRET) technology, allows the detection and quantitation of specific PCR products through the use of nucleic acid probes. FRET enhances real-time PCR monitoring, as it can be used with fluorescent dyes that have various emission spectra. Different wavelengths of fluorescent light can be simultaneously monitored, and thus, the presence of one or more sequence-specific targets can be detected (121, 122).

The FRET technology is based on the coupling of fluorophore and quencher pairs that have overlapping absorption and emission spectra. As PCR amplifies the DNA strands, multicolor fluorophores coupled to DNA probes are brought into or removed from juxtapositions so that stored energy can be transferred from one to another. Detection of PCR products is done by scanning for signals produced by fluorescent donor and acceptor dyes when energy transfers occur in fluorescent molecules concurrent with PCR amplification or hybridization. An example of a donor-acceptor dye combination is depicted in Fig. 5.

Two modes commonly used for FRET fluorescence scanning are emission scanning and synchronous scanning. Emission scanning is performed with a laser source of fixed excitation wavelength. This type of scanning allows the detection of an extremely low concentration of analytes. Synchronous scanning occurs when both excitation and emission wavelengths are scanned simultaneously, allowing the detection of multiple reporter dyes with widely spaced excitation wavelengths. Using this method, multiplex PCR products can be detected, creating the potential for increased sample throughput and cost savings, as multiple targets are amplified within the same reaction; however, the

relative sensitivity of the method suggests that it is best used when detection of low analyte concentrations is not required.

Molecular Beacons

Molecular beacons are oligonucleotide probes produced in the shape of a hairpin rather than the traditional linear probe conformation. The probes contain various fluorophores and a quencher molecule, DABCYL (4-[4'-dimethylaminophenylazol] benzoic acid). In closed formation, they act as switches, bringing the fluorophore and quencher molecules together and turning the fluorescence off. When the probes bind to their targets in the PCR product, the structure changes. The fluorophore is separated from the quencher molecule, and fluorescence is turned on (Fig. 6). Fluorescence is measured with a spectrophotometer or luminometer in the PCR annealing step when the beacon is bound to its complementary target (113, 114). Molecular beacons have already been used for the detection of drug-resistant tuberculosis (87) and are likely to play an important role in the detection of other important infectious diseases.

Like FRET technology, molecular beacons can be used to monitor endpoint, real-time, and quantitative PCRs. A variety of fluorophores can be used to distinguish among multiple targets in the same reaction, even if the targets differ by only one nucleotide (113). Probes designed with beacons differ from those designed as linear hydrolysis probes, making them suitable for allele discrimination, since the hairpin structures and their hybrids are less stable than linear-probe hybrids. The beacon's stem-loop structure imparts an increased ability to detect single-base-pair mismatches and different alleles. The instability of the hybrids allows different populations of PCR products to be distinguished by their differences in melting parameters and might allow for enhanced specificity of PCR designed with molecular beacon probes under conditions of high stringency.

Several other techniques may be enhanced by the use of molecular beacons. Since the quenching of molecular beacons occurs through direct transfer of energy from a reporter fluorophore dye to a quencher molecule, a common quencher can be used, thereby allowing an increased number of fluorophores to be combined as reporter molecules

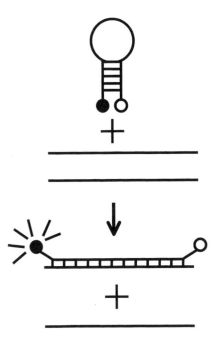

FIGURE 6 Schematic of a molecular-beacon probe binding to denatured DNA template. As the beacon probe binds, the quencher molecule is removed from proximity to the fluorophore, allowing the dye to fluoresce.

(113). Multiple reporter dyes may prove advantageous when designing multiplex PCR. In addition, primer and probe design are simplified, allowing thermal-cycling parameters to be more flexible than those required for linear hydrolysis probes.

Real-Time PCR Instruments

Several automated systems are able to use FRET technology to perform real-time PCR, allowing concurrent detection and identification of products within one sealed reaction tube. These closed PCR systems are more easily adapted to the routine clinical laboratory workplace, since the risk of environmental contamination with PCR amplicons is lessened; however, typical unidirectional workflow and amplicon inactivation protocols are still a necessity. Even with closed-system real-time PCR procedures, careful attention to amplicon control is essential. The liberation of amplicons through carelessness or broken reaction tubes can lead to contamination events. Unfortunately, since many published real-time PCR protocols rely on the amplification of short products, the amplicons may not be efficiently inactivated by any of the available enzymatic and/or photochemical techniques (22).

ABI Prism 5700 and 7700 Sequence Detection Systems

The ABI PRISM 5700 and 7700 sequence detection systems are high-throughput 96-well instruments that combine traditional thermal cycling with TaqMan chemistry, a combination of FRET technology and the fluorogenic 5′-nuclease activity of DNA polymerase. The 5′-nuclease assay (39) makes use of nonextendable linear oligonucleotide hybridization probes. The hybridization probes are labeled with a fluorescent reporter dye, such as FAM (6-car-

boxy-fluorescein), at the 5′ end and a nearby quencher molecule, such as TAMRA (6-carboxy-tetramethylrhodamine). During denaturation of the PCR product, the probe binds to the denatured target DNA strands, and the fluorescent emission of the reporter dye is quenched due to its proximity to the quencher molecule. After each annealing cycle, DNA polymerase, with 5′ exonuclease activity, will extend the specific target strand and concurrently hydrolyze and cleave the hybridization probe from the strand. As the probe is cleaved, the quencher molecule is removed from its proximity to the reporter dye (Fig. 7). A laser light source stimulates the two dyes and, if specific amplification of target DNA has occurred, the reporter dye fluorescence will increase and accumulate with each PCR cycle (39, 59).

Emission spectra are measured in real time by the ABI 5700 or 7700 sequence detector, which can simultaneously detect emitted light from all 96 wells (Fig. 8). Fluorescent light is monitored within a closed system through the sealed caps of the microwells on the plate. Ninety-six individual fiber optic lines send the light images to a spectrograph equipped with a charge-coupled device (CCD) (28, 32). Software converts the emission data to amplification plots that can be used to display the real-time data acquired during the PCR process or to represent the end product of the PCR (Fig. 9).

Microwell plates containing the dehydrated reaction components for TaqMan chemistry can be designed, prepared, shipped, and stored as ready-to-use plates. The plates are rehydrated, template is added, and 96 tests can be performed in approximately 1 h and 45 min. TaqMan reactions are centered on PCR primers and probes designed to function optimally at a fixed temperature, and Primer Express software facilitates the design of both primers and probes.

The advantages of ABI PRISM systems include high throughput, the ability to collect data from real-time PCR amplification, and confirmation via probe hybridization in a closed system. By changing the emission spectra, multiple primers and probes can be used at once, making it easier to design a battery of molecular tests. Colorimetric detection can be adjusted to incorporate as many as seven different colors for emission detection and up to three or four probes per reaction well (57). The system is extremely versatile in that it can detect, quantitate, and confirm multiple PCR targets at one time. Use of the default amplification programs requires that all thermal parameters for annealing occur at approximately 59°C, but the default temperatures can be altered for specific applications. Furthermore, PCR products can be removed from the system, directly sequenced, and compared to the ABI Microseq 16S rRNA database or a nonproprietary database to determine the sequence identities of unknown organisms. The passive reference, included in the TaqMan PCR reagent kit, allows the software to normalize the reporter dye to ensure reproducibility. Positive internal controls can be used to ensure the lack of PCR inhibitors in the specimen and further increase confidence in the PCR results. The plastic closed plate system limits the potential for environmental contamination by PCR amplicons. The system is user friendly and has computer programs for monitoring reactions and storing information.

Routine clinical use of this instrument may be forthcoming. Food and Drug Administration (FDA)-approved diagnostic kits are planned for the instrument; however, the ABI technology and similar technologies are subject to royalties, and therefore, the cost-effectiveness of routine testing with this system will depend on the limits of those royalties. TaqMan technology has already been used for rapid

Polymerization

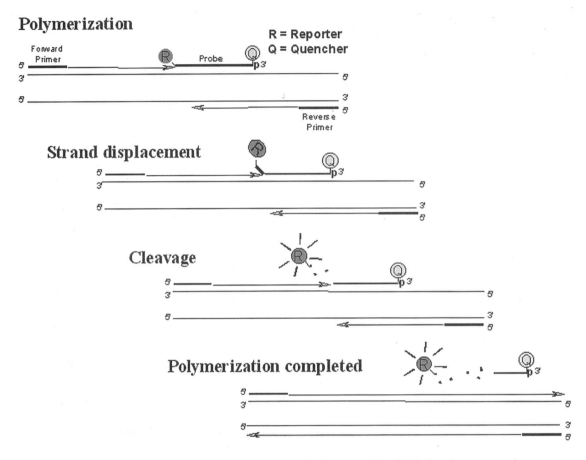

R = Reporter
Q = Quencher

Strand displacement

Cleavage

Polymerization completed

FIGURE 7 Schematic of TaqMan chemistry. Hybridization probes, labeled with a reporter dye and a fluorescent quencher molecule, are bound to the denatured target DNA strand. As PCR occurs and the opposing strand is generated, the 5'-nuclease activity of DNA polymerase will hydrolyze the probe, removing the reporter from its close proximity to the quencher dye and allowing measurable fluorescent signal to accumulate. (Courtesy of ABI.)

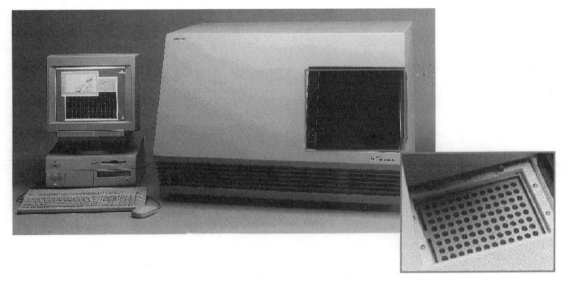

FIGURE 8 ABI PRISM 7700 sequence detection system. (Courtesy of ABI.)

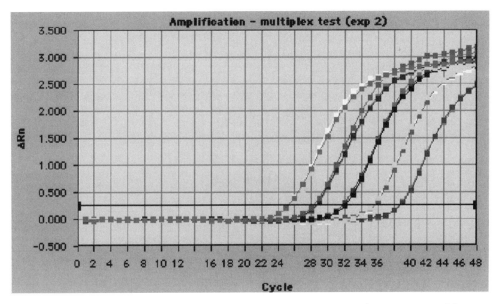

FIGURE 9 Amplification plots for the ABI PRISM 7700 sequence detection system. Negative controls are used to establish fluorescent-signal background thresholds. As exponential amplification of a specific target occurs, the amplification curve for a positive sample will cross those thresholds. The PCR cycle number at which a curve crosses the threshold is related to the original concentration of nucleic acid in the sample and can provide a measure of quantitation to the assay. (Courtesy of ABI.) ΔRn, change in relative fluorescence; exp, experiment.

detection of *Escherichia coli* O157:H7, dengue virus (55, 101), *Ehrlichia* spp. (7, 19), *Yersinia pestis* (34), hepatitis B virus (2), cytomegalovirus (66), and the *mecA* gene of staphylococci (48).

LightCycler

The LightCycler (Idaho Technology, Idaho Falls, Idaho, and Roche Molecular Biochemicals) uses both heated and ambient air to provide rapid cycling parameters for PCR

(Fig. 10). The rapid cycling is combined with light-emitting diode excitation of fluorophores, such as SYBR-Green I, which binds to newly synthesized double-stranded DNA in the PCR. Increases in fluorescence are proportional to the DNA concentration, allowing for simultaneous detection and quantitation of DNA as the PCR takes place in glass reaction tubes (Fig. 11), which are seated in the LightCycler rotor. Synchronous scanning and detection of fluorescent PCR products is achieved by the use of a fluorimeter that monitors characteristic increases in emission spectra (Fig. 12). Emission signals are read once per PCR cycle and can be monitored continuously on a computer screen. Confirmation of amplified sequences can be done by direct

FIGURE 10 LightCycler system for real-time PCR. (Courtesy of Roche Molecular Biochemicals.)

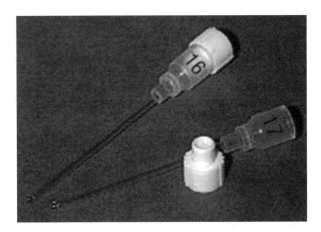

FIGURE 11 Sealed glass capillaries house the PCR mix and the sample. The reaction tubes are placed in a circular rotor within the LightCycler system for rapid thermal cycling and amplicon detection. (Courtesy of Roche Molecular Biochemicals.)

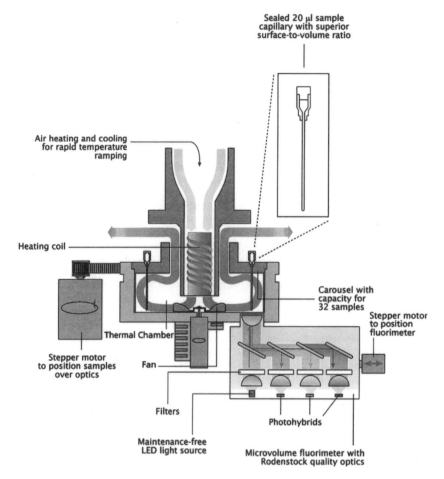

FIGURE 12 The LightCycler fluorimeter monitors fluorescent changes as they occur when DNA is amplified in glass reaction tubes seated in the circular rotor. (Courtesy of Roche Molecular Biochemicals.)

sequencing of the amplified product or through specific hybridization of probes to the target sequence (121, 122).

Specific hybridization of DNA products can occur simultaneously with amplification and can be monitored by the use of various fluorescent dyes, which are bound to hybridization probes and have different emission spectra. The probes are designed to be in close proximity to each other (1 to 5 bp apart). When the probes hybridize to the amplified DNA strand, FRET occurs and causes fluorescence to be emitted (Fig. 13) (121).

Additionally, after PCR is performed, detection of the diminished fluorescent signal, observed when the amplicon is heated to its denaturation temperature, enables a melting curve analysis to be performed on PCR products (Fig. 14). Because mismatches, in the form of point mutations and polymorphisms, combine with other characteristics of the amplicon to affect the melting of the DNA strands, melting temperature curves can be used to distinguish between PCR products with as little as 1 bp difference (88, 95, 121).

The LightCycler system's advantages include a relatively high throughput at 32 tests per run. The LightCycler performs real-time PCR with real-time data acquisition and real-time display of data. Rapid cycling results in reduction of nonspecific PCR products and can produce both reaction and analysis in as little as 30 min. Sealed capillary reaction

tubes reduce the possibility of cross contamination of the template and amplicons (122); however, care must be taken when handling the glass reaction tubes so that no breakage

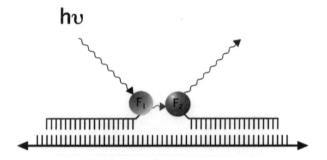

FIGURE 13 FRET. A pair of oligonucleotide probes is designed to hybridize in close proximity to one another on the target DNA strand. As oligonucleotide probes, synthesized to contain a fluorescent dye, are bound to the target DNA strand, energy emitted from fluor 1 (F_1) will excite fluor 2 (F_2) to emit a specific wavelength that is monitored by the LightCycler instrument. Fluorescent signals provide a measure of real-time PCR and specific hybridization of probes. (Courtesy of Roche Molecular Biochemicals.) hυ, light.

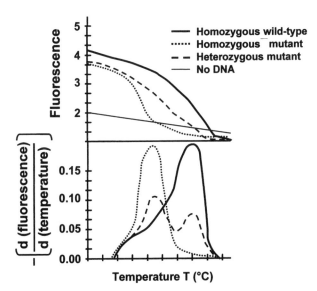

FIGURE 14 Melting curve analysis for FRET-based probes. Fluorescence will diminish as the temperature is increased, since probes will be removed from the target DNA strands. Mismatches in the target sequence (mutations) will allow the probe to be removed more easily than the same probe bound to a target sequence that is an exact match for the probe (wild type). By plotting the ratio of fluorescent signal to temperature, the temperature at which most of the melting takes place can be monitored and wild-type sequences can be distinguished from those of mutants. (Courtesy of Roche Molecular Biochemicals.)

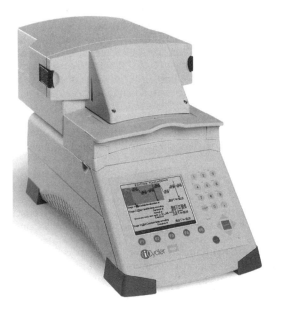

FIGURE 15 iCycler. (Courtesy of Bio-Rad.)

occurs, since liberation of the amplicons can lead to contamination events. As for the TaqMan chemistry system, use of the LightCycler is subject to royalties. Home brew assays have been designed (23, 60, 75), and diagnostic methods for the clinical microbiology laboratory have been published (23, 47, 91).

iCycler

As an alternative to the systems described above, the iCycler (Bio-Rad, Hercules, Calif.) may hold promise as a lower-priced real-time PCR instrument (Fig. 15). Peltier heating and cooling couple with National Institute of Standards and Technology (NIST) traceable temperature performance to make the iCycler adaptable for use in routine clinical laboratories, although this use is not yet licensed. Guidelines are available for the conversion of traditional PCR protocols and should enhance the applicability of this real-time system for certain clinical laboratory applications. The iCycler features interchangeable sample blocks, which allow flexibility between 0.5 and 0.2 ml, and 96- to 384-well PCR plates. Dual sample blocks allow independent protocols to be run simultaneously. As PCR occurs, light is transmitted from all tubes or wells and is simultaneously monitored by a CCD detector. Five excitation and emission filters combine with light intensifier technology to enhance the sensitivity of emission detection.

Smart Cycler

The Cepheid Smart Cycler, distributed by Fisher Scientific (Pittsburgh, Pa.), has a unique heating and cooling protocol

which uses ceramic heater plates coupled with a high-efficiency cooling fan. Real-time monitoring of the fluorescent signal occurs through optical detection windows in the proprietary reaction tubes. High-intensity light-emitting diodes, silicon photodetectors, and multiple filters are reported to allow multiplexed targets with simultaneous detection of up to four different spectral bands. The reaction tubes prevent sample contamination and can be placed in each of 16 independently programmable reaction sites, thereby allowing multiple low-sample-number tests to be performed simultaneously. Alternatively, the sample blocks can be linked together for higher throughput.

Rotor-Gene

The Corbett Research Rotor-Gene (distributed by Phenix Research Products, Hayward, Calif.) allows real-time fluorescent detection of amplicons to take place through traditional 0.2-ml thin-walled PCR tubes. The system may hold promise for conversion of existing protocols to real-time PCR methods.

Systems Using Molecular Beacons

Several systems can incorporate the use of molecular beacons as probes. Some, like those from Stratagene (La Jolla, Calif.) and Intergen (Purchase, N.Y.), are designed specifically with beacon technology in mind. Others, such as the ABI PRISM, LightCycler, and iCycler, are designed for adaptations of FRET technology but can also be used with molecular-beacon probes.

Stratagene

Stratagene offers custom synthesis of molecular beacons coupled to a variety of fluorophores. Beacons are not yet subject to the high proprietary fees associated with some technologies, and their potential use in clinical microbiology is very promising. Beacons can be monitored in a variety of different nonproprietary fluorescence readers; however,

Stratagene's Mx4000 system has been designed to incorporate a 96-well microplate fluorescence reader with a PCR thermal cycler to detect PCR products tagged with molecular beacons. The system can accommodate eight strip tubes or individual tubes with up to four different probes per reaction and can detect wavelengths in the range of 350 to 830 nm. Real-time quantitative PCR data can be viewed in a variety of formats. Control kits with beacon-based probes for human housekeeping genes are available through Stratagene. Coupled with Stratagene's reverse transcription-PCR reagent kit, the system can also be used to quantify gene expression.

Amplifluor

The Amplifluor Universal Amplification and Detection System (Intergen) is based on the use of molecular beacons, but in this system, a traditional PCR primer combines with a fluorescent hairpin oligonucleotide and is called a Uniprimer. The Uniprimers function as both amplification and detection primers and incorporate into the PCR product, enabling direct detection and quantification (Fig. 16). Two specific primers and one labeled universal primer are used to amplify DNA. Fluorescent signal is generated when a Uniprimer unfolds during incorporation into an amplification product.

One advantage of this system is that any traditional PCR can be adapted for use by the addition of a tail sequence to the 5' end of an existing target-specific primer. Furthermore, the system is flexible in that fluorescent endpoints can be determined with any fluorescent-plate reader. A

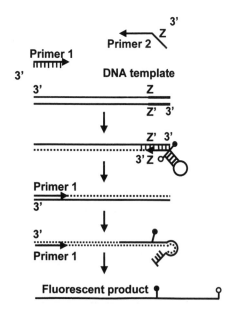

FIGURE 16 The Amplifluor Universal Amplification and Detection System uses traditional PCR primers, one of which synthesizes with a tail (the Z region). The Z region is replicated as the opposing strand is generated. Primers are combined with a fluorescent hairpin oligonucleotide, called a Uniprimer, that binds specifically to the Z regions. Fluorescent signal is generated when a Uniprimer unfolds during incorporation into an amplification product. (Courtesy of Intergen.)

potential disadvantage is that nonspecifically amplified products will also generate a signal. Intergen offers kits for human papillomavirus, and its Universal Amplification System has been successfully used with the ABI PRISM 7770, the LightCycler, and other systems for real-time PCR. The system can also be adapted for genotyping of infectious disease agents, and an in situ PCR kit format is offered for human papillomavirus type 16.

OTHER EMERGING DIAGNOSTIC TECHNOLOGIES
Signal Amplification Technology

In order to avoid the risks involved with PCR performed in open systems, several manufacturers have designed systems that detect DNA by hybridization of probes to targets. In these systems, only the signal generated from the binding is amplified; DNA and RNA are not amplified. Various protocols have been developed using signal amplification platforms (8, 13, 53, 76, 77, 100, 105, 106). These systems provide benefit to laboratories that do not have properly engineered space to allow the control of PCR amplicons. Two new commercial signal amplification systems, discussed here, appear to have the potential to be considered for use in clinical laboratories.

QuantiGene

The Bayer (Tarrytown, N.Y.) QuantiGene assay system is one example of a new signal amplification system. With this system, Probe Designer Software enables the design of oligonucleotide probes for use in signal amplification protocols. QuantiGene can perform quantitation of mRNA from lysed cells and whole tissue. Detection limits are as low as two or three copies/ml, and the system is expected to have many practical applications in diagnostic microbiology, pharmacology, and drug discovery.

Invader Assay

The homogenous Invader assay (Third Wave Technologies, Inc., Madison, Wis.) is a novel process that can be used to detect and analyze DNA and RNA without the thermal cycling of a traditional PCR-based amplification process. Invader technology can be used by small-volume users and as a high-throughput system for genetic analysis of single-nucleotide polymorphisms (SNPs) and other genetic variations associated with specific diseases. The assays can also be used for genotyping and gene expression analysis. The technology is reported to be more robust than DNA sequencing and does not suffer from PCR amplicon carry-over problems because the target is not amplified. In this process, only a signal is amplified, making it a reasonable alternative to PCR for laboratories with limited space and engineering controls for amplicon control. The system produces fluorescent signals that are adaptable to detection by a variety of automated platforms and microplate formats from 92 to 1,536 wells per plate (53, 97).

The system uses thermostable structure-specific 5' nucleases (flap endonucleases) called Cleavase enzymes, members of a family of enzymes that cleave nucleic acid molecules at specific sites based on structure rather than sequence (45, 65). As illustrated in Fig. 17, the Invader DNA reaction uses a two-step cascade to produce a signal. In the first step of the

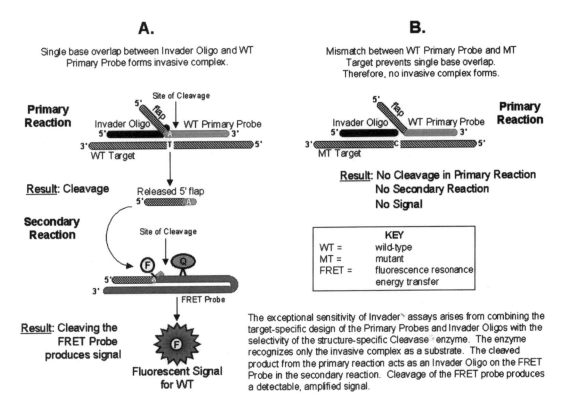

FIGURE 17 Schematic of the Invader assay. Image created by Jodi Hoeser. (Courtesy of Third Wave Technologies, Inc.)

cascade (the primary reaction), multiple copies of a synthetic oligonucleotide are cleaved, but only in the presence of the target of interest. In the second step (the secondary reaction), the cleavage products of the first step anneal to a hairpin-shaped probe, which contains a self-complementary region that is labeled with two dyes: a fluorescent donor dye and an acceptor dye that uses a FRET mechanism to quench the signal from the donor dye (11, 108). When the cleavage product binds to the hairpin-shaped probe and the enzyme cleaves the probe, a detectable fluorescent signal is produced and amplified (Fig. 17A). Significantly, if the sequences of the primary probe and the target are not complementary at the base of interest, just upstream from the cleavage site, no invasive structure will form, cleavage will be suppressed, and no secondary reaction will occur (Fig. 17B). The Cleavase enzyme cleaves the overlapped invasive cleavage structure at least 340-fold more efficiently than the nonoverlapping structure and allows excellent discrimination between nucleotide sequences that differ by only a single base (45, 63, 65). The assay follows the same general format for both DNA and RNA targets, but the enzyme and some reaction components are specific for DNA or RNA. (Contact the manufacturer for recent, minor modifications to the Invader technology.)

The isothermal Invader reaction runs near the melting temperature of the primary probes, which are provided in excess, but below the melting temperature of the Invader oligonucleotide. Consequently, cleaved and uncleaved primary probes cycle rapidly on and off the target, and the Invader oligonucleotide remains hybridized to the target.

The two-step cascade of the Invader assay amplifies the signal 1- to 10-million-fold per target strand per h (53).

One advantage of Invader technology is the reported ease and speed of developing assays for different targets. Due to the adaptability of oligonucleotide design and the power of Third Wave's proprietary oligonucleotide design software, InvaderCreator, oligonucleotides can be made to "fit" specified reaction parameters so that reaction conditions may also remain constant from target to target. Consequently, Invader assays for multiple targets can be run simultaneously in different wells on the same microtiter plate. Alternatively, in high-throughput settings, each well of a single plate can be used to detect the same target. The Invader assay can be used in combination with a variety of automated platforms, including capillary electrophoresis, mass spectrometry, and the ABI PRISM systems.

Invader technology has already been used successfully in clinical reference laboratories and in genome centers to detect SNPs and mutations from both genomic DNA (63, 65) and PCR-amplified fragments (17, 73). Other clinical assays are in preparation. Furthermore, Invader assays have been used to detect and quantify gene expression (53) and to genotype or test for drug resistance mutations in S. aureus and M. tuberculosis (17).

Host Response Profiling via Microarray Technology

Genomic analysis of human pathogens has been the historical focus of most molecular diagnostics efforts; however, DNA sequence information provides only a static snapshot of all of the possible ways a cell might use its genes. As a pathogenic organism reacts to its environment, changes occur that may affect the outcome of the organism's defenses against antimicrobial therapy or the host immune response. Likewise, study of the human cell response to infection, which may vary because of multifactorial genetic or environmental factors, could provide key information that relates to variability in human susceptibility or resistance to infectious disease (86).

Because of the concurrent and often interdependent events involved in pathogenesis and infection, emphasis is now being placed on the study of both pathogen and host responses to disease via gene expression analysis (9, 72, 83, 85, 104). At present, at least four infectious diseases (Puumala hantavirus infection, tuberculosis, Lyme disease, and AIDS) are known to have corresponding host genes that influence disease susceptibility. More examples are sure to follow. Linking gene expression analysis to the pathogenic changes associated with infectious-disease conditions will enable identification of relevant disease targets, both of the infectious agent and the host, bridging the gaps among genome sequence, cellular behavior, and disease (52, 111, 123). This host profiling should allow the identification of molecular targets in infected hosts prior to culture positivity of their samples. Targets like these may also be useful as predictors of disease outcome or therapeutic success.

In addition, the science of pharmacogenomics promises to assess variables related to the human host's capability to metabolize drugs. For example, degradative liver enzymes, such as cytochrome P-450, are being studied. SNP patterns, found in a variety of genes, may also prove useful for the creation of SNP profiles to predict an individual's capacity to effectively use a given drug of choice and to avoid toxicity (9, 72). Of particular interest for these various endeavors is the use of genetic microarrays. Microarray technology enables high-throughput gene expression analysis and holds promise as a discovery technique to identify new genetic targets, in both host and pathogen, which may be used to design new diagnostic methods.

Microarrays

Diagnostic research laboratories will find that microarray (DNA chip or DNA array) technology offers various applications, such as gene discovery and mapping, genotyping, gene expression, and detection of mutations or polymorphisms, including SNPs. This technology will eventually provide a variety of applications for disease management and will be applicable to clinical microbiology laboratories, allowing testing that may include identification of host gene expression or detection of multiple gene targets for organism identification or drug resistance patterns (21, 52, 111). Already, genomic DNA from yeast has been placed on a microarray (83), and human genes were screened to identify mRNAs that change in relation to cytomegalovirus infection (124). RNA from a streptococcus was hybridized to individual chips, allowing quantitative measurement of transcriptional activities and providing clues to developing drug resistance and other pathogenic events (98, 112). Recently, in a collaborative study with Biomerieux and Affymetrix, high-density DNA arrays were successfully used to simultaneously identify Mycobacterium spp. as well as to screen for rifampin resistance (112).

Several types of chips exist that are used for a number of different applications. Chips can be used for specimen preparation, size separation of nucleic acids, modified PCR, sequencing, and hybridization and detection assays. It is the last two which will be discussed in the most detail.

Hybridization arrays are composed of various forms of single-stranded DNA, such as cDNA or oligonucleotides, which are dotted or synthesized and subsequently immobilized in a specific location on a substrate, such as glass, nylon, or silicon. The DNA can represent a variety of known genes or gene segments. When the chip is exposed to (i.e., probed with) cDNAs that are made from an unknown sample of RNA from a tissue or cell, complementary sequences will hybridize with various sequences already positioned on the chip (Fig. 18). By monitoring the hybridization pattern of the unknown cDNA, the identity of the hybridized nucleic acids can be determined and will indicate which RNAs were in the tissue and, in general, which genes are being expressed. The chips can depict the expression patterns of thousands of genes at one time.

The chips can be processed and analyzed with a variety of different instruments. For the hybridization arrays, chips are usually exposed to fluorescently labeled sample DNA, placed in a microfluidics wash station, and hybridized in a special hybridization chamber. Once the complementary sequence is hybridized, the chip is scanned by confocal epifluorescence microscopy to measure emitted light, which is proportional to the amount of labeled DNA target bound at each location on the probe array. Fluorescent images are digitized, color hybridization displays are produced by computer, and semiquantitative hybridization information is obtained (Fig. 19). A stronger signal is produced when there is an exact match between the probe and the target. Therefore, the resulting signal is a measure of both sequence homology and the quantity of DNA bound to the chip.

Chips from several manufacturers are presently available. Most notable is the GeneChip array, produced by Affymetrix (Santa Clara, Calif.), which adapts semiconductor industry photolithographic processes and solid-phase chemistry to synthesize oligonucleotide probe sequences directly onto a glass chip substrate of approximately 1 cm^2. Each characterized probe is located in a probe feature, a specific and identifiable area on the probe array (Fig. 19). Each probe feature contains millions of copies of a given probe. The GeneChip instrument system can be purchased with a fluidics station, a hybridization oven, an Agilent GeneArray to measure emitted light, a PC workstation, and GeneChip Data Analysis Suite software (Fig. 20). The Affymetrix GeneChip system has also been used to monitor the antiretroviral drug resistance of human immunodeficiency virus type 1 (HIV-1) (115). Human gene expression arrays are also available.

Alternatively, oligonucleotides can be placed on chips through a variety of technologies. One technology, analogous to that of an ink jet printer, deposits nucleotides instead of ink (Incyte Pharmaceuticals, Palo Alto, Calif., and Rosetta Informatics, Kirkland, Wash.). Variations of these methods exist, including hybridization of total RNA or double-stranded DNA to oligonucleotide arrays.

Affymetrix, HySeq (Sunnyvale, Calif.), and Genomic Solutions (Ann Arbor, Mich.) have increased manufacture to an industrial scale (68). Several other companies also offer production capability (84). Among them, NEN Life Science Products, Inc. (Boston, Mass.) offers the MICRO-MAX chip, prespotted with human genes of various derivations for differential gene expression analysis. Stratagene also offers human cDNA microarrays for host expression profiling and gene discovery, and Becton Dickinson has

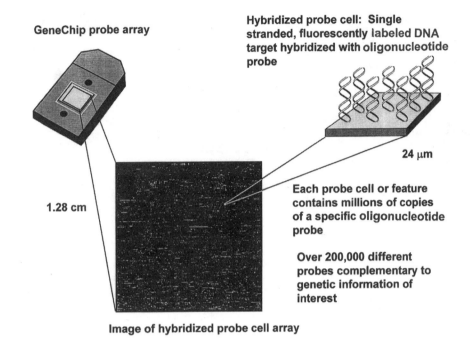

GeneChip probe array

1.28 cm

Hybridized probe cell: Single stranded, fluorescently labeled DNA target hybridized with oligonucleotide probe

24 μm

Each probe cell or feature contains millions of copies of a specific oligonucleotide probe

Over 200,000 different probes complementary to genetic information of interest

Image of hybridized probe cell array

FIGURE 18 Schematic of a hybridization experiment using the GeneChip probe array. (Courtesy of Affymetrix.)

plans to market the APEX chip, which uses electrical potential to enhance hybridization of DNA to the capture probes on the microarrays. In addition, several companies offer kit formats or contracted services for microarray expression analysis. Genomic Solution's GeneTAC 2000 Biochip Analyzer provides automated imaging and analysis of up to 24 fluor-labeled biochips at once and offers contract services for chip production.

Another possibility for the use of microarrays lies in the fact that DNA chip technology has the potential to combine DNA preparation, detection, and identification. Because

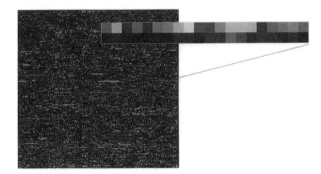

FIGURE 19 Close up of probe array surface. Various fluorescent-signal intensities, depicted by color hybridization displays, are generated and digitized. A stronger signal is produced when there is an exact match between the probe and the target. Therefore, the resulting signal is a measure of both sequence homology and the quantity of DNA bound to the chip. Semiquantitative hybridization information is obtained. (Courtesy of Affymetrix.)

chip technology requires that sample DNA be amplified by PCR in such a way that a fluorescent label is inserted into the PCR product, it has some of the sample preparation limitations of any PCR. Optimal use of this technology will require that processes be developed and integrated with the chip technology to preclude the interference of PCR inhibitors and minimize loss of sample during preparation.

Current limitations of microarray technology include the high costs of instrumentation and disposables. In addition, our current inability to easily analyze the enormous amount of information potentially available through this technology will present challenges. Fortunately, production costs are likely to decline steeply over the next few years. As economies of scale are achieved, less expensive diagnostic arrays will become available. In addition, arrays that are limited to a specific purpose require less genetic material per chip and may provide a more cost-effective approach, allowing more widespread use of array technology. Although this technology is not likely to play a significant role in clinical microbiology for several years, gene expression analysis promises to provide a better overall understanding of disease progression and may eventually allow the identification of more timely and relevant targets for infectious-disease detection. Moreover, the possibility of miniaturization gives chip technology potential as a point-of-care testing method (38). New developments are likely to continue the evolution of array technology to make it one of the most powerful tools available for molecular diagnostics. For a thorough review of suppliers of microarrays, see reference 18a.

DNA Sequencing

Gel-Based Automation

A few systems have been designed for post-PCR product sequencing via gel electrophoresis. ABI manufactures the

FIGURE 20 The GeneChip instrument system with fluidics station, hybridization oven, Agilent GeneArray to measure emitted light, PC workstation, and GeneChip Data Analysis Suite software. (Courtesy of Affymetrix.)

PRISM 377 DNA sequencer, which provides automated sequence data, based on the Sanger enzymatic method for DNA sequencing. This instrument uses four fluor-labeled oligonucleotide primers to label DNA fragments produced by PCR (102). Labeled fragments are coelectrophoresed and identified by the colors of their different emission spectra. The system is designed for mid- to high throughput applications with 36 to 576 electrophoresis lanes per day.

Another slab gel system, LI-COR-NEN Global IR[2] System (LI-COR, Lincoln, Nebr.), uses two dyes, near-infrared technology, and a scanning fluorescent microscope with two laser diodes to excite the fluorophores. The system provides relatively low output; however, it provides extreme sequence accuracy and is able to read long regions of DNA. Moreover, it is the only system that is internet compatible and provides software for cross-platform operation with both Windows and Macintosh operating systems.

The OpenGene system is available from Visible Genetics (Athens, Ga.) and includes the capability for both sequencing and fragment analysis. It uses a GelToaster system to easily prepare electrophoresis gels in less than 5 min. MicroCel cassette gels are then loaded with a PCR product and placed into a gel sequencer tower that can electrophorese multiple gels and detect fluorescent emissions of tagged nucleic acids within the PCR product. The system is reported to sequence PCR products as large as 1,000 bp. Sequence data is stored, analyzed, and archived in the GeneLibrarian software system to monitor a specific genome or to detect changes associated with drug resistance. Databases for hepatitis C virus, hepatitis B virus, and M. *tuberculosis* are in process, and the TrueGene assay is available for HIV-1 genotyping and analysis of HIV drug resistance. In addition, the Vigilance II study is in progress using this methodology to accumulate HIV genotype results from patients across the United States. Fifty test sites are anticipated, and the study will monitor the prevalence of HIV mutations and clinical outcomes associated with the data.

Capillary Electrophoresis

The ABI PRISM 3700 DNA analyzer is a capillary electrophoresis system that automates sample loading, electrophoresis, data collection, and analysis. Capillary electrophoresis has several advantages over slab gel electrophoresis. Manual gel pouring can be avoided, small sample volumes can be tested, and run times are accelerated. The system can provide production scale sequencing

with up to 96 samples run in 3 h with minimal operator intervention. For lower-throughput applications, ABI offers the PRISM 310 genetic analyzer, which can perform 10 to 24 sequences per day.

Electrophoresis occurs in a fluid gel matrix called POP (performance-optimized polymer), which is pumped into the capillaries prior to electrophoresis. When a capillary and a voltage-generating electrode are placed in a sample and voltage is applied, negatively charged DNA molecules enter the capillary and migrate to a positive charge located at the opposite end. Product detection occurs as argon lasers, focused near the ends of the silica capillaries, excite fluor-labeled DNA fragments in the POP6 solution. Emitted light is collected, separated into different spectra, and imaged on a CCD. Data are processed and stored as electropherograms, graphical displays of DNA sequences (Fig. 21).

The Amersham Pharmacia Biotech (Sunnyvale, Calif.) MEGABASE 1000 DNA sequencing system differs from the ABI sequencers in that it uses linear polyacrylamide Long Read Matrix to fill the capillaries. Confocal optics combined with multiple filter sets allows the detection of sequencing products to occur at fixed wavelengths.

Beckman Coulter's CEQ 2000 DNA analysis system is an eight-capillary system that uses laser-induced fluorescence of CEQ Well RED dyes. These dyes are detected in the red spectrum and provide high sensitivity and low background. Fluorescence detection can be coupled with automated workstations, such as the Biomek 2000.

Pyrosequencing

Pyrosequencing (Pyrosequencing, Westborough, Mass. and Uppsala, Sweden) is a process by which DNA is sequenced via the synthesis of short to medium-length DNA fragments. With this technique, used in the PSQ96 system, a sequencing primer is hybridized to a DNA template and incubated with DNA polymerase, ATP sulfurylase, luciferase, apyrase, and the substrates adenosine 5′-phosphosulfate and luciferin (Fig. 22). As complementary deoxynucleoside triphosphates are added to the reaction mixture and incorporated into the strand, pyrophosphate is released in a quantity equal to that of the original incorporated nucleotides. ATP is generated and drives the luciferase reaction so that light is produced, monitored by a CCD, and converted to graphed data of peaks called a pyrogram. Pyrosequencing may be used for SNP detection and

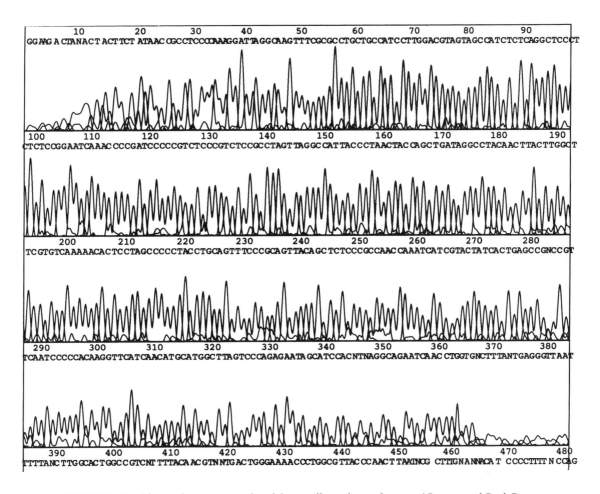

```
        10         20         30         40         50         60         70         80         90
GGAAGA CTANACT ACT TCT ATAAC CGCCTCCCCAAAGGATTAGGCAAGTTTCGCGCCTGCTGCCATCCTTGGACGTAGTAGCCATCTCTCAGGCTCCCT

   100        110        120        130        140        150        160        170        180        190
CTCTCCGGAATCAAAC CCCGATCCCCCGTCTCCCGTCTCCGCCTAGT TAGGC CAT TACCCTAACTAC CAGCTGATAGGCCTACAAC TTACTTGGCT

      200        210        220        230        240        250        260        270        280
TCGTGTCAAAAACACTCCTAGCCCCCTACCTGCAGTTTCCCGCAGTTACAGC TCTCCCGCCAACCAAATCATCGTACTATCACTGAGCCGNCCGT

   290        300        310        320        330        340        350        360        370        380
TCAATCCCCCACAAGGTTCATCAACATGCATGGCTTAGTCCCAGAGAATAGCATCCACNTNAGGCAGAATCAAC CTGGTGNCTTTANTGAGGGTTAAT

      390        400        410        420        430        440        450        460        470        480
TTTTANC TTGGCACTGGCCGTCNT TTTACAACGTNNTGACTGGGAAAACCCTGGCGTTACCCAACT TAATNCG CTTTGNANNACAT CCCCTTTT N CCAG
```

FIGURE 21 Electropherogram produced by capillary electrophoresis. (Courtesy of Paul Rys, Mayo Clinic, Rochester, Minn.)

for sequencing small PCR products and is reported to yield accurate and consistent analysis equal to that of sequencing via mass spectrometry.

Mass Spectrometry

The use of mass spectrometry in the clinical microbiology laboratory has been limited to urea breath tests in the diagnosis of *Helicobacter pylori* infections (61); however, mass spectrometry may eventually provide greater accuracy and precision for DNA sequencing than either gel-based or hybridization sequencing methods. The technique provides several technical advantages. The platform is flexible in that it can also be coupled with two-dimensional protein electrophoresis for proteomics protocols. Sequenom (San Diego, Calif.) has plans to combine microarray chip technology with mass spectroscopy to provide an alternative approach to the analysis of multiplex PCR products. In addition, direct analysis of products by mass spectrometry eliminates the need for radioactive or fluorescent probe labels and avoids the pitfalls associated with nonspecific probe binding. This combination of technology appears to be useful for SNP genotyping and identification of polymorphisms but can also be used to sequence PCR products (42, 54).

Mass spectrometry and its variations have already been used to analyze 16S ribosomal DNA (rDNA) (31) and as a means to identify viral infections in cell culture (1). Through the use of mass spectrometry, a protein indicative of human M. *tuberculosis* infection has been identified (33) and muramic acid has been detected in septic synovial fluids (25). The epidemiologies of influenza virus and a variety of other pathogens have been improved by the use of this technology (30, 43, 49, 67, 80).

Chromatography

The WAVE DNA fragment analysis system (Transgenomic, Inc., Omaha, Nebr.) makes mutation detection faster and more cost-effective than traditional gel-based sequencing systems. This system can perform polymorphic marker mapping, nucleic acid sizing and quantitation, and SNP analysis. The system also automates screening for mutations by coupling heteroduplex analysis with chromatography. PCR is used to produce amplified DNA from wild-type and mutant genes. PCR samples are loaded into 96-well plates, and an autosampler injects the samples into the DNASEP cartridge. Using the DNASEP column matrix, chromatography is run at a temperature that will partially denature the heteroduplexes and keep the perfectly matched strands of

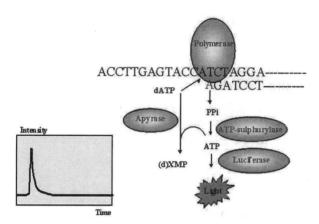

FIGURE 22 Pyrosequencing chemistry. A sequencing primer is hybridized to a DNA template and incubated with DNA polymerase, ATP sulfurylase, luciferase, apyrase, and the substrates adenosine 5′-phosphosulfate and luciferin. As complementary deoxynucleoside triphosphates are added individually to the reaction mixture, they are incorporated into short to medium-length DNA strands, and pyrophosphate is released in a quantity equal to that of the original incorporated nucleotides. ATP is produced and drives the luciferase reaction so that light is generated. (Courtesy of Pyrosequencing.)

the homoduplex intact and completely annealed. The mix of homo- and heteroduplexes is then separated by the use of both temperature-modulated heteroduplex analysis and ion pair reverse-phase column-based liquid chromatography. Heteroduplexes containing the gene mutations have reduced retention time in the column and can be eluted and directly sequenced to identify the mutation.

The WAVE-HS system offers the increased sensitivity of fluorescent detection and can analyze one sample in as little as 5 to 7 min. The system's fragment collector allows fragments to be saved for further analysis and sequencing. Computer software can produce chromatographylike traces or familiar gel-like bands as data output.

PNAs

Peptide nucleic acids (PNAs) are novel synthetic DNA-like compounds with nucleotide bases attached to a peptide backbone. PNAs can be used as both PCR primers and hybridization probes for PCR, in situ hybridization, and other traditional oligonucleotide applications. Probes can be designed to selectively amplify rare single-base-pair mutations in a mixed population or as a sequence-specific capture probe for nucleic acid purification. The advantages of their use as replacements for oligonucleotides include (i) superior hybridization affinity for RNA and DNA, (ii) high stability, and (iii) convenient solid-phase synthesis (79).

PNAs have been used to distinguish between tuberculous and nontuberculous mycobacteria in liquid cultures (81), to detect M. *tuberculosis* in sputum (107), and to detect specific PCR products (99). It is quite possible that PNAs will have practical applications in molecular diagnostics.

Flow Cytometry

Flow cytometry provides rapid measurement of light scattering and fluorescent emission of cells or particles in liquid as they pass though a light beam that excites fluorochromes on

or within the cells. The light and emissions are collected and sent to a computer, where the size and distribution of light scatter is displayed in two-dimensional or three-dimensional plots. Although traditionally used in the areas of immunology and cell biology, this technology could be adapted for rapid detection of organisms and antimicrobial susceptibility patterns, especially those of mixed populations, or for identification of organisms in normally sterile body fluids (3).

Applications for diagnostic microbiology include direct detection of bacteria, fungi, parasites, virus-infected cells, and viral antigens and nucleic acids (3, 10, 12, 20, 24, 29, 40, 41, 44, 74, 103). Furthermore, serologic diagnosis can be performed by using serum and monitoring specific antibody binding to antigen-coated microspheres (3). Monitoring of bactericidal and bacterostatic effects by flow cytometry has been shown to reduce turnaround time for susceptibility testing and can provide a method by which the effects of antimicrobial therapy can be monitored (3, 16, 69, 96). Commercial kits are available (Molecular Probes) to monitor bacterial viability. The limitations of flow cytometry include relatively high costs and low throughput.

INFORMATION MANAGEMENT

Bioinformatics is a science that allows a convergence of diagnostic medicine and computational biology. As technology continues to provide us with information at a rate that outpaces our ability to understand it, bioinformatics will become an increasingly important tool for infectious-disease diagnosis. Current information and genomic World Wide Web-based databases, such as GenBank, are available from the National Center for Biotechnology Information (http://www.ncbi.nlm.nih.gov/) and will become more important to clinical laboratories in the future. The National Center for Biotechnology Information, established in 1988 as a national resource for molecular biology information, creates public databases, conducts research in computational biology, develops software tools for analyzing genome data, and disseminates biomedical information to facilitate the understanding of molecular processes affecting human health and disease.

Related databases, such as the rRNA WWW Server (http://www-rrna.uia.ac.be/), the Ribosomal Database Project II (http://www.cme.msu.edu/RDP/html/index.html), The Institute for Genomic Research (http://www.tigr.org), and the Ribosomal Differentiation of Medical Microorganisms (http://www.ridom.hygiene.uni-wuerzburg.de/), also provide much needed nonproprietary sources of genomic information. Each will likely continue to incorporate improved database management and standardization, more user-friendly graphical interfaces, and relational databases with intuitive multidimensional queries in order to facilitate data analysis. Commercially, the MicroSeq 16S rRNA gene kit (ABI) combines a PCR kit format with an rDNA database (available for purchase) composed of specific and conserved rRNA gene sequences used for organism identification, but it is still somewhat limited in its scope and the manufacturer's commitment to product support for clinical laboratories. In addition, BioSystematica (Tavistock, Devon, United Kingdom) offers BioNumerics software, a relatively user-friendly system designed to store and analyze both the molecular sequence data and the phenotypic characteristics of an organism. This software will improve current organism identification schemes by allowing the routine use of polyphasic algorithms for identification.

Analysis comparing human and nonhuman genomes will enable both the genetic profiling of disease agents and determination of human genetic response to disease. As human disease management protocols evolve to include genotypic patterns for disease susceptibility and the pharmacogenomics related to the relevant therapeutic options, clinical laboratories may find themselves screening for both pathogen- and host-related genomic markers. Practices and algorithms like these will be the logical extension of the Human Genome Project and will allow successful integration of host-pathogen interactions, but they will require innovations in data management and improved data query capabilities. Algorithms for data mining (118) and improved encryption will also be necessary to gain full advantage of existing clinical databases and to protect the sensitive patient genotype data present in those databases.

To keep pace with the molecular information overload, technology companies that have unique software interfaces for high-throughput technologies may discover the need to integrate with other high-throughput instrumentation. The American Society for Testing and Materials is in the process of adopting the Laboratory Equipment Control Interface Specification. This specification will enable laboratory automation and robotics to have a single interface and connectivity to all types of commercial hardware. In addition, industry standards are being considered. Molecular Dynamics and Affymetrix have formed the Genetic Analysis Technology Consortium in an attempt to standardize the rapidly growing field of array-based genetic analysis. The consortium was created to provide a unified technology platform to design, process, read, and analyze DNA arrays, eliminating the need for redundant equipment and software.

QUALITY CONTROL AND STANDARDIZATION

In our collective struggle to maintain cutting-edge technology in our practices, we must not forget the roots of quality in clinical laboratory diagnostics. Home brew molecular assays have been offered for both research and diagnostic purposes and are often self-validated. To a greater or lesser extent, we have created what we consider to be reasonable and cost-effective validation and quality control practices. Some of these practices are included in the checklists and guidelines put forth by the Clinical Laboratory Improvement Act, the College of American Pathologists, the NCCLS, and the International Organization for Standardization standards. Unfortunately for microbiologists, most of the current guidelines have focused on the practices of forensics and molecular pathology. Specifics that may be relevant to infectious-disease diagnostics are intertwined in those venues. Not surprisingly, standardization is not the norm.

Not all molecular infectious-disease-testing protocols hold to the same level of precision, accuracy, and reproducibility as that which is expected of the FDA-approved tests most often used in our practices. Molecular testing laboratories and manufacturers of molecular reagents and oligonucleotides are not bound to the same validation criteria necessary for most commercially available laboratory assays. Measures of sensitivity and specificity, predictive value, contamination events, reagent purity, and the like are not public knowledge. Changes in methodology are occurring at a rapid pace, and our clients may have various levels of understanding of the limitations of molecular testing methodologies.

It is the responsibility of clinical microbiologists to carefully and fastidiously interpret the existing guidelines and extrapolate them to our practices. Additions to the current regulatory standards and recommendations will be necessary and, in some cases, are under way to enhance the quality control systems and method verification and validation requirements for home brew molecular infectious-disease testing. Clinical chemists and blood bank practitioners, who are well versed in validation studies and quality systems practices, have information that is relevant to our endeavors. We must make use of all resources, advancing our training efforts to include topics such as biochemistry, biorobotics, mass spectroscopy, and the like. New technology has placed PCR and other molecular protocols within our reach. Old boundaries no longer exist, and we must prepare ourselves to meet the challenges of quality molecular testing in our laboratories.

COST-BENEFIT

Without a standardized cost-effective approach, improved standards of patient care, facilitated by the use of molecular technologies, may be available only to those patients who can most readily pay for the technology. Research scientists, clinical microbiologists, infectious-disease physicians, and public health officials must maintain a closer cooperation in order to achieve a balance between what is possible and what is practical. Ethical cost-benefit schemes must be devised in order to gain benefit from new advances in molecular infectious-disease diagnostics. Algorithms such as those used for prescreening laboratory requests for herpes simplex virus PCR from cerebrospinal fluid samples (109) must continue to be developed so that improved standards of care may be realized without a large cost detriment. Leadership must exist at the national and local levels to prioritize public health risks and to assess which infectious diseases are most likely to be benefited by a molecular diagnosis.

PREDICTIONS

To date, economic constraints of the current health care climate have limited molecular diagnostic approaches for infectious disease to a few select circumstances: (i) the identification of uncultivable or difficult-to-cultivate organisms, (ii) situations where faster diagnosis and intervention prevents significant morbidity, (iii) screening for new infectious-disease causes and links to related diseases, (iv) situations in which multiplexed targets or universal primers allow for identification of multiple pathogens, (v) epidemiological studies, (vi) signal amplification technology to avoid amplicon contamination or PCR performed in specially designed facilites, and (vii) FDA-approved molecular tests preadapted to meet the clinical laboratories' expectations of quality control and amplicon control. Changes are on the immediate horizon. Novel technology and advances in automated, closed-system, integrated, and parallel platforms will enable molecular technology to move from research laboratories to diagnostic clinical laboratories. Changes in our strategic thinking will occur, as more molecular diagnostic techniques are introduced into the health care workplace. It is clear that advances in the PCR paradigm and in the automation of molecular technology will eventually allow routine molecular testing to occur in even the smallest laboratories.

Automation will occur in all areas of testing, including specimen processing and nucleic acid extraction, amplification of nucleic acid targets or their facsimiles, and detection of nucleic acids or their tagged signals. Gel-based technology will remain, but it will also be improved through the use of automation and digital imaging techniques. Advances in nonisotopic, non-gel-based detection and identification of PCR products may allow for cost-effective integration into the clinical microbiology workplace.

PCR is a technology on the verge of widespread commercialization, but it is nevertheless a technology without a widespread infrastructure to support basic standard reference materials and controls. While there have been a few efforts to standardize quality control and proficiency testing for the new discipline of molecular diagnostics, much more work is needed. Improved interdisciplinary collaboration of microbiologists, molecular biologists, epidemiologists, cell biologists, clinical chemists, pathologists, laboratory managers, physicians, educators, and manufacturers will be necessary to properly utilize the strengths of molecular testing, unravel the complex interactions of host and pathogen, and allow improved disease detection and identification in a cost-effective manner.

PCR combined with rDNA sequencing will continue to play a role in the identification and phylogenetic placement of both prokaryotes and eukaryotic organisms; however, its use may best serve the purposes of clinical microbiology when combined with polyphasic approaches to organism identification. There are some limitations of this technique as a standalone tool. For some pathogens, it will be necessary to combine genotypic analysis with phenotype, gene expression, or protein analysis for accurate identification (51, 62).

Identification of unknown infectious-disease agents, atypical presentations of an infection, or infectious diseases associated with chronic diseases of multifactorial origin will occur through the combination of molecular diagnostic tools and the epidemiological aspects of disease. Moreover, sequential analysis of gene expression will allow the assessment of pathogenicity and drug resistance by analyzing the pathogen's current metabolic, genetic, or environmental state.

Although analysis of genomic information has been the historical focus of molecular diagnostics efforts, DNA sequence information provides only a static snapshot of all of the possible ways a cell might use its genes. As a cell reacts to its environment, dynamic events may affect the outcome of the organism's defenses against antimicrobial therapy or the host immune response. Likewise, study of the human cell response to infection may determine susceptibility to disease. Because of these concurrent and often interdependent events, emphasis is being given to the study of the proteins produced by the genes present in an organism; this discipline is called proteomics. Proteomic technology may be used as a discovery technique to identify new targets for diagnosis, but it is not likely to play a significant role in clinical microbiology for several years.

Molecular strategies of the future must encompass high throughput, rapid turnaround time, cost-effectiveness, and compatible or integrated analysis platforms. As molecular testing becomes more widely available, molecular diagnostic strategies will shift from routine isolation of the nucleic acids of a single pathogen to multiplex real-time PCR, parallel testing, DNA chip, and sequencing technologies that allow detection of multiple targets and pathogens. Well-designed molecular protocols ideally should improve the standards of patient care and provide an overall cost benefit. In order to achieve these goals, major advances will have to occur and will likely do so through the combined use of robotics and automation for specimen preparation,

high-throughput real-time PCR, and parallel testing protocols for multiple targets or organisms and by the efforts of clinical microbiologists to adapt technology to their workplaces. Collaborations with clinical chemists, molecular geneticists, and pathologists will be necessary if overlapping technologies are to be shared and used most effectively.

Improvements in nucleic acid detection and identification using fluorescent chemistry, microarray DNA chip technology, mass spectrometry, and chromatography will combine with advances in bioinformatics to facilitate post-PCR detection and product analysis via connective and interface-ready automated platforms. Finally, standardized algorithms for molecular testing protocols will be needed in order to advance the technology without overutilization of health care resources. As microbiologists, we must responsibly prepare molecular algorithms to include the ability to identify new pathogens or new presentations of a disease, which may be overlooked if we focus too closely on only a certain set of molecular parameters for disease diagnosis.

The opportunity to participate in shaping the future of infectious-disease diagnostics may well be the greatest challenge clinical microbiologists face in this decade; the possibilities seem almost endless. Progress must be tempered with common sense so that optimal patient care benefit is derived from our efforts. In our lifetimes, technology will be only as good as the collective efforts, ethics, and judgement of the scientists who design and provide molecular diagnostic services. It seems apparent that we will educate ourselves in ways we may never have imagined so that, as clinical microbiologists, we can continue to provide the best diagnostic care possible. Working alone, we may not be able to achieve the goals for effective integration of these technologies; working together, it seems that we can.

REFERENCES

1. **Abbott, R., M. Anbar, H. Faden, J. McReynolds, W. Rieth, M. Scanlon, L. Verkh, and B. Wolff.** 1980. Diagnosis of viral infections by multicomponent mass spectrometric analysis. *Clin. Chem.* **26:**1443–1449.

2. **Abe, A., K. Inoue, T. Tanaka, J. Kato, N. Kajiyama, R. Kawaguchi, S. Tanaka, M. Yoshiba, and M. Kohara.** 1999. Quantitation of hepatitis B virus genomic DNA by real-time detection PCR. *J. Clin. Microbiol.* **37:**2899–2903.

3. **Alvarez-Barrientos, A., J. Arroyo, R. Canton, C. Nombela, and M. Sanchez-Perez.** 2000. Applications of flow cytometry to clinical microbiology. *Clin. Microbiol. Rev.* **13:**167–195.

4. **Anderson, B.** 1998. Identifying novel bacteria using broad-range polymerase chain reaction, p. 117–127. *In* S. Specter, M. Bendinelli, and H. Friedman (ed.), *Rapid Detection of Infectious Agents.* Plenum Press, New York, N.Y.

5. **Anderson, B., C. Kelly, R. Threlkel, and K. Edwards.** 1993. Detection of Rochalimaea henselae in cat-scratch disease. *J. Infect. Dis.* **168:**1034–1036.

6. **Anderson, B., K. Sims, R. Regnery, L. Robinson, M. J. Schmidt, S. Goral, C. Hager, and K. Edwards.** 1994. Detection of *Rochalimaea henselae* DNA in specimens from cat scratch disease patients by PCR. *J. Clin. Microbiol.* **32:**942–948.

7. **Anderson, B. E., J. E. Dawson, D. C. Jones, and K. H. Wilson.** 1991. Ehrlichia chaffeensis, a new species associated with human ehrlichiosis. *J. Clin. Microbiol.* **29:**2838–2842.

8. **Bhatt, R., B. Scott, S. Whitney, R. N. Bryan, L. Cloney, and A. Lebedev.** 1999. Detection of nucleic acids by cycling probe technology on magnetic particles: high sensitivity and ease of separation. *Nucleosides Nucleotides* **18:**1297–1299.

9. **Brahic, M., and J. F. Bureau.** 1997. Of genes and germs. *Trends Microbiol.* **48:**2–48.

10. **Brussaard, C. P., D. Marie, and G. Bratbak.** 2000. Flow cytometric detection of viruses. *J. Virol. Methods* **85:**175–182.

11. **Cardullo, R. A., S. Agrawal, C. Flores, P. C. Zamecnik, and D. E. Wolf.** 1988. Detection of nucleic acid hybridization by nonradiative fluorescence resonance energy transfer. *Proc. Natl. Acad. Sci. USA* **85:**8790–8794.

12. **Chang, W. L., H. C. van der Heyde, and B. S. Klein.** 1998. Flow cytometric quantitation of yeast a novel technique for use in animal model work and in vitro immunologic assays. *J. Immunol. Methods* **211:**51–63.

13. **Chernesky, M. A.** 1999. Nucleic acid tests for the diagnosis of sexually transmitted diseases. *FEMS Immunol. Med. Microbiol.* **24:**437–446.

14. **Cinque, P., G. M. Cleator, T. Weber, P. Monteyne, C. J. Sindic, and A. M. van Loon.** 1996. The role of laboratory investigation in the diagnosis and management of patients with suspected herpes simplex encephalitis: a consensus report. The EU Concerted Action on Virus Meningitis and Encephalitis. *J. Neurol. Neurosurg. Psychiatry* **61:**339–345.

15. **Cockerill, F. R.** 1999. Genetic methods for assessing antimicrobial resistance. *Antimicrob. Agents Chemother.* **43:**199–212.

16. **Comas, J., and J. Vives-Rego.** 1997. Assessment of the effects of gramicidin, formaldehyde, and surfactants on Escherichia coli by flow cytometry using nucleic acid and membrane potential dyes. *Cytometry* **29:**58–64.

17. **Cooksey, R. C., B. P. Holloway, M. C. Oldenburg, S. Listenbee, and C. W. Miller.** 2000. Evaluation of the invader assay, a linear signal amplification method, for identification of mutations associated with resistance to rifampin and isoniazid in *Mycobacterium tuberculosis*. *Antimicrob. Agents Chemother.* **44:**1296–1301.

18. **Cooksey, R. C., G. P. Morlock, S. Glickman, and J. T. Crawford.** 1997. Evaluation of a line probe assay kit for characterization of *rpoB* mutations in rifampin-resistant *Mycobacterium tuberculosis* isolates from New York City. *J. Clin. Microbiol.* **35:**1281–1283.

18a.**Cummings, C. A., and D. A. Relman.** 2000. Using DNA microarrays to study host-microbe interactions. *Emerg. Infect. Dis.* **6:**513–525.

19. **Dawson, J. E., B. E. Anderson, D. B. Fishbein, J. L. Sanchez, C. S. Goldsmith, K. H. Wilson, and H. L. Dupont.** 1991. Isolation and characterization of an *Ehrlichia* sp. from a patient diagnosed with human ehrlichiosis. *J. Clin. Microbiol.* **29:**2741–2745.

20. **Detrick, B., J. J. Hooks, J. Keiser, and I. Tabbara.** 1999. Detection of cytomegalovirus proteins by flow cytometry in the blood of patients undergoing hematopoietic stem cell transplantation. *Exp. Hematol.* **27:**569–575.

21. **Diehn, M., A. A. Alizadeh, and P. O. Brown.** 2000. Examining the living genome in health and disease with DNA microarrays. *JAMA* **283:**2298–2299.

22. **Espy, M. J., T. F. Smith, and D. H. Persing.** 1993. Dependence of polymerase chain reaction product inactivation protocols on amplicon length and sequence composition. *J. Clin. Microbiol.* **31:**2361–2365.

23. **Espy, M. J., J. R. Uhl, P. S. Mitchell, J. N. Thorvilson, K. A. Svien, A. D. Wold, and T. F. Smith.** 2000. Diagnosis of herpes simplex virus infections in the clinical laboratory by LightCycler PCR. *J. Clin. Microbiol.* **38:**795–799.

24. **Fenili, D., and B. Pirovano.** 1998. The automation of sediment urinalysis using a new urine flow cytometer (UF-100). *Clin. Chem. Lab. Med.* **36:**909–917.

25. **Fox, A., K. Fox, B. Christensson, D. Harrelson, and M. Krahmer.** 1996. Absolute identification of muramic acid, at trace levels, in human septic synovial fluids in vivo and absence in aseptic fluids. *Infect. Immun.* **64:**3911–3915.

26. **Fredericks, D. N., and D. A. Relman.** 1999. Application of polymerase chain reaction to the diagnosis of infectious diseases. *Clin. Infect. Dis.* **29:**475–488.

27. **Gaydos, C. A., and T. C. Quinn.** 1998. Ligase chain reaction for detecting sexually transmitted diseases, p. 83–92. *In* S. Specter, M. Bendinelli, and H. Friedman (ed.), *Rapid Detection of Infectious Agents.* Plenum Press, New York, N.Y.

28. **Gibson, U. E. M., C. A. Heid, and P. M. Williams.** 1996. A novel method for real time quantitative RT-PCR. *Genome Res.* **6:**995–1001.

29. **Guinet, F., A. Louise, H. Jouin, J. C. Antoine, and C. W. Roth.** 2000. Accurate quantitation of Leishmania infection in cultured cells by flow cytometry. *Cytometry* **39:**235–240.

30. **Haag, A. M., S. N. Taylor, K. H. Johnston, and R. B. Cole.** 1998. Rapid identification and speciation of Haemophilus bacteria by matrix-assisted laser desorption/ionization time-of-flight mass spectrometry. *J. Mass Spectrom.* **33:**750–756.

31. **Hall, V., G. L. O'Neill, J. T. Magee, and B. I. Duerden.** 1999. Development of amplified 16S ribosomal DNA restriction analysis for identification of *Actinomyces* species and comparison with pyrolysis-mass spectrometry and conventional biochemical tests. *J. Clin. Microbiol.* **37:**2255–2261.

32. **Heid, C. A., J. Stevens, K. J. Livak, and P. M. Williams.** 1996. Real time quantitative PCR. *Genome Res.* **6:**986–994.

33. **Hendrickson, R. C., J. F. Douglass, L. D. Reynolds, P. D. McNeill, D. Carter, S. G. Reed, and R. L. Houghton.** 2000. Mass spectrometric identification of Mtb81, a novel serological marker for tuberculosis. *J. Clin. Microbiol.* **38:**2354–2361.

34. **Higgins, J. A.** 1999. 5′ nuclease PCR assay to detect *Yersinia pestis*. *J. Clin. Microbiol.* **36:**2284–2288.

35. **Higuchi, R., G. Dollinger, P. S. Walsh, and R. Griffith.** 1992. Simultaneous amplification and detection of specific DNA sequences. *Bio/Technology* **10:**413–417.

36. **Higuchi, R., C. Fockler, G. Dollinger, and R. Watson.** 1993. Kinetic PCR: real time monitoring of DNA amplification reactions. *Bio/Technology* **11:**1026–1030.

37. **Hirano, K., C. Abe, and M. Takahashi.** 1999. Mutations in the *rpoB* gene of rifampin-resistant *Mycobacterium tuberculosis* strains isolated mostly in Asian countries and their rapid detection by line probe assay. *J. Clin. Microbiol.* **37:**2663–2666.

38. **Hodgson, J.** 1998. Shrinking DNA diagnostics to fill the markets of the future. *Nat. Biotechnol.* **16:**725–727.

39. **Holland, P. M., R. D. Abramson, R. Watson, and D. H. Gelfand.** 1991. Detection of specific polymerase chain reaction product by utilizing the 5′ to 3′ exonuclease activity of *Thermus aquaticus* DNA polymerase. *Proc. Natl. Acad. Sci. USA* **88:**7276–7280.

40. Honda, J., Y. Okubo, K. Arikawa, M. Kumagai, M. Kuwamoto, and K. Oizumi. 1996. Detection of human cytomegalovirus-antigen in peripheral blood cells by using a flow-cytometer. *Kansenshogaku Zasshi* **70**:923–930.

41. Ibrahim, P., A. S. Whiteley, and M. R. Barer. 1997. SYTO16 labelling and flow cytometry of Mycobacterium avium. *Lett. Appl. Microbiol.* **25**:437–441.

42. Jackson, P. E., P. F. Scholl, and J. D. Groopman. 2000. Mass spectrometry for genotyping: an emerging tool for molecular medicine. *Mol. Med. Today* **6**:271–276.

43. Jackson, R. M., M. L. Heginbothom, and J. T. Magee. 1997. Epidemiological typing of Klebsiella pneumoniae by pyrolysis mass spectrometry. *Zentbl. Bakteriol.* **285**:252–257.

44. Jiwa, N. M., W. E. Mesker, J. J. Ploem-Zaaijer, W. van Dorp, T. H. The, and A. K. Raap. 1994. Quantification of low frequency white blood cells expressing human cytomegalovirus antigen by image cytometry. *Cytometry* **16**:69–73.

45. Kaiser, M. W., N. Lyamicheva, W. Ma, C. Miller, B. Neri, L. Fors, and V. I. Lyamichev. 1999. A comparison of eubacterial and archaeal structure-specific 5'-exonucleases. *J. Biol. Chem.* **274**:21387–21394.

46. Kellam, P. 1998. Molecular identification of novel viruses. *Trends Microbiol.* **6**:160–166.

47. Kessler, H. H., G. Muhlbauer, B. Rinner, E. Stelzl, A. Berger, H. W. Dorr, B. Santner, E. Marth, and H. Rabenau. 2000. Detection of herpes simplex virus DNA by real-time PCR. *J. Clin. Microbiol.* **38**:2638–2642.

48. Killgore, G. E., B. Holloway, and F. C. Tenover. 2000. A 5' nuclease PCR (TaqMan) high-throughput assay for detection of the *mecA* gene in staphylococci. *J. Clin. Microbiol.* **38**:2516–2519.

49. Kiselar, J. G., and K. M. Downard. 1999. Antigenic surveillance of the influenza virus by mass spectrometry. *Biochemistry* **38**:14185–14191.

50. Koehler, J. E., F. D. Quinn, T. G. Berger, and J. W. Tappero. 1992. Isolation of Rochalimaea species from cutaneous and osseous lesions of bacillary angiomatosis. *N. Engl. J. Med.* **327**:1625–1631.

51. Kolbert, C. P., and D. H. Persing. 1999. Ribosomal DNA sequencing as a tool for identification of bacterial pathogens. *Curr. Opin. Microbiol.* **2**:299–305.

52. Kozian, D. H., and B. J. Kirschbaum. 1999. Comparative gene-expression analysis. *Trends Biotechnol.* **17**:77.

53. Kwiatkowski, R. W., V. Lyamichev, M. de Arruda, and B. Neri. 1999. Clinical, genetic, and pharmacogenetic applications of the Invader assay. *Mol. Diagn.* **4**:353–364.

54. Laken, S. J., P. E. Jackson, K. W. Kinzler, B. Vogelstein, P. T. Strickland, J. D. Groopman, and M. D. Friesen. 1998. Genotyping by mass spectrometric analysis of short DNA fragments. *Nat. Biotechnol.* **16**:1352–1356.

55. Laue, T., P. Emmerich, and H. Schmitz. 1999. Detection of dengue virus RNA in patients after primary or secondary infection by using TaqMan automated amplification system. *J. Clin. Microbiol.* **37**:2543–2547.

56. Lee, L. G., C. R. Connell, and W. Bloch. 1993. Allelic discrimination by nick-translation PCR with fluorogenic probes. *Nucleic Acids Res.* **21**:3761–3766.

57. Lee, L. G., K. J. Livak, B. Mullah, R. J. Graham, R. S. Vinayak, and T. M. Woudenberg. 1999. Sevencolor, homogeneous detection of six PCR products. *BioTechniques* **27**:342–349.

58. Lee, N. Y., Y. Tang, M. J. Espy, C. P. Kolbert, P. N. Rys, P. S. Mitchell, H. L. Henry, D. H. Persing, and T. F. Smith. 1999. Role of genetic analysis of thymidine kinase gene of herpes simplex virus for determination of neurovirulence and resistance to acyclovir. *J. Clin. Microbiol.* **37**:3171–3174.

59. Livak, K. J., S. J. A. Flood, J. Marmaro, W. Giusti, and K. Deetz. 1995. Oligonucleotides with fluorescent dyes at opposite ends provide a quenched probe system useful for detecting PCR product and nucleic acid hybridization. *PCR Methods Appl.* **4**:357–362.

60. Loeffler, J., N. Henke, H. Hebart, D. Schmidt, L. Hagmeyer, U. Schumacher, and H. Einsele. 2000. Quantification of fungal DNA by using fluorescence resonance energy transfer and the light cycler system. *J. Clin. Microbiol.* **38**:586–590.

61. Logan, R. P. 1998. Urea breath tests in the management of Helicobacter pylori infection. *Gut* **43**(Suppl. 1):S47–S50.

62. Ludwig, W. 1999. Phylogeny of bacteria beyond the 16S rRNA standard. *ASM News* **65**:752–757.

63. Lyamichev, V. I., M. W. Kaiser, N. E. Lyamichev, A. V. Vologodskii, J. G. Hall, W. P. Ma, H. T. Allawi, and B. P. Neri. 2000. Experimental and theoretical analysis of the invasive signal amplification reaction. *Biochemistry* **39**:9523–9532.

64. Lyamichev, V., M. A. D. Brow, and J. E. Dahlberg. 1993. Structure-specific endonucleolytic cleavage of nucleic acids by eubacterial DNA polymerases. *Science* **260**:778–783.

65. Lyamichev, V., A. L. Mast, J. G. Hall, J. R. Prudent, M. W. Kaiser, T. Takova, R. W. Kwiatkowski, T. J. Sander, M. de Arruda, D. A. Arco, B. P. Neri, and M. A. Brow. 1999. Polymorphism identification and quantitative detection of genomic DNA by invasive cleavage of oligonucleotide probes. *Nat. Biotechnol.* **17**:292–296.

66. Machida, U., M. Kami, T. Fukui, Y. Kazuyama, M. Kinoshita, Y. Tanaka, Y. Kanda, S. Ogawa, H. Honda, S. Chiba, K. Mitani, Y. Muto, K. Osumi, S. Kimura, and H. Hirai. 2000. Real-time automated PCR for early diagnosis and monitoring of cytomegalovirus infection after bone marrow transplantation. *J. Clin. Microbiol.* **38**:2536–2542.

67. Magee, J. T., E. A. Randle, S. J. Gray, and S. K. Jackson. 1993. Pyrolysis mass spectrometry characterisation and numerical taxonomy of Aeromonas spp. *Antonie Leeuwenhoek* **64**:315–323.

68. Marshall, A., and J. Hodgson. 1998. DNA chips: an array of possibilities. *Nat. Biotechnol.* **16**:27–31.

69. Martinez, O. V., H. G. Gratzner, T. I. Malinin, and M. Ingram. 1982. The effect of some beta-lactam antibiotics on Escherichia coli studied by flow cytometry. *Cytometry* **3**:129–133.

70. Matsiota-Bernard, P., G. Vrioni, and E. Marinis. 1998. Characterization of rpoB mutations in rifampin-resistant clinical *Mycobacterium tuberculosis* isolates from Greece. *J. Clin. Microbiol.* **36**:20–23.

71. Matthias, F., J. Reidl, and U. Vogel. 1998. Genomics and infectious diseases: approaching the pathogens. *Trends Microbiol.* **6**:346–349.

72. McNicholl, J. 1998. Host genes and infectious diseases. *Emerg. Infect. Dis.* **4**:423–426.

73. Mein, C. A., B. J. Barratt, M. G. Dunn, T. Siegmund, A. N. Smith, L. Esposito, S. Nutland, H. E. Stevens, A. J. Wilson, M. S. Phillips, N. Jarvis, S. Law, M. de Arruda, and J. A. Todd. 2000. Evaluation

of single nucleotide polymorphism typing with invader on PCR amplicons and its automation. *Genome Res.* **10:**330–343.

74. **Moss, D. M., G. P. Croppo, S. Wallace, and G. S. Visvesvara.** 1999. Flow cytometric analysis of microsporidia belonging to the genus *Encephalitozoon. J. Clin. Microbiol.* **37:**371–375.

75. **Niesters, H. G., J. van Esser, E. Fries, K. C. Wolthers, J. Cornelissen, and A. D. Osterhaus.** 2000. Development of a real-time quantitative assay for detection of Epstein-Barr virus. *J. Clin. Microbiol.* **38:**712–715.

76. **Nolte, F. S.** 1998. Branched DNA signal amplification for direct quantitation of nucleic acid sequences in clinical specimens. *Adv. Clin. Chem.* **33:**201–235.

77. **Nurmi, J., A. Ylikoski, T. Soukka, M. Karp, and T. Lovgren.** 2000. A new label technology for the detection of specific polymerase chain reaction products in a closed tube. *Nucleic Acids Res.* **28:**E28.

78. **O'Duffy, J. D., W. L. Griffing, C. Li, M. F. Abdel-malek, and D. H. Persing.** 1999. Whipple's arthritis. *Arthritis Rheum.* **42:**812–817.

79. **Orum, H., C. Kessler, and T. Koch.** 1997. Peptide nucleic acid, p. 29–48. *In* H. Lee, S. Morse, and O. Olsvik (ed.), *Nucleic Acid Amplification Technologies: Application to Disease Diagnosis.* Biotechniques Book/Eaton Publishing, Natick, Mass.

80. **Osipov, G. A., and A. M. Demina.** 1996. Chromatographic mass spectrometry detection of microorganisms in anaerobic infectious processes. *Vestn. Ross. Akad. Med. Nauk* **2:**52–59.

81. **Padilla, E., J. M. Manterola, O. F. Rasmussen, J. Lonca, J. Dominguez, L. Matas, A. Hernandez, and V. Ausina.** 2000. Evaluation of a fluorescence hybridisation assay using peptide nucleic acid probes for identification and differentiation of tuberculous and non-tuberculous mycobacteria in liquid cultures. *Eur. J. Clin. Microbiol. Infect. Dis.* **19:**140–145.

82. **Palleroni, N. J.** 1997. Prokaryotic diversity and the importance of culturing. *Antonie Leeuwenhoek* **72:**3–19.

83. **Perkins, B. A.** 1998. Explaining the unexplained in clinical infectious diseases: looking forward. *Emerg. Infect. Dis.* **4:**395–397.

84. **Persidis, A.** 1998. Biochips. *Nat. Biotechnol.* **16:**981–983.

85. **Persidis, A.** 1998. Proteomics. *Nat. Biotechnol.* **16:**393–394.

86. **Petrich, A. K., K. E. Luinstra, D. Groves, M. A. Chernesky, and J. B. Mahony.** 1999. Direct detection of vanA and vanB genes in clinical specimens for rapid identification of vancomycin resistant enterococci (VRE) using multiplex PCR. *Mol. Cell. Probes* **13:**275–281.

87. **Piatek, A. S., et al.** 1998. Detecting drug-resistant tuberculosis: beacons in the dark. *Nat. Biotechnol.* **16:**359–363.

88. **Pietila, J., Q. He, J. Oksi, and M. K. Viljanen.** 2000. Rapid differentiation of *Borrelia garinii* from *Borrelia afzelii* and *Borrelia burgdorferi* sensu stricto by LightCycler fluorescence melting curve analysis of a PCR product of the *recA* gene. *J. Clin. Microbiol.* **38:**2756–2759.

89. **Pozzi, G., M. Meloni, E. Iona, G. Orru, O. F. Thoresen, M. L. Ricci, M. R. Oggioni, L. Fattorini, and G. Orefici.** 1999. *rpoB* mutations in multidrug-resistant strains of *Mycobacterium tuberculosis* isolated in Italy. *J. Clin. Microbiol.* **37:**1197–1199.

90. **Ramers, C., G. Billman, M. Hartin, S. Ho, and M. H. Sawyer.** 2000. Impact of a diagnostic cerebrospinal fluid enterovirus polymerase chain reaction test on patient management. *JAMA* **283:**2680–2685.

91. **Reischl, U., H. J. Linde, M. Metz, B. Leppmeier, and N. Lehn.** 2000. Rapid identification of methicillin-resistant *Staphylococcus aureus* and simultaneous species confirmation using real-time fluorescence pCR. *J. Clin. Microbiol.* **38:**2429–2433.

92. **Relman, D. A.** 1998. Detection and identification of previously unrecognized microbial pathogens. *Emerg. Infect. Dis.* **4:**382–388.

93. **Relman, D. A., J. S. Loutit, T. M. Schmidt, S. Falkow, and L. S. Tompkins.** 1990. The agent of bacillary angiomatosis. An approach to the identification of uncultured pathogens. *N. Engl. J. Med.* **323:**1573–1580.

94. **Relman, D. A., T. M. Schmidt, R. P. MacDermott, and S. Falkow.** 1992. Identification of the uncultured bacillus of Whipple's disease. *N. Engl. J. Med.* **327:**293–301.

95. **Ririe, K. M., R. P. Rasmussen, and C. T. Wittwer.** 1997. Product differentiation by analysis of DNA melting curves during polymerase chain reaction. *Anal. Biochem.* **245:**154–160.

96. **Roth, B. L., M. Poot, S. T. Yue, and P. J. Millard.** 1997. Bacterial viability and antibiotic susceptibility testing with SYTOX green nucleic acid stain. *Appl. Environ. Microbiol.* **63:**2421–2431.

97. **Ryan, D., B. Nuccie, and D. Arvan.** 1999. Non-PCR-dependent detection of the factor V Leiden mutation from genomic DNA using a homogeneous invader microtiter plate assay. *Mol. Diagn.* **4:**135–144.

98. **Saizieu, A., U. Certa, J. Warrington, C. Gray, W. Keck, and J. Mous.** 1998. Bacterial transcript imaging by hybridization of total RNA to oligonucleotide arrays. *Nat. Biotechnol.* **16:**45–48.

99. **Sawata, S., E. Kai, K. Ikebukuro, T. Iida, T. Honda, and I. Karube.** 1999. Application of peptide nucleic acid to the direct detection of deoxyribonucleic acid amplified by polymerase chain reaction. *Biosens. Bioelectron.* **14:**397–404.

100. **Shah, J. S., J. Liu, D. Buxton, A. Hendricks, L. Robinson, G. Radcliffe, W. King, D. Lane, D. M. Olive, and J. D. Klinger.** 1995. Q-beta replicase-amplified assay for detection of *Mycobacterium tuberculosis* directly from clinical specimens. *J. Clin. Microbiol.* **33:**1435–1441.

101. **Sharma, V. K., E. A. Dean-Nystrom, and T. A. Casey.** 1999. Semi-automated fluorogenic PCR assays (TaqMan) for rapid detection of *Eschericia coli* O157:H7 and other shiga toxigenic *E. coli. Mol. Cell. Probes* **13:**291–302.

102. **Smith, L. M., J. Z. Sanders, R. J. Kaiser, P. Hughes, C. Dodd, C. R. Connell, C. Heiner, S. B. H. Kent, and L. E. Hood.** 1986. Fluorescence detection in automated DNA sequence analysis. *Nature* **321:**674–679.

103. **Smith, P. L., C. R. Walker Peach, R. J. Fulton, and D. B. DuBois.** 1998. A rapid, sensitive, multiplexed assay for detection of viral nucleic acids using the FlowMetrix system. *Clin. Chem.* **44:**2054–2056.

104. **Snewin, V. A., and D. W. Holden.** 1999. Signs and portents: molecular signals and infectious diseases. *Trends Microbiol.* **7:**62–63.

105. **Speel, E. J.** 1999. Robert Feulgen Prize Lecture 1999. Detection and amplification systems for sensitive, multiple-target DNA and RNA in situ hybridization:

looking inside cells with a spectrum of colors. *Histochem. Cell Biol.* **112**:89–113.

106. **Speel, E. J., A. H. Hopman, and P. Komminoth.** 1999. Amplification methods to increase the sensitivity of in situ hybridization: play card(s). *J. Histochem. Cytochem.* **47**:281–288.

107. **Stender, H., T. A. Mollerup, K. Lund, K. H. Petersen, P. Hongmanee, and S. E. Godtfredsen.** 1999. Direct detection and identification of *Mycobacterium tuberculosis* in smear-positive sputum samples by fluorescence in situ hybridization (FISH) using peptide nucleic acid (PNA) probes. *Int. J. Tuberc. Lung Dis.* **3**:830–837.

108. **Stryer, L.** 1978. Fluorescence energy transfer as a spectroscopic ruler. *Annu. Rev. Biochem.* **47**:819–846.

109. **Tang, Y., J. R. Hibbs, K. R. Tau, Q. Qian, H. A. Skarhus, T. F. Smith, and D. H. Persing.** 1999. Effective use of polymerase chain reaction for the diagnosis of central nervous system infections. *Clin. Infect. Dis.* **29**:803–806.

110. **Tang, Y., and D. H. Persing.** 1999. Molecular detection and identification of microorganisms, p. 215–244. *In* P. R. Murray, E. J. Baron, M. A. Pfaller, F. C. Tenover, and R. H. Yolken (ed.), *Manual of Clinical Microbiology*, 7th ed. ASM Press, Washington, D.C.

111. **Tomb, J.-F.** 1998. A panoramic view of bacterial transcription. *Nat. Biotechnol.* **16**:23.

112. **Troesch, A., H. Nguyen, C. G. Miyada, S. Desvarenne, T. R. Gingeras, P. M. Kaplan, P. Cros, and C. Mabilat.** 1999. Mycobacterium species identification and rifampin resistance testing with high-density DNA probe arrays. *J. Clin. Microbiol.* **37**:49–55.

113. **Tyagi, S., D. P. Bratu, and R. F. Kramer.** 1998. Multicolor molecular beacons for allele discrimination. *Nat. Biotechnol.* **16**:49–53.

114. **Tyagi, S., and R. F. Kramer.** 1996. Molecular beacons: probes that fluoresce upon hybridization. *Nat. Biotechnol.* **14**:303–308.

115. **Vahey, M., M. E. Nau, S. Barrick, J. D. Cooley, R. Sawyer, A. A. Sleeker, P. Vickerman, S. Bloor, B. Larder, N. L. Michael, and S. A. Wegner.** 1999. Performance of the Affymetrix GeneChip HIV PRT 440 Platform for antiretroviral drug resistance genotyping of human immunodeficiency virus type 1 clades and viral isolates with length polymorphisms. *J. Clin. Microbiol.* **37**:2533–2537.

116. **Vandamme, P., B. Pot, M. Gillis, P. de Vos, K. Kersters, and J. Swings.** 1996. Polyphasic taxonomy, a consensus approach to bacterial systematics. *Microbiol. Rev.* **60**:407–438.

117. **Wayne, L. G., R. C. Good, E. C. Bottger, R. Butler, M. Dorsch, T. Ezaki, W. Gross, V. Jonas, J. Kilburn, P. Kirschner, M. I. Krichevsky, M. Ridell, T. M. Shinnick, B. Springer, E. Stackebrandt, I. Tarnok, Z. Tarnok, H. Tasaka, V. Vincent, N. G. Warren, C. A. Knott, and R. Johnson.** 1996. Semantide- and chemotaxonomy-based analyses of some problematic phenotypic clusters of slowly growing mycobacteria, a cooperative study of the International Working Group on Mycobacterial Taxonomy. *Int. J. Syst. Bacteriol.* **46**:280–297.

118. **Wedin, R.** 1999. Visual data mining speeds drug discovery. *Mod. Drug Discov.* **2**:39–47.

119. **Wiedbrauk, D. L., and R. L. Hodinka.** 1998. Applications of the polymerase chain reaction, p. 97–112. *In* S. Specter, M. Bendinelli, and H. Friedman (ed.), *Rapid Detection of Infectious Agents.* Plenum Press, New York, N.Y.

120. **Wilson, K. H.** 1994. Detection of culture-resistant bacterial pathogens by amplification and sequencing of ribosomal DNA. *Clin. Infect. Dis.* **18**:958–962.

121. **Wittwer, C. T., M. G. Herrmann, A. A. Moss, and R. P. Rasmussen.** 1997. Continuous fluorescence monitoring of rapid cycle DNA amplification. *Bio/Techniques* **22**:130–138.

122. **Wittwer, C. T., K. M. Ririe, R. V. Andrew, D. A. David, R. A. Gundry, and U. J. Bali.** 1997. The Lightcycler: a microvolume multisample fluorimiter with rapid temperature control. *Bio/Techniques* **22**:176–181.

123. **Zheng, X., and D. H. Persing.** 1998. Genetic amplification techniques for diagnosing infectious diseases, p. 69–81. *In* S. Specter, M. Bendinelli, and H. Friedman (ed.), *Rapid Detection of Infectious Agents.* Plenum Press, New York, N.Y.

124. **Zhu, H., J. P. Cong, G. Matmora, T. Gingeras, and T. Shenk.** 1998. Cellular gene expression altered by human cytomegalovirus: global monitoring with oligonucleotide arrays. *Proc. Natl. Acad. Sci. USA* **95**:14470–14475.

APPENDIX

Manufacturers and Distributors

JAMSHID MOGHADDAS AND ALLAN L. TRUANT

In this section are listed many of the vendors described in the previous chapters. We have listed them in alphabetical order and have not grouped them with respect to the products they offer, since many companies offer a wide array of products, which would cause much duplication.

When reviewing and evaluating vendors for selected products, tests, or instruments, the reader is referred to the specific chapter for a discussion of which vendors offer the product in which the reader is interested. We attempt here to list the vendor address, telephone number, toll-free telephone number (if available), fax number, and website and e-mail addresses (when available). Only those companies which offer clinical microbiology products are listed here.

We encourage the reader to contact the editor or publisher with any errors, changes, or additional vendors which should be added or corrected in future versions of this manual.

As emphasized in other parts of this manual, the users of any kit, reagent, instrument, or product must evaluate the performance of each product in their laboratory or setting, investigate the U.S. Food and Drug Administration review status, and use the product within its stated shelf life and for the recommended clinical uses.

AATI (Advanced Analytical Technologies, Inc.)
2901 S. Loop Dr.
Ames, IA 50010
Phone: (515) 296-5315
Fax: (515) 296-6789
Web: http://www.advanced-analytical.com
E-mail: dpawelko@aati-us.com

AB Applied Biosystems
850 Lincoln Center Dr.
Foster City, CA 94404
Phone: (800) 248-0281, (800) 545-7547
Fax: (650) 638-6274
Web: http://www.appliedbiosystems.com

AB BIODISK North America, Inc.
200 Centennial Ave., Suite 206
Piscataway, NJ 08854-3910
Phone: (908) 457-8408, (800) 874-8814
Fax: (908) 457-8980, (732) 457-8980
E-mail: Etest@abbiodisk.se

Abbott Laboratories
100 Abbott Park Rd.
Abbott Park, IL 60064
Phone: (800) 323-9100 (U.S.),
  (800) 268-2349 (Canada)
Fax: (847) 938-3616
Web: http://www.abbott.com

ABI Biotechnologies
Rivers Park II
9108 Guilford Rd.
Columbia, MD 21046
Phone: (800) 426-0764
Fax: (301) 497-9773
Web: http://www.abionline.com

AccuDx, Inc.
9466 Black Mountain Rd.
Suite 130
San Diego, CA 92126
Phone: (858) 271-7429
Fax: (858) 777-3600
Web: http://www.accudx.com
E-mail: sales@accudx.com

ACGT, Inc.
1955 Raymond Dr.
Suite 104
Northbrook, IL 60062
Phone: (847) 559-8631
Fax: (847) 559-8632
Web: http://www.acgtinc.com
E-mail: dnaseq@acqt.com

ACON Laboratories, Inc.
11175 Flintkote Ave., Suite F
San Diego, CA 92121
Phone: (858) 535-2030
Fax: (858) 535-2035

Web: http://www.aconlabs.com
E-mail: info@aconlabs.com

A. Daigger & Company
620 Lakeview Pkwy.
Vernon Hills, IL 60061
Phone: (800) 621-7193
Fax: (800) 320-7200
Web: http://www.daigger.com
E-mail: daigger@daigger.com

Affymetrix, Inc.
3380 Central Expwy.
Santa Clara, CA 95051
Phone: (408) 731-5503
Fax: (408) 481-9442
Web: http://www.affymetrix.com
E-mail: sales@affymetrix.com

Agen Biomedical Limited
11 Durbell St.
Acacia Ridge
Brisbane, Queensland 4110
Australia
Phone: 61-7-3370-6300
Fax: 61-7-3370-6370
Web: http://www.agen.com.au
E-mail: mail@agen.com.au

Agilent Technologies
P.O. Box 10395
Palo Alto, CA 94306

Phone: (800) 227-9770
Web: http://www.agilent.com
E-mail: webmaster@agilent.com

Air Sea Atlanta, Incorporated
1234 Logan Cir.
Atlanta, GA 30318
Phone: (404) 351-8600
Fax: (404) 351-4005
Web: http://www.airseaatlanta.com
E-mail: proane@airseaatlanta.com

Alexon-Trend
14000 Unity St. N.W.
Ramsey, MN 55303-9115
Phone: (800) 366-0096, (612) 323-7800
Fax: (612) 712-2371
Web: http://www.alexon-trend.com
E-mail: enutter@alexon-trend.com

Allegiance Healthcare Corp.
1450 Waukegan Rd.
McGaw Park, IL 60085-6786
Phone: (800) 964-5227
Fax: (856) 339-0443
Web: http://www.allegiance.net

Alliance for the Prudent Use of Antibiotics
P.O. Box 1372
Boston, MA 02117-1372
Phone: (617) 636-0966
Fax: (617) 636-3999
Web: http://www.apua.org
E-mail: apua@opal.tufts.edu

Alpha Innotech Corporation
14743 Catalina St.
San Leonard, CA 94577
Phone: (800) 795-5556
Fax: (510) 483-3227
Web: http://www.alphainnotech.com
E-mail: info@aicemail.com

Alpha-Tec Systems, Incorporated
P.O. Box 5435
Vancouver, WA 98668-5435
Phone: (360) 260-2779, (800) 221-6058
Fax: (360) 260-3277
E-mail: info@alphatecsystems.com

Ambion, Incorporated
2130 Woodward St.
Austin, TX 78744-1832
Phone: (512) 651-0200, (800) 888-8804 (U.S.); (800) 445-1161 (Canada)
Fax: (512) 651-0201
Web: http://www.ambion.com

Amerex Instruments, Inc.
P.O. Box 787
Lafayette, CA 94549
Phone: (510) 937-0182
Fax: (510) 937-0950
Web: http://www.amerexinst.com
E-mail: marketing@amerexinst.com

American Laboratory Products Company (ALPCO)
P.O. Box 451
Windham, NH 03087
Phone: (800) 592-5726
Fax: (603) 898-6854
Web: http://www.alpco.com
E-mail: email@alpco.com

American Radiolabeled Chemical Inc. (ARC)
11624 Bowling Green Dr.
St. Louis, MO 63146
Phone: (314) 991-4545, (800) 331-6661
Fax: (314) 991-4692, (800) 999-9925
Web: http://www.arc-inc.com
E-mail: arinc@arc-inc.com

American Type Culture Collection (ATCC)
10801 University Blvd.
Manassas, VA 20110-2209
Phone: (703) 365-2700
Fax: (703) 365-2701
Web: http://www.atcc.org

Amersham Pharmacia Biotech, Inc.
800 Centennial Ave.
P.O. Box 1327
Piscataway, NJ 08855-1327
Phone: (732) 457-8000
Fax: (732) 457-0557
Web: http://www.apbiotech.com

AMSCO International, Inc.
See Steris Corporation

Anaerobe Systems
15906 Concord Cir.
Morgan Hill, CA 95037
Phone: (408) 782-7557, (800) 443-3108
Fax: (408) 782-3031
Web: http://www.anaerobesystems.com
E-mail: mikecox@anaerobesystems.com

Antech Diagnostics
10 Executive Blvd.
Farmingdale, NY 11735
Phone: (800) 745-4725 (West), (800) 872-1001 (East)
Web: http://www.antechdiagnostics.com

Applied Biosystems
PE Biosystems Division Headquarters
850 Lincoln Center Dr.
Foster City, CA 94404
Phone: (800) 345-5224, (650) 638-5800
Fax: (650) 638-5884
Web: http://www.appliedbiosystems.com

Applied Precision, Inc.
1040 12th Ave., N.W.
Issaquah, WA 98027
Phone: (425) 313-4557
Web: http://www.appliedprecision.com
E-mail: hotline@api.com

Appropriate Technical Resources, Inc. (ATR)
P.O. Box 460
Laurel, MD 20725-0460
Phone: (800) 827-5931
Web: http://www.atrbiotech.com
E-mail: info@atrbiotech.com

Argene Inc.
198 N. Queens Ave.
North Massapequa, NY 11758
Phone: (516) 795-5583, (888) 4ARGENE
Fax: (516) 795-4942
Web: http://www.argenbiosoft.com
E-mail: argeninc@argenbiosoft.com

Associates of Cape Cod, Inc.
704 Main St.
Falmouth, MA 02540
Phone: (888) 395-ACC1, (508) 540-3444
Fax: (508) 540-8680
Web: http://www.acciusa.com
E-mail: accinc@acciusa.com

Association of Public Health Laboratories (APHL)
1211 Connecticut Ave., N.W.
Suite 608
Washington, DC 20036
Phone: (202) 822-5227
Fax: (202) 887-5098
Web: http://www.aphl.org

Atlanta Biological
1425 Oakbrook Dr., Suite 400
Norcross, GA 30093
Phone: (800) 780-7788
Fax: (800) 780-7374
Web: http://www.atlantabio.com
E-mail: service@atlantabio.com

Autogen
35 Loring Dr.
Framingham, MA 01702
Phone: (800) 292-5678
Fax: (508) 875-5329
Web: http://www.autogen.com

Avanti Polar Lipids, Incorporated
700 Industrial Park Dr.
Alabaster, AL 35007-9105
Phone: (800) 227-0651
Fax: (800) 229-1004
Web: http://www.avantilipids.com

Avecon Diagnostics, Inc.
405 South Main St.
Coopersburg, PA 18036
Phone: (800) 249-5875

Bacterial Barcodes, Incorporated
8080 North Stadium Dr.
Suite 1200
Houston, TX 77054
Phone: (866) 473-7727

Fax: (713) 467-7766
Web: http://www.bacbarcodes.com
E-mail: info@bacterialbarcodes.com

Baker Company
P.O. Drawer E
Sanford Airport Rd.
Sanford, ME 04073
Phone: (800) 992-2537, (207) 324-8773
Fax: (207) 324-3869
Web: http://www.bakerco.com

Barnstead/Thermolyne Corp.
2555 Kerper Blvd.
Dubuque, IA 52001
Phone: (800) 553-0039
Fax: (847) 537-9228
Web:
   http://www.barnsteadthermolyne.com
E-mail: robertepf@aol.com

Bartels
(a Trinity Biotech Company)
5919 Farnsworth Ct.
Carlsbad, CA 92008
Phone: (800) 331-2291
Fax: (760) 929-0124

Bayer Corporation Diagnostic Division
511 Benedict Ave.
Tarrytown, NY 10591
Phone: (800) 348-8100, (800) 255-3232
Fax: (914) 524-2132
Web: http://www.bayerdiag.com

BBI Diagnostics
375 West St.
West Bridgewater, MA 02379
Phone: (800) 676-1881
Fax: (508) 580-2202
E-mail: jbarron@bbii.com

B. Brown Biotech Inc. USA
999 Postal Rd.
Allentown, PA 18103
Phone: (800) 258-9000, (610) 266-6262
Web: http://www.bbraunbiotech.com

BD Biosciences
7 Loveton Cir.
Sparks, MD 21152
Phone: (800) 638-8663, (410) 316-4000
Phone: (877) 362-2700 (Industrial
   Microbiology)
Web: http://www.bdms.com

Beckman Coulter, Inc.
4300 N. Harbor Blvd.
P.O. Box 3100
Fullerton, CA 92838-3100
Phone: (714) 871-4848
Fax: (714) 773-8283
Web: http://www.beckmancoulter.com

Becton Dickinson Vacutainer Systems
1 Becton Dr.
Franklin Lakes, NJ 07417-1885

Phone: (888) 237-2762
Fax: (800) 847-2220
Web: http://www.bd.com

Binax, Inc.
217 Read St.
Portland, ME 04103
Phone: (800) 323-3199
Fax: (207) 761-2074
Web: http://www.binax.com
E-mail: johno@binax.com

The Binding Site, Incorporated
5889 Oberlin Dr., #101
San Diego, CA 92121
Phone: (858) 453-9177
Fax: (858) 453-9189
Web: http://www.bindingsite.com
E-mail: info@thebindingsite.com

Bio 101, Inc.
1070 Joshua Way
Vista, CA 92083
Phone: (760) 598-7299
Fax: (760) 598-0116
Web: http://www.bio101.com

BioChem ImmunoSystems
754 Roble Rd., Suite 70
Allentown, PA 18109
Phone: (800) 549-7423
Fax: (610) 264-8102

BioControl Systems, Inc.
12822 S.E. 32nd St.
Bellevue, WA 98005
Phone: (425) 603-1123
Fax: (425) 603-0080
Web: http://www.rapidmethods.com

Biodesign International
105 York St.
Kennebunk, ME 04043
Phone: (207) 985-1944
Fax: (207) 985-6322
Web: http://www.biodesign.com
E-mail: info@biodesign.com

BioFx
BioFX Laboratories, L.L.C.
9633 Liberty Rd., Suite S
Randallstown, MD 21133
Phone: (410) 496-6006, (800) 445-6447
Fax: (410) 496-6008
Web: http://www.biofx.com
E-mail: biofx@biofx.com

Biogenesis
P.O. Box 1016
Kingston, NH 03848
Phone: (603) 642-8302
Fax: (603) 642-8322
Web: http://www.biogenesis.co.uk
E-mail: biogenesis@sprintmail.com

Biokit
08186 Llissa d'Amunt
Barcelona
Spain
Phone: 34-93-860-90-00
Fax: 34-93-860-90-17
Web: http://www.biokit.com

BioLabs Inc.
32 Tozer Rd.
Beverly, MA 01915-5510
Phone: (978) 927-5054
Fax: (978) 922-7085
Web: http://www.neb.com
E-mail: info@neb.com

Biolog, Inc.
3938 Trust Way
Hayward, CA 94545
Phone: (510) 785-2564
Fax: (510) 782-4639
Web: http://www.biolog.com
E-mail: info@biolog.com

BioMed Diagnostics, Inc.
1430 Koll Cir., Suite 101
San Jose, CA 95112
Phone: (408) 451-0400, (800) 964-6466
Fax: (408) 456-0409
Web: http://www.biomed1.com
E-mail: info@biomed1.com

Bio-Medical Products Corp.
10 Halstead Rd.
Mendham, NJ 07945
Phone: (973) 292-5100
Fax: (973) 539-3476
Web: http://www.biomedicalproducts.com
E-mail: Biomedicalproducts@msn.com

BioMérieux SA
69280 Marcy L'Etoile
France
Phone: 33 4 78 87 20 00
Fax: 33 4 78 87 20 90
Web: http://www.biomerieux.fr

BioMerieux Vitek, Inc.
595 Anglum Rd.
Hazelwood, MO 63042-2320
Phone: (314) 731-8500, (800) 638-4835
Fax: (314) 731-8700, (800) 325-1598
Web: http://www.biomerieux-vitek.com
   http://www.biomerieux.com
E-mail: bmxvitek@vitek.com

Bionor AS
Stromdaljordet 4
P.O. Box 1868
N-3703 Skien
Norway
Phone: 47-35-50-57-50
Fax: 47-35-50-57-01
Web: http://www.bionor.com

Bio PLAS, Inc.
Corporate Office
4200 California St.
Suite 100
San Francisco, CA 94118
Phone: (415) 668-8917
Fax: (415) 668-8354
Canadian Sales Office
4659 Albion Rd.
Gloucester, Ontario K1X 1A4
Canada
Phone: (613) 822-3107
Fax: (613) 822-3109
Web: http://www.bioplas.com
E-mail: bioplas@aol.com

Bio-Rad Laboratories, Inc.
Clinical Diagnostics Group
4000 Alfred Nobel Dr.
Hercules, CA 94547
Phone: (800) 424-6723, (800) 2BIORAD
Fax: (510) 741-5824
Web: http://www.bio-rad.com

Biosite Diagnostics, Inc.
11030 Roselle St.
San Diego, CA 92121
Phone: (800) 745-8026, (858) 455-4808
Fax: (858) 455-4815
Web: http://www.biosite.com

BioSource International
542 Flynn Rd.
Camarillo, CA 93012
Phone: (800) 242-0607
Fax: (805) 987-3385
Web: http://www.biosource.com
E-mail: tech.support.@Biosource.com

BioStar, Inc.
6655 Lookout Rd.
Boulder, CO 80301
Phone: (800) 637-3717, (303) 530-3888
Fax: (303) 530-6601
Web: http://www.biostar.com

Biosynth International, Inc.
1665 W. Quincy, Suite 155
Naperville, IL 60540
Phone: (800) 270-2436
Web: http://www.biosynth.com
E-mail: Welcome@biosynth.ch

Bio-Tek Instruments
Highland Park
Box 998
Winooski, VT 05404-0998
Phone: (802) 655-4040,
(800) 451-5172
Fax: (802) 655-7941
Web: http://www.biotek.com

Biotest Diagnostic Corporation
66 Ford Rd., Suite 131
Danville, NJ 07834
Phone: (800) 522-0090,
(973) 625-1300

Fax: (973) 625-9454
Web: http://www.biotest.com
E-mail: CustomerService@biotest.com

Biotrace Ltd.
The Science Park
Bridgend CF31 3NA
Wales
United Kingdom
Phone: 44 1656 64 14 00
Fax: 44 1656 76 88 35
Web: http://www.biotrace.com,
http://www.biotrace.co.uk

Biotrin
The Rise
Mount Merrion
County Dublin
Ireland
Phone: 353 1 283 1166
Fax: 353 1 283 1232
Web: http://www.biotrin.ie

Biotronics
50 Stedman St.
Lowell, MA 01850
Phone: (800) 722-8436
Fax: (978) 459-7809
Web: http://www.biotronics.com

BioWhittaker, Inc.
8830 Biggs Ford Rd.
Walkersville, MD 21793
Phone: (301) 898-7025, (800) 638-8174
Fax: (301) 845- 4024
Web: http://www.clonetics.com
E-mail: inet@biowhittaker.com

Blackhawk BioSystems, Inc.
12945 Alcosta Blvd.
San Ramon, CA 94583
Phone: (510) 866-1458
Fax: (510) 866-2941
Web:
http://www.blackhawkbiosystems.com
E-mail:
webmaster@blackhawkbiosystems.com

Boston Biomedica, Inc.
375 West St.
West Bridgewater, MA 02379
Phone: (508) 580-1900, (800) 676-1881
Fax: (508) 580-2202
Web: http://www.bbii.com
E-mail: bmatevich@bbii.com

Brinkmann Instruments, Incorporated
One Cantiague Rd.
P.O Box 1019
Westbury, NY 11590-0207
Phone: (800) 645-3030, (516) 334-7500
Fax: (516) 334-7506
Web: http://www.brinkmann.com
E-mail: info@brinkman.com

Bristol-Myers Squibb Company
777 Scudders Mill Rd.

Plainsboro, NJ 08536
Web: http://www.bms.com

Butler
5000 Bradenton Ave.
Dublin, OH 43017
Phone: (800) 258-2148
Fax: (614) 761-9096

Calbiochem-Novabiochem International, Inc.
P.O. Box 12087
La Jolla, CA 92039-2087
Phone: (800) 854-3417
Fax: (614) 761-9096
Web: http://www.calbiochem.com

Calypte Biomedical Corporation
1265 Harbor Bay Pkwy.
Alameda, CA 94502
Phone: (510) 749-5100, (877) 225-9783
Fax: (510) 814-8408
Web: http://www.calypte.com

Capricorn Products, Inc.
301 U.S. Route 1
Scarborough, ME 04074
Phone: (207) 885-0480
Fax: (207) 885-0494
Web: http://www.capricornproducts.com
E-mail: info@capricornproducts.com

Carl Zeiss, Inc., Microscopy Division
One Zeiss Dr.
Thornwood, NY 10594
Phone: (800) 233-2343 (U.S.),
(416) 449-4660 (Canada)
Fax: (914) 681-7446 (U.S.),
(914) 681-7465 (Canada)
Web: http://www.zeiss.com
E-mail: micro@zeiss.com \ micro

C.B.S. Scientific Company, Inc.
P.O. Box 856
Del Mar, CA 92014
Phone: (858) 755-4959
Fax: (858) 755-0733
Web: http://www.cbsscientific.com

Cell Press
1050 Massachusetts Ave.
Cambridge, MA 02138
Phone: (617) 661-7057
Fax: (617) 661-7061
Web: http://www.cellpress.com
E-mail: mark@cell.com

Centaur, Inc.
P.O. Box 25667
Overland Park, KS 66225-5667

CENTERCHEM, Inc.
High Ridge Rd.
Stamford, CT 06905
Phone: (203) 975-9800
Fax: (203) 975-8777
E-mail: Salley@centerchem.com

Center for Scientific Review (CSR)
National Institutes of Health
6701 Rockledge Dr.
Bethesda, Maryland 20892
Phone: (301) 435-0715
Fax: (301) 480-1987
Web: http://www.csr.nih.gov

Cepheid
1190 Borregas Ave.
Sunnyvale, CA 94089
Phone: (408) 541-4191
Fax: (408) 541-4192
Web: http://www.cepheid.com

Chembio Diagnostic Systems, Inc.
3661 Horseblock Rd.
Medford, NY 11763
Phone: (631) 924-1135
Fax: (631) 924-6033
Web: http://www.chembio.com

Chemicon International, Inc.
28835 Single Oak Dr.
Temecula, CA 92590
Phone: (800) 437-7500, (909) 676-8080
Fax: (909) 676-9209
Web: http://www.chemicon.com
E-mail: custserv@chemicon.com

Chemunex, Inc
1 Deer Park Dr., Suite H-2
Monmouth Junction, NJ 08852
Phone: (800) 411-6734, (732) 329-1153
Fax: (732) 329-1192
Web: http://www.chemunex.com

Chiron Diagnostics
4560 Horton St.
Emeryville, CA 94608
Phone: (510) 655-8730
Fax: (510) 655-9910
Web: http://www.chiron.com
E-mail: corpcomm@cc.chiron.com

Chromagar
267 Rue Lecourbe
Paris
France F-75015
Web: http://www.chromagar.com

Cogent Technologies, Limited
11140 Luschek Dr.
Cincinnati, OH 45241

Combact Diagnostic Systems, Limited
60 Medinat Hayehudim
P.O.B. Herzliya 46120
Israel

Copan Diagnostics Inc.
2175 Sampson Ave., Suite 124
Corona, CA 91719
Phone: (800) 216-4016, (909) 549-8793
Fax: (909) 549-8850

Web: http://www.copan.com
E-mail: info@copan.com

Corning Inc., Science Products Division
45 Nagog Park
Acton, MA 01720
Web: http://www.corning.com

CORTEX BIOCHEM
1933 Davis St., Suite 321
San Leandro, CA 94577
Phone: (800) 888-7713, (510) 568-2228
Fax: (510) 568-2467
Web: http://www.cortex-biochem.com
E-mail: cortbioc@ix.netcom.com

Covance Research Products, Inc.
310 Swampbridge Rd.
Denver, PA 17517

Coy Laboratory Products, Inc.
14500 Coy Dr.
Grass Lake, MI 49137
Phone: (734) 475-2200
Fax: (734) 475-1846
Web: http://www.coylab.com
E-mail: sales@coylab.com

CSL Limited
45 Poplar Rd.
Parkville, Victoria 3052
Australia
Phone: 61 3 9389 1911
Fax: 61 3 9389 1434

Current Technologies, Inc.
P.O. Box 21
Crawfordsville, IN 47933
Phone: (800) 456-4022, (765) 364-0490
Fax: (765) 364-1607
E-mail: currtech@iserve.net

Cytovax Biotechnologies Inc.
8925 51st Ave., Suite 308
Edmonton, Alberta T6E 5J3
Canada
Phone: (780) 448-0621, (800) 661-1426
Fax: (780) 448-0624
Web: http://www.cytovax.com
E-mail: biotech@cytovax.com

Dade Behring Microscan Microbiology
Systems
1717 Deerfield Rd.
West Sacramento, CA 95691
Phone: (800) 393-9362, (847) 267-5300
Web: http://www.dadebehring.com

DAKO Corporation
6392 Via Real
Carpinteria, CA 93013
Phone: (805) 566-6655
Fax: (805) 566-6688
Web: http://www.dakousa.com

Diagnostic Products Corporation (DPC)
5700 West 96th St.
Los Angeles, CA 90045-5597
Phone: (800) 986-1001, (310) 642-5180
Fax: (310) 642-0192
Web: http://www.dpcweb.com
E-mail: info@dpconline.com

DiaMedix Corporation
2140 N. Miami Ave.
Miami, FL 33127
Phone: (305) 324-2300, (800) 327-4565
Fax: (305) 324-2395
Web: http://www.diamedix.com

DiaSorin
1990 Industrial Blvd.
Stillwater, MN 55082
Phone: (800) 328-1482, (612) 439-9710
Fax: (651) 779-7847
Web: http://www.diasorin.com

Digene Corporation
1201 Clopper Rd.
Gaithersburg, MD 20878
Phone: (800) 344-3631, (301) 944-7000
Fax: (301) 944-7199
Web: http://www.digene.com
E-mail: inforeg@digene.com

Digital Instruments
112 Robin Hill Rd.
Santa Barbara, CA 93117
Phone: (800) 873-9750, (800) 967-1400
Fax: (805) 967-7717
Web: http://www.di.com
E-mail: info@di.com

DIOMED
Sanayi cad no: 19,81260
Dudullu
Istanbul
Turkey
Phone: 90.216.420 03 73
Fax: 90.216.466 63 01
Web: http://www.diomed.com.tr
E-mail: nyildiz@superonline.com.tr

Drummond Scientific
500 Parkway, Box 700
Broomall, PA 19008
Phone: (800) 523-7480, (610) 353-0200
Fax: (610) 353-6204
Web: http://www.drummondsci.com
E-mail: info@drummondsci.com

Dynex Technologies, Inc.
See ThermoLabsystems

Epicentre Technologies Corporation
1202 Ann St.
Madison, WI 53716
Phone: (608) 258-3080
Fax: (608) 258-3099
Web: http://www.epicentre.com

Eppendorf
5603 Arapahoe
Boulder, CO 80303
Phone: (303) 583-7007, (800) 533-5703
Fax: (303) 440-0835
Web: http://www.eppendorf.com

EQUITECH-BIO
306 Hwy. 27W
Ingram, TX 78025
Phone: (830) 367-2027, (800) 259-0591
Fax: (830) 367-5292
Web: http://www.equitech-bio.com
E-mail: equitechbio@msn.com

European Collection of Cell Cultures
See United Kingdom National Culture
  Collection

Evergreen Scientific
2254 E. 49th St.
P.O. Box 58248
Los Angeles, CA 90058
Phone: (323) 583-1331, (800) 421-6261
Fax: (323) 581-2503
Web: http://www.evergreensci.com

Ever Ready Thermometer Company Inc.
228 Lackawanna Ave.
West Paterson, NJ 07424

Evsco Pharmaceuticals
P.O. Box 685
Harding Hwy.
Buena, NJ 06310
Phone: (800) 387-2607
Fax: (856) 697-9711
Web: http://www.evscopharm.com

EY Laboratories, Inc.
107 North Amphlett Blvd.
San Mateo, CA 94401
Phone: (650) 342-3296, (800) 821-0044
Fax: (650) 342-2648
Web: http://www.eylabs.com

Fisher Scientific Company L.L.C.
2000 Park Lane Dr.
Pittsburgh, PA 15275-1126
Phone: (800) 640-0640
Fax: (800) 290-0290
Web: http://www.fishersci.com

Forma Scientific, Inc.
P.O. Box 649
Marietta, OH 45750
Phone: (740) 373-4763, (800) 848-3080
Fax: (740) 373-6770
Web: http://www.forma.com

Fotodyne, Inc.
950 Walnut Ridge Dr.
Heartland, WI 53029-9388
Phone: (800) 362-3686
Fax: (800) 362-3642
Web: http://www.fotodyne.com
E-mail: customer-service@fotodyne.com

FTS Systems, Inc.
3538 Main St.
P.O. Box 158
Stone Ridge, NY 12484
Phone: (800) 824-0400
Fax: (914) 687-7481
Web: http://www.ftssystems.com
E-mail: info@ftssystems.com

Fujirebio America, Inc.
30 Two Bridges Rd., Suite 250
Fairfield, NJ 07004-1550
Phone: (973) 227-8888, (888) 499-9998
Fax: (973) 227-8585
Web: http://www.fujirebioamerica.com

Gelman Science
600 South Wagner Rd.
Ann Arbor, MI 48103-9019
Phone: (800) 521-1520
Fax: (734) 913-6114
Web: http://www.pall.gelman.com

Genelabs Diagnostics, Inc.
505 Penobscot Dr.
Redwood City, CA 94063
Phone: (650) 369-9500
Fax: (650) 369-6154
Web: http://www.genelabs.com

Genomic Solutions
525 Avis Dr.
Ann Arbor, MI 48108
Phone: (877) 436-6642
Fax: (734) 975-4804
Web: http://www.genomicsolutions.com
E-mail: info@genomicsolutions.com

Gen-Probe, Inc.
10210 Genetic Center Dr.
San Diego, CA 92121-4362
Phone: (800) 523-5001 (U.S.), (800)
  342-7441 (Canada)
Fax: (800) 288-3141
Web: http://www.gen-probe.com

Gentra System
13355 10th Ave. N., Suite 120
Minneapolis, MN 55441
Phone: (800) 866-3039
Fax: (763) 543-0699
Web: http://www.gentra.com
E-mail: info@gentra.com

Genzyme Diagnostics
1531 Industrial Rd.
San Carlos, CA 94070
Phone: (800) 717-6314
Fax: (650) 594-0571
Web: http://www.genzyme.com
E-mail: Medixbio@aol.com

Genzyme Virotech GmbH
Löwenplatz 5
D-65428 Rüsselsheim
Germany

Phone: 49 6142-69090
Fax: 49 6142-82621

Getinge/Castle Inc.
1777 East Henrietta Rd.
Rochester, NY 14623-3133
Phone: (716) 475-1400
Fax: (716) 272-5033
Web: http://www.getinge.com
E-mail: info@getingecastle.com

G.F.M.D. Limited
22650 Heslip
Novi, MI 48375
Phone: (800) 405-3600
Fax: (248) 305-6110

Gibson Laboratories, Inc.
1040 Manchester St.
Lexington, KY 40508
Phone: (800) 477-4763
Fax: (859) 253-1476
Web: http://www.gibsonlabs.com

Glycosynth
14 Craven Ct.
Winwick Quay
Warrington, Cheshire WA2 8QU
England
Phone: 44-1925 575075
Fax: 44-1925 575121
E-mail: info@glycosynth.co.uk

Gow-Mac Instrument Co.
277 Brodhead Rd.
Bethlehem, PA 18017
Phone: (610) 954-9000
Fax: (610) 954-0599
Web: http://www.gow-mac.com
E-mail: sales@GOW-MAC.com

Granbio, Inc.
P.O. Box 892140
Temecula, CA 92589-2140
Phone: (909) 676-0049

Greiner America, Inc.
P.O. Box 953279
Lake Mary, FL 32795-3279
Phone: (800) 884-4703, (407) 333-2800
Fax: (407) 333-3001
Web: http://www.greineramerica.com
E-mail: info@greineramerica.com

Grifols-Quest
1980 N.E. 147th St.
North Miami, FL 33181
Phone: (800) 379-0957
Fax: (703) 631-7816
Web: http://www.grifols.com

Hach Company
P.O. Box 389
Loveland, CO 80539
Phone: (800) 227-4224
Fax: (970) 207-1088
Web: http://www.hach.com

Hamilton Company
4970 Energy Way
Reno, NV 89502
Phone: (800) 648-5950, (775) 858-3000
Fax: (775) 856-7259
Web: http://www.hamiltoncompany.com
E-mail: rlund@hamiltoncompany.com

Hardy Diagnostics
1430 W. McCoy
Santa Monica, CA 93455
Phone: (805) 346-2766
Fax: (805) 346-2760
Web: http://www.hardydiagnostics.com
E-mail: sales@hardydiagnostics.com

Harris Manufacturing
275 Aiken Rd.
Asheville, NC 28804
Phone: (800) 221-4201
Fax: (828) 658-0363
Web: http://www.harrisphq.com
E-mail: mary.ramaglia@glse.gensig.com

Heathrow Scientific
620 Lakeview Pkwy.
Vernon Hills, IL 60061
Phone: (847) 478-9020, (800) 741-4597
Fax: (847) 478-9026
Web: http://www.heathrowscientific.com
E-mail:
    heathrow@heathrowscientific.com

Helica Biosystems, Inc.
223 Imperial Hwy., Suite 165
Fullerton, CA 92835
Phone: (877) 9HELICA
Fax: (714) 578-7831
Web: http://www.helica.com
E-mail: info@helica.com

Hemagen Diagnostics, Inc.
40 Bear Hill Rd.
Waltham, MA 02154
Phone: (800) 436-2436
Fax: (781) 890-3748
Web: http://www.hemagen.com
E-mail: cwilland@hemagen.com

Heska
1613 Prospect Pkwy.
Fort Collins, CO 80525
Phone: (970) 493-7272
Fax: (970) 484-9505
Web: http://www.heska.com

Hitachi Genetic Systems
1201 Harbor Bay Pkwy.
South San Francisco, CA 94080
Phone: (800) 624-6174, (650) 615-9600
Fax: (650) 615-7699
Web: http://www.hitachi-soft.com/gs
E-mail: gene@hitsoft.com

HTI Bio-Product, Inc.
P.O. Box 1319

Ramona, CA 92065
Phone: (800) 481-9737, (760) 788-9691
Fax: (760) 788-9694

ID Biomedical Corp.
8855 Northbrook Ct.
Burnaby, British Columbia V5J 5J1
Canada

IDEXX Laboratories, Inc.
One IDEXX Dr.
Westbrook, ME 04092
Phone: (207) 856-0300
Fax: (207) 856-0346
Web: http://www.idexx.com

IIT Research Institute
10 West 35th St.
Chicago, IL 60616
Phone: (312) 567-4000
Fax: (312) 567-4852
Web: http://www.iitri.org
E-mail: bgingras@iitri.org

IKA Works, Inc.
2635 North Chase Pkwy.
Wilmington, NC 28405
Phone (910) 452-7059
Fax: (910) 452-7693
Web: http://www.ika.net
E-mail: ysa@ika.net

IMMCO Diagnostics
60 Pineview Dr.
Buffalo, NY 14228-2120
Phone: (800) 537-TEST, (716) 691-0466
Fax: (716) 691-0466
Web: http://www.immcodiagnostics.com

ImmunCell
65 Evergreen Dr.
Portland, ME 04103
Phone: (800) 387-2607
Fax: (207) 878-2117
Web: http://www.bioportfolio.com

Immunetics, Inc.
63 Rogers St.
Cambridge, MA 02142

Immuno Concepts N.A., Limited
9779 Business Park Dr.
Sacramento, CA 95827
Phone: (916) 363-2649, (800) 251-5115
Web: http://www.immunoconcepts.com

Inmark, Inc.
220 Fisk Dr., S.W.
Atlanta, GA 30336

Innogenetics
2580 Westside Pkwy.
Suite 400
Alpharetta, GA 30004
Phone: (678) 393-1672
Fax: (678) 393-1673

Web: http://www.innogenetics.com
E-mail: brian_brody@innogenetics.com

INOVA Diagnostics, Inc.
10180 Scripps Ranch Blvd.
San Diego, CA 92131-1234
Phone: (619) 586-9900, (800) 545-9495
Fax: (619) 586-9911
Web: http://www.inovadx.com

The Institute for Genomic Research
9712 Medical Center Dr.
Rockville, MD 20850
Phone: (301) 838-0200
Fax: (301) 838-0208
Web: http://www.tigr.org

Institute POURQUIER
326, rue de la Galéra
Parc Euromédecine
34090 Montpellier
France
Phone: 33 4 99 23 24 25
Fax: 33 4 67 04 20 25
Web: http://www.institut-pourquier.fr
E-mail: institut.pourquier@wanadoo.fr

Institute Virion-Serion GmbH
Konradstrasse 1
D-97072 Wurzburg
Germany
Phone: 49 931 309860
Fax: 49 931-52650

Integra Biosciences, Inc.
10097 Tyler Pl., Suite 10
Ijamsville, MD 21754

Integrated Diagnostics, Inc.
1756 Sulphur Spring Rd.
Baltimore, MD 21227
Phone: (410) 737-8500, (800) TEC-INDX
Fax: (410) 536-1212
Web: http://www.indxdi.com

INTERGEN
202 Perry Pkwy.
Gaithersburg, MD 20877
Phone: (800) 431-4505, (301) 519-2170
Fax: (800) 468-7436, (301) 963-5017
Web: http://www.intergenco.com

International Dynamics Corporation (IDC)
17300 SW Upper Boones Ferry Rd.
Suite 120
Portland, OR 97224
Phone: (503) 684-8008, (800) 323-4810
Fax: (503) 684-9559
Web: http://www.teleport.com/~idclatex/
E-mail: idclatex@teleport.com

International Equipment Company (IEC)
300 Second Ave.
Needham Heights, MA 02494
Phone: (800) 843-1113, (781) 455-9729
Fax: (781) 444-6743

Web: http://www.labcentrifuge.com
E-mail: kathleen.dunleavy@iec-centrifuge.com

International Microbio
BP 705
83030 Toulon Cedex 9
France
Phone: 33 4 94 88 5 00
Fax: 33 4 94 88 55 22
Web: http://www.int-microbio.com

INTERSEP
205 North Collier Blvd.
Marco Island, FL 34145
Phone: (941) 642-1010
Fax: (941) 642-1040
Web: http://www.intersep.com

INTRACEL Corporation
Corporate Headquarters:
1330 Piccard Dr.
Rockville, MD 20850
Phone: (301) 258-5200
Fax: (301) 296-0082
Diagnostic Products and Sales:
2005 NW Sammamish Rd., Suite 107
Issaquah, WA 98027
Phone: (425) 392-2992
Fax: (425) 557-1894
Web: http://www.intracel.com
E-mail: info@intracel.com

Invitrogen
1600 Faraday Ave.
Carlsbad, CA 92008-9930
Phone: (800) 533-4363
Web: http://www.resgen.com

Irving Scientific
2511 Daimler St.
Santa Ana, CA 92705
Phone: (800) 437-5706
Fax: (949) 261-6533
Web: http://www.irvinesci.com

ISC Bioexpress
420 N. Kays Dr.
Kaysville, UT 84037
Phone: (800) 999-2901, (800) 288-2901
Fax: (801) 547-5051
Web: http://www.bioexpress.com
E-mail: isc@bioexpress.com

Jackson ImmunoResearch Laboratories, Inc.
872 West Baltimore Pike
P.O. Box 9
West Grove, PA 19390
Phone: (800) 367-5296, (610) 869-4067
Fax: (610) 869-0171
Web: http://www.jacksonimmuno.com
E-mail: stegeman@jacksonimmuno.com

Jencons Scientific
800 Bursca Dr.
Bridgeville, PA 15017
Phone: (412) 257-8861
Fax: (412) 257-8809
Web: http://www.jencons.co.uk
E-mail: mwilliams@jencons.com

Jordon Scientific
2200 Kennedy St.
Philadelphia, PA 19137
Phone: (800) 523-0171, (215) 535-8300
Fax: (215) 289-1597
Web: http://www.jordonscientific.com
E-mail: jordon@jordonscientific.com

Jouan, Inc.
170 Marcel Dr.
Winchester, VA 22602
Phone: (800) 662-7477, (540) 869-8623
Fax: (540) 869-8626
Web: http://www.jouaninc.com
E-mail: ibrown@jouaninc.com

Key Scientific Products
12401 Washington Ave.
Rockville, MD 20852
Phone: (301) 881-2045
Fax: (301) 881-8306
Web: http://www.keyscr.com
E-mail: info@keysci.com

Kirkegaard & Perry Laboratories, Inc.
2 Cessna Ct.
Gaithersburg, MD 20879-4174
Phone: (800) 638-3167

KMI Diagnostics
818 51st Ave., N.E.
Minneapolis, MN 55421
Phone: (612) 572-9354
Fax: (612) 586-0748
Web: http://www.kmidiagnostics.com
E-mail: mkowal@kmidiagnostics.com

Labconco Corp.
8811 Prospect Ave.
Kansas City, MO 64132
Phone: (800) 732-0031
Fax: (816) 363-0130
Web: http://www.labconco.com
E-mail: labconco@labconco.com

Lab-Line Instruments, Inc.
15th and Bloomingdale Ave.
Melrose Park, IL 60160

LabPlas, Inc.
1950 Rue Bombardier
Ste-Julie, Quebec J3E 2J9
Canada
Phone: (450) 649-7343
Fax: (450) 649-3113
Web: http://www.labplas.com
E-mail: rmeijer@labplas.com

LabSource
319 W. Ontario
Chicago, IL 60610
Phone: (800) 545-8823
Web: http://www.labsource.com

Labsystems
8 East Forge Pkwy.
Franklin, MA 02038
Phone: (800) LAB-PROD,
(508) 520-2229
Fax: (508) 520-2229
Web: http://www.affinity-sensors.com
E-mail: ron.gulka@thermobio.com

Leica Microsystems, Inc.
111 Deer Lake Rd.
Deerfield, IL 60015
Phone: (800) 248-0123
Fax: (847) 405-0147

Life Science Products
P.O. Box 8098
St. Joseph, MO 64508-8098
Phone: (800) 825-0341
Fax: (816) 279-4725

Life Technologies, Inc.
9800 Medical Center Dr.
P.O. Box 6482
Rockville, MD
Phone: (800) 338-5772
Fax: (800) 331-2286

Litmus Concepts, Inc.
2981 Copper Rd.
Santa Clara, CA 95051
Phone: (408) 245-5525
Fax: (408) 245-3301
Web: http://www.litmusconcepts.com
E-mail: bleiva@litmusconcepts.com

MarDx Diagnostic, Inc.
5919 Farnsworth Ct.
Carlsbad, CA 92008
Phone: (800) 331-2291,
(760) 929-0124
Fax: (760) 929-0124

Medical Packaging Corp.
941 Avenida Acaso
Camarillo, CA 93012
Phone: (800) 792-0600, (800) 672-0025,
(805) 388-2383
Fax: (805) 288-5531

MedMira Laboratories
200 Ronson Dr., Suite 101
Toronto, Ontario M9W 5Z9
Canada
Phone: (416) 644-0011
Fax: (416) 644-0007
Web: http://www.medmira.com
E-mail: sales@medmira.com

Meridian Diagnostics, Inc.
3471 River Hills Dr.
Cincinnati, OH 45244
Phone: (800) 543-1980, (800) 343-3858
Fax: (513) 271-3762
Web: http://www.meridiandiagnostics.com

Merlin
Keinstraße 14
D-53332 Bornheim-Hersel
Germany
Phone: 49 22 22-96 31 0
Fax: 49 22 22-96 31 9 0

Merlin Diagnostics
163 Cabot St.
Beverly, MA 01915
Web: http://www.merlindiagnostics.com

MicroBiologics, Inc.
217 Osseo Ave., North
St. Cloud, MN 56303-4455
Phone: (320) 253-1640,
(800) 599-BUGS
Fax: (320) 253-6250
Web: http://www.microbiologics.com

Microbiology International
10242 Little Rock Ln.
Frederick, MD 21702
Phone: (800) EZ-MICRO,
(301) 662-8096
Web: http://www.microbiology-intl.com
E-mail: gina.mcjonathan@microbiology-intl.com

Micro Test, Inc.
4325 Business Park Ct.
Lilburn, GA 30047
Phone: (800) 646-6678
Fax: (770) 935-4277
E-mail: microtest@transport-media.com

MIDI, Inc.
125 Sandy Dr.
Newark, DE 19713
Phone: (800) 276-8068, (302) 737-4297
Fax: (302) 737-7781
Web: http://www.midilabs.com
E-mail: info@midilabs.com

Millipore Corp.
80 Ashley Rd.
Bedford, MA 07130
Phone: (800) 221-1975
Fax: (781) 533-8887
Web: http://www.millipore.com

Miltenyi Biotec, Inc.
12740 Earhart Ave.
Auburn, CA 95602
Phone: (800) 367-6227
Fax: (530) 887-5348
Web: http://www.miltenyibiotec.com

Mitsubishi Gas Chemical America, Inc.
520 Madison Ave., 17th Floor

New York, NY 10022
Phone: (212) 752-4620
Fax: (212) 758-4012
Web: http://www.mgc-a.com
E-mail: henry@mgc-a.com

MLA Systems
270 Marble Ave.
Pleasantville, NY 10570-3448
Phone: (914) 747-3020
Fax: (914) 747-0498
Web: http://www.mlasystems.com

Molecular Biology Insights, Inc.
8685 U.S. Hwy. 24
Cascade, CO 80809-1333
Phone: (800) 747-4362
Fax: (719) 684-7989

Molecular Bio-Products
9880 Mesa Rim Rd.
San Diego, CA 92121
Phone: (800) 995-2787
Fax: (858) 453-4367
Web: http://www.mbpinc.com
E-mail: info@mbpinc.com

Molecular Probes, Inc.
4849 Pitchford Ave.
Eugene, OR 97402-0469
Phone: (541) 465-8300
Fax: (541) 344-6504
Web: http://www.probes.com

Molecular Probes Europe BV
PoortGebouw
Rijnsburgerweg 10
2333 AA Leiden
The Netherlands
Phone: 31-71-5233378
Fax: 31-71-5233419
Web: http://www.probes.com

Molecular Technologies
280 Vance Rd.
Valley Park, MO 63088
Phone: (800) 227-9997
Fax: (636) 225-9998
Web: http://www.moltec.com
E-mail: custserv@midsci.com

Molecular Toxicology, Inc.
157 Industrial Park Dr.
Boone, NC 28607

Monoclonal Technologies, Inc. (M-Tech)
16230 Birmingham Hwy.
Alpharetta, GA 30004
Phone: (888) 683-2414, (770) 277-1911
Fax: (770) 277-7988
Web: http://www.4m-tech.com
E-mail: m-tech@mindspring.com

Moss Inc.
P.O. Box 189
Pasadena, MD 22123-0189
Phone: (800) 932-6677

Fax: (410) 768-3971
Web: http://www.mosssubstrates.com
E-mail: bmoss2@erols.com

Mycoplasma Experience
1 Norbury Rd.
Reigate, Surrey RH2 9BY
England
Phone: 44 1737 226662
Fax: 44 1737 224751
Web: http://www.mycoplasma-exp.com
E-mail: mexp@mycoplasma-exp.com

Nalge Nunc International
2000 North Aurora Rd.
Naperville, IL 60563-1796
Phone: (800) 288-6862
Fax: (630) 416-2556
Web: http://www.nalgenunc.com
E-mail: bsylvester@nalgenunc.com

Namsa
2261 Tracy Rd.
Northwood, OH 43619
Phone: (419) 662-4319
Fax: (419) 666-2954
Web: http://www.namsa.com
E-mail: info@namsa.com

NAPCO
170 Marcel Dr.
Winchester, VA 22602
Phone: (800) 621-8820
Fax: (540) 869-0130
Web: http://www.napco2.com
E-mail: info@napco2.com

Nasco
901 Janesville Ave.
Fort Atkinson, WI 53538
Phone: (800) 558-9595
Fax: (920) 563-8296
Web: http://www.nascofa.com
E-mail: expert@nascofa.com

National Laboratory Training Network
Eastern Office
Delaware Public Health Laboratory
P.O. Box 1047
Smyrna, DE 19977-1047
Phone: (302) 653-2841
Fax: (302) 653-2844

NEN Life Science Products, Inc.
549 Albany St.
Boston, MA 02118-2512
Phone: (800) 551-2121
Fax: (617) 482-1380
Web: http://www.nen.com
E-mail: techsupport@nenlifesci.com

Neogen Corporation
620 Lesher Pl.
Lansing, MI 48912
Phone: (517) 372-9200
Fax: (517) 372-0108

Web : http://www.neogen.com
E-mail: neogen-info@neogen.com

New Brunswick Scientific
44 Talmadge Rd., Box 4005
Edison, NJ 08818-4005
Phone: (800) 631-5417
Fax: (800) 489-1400
Web: http://www.nbsc.com
E-mail: culotta@nbsc.com

New England Biolabs
32 Tozer Rd.
Beverly, MA 01915
Phone: (800) 632-5227
Fax: (800) 632-7440
Web: http://www.neb.com
E-mail: orders@neb.com

New Horizons Diagnostics Corp.
9110 Red Branch Rd.
Columbia, MD 21045
Phone: (800) 888-5015, (410) 992-9357
Fax: (410) 992-0328
Web: http://www.NHDiag.com
E-mail: NHDiag@aol.com

Nikon, Inc.
1300 Walt Whitman Rd.
Melville, NY 11747
Phone: (800) 52-NIKON
Web: http://www.nikon.com
E-mail: biosales@nikonincmail.com

Nor-Lake Scientific
727 Second St.
Hudson, WI 54016
Phone: (800) 241-1734
Fax (715) 386-6149
Web: http://www.norlake.com

**NOVEL EXPERIMENTAL
TECHNOLOGY**
11040 Roselle St.
San Diego, CA 92121
Phone: (800) 403-5024, (800) 456-6839
Fax: (619) 452-6635
E-mail: cindyh@novex.com

NuAire, Inc.
2100 Fernbrook Ln.
Plymouth, MN 55447
Phone: (612) 553-1270
Web: http://www.nuaire.com

NVSL-Ames
1800 Dayton Ave.
P.O. Box 844
Ames, IA 50010
Phone: (515) 663-7266
Fax: (515) 663-7397

NVSL-FADDL
P.O. Box 848
Greenport, NY 11944-0848

Phone: (631) 323-2500
Fax: (631) 323-2798

OEM Concepts Inc.
Bldg. 25 Unit 96
1889 Route A
Toms River, NJ 08755
Phone: (732) 341-3570, (877) 341-3570
Fax: (732) 286-3173
Web: http://www.oemconcepts.com
E-mail: info@oemconcepts.com

Olympus America, Inc.
Two Corporate Center Dr.
Melville, NY 11747-3157
Phone: (800) 223-0125
Fax: (516) 844-5112
Web: http://www.olympus.com

Omni International, Inc.
6530 Commerce Ct., Suite 100
Warrenton, VA 20187
Phone: (800) 776-4431, (540) 347-5331
Fax: (540) 347-5352
Web: http://www.omni-inc.com

Online Engineering, Inc.
3802 Industrial Blvd., Unit 5
Bloomington, IN 47403
Phone: (812) 339-9511
Fax: (812) 339-9512
Web: http://www.online-engineering.com
E-mail: info@online-engineering.com

OPERON
1000 Atlantic Ave., Suite 108
Alameda, CA 94501
Phone: (800) 688-2248
Fax: (510) 865-5255
Web: http://www.operon.com
E-mail: dna@operon.com

OraSure Technologies, Inc.
150 Webster St.
Bethlehem, PA. 18015
Phone: (610) 882-1820
Fax: (610) 882-1830
Web: http://www.stctech.com

Organon-Teknika
100 Akzo Ave.
Durham, NC 27712
Phone: (919) 620-2000, (800) 682-2666
Fax: (732) 577-7609
Web: http://www.organonteknika.com

Orgenics, Ltd.
North Industrial Zone
70650 Yavne
Israel
Phone: 972-8-9429201
Fax: 972-8-9438758
Web: http://www.orgenics.com

Orion Diagnostica
P.O. Box 83
02101 Espoo
Finland
Phone: 358 10 42 995
Fax: 358 10 429 2794
Web: http://www.oriondiagnostica.fi

Ortho-Clinical Diagnostics
A Johnson & Johnson Company
100 Indigo Creek Dr.
Rochester, NY 14626
Phone: (800) 828-6316
Fax: (716) 453-3660
Web: http://www.orthoclinical.com

Owl Separation Systems
55 Heritage Ave.
Portsmouth, NH 03801
Phone: (800) 242-5560, (603) 559-9297
Fax: (603) 559-9258
Web: http://www.owlsci.com

Oxford Molecular Group, Inc.
11350 McCormick Rd.
Executive Plaza III, Suite 1100
Hunt Valley, MD 21030
Phone: (800) 876-9994
Fax: (410) 527-4599

Oxoid, Inc.
217 Colonnade Rd.
Nepean, Ontario K2E 7K3
Canada
Phone: (800) 267-6391
Fax: (613) 226-3728
Web: http://www.oxoid.ca
E-mail: sales@oxoid.ca

Oxyrase, Inc.
P.O. Box 1345
Mansfield, OH 44901
Phone: (419) 589-8800
Fax: (419) 589-9919
Web: http://www.oxyrase.com

Pall Corporation
600 South Wagner Rd.
Ann Arbor, MI 48103-9019
Phone: (734) 665-0651
Fax: (734) 913-6114
Web: http://www.pall.com

PanBio Pty., Ltd.
116 Lutwyche Rd.
Windsor
Brisbane, Queensland 4030
Australia
Phone: 61 7 3357 1177
Fax: 61 7 3357 1222
Web: http://www.panbio.com.au
E-mail: carl_Stubbings@PanBio.com.au

PBS Orgenics
19, rue de Lambrechts
92404 Courbevoie Cedex
France
Phone: 33 1 41 99 92 92

PE Biosystems
850 Lincoln Center Dr.
Foster City, CA 94404
Phone: (650) 638-6640, (800) 345-5224
Fax: (650) 638-5884
E-mail: wakidah@pebio.com

Perkin-Elmer Analytical Instruments
761 Main Ave.
Norwalk, CT 06859
Phone: (800) 762-4003
Fax: (203) 762-4054
Web: http://www.nen.com
E-mail: info@perkin-elmer.com

Pfizer USA Pharmaceuticals
235 East 42nd St.
New York, NY 10017
Web: http://www.pfizer.com

Pierce Chemical Company
3747 North Meridian Rd.
Rockford, IL 61101
Phone: (800) 874-3723, (815) 987-4603
Fax: (815) 968-3556
E-mail: tim.brennan@mail.piercenet.com

Piramoon Technologies, Inc.
Santa Clara, CA 95054
Phone: (408) 988-1103
Fax: (408) 988-1196
Web: http://www.piramoon.com
E-mail: mp@piramoon.com

PML Microbiologicals, Inc.
27120 S.W. 95th Ave.
Wilsonville, OR 97070
Phone: (800) 547-0659, (503) 570-2500
Fax: (503) 570-2501
Web: http://www.pmlmicro.com
E-mail: inquiry@pmlmicro.com

Precision Scientific
170 Marcel Dr.
Winchester, VA 22602
Phone: (540) 869-9892, (800) 621-8820
Fax: (540) 869-0130
Web: http://www.precisionsci.com
E-mail: info@precisionsci.com

Program for Appropriate Technology in Health (PATH)
4 Nickerson St.
Suite 300
Seattle, WA 98109-1699
Phone: (206) 285-3500
Fax: (206) 285-6619
Web: http://www.path.org

Pro-lab Diagnostics
B2100 Kramer Ln.
Austin, TX 78758
Phone: (800) 522-7740
Fax: (800) 332-0450
Web: http://www.pro-lab.com
E-mail: support@pro-lab.com

Promega Corp.
2800 Wood Hollow Rd.
Madison, WI 53711
Phone: (800) 356-9526
Fax: (800) 356-1970
Web: http://www.promega.com
E-mail: custserv@promega.com

Pyrosequencing, Inc.
2200 West Park Dr., Suite 320
Westborough, MA 01581
Phone: (877) 797-6767
Fax: (508) 898-3306
Web: http://www.pyrosequencing.com

Qiagen, Inc.
28159 Stanford Ave.
Valencia, CA 91355
Phone: (800) 426-8157
Fax: (800) 718-2056
Web: http://www.qiagen.com

Qualicon, Inc.
P.O. Box 80357/1024A
Route 141 and Henry Clay Rd.
Wilmington, DE 19880-0357
Phone: (800) 863-6842, (302) 695-9400
Fax: (302) 695-9027
Web: http://www.qualicon.com

Quantum Design
11578 Sorrento Valley Rd.
San Diego, CA 92121-1311
Phone: (619) 481-4400
Fax: (619) 481-7410
E-mail: info@quandsn.com

QueLab Laboratories, Inc.
2331 Dandurand
Montreal, Quebec H2G 3C5
Canada
Phone: (514) 277-2558
Fax: (514) 277-4714
E-mail: marga@quelab.qc.ca

Quidel Corporation
10165 Mckellar Ct.
San Diego, CA 92121
Phone: (800) 874-1517
Web: http://www.quidel.com
E-mail: webguru@quidel.com

Quintiles
1300 17th St. N., Suite 300
Arlington, VA 22209-3801
Phone: (703) 276-0400
Fax: (703) 526-8399
Web: http://www.cro.quintiles.com
E-mail: pohanley@qarl.quintiles.com

Rainin Instrument Company
5400 Hollis St.
Emeryville, CA 94608
Phone: (800) 472-4646, (781) 935-3050
Web: http://www.rainin.com

Ramco Laboratories, Inc.
4507 Mt. Vernon
Houston, TX 77006
Phone: (800) 231-6238, (713) 526-1528
Web: http://www.ramco.com
E-mail: ramcolab@aol.com

Randolph Biomedical
21 McElroy St. West
Warwick, RI 02893
Phone: (401) 826-1407
Fax: (954) 697-2624

Raven Biological Laboratories
8607 Park Dr.
Omaha, NE 68106-1428
Phone: (800) 728-5702
Fax: (402) 593-0921
Web: http://www.ravenlabs.com
E-mail: sales@ravenlabs.com

RELA Medical Software
6175 Longbow Dr.
Boulder, CO 80301
Phone: (303) 530-2626, (800) 866-3716
Fax: (303) 530-2866
Web: http://www.cmed.com
E-mail: relainfo@cmed.com

Remel
12076 Santa Fe Dr.
Lenexa, KS 66215
Phone: (800) 447-3635
Fax: (913) 888-5884
Web: http://www.remelinc.com
E-mail: remel@remelinc.com

Research Organics
4353 East 49th St.
Cleveland, OH 44125
Phone: (800) 321-0570,
   (216) 883-8025
Phone (West Coast):   (619) 259-1534
Fax: (216) 883-1576
E-mail: amiller@resorg.com

Revco
275 Aiken Rd.
Asheville, NC 28804
Phone: (800) 252-7100
Fax: (828) 645-3368
Web: http://www.revco-sci.com
E-mail: sales@revco-sci.com

Robbins Scientific Corp.
1250 Elko Dr.
Sunnyvale, CA 94089-2213
Phone: (800) 752-8585

Fax: (408) 734-0300
Web: http://www.robsci.com
E-mail: custerv@robsci.com

Roche
9115 Hague Rd.
Indianapolis, IN 46250
Phone: (800) 526-1247, (800) 428-5074
Fax: (317) 521-3116
Web: http://www.roche.com
E-mail: diagnostics.webmaster@roche.com

Rockland Immunochemicals
Box 316
Gilbertsville, PA 19525
Phone: (800) 656-ROCK
Web: http://www.rockland-inc.com

Rosys Anthos
Churchmans Center
11A Parkway Cir.
New Castle, DE 19720
Phone: (302) 326-0433
Fax: (302) 326-0492
Web: http://www.rosys-anthos.com
E-mail: info@rosys-anthos.com

SafePath Laboratories, L.L.C.
1400 Energy Park Dr., Suite 20
St. Paul, MN 55108
Phone: (651) 659-9093
Fax: (801) 912-6697
Web: http://www.safepath.com

Saf-T-Pak, Inc.
10807 182nd St.
Edmonton, Alberta T8N 1J5
Canada
Phone: (780) 486-0211, (800) 814-7484
Fax: (780) 486-0235
Web: http://www.saftpak.com
E-mail: saftpak@compuserve.com

Saliva Diagnostic System
3661 Horseblock Rd., Suite E
Medford, NY 11763
Phone: (631) 205-0700
Fax: (212) 937-3801
Web: http://www.salv.com

Sanochemia Pharmazeutika AG
Boltzmanngasse 11
A-1090 Vienna
Austria
Phone: 43-1-319-1456-35
Fax: 43-1-319-1456-44
Web: http://www.sanochemia.at

Sanofi Diagnostics Pasteur
6565 185th Ave., N.E.
Redmond, WA 95082
Phone: (800) 666-5111, (800) 666-2111
Fax: (425) 861-5182

Sanofi Diagnostics Pasteur
3 boulevard Raymond Poincaré
92430 Marnes-La-Coquette
France

Phone: 33 1 47 95 60 00
Fax: 33 1 47 41 91 33

Sanyo Scientific
900 North Arlington Heights Rd.
Itasca, IL 60143
Phone: (800) 858-8442
Fax: (708) 775-0044

Sarstedt, Inc.
1025 St. James Church Rd.
Newton, NC 28658
Phone: (800) 257-5101
Fax: (828) 465-4003
Web: http://www.sarstedt.com

SA Scientific, Inc.
4919 Golden Quail
San Antonio, TX 78240
Phone: (210) 699-8800
Fax: (210) 699-6545
E-mail: sas@sascientific.com

SAS Institute Inc.
SAS Campus Dr.
Cary, NC 27513-2414
Phone: (800) 727-3228
Fax: (919) 677-8166
Web: http://www.statview.com
E-mail: statview@sas.com

Savant Instruments/E-C Apparatus
100 Colin Dr.
Holbrook, NY 11741-4306
Phone: (800) 634-8886, (516) 244-2929
Fax: (516) 244-0606
E-mail: charlotte.simpson@tmquest.com

Savyon Diagnostics, Ltd.
(A subsidiary of Healthcare Technologies, Ltd.)
Habosem 3
Ashdod
Israel 77101
Phone: 972-8-8562920
Fax: 972-8-8563258
Web: http://www.hctech.com

Scanalytics Inc.
8550 Lee Hwy., Suite 400
Fairfax, VA 22031-1515
Phone: (703) 208-2230
Fax: (703) 208-1960
Web: http://www.scanalytics.com
E-mail: sales@scanalytics.com

Scientific Industries, Inc.
70 Orville Dr.
Bohemia, NY 11716
Phone: (888) 850-6208
Fax: (631) 567-5896
Web: http://www.scind.com
E-mail: info@scientificindustries.com

SciQuest.com
P.O. Box 12156
Research Triangle Park, NC 27709

Phone: (919) 281-2130, (800) 233-1121
Fax: (919) 281-2199
Web: http://www.sciquest.com
E-mail: customercare@sciquest.com

Sequenom
Corporate Headquarters
3595 John Hopkins Ct.
San Diego, CA 92121-1331
Phone: (858) 202-9000
Fax: (858) 202-9001
U.S. East Coast Office
142-F North Rd., Suite 150
Sudbury, MA 01776
Phone: (978) 371-9830
Fax: (978) 371-9844
Web: http://www.sequenom.com

Seradyn, Inc.
1200 Madison Ave.
Indianapolis, IN 46225
Phone: (800) 428-4007
Fax: (317) 266-2991
Web: http://www.seradyn.com

Sera Quest
1938 N.E. 148th Terrace
North Miami, FL 33181

Shamrock Scientific Specialty Systems, Inc.
34 Davis Dr.
Bellwood, IL 60104
Phone: (800) 323-0249
Fax: (800) 248-1907

Shared Systems, Inc.
3961 Columbia Rd.
P.O. Box 211587
Martinez, GA 30917
Phone: (706) 868-0408
Fax: (706) 868-6588

Sheldon Manufacturing, Inc.
300 North 26th Ave.
Comelius, OR 97113
Phone: (800) 322-4897
Fax: (503) 640-1366
Web: http://www.shellab.com
E-mail: webmaster@shellab.com

Shimadzu Scientific Instruments, Inc.
7102 Riverwood Dr.
Columbia, MD 21046
Phone: (800) 477-1227
Fax: (410) 381-1222
Web: http://www.ssi.shimadzu.com
E-mail: webmaster@shimadzu.com

Shire
275 Blvd. Armand-Frappier
Laval, Quebec H7V 4A7
Canada
Phone: (450) 681-1744

Fax: (450) 978-7755
Web: http://www.shire.com

SIERRA Diagnostics, Inc.
21109 Longeway #C
Sonora, CA 95370
Phone: (888) 807-0900
Fax: (209) 536-0853
Web: http://www.sierradiagnostics.com
E-mail: info@sierradiagnostics.com

Sigma Diagnostics
545 South Ewing Ave.
St. Louis MO 63103
Phone: (800) 325-3010, (314) 771-5750
(U.S.); (800) 565-1400, (905) 829-9500
(Canada)
Fax: (314) 286-7819
Web: http://www.sigma-aldrich.com

Simport Plastic, Limited
176 Rue du Parc
Beloeil, Quebec J3G 4S5
Canada
Phone: (450) 464-1723
Fax: (450) 464-3394
Web : http://www.simport.com
E-mail: simport@mlink.net

SIR SCAN
I2a B.P. 42 Parc de la Méditerranée
District de Montpellier
34472 Pérols Cedex
France
Phone: 33 4 67 50 48 05
Fax: 33 4 67 17 09 06
E-mail: i2a@mnet.fr

SI Scientific Industries, Inc.
Airport International Plaza
70 Orville Dr.
Bohemia, NY 11716
Phone: (516) 567-4700
Fax: (516) 567-5896
Web: http://www.scind.com
E-mail: info@scind.com

Sonics & Materials, Inc.
53 Church Hill Rd.
Newtown, CT 06470
Phone: (203) 270-4600
Fax: (203) 270-4610
Web: http://www.sonicsandmaterials.com

Sooner Scientific, Inc.
P.O. Box 180
Garvin, OK 74736
Phone: (580) 286-9408
Fax: (580) 286-7047
Web: http://www.soonersci.com
E-mail: sonrsci@ionet.net

Span Diagnostics Ltd.
173-B, New Industrial Estate
Udhna, Surat-394210
India

Phone: 91-261-677-211
Fax: 91-261-665-757
Web: http://www.spandiag.com

Spectronics Corporation
956 Brush Hollow Rd.
P.O. Box 483
Westbury, NY 11590
Phone: (800) 274-8888, (516) 333-4840
Fax: (800) 491-6868, (516) 333-4859
Web: http://www.spectroline.com
E-mail: uvuv@aol.com

SRI Instrument
20720 Earl St.
Torrance, CA 90503-2162
Phone: (310) 214-5092
Fax: (310) 214-5097
Web: http://www.srigc.com
E-mail: webmaster@srigc.com

Starplex Scientific
50 Steinway Blvd.
Etobicoke, Ontario M9W 6Y3
Canada
Phone: (416) 674-7474, (800) 665-0954
Fax: (416) 674-6067

Stellar Bio Systems, Inc.
9075 Guilford Rd.
Columbia, MD 21046
Phone: (800) 962-6790
Fax: (410) 381-8984
Web: http:http://www.stellarbio.com
E-mail: info@stellarbio.com

**Steris Corporation (AMSCO
International, Inc.)**
5960 Heisley Rd.
Mentor, OH 44060
Phone: (216) 354-2600
Fax: (216) 639-4459

Stratagene
11011 North Torrey Pines Rd.
La Jolla, CA 92037
Phone: (800) 424-5444
Fax: (512) 321-3128
Web: http://www.stratagene.com
European Corporate Office
Gebouw California
Hogehilweg 15
1101 CB Amsterdam Zuidoost
The Netherlands
Phone: 00800-7400-7400
Fax: 00800-7001-7001

Svanova Biotech
National Veterinary Institute
SE-751 89 Uppsala
Sweden
Phone: 46 18 67 40 00
Fax: 46 18 30 91 62

U.S. distributor:
Diagnostic Chemical Limited
Marketing and Product Manager
Veterinary Diagnostics
160 Christian St.
Oxford, CT 06478
Web: http://www.sva.se/emeny.html
E-mail: sva@sva.se

Synbiotics Corp.
11011 Via Frontera
San Diego, CA 92127
Phone: (800) 247-1725
Web: http://www.synbiotics.com

Syngene
97H Monocacy Blvd.
Frederick, MD 21701
Phone: (301) 662-7144,
(877) GELDOCS
Fax: (301) 662-8096
Web: http://www.syngene.com
E-mail: ussales@syngene.com

SynPep Corporation
P.O. Box 2999
Dublin, CA 94568
Phone: (800) 899-3436,
(925) 803-9250
Fax: (800) 899-6534, (925) 803-0786
Web: http://www.synpep.com
E-mail: peptide@synpep.com

Syracuse Bioanalytical, Inc.
Langmuir Laboratory Box 1013
95 Brown Rd., Suite 144
Ithaca, NY 14850

Tecan U.S.
P.O. Box 13953
Research Triangle Park, NC 27709
Phone: (919) 361-5200
Fax: (919) 361-5201
Web: http://www.tecan-us.com

TechLab, Inc.
1861 Pratt Dr., Suite 1030
Blacksburg, VA 24060
Phone: (800) TECHLAB
Fax: (540) 953-1665

Thermo Hybaid
8 East Forge Pkwy.
Franklin, MA 02038
Phone: (888) 4-HYBAID
Fax: (508) 541-3041
Web: http://www.thermohybaid.com
E-mail: info-us@thermohybaid

**ThermoLabsystems (formerly Dynex
Technologies, Inc.)**
14340 Sullyfield Cir.
Chantilly, VA 20151-1683
Phone: (800) 336-4543,
(703) 631-7800

Fax: (703) 631-7816
Web: http://www.dynextechnologies.com
E-mail: info@dynextechnologies.com

Thermo Labsystems Oy
Sorvaajankatu 15
P.O. Box 208
FIN-00811
Helsinki
Finland
Phone: 358-9-329 100
Fax: 358-9-3291 0312
Web: http://www.labsystems.fi

Third Wave Technologies Inc.
502 South Rosa Rd.
Madison, WI 53719
Phone: (608) 273-8933, (888) 898-2357
Fax: (608) 273-8618
Web: http://www.twt.com
E-mail info@twt.com

Tomy Tech USA, Inc.
40479 Encyclopedia Cir.
Fremont, CA 94538
Phone: (510) 440-1976,
   (800) 545-TOMY
Fax: (510) 440-1975

Tosh Medics, Inc.
347 Oyster Point Blvd., Suite 201
South San Francisco, CA 94080
Phone: (800) 248-6764
Fax: (650) 615-0415
Web: http://www.tosohm.com

Toucan Technologies, Inc.
1158 Altadena Dr.
Cincinnati, OH 45230
Phone: (800) 506-2266

Transgenomic
5600 S. 42nd St.
Omaha, NE 68107
Phone: (888) 233-9283
Fax (402) 733-1932
Web: http://www.transgenomic.com
E-mail: info@transgenomic.com

TREK International Inc.
25760 First St.
Westlake, OH 44145
Phone: (800) 871-8909, (440) 808-0000
Fax: (440) 808-0400
Web: http://www.trekds.com/onsite
E-mail: jbarkley@trekds.com

Tridelta Development Ltd.
P.O. Box 14
Greystones
County Wicklow
Ireland
Phone: 353-1-276-5105
Web: http://www.tridaltaltd.com

Trinity Biotech
IDA Business Park
Bray, County Wicklow
Ireland
Phone: (800) 603-8076
Fax: 353-1-276-9888
Web: http://www.trinitybiotech.com

Triple G Corporation
3100 Steeles Ave. East
Suite 600
Markham, Ontario L3R 8T3
Canada
Phone: (888) 874-7534
Fax: (905) 305-0046
Web: http://www.tripleg.com

Tropical Biological
Medical Chemical Corporation
19430 Van Ness Ave.
Torrance, CA 90501
Phone: (800) 424-9394

Troy Biologicals Inc.
1238 Rankin
Troy, MI 48083
Phone: (800) 521-0445
Fax: (248) 585-2490
E-mail: tom@troybio.com

Turner Designs
845 W. Maude Ave.
Sunnyvale, CA 94086
Phone: (408) 749-0994
Fax: (408) 749-0998
Web: http://www.turnerdesigns.com

United Biomedical, Inc.
25 Davids Dr.
Happauge, NY 11788
Phone: (631) 273-2828
Fax: (631) 273-1717
Web: http://www.unitedbiomedical.com

United Kingdom National Culture Collection

Animal, Human Cell Line, or Viruses
European Collection of Cell Cultures
Center for Applied Microbiology and
   Research
Porton Down, Salisbury, Wiltshire SP4 0JG
United Kingdom
Phone: 44 1980 612512
Fax: 44 1980 611315
Web: http://www.ukncc.co.uk
E-mail: ecacc@camr.org.uk

Medically Important Fungi
National Collection of Pathogenic Fungi
Public Health Laboratory
Mycology Reference Laboratory
Myrtle Rd.
Kingstown
Bristol BS2 8EL
United Kingdom

Phone: 44 1179 291326
Fax: 44 1179 226611
Web: http://www.ukncc.co.uk

Medically Important Bacteria
National Collection of Type Cultures
Central Public Health Laboratory
61 Colindale Ave.
London NW9 5HT
United Kingdom
Phone: 44 181 2004400
Fax: 44 181 2057483
E-mail: bholemes@phls.nhs.uk

Universal Healthwatch
8990-E Oakland Center
Route 108
Columbia, MD 21045

U.S. Department of Agriculture Animal and Plant Health Inspection Service
4700 River Rd.
Riverdale, MD 20737
Phone: (301) 734-8695

UVP, Inc.
2066 W. 11th St.
Upland, CA 91786
Phone: (800) 452-6788, (909) 946-3197
Fax: (909) 946-3597
Web: http://www.uvp.com
E-mail: uvp@uvp.com

Varian Instruments
2700 Mitchell Dr.
Walnut Creek, CA 94598
Phone: (800) 926-3000
Web: http://www.varianinc.com
E-mail: webmaster@varianinc.com

Vedco Inc.
2121 S.E. Bush Rd.
St. Joseph, MO 64504
Phone: (816) 238-8840
Web: http://www.vedco.com

Veterinary Diagnostic Technology, Inc.
4890 Van Gordon St.
Suite 101
Wheat Ridge, CO 80033

Vétoquinol N-A Inc.
2000 Chemin Georges
Lavaltrie, Québec J0K 1H0
Canada
Phone: (450) 586-2252, (800) 363-1700
Fax: (450) 586-4649
Web: http://www.vetoquinol.ca
E-mail: info@vetoquinol.ca

Viral Antigens, Inc.
5171 Wilfong Rd.
Memphis, TN 38134

VircoLab, Inc.
Alpha Center
Johns Hopkins Bayview Research Campus
5210 Eastern Ave.
Baltimore, MD 21224
Phone: (410) 558-7031
Fax: (410) 558-7071
Web: http://www.vircolab.com
E-mail: info@vircolab.com

Viridae Clinical Sciences Inc.
1134 Burrard St.
Vancouver, British Columbia V6Z 1Y8
Canada
Phone: (604) 689-9404
Fax: (604) 689-5153
Web: http://www.viridae.com

ViroLogic, Inc.
270 East Grand Ave.
South San Francisco, CA 94080
Phone: (800) 777-0177
Fax: (650) 615-0177
Web: http://www.virologic.com

ViroMED
6101 Blue Dr.
Minneapolis, MN 55343-9108
Phone: (800) 582-0077, (612) 931-0077
Fax: (612) 939-4215
E-mail: csiegel@viromedlabs.com

Visible Genetics, Inc.
700 Bay St., Suite 1000
Toronto, Ontario M5G 1Z6
Canada
Phone: (888) 463-6844
Fax: (416) 813-3262
Web: http://www.visgen.com
E-mail pqula@visgen.com

Vista Technology, Inc.
2316 Delaware Ave., #333
Buffalo, NY 14216-2606
Phone: (416) 798-4988, (800) 667-3411
Fax: (905) 475-7309
Web: http://www.vita-tech.com
E-mail: info@vita-tech.com

VMRD, Inc.
P.O. Box 502
4641 Pullman-Albion Rd.
Pullman, WA 99163
Phone: (800) 222-8673
Fax: (509) 332-5356
Web: http://www.vmrd.com

VWR Scientific Products
1310 Goshen Pkwy.
West Chester, PA 19380
Phone: (800) 932-5000, (847) 726-9633
Fax: (847) 726-9612
Web: http://www.vwrsp.com

Wampole Laboratories
Division of Carter-Wallace
Half Acre Rd.
P.O. Box 1001
Cranbury, NJ 08512
Phone: (800) 257-9525
Web: http://www.wampolelabs.com

Waring Products Division
283 Main St.
New Hartford, CT 06057
Phone: (860) 379-0731
Fax: (860) 738-0249

Wescor, Inc.
459 South Main St.
Logan, UT 84321
Phone: (800) 453-2725

Fax: (435) 752-4127
Web: http://www.wescor.com
E-mail: biomed@wescor.com

Wheaton Science Products
1501 N. 10th St.
Millville, NJ 08332
Phone: (800) 225-1437, (609) 825-1100
Fax: (609) 293-6374
E-mail: Ned.morgan@wheaton.com

ZeptoMetrix Corporation
872 Main St.
Buffalo, NY 14202
Phone: (800) 274-5487
Fax: (716) 882-0959
Web: http://www.zeptometrix.com

Zeus Scientific, Inc.
351 W. Camden St.
Baltimore, MD 21201
Phone: (800) 526-3874, (800) 286-2111
Fax: (908) 526-2058
Web: http://www.zeusscientific.com

Zylux Corp.
144 Ridgeway Square
Oak Ridge, TN 37830
Phone: (865) 481-8181
Fax: (865) 481-8182
Web: http://www.zylux.com
E-mail: contact@zylux.com

ZymeTx, Inc.
800 Research Pkwy., Suite 100
Oklahoma City, OK 73104-3600
Phone: (405) 271-1314, (888) 817-1314
Fax: (405) 271-1038
Web: http://www.zymetx.com
E-mail: pederseng@zymetx.com

Index